GLOSSARY OF SYMBOLS

SYMBOL	MEANING	PAGE
$\sin \theta$; $\cos \theta$	Sine and cosine of θ	226, 228
$\tan \theta$, $\cot \theta$, $\sec \theta$, $\mathrm{cosec}\, \theta$	Tangent, cotangent, secant, and cosecant of θ	226, 230
Sin^{-1}, Cos^{-1}, Tan^{-1}	Principal inverse sine, cosine, and tangent	249, 251, 252
T; ω	Period; angular frequency	256, 258, 521
$\int f(x)\, dx$	Indefinite integral or antiderivative of $f(x)$	269
$F(x)$, $G(x)$	Antiderivatives of $f(x)$, $g(x)$	269
$\sum\limits_{k=m}^{n} \cdots$	The sum from $k = m$ to $k = n$ of $\ldots$	309
$\int_{a}^{b} f(x)\, dx$	The definite integral of $f(x)$ from a to b	322
$[F(x)]_{a}^{b}$	$F(b) - F(a)$	323
$\bar{x}$, $\bar{y}$	Coordinates of center of mass	346, 350
S	Sample space	360
$E_1 \cup E_2$	Union of two events E_1 and E_2	363
$E_1 \cap E_2$	Intersection of two events E_1 and E_2	364
E'	Complementary event to E	365
$P(E)$	Probability of the event E	368
$n!$	Factorial n	379
$_{n}P_{r}$	Permutations of r objects from among n	381
$\binom{n}{r}$	Combinations of r objects from among n	382
A, a	Alleles	390
AA, Aa, aa	Genotypes	390
D, R	Dominant and recessive genotypes	392
$F(x)$; $f(x)$	Probability distribution function; density function	405, 408
μ; σ^2	Mean; variance	412, 413
Z	Standard normal random variable	417
(x, y, z)	Coordinates in three dimensions	435
$\dfrac{\partial f}{\partial x}, \dfrac{\partial f}{\partial y}, f_x, f_y, z_x, z_y$	Partial derivatives	452, 457
$\dfrac{\partial^2 f}{\partial x^2}, \dfrac{\partial^2 f}{\partial x\, \partial y}, f_{xx}, f_{yy}$	Second partial derivatives	455, 457
$\Delta(x, y)$	$f_{xx} f_{yy} - f_{xy}^2$	469
$\overrightarrow{PQ}$; $\lvert \overrightarrow{PQ} \rvert$	Directed line segment; length of $\overrightarrow{PQ}$	593
$\mathbf{A}$; $\hat{\mathbf{A}}$	Vector; unit vector parallel to $\mathbf{A}$	594
$\mathbf{i}, \mathbf{j}$	Unit vectors in x and y coordinate directions	596
(x, y)	$x\mathbf{i} + y\mathbf{j}$ or position vector of point (x, y)	596
$\mathbf{A} \cdot \mathbf{B}$	Inner product of $\mathbf{A}$ and $\mathbf{B}$	604
$m \times n$	Size of a matrix (m rows, n columns)	627
A_{ij}; (A_{ij})	ij-element of the matrix $\mathbf{A}$; $(A_{ij}) = \mathbf{A}$	628

MATHEMATICS
for the Biological Sciences

MATH

for the

Jagdish C. Arya
Robin W. Lardner

Department of Mathematics
Simon Fraser University

EMATICS

Biological Sciences

PRENTICE-HALL, INC., Englewood Cliffs, New Jersey 07632

Library of Congress Cataloging in Publication Data

Arya, Jagdish C. (date)
    Mathematics for the biological sciences.

    Includes index.
    1. Biomathematics. I. Lardner, Robin W., (date) joint
author. II. Title.
QH323.5.A79   510   78-13424
ISBN 0-13-562439-8

MATHEMATICS FOR THE BIOLOGICAL SCIENCES
Jagdish C. Arya and *Robin W. Lardner*

©1979 by PRENTICE-HALL, INC.,
Englewood Cliffs, New Jersey 07632

Printed in the United States of America

10 9 8 7 6 5

Editorial/production supervision by Linda Mihatov
Cover and interior design by Virginia M. Soulé
Manufacturing buyer: Phil Galea
Cover photograph: DNA model by Fritz Goro. Courtesy
    Life Picture Service (1963).

PRENTICE-HALL INTERNATIONAL, INC., *London*
PRENTICE-HALL OF AUSTRALIA PTY. LIMITED, *Sydney*
PRENTICE-HALL OF CANADA, LTD., *Toronto*
PRENTICE-HALL OF INDIA PRIVATE LIMITED, *New Delhi*
PRENTICE-HALL OF JAPAN, INC., *Tokyo*
PRENTICE-HALL OF SOUTHEAST ASIA PTE. LTD., *Singapore*
WHITEHALL BOOKS LIMITED, *Wellington, New Zealand*

To Our Parents

Contents

FUNCTIONS 1

THE DERIVATIVE 60

EXPONENTIAL AND LOGARITHM FUNCTIONS 117

APPLICATIONS OF DERIVATIVES 160

TRIGONOMETRIC FUNCTIONS   224

INTEGRATION   268

THE DEFINITE INTEGRAL   308

PROBABILITY 359

FUNCTIONS OF
SEVERAL VARIABLES 428

DIFFERENTIAL EQUATIONS 486

DIFFERENCE EQUATIONS 553

VECTORS AND MATRICES 592

APPENDICES

ANSWERS TO ODD-NUMBERED EXERCISES 666

Preface

A number of years ago we decided at Simon Fraser University to separate those students taking our calculus courses whose specialization was in the biological sciences from the main stream of students taking first-year mathematics courses. Our aim was to offer them a special two-semester course in calculus which would be oriented towards their own field of interest and which would concentrate on teaching the applications of calculus rather than on teaching pure mathematics for its own sake. To a large extent, this aim has been achieved and the course has proved to be extremely successful.

In the initial years we faced the problem of the lack of a suitable textbook covering the material we wished to include and written specifically for students majoring in the life sciences. Eventually we were driven to the length of preparing our own classroom notes for the course, from which this text evolved. It has been used, in manuscript form, as the basis for our course for the last two years, in the process of which it has undergone a number of refinements.

In writing the book, our purpose was to emphasize those portions of calculus which are of prime concern in the life sciences and to suppress or even ignore topics which are of peripheral interest. We have concentrated on teaching the student how to use calculus as a technique, and we have avoided the rigorous mathematical proofs of most of the results. It is our experience that the student who learns to master the techniques usually develops a reasonably sound intuitive grasp in the process, and the lack of rigorous proofs is not a serious deficiency. We have also illustrated the various mathematical techniques with examples and applications from the life sciences throughout.

The result of this approach has been a book which in many ways differs quite substantially from existing textbooks in first-year calculus. Much of this difference appears in the way in which examples from the biological sciences are used to motivate or to illustrate the various mathematical developments. However, it also shows in the choice of and emphasis on the various topics considered. The following brief review of the contents of the book will demonstrate this point.

Chapters 1 and 2 contain a discussion of functions, limits, derivatives, and techniques of differentiation. Apart from the emphasis on teaching by explanation and example rather than by proof of theorems, most of the material in these chapters is fairly standard. However, we have included a section on linear inequalities which brings the student to the edge (but not over it) of linear programming.

Chapter 3 provides an early introduction to the exponential and logarithm functions, which are so important in the biological sciences. In most calculus books this material is postponed until it can be given a rigorous foundation after integration has been developed. We have preferred to introduce the topic early, on a nonrigorous but intuitively acceptable basis. We have given in Section 3.3 a discussion of logarithmic plots and a more detailed than usual discussion in Sections 3.1 and 3.5 of exponential growth and decay processes.

Chapter 4 gives applications of differential calculus to curve sketching, optimization, to Newton's method, and to approximation of functions using differentials. Great emphasis is placed in Section 4.3 on optimization problems in bioscience.

Chapter 5 provides a discussion of the trigonometric functions. We find that a substantial number of incoming students are weak in this area, and in Section 5.1 have given a short review of elementary trigonometry. The following sections extend the differential calculus to the trigonometric functions and in Section 5.5 periodic functions are discussed. Periodic phenomena are so important in the life sciences that this subject, although not usually covered in mathematics courses until the third year, should be introduced to students in this area. The section leads up to Fourier series.

It is not uncommon in first-year calculus courses to spend half of the second semester course teaching techniques of integration which, on the whole, are quite difficult to learn. For students in a nonmathematical area like the life sciences this is often a waste of effort in our opinion. Our primary approach to integration in Chapter 6 is through the use of tables of integrals, and we proceed to introduce such tables in Section 6.3 immediately after covering the substitution method. However, the instructor who prefers to dwell at some length on techniques of integration can, if he wishes, postpone coverage of Section 6.3 until after the sections on partial fractions, trigonometric substitutions, and integration by parts.

Chapter 7 provides the application of the definite integral to the calculation of areas, volumes, and centers of mass.

It is likely that almost all of the material in the first seven chapters would be included in any two-semester course on calculus. The remaining five chapters contain material which would be less standard in such courses. The book is 20–25% longer than could normally be covered in two semesters, so that an instructor has a certain degree of choice in his selection of material. It is envisaged that typically only two or three of Chapters 8–12 would be covered, depending on the frequency with which the class meets. These five chapters are independent of one another, so that any choice of the chapters to be covered is feasible.

Chapter 8 gives an introduction to probability. It includes counting methods, some discrete probability models of particular concern in the life sciences, and continuous distributions (the normal distribution, of course, is emphasized). A whole section is devoted to applications in genetics.

Chapter 9 gives an introduction to functions of several variables leading up to approximations by differentials, optimization of functions of two variables, and the least-squares method.

Chapters 10 and 11 provide an introduction to differential and difference equations. Particular emphasis is given to the application of these equations to model-building in ecology, for example, to problems in population dynamics and to survival problems. Among other areas of application are chemical kinetics and the spread of epidemics. The instructor who wishes to introduce model-building at an earlier stage in his course might consider the possibility of using Sections 11.1 and 11.2 as the basis for doing this, since the level of mathematics in these articles is quite elementary.

Chapter 12 contains an introduction to vectors and matrices and their applications to relative velocities, forces, systems of linear equations, and some models of interacting systems.

Throughout the book we have illustrated the various mathematical techniques with a large number of worked examples. Usually we have first given examples of a purely mathematical nature, then where appropriate have followed these with worked problems showing the application in some area of bio-science. A large number (usually 20–30) of exercises are given at the end of each section. Again these are divided into purely mathematical exercises which help the student to learn the basic mathematical technique, and then problems from the bio-sciences which permit him to apply the techniques he has learned.

The more difficult theorems in the subject are stated without proof. Many of the simpler theorems are proved. However, we have found that many students in this type of course become bored by too many proofs, even relatively easy ones, and the worked examples given are such as to permit the student to omit many of the proofs of the theorems if he wishes without losing too much.

It is a pleasure to acknowledge our indebtedness to those colleagues and students who have made suggestions for improving the book. Our thanks go to Professors Herbert W. Hethcote, University of Iowa, Jack L. Goldberg, University of Michigan, and Frank A. Wattenburg, University of Massachusetts, Amherst. We particularly appreciate the detailed, line-by-line review prepared on the final manuscript by Professor Wattenburg. The reviewers' comments have had a substantial effect on the final form which the book has assumed.

JAGDISH C. ARYA
ROBIN W. LARDNER

Applications in the Life Sciences

NOTE: Text discussions are coded T; Examples, Exp; and Exercises, E, in the listing given below.

MATHEMATICS
for the Biological Sciences

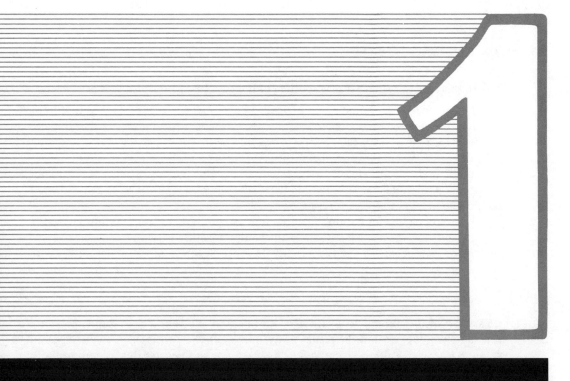

Functions

1.1 REAL NUMBER SYSTEM

In this section we give a brief outline of the structure of the real number system.

The numbers 1, 2, 3, and so on are called *natural numbers*. If we add or multiply any two natural numbers, the result is always a natural number. For example: $8 + 5 = 13$, $8 \times 5 = 40$; the results 13 and 40 are natural numbers. But if we subtract or divide two natural numbers, the result is *not* always a natural number. For example, $8 - 5 = 3$ and $8 \div 2 = 4$ are natural numbers but $5 - 8$ and $2 \div 7$ do not result in natural numbers. Thus, within the system of natural numbers we can add and multiply, but cannot always subtract or divide.

To overcome the limitation of subtraction, we extend the natural number system to the system of *integers*. This we do by including, together with all the natural numbers, all of their negatives and the number 0 (zero). Thus, we can represent the system of integers in the form

$$\ldots, -3, -2, -1, 0, 1, 2, 3, \ldots$$

Clearly all of the natural numbers are also integers. We can add, multiply, or subtract any two integers and the result always turns out to be an integer. For example: $-3 + 8 = 5, (-3)(5) = -15$, and $3 - 8 = -5$ are all integers. But we still cannot divide (division by zero excluded) any two integers to get an integer as a result. For example: $8 \div (-2) = -4$ is an integer whereas $-8 \div 3$ is not an integer. Thus, within the system of integers we can add, multiply, and subtract but cannot divide.

To overcome this limitation of division, we extend the system of integers to the system of *rational numbers*.

We define a number as a rational number if it can be expressed as a ratio of two integers (the denominator being nonzero). Thus $\frac{8}{3}, -\frac{5}{7}, \frac{0}{3}, 6 = \frac{6}{1}$, etc., are all rational numbers. We can add, multiply, subtract, and divide any two rational numbers (except that division by zero is excluded) and the result is always a rational number. Thus all the four fundamental operations of arithmetic, addition, multiplication, subtraction, and division, are possible within the system of rational numbers.

There also exist some numbers in everyday use that are not rational, that is, they cannot be expressed as the ratio of two integers. For example: $\sqrt{2}, \sqrt{3}, \pi$, etc., are not rational numbers; such numbers are called *irrational numbers*.

The term *real number* is used to describe a number that is either rational or irrational. To give a complete definition of real numbers would involve the introduction of a number of new ideas, and we shall not embark on this task. However, it is possible to get a good idea of what real numbers are by thinking of them in terms of decimals. Any rational number can be expressed as a decimal simply by dividing the denominator into the numerator by long division. It is found in every case that the decimal either terminates or develops a pattern that repeats indefinitely. For example, $\frac{1}{4} = 0.25$ and $\frac{93}{80} = 1.1625$ both correspond to decimals that terminate, whereas $\frac{1}{6} = 0.1616161\ldots$ and $\frac{4}{7} = 0.5714285714285\ldots$ correspond to decimals with repeating patterns.

2

An irrational number can also be represented by a decimal, but in this case the decimal continues indefinitely without developing any recurrent pattern, for example, $\sqrt{2} = 1.4142135623\ldots$ and $\pi = 3.1415926535\ldots$ to ten decimal places. No matter how many decimal places we express these numbers to, we will never find that they develop a repeating pattern.

Thus we can regard the system of real numbers as being equivalent to all possible decimals. Those decimals that terminate or are recurrent correspond to the rational numbers, while the rest correspond to the irrational numbers.

The square of any real number (positive, zero, or negative) is always nonnegative. Thus $\sqrt{-3}$, $\sqrt{-2/5}$, etc. are not real numbers. As long as we consider only real numbers, no meaning can be attached to the square root of a negative number. (Those students who have encountered the term will recall that these numbers do exist in a certain sense and are called imaginary numbers. We shall not consider such numbers at all in this book.)

A NOTE ON DIVISION BY ZERO. Division and multiplication are two inverse operations in mathematics. The statement $\frac{a}{b} = c$ is true if and only if the inverse statement $b \times c = a$ is true. Now consider $\frac{3}{0}$; this is not a number because there is no number c for which the inverse statement $0 \times c = 3$ is true. Again, $\frac{0}{0}$ is also not a number because the inverse statement $0 \times c = 0$ is satisfied by any arbitrary number c. Thus we conclude that division by zero is meaningless and in any fraction we cannot allow the denominator to be zero.

Inequalities

If a and b are two real numbers, the notation $a > b$ is used to denote that *a is greater than b* and the notation $a < b$ is used to denote that *a is less than b*. The statement $a < b$ means the same as the statement $b > a$. For example, the following statements are all true: $8 > 7$; $-1 > -\pi$; $0 < 2.5$; $-4 < -3$. If x is a variable, the statement $x > 3$ means that the quantity denoted by the letter x is restricted to values that are greater than 3.

The notation $a \geq b$ is used to denote that either $a > b$ or $a = b$. Thus we can write $8 \geq 7$ and $-1 \geq -\pi$, both of which are true. In addition, the statement $7 \geq 7$ is also true. In a similar way we write $a \leq b$ to mean that either $a < b$ or $a = b$. For example, $-1 \leq 2$ and $-3 \leq -3$ are true statements.

Inequalities are often written in combined form. For example, the inequality $1 \leq x < 3$ means that both $1 \leq x$ and $x < 3$. In other words, the quantity x is greater than or equal to 1 and is less than 3 in value.

There are certain rules by means of which inequalities can be manipulated. The first rule is that *any real number can be added to or subtracted from both sides of an inequality and the new inequality will still be true.*

EXAMPLES   **(a)** $8 > 7$ is true. If 4 is added to both sides, we obtain $8 + 4 > 7 + 4$, or $12 > 11$, which is still true. If 9 is subtracted from both sides, we obtain $8 - 9 > 7 - 9$, or $-1 > -2$, which again is true.

(b) If $x - 1 \geq 2$ then $x \geq 3$. The second inequality is obtained by adding 1 to both sides of the first:

$$x - 1 + 1 \geq 2 + 1, \quad \text{or} \quad x \geq 3.$$

The second rule is that *both sides of an inequality can be multiplied by any positive real number and the inequality so obtained is still true. However, if an inequality is multiplied by a negative real number, the direction of the inequality sign must be reversed.*

EXAMPLES   **(a)** $8 > 7$ is true. If both sides are multiplied by 3, we get $3(8) > 3(7)$, or $24 > 21$, which is true. However if both sides are multiplied by -3, the inequality sign must be changed from $>$ to $<$: $(-3)8 < (-3)7$, or $-24 < -21$, which is again true.

(b) If $\frac{1}{2}x \geq 1$, then multiplying by 2, we get

$$2(\tfrac{1}{2}x) \geq 2(1), \quad \text{or} \quad x \geq 2.$$

If $-2x \leq 6$, then multiplying by $(-\frac{1}{2})$ and reversing the inequality, we get

$$(-\tfrac{1}{2})(-2x) \geq (-\tfrac{1}{2})(6).$$

After simplification this becomes $x \geq -3$. In other words, if x is such that $(-2x)$ is less than or equal to 6, then x itself must be greater than or equal to -3 in value.

EXAMPLE   For what values of x does the inequality $2 - x < 3x + 10$ hold?

SOLUTION   We must move all the x-terms to one side of the inequality and all the numbers to the other side. So we subtract $(3x + 2)$ from both sides:

$$2 - x - 3x - 2 < 3x + 10 - 3x - 2, \quad \text{or} \quad -4x < 8.$$

We now multiply by $(-\frac{1}{4})$ (which is negative, so we reverse the inequality):

$$(-\tfrac{1}{4})(-4x) > (-\tfrac{1}{4})(8), \quad \text{or} \quad x > -2.$$

The given inequality holds for all values of x greater than -2.

EXAMPLE   Find the values of x that satisfy the following inequality:

$$2x - 1 \leq 2 - x < 3x + 10.$$

SOLUTION   In problems such as this one we must split the given double inequality into its two component parts. In this case, the inequalities $2x - 1 \leq 2 - x$ and $2 - x < 3x + 10$ must *both* be satisfied. We know from the preceding example that the second of these is equivalent to the condition $x > -2$. We must consider the other inequality: $2x - 1 \leq 2 - x$. Adding $x + 1$ to both sides we move all the x's to the left and all the numbers to the right:

$$2x - 1 + x + 1 \leq 2 - x + x + 1, \quad \text{or} \quad 3x \leq 3.$$

Multiplying by $\frac{1}{3}$, we then get the result that $x \le 1$. Thus the given inequality is satisfied if both $x > -2$ and $x \le 1$. We can combine these in the form $-2 < x \le 1$: x must be greater than -2 but less than or equal to 1.

The following examples illustrate a different type of inequality involving a variable x.

EXAMPLE   Find the values of x for which $(x - 1)(x - 2) > 0$ and $(x - 1)(x - 2) < 0$.

SOLUTION   The given expression $(x - 1)(x - 2)$ is the product of two numbers $(x - 1)$ and $(x - 2)$. It is positive when both of these numbers are positive and when both of them are negative. When one of the numbers $(x - 1)$ and $(x - 2)$ is positive and the other is negative, their product, $(x - 1)(x - 2)$, is negative.

Now $x - 1$ is positive for $x > 1$ and is negative for $x < 1$. And similarly $x - 2$ is positive for $x > 2$ and is negative for $x < 2$. Thus we can make the following table, which gives the signs of these two factors for $x < 1$, $1 < x < 2$, and for $x > 2$.

	$x < 1$	$1 < x < 2$	$x > 2$
$x - 1$	$-$ve	$+$ve	$+$ve
$x - 2$	$-$ve	$-$ve	$+$ve
$(x - 1)(x - 2)$	$+$ve	$-$ve	$+$ve

Then, for example, for $x < 1$, both $(x - 1)$ and $(x - 2)$ are negative, so their product $(x - 1)(x - 2)$ is positive. For $1 < x < 2$ the two factors are of opposite sign, so $(x - 1)(x - 2) < 0$. And for $x > 2$, both factors are positive, so $(x - 1)(x - 2) > 0$. We can summarize these results as follows:

$$(x - 1)(x - 2) > 0 \quad \text{for } x < 1 \text{ and for } x > 2;$$
$$(x - 1)(x - 2) < 0 \quad \text{for } 1 < x < 2.$$

EXAMPLE   For what values of x is $\sqrt{4 - x^2}$ a real number?

SOLUTION   In order that the quantity given should be a real number, the expression under the square-root sign must be greater than or equal to zero. Thus the condition is

$$4 - x^2 \ge 0, \quad \text{or} \quad x^2 - 4 \le 0.$$

The expression here can be factored since it is the difference of two squares:

$$x^2 - 4 = x^2 - 2^2 = (x - 2)(x + 2).$$

So the condition on x is that $(x - 2)(x + 2) \le 0$. In this form, the present example closely resembles the last one. We must find for what values of x the two factors $(x - 2)$ and $(x + 2)$ are of opposite sign.

Now $(x - 2)$ is positive for $x > 2$ and negative for $x < 2$, and $(x + 2)$ is positive for $x > -2$ and negative for $x < -2$. Thus for $-2 < x < 2$ the factor $(x + 2)$ is positive while the factor $(x - 2)$ is negative. So we have the result for their product: $(x - 2)(x + 2) \leq 0$ for $-2 \leq x \leq 2$.

EXERCISES 1.1

Solve the following inequalities for the variable x.

1. $x + 1 \leq 2$.

2. $2x - 1 > 5$.

3. $1 - 4x > -3$.

4. $3 - 2x \leq -5$.

5. $x + 1 > 3 - x$.

6. $4 - 2x < 3x - 6$.

7. $4 + 3x \geq x - 6$.

8. $1 - x \geq x - 7$.

9. $x + 1 \leq 2x + 3 \leq 8 - 3x$.

10. $1 - x \leq x + 3 \leq 9 - 2x$.

11. $(x + 1)(x - 2) > 0$.

12. $(x - 1)(x - 3) > 0$.

13. $x^2 - 4x + 3 < 0$.

14. $x^2 + 2x - 3 \leq 0$.

15. $x^2 - 3 \geq 0$.

16. $9 - x^2 \geq 0$.

1.2 SETS AND THEIR REPRESENTATION

Any collection of objects is referred to as a *set*. The objects constituting the set are called the *members* or *elements* of the set. A set can be specified in two ways, either by making a list of all of its members or by stating a rule for membership in the set. Let us examine these two methods in turn.

i. *Listing Method*. If it is possible to specify all of the elements of a set, the set can be described by listing all the elements and enclosing the list inside braces. For example, $\{1, 2, 5\}$ denotes the set consisting of the three numbers 1, 2, and 5; $\{p, q\}$ denotes the set whose only members are the two letters p and q.

In cases where the set contains a large number of elements it is often possible to employ what is called a *partial listing*. For example, $\{2, 4, 6, \ldots, 100\}$ denotes the set of all the even integers from 2 to 100. The ellipsis "..." is used to imply that the sequence of elements continues in a manner that is clear from the first few members listed. The sequence terminates at 100.

By use of the ellipsis, the listing method can be employed in some cases when the set in question contains infinitely many members. For example, $\{1, 3, 5, \ldots\}$ denotes the set of all odd natural numbers. The absence of any number following the ellipsis indicates that in this case the sequence does not terminate, but carries on indefinitely.

ii. *Rule Method.* There are many cases in which it is not possible or in which it would be inconvenient to list all of the members of a particular set. In such a case the set can be specified by stating a rule for membership. For example, consider the set of all people living in Canada at this present moment. To specify this set by listing all the members by name would clearly be a prodigious task. Instead we can denote it as follows:

$$\{x \mid x \text{ is a person, currently resident in Canada}\}.$$

The symbol | stands for "such that," and this expression should be read "the set of all x such that x is a person currently resident in Canada." The statement that follows | inside the braces is the rule that specifies the membership in the set.

As a second example, consider the set

$$\{x \mid x \text{ is a point on this page}\},$$

which denotes the set of all of the points on this page. This is an example of a set that could not be specified by the listing method even if we wanted to do so.

Many sets can be specified either by listing or by stating a rule, and we can choose whichever of the two methods we like. We shall give several examples of sets specified using both methods.

1. If N denotes the set of all natural numbers, then we can write

$$N = \{1, 2, 3, \ldots\}$$
$$= \{k \mid k \text{ is a natural number}\}.$$

2. If P denotes the set of integers between -2 and $+3$ including both -2 and $+3$ themselves, then

$$P = \{-2, -1, 0, 1, 2, 3\}$$
$$= \{x \mid x \text{ is an integer}, -2 \le x \le 3\}.$$

We observe that in this example, the membership rule consists of two conditions separated by a comma. Both conditions must be satisfied by any member of the set.

3.
$$Q = \{1, 4, 7, \ldots, 37\}$$
$$= \{x \mid x = 3k + 1 \text{ where } k \text{ is an integer}, 0 \le k \le 12\}.$$

4. The set of all students currently taking this course can be represented in the form

$$S = \{x \mid x \text{ is a student currently enrolled in Math 154}$$
$$\text{at Simon Fraser University}\}.$$

This set could also be specified by listing the names of all of the students involved.

Further examples of sets are:

5. The set of all real numbers that are greater than 1 and less than 2.
$$T = \{x \,|\, x \text{ is a real number, } 1 < x < 2\}.$$
This set cannot be listed because it has too many elements. (The student at this stage is not expected to understand why this is so.)
6. The set of all fresh-water species of fish. This set can be listed, although the list would be quite long.
7. The set of all fresh-water fish. Note the distinction between this example and the previous one. In item 6, the elements of the set are the species of fish such as "rainbow trout," "pike," and so on. In item 7, the elements of the set are the individual fish themselves.
8. The set of all chemical elements can be represented in the form
$$C = \{x \,|\, x \text{ is a chemical element}\}.$$

A set is said to be *finite* if the number of elements belonging to it is finite. A set containing an infinite number of elements is called an *infinite* set. In the above examples, the sets in examples 2, 3, 4, 6, 7, and 8 are all finite, while those in examples 1 and 5 are infinite.

It is usual to use capital letters to denote sets and lower-case letters to denote the elements in the sets. It will be noted that we have followed this convention in the preceding examples. If A is any set and x any object, the notation $x \in A$ is used to denote the fact that x is a member of A. The statement "$x \in A$" is read "x belongs to A" or "x is an element of A." The converse statement "x is not an element of A" is denoted by writing $x \notin A$.

Referring to example 2 above, we have the results that $2 \in P$, but $6 \notin P$. For the set in example 5, $\sqrt{2} \in T$, $\frac{3}{2} \in T$, but $2 \notin T$ and $\pi \notin T$.

A set that contains no elements is called an *empty set*. (The terms *null set* and *void set* are also used.) The symbol $\varnothing$ is used to denote a set that is empty, and the statement $A = \varnothing$ means that the set A contains no members. Examples of empty sets are as follows:

$$\{x \,|\, x \text{ is an integer and } 3x = 2\} = \varnothing.$$

$$\{x \,|\, x \text{ is a real number and } x^2 + 1 = 0\} = \varnothing.$$

The set of all insects having eight legs $= \varnothing$.

The set of all magnets having only one pole $= \varnothing$.

An important relation defined on sets is the *subset* relation. We say that set A is a subset of set B if every element of A is also an element of set B. This relation is written as "$A \subseteq B$," i.e., A is a subset of B. For example:

   i. If $\qquad\qquad A = \{2, 4, 6\}, \qquad B = \{1, 2, 3, 4, 5, 6, 7, 8\},$
      then

$$A \subseteq B.$$

ii. If

 N is the set of all natural numbers,

 I is the set of all integers,

 Q is the set of all rational numbers, and

 R is the set of all real numbers,

then

$$N \subseteq I \subseteq Q \subseteq R.$$

iii. The set of all girl students at this university is a subset of the set of all students at this university.

iv. Every set is a subset of itself, i.e.,

$$A \subseteq A \text{ for any set } A.$$

v. The empty set is a subset of every set, that is,

$$\varnothing \subseteq A \text{ for any set } A.$$

vi. The set of all mammals is a subset of the set of all vertebrates, which in turn is a subset of the set of all animals.

EXERCISES 1.2

Use the listing method to describe the following sets:

1. The set of all integers less than 5 and greater than -2.

2. The set of all natural numbers less than 50.

3. The set of all prime numbers less than 20.

4. $\{y \mid y = 1/(h + 2), h \text{ is a natural number}\}$.

5. $\{x \mid x \text{ is a prime factor of } 36\}$.

6. $\{p \mid p = 1/(n - 1), n \text{ is a prime number less than } 20\}$.

Use the rule method to describe the following sets:

7. The set of all even numbers less than 100.

8. The set of all prime numbers less than 30.

9. $\{1, 3, 5, 7, 9, \ldots, 19\}$.

10. $\{\ldots, -4, -2, 0, 2, 4, 6, \ldots\}$.

11. $\{3, 6, 9, \ldots\}$.

12. $\{1, \frac{1}{2}, \frac{1}{3}, \frac{1}{4}, \ldots\}$.

Which of the following statements 13–21 are true and which are false?

13. $2 \in \{1, 2, 3\}$.

14. $3 \subseteq \{1, 2, 3, 4\}$.

15. $\{5\} \subseteq \{1, 2, 3, 5, 7\}$.

16. $4 \in \{1, 2, 5, 7\}$.

17. $\{a, b\} \in \{a, b, c\}$.

18. $\{p, q, r\} \subseteq \{a, b, c, \dots, x, y, z\}$.

19. $\{x \mid x$ is real and $x + 1 = 2\} \subseteq \{1, 2, 3\}$.

20. The set of all rectangles in a plane is a subset of the set of all squares in a plane.

21. The set of all equilateral triangles is a subset of all triangles.

22. If A is the set of all squares in a plane, B is the set of all rectangles in a plane, and C is the set of all quadrilaterals in a plane, which of these sets are subsets of the others?

23. A is the set $\{a, b, c\}$ containing three elements. List all of the subsets of A. How many different subsets does A have? (*Note:* Do not forget to count $\varnothing$ and A.)

*24. Set B has 4 elements. How many different subsets does B have? Repeat for set C, which contains five elements.

*25. Set D has n elements. How many different subsets does D have?

*26. A, B, and C are sets of real numbers defined as follows:

$$A = \{x \mid 2 \leq x \leq 4\}.$$
$$B = \{x \mid x^2 - 6x + 9 < 0\}.$$
$$C = \{x \mid 2x - 9 < 3 - x \leq x - 1\}.$$

Determine which of A, B, and C are subsets of one another.

1.3 FUNCTIONS

The idea of a function is one of the most basic concepts in mathematics. A function expresses the idea of one quantity depending on or being determined by another. For example:

    a. The area of a square depends on the length of its side; if the length of the side of a square is given then its area can be determined.

    b. The volume of a sphere depends on its radius.

    c. The average yield of crops per acre depends on the amount of fertilizer they have received.

    d. The average height of a certain species of plant depends on the age of the plant.

    e. The response of a nerve depends on the magnitude of the stimulus applied to it.

We shall begin by giving a formal definition of a function.

---

*An asterisk is used to denote exercises whose level of difficulty is higher than the others.

DEFINITION   Let X and Y be two non-empty sets. Then a function from X to Y is a *rule* that assigns to each element $x \in X$ a unique $y \in Y$.

If a function assigns y to a particular $x \in X$, then we speak of y as being the *value* of the function at x. A function is generally denoted by a single letter, say f or g, or F or G, etc.

Let f denote a given function. The set X for which f assigns a unique $y \in Y$ is called the *domain* of the function f. It is often denoted by D_f. The corresponding set of values $y \in Y$ is called the *range* of the function and is often denoted by R_f.

EXAMPLES   **(a)**   Let X be the set of all persons currently residing in Canada. Let f be the rule that assigns to each person in the set his or her age (in years). Since every person has a unique age, this rule does define a function. In this case, the domain is the set of all persons residing in Canada and the range is the subset of natural numbers $\{0, 1, 2, \ldots, 105\}$. (The oldest person in Canada is currently 105 years of age.)

(b)   Let X denote the set of all circles in a plane, and let f be the rule that assigns to each circle belonging to X, a unique number equal to the area of the circle. Clearly f is a function

D_f is the set of all circles in the plane, and

R_f is the set of all nonnegative real numbers.

(c)   The growth of a culture of bacteria is a function of time. Here the domain is a certain set of values of time and the range of the function is the set of values of the size of the culture.

If a function f assigns a value y in the range to a certain x in the domain, then we write

$$y = f(x);$$

$f(x)$ is read as "f of x," and is called the value of f at x. Note that $f(x)$ is not the product of f and x.

If a function f is expressed by a relation of the type $y = f(x)$, then x is called the *independent variable* or *argument* of f, and y is called the *dependent variable*.

Generally we shall come across functions that are expressed by stating the value of the function by means of an algebraic formula in terms of the independent variable involved. For example:

$$f(x) = x^2 + 3x + 7,$$

$$g(t) = 2t^2 + \frac{3}{t-2}, \quad \text{etc.}$$

EXAMPLE   Given $f(x) = 3x^2 - 7x + 2$. Find the values of f when $x = a$, $x = -3$, $x = \frac{1}{2}$, and $x = -\frac{2}{3}$, i.e., find $f(a), f(-3), f(\frac{1}{2})$ and $f(-\frac{2}{3})$.

SOLUTION We have $f(x) = 3x^2 - 7x + 2$. To find $f(a)$ we replace x by a in the given equation and get
$$f(a) = 3a^2 - 7a + 2.$$

To evaluate $f(-3)$, we substitute -3 for x on both sides of the given equation:
$$f(-3) = 3(-3)^2 - 7(-3) + 2$$
$$= 27 + 21 + 2 = 50.$$

Similarly,
$$f(\tfrac{1}{2}) = 3(\tfrac{1}{2})^2 - 7(\tfrac{1}{2}) + 2$$
$$= \tfrac{3}{4} - \tfrac{7}{2} + 2 = -\tfrac{3}{4},$$

and
$$f(-\tfrac{2}{3}) = 3(-\tfrac{2}{3})^2 - 7(-\tfrac{2}{3}) + 2$$
$$= \tfrac{4}{3} + \tfrac{14}{3} + 2 = 8.$$

EXAMPLE If $f(x) = 2x^2 - 3x$, evaluate $f(1 + h)$ and $f(x + h) - f(x)$.

SOLUTION To evaluate $f(1 + h)$ we must replace the argument x by $1 + h$:
$$f(x) = 2x^2 - 3x$$
$$f(1 + h) = 2(1 + h)^2 - 3(1 + h)$$
$$= 2(1 + 2h + h^2) - 3(1 + h)$$
$$= 2 + 4h + 2h^2 - 3 - 3h$$
$$= -1 + h + 2h^2.$$

Similarly, to evaluate $f(x + h)$ we use $x + h$ as the argument of f:
$$f(x + h) = 2(x + h)^2 - 3(x + h)$$
$$= 2(x^2 + 2xh + h^2) - 3(x + h)$$
$$= 2x^2 + 4xh + 2h^2 - 3x - 3h.$$

Therefore
$$f(x + h) - f(x) = 2x^2 + 4xh + 2h^2 - 3x - 3h - (2x^2 - 3x)$$
$$= 4xh + 2h^2 - 3h.$$

In most cases the domains and ranges of the function we are concerned with are subsets of the set of real numbers. In such cases the function is commonly represented by its *graph*. The graph of a function f is obtained by plotting all of the points (x, y) where x belongs to the domain of f and $y = f(x)$, treating x and y as Cartesian coordinates.

EXAMPLE Consider $f(x) = x^2$. The domain of f is the set of all real numbers and the range is the set of all nonnegative real numbers. Some of the values of this function are set out in the following table, in which certain values of x are listed in the top

12

row, and the corresponding values of $y = f(x)$ are given beneath each value of x.

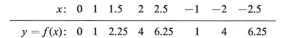

x:	0	1	1.5	2	2.5	-1	-2	-2.5
$y = f(x)$:	0	1	2.25	4	6.25	1	4	6.25

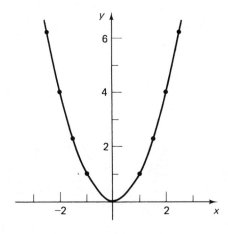

Figure 1.1

The points corresponding to the values of x and y in this table are plotted as heavy dots in Fig. 1.1. The graph of the function $f(x) = x^2$ is shown as the U-shaped curve (parabola), passing through the heavy dots.

Any given curve (or set of points) in the xy-plane is the graph of some function, provided that any vertical line meets the graph in at most one point. For example, the graphs in Fig. 1.2 all represent functions. [Note that in Fig. 1.2(c) the domain of the function is the set of integers $\{1, 2, 3, 4, 5\}$ so that the graph consists of five points rather than a curve.]

On the other hand, the graphs in Fig. 1.3 do not represent functions. The reason for this is that

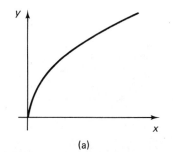

(a)

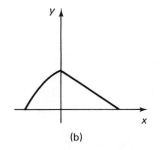

(b)

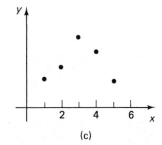

(c)

Figure 1.2

there are vertical lines that meet the graphs in more than one point. Thus, corresponding to the value $x = x_0$ on the first graph there are two values y_1 and y_2 for y. In such a case the value of x does not determine a unique value of y.

On the graph of a function, the values along the x-axis at which the graph is defined constitute the domain of the function. Correspondingly, the values along the y-axis at which the graph has points constitute the range of the function. This is illustrated in Fig. 1.4.

Often the domain of a function is not stated explicitly. In such cases it is understood to be the set of all values of the argument for which the given rule makes sense. For a function f defined by an algebraic expression, the domain of f is the set of all real numbers x for which $f(x)$ is a well-defined real number. For example, the domain

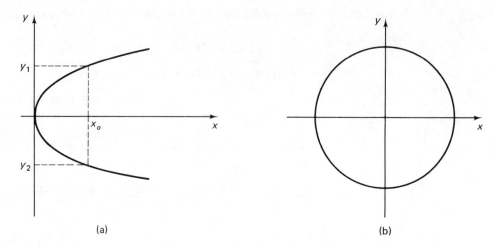

(a)   (b)

Figure 1.3

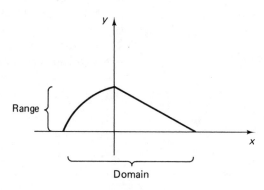

Figure 1.4

of the function $f(x) = \sqrt{x}$ is the set of nonnegative real numbers, since the square root makes sense only for $x \geq 0$. Similarly, in the case of the function $g(x) = x^2/(x - 3)$, the domain is the set of all real numbers except $x = 3$, since when $x = 3$ the denominator becomes zero and $g(3)$ is not defined.

In general, when finding the domain of a function we must bear these two conditions in mind: Any expression under a square-root sign cannot be negative, and the denominator of any fraction cannot be zero.

EXAMPLE   Find the domain and range of g, where

$$g(x) = \frac{x + 3}{x - 2}.$$

SOLUTION   Clearly $g(x)$ is not a well-defined real number for $x = 2$. For any other value of x (except 2), $g(x)$ is a well-defined real number. Thus the domain of f is the set of all real numbers except 2.

To find the range of g, we replace $g(x)$ by y to get

$$y = \frac{x + 3}{x - 2}$$

and solve this equation for x in terms of y. Multiply both sides of this equation by $(x - 2)$:

$$y(x - 2) = x + 3$$
$$xy - 2y = x + 3$$
$$xy - x = 3 + 2y$$
$$x(y - 1) = 3 + 2y$$
$$x = \frac{3 + 2y}{y - 1}.$$

From this equation we see that y can have any value except 1. Thus the range of g is the set of all real numbers except 1.

The graph of g is shown in Fig. 1.5. Observe that for $x = 2$, y is not defined and that there is no value of x for which $y = 1$.

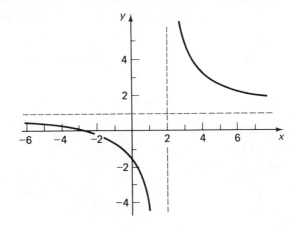

Figure 1.5

EXAMPLE   Find the domain and range of f where

$$f(x) = \sqrt{x - 4}.$$

SOLUTION   The domain of f is the set of all values of x for which the expression under the radical sign is nonnegative, i.e.,

$$x - 4 \geq 0 \quad \text{or} \quad x \geq 4.$$

For $x < 4$, the value of $f(x)$ is not a real number, since the quantity $(x - 4)$ under the square-root sign is negative.

From the given functional equation, $f(x)$ is always nonnegative for each value of x in the domain of f. Thus the range is the set of all nonnegative real numbers. The graph of f is as shown in Fig. 1.6.

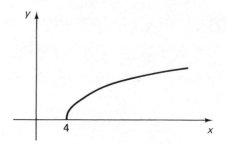

Figure 1.6

EXAMPLE   Find the domain of $f(x) = \sqrt{(x-2)(x-3)}$.

SOLUTION   The domain is the set of x for which the quantity under the square-root sign is greater than or equal to zero:

$$D_f = \{x \mid (x-2)(x-3) \geq 0\}.$$

This condition is met when both factors $(x-2)$ and $(x-3)$ are of the same sign or when one of them is zero. Both $(x-2)$ and $(x-3)$ are positive when $x > 3$, and they are both negative when $x < 2$. Therefore,

$$D_f = \{x \mid x \leq 2 \quad \text{or} \quad x \geq 3\}.$$

The domain consists of the real numbers less than or equal to 2, together with those greater than or equal to 3.

   In the preceding examples we have been concerned with functions that are defined by a single algebraic expression for all values of the independent variable throughout the domain of the function. It sometimes happens that we need to use functions that are defined by more than one expression.

EXAMPLE   Electricity is charged to consumers at the rate of 10¢ per unit for the first 50 units and 3¢ per unit for amounts in excess of this. Find the function $c(x)$ that gives the cost of using x units of electricity.

SOLUTION   For $x \leq 50$, each unit costs 10¢, so the total cost of x units is $10x$ cents. So $c(x) = 10x$ for $x \leq 50$. When $x = 50$, we get $c(50) = 500$; the cost of the first 50 units is equal to 500¢.

   When $x > 50$, the total cost is equal to the cost (500¢) for the first 50 units plus the cost of the rest of the units used. The number of these excess units is $(x - 50)$, and they cost 3¢ each, so their total cost is $3(x - 50)$ cents. Thus the total bill when $x > 50$ comes to

$$c(x) = 500 + 3(x - 50) = 500 + 3x - 150$$
$$= 350 + 3x.$$

We can write $c(x)$ in the form

$$c(x) = \begin{cases} 10x & (x \leq 50) \\ 350 + 3x & (x > 50). \end{cases}$$

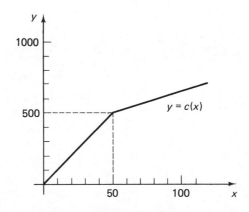

Figure 1.7

The graph of $y = c(x)$ is shown in Fig. 1.7. Note the way in which the nature of the graph changes at $x = 50$, where one formula takes over from the other.

EXAMPLE   Consider

$$f(x) = \begin{cases} 4 - x & (0 \leq x \leq 4) \\ \sqrt{x - 4} & (x > 4). \end{cases}$$

The domain of this function is the set of all nonnegative real numbers. For $0 \leq x \leq 4$ the function is defined by the algebraic expression $f(x) = 4 - x$, while for $x > 4$ it is defined by the expression $f(x) = \sqrt{x - 4}$.

The graph of this function is shown in Fig. 1.8. It consists of two segments. For x between 0 and 4 the graph consists of a straight line segment whose equation is $y = 4 - x$. The graph of $f(x)$ for $x > 4$ is the same as the graph in Fig. 1.6. The range of $f(x)$ in this example is the set $\{y \mid y \geq 0\}$.

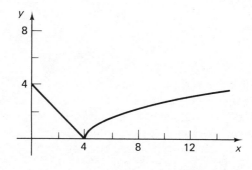

Figure 1.8

In these examples the function under consideration has been defined by two algebraic expressions. It is sometimes necessary to consider functions defined by three or even more different expressions.

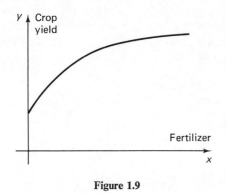

Figure 1.9

There exist many cases in the empirical sciences, and in particular in biology, where a function is defined by empirical data rather than by an algebraic expression such as those expressions that specified the functions in the preceding examples.

For example, let y denote the average yield in bushels per acre of a certain corn crop and let x denote the number of pounds per acre of fertilizer. Certainly y depends on x, and the graph of the function relating y to x would have the form indicated in Fig. 1.9. This function is not defined by an algebraic expression, but could be determined by conducting a suitable experiment.

The following example illustrates this point more fully.

EXAMPLE   The size of any biological population changes with time. In experiments on a colony of fruit flies (*Drosophila*) growing in a certain environment in which the food supply was restricted, Pearl and Parker found the results given in Fig. 1.10. The experimental population counts obtained at intervals of 3 days are indicated by circles on the graph. A smooth curve has been drawn, which approximately passes through all of the experimental points.

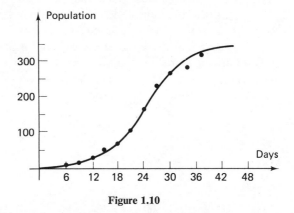

Figure 1.10

In this example, as with most functions determined by experimental data, the complete graph of the function is not known. The experiment determines only a finite number of points on the graph. In such cases we draw a smooth curve, as shown in the figure, which passes close to the points obtained from the experiment. The smooth curve then provides an approximate representation of the experimental results.

Usually a set of experimental measurements will be found not to lie on a very smooth curve. This is clear in the preceding example, where some of the experimental points lie above and others lie below the curve that has been drawn. Such departures from smoothness can, as a rule, be ascribed either to experimental errors, i.e., errors in the measuring apparatus, or to random influences on the experiment that have affected some measurements differently from others. In drawing a smooth curve through the measured points, these errors and random fluctuations are averaged out.

The process by which the smooth curve is found belongs to an area of calculus known as "curve-fitting." We shall discuss some of the more elementary curve-fitting procedures in Chapter 9.

EXERCISES 1.3

1. Given $f(x) = 5x^2 - 7x + 3$, find:
   (a) $f(0)$.   (b) $f(3)$.
   (c) $f(-1)$.   (d) $f(-\frac{1}{2})$.
   (e) $f(a)$.   (f) $f(a + h)$.
   (g) $\dfrac{f(a + h) - f(a)}{h}$.

2. Given $f(x) = \frac{1}{2}x^3 - 3x + 7$, find:
   (a) $f(1)$.   (b) $f(-1)$.
   (c) $f(2)$.   (d) $f(-\frac{2}{3})$.
   (e) $f(c)$.   (f) $f(c + h)$.
   (g) $\dfrac{f(c + h) - f(c)}{h}$.

   Given $f(x) = 3x + 5 \quad \text{if } x \geq 3$
   $\qquad\qquad\quad = 2x - 7 \quad \text{if } x < 3,$

   find:

3. (a) $f(2)$.   (b) $f(5)$.

4. (a) $f(3)$.   (b) $f(-2)$.

   Find the domain and range of the following functions:

5. $f(x) = x^2 + 2$.

6. $g(x) = \dfrac{1}{x - 2}$.

7. $F(x) = \dfrac{x + 1}{x - 1}$.

8. $f(x) = \sqrt{x - 3}$.

9. $g(x) = \dfrac{1}{\sqrt{1 - x}}$.

10. $f(x) = -\sqrt{2 - 3x}$.

11. $f(x) = 1 + \sqrt{1 - x^2}$.   ←

12. $H(x) = -\sqrt{x(1 - x)}$.

13. The size of an insect population at time t (measured in days) is given by:

$$p(t) = 3000 - \frac{2000}{1 + t^2}.$$

Determine the initial population $p(0)$ and the population sizes after 1 and 2 days.

14. The size of a bacteria population at time t (measured in hours) is given by

$$p(t) = 5000 + 3000t - 2000t^2.$$

Find the size of (a) the initial population $p(0)$, and (b) the population after 1 hour (hr). When will the population size be zero?

15. In a test for blood sugar metabolism, conducted over a time interval, the amount of sugar in the blood was a function of time t (measured in hours) and given by:

$$A(t) = 3.9 + 0.2t - 0.1t^2.$$

Find the amount of sugar in the blood (a) at the beginning of the test, (b) 1 hr. after the beginning, and (c) $2\frac{1}{2}$ hr after the beginning.

State whether or not the following graphs represent functions.

16.

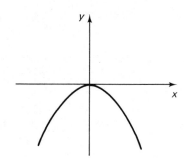

17.

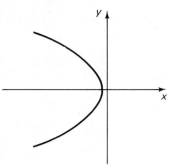

18.

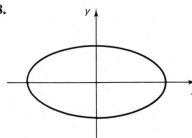

19.

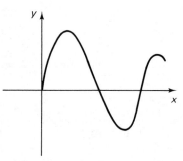

20.

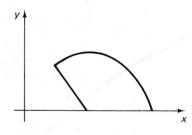

Find the domains of the following functions:

21. $f(x) = \sqrt{(x-1)(2-x)}$.   **22.** $f(x) = \sqrt{x^2 - 5x - 6}$.

23. $f(x) = \sqrt{x^2 - 16}$.   **24.** $f(x) = \sqrt{(x+2)(4-x)}$.

25. Hamburger costs 79¢ per pound for amounts less than 5 pounds (lb) and 69c per pound for amounts greater than 5 lb. $F(x)$ denotes the cost in cents of x pounds of hamburger. Express $F(x)$ by means of suitable algebraic expressions and draw its graph.

26. Sugar cost 50¢/kilogram (kg) for amounts up to 10 kg and 40¢/kg for amounts in excess of 10 kg. If $c(x)$ denotes the cost of x kilograms of sugar, express $c(x)$ by means of algebraic formulas and sketch its graph.

1.4   LINEAR FUNCTIONS

In this section we shall consider functions like $f(x) = 2x$, $f(x) = 3x - 1$, or $f(x) = -x + 4$, whose graphs are shown in Fig. 1.11. All three of these functions have graphs that are straight lines. They are examples of linear functions. In general, a *linear function* is a function whose graph is a straight line.

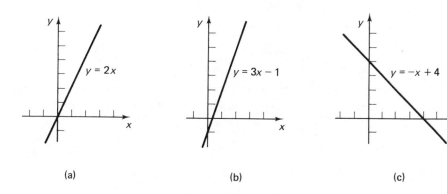

(a)   (b)   (c)

Figure 1.11

EXAMPLE   Hemoglobin (the red coloring matter of the red blood corpuscles) transfers oxygen from the lungs to the other parts of the body. The rate of combination y of hemoglobin and oxygen in the lungs is found to be related to the concentration x of oxygen in the air by the equation

$$y = mx,$$

where m is a constant that depends on the pressure of the oxygen and other factors. The graph of this relation is a straight line. Thus y is a linear function of x.

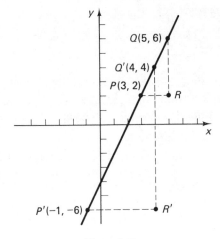

Figure 1.12

One of the most important properties of a straight line is how steeply it rises or falls, and we wish to introduce a quantity that will measure the steepness of a given line. Let us begin by considering an example. The equation $y = 2x - 4$ has as its graph the straight line shown in Fig. 1.12. Let us choose two points on this line, for example, the points $(3, 2)$ and $(5, 6)$, which are denoted respectively by P and Q in the figure. The difference between the x-coordinates of these two points, denoted by PR in the figure, is called the *run* from P to Q:

$$\text{run} = PR = 5 - 3 = 2.$$

The difference between the y-coordinates of P and Q, equal to the distance QR, is called the *rise* from P to Q:

$$\text{rise} = QR = 6 - 2 = 4.$$

We note that the rise is equal to twice the run. This would have turned out to be the case no matter which pair of points we had chosen on the given graph. For example, let us take the two points $P'(-1, -6)$ and $Q'(4, 4)$ (Fig. 1.12). Then

$$\text{run} = P'R' = 4 - (-1) = 5; \quad \text{rise} = Q'R' = 4 - (-6) = 10.$$

Again we see that the ratio rise/run is equal to 2.

The reason why the same ratio of rise to run is obtained in the two cases is that the two triangles PQR and $P'Q'R'$ are similar to one another. Therefore the ratios of corresponding sides are equal: $QR/PR = Q'R'/P'R'$. This ratio is called the *slope* of the given straight line. The line in the preceding figure has a slope equal to 2.

The slope of a general straight line is defined similarly. Let P and Q be any two points on the given line (Fig. 1.13). Let them have coordinates (x_1, y_1) and (x_2, y_2)

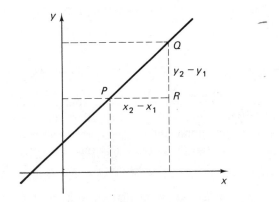

Figure 1.13

respectively. Let R be the intersection of the horizontal line through P and the vertical line through Q. The distance PR is called the *run* between P and Q and the distance QR is called the *rise* between P and Q.

In terms of the coordinates,

$$\text{rise} = QR = y_2 - y_1$$
$$\text{run} = PR = x_2 - x_1.$$

(Note that if Q turns out to lie below R, which happens when the line slopes downwards to the right, the rise is negative.)

The *slope* of the line is defined to be the ratio of rise to run. It is usually denoted by the letter m. Hence

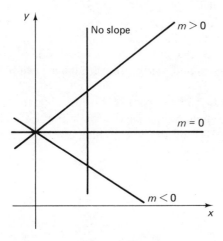

Figure 1.14

$$m = \frac{\text{rise}}{\text{run}} = \frac{y_2 - y_1}{x_2 - x_1} \qquad (1)$$

Note that this expression for slope is a meaningful one as long as $x_2 - x_1 \neq 0$ (that is, provided that the line is nonvertical).

It should be noted that the slope of a line remains the same, no matter how we choose the positions of the two points P and Q on the line.

If the slope m of a line is positive, the line ascends to the right (Fig. 1.14). The larger the value of m, the more steeply the line is inclined to the horizontal. If m is negative, then the line descends to the right. If $m = 0$, then the line is a horizontal one. The slope of a vertical line is not defined.

EXAMPLE Find the slope of the line joining the two points $(1, -3)$ and $(3, 7)$.

SOLUTION Using formula (1), the slope is

$$m = \frac{7 - (-3)}{3 - 1} = \frac{10}{2} = 5.$$

EXAMPLE The slope of the line joining the two points $(3, 2)$ and $(5, 2)$ is

$$m = \frac{2 - 2}{5 - 3} = 0.$$

Thus the line joining these two points is a horizontal one.

EXAMPLE The slope of the line joining $P(2, 3)$ and $Q(2, 6)$ is given by

$$m = \frac{6 - 3}{2 - 2} = \frac{3}{0},$$

which is undefined. Thus the line joining P and Q has *no* slope. In this case the line PQ is a vertical one.

Let us ask the question: What information do we need to be given in order to be able to draw a particular straight line? One way in which a line can be specified is by giving two points that lie on it. Once two points are specified, the whole line is determined, since there is only one straight line through two given points.

Through any *one* point there are of course many different straight lines with slopes ranging from large to small, positive or negative. However, if the slope is given, then there is only one line through the point in question. Thus a second way in which a straight line can be specified is by giving one point on it and the value of its slope.

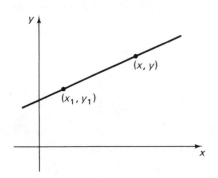

Our immediate task will be to determine the equation of the straight line whose slope is given (equal to m) and that passes through a given point, (x_1, y_1).

Let (x, y) be a point on the line different from the given point (x_1, y_1), as shown in Fig. 1.15. Then the slope m of the line joining the two points (x_1, y_1) and (x, y) is given by

Figure 1.15

$$m = \frac{y - y_1}{x - x_1}.$$

It follows therefore that

$$y - y_1 = m(x - x_1). \tag{2}$$

This is called the *point–slope* formula for the line.

Let us take (x_1, y_1) to be $(0, b)$, as shown in Fig. 1.16. Then equation (2) becomes

$$y - b = m(x - 0)$$

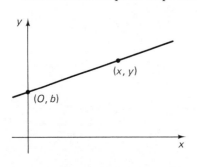

or

$$y = mx + b. \tag{3}$$

The quantity b, which gives the distance along the y-axis, which is cut off by the straight line, is called the *y-intercept* of the line. Equation (3) is called the *slope–intercept* formula.

Figure 1.16

It follows from this result that the graph of any function of the form $f(x) = mx + b$ is a straight line of slope m and y-intercept b. For example, the function $f(x) = 2x - 4$ has a straight-line graph with $m = 2$ and y-intercept $b = -4$.

EXAMPLE   Find the equation of the line through the point $(5, -3)$ whose slope is -2.

SOLUTION   Using (2) with $m = -2$ and $(x_1, y_1) = (5, -3)$, we find that the required equation of the straight line is

$$y - (-3) = -2(x - 5)$$
$$y + 3 = -2x + 10$$
$$y = -2x + 7.$$

EXAMPLE   Given the linear equation $2x + 3y = 6$, find the slope and y-intercept of its graph.

SOLUTION   To find the slope and y-intercept of the line, we must express the given equation in the form

$$y = mx + b$$

(i.e., we must solve the equation for y in terms of x). We have:

$$2x + 3y = 6$$
$$3y = -2x + 6$$
$$y = -\tfrac{2}{3}x + 2.$$

Comparing with the general form $y = mx + b$, we have

$$m = -\tfrac{2}{3} \quad \text{and} \quad b = 2.$$

Thus the slope is equal to $-\tfrac{2}{3}$ and the y-intercept is equal to 2.

EXAMPLE   Find the equation of the straight line passing through the two points $(1, -2)$ and $(5, 6)$.

SOLUTION   The slope of the line joining $(1, -2)$ and $(5, 6)$ is

$$m = \frac{6 - (-2)}{5 - 1} = \frac{8}{4} = 2.$$

Thus from the point–slope formula, the equation of the straight line through $(1, -2)$ with slope $m = 2$ is

$$y - (-2) = 2(x - 1)$$
$$y + 2 = 2x - 2$$
$$y = 2x - 4.$$

EXAMPLE   Suppose that only dried lentils and dried soybeans are available to satisfy a person's daily requirement for protein, which is 75 grams (gm). One gram of lentils contains 0.26 gm of protein and 1 gm of soybeans contains 0.35 gm of

25

protein. Let his daily consumption be x grams of lentils and y gm of soybeans. What is the relationship between x and y that exactly satisfies his protein requirement?

SOLUTION   The amount of protein in grams obtained from x grams of lentils is $0.26x$ and from y grams of soybeans is $0.35y$. The total protein is therefore

$$(0.26x + 0.35y) \text{ gm.}$$

This must be equal to his daily need, which is 75 gm. Thus,

$$0.26x + 0.35y = 75,$$

or multiplying through by 100, we have

$$26x + 35y = 7500.$$

Therefore

$$y = \frac{-26}{35}x + \frac{7500}{35}.$$

We see that x and y satisfy a linear relation.

Linear relations are often used in the analysis of experimental data. The following illustrates a typical case in point.

EXAMPLE   The number of eggs laid by a fish depends upon the size of the fish: For any species, the bigger the fish, the larger the number of eggs. Figure 1.17 shows data for salmon, the number of eggs in thousands being plotted against the length of fish in centimeters. The experimental points are shown as small circles.

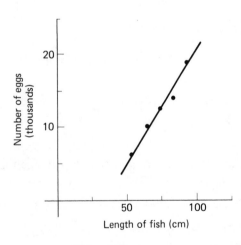

It is clear that the experimental points lie close to a straight line, and the straight line that lies closest to the set of points has been drawn in the figure. Because of this, we would say that the number of eggs is a linear function of the length of the salmon, even though the experimental points do not lie precisely on a straight line.

Figure 1.17

Thus when a biologist uses a linear function to describe a relationship between a dependent and an independent variable, he commonly does not assert that the true relationship is linear, but rather that a linear function is a good approximation to the experimental data over the range of interest.

EXERCISES 1.4

Find the slopes of the lines joining the pairs of points given below:

1. (3, 1) and (5, 7).

2. (6, −2) and (2, −6).

3. (1, 3) and (3, −7).

4. (2, 5) and (3, 5).

5. (−2, 4) and (3, 4).

Find the equation of the straight lines satisfying the conditions in each of the following exercises. Sketch the graph in each case.

6. Passing through (0, 0) with slope 3.

7. Passing through (1, −2) with slope −3.

8. Passing through (3, 4) with zero slope.

9. Passing through (2, −3) and (4, 5).

10. Passing through (2, 1) and (3, 4).

11. With slope −2 and y-intercept 3.

12. With slope 3 and y-intercept −2.

Find the slope and the y-intercept for each of the following linear relations.

13. $3x + 5y = 15$.

14. $2x = 13 - 4y$.

15. $y + 2x + 6 = 0$.

16. $\dfrac{x}{2} + \dfrac{y}{3} = 1$.

17. A patient in the hospital who is on a liquid diet has the choice of prune juice and orange juice to satisy his daily requirement of thiamine, which is 1 milligram (mg). One ounce (oz) of prune juice contains 0.05 mg of thiamine, and 1 oz of orange juice contains 0.08 mg of thiamine. Let his daily consumption be x oz of prune juice and y oz of orange juice. What is the relationship between x and y that exactly satisfies his thiamine requirement?

18. An individual on a strict diet plans to breakfast on cornflakes, milk, and a boiled egg. After allowing for the egg, his diet allows a further 300 calories (cal) for this meal. One ounce of milk contains 20 cal and 1 oz (about one cupful) of cornflakes (plus sugar) contains 160 cal. What is the relation between the number of ounces of milk and of cornflakes that can be consumed?

19. Substance *A* contains 5 mg of niacin per ounce and substance *B* contains·2 mg of niacin per ounce. In what proportions should *A* and *B* be mixed so that the resulting mixture contains 4 mg of niacin per ounce?

20. A crop of potatoes yields an average 16 metric tons of protein per square kilometer of planted area, while corn yields 24 metric tons/km². In what proportions must potatoes and corn be planted in order to yield 21 tons of protein per km² from the combined crop?

1.5 LINEAR INEQUALITIES

The inequality $y > 2x - 4$ connecting the two variables x and y is an example of what are called *linear inequalities*. Let us begin by examining this particular example in terms of a graph.

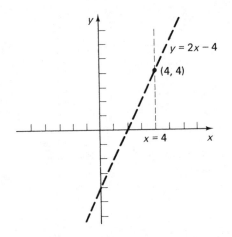

Figure 1.18

The equation $y = 2x - 4$ has as its graph a straight line whose slope is 2 and whose y-intercept is -4. It is shown as a dashed line in Fig. 1.18. As an example, when $x = 4$, $y = 2(4) - 4 = 4$, so the point $(4, 4)$ lies on the line, as shown in the figure.

Now consider the inequality $y > 2x - 4$. When $x = 4$, this takes the form $y > 2(4) - 4$, or $y > 4$. Thus the inequality is satisfied at all of the points $(4, y)$ where $y > 4$. Graphically, this means that on the vertical line $x = 4$, the inequality $y > 2x - 4$ is satisfied at all points that lie *above* the point $(4, 4)$.

Similarly we can take *any* vertical line, say $x = x_1$. Then the inequality $y > 2x - 4$ is satisfied by all of the points (x_1, y) that lie on this vertical line and that lie above the point (x_1, y_1) where the vertical line meets the line $y = 2x - 4$. (That is, $y_1 = 2x_1 - 4$, and the inequality is met provided that $y > y_1$.)

We conclude from this therefore that the inequality $y > 2x - 4$ is satisfied at all of the points (x, y) that lie *above* the straight line $y = 2x - 4$. This region in the xy-plane is said to be the *graph* of the given inequality.

A linear inequality between two variables x and y is a relationship of the form $ax + by + c > 0$ or $ax + by + c \geq 0$. The graph of a linear inequality consists of all those points (x, y) that satisfy the inequality. It consists of a region in the xy-plane, not simply a line or a curve.

The graph of the inequality $ax + by + c > 0$ is a half-plane bounded by the straight line whose equation is $ax + by + c = 0$. Figure 1.19 shows a few graphs which illustrate some linear inequalities.

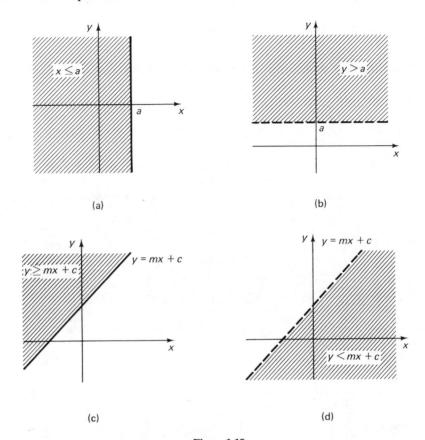

Figure 1.19

In each case the half-plane of points that satisfy the inequality is shaded. The graph of $y > mx + c$ is the half-plane above the line $y = mx + c$, and the graph of $y < mx + c$ is the half-plane below the line $y = mx + c$. If the inequality includes the line, we show the graph by a solid line; otherwise we use a dashed line. A dashed line always corresponds to a strict inequality ($>$ or $<$) and a solid line corresponds to a weak inequality ($\geq$ or $\leq$).

EXAMPLE   Sketch the graph of the linear inequality,

$$2x - 3y < 6.$$

SOLUTION   First we solve the given inequality for y in terms of x (i.e., express it in one of the forms $y > mx + c$ or $y < mx + c$). We have:

$$2x - 3y < 6 \quad \text{or} \quad -3y < -2x + 6.$$

We now divide both sides by -3 (note: when we divide by a negative number the direction of the inequality changes):

$$y > \tfrac{2}{3}x - 2.$$

Next we plot the line $y = \tfrac{2}{3}x - 2$. For $x = 0$, we have $y = -2$. Thus $(0, -2)$ is a point on this line. Again, when $y = 0$, we have $\tfrac{2}{3}x - 2 = 0$ or $x = 3$. Thus $(3, 0)$ is another point on the line. We plot these two points in Fig. 1.20 and join

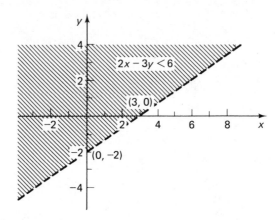

Figure 1.20

them by a dashed straight line (dashed because we are given a strict inequality). Since the given inequality, when solved for y, involves the "greater than" sign, the graph is the half-plane above the dashed line.

The following examples illustrate the application of linear inequalities to some problems of practical interest.

EXAMPLE One oz of whole egg contains 165 mg of cholesterol and 1 oz of liver contains 90 mg of cholesterol. A person who is on a certain diet should consume an average of less than 300 mg of cholesterol per day. Find the relationship between the quantities of egg and liver that can be allowed in the diet, assuming that these are the person's major sources of cholesterol. Sketch the graph of this relation.

SOLUTION Let x oz of egg and y oz of liver be consumed on average per day. The amount of cholesterol contained in the eggs and liver will be

$$(165x + 90y) \text{ mg}$$

and this should be less than 300 mg. Thus,

$$165x + 90y < 300$$
$$90y < -165x + 300$$
$$y < -\tfrac{165}{90}x + \tfrac{300}{90}$$
$$y < -\tfrac{11}{6}x + \tfrac{10}{3}.$$

The graph of this inequality is illustrated in Fig. 1.21. (Note that in this example only the region for which $x \geq 0$ and $y \geq 0$ has any significance, so the shaded region is a triangle rather than a half-plane.)

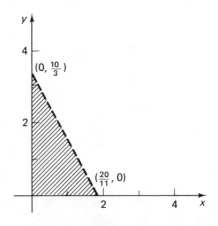

Figure 1.21

In many practical situations, problems arise involving more than one linear inequality. The following example illustrates a case in which two such inequalities occur.

EXAMPLE   A hospital dietician plans to meet a patient's daily requirement of 70 mg of vitamin C by use of a combination of prune and orange juices. It is also required that no more than 300 cal should be contained in the combined juice diet. Prune juice contains 9 mg of vitamin C and 77 cal/oz, and orange juice contains 44 mg of vitamin C and 46 cal per/oz. Find the relationships between the quantities of prune juice and orange juice in the diet necessary to meet the requirements of calorie and vitamin C contents. Illustrate these relationships graphically.

SOLUTION   Let the diet contain x ounces of prune juice and y ounces of orange juice per day. The daily consumption of vitamin C is $(9x + 44y)$ mg. Since this must be at least 70 mg, we arrive at the inequality

$$9x + 44y \geq 70.$$

Secondly, the number of calories in such a diet is equal to $(77x + 46y)$. Since this must not exceed 300, we have

$$77x + 46y \leq 300.$$

In Fig. 1.22, the two straight lines whose equation are $9x + 44y = 70$ and $77x + 46y = 300$ have been drawn. The inequalities above are satisfied by any point (x, y) which lies above the first of the two lines and below the second of them. Thus, in order to satisfy *both* inequalities, the point (x, y) must lie within the triangle which has been shaded.

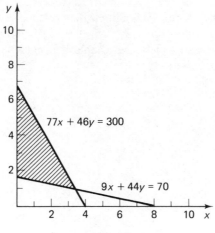

Figure 1.22

EXERCISES 1.5

Sketch the graphs of the following inequalities in the xy-plane.

1. $x + y > 1$.

2. $2x + 3y < 6$.

3. $2x - y \leq 4$.

4. $3x \geq y - 6$.

5. $2x + 3 > 0$.

6. $4 - 3y \leq 0$.

Sketch the graphs of the following sets of inequalities:

7. $x + y > 2$ and $3x + y < 3$.

8. $2x + y > 4$; $x + 2y < 4$ and $2x - 3y < 3$.

9. A mother decides that her child should obtain *at least* 400 cal from his breakfast foods. The child has a choice of eating either of two cereals whose brand names are Cereal-One and Cereal-Two, or a mixture of the two. One oz of Cereal-One contains 100 cal and 1 oz of Cereal-Two contains 120 cal. Find the possible values of x and y if a child consumes x oz of Cereal-One and y oz of Cereal-Two.

10. Miss X has been informed by her doctor that she would be less depressed if she obtained at least the minimum adult requirement of thiamine, which is 1 mg/day. The doctor suggests that she get half of this from breakfast cereal. The cereal A contains 0.12 mg of thiamin per ounce and the cereal B contains 0.08 mg of thiamin per ounce. Determine the possible amounts of these cereals to provide her with at least one-half of the adult daily requirement of thiamine.

11. The storeroom of a chemistry department stocks at least 300 beakers of one size and at least 400 beakers of a second size. It is decided that the total number of beakers stored should not exceed 1200. Determine the possible numbers of the two kinds of beakers that can be stored, and show this by a graph.

12. In the preceding exercise, assume that beakers of the first size occupy 9 sq in. of shelf-space and those of the second size occupy 6 sq in. The total area of shelf-space available for storage is at most 62.5 sq ft. Determine the possible numbers of the two beakers and show this by a graph.

13. A person is considering replacing part of the meat in his diet by soybeans. One oz of meat contains on average about 7 gm of protein while 1 oz of soybeans (un-dried) contains about 3 gm of protein. If he demands that his daily protein intake from meat and soybeans together should be at least 50 gm, what combination of these two would form an acceptable diet?

14. Sirloin steak costs 15¢/oz, and each ounce contains 110 cal and 7 gm of protein. Roast chicken costs 8¢/oz, and each ounce contains 83 cal and 7 gm of protein. Represent graphically the combinations of x oz of steak and y oz of chicken that do not exceed $1.00 in cost and that contain less than 900 cal and at least 60 gm of protein.

15. A fish pool is stocked each spring with two species of fish S and T. The average weight of the fish stocked is 3 lb for S and 2 lb for T. Two foods, F_1 and F_2, are available in the pool. The average daily requirement of a fish of species S is 2 units of F_1 and 3 units of F_2 whereas for species T, it is 3 units of F_1 and 1 unit of F_2. If at most 600 units of F_1 and 300 units of F_2 are available each day, how should the pool be stocked so that the total weight of the fish in the pool is at least 400 lb? Illustrate by graph.

1.6   OTHER SIMPLE FUNCTIONS

We often come across situations where the functional relationship between the dependent and independent variables is *not* linear. As an example from the biological sciences, for *very slow flow* of blood through a vessel, the rate of flow is approximately a linear function of the drop in pressure between two chosen points along the vessel. But even at moderate rates of flow, the flow is a *nonlinear* function of the pressure drop. In this section, we shall discuss a number of simple functions that are of common use and interest.

Quadratic Functions

A function of the form

$$f(x) = ax^2 + bx + c$$

where $a, b,$ and c are given constants ($a \neq 0$) is called a *quadratic function*. The domain of $f(x)$ is the set of all real numbers. Let us consider first the case when b and c are both zero, in which case the function reduces to $f(x) = ax^2$.

The graph of the equation $y = ax^2$ is a parabola whose vertex is at the origin. If $a > 0$, the parabola opens upwards, while if $a < 0$ the parabola opens downwards, as shown in Fig. 1.23. For example, $y = 3x^2$ gives a parabola opening upwards as its

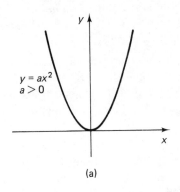

(a)

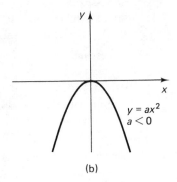

(b)

Figure 1.23

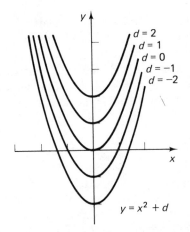

Figure 1.24

graph while $y = -(\frac{1}{2})x^2$ gives a parabola opening downwards.

Next let us consider the equation $y = x^2 + d$ for different values of the constant d. The graphs of this equation for $d = -2, -1, 0, 1, 2$ are shown in Fig. 1.24. All of the graphs are the same shape (parabolas that open upwards) and are the same size, but are shifted relative to one another in the vertical direction. $y = x^2 + 1$ has its vertex at the point $(0, 1)$; $y = x^2 - 2$ has its vertex at $(0, -2)$; and so on. In general $y = x^2 + d$ has its vertex at the point $(0, d)$.

This property extends directly to the equation $y = ax^2 + d$ for any value of a. The graph of $y = ax^2 + d$ is the same as the graph of $y = ax^2$ but is displaced by an amount d in the y-direction.

Thus the graph of $y = ax^2 + d$ is a parabola whose vertex is at $(0, d)$ and which opens upwards if $a > 0$ and downwards if $a < 0$.

EXAMPLES (a) $y = 2x^2 - 3$ is a parabola opening upwards (since $a = 2 > 0$) with its vertex at the point $(0, -3)$ (since $d = -3$). Its graph is shown in Fig. 1.25.

(b) $y = -\frac{1}{2}x^2 + 1$ is a parabola opening downwards ($a = -\frac{1}{2} < 0$) with its vertex at the point $(0, 1)$.

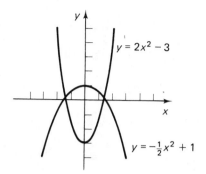

Figure 1.25

Consider now the equation $y = (x - k)^2$, where k is a constant. This is still an example of a quadratic function since, upon expanding the square, we obtain that $y = x^2 - 2kx + k^2$. Comparing with the general quadratic function $y = ax^2 + bx + c$, we see that for this particular case, $a = 1, b = -2k$ and $c = k^2$.

The graph of $y = (x - k)^2$ is exactly the same as the graph of $y = x^2$ except that it is shifted in the direction of the x-axis by an amount k. Thus $y = (x - k)^2$ has a graph that is a parabola opening upwards with its vertex at the point $(k, 0)$. Figure 1.26 shows the graphs of $y = x^2$ and $y = (x - 2)^2$, the latter having its vertex at $(2, 0)$.

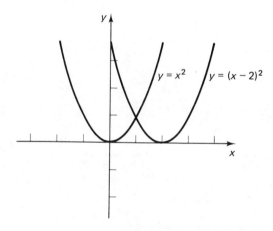

Figure 1.26

In a similar way, the equation $y = a(x - k)^2$ has exactly the same graph as $y = ax^2$ except that it is displaced along the x-axis by an amount k. Therefore $y = a(x - k)^2$ represents a parabola with its vertex at $(k, 0)$ which opens upwards if $a > 0$ and downwards if $a < 0$. These graphs are illustrated in Fig. 1.27.

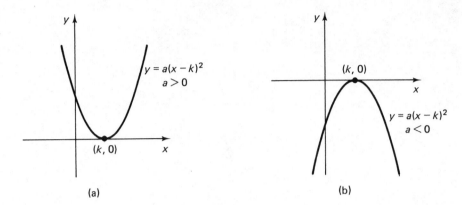

Figure 1.27

We see from these examples that the equations $y = ax^2 + d$ and $y = a(x - k)^2$ both have graphs that are identical in shape and size to the parabola $y = ax^2$. The first has its vertex on the y-axis at $(0, d)$ while the second has its vertex on the x-axis at $(k, 0)$. It seems reasonable to anticipate therefore that if we include both d and k and consider the equation $y = a(x - k)^2 + d$, we shall obtain a graph that is still the same parabola, but whose vertex is now at the point (k, d) in the xy-plane. This is indeed the case, and the graph of this equation is shown in Fig. 1.28 for $a > 0$ and in Fig. 1.29 for $a < 0$.

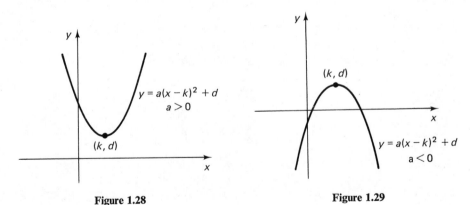

Figure 1.28   **Figure 1.29**

The general quadratic function
$$y = ax^2 + bx + c$$
can always be written in the form $y = a(x - k)^2 + d$ for certain values of k and d. Let us see how this can be done. First of all, we write

$$y = ax^2 + bx + c = a\left(x^2 + \frac{b}{a}x\right) + c. \tag{1}$$

Then we observe the following identity:

$$\left(x + \frac{b}{2a}\right)^2 = x^2 + 2x\left(\frac{b}{2a}\right) + \left(\frac{b}{2a}\right)^2$$

$$= x^2 + \frac{b}{a}x + \frac{b^2}{4a^2},$$

from which it follows that

$$x^2 + \frac{b}{a}x = \left(x + \frac{b}{2a}\right)^2 - \frac{b^2}{4a^2}.$$

Substituting this into Eq. (1) above, we find that the quadratic function takes the form

$$y = a\left(x + \frac{b}{2a}\right)^2 - \frac{b^2}{4a} + c.$$

This is of the form $y = a(x - k)^2 + d$, provided that we identify

$$k = -\frac{b}{2a}, \qquad d = \frac{4ac - b^2}{4a}.$$

We conclude from this that the equation $y = ax^2 + bx + c$ always represents a parabola, opening upwards if $a > 0$ and downwards if $a < 0$. The vertex of the parabola is at the point

$$\left(-\frac{b}{2a}, \frac{4ac - b^2}{4a}\right).$$

EXAMPLE   Find the vertex of the parabola whose equation is

$$y = -2x^2 + 12x.$$

SOLUTION   The answer can be found most directly by substituting values $a = -2, b = 12$ and $c = 0$ into the following formulas:

$$-\frac{b}{2a} = -\frac{12}{2(-2)} = 3,$$

$$\frac{4ac - b^2}{4a} = \frac{4(-2)(0) - 12^2}{4(-2)} = 18.$$

The vertex is therefore at the point $(3, 18)$. However, we could alternatively follow through the steps by which this formula for the vertex was derived:

$$y = -2x^2 + 12x$$
$$= -2(x^2 - 6x)$$
$$= -2(x^2 - 6x + 9 - 9)$$
$$= -2[(x - 3)^2 - 9]$$
$$= -2(x - 3)^2 + 18.$$

This is of the form $y = a(x - k)^2 + d$, with $a = -2, k = 3$ and $d = 18$. The vertex is therefore at $(k, d) = (3, 18)$.

The vertex of a parabola represents the lowest point when $a > 0$ or the highest point when $a < 0$. It follows therefore that for $a > 0$, the function $f(x) = ax^2 + bx + c$ takes its minimum value at the vertex of the corresponding parabola. That is, $f(x)$ is smallest when $x = -b/2a$ and this smallest value of $f(x)$ is equal to $(4ac - b^2)/4a$. Correspondingly when $a < 0$, the function $f(x) = ax^2 + bx + c$ takes its largest value when $x = -b/2a$, and the maximum value of $f(x)$ is $(4ac - b^2)/4a$.

Problems in which we are required to calculate the maximum and minimum values of certain functions arise very frequently in applications. We shall study them at some length in Chapter 4. However some of these problems can be solved by making use of the properties of parabolas. The following example belongs to this category.

EXAMPLE   A farmer has 200 yards (yd) of fencing with which he wishes to enclose a rectangular field. One side of the field can make use of a fence that already exists. What is the maximum area he can enclose?

SOLUTION   Let the sides of the field be denoted by x and y, as shown in Fig. 1.30, the y-side being parallel to the fence that already exists. Then the length of new fence is $2x + y$, which must equal the available 200 yd:

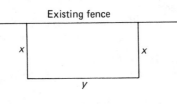

Figure 1.30

$$2x + y = 200.$$

The area enclosed is $A = xy$. But $y = 200 - 2x$; therefore,

$$A = x(200 - 2x) = 200x - 2x^2.$$

Thus we see that A is a quadratic function of x, the coefficients being $a = -2$, $b = 200$, $c = 0$. Therefore, since $a < 0$, the quadratic function has a maximum value at the vertex, that is when

$$x = -\frac{b}{2a} = -\frac{200}{2(-2)} = 50.$$

The maximum value of A is given by

$$\frac{4ac - b^2}{4a} = \frac{4(-2)(0) - 200^2}{4(-2)} = 5{,}000.$$

So the maximum area that can be enclosed is 5,000 sq yd. The dimensions of this largest area are $x = 50$ yd and $y = 100$ yd.

Power Functions

A function of the form

$$f(x) = ax^n,$$

where a and n are nonzero constants, is called a *power function*. We shall consider some special cases of functions of this type.

$n = 2$. In this case $f(x) = ax^2$, and we have a special case of the quadratic functions discussed earlier. The graph of $y = ax^2$ is a parabola with its vertex at the origin, opening upwards if $a > 0$ and downwards if $a < 0$ (Fig. 1.23).

$n = \frac{1}{2}$. In this case, $f(x) = ax^{1/2}$. The graph of this function is one half of a parabola that opens towards the right (Fig. 1.31). If $a > 0$, the graph is the upper half of

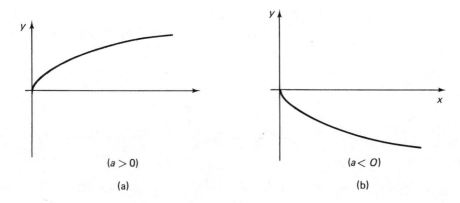

(a > 0)

(a)

(a< 0)

(b)

Figure 1.31

the parabola, while if $a < 0$, it is the lower half. Thus, the graph rises or falls to the right as $a > 0$, or $a < 0$.

The domain of f is the set of all nonnegative real numbers and the range of f is the set of all nonnegative or nonpositive real numbers according as $a > 0$ or $a < 0$.

$n = -1$. In this case, $f(x) = a/x$. The graph of $f(x)$ is a rectangular hyperbola (Fig. 1.32), which has its branches in the first and third quadrants if $a > 0$. If $a < 0$,

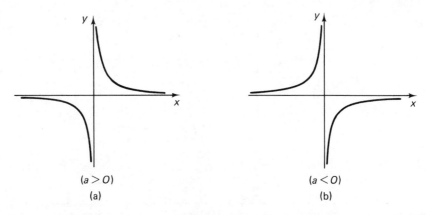

(a > 0)

(a)

(a < 0)

(b)

Figure 1.32

the two branches of the hyperbola are in the second and fourth quadrants. As x increases numerically, $f(x)$ gets closer and closer to zero but is never zero. Similarly, when x moves nearer and nearer to zero, $f(x)$ becomes numerically larger and larger. The two coordinate axes are the asymptotes of the hyperbola.

The domain and range of $f(x)$ are both equal to the set of all real numbers except zero.

$n = 3$. In this case, $f(x) = ax^3$. The graph of $f(x)$ is a cubic parabola as shown in Fig. 1.33. The domain and range of f are both equal to the set of all real numbers.

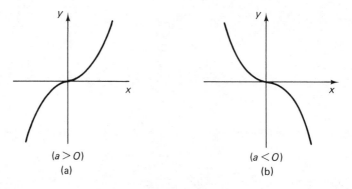

$(a > 0)$

(a)

$(a < 0)$

(b)

Figure 1.33

General n. Figure 1.34 provides a comparison of the graphs of the function $y = ax^n$ for various values of n. The case $a > 0$ is shown, and the graphs are only drawn for the quadrant in which x and y are nonnegative. (Commonly in biological applications we are concerned with variables that take only nonnegative values.)

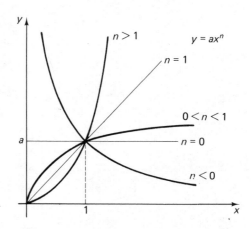

Figure 1.34

We see that all of the graphs pass through the point $(1, a)$. When $n > 1$, the graph has a positive slope, which increases as x increases. The functions $y = ax^2$ and $y = ax^3$, which were encountered previously, are examples that fall into this category. The case $n = 1$ corresponds to the straight line $y = ax$ passing through the origin and the point $(1, a)$.

When $0 < n < 1$, the graph of $y = ax^n$ still has positive slope, but its slope decreases as x increases. An example of this type is the function $y = ax^{1/2}$, whose graph is half of a parabola, as shown in Fig. 1.31.

The case $n = 0$ corresponds to a horizontal straight line. When $n < 0$, the function $y = ax^n$ has a graph whose slope is negative and that is asymptotic to the x- and y-axes. The rectangular hyperbola, whose equation is $y = ax^{-1}$, is an example of such a graph.

Circle

A circle is the set of all points that lie at a *constant distance* (called the radius) from a *given point* (called the center).

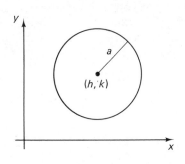

Figure 1.35

Let us find the equation of the circle with its center at the point (h, k) and its radius equal to a (Fig. 1.35). Let (x, y) be any point on the circle. Then the distance between this point (x, y) and the center (h, k) is given by the distance formula to be

$$\sqrt{(x - h)^2 + (y - k)^2}.$$

Setting this equal to the given radius a, we obtain the equation

$$\sqrt{(x - h)^2 + (y - k)^2} = a,$$

which on squaring gives,

$$\boxed{(x - h)^2 + (y - k)^2 = a^2.}$$

This is the standard equation of the circle whose center is (h, k) and radius is a. In particular, if the center is at the origin, $h = k = 0$ and the equation of the circle reduces to

$$\boxed{x^2 + y^2 = a^2.}$$

EXAMPLE   Find the equation of a circle with its center at $(2, -3)$ and its radius equal to 5.

SOLUTION   Here $h = 2$, $k = -3$ and $a = 5$. Thus, using the standard equation of the circle, we have

$$(x - 2)^2 + [y - (-3)]^2 = 5^2$$

or

$$(x - 2)^2 + (y + 3)^2 = 25.$$

On expanding the squares this reduces to

$$x^2 + y^2 - 4x + 6y - 12 = 0.$$

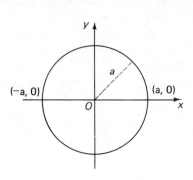

Figure 1.36

The set of points (x, y) that satisfy the relation $x^2 + y^2 = a^2$ consists of the points on the circle whose center is the origin and whose radius is a. We can speak of this circle as being the graph of the relation $x^2 + y^2 = a^2$ (Fig. 1.36). Clearly this circle cannot represent a function because for any value of x lying between $-a$ and $+a$ (except for $x = \pm a$) there are two values of y. We can see this algebraically by solving the equation $x^2 + y^2 = a^2$ for y, in which case we obtain

$$y = \pm\sqrt{a^2 - x^2},$$

showing that there are two values of y corresponding to the choice of the $+$ or $-$ sign.

The complete circle represents two functions (Fig. 1.37). The upper semi-circle is the graph of the function $y = +\sqrt{a^2 - x^2}$, in which the positive square root is taken for y; the lower semi-circle is the graph of the function $y = -\sqrt{a^2 - x^2}$, in which the negative square root is taken.

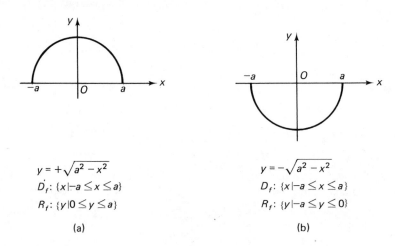

Figure 1.37

Absolute Value Functions

If x is a real number, the absolute value of x, denoted by $|x|$, is defined as

$$|x| = x \quad \text{if } x \geq 0$$
$$= -x \quad \text{if } x \leq 0.$$

For example,

$$|3| = 3, \qquad |0| = 0, \qquad |-2| = -(-2) = 2.$$

Clearly $|x| \geq 0$; thus the absolute value of a real number is always nonnegative. The graph of the absolute value function is shown in Fig. 1.38.

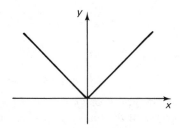

Figure 1.38

EXAMPLE   Consider the function

$$f(x) = |x - 2|.$$

The domain of f is the set of all real numbers and the range is the set of all nonnegative real numbers.

Let us draw the graph of $f(x)$. Setting $y = f(x)$, we have:

$$y = |x - 2|,$$

or, using the above definition of absolute value,

$$y = x - 2 \text{ if } x - 2 \geq 0 \text{ (i.e., if } x \geq 2)$$

and

$$y = -(x - 2) \text{ if } x - 2 \leq 0 \text{ (i.e., if } x \leq 2).$$

Since the range of f is the set of all nonnegative real numbers (that is, $y \geq 0$), the graph of $f(x)$ consists of those portions of the two straight lines

$$y = x - 2 \quad \text{and} \quad y = -(x - 2) = 2 - x$$

for which $y \geq 0$. Thus the graph is as shown in Fig. 1.39.

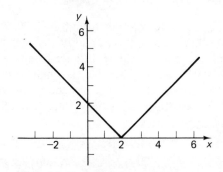

Figure 1.39

EXAMPLE   Consider the function
$$f(x) = |x| - 2.$$

The domain of f is the set of all real numbers and the range of f is the set of all real numbers greater than or equal to -2.

Setting $y = f(x)$, we have
$$y = |x| - 2$$
or
$$y = x - 2 \quad \text{if } x \geq 0$$
and
$$y = -x - 2 \text{ if } x \leq 0.$$

As before, the graph of f consists of portions of two straight lines, in this case those portions of the lines $y = x - 2$ and $y = -x - 2$ for which $y \geq -2$. The graph is shown in Fig. 1.40.

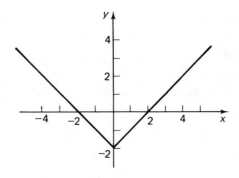

Figure 1.40

EXAMPLE   Let x be the distance in miles along a certain bird migration route. Along the route there are food sources at the start $(x = 0)$, at $x = 400$, and at the end, which is at $x = 1000$. The function $f(x)$ is the distance of the point x from the nearest source of food. Graph $f(x)$. What is the greatest distance of any point on the route from a food source?

SOLUTION   Along the route, x varies from 0 to 1000. The distance of the point x from the food source at $x = 0$ is x, and its distance from the food source at $x = 1000$ is equal to $(1000 - x)$. The distance from the food source at $x = 400$ is equal to $|x - 400|$. The function $f(x)$ is equal to the smallest of these three distances. The graphs of these three functions are shown in Fig. 1.41. The graph of $y = f(x)$ is drawn as a heavy line in the figure. We can see that $f(x)$ is given explicitly as follows:

$$f(x) = \begin{cases} x & \text{if } 0 \leq x \leq 200 \\ |x - 400| & \text{if } 200 \leq x \leq 700 \\ (1000 - x) & \text{if } 700 \leq x \leq 1000. \end{cases}$$

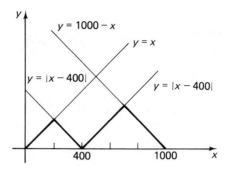

Figure 1.41

The maximum value of $f(x)$ occurs at $x = 700$, at which point $f(x) = 300$; the maximum distance from a food source is 300 miles (mi).

EXAMPLE   Consider the function

$$f(x) = \frac{|x|}{x}.$$

Clearly the function is not defined for $x = 0$, since for this value of x the denominator becomes zero. Thus, the domain of f is the set of all real numbers except zero.

$$\text{If } x > 0, \quad f(x) = \frac{|x|}{x} = \frac{x}{x} = 1,$$

and

$$\text{if } x < 0, \quad f(x) = \frac{|x|}{x} = \frac{-x}{x} = -1.$$

Thus the range consists of only two numbers: 1 and -1.

The graph of f consists of two straight lines (one above and one below the x-axis) that are parallel to the x-axis and at a distance 1 (one) from it. This is shown in Fig. 1.42. Note the use of small circles at the ends of the two lines to indicate that the end points $(0, 1)$ and $(0, -1)$ do not belong to the graph.

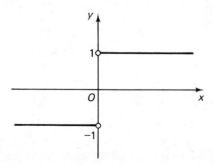

Figure 1.42

EXERCISES 1.6

Find the vertices of the parabolas given by the following equations.

1. $y = 2x^2 + 3x + 1$.

2. $y = 4x - x^2$.

3. $y = 3 - x - 3x^2$.

4. $y = 4x^2 + 16x + 4$.

Find the domains and ranges of the following functions and sketch their graphs:

5. $f(x) = \sqrt{4 - x^2}$.

6. $f(x) = 2 - \sqrt{9 - x^2}$.

7. $g(x) = -\sqrt{3 - x}$.

8. $f(x) = \sqrt{x - 2}$.

9. $f(x) = \dfrac{1}{x}$.

10. $f(x) = \dfrac{-3}{x - 2}$.

11. $f(x) = x^3$.

12. $f(x) = 1 - x^3$.

13. $f(x) = 2 - |x|$.

14. $g(x) = |x| + 3$.

15. $f(x) = |x + 3|$.

16. $F(x) = -|x - 2|$.

17. $f(x) = \dfrac{|x - 3|}{x - 3}$.

18. $G(x) = \dfrac{2 - x}{|x - 2|}$.

19. Which of the following half circles represent the graphs of functions? In each case where the answer is positive, determine the equation for the function from the graph.

(a)

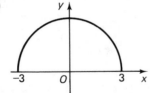

(b)

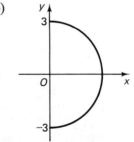

(c)

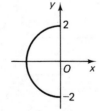

(d)

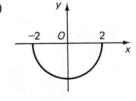

Determine the equation of each of the circles whose radii and centers are given below:

20. Radius 5, center $(0, -2)$.

21. Radius 3, center $(2, 5)$.

22. Radius 2, center $(0, 0)$.

23. Radius 4, center $(-3, 0)$.

24. Find the equation of a circle that lies in the first quadrant, has a radius of 2 units, and touches the coordinate axes.

25. Find the equation of a circle with its center at $(3, -3)$, and that touches the coordinate axes.

26. Find the two numbers x and y that satisfy the condition that $3x - y = 18$ and for which their product xy is as small as possible.

27. A farmer has 400 yd of fencing with which to build a rectangular paddock. What is the largest area he can enclose?

28. The yield of apples from each tree in an orchard is $(500 - 5x)$ lb, where x is the density with which the trees are planted (i.e., number of trees per acre). Find the value of x that makes the total yield per acre a maximum.

29. If rice plants are sown at a density of x plants per square foot, the yield of rice from a certain location is $x(10 - 0.5x)$ bushels per acre. What value of x maximizes the yield per acre?

30. A bird flies a distance of 1000 mi over the ocean, its route passing over two islands, one after 200 mi and the other after 700 mi. If x is the distance along the route to a given point $(0 \leq x \leq 1000)$, determine the function $f(x)$ that is equal to the distance of that point from the nearest land. Sketch its graph.

31. In the previous exercise the function $g(x)$ is equal to the distance of the point x from the nearest land *ahead* of the bird. Write algebraic expressions for $g(x)$.

1.7 MORE ON FUNCTIONS

Let $f(t)$ and $g(t)$ be the sizes of two neighboring populations of the same species as functions of time t. We may under certain circumstances wish to consider the combined population of the two groups. Its size is, of course, equal to $f(t) + g(t)$. From the two functions f and g we have in this way obtained a third function, the sum of f and g.

This kind of example leads us to the following abstract definition. Given two functions f and g, the sum function $f + g$ and the difference function $f - g$ are defined by the equations

$$(f + g)(x) = f(x) + g(x)$$
$$(f - g)(x) = f(x) - g(x).$$

The domains of the sum and difference functions are equal to the common part of the domains of f and g.

We can also define a product function fg and a quotient function f/g in the obvious way:

$$(fg)(x) = f(x)g(x)$$

$$\frac{f}{g}(x) = \frac{f(x)}{g(x)}.$$

The domain of fg is again the common part of the domains of f and g, but for the quotient function we must exclude from this common part any values of x for which $g(x) = 0$ in order to get the domain.

An illustration of the quotient function is provided by the following example. Let $g(t)$ be the size of a certain population at time t, and let $f(t)$ be the total food supply available to the whole population. Then the food supply per individual is equal to the quotient $f(t)/g(t)$ at time t.

EXAMPLE   Let $f(x) = 1/(x - 1)$ and $g(x) = \sqrt{x}$. Find $f + g$, $f - g$, fg and f/g. Determine also their domains.

SOLUTION   We have

$$f(x) = \frac{1}{x - 1}, \qquad g(x) = \sqrt{x}.$$

$$(f + g)(x) = f(x) + g(x) = \frac{1}{x - 1} + \sqrt{x}$$

$$(f - g)(x) = f(x) - g(x) = \frac{1}{x - 1} - \sqrt{x}$$

$$(fg)(x) = f(x)g(x) = \frac{1}{x - 1} \cdot \sqrt{x} = \frac{\sqrt{x}}{x - 1}$$

and

$$\left(\frac{f}{g}\right)(x) = \frac{f(x)}{g(x)} = \frac{1}{x - 1}\bigg/ \sqrt{x} = \frac{1}{\sqrt{x}(x - 1)}.$$

$f(x)$ is not defined for $x = 1$ since for this value of x the denominator is zero, thus the domain of f is the set of all real numbers except 1.

Similarly, $g(x)$ is defined for values of x for which the expression under the radical sign is nonnegative, i.e., $x \geq 0$. Thus

$$D_f : \{x \mid x \neq 1\}$$

$$D_g : \{x \mid x \geq 0\}.$$

The common part of D_f and D_g is

$$\{x \mid x \geq 0 \text{ and } x \neq 1\},$$

and this set provides the domain of $f + g$, $f - g$ and fg.

$g(x) = \sqrt{x}$ is zero when $x = 0$, so this point must be excluded from the domain of f/g. Thus the domain of f/g is

$$\{x \mid x > 0 \quad \text{and} \quad x \neq 1\}.$$

A rather different way in which two functions can be combined in order to yield a third function is called the *composition* of functions. Consider the following situation.

The rate R at which a certain compound is formed in the course of a chemical reaction depends on the temperature T at which the reactants are maintained. In general we can write $R = f(T)$, but let us consider as an example that $R = 2T^3 + 3T$. Now suppose that the temperature is changed as a function of time, so that $T = g(t)$ for a certain function g, but let us again assume a specific form for g, namely $T = 4t - 1$. Then because R is a function of T and because T varies with time t, R also must vary with t. In fact, we can write

$$R = 2T^3 + 3T = 2(4t - 1)^3 + 3(4t - 1),$$

thus expressing R explicitly as a function of t.

Observe the way in which R is obtained as a function of t. R is given initially as a function of T and T is replaced by $g(t)$ [which is $(4t - 1)$ in this example] in the original formula expressing R in terms of T. Thus in general we have $R = f(T)$ and $T = g(t)$, and after replacing T by $g(t)$ we obtain $R = f[g(t)]$, expressing R as a function of t. This leads to the following formal definition.

DEFINITION   Let f and g be two functions. Let x belong to the domain of g and be such that $g(x)$ belongs to the domain of f. The *composite function* $f \circ g$ (read as "f circle g") is defined by

$$(f \circ g)(x) = f[g(x)].$$

The domain of $f \circ g$ is

$$D_{f \circ g} = \{x \mid x \in D_g \quad \text{and} \quad g(x) \in D_f\}.$$

EXAMPLE   Let $f(x) = 1/(x - 2)$ and $g(x) = \sqrt{x}$. Evaluate

$$(f \circ g)(9), \qquad (f \circ g)(4), \qquad (f \circ g)(x)$$
$$(g \circ f)(6), \qquad (g \circ f)(1), \qquad (g \circ f)(x).$$

Determine the domains of $f \circ g$ and $g \circ f$.

SOLUTION   **i.** $g(9) = \sqrt{9} = 3$. Therefore,

$$(f \circ g)(9) = f[g(9)] = f(3) = \frac{1}{3 - 2} = 1.$$

ii. $g(4) = \sqrt{4} = 2$.

$$(f \circ g)(4) = f[g(4)] = f(2) = \frac{1}{2 - 2},$$

which is not defined. The value $x = 4$ does not belong to the domain of $f \circ g$, so that $(f \circ g)(4)$ cannot be found.

iii. $g(x) = \sqrt{x}$.

$$(f \circ g)(x) = f[g(x)] = \frac{1}{g(x) - 2} = \frac{1}{\sqrt{x} - 2}.$$

Clearly $f \circ g$ is defined for nonnegative values of x, except for $x = 4$, since when $x = 4$ the denominator becomes zero. Thus

$$D_{f \circ g} = \{x \mid x \geq 0 \quad \text{and} \quad x \neq 4\}.$$

iv. $f(6) = 1/(6 - 2) = \frac{1}{4}.$

$$(g \circ f)(6) = g[f(6)] = g(\tfrac{1}{4}) = \sqrt{\tfrac{1}{4}} = \tfrac{1}{2}.$$

v. $f(1) = 1/(1 - 2) = -1.$

$$(g \circ f)(1) = g[f(1)] = g(-1) = \sqrt{-1},$$

which is not a real number. We cannot evaluate $(g \circ f)(1)$ as 1 does not belong to the domain of $g \circ f$.

vi. $f(x) = 1/(x - 2).$

$$(g \circ f)(x) = g[f(x)] = \sqrt{f(x)} = \sqrt{\frac{1}{x - 2}} = \frac{1}{\sqrt{x - 2}}.$$

Clearly

$$D_{g \circ f} = \{x \mid x > 2\}.$$

EXAMPLE   Given $f(x) = \sqrt{x}$ and $g(x) = x^2$. Determine $(f \circ g)(x)$ and $(g \circ f)(x)$. Also find their domains.

SOLUTION

$$(f \circ g)(x) = f[g(x)] = f(x^2) = \sqrt{x^2} = |x|.$$

The domain of $f \circ g$ is the set of all real numbers.

$$(g \circ f)(x) = g[f(x)] = g(\sqrt{x}) = (\sqrt{x})^2 = x.$$

Note that in this case, the domain is *not* the set of all real numbers because for negative values of x, $f(x)$ is not defined.

The domain of $g \circ f$ is given by

$$\begin{aligned}
D_{g \circ f} &= \{x \mid x \in D_f \quad \text{and} \quad f(x) \in D_g\} \\
&= \{x \mid x \geq 0 \quad \text{and} \quad f(x) \text{ is a real number}\} \\
&= \{x \mid x \geq 0\}.
\end{aligned}$$

EXAMPLE   The rate R at which a certain chemical compound is formed during a chemical reaction is a function of the temperature T of the reactants. It is observed that, as a function of time t, $R = (2 + t)/(1 + t)$ and $T = 1/(1 + t)^2$. How does R depend on T?

SOLUTION   If $R = f(T)$ and $T = g(t)$, then R is given as a composite function of t: $R = (f \circ g)(t)$. Thus we are given that $(f \circ g)(t) = (2 + t)/(1 + t)$ and $T = g(t) = 1/(1 + t)^2$, and we are required to find the function f. The simplest procedure is to solve the given equation $T = g(t)$ in order to express t as a function of T, and then substitute this into the given expression for R as a function of t. We get

$$(1 + t)^2 = T^{-1},$$

and therefore

$$1 + t = T^{-1/2} \quad \text{or} \quad t = T^{-1/2} - 1.$$

Thus

$$R = \frac{2 + t}{1 + t} = \frac{2 + T^{-1/2} - 1}{T^{-1/2}}$$

$$= (T^{-1/2} + 1)T^{1/2}$$

$$= 1 + T^{1/2}.$$

This is the required result, expressing the rate of reaction as a function of temperature.

EXERCISES 1.7

Find the sum, difference, product, and quotient of the two functions f and g in each of the following examples. Determine the domains of the resulting functions.

1. $f(x) = x^2,$ $\qquad g(x) = \dfrac{1}{x - 1}.$

2. $f(x) = x^2 + 1,$ $\quad g(x) = \sqrt{x}.$

3. $f(x) = \sqrt{x - 1},$ $g(x) = \dfrac{1}{x + 2}.$

4. $f(x) = 1 + \sqrt{x}, g(x) = \dfrac{2x + 1}{x + 2}.$

5. $f(x) = (x + 1)^2,$ $g(x) = \dfrac{1}{x^2 - 1}.$

Given $f(x) = x^2$ and $g(x) = \sqrt{x - 1}$, evaluate

6. $(f \circ g)(5).$ $\qquad\qquad$ **7.** $(g \circ f)(3).$

8. $(f \circ g)(\frac{5}{4}).$ $\qquad\qquad$ **9.** $(g \circ f)(-2).$

10. $(f \circ g)(\frac{1}{2}).$ $\qquad\qquad$ **11.** $(g \circ f)(\frac{1}{3}).$

12. $(f \circ g)(2).$ $\qquad\qquad$ **13.** $(g \circ f)(1).$

Determine $(f \circ g)(x)$ and $(g \circ f)(x)$ in the following examples. In each case determine the domains of $f \circ g$ and $g \circ f$.

14. $f(x) = x^2,$ $\qquad g(x) = 1 + x.$

15. $f(x) = \sqrt{x} + 1, g(x) = x^2.$

16. $f(x) = \dfrac{1}{x + 1},$ $\quad g(x) = \sqrt{x} + 1.$

17. $f(x) = 2 + \sqrt{x}, g(x) = (x - 2)^2.$

Determine $g(x)$ if

18. $f(x) = x^2$ and $(f \circ g)(x) = (1 + x)^2$.

19. $f(x) = \sqrt{x - 1}$ and $(f \circ g)(x) = x^2$.

20. $f(x) = \dfrac{1}{x + 2}$ and $(g \circ f)(x) = x + 2$.

21. $f(x) = \dfrac{x + 1}{x - 2}$ and $(g \circ f)(x) = \dfrac{3}{x - 2}$.

22. $f(x) = \sqrt{x}$ and $(f \circ g)(x) = |x|$.

Determine $f(x)$ and $g(x)$ if the composite function $f \circ g$ is as follows: (The answer is not unique.)

23. $(f \circ g)(x) = (x^2 + 1)^3$.

24. $(f \circ g)(x) = \sqrt{2x + 3}$.

1.8   LIMITS AS $x \to \infty$

Often in the application of mathematics to the biological sciences we are concerned with using mathematical equations to describe processes that evolve in time. In such cases, the independent variable x is identified with "time" (usually t is used in place of x). The state of the system at any instant of time is described by a certain function of x (or in more complex systems, it may be necessary to use several functions of x in order to specify the state of the system).

For example, we may be concerned with an experiment to inves:igate the variations in the amount of sugar in an individual's blood, or with an experiment to measure the growth of a population of microorganisms. In the first case we would use the blood sugar level as a function of time in order to describe the system; in the second we would use the population size (either the number of organisms or the total weight of the population) as a function of time.

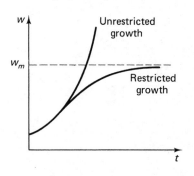

Figure 1.43

When we are investigating systems that evolve in time, we are often interested in the behavior of the system for large values of t. While the behavior of the system may in general be very complex, it often happens that after a while it settles down into a much simpler pattern. This so-called *asymptotic behavior* of a system is in many cases very important.

As an illustration, consider a culture of bacteria growing in a laboratory. The size of the culture (measured, say, by its weight w) is a function of t, as shown in Fig. 1.43. The growth of the culture may be unrestricted, that is, the weight may continue to increase indefinitely with time. However if the rate of supply of food to the culture is limited, the growth of the culture will slow down. For large values of t, w approaches a certain value w_m, which is the maximum population weight that can be sustained by the given food supply. In this case, the asymptotic behavior of the system is that w approaches the constant value w_m. We call w_m the *limiting value* of w as t approaches infinity, and we write

$$w \longrightarrow w_m \text{ as } t \longrightarrow \infty \quad \text{or} \quad \lim_{t \to \infty} w = w_m.$$

(The notation $x \longrightarrow y$ is read as "x approaches y" or "x gets closer and closer to y.")

With this type of example in mind, we shall now study limits as $x \longrightarrow \infty$ in a more abstract way.

Consider the function f defined by

$$f(x) = \frac{2x + 1}{x}.$$

Let us determine the behavior of f as x gets larger and larger without bound, that is, as x approaches infinity. Let x take the values 1, 10, 100, 1000, 10,000, ..., etc. The corresponding values of $f(x)$ are given in the following table:

x:	1	10	100	1000	10,000	...
$f(x)$:	3	2.1	2.01	2.001	2.0001	...

We observe from the above table that as x gets larger and larger without bound, $f(x)$ gets closer and closer to 2. In other words, the difference between $f(x)$ and 2 can be made as small as we please by choosing x sufficiently large. In terms of limits we say that $f(x) \longrightarrow 2$ as $x \longrightarrow \infty$ and write

$$\lim_{x \to \infty} f(x) = 2.$$

The graph of $y = (2x + 1)/x$ is shown in Fig. 1.44 for $x > 0$.

Consider another example: $f(x) = 1/x$. The behavior of $f(x)$ as x gets larger and larger is shown in the following table:

x:	1	10	100	1000	10,000	100,000	...	$\longrightarrow \infty$
$f(x)$:	1	0.1	0.01	0.001	0.0001	0.00001	...	$\longrightarrow 0$

It is clear from the above table that as x becomes larger and larger, $f(x)$ gets closer and closer to zero; that is, the difference between $f(x)$ and zero can be made as small as we please by taking x sufficiently large (Fig. 1.45). Thus

$$\lim_{x \to \infty} f(x) = 0 \quad \text{or} \quad \lim_{x \to \infty} \frac{1}{x} = 0.$$

Now we give a formal definition.

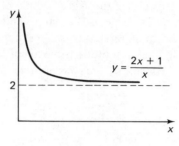

Figure 1.44   **Figure 1.45**

DEFINITION   Let $f(x)$ be defined for all sufficiently large real numbers. Then $f(x)$ is said to approach the *limiting value L* as x approaches infinity, provided the difference between $f(x)$ and L can be made as small as we please by taking any value of x that is sufficiently large. We write this

$$\lim_{x \to \infty} f(x) = L.$$

If $f(x)$ also gets larger and larger with the increase of x and becomes unbounded, then as a matter of convention, we write

$$\lim_{x \to \infty} f(x) = \infty.$$

We know from the example above that

$$\lim_{x \to \infty} \frac{1}{x} = 0.$$

In fact, we have a more general result, which is that

$$\lim_{x \to \infty} \frac{1}{x^n} = 0 \quad \text{for all} \quad n > 0.$$

For example,

$$\lim_{x \to \infty} \frac{1}{x^2} = 0, \quad \lim_{x \to \infty} \frac{1}{\sqrt{x}} = 0, \quad \text{etc.}$$

We shall use this standard result in determining the limits of rational algebraic functions as the independent variable approaches infinity. The method is to divide the numerator and denominator of any fraction that occurs in the function by the highest power of the variable occurring in the denominator, and then to use the above standard results to obtain the limit. This is illustrated in the following examples.

EXAMPLE   Evaluate

$$\lim_{x \to \infty} \frac{2x^2 - 3x + 7}{3x^2 + 5x - 1}.$$

SOLUTION   We divide the numerator and denominator by x^2 (the highest power of x in the denominator) to get

$$\lim_{x \to \infty} \frac{2x^2 - 3x + 7}{3x^2 + 5x - 1} = \lim_{x \to \infty} \frac{\dfrac{2x^2 - 3x + 7}{x^2}}{\dfrac{3x^2 + 5x - 1}{x^2}}$$

$$= \lim_{x \to \infty} \frac{2 - \dfrac{3}{x} + \dfrac{7}{x^2}}{3 + \dfrac{5}{x} - \dfrac{1}{x^2}}.$$

As x gets larger and larger, the terms $-3/x$ and $7/x^2$ in the numerator get smaller and smaller, and the numerator itself approaches the value 2. Similarly, the denominator approaches the limiting value 3 as $x \to \infty$ since the terms $5/x$ and $-1/x^2$ approach zero. Thus the given expression approaches a limiting value of $\frac{2}{3}$.

EXAMPLE   Evaluate

$$\lim_{x \to \infty} \frac{5x^2 + 3}{7x^3 - 1}.$$

SOLUTION   Divide the numerator and denominator by x^3 (the highest power of x in the denominator) to get:

$$\lim_{x \to \infty} \frac{5x^2 + 3}{7x^3 - 1} = \lim_{x \to \infty} \frac{\dfrac{5x^2 + 3}{x^3}}{\dfrac{7x^3 - 1}{x^3}}$$

$$= \lim_{x \to \infty} \frac{\dfrac{5}{x} + \dfrac{3}{x^3}}{7 - \dfrac{1}{x^3}}.$$

As $x \to \infty$, the numerator in this final fraction approaches zero since both of the terms $5/x$ and $3/x^3$ become smaller and smaller. The denominator, on the other hand, approaches the finite limiting value 7. So the given function approaches the limiting value $\frac{0}{7}$, or 0.

EXAMPLE   Evaluate

$$\lim_{x \to \infty} \frac{2x^3 - 3x + 7}{5x^2 + x - 1}.$$

SOLUTION   Divide the numerator and denominator by x^2 (the highest power of x in the denominator) to get

$$\lim_{x \to \infty} \frac{2x^3 - 3x + 7}{5x^2 + x - 1} = \lim_{x \to \infty} \frac{2x - \dfrac{3}{x} + \dfrac{7}{x^2}}{5 + \dfrac{1}{x} - \dfrac{1}{x^2}}.$$

In this second fraction, the denominator approaches the finite limiting value 5. In the numerator the last two terms, $-3/x$ and $7/x^2$, approach zero, but the first term, $2x$, gets larger and larger as $x \longrightarrow \infty$. Thus the fraction as a whole gets larger and larger and we write

$$\lim_{x \to \infty} \frac{2x^3 - 3x + 7}{5x^2 + x - 1} = \infty.$$

EXAMPLE   During a chemical reaction, the amount of a certain compound formed at time x is given by

$$C(x) = \frac{3x + 7}{\sqrt{4x^2 + 5}}.$$

What is the limiting value of $C(x)$ over very large periods of time (i.e., as $x \longrightarrow \infty$)?

SOLUTION   The highest power of x in the denominator is x^2, which occurs in the radical sign. Thus we divide the numerator and denominator by $\sqrt{x^2} = x$ to get

$$\lim_{x \to \infty} \frac{3x + 7}{\sqrt{4x^2 + 5}} = \lim_{x \to \infty} \frac{\dfrac{3x + 7}{x}}{\dfrac{\sqrt{4x^2 + 5}}{x}}$$

$$= \lim_{x \to \infty} \frac{3 + \dfrac{7}{x}}{\sqrt{\dfrac{4x^2 + 5}{x^2}}}$$

$$= \lim_{x \to \infty} \frac{3 + \dfrac{7}{x}}{\sqrt{4 + \dfrac{5}{x^2}}}$$

$$= \frac{3 + 0}{\sqrt{4 + 0}} = \frac{3}{2}.$$

EXAMPLE   The weight of a culture of bacteria as a function of time t is given by

$$w(t) = \frac{\sqrt{t^3 - 3t^2}}{\sqrt[4]{t^6 + 2t}}.$$

Find $\lim_{t \to \infty} w(t)$.

SOLUTION   The highest power of t in the denominator is $t^{3/2}$ (i.e., t^6 under the fourth root sign). So dividing top and bottom by this power of t, we get

$$w(t) = \frac{\sqrt{\dfrac{t^3 - 3t^2}{t^3}}}{\sqrt[4]{\dfrac{t^6 + 2t}{t^6}}}$$

$$= \frac{\sqrt{1 - \dfrac{3}{t}}}{\sqrt[4]{1 + \dfrac{2}{t^5}}}.$$

As $t \to \infty$, the terms $-3/t$ and $2/t^5$ approach zero, and we get

$$\lim_{t \to \infty} w(t) = \frac{\sqrt{1-0}}{\sqrt[4]{1+0}} = 1.$$

Limits of functions as $x \to -\infty$ can be evaluated in a similar way: by dividing the numerator and denominator by the highest power of x that occurs in the denominator. This is quite straightforward, but a little care is needed when the function involves radicals. The type of consideration that can arise is illustrated by the following example.

EXAMPLE   Evaluate

$$\lim_{x \to -\infty} \frac{3x + 7}{\sqrt{4x^2 + 5}}.$$

SOLUTION   We must divide the numerator and denominator by $\sqrt{x^2}$, the highest power in the denominator. However when x is negative, $\sqrt{x^2}$ is not equal to x but rather to $-x$. Therefore,

$$\lim_{x \to -\infty} \frac{3x + 7}{\sqrt{4x^2 + 5}} = \lim_{x \to -\infty} \frac{\dfrac{3x + 7}{-x}}{\sqrt{\dfrac{4x^2 + 5}{x^2}}}$$

$$= \lim_{x \to -\infty} \frac{-3 - \dfrac{7}{x}}{\sqrt{4 + \dfrac{5}{x^2}}}$$

$$= \frac{-3 - 0}{\sqrt{4 + 0}} = -\frac{3}{2}.$$

EXERCISES 1.8

Evaluate the following limits:

1. $\displaystyle\lim_{x \to \infty} \frac{3x^2 - 5x + 1}{x^2 + x + 2}.$

2. $\displaystyle\lim_{x \to \infty} \frac{2x + 5}{3x - 1}.$

3. $\displaystyle\lim_{x \to \infty} \frac{x^2 + 1}{2x^3 - x}.$

4. $\displaystyle\lim_{x \to -\infty} \frac{5x + 1}{x^2 + 1}.$

5. $\displaystyle\lim_{x \to \infty} \frac{x^2 - 1}{x - 1}.$

6. $\displaystyle\lim_{x \to \infty} \frac{x^3 - 8}{x^2 - 4}.$

7. $\displaystyle\lim_{x \to -\infty} \frac{x^2 + 4}{2x^2 + 5}.$

8. $\displaystyle\lim_{x \to -\infty} \frac{x^3 + 7x}{5x^2 + 1}.$

9. $\displaystyle\lim_{x \to \infty} \frac{2x + 1}{\sqrt{x^2 - 1}}.$

10. $\displaystyle\lim_{x \to \infty} \frac{3x^2 + 1}{\sqrt{9x^4 + 7}}.$

11. $\lim\limits_{x\to\infty} \dfrac{2x+3}{\sqrt{4x^3-1}}$.

12. $\lim\limits_{x\to-\infty} \dfrac{2x+3}{\sqrt{9x^2-1}}$.

13. $\lim\limits_{x\to-\infty} \dfrac{5x^2+1}{\sqrt{4x^2-1}}$.

14. $\lim\limits_{x\to-\infty} \dfrac{|x-2|}{x-2}$.

15. $\lim\limits_{x\to\infty} \dfrac{|x|+|x-2|}{x}$.

REVIEW EXERCISES FOR CHAPTER 1

1. Are the following statements true or false? If false, give the corresponding correct statements.
   (a) Every real number is a rational number.
   (b) Every natural number is a real number.
   (c) All integers are rational numbers.
   (d) A given curve is the graph of a function if any vertical line meets the curve in at least one point.
   (e) If f and g are two functions such that the composite functions $f \circ g$ and $g \circ f$ are defined, then $f \circ g = g \circ f$.
   (f) Every equation in x and y expresses y as a function of x.
   (g) The graph of a linear equation $ax + by + c = 0$ is a straight line for all values of the constants a, b, and c.
   (h) The graph of $ax + by + c > 0$ $(b \neq 0)$ is the half-plane above the line $ax + by + c = 0$.
   (i) The domain of $f(x) = |x - 2|$ is the set of all real numbers greater than or equal to 2.
   (j) $(x^2 - 9)/(x - 3) = x + 3$ for all real values of x.
   (k) If f, g are two functions, then $f + g, f \cdot g$ and f/g have the same domain.
   (l) $-2y > 4x - 6$ is equivalent to $y > -2x + 3$.
   (m) $\sqrt{x^2} = |x|$ for all real numbers x.

2. For each of the following, give an example of a function f that satisfies the following property for all values of x and y.
   (a) $f(x) = f(-x)$ [such a function is called an even function].
   (b) $f(-x) = -f(x)$ [such a function is called an odd function].
   (c) $f(x + y) = f(x) + f(y)$ [a function with this property is called a linear function].

3. If $f(x) = |x|$ and $g(x) = x^2$, determine $f \circ g$ and $g \circ f$ and their domains. Is $f \circ g = g \circ f$?

4. Two functions f and g are said to be equal if $f(x) = g(x)$ for all x in the domain and $D_f = D_g$. Use this criterion to determine which of the following functions are equal to $f(x) = (2x^2 + x)/x$.

(a) $g(x) = 2x + 1.$

(b) $h(x) = \sqrt{1 + 4x + 4x^2}.$

(c) $F(x) = \dfrac{2x^3 + x^2}{x^2}.$

(d) $G(x) = \dfrac{(x^3 + 2x)(1 + 2x)}{x(x^2 + 2)}.$

5. Determine the equation of a straight line
   (a) whose slope is 3 and y-intercept is 4 units.
   (b) whose slope is 3 and passes through the point $(7, 5)$.

6. Find the equation of the straight line passing through the points $(3, 9)$ and $(4, 8)$; show that the point $(5, 7)$ lies on this line.

7. Determine the equation of the circle whose center is $(-1, 2)$ and whose radius is 4.

8. Determine the equation of the circle whose center is $(-2, 4)$ and that passes through the point $(1, 0)$.

9. Find the vertex of the parabola $y = 3x^2 + 2x + 1$.

10. Find the vertex of the parabola $y = 4 - 2x - 2x^2$ and sketch its graph.

11. Draw the graph of the region in the xy-plane that satisfies the inequalities
$$x + y < 5, \qquad 2x + y \geq 6, \qquad x - y \leq 2.$$

Evaluate the limits

12. $\displaystyle\lim_{x \to \infty} \frac{(x + 1)(2x + 3)}{(x + 2)(3x + 4)}.$

13. $\displaystyle\lim_{x \to \infty} \sqrt{\frac{2 + 3x}{6x - 1}}.$

14. If $f(x) = x/(1 + x)$ and $g(x) = 1/(x - 1)$, evaluate $f \circ g(x)$ and find the domain of this function.

15. In exercise 14, evaluate $g \circ f(x)$ and find the domain of this function.

2

The
Derivative

2.1 INCREMENTS AND RATES

Differential calculus is the study of the changes that occur in one quantity when other quantities on which it depends change. Some examples drawn from the biological sciences are:

a. the change in crop yield that occurs with each additional pound of fertilizer used;
b. the change in the blood pressure of a patient produced by each additional milligram of a certain drug administered to him;
c. the change in the growth of a culture of bacteria with each additional hour.

DEFINITION Let a variable x have a first value x_1 and then a second value x_2. The change in the value of x, which is $(x_2 - x_1)$, is called the *increment* in x and is denoted by Δx.

Δ (delta) is a Greek letter that is used to denote a change or increment in any variable:

Δx denotes the change in the variable x;
Δt denotes the change in the variable t;
Δu denotes the change in the variable u; and so on.

Let $y = f(x)$ be a variable dependent on x. When x has the value x_1, y has the value $y_1 = f(x_1)$. Similarly, when $x = x_2$, y has the value $y_2 = f(x_2)$. The increment in y is then

$$\Delta y = y_2 - y_1 = f(x_2) - f(x_1).$$

Solving $\Delta x = x_2 - x_1$ for x_2, we have $x_2 = x_1 + \Delta x$. Using this value of x_2 in Δy, we get:

$$\Delta y = f(x_1 + \Delta x) - f(x_1).$$

Since x_1 can be any arbitrary value of x, we can drop the subscript and write

$$\boxed{\Delta y = f(x + \Delta x) - f(x).}$$

Alternatively, since $f(x) = y$, we can write

$$\boxed{y + \Delta y = f(x + \Delta x).}$$

Let P be the point (x_1, y_1) and Q be the point (x_2, y_2), both of which lie on the graph of the function $y = f(x)$ (Fig. 2.1). Then the increment Δx is equal to the horizontal distance between P and Q while Δy is equal to the vertical distance from P to Q. In other words, Δx is the *run* and Δy is the *rise* from P to Q.

In the case illustrated in Fig. 2.1, both Δx and Δy are positive. It is possible for either or both Δx and Δy to be negative, and Δy can also be zero. A typical example of a case when $\Delta x > 0$ and $\Delta y < 0$ is illustrated in Fig. 2.2.

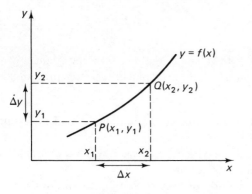

Figure 2.1   **Figure 2.2**

In some of the applications later on, we shall want to think of the increment Δx as being small; that is, we shall want to consider only small changes in the independent variable. It is in fact often understood that Δx means a small increment in x rather than just any increment. As far as this present section is concerned, however, no restriction is placed on the size of increments considered; they can be as small or as large as we like.

EXAMPLE   Given $f(x) = x^2$, find Δy if $x = 1$ and $\Delta x = 0.2$.

SOLUTION   Substituting the values of x and Δx in the formula for Δy, we have

$$\begin{aligned}
\Delta y &= f(x + \Delta x) - f(x) \\
&= f(1 + 0.2) - f(1) \\
&= f(1.2) - f(1) \\
&= (1.2)^2 - (1)^2 \\
&= 1.44 - 1 = 0.44.
\end{aligned}$$

Thus a change of 0.2 in the value of x results in a change in y of 0.44. This is illustrated graphically in Fig. 2.3.

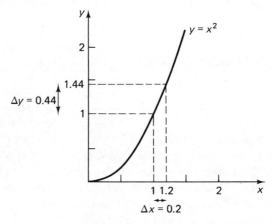

Figure 2.3

EXAMPLE   For the function $y = x^2$, find Δy when $x = 1$ for any increment Δx.

SOLUTION
$$\begin{aligned}
\Delta y &= f(x + \Delta x) - f(x) \\
&= f(1 + \Delta x) - f(1) \\
&= (1 + \Delta x)^2 - (1)^2 \\
&= 1 + 2\,\Delta x + \Delta x^2 - 1 \\
&= 2\,\Delta x + \Delta x^2.
\end{aligned}$$

Since this expression for Δy holds for all increments Δx, we can recover the result of the preceding example by substituting $\Delta x = 0.2$. We get
$$\Delta y = 2(0.2) + (0.2)^2 = 0.4 + 0.04 = 0.44,$$
as before.

EXAMPLE   For the function $y = x^2$, find Δy for general values of x and Δx.

SOLUTION
$$\begin{aligned}
\Delta y &= f(x + \Delta x) - f(x) \\
&= (x + \Delta x)^2 - x^2 \\
&= x^2 + 2x\,\Delta x + \Delta x^2 - x^2 \\
&= 2x\,\Delta x + \Delta x^2.
\end{aligned}$$

Again it is clear that we recover the result of the preceding example by substituting $x = 1$. The expression in this last example, however, provides the increment in y for any values of x and Δx.

EXAMPLE   The size of an insect population at time t (measured in days) is given by
$$f(t) = 5000 - \frac{3000}{1 + t}.$$

Determine the change in the population when $t = 2$ and $\Delta t = 3$ (i.e., the change in the population between the second and the fifth day).

SOLUTION   Here x is replaced by t, so that the formula for Δy becomes
$$\Delta y = f(t + \Delta t) - f(t).$$
Substituting the given values of t, and Δt, we have:
$$\begin{aligned}
\Delta y &= f(2 + 3) - f(2) \\
&= f(5) - f(2) \\
&= 5000 - \frac{3000}{1 + 5} - \left(5000 - \frac{3000}{1 + 2}\right) \\
&= (5000 - 500) - (5000 - 1000) \\
&= 4500 - 4000 = 500.
\end{aligned}$$

Thus the population increases by 500 during the 3 days that follow the second day.

The changes in the dependent variable when stated in absolute terms as in the above examples are less informative than they would be if stated in relative terms. For example, absolute statements such as "the temperature dropped by 8°C" or "the population increased by 800" are less informative than relative statements such as "the temperature dropped by 8°C in the last 4 hr" or "the population increased by 800 in the last 5 days." From these last statements not only do we know by how much the variable (temperature or population) changed but we can also calculate the average *rate* at which it is changing during the given period of time. In these two examples, the average drop in temperature during the last 4 hr is $\frac{8}{4} = 2°C/hr$, and the average increase in population during the last 5 days is $\frac{800}{5} = 160$ per day.

DEFINITION   The *average rate of change* of a function f over an interval x to $x + \Delta x$ is defined by the ratio $\Delta y/\Delta x$. Thus average rate of change of y with respect to x is described in the following equation:

$$\frac{\Delta y}{\Delta x} = \frac{f(x + \Delta x) - f(x)}{\Delta x}.$$

NOTE.   It is necessary that the whole interval from x to $x + \Delta x$ belong to the domain of f.

Graphically if P is the point $[x, f(x)]$ and Q the point $[x + \Delta x, f(x + \Delta x)]$ on the graph of $y = f(x)$, then $\Delta y = f(x + \Delta x) - f(x)$ is the rise and Δx the run from P to Q (Fig. 2.4). From the definition of slope, we can say that $\Delta y/\Delta x$ is the slope of the

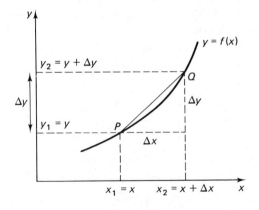

Figure 2.4

straight line segment PQ. Thus the average rate of change of y with respect to x is equal to the slope of the chord PQ joining the two points P and Q on the graph of $y = f(x)$. These points correspond to the values x and $x + \Delta x$ of the independent variable.

EXAMPLE   Find the average rate of change of the volume of a sphere when its radius changes from 3 cm to 5 cm.

SOLUTION If V denotes the volume of a sphere of radius r, then
$$V = \tfrac{4}{3}\pi r^3.$$
Here $r = 3$, $r + \Delta r = 5$; therefore, $\Delta r = 5 - 3 = 2$.
$$\Delta V = f(r + \Delta r) - f(r)$$
$$= f(5) - f(3)$$
$$= \tfrac{4}{3}\pi(5^3) - \tfrac{4}{3}\pi(3^3)$$
$$= \tfrac{4}{3}\pi(125 - 27) = \tfrac{4}{3}\pi(98) = \tfrac{392}{3}\pi.$$
Thus the average rate of change in volume is
$$\frac{\Delta V}{\Delta r} = \frac{\frac{392}{3}\pi}{2} = \frac{392\pi}{6} = \frac{196\pi}{3} \text{ cm}^3/\text{cm}.$$

EXAMPLE A population of bacteria is introduced to a nutrient medium. Suppose that the weight of the population in milligrams changes according to the formula
$$P(t) = 50 + \frac{100t}{21 + t^2},$$
where t is the time measured in hours. Determine the average rate of growth of the population during the five-hour period starting at $t = 2$ hr.

SOLUTION Here $t = 2$ and $\Delta t = 5$. The increment in the population size corresponding to this increment in time is
$$\Delta P = P(t + \Delta t) - P(t)$$
$$= P(2 + 5) - P(2)$$
$$= \left(50 + \frac{700}{21 + 49}\right) - \left(50 + \frac{200}{21 + 4}\right)$$
$$= 60 - 58 = 2.$$
Thus the average rate of change of population during the interval from $t = 2$ hr to $t = 7$ hr is
$$\frac{\Delta P}{\Delta t} = \frac{2}{5} = 0.4 \text{ mg/hr}.$$

EXAMPLE When any object is released from rest and allowed to fall freely under the force of gravity, the distance s (in feet) traveled in t seconds is given by
$$s(t) = 16t^2.$$
Determine the average speed of the object during the following intervals of time:

(a) the time interval from 3 to 5 sec;
(b) the fourth second (i.e., from $t = 3$ to $t = 4$ sec);
(c) the interval between 3 and $3\frac{1}{2}$ sec;
(d) the time interval from t to $t + \Delta t$.

65

SOLUTION   The average speed of any moving object is equal to the distance traveled divided by the interval of time concerned. During the interval of time from t to $t + \Delta t$, the distance traveled is the increment Δs, and the average speed is the ratio $\Delta s/\Delta t$.

(a)   Here $t = 3$ and $t + \Delta t = 5$. Therefore

$$\frac{\Delta s}{\Delta t} = \frac{s(t + \Delta t) - s(t)}{\Delta t} = \frac{s(5) - s(3)}{5 - 3}$$

$$= \frac{16(5^2) - 16(3^2)}{2} = \frac{400 - 144}{2}$$

$$= \frac{256}{2} = 128 \text{ ft/sec.}$$

Thus during the interval of time $t = 3$ to $t = 5$, the body falls an increment of 256 ft and has an average speed of 128 ft/sec.

(b)   Here $t = 3$ and $t + \Delta t = 4$. Therefore

$$\frac{\Delta s}{\Delta t} = \frac{s(t + \Delta t) - s(t)}{\Delta t} = \frac{s(4) - s(3)}{4 - 3}$$

$$= \frac{16(4^2) - 16(3^2)}{1} = 256 - 144 = 112 \text{ ft/sec.}$$

The body has an average speed of 112 ft/sec during the fourth second of fall.

(c)   Here $t = 3$ and $\Delta t = 3\frac{1}{2} - 3 = \frac{1}{2}$. Therefore

$$\frac{\Delta s}{\Delta t} = \frac{s(t + \Delta t) - s(t)}{\Delta t} = \frac{16(3\frac{1}{2})^2 - 16(3^2)}{\frac{1}{2}}$$

$$= \frac{196 - 144}{\frac{1}{2}} = \frac{52}{\frac{1}{2}} = 104 \text{ ft/sec.}$$

Thus the body has an average speed of 104 ft/sec during the time interval 3 to $3\frac{1}{2}$ sec.

(d)   In the general case,

$$\frac{\Delta s}{\Delta t} = \frac{s(t + \Delta t) - s(t)}{\Delta t}$$

$$= \frac{16(t + \Delta t)^2 - 16t^2}{\Delta t}$$

$$= \frac{16[t^2 + 2t \cdot \Delta t + (\Delta t)^2] - 16t^2}{\Delta t}$$

$$= \frac{32t \cdot \Delta t + 16(\Delta t)^2}{\Delta t}$$

$$= 32t + 16(\Delta t),$$

66   which is the required average speed during the interval from t to $t + \Delta t$.

From this last result (d), all of the previous results in this example can be obtained as special cases by putting t and Δt equal to appropriate values. For example, the result of (a) is obtained by setting $t = 3$ and $\Delta t = 2$:

$$\frac{\Delta s}{\Delta t} = 32t + 16(\Delta t) = 32(3) + 16(2) = 96 + 32 = 128.$$

EXERCISES 2.1

Find the increments of the following functions for the given intervals:

1. $f(x) = 2x + 7$, $x = 3$, $\Delta x = 0.2$.

2. $f(x) = 2x^2 + 3x - 5$, $x = 2$, $\Delta x = 0.5$.

3. $g(x) = \dfrac{x^2 - 4}{x - 2}$, $x = 1$, $\Delta x = 2$.

4. $f(t) = \dfrac{900}{t}$, $t = 25$, $\Delta t = 5$.

5. $p(t) = 2000 + \dfrac{500}{1 + t^2}$, $t = 2$, $\Delta t = 1$.

6. $h(x) = ax^2 + bx + c$, x to $x + \Delta x$.

7. $F(x) = x + \dfrac{2}{x}$, x to $x + \Delta x$.

8. $G(t) = 300 + \dfrac{5}{t + 1}$, t to $t + \Delta t$.

Determine the average rates of change of the following functions for the given intervals:

9. $f(x) = 3 - 7x$, $x = 2$, $\Delta x = 0.5$.

10. $f(x) = 3x^2 - 5x + 1$, $x = 3$, $\Delta x = 0.2$.

11. $g(x) = \dfrac{x^2 - 9}{x - 3}$, $x = 2$, $\Delta x = 0.5$.

12. $h(x) = \dfrac{3x^2 + 1}{x}$, $x = 5$, $\Delta x = 0.3$.

13. $f(t) = \sqrt{4 + t}$, $t = 5$, $\Delta t = 1.24$.

14. $F(x) = \dfrac{3}{x}$, x to $x + \Delta x$.

15. $G(t) = t^3 + t$, $t = a$ to $a + h$.

16. $f(x) = \dfrac{3}{2x + 1}$, x to $x + \Delta x$.

17. The size of a bacteria population at time t (measured in hours) is given by
$$p(t) = 10,000 + 1000t - 120t^2.$$
Determine the average rate of growth between times
(a) $t = 3$ hr and $t = 5$ hr.
(b) $t = 3$ hr and $t = 4$ hr.
(c) $t = 3$ hr and $t = 3\frac{1}{2}$ hr.
(d) $t = 3$ hr and $t = 3\frac{1}{4}$ hr.
(e) t and $t + \Delta t$.

18. Suppose the size of a population of salmon fingerlings in a hatchery tank at time t (measured in days) is given by
$$p(t) = 100(t + 3)^2.$$
Find the average rate of change in the size of population between the days
(a) $t = 0$ and $t = 1$.
(b) $t = 1$ and $t = 2$.
(c) $t = 1$ and $t = 3$.
(d) $t = 2$ and $t = 3$.

19. When a certain drug is given to a person, its reaction is measured by noting the change in blood pressure, change in body temperature, pulse rate, and other physiological changes. The strength S of the reaction depends on the amount x of drug administered, and is given by
$$S(x) = x^2(5 - x).$$
Determine the average rate of change in the strength of reaction when the amount of drug used changes from $x = 1$ unit to $x = 3$ units.

20. The number of pounds of good quality peaches, P, produced by an average tree in a certain orchard depends on the number of pounds of insecticide x with which the tree is sprayed, according to the formula
$$P = 300 - \frac{100}{1 + x}.$$
Calculate the average rate of increase of P when x changes from 0 to 3.

2.2 LIMITS

In the last example of Sec. 2.1, we discussed the average speeds of a falling body during a number of different time intervals. However, in many instances, both in science and in everyday life, the average speed of a moving object does not provide the information that is of most importance. For example, if a person traveling in an automobile hits a concrete wall, it is not his average speed from the start to the point where he hits the wall, but the speed at the *instant of collision* that determines whether he will survive the accident.

What do we mean by the speed of a moving object at a certain instant of time (or *instantaneous speed* as it is usually termed)? Most people would accept that there is such a thing as instantaneous speed—it is precisely the quantity that is measured by the speedometer of an automobile—but the definition of instantaneous speed presents a difficulty. Speed is defined as the distance traveled in a certain interval of time divided by the length of time. But if we are concerned with the speed at a particular instant of time, we ought to consider an interval of time of zero duration. However, during such an interval, the distance traveled would be zero, and for the speed, distance divided by time, we would obtain $\frac{0}{0}$, a meaningless quantity.

In order to define the instantaneous speed of a moving object at a certain time t, we proceed as follows. During any interval of time from t to $t + \Delta t$, an increment of distance Δs is traveled. The average speed is $\Delta s / \Delta t$. Now let us imagine that the increment Δt is taken smaller and smaller, so that the corresponding interval of time is very short. Then it is reasonable to suppose that the average speed $\Delta s / \Delta t$ over such a very short interval will be very close to the instantaneous speed at time t. Furthermore, the shorter we make the interval Δt, the better will the average speed approximate the instantaneous speed. In fact we can imagine that Δt is allowed to get arbitrarily close to zero, so that the average speed $\Delta s / \Delta t$ can be made as close as we like to the instantaneous speed.

In the last example of Sec. 2.1 we saw that for a body falling under gravity, the average speed during the time interval from t to $t + \Delta t$ is given by

$$\frac{\Delta s}{\Delta t} = 32t + 16(\Delta t).$$

Setting $t = 3$ we obtain the average speed during a time interval of length Δt following 3 sec of fall:

$$\frac{\Delta s}{\Delta t} = 96 + 16(\Delta t).$$

Some values of this velocity are set out in the following table for different values of the increment Δt. For example, the average velocity between 3 and 3.1 sec is obtained by setting $\Delta t = 0.1$: $\Delta s / \Delta t = 96 + 16(0.1) = 96 + 1.6 = 97.6$ ft/sec.

Δt:	0.5	0.25	0.1	0.01	0.001
$\Delta s / \Delta t$:	104	100	97.6	96.16	96.016

It is clear from the values in this table that as Δt gets smaller and smaller, the average velocity gets closer and closer to 96 ft/sec. We can reasonably conclude, therefore, that this figure gives the instantaneous speed at $t = 3$.

This example is typical of a whole class of problems in which we need to examine the behavior of a certain function as its argument gets closer and closer to a particular value. In this case we are concerned with the behavior of the average speed $\Delta s / \Delta t$ as Δt gets closer and closer to 0. In general we may be interested in the behavior of a function $f(x)$ of a variable x as x approaches a particular value, for example, c. When we say that x approaches c, we mean that x takes a succession of values that get arbitrarily close to the value c, but x never equals c. (Note that the average speed $\Delta s/\Delta t$ is not defined for $\Delta t = 0$. We can only take a very, very small value of Δt but never a zero value.) We write "$x \longrightarrow c$" to mean "x approaches c"; for example, we would write $\Delta t \longrightarrow 0$ in the above example.

As an example, let us consider the function $f(x) = 2x + 3$, and let us suppose that $x \longrightarrow 1$. We shall allow x to take the succession of values 0.8, 0.9, 0.99, 0.999, 0.9999, etc., which clearly are getting closer and closer to 1. The corresponding values of $f(x)$ are given in the following table:

x:	0.8	0.9	0.99	0.999	0.9999	$\ldots$
$f(x)$:	4.6	4.8	4.98	4.998	4.9998	$\ldots$

It is clear from the table that as x gets closer to 1, $f(x)$ gets closer to the value 5. We write $f(x) \longrightarrow 5$ as $x \longrightarrow 1$.

The values of x considered in the above table were all less than 1. In such a case we say that x approaches 1 from below. We can also consider the alternative case in which x approaches 1 from above, that is, x takes a succession of values getting closer and closer to 1, but always remaining greater than 1. For example, we might allow x to take the sequence of values 1.5, 1.1, 1.01, 1.001, 1.0001, etc. The corresponding values of $f(x)$ are given below:

x:	1.5	1.1	1.01	1.001	1.0001	$\ldots$
$f(x)$:	6	5.2	5.02	5.002	5.0002	$\ldots$

Again it is clear that $f(x)$ gets closer and closer to 5 as x approaches 1 from above.

Thus as x approaches 1 either from below or from above, $f(x) = 2x + 3$ approaches 5. We say that the *limit* (or *limiting value*) of $f(x)$ as x approaches 1 is equal to 5. This is written

$$\lim_{x \to 1} (2x + 3) = 5.$$

We now give a formal definition of a limiting value.

DEFINITION   Let $f(x)$ be a function that is defined for all values of x close to c, except possibly at the point c itself. Then *L is said to be the limiting value of $f(x)$ as x approaches c if the difference between $f(x)$ and L can be made as small as we please by taking any value of x that is sufficiently close to c.* In symbols we write

$$\lim_{x \to c} f(x) = L.$$

or

$$f(x) \longrightarrow L \text{ as } x \longrightarrow c.$$

In the above example $f(x) = 2x + 3$, $c = 1$, and $L = 5$. We can make the values of the function $(2x + 3)$ as close to 5 as we like by choosing x sufficiently close to 1.

In the example given, the limiting value of the function $f(x) = 2x + 3$ as $x \to 1$ can be obtained simply by substituting $x = 1$ into the formula $2x + 3$, which defines the function. The question arises as to whether limits can always be found by substituting the value of x. The answer to this question is: sometimes but not always. The following example illustrates a case when direct substitution does not work.

EXAMPLE   If

$$f(x) = \frac{x^2 - 9}{x - 3},$$

evaluate

$$\lim_{x \to 3} f(x).$$

SOLUTION   If we substitute $x = 3$ into $f(x)$, we obtain $\frac{0}{0}$, and we conclude that $f(x)$ is not defined for $x = 3$. However, $\lim_{x \to 3} f(x)$ does exist, since we can write

$$f(x) = \frac{x^2 - 9}{x - 3} = \frac{(x - 3)(x + 3)}{x - 3} = x + 3.$$

The cancellation of the factors $(x - 3)$ is valid for all $x \neq 3$ (but of course is not valid for $x = 3$ itself). It is readily seen that as x approaches 3, the function $(x + 3)$ gets closer and closer to 6 in value. Consequently,

$$\lim_{x \to 3} f(x) = \lim_{x \to 3} (x + 3) = 3 + 3 = 6.$$

When evaluating $\lim_{x \to c} f(x)$ it is quite legitimate to cancel factors of $(x - c)$, as we have done in the preceding example, in spite of the fact that when $x = c$ these factors are zero. This is because the limit is concerned with the behavior of $f(x)$ close to $x = c$, but is not concerned at all with value of f at $x = c$ itself. As long as $x \neq c$, factors of $(x - c)$ can be cancelled. In fact, the preceding example illustrates a case where $f(x)$ is not even defined for $x = c$, and yet $\lim_{x \to c} f(x)$ exists.

Let us examine the idea of limits from the point of view of the graph of the function involved. We shall first consider our initial example in which $f(x) = 2x + 3$. The graph of this function (Fig. 2.5) is a straight line of slope 2 and intercept 3. When $x = 1$ on the graph, $y = 5$. Consider any sequence of points P_1, P_2, P_3, etc., on the graph that are such that their x-coordinates are getting closer to 1. Clearly, the points themselves must get closer to the point $(1, 5)$ on the graph and their y-coordinates approach the limiting value 5. This corresponds to our earlier statement that $\lim_{x \to 1} (2x + 3) = 5$.

The example $f(x) = (x^2 - 9)/(x - 3)$ is a little different. We saw before that, as long as $x \neq 3$, we can write $f(x) = x + 3$. So this function also has a straight line as its graph, the slope being 1 and intercept 3. However $f(x)$ is not defined for $x = 3$, so the point $(3, 6)$ is missing from the graph. This fact is indicated in Fig. 2.6 by the use of a small circle at this point on the straight line. Again if we consider a sequence

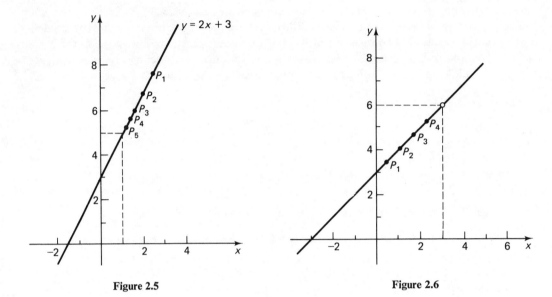

Figure 2.5   Figure 2.6

of points P_1, P_2, P_3, etc., on the graph for which the x-coordinates approach the value 3, then the points themselves must approach the point $(3, 6)$, even though this point is missing from the graph. Thus, in spite of the fact that $f(3)$ does not exist, the limit of $f(x)$ as $x \rightarrow 3$ does exist and is equal to 6.

The calculation of the limiting values of functions rests on a number of theorems concerning limits. We shall now state these theorems and illustrate their significance with a number of examples, but shall not give proofs of them.

THEOREM 2.2.1

If m, b, and c are any three constants, then

$$\lim_{x \to c} (mx + b) = mc + b.$$

We note that the function $y = mx + b$ has as its graph a straight line with slope m and y-intercept b. When $x = c$, y is always defined and $y = mc + b$. As x approaches c, the point (x, y) on the graph of this function gets closer and closer to the point $(c, mc + b)$. That is, the value of y gets closer and closer to $mc + b$, as stated in the theorem.

EXAMPLES   (a)   Taking $m = 2, b = 3$, and $c = 1$ we get the result

$$\lim_{x \to 1} (2x + 3) = 2(1) + 3 = 5,$$

which was discussed above.

(b)   Taking $m = 1, b = 3$, and $c = 3$, we get

$$\lim_{x \to 3} (x + 3) = 3 + 3 = 6,$$

again reproducing a result stated earlier.

THEOREM 2.2.2

(a) $\lim\limits_{x \to c} b\, f(x) = b \lim\limits_{x \to c} f(x)$

(b) $\lim\limits_{x \to c} [f(x)]^n = [\lim\limits_{x \to c} f(x)]^n$ provided that $[f(x)]^n$ is defined for x close to c.

EXAMPLES

(a) $\lim\limits_{x \to 3} x^2 = [\lim\limits_{x \to 3} x]^2$ by Theorem 2.2.2(b)

$\qquad\qquad = 3^2$ by Theorem 2.2.1

$\qquad\qquad = 9.$

(b) $\lim\limits_{x \to 1} 5(2x + 3)^{-1} = 5 \lim\limits_{x \to 1} (2x + 3)^{-1}$ by Theorem 2.2.2(a)

$\qquad\qquad\qquad\quad = 5[\lim\limits_{x \to 1} (2x + 3)]^{-1}$ by Theorem 2.2.2(b)

$\qquad\qquad\qquad\quad = 5[2(1) + 3]^{-1}$ by Theorem 2.2.1

$\qquad\qquad\qquad\quad = 5(5)^{-1} = 1.$

(c) $\lim\limits_{x \to 3} \dfrac{(x^2 - 9)^3}{12(x - 3)^3} = \dfrac{1}{12} \lim\limits_{x \to 3} \left(\dfrac{x^2 - 9}{x - 3}\right)^3$ by Theorem 2.2.2(a)

$\qquad\qquad\qquad = \dfrac{1}{12}\left[\lim\limits_{x \to 3} \left(\dfrac{x^2 - 9}{x - 3}\right)\right]^3$

$\qquad\qquad\qquad = \tfrac{1}{12}(6)^3$ by the result of a preceding example

$\qquad\qquad\qquad = 18.$

THEOREM 2.2.3

(a) $\lim\limits_{x \to c} [f(x) + g(x)] = \lim\limits_{x \to c} f(x) + \lim\limits_{x \to c} g(x).$

(b) $\lim\limits_{x \to c} [f(x) - g(x)] = \lim\limits_{x \to c} f(x) - \lim\limits_{x \to c} g(x).$

(c) $\lim\limits_{x \to c} [f(x)\, g(x)] = [\lim\limits_{x \to c} f(x)][\lim\limits_{x \to c} g(x)].$

(d) $\lim\limits_{x \to c} [f(x)/g(x)] = [\lim\limits_{x \to c} f(x)]/[\lim\limits_{x \to c} g(x)]$ provided that the denominator on the right-hand side does not equal zero.

EXAMPLES

(a) $\lim\limits_{x \to 3} (x^2 + 2x) = \lim\limits_{x \to 3} x^2 + \lim\limits_{x \to 3} 2x$ by Theorem 2.2.3(a)

$\qquad\qquad\qquad = 3^2 + 2(3)$ by a preceding example and Theorem 2.2.1

$\qquad\qquad\qquad = 9 + 6 = 15.$

(b) $\lim\limits_{x \to -1} \left(2x^3 - \dfrac{3}{x - 1}\right) = \lim\limits_{x \to -1} (2x^3) - \lim\limits_{x \to -1} \left(\dfrac{3}{x - 1}\right)$ by Theorem 2.2.3(b)

$\qquad\qquad\qquad\quad = 2 \lim\limits_{x \to -1} x^3 - 3 \lim\limits_{x \to -1} (x - 1)^{-1}$ by Theorem 2.2.2(a)

$\qquad\qquad\qquad\quad = 2[\lim\limits_{x \to -1} x]^3 - 3[\lim\limits_{x \to -1} (x - 1)]^{-1}$ by Theorem 2.2.2(b)

$\qquad\qquad\qquad\quad = 2(-1)^3 - 3(-1 - 1)^{-1}$ by Theorem 2.2.1

$\qquad\qquad\qquad\quad = -2 + \tfrac{3}{2} = -\tfrac{1}{2}.$

(c) $\lim_{x \to 3} \dfrac{(x-1)(x^2-9)}{(x-3)} = \lim_{x \to 3} (x-1) \lim_{x \to 3} \left(\dfrac{x^2-9}{x-3}\right)$ by Theorem 2.2.3(c)

$$= \lim_{x \to 3} (x-1) \lim_{x \to 3} (x+3)$$

$$= (3-1)(3+3) = 12.$$

(d) $\lim_{x \to -2} \left(\dfrac{x^2}{x-1}\right) = \dfrac{\lim_{x \to -2} x^2}{\lim_{x \to -2} (x-1)}$ by Theorem 2.2.3(d)

$$= \dfrac{\left(\lim_{x \to -2} x\right)^2}{(-2-1)}$$ by Theorems 2.2.2(b) and 2.2.1

$$= \dfrac{(-2)^2}{-3}$$ by Theorem 2.2.1

$$= -\tfrac{4}{3}.$$

It will probably not have escaped the reader's notice that in most of these examples the limiting value of the function involved could have been obtained simply by substituting the limiting value of x into the given function. This method of simple substitution will in fact always produce the right answer in the case of rational algebraic functions, provided that upon substitution we get a well-defined real number, that is, provided that the result is not of the form $\dfrac{0}{0}$ or constant/0. However it is recommended that the student do a number of exercises using the above limit theorems in the manner illustrated by the preceding examples. The reason for this is that we shall in later chapters encounter cases in which use of the theorems plays an essential role, and the limits will not be capable of evaluation by substitution. Only after having mastered the use of the theorems should the student adopt the method of substitution as a means of evaluating limits.

It may happen that upon substituting $x = c$ into $f(x)$, we obtain a result of the type constant/0. For example, suppose we tried to evaluate $\lim_{x \to 0} (1/x)$. Substituting $x = 0$ we obtain the result $1/0$, which is not defined. In such a case we would say that the limit does not exist. The function $1/x$ becomes indefinitely large as x approaches zero, and does not approach any limiting value.

A further very important case that can arise is that upon substituting $x = c$ into $f(x)$ we obtain the result $\dfrac{0}{0}$, which is undefined. Limits of this type can often be evaluated by cancelling factors of $(x - c)$ from the numerator and denominator of fractions that occur in $f(x)$. This technique was illustrated earlier in the section, and further examples will now be given.

EXAMPLE   Evaluate

$$\lim_{x \to -1} \frac{x^2 + 3x + 2}{1 - x^2}.$$

SOLUTION   Putting $x = -1$, we get

$$\frac{(-1)^2 + 3(-1) + 2}{1 - (-1)^2} = \frac{1 - 3 + 2}{1 - 1} = \frac{0}{0}.$$

Consequently, we factor the numerator and denominator and cancel out $(x + 1)$ before substituting $x = -1$:

$$\lim_{x \to -1} \frac{x^2 + 3x + 2}{1 - x^2} = \lim_{x \to -1} \frac{(x + 1)(x + 2)}{(1 - x)(1 + x)}$$

$$= \lim_{x \to -1} \frac{x + 2}{1 - x} = \frac{-1 + 2}{1 - (-1)} = \frac{1}{2}.$$

EXAMPLE   Evaluate

$$\lim_{x \to 0} \frac{\sqrt{1 + x} - 1}{x}.$$

SOLUTION   When we substitute $x = 0$, we get

$$\frac{\sqrt{1 + 0} - 1}{0} = \frac{0}{0}.$$

In this case, we cannot factor the numerator directly to get the x that we need to cancel the x in the denominator. We overcome this difficulty by rationalizing the numerator, which is accomplished by multiplying the numerator and denominator by $(\sqrt{1 + x} + 1)$. Thus

$$\lim_{x \to 0} \frac{\sqrt{1 + x} - 1}{x} = \lim_{x \to 0} \frac{\sqrt{1 + x} - 1}{x} \cdot \frac{\sqrt{1 + x} + 1}{\sqrt{1 + x} + 1}$$

$$= \lim_{x \to 0} \frac{(\sqrt{1 + x})^2 - 1^2}{x(\sqrt{1 + x} + 1)} = \lim_{x \to 0} \frac{(1 + x) - 1}{x(\sqrt{1 + x} + 1)}$$

$$= \lim_{x \to 0} \frac{x}{x(\sqrt{1 + x} + 1)}$$

$$= \lim_{x \to 0} \frac{1}{\sqrt{1 + x} + 1}$$

$$= \frac{1}{\sqrt{1 + 0} + 1} = \frac{1}{2}.$$

Note that in these examples, the final limit has been evaluated by substitution. In reality, the theorems on limits underlie this substitution procedure.

EXAMPLE   Given $f(x) = 2x^2 + 3x + 7$. Evaluate

$$\lim_{h \to 0} \frac{f(a + h) - f(a)}{h}.$$

SOLUTION   Substituting $x = a$ and $x = a + h$ respectively, we have

$$f(a) = 2a^2 + 3a + 7$$

and

$$f(a + h) = 2(a + h)^2 + 3(a + h) + 7$$
$$= 2(a^2 + 2ah + h^2) + 3a + 3h + 7$$
$$= 2a^2 + 4ah + 2h^2 + 3a + 3h + 7.$$

Thus

$$\lim_{h \to 0} \frac{f(a + h) - f(a)}{h}$$

$$= \lim_{h \to 0} \frac{(2a^2 + 4ah + 2h^2 + 3a + 3h + 7) - (2a^2 + 3a + 7)}{h}$$

$$= \lim_{h \to 0} \frac{2h^2 + 4ah + 3h}{h}$$

$$= \lim_{h \to 0} (2h + 4a + 3)$$

$$= 2(0) + 4a + 3 = 4a + 3.$$

EXERCISES 2.2

Evaluate the following limits:

1. $\lim\limits_{x \to 2} (3x^2 + 7x - 1).$

2. $\lim\limits_{x \to -1} (2x^2 + 3x + 1).$

3. $\lim\limits_{x \to 3} \dfrac{x + 1}{x - 2}.$

4. $\lim\limits_{x \to 3} \dfrac{x^2 + 1}{x + 3}.$

5. $\lim\limits_{x \to 5} \dfrac{x^2 - 25}{\sqrt{x^2 + 11}}.$

6. $\lim\limits_{x \to 4} \dfrac{x^2 - 16}{x - 4}.$

7. $\lim\limits_{x \to -2} \dfrac{x^2 - 4}{x^2 + 3x + 2}.$

8. $\lim\limits_{x \to 1} \dfrac{x^2 - 1}{x^2 + x - 2}.$

9. $\lim\limits_{x \to 3} \dfrac{x^2 - 5x + 6}{x - 3}.$

10. $\lim\limits_{x \to 1} \dfrac{x^3 - 1}{x^2 - 1}.$

11. $\lim\limits_{x \to 2} \dfrac{x + 1}{x - 2}.$

12. $\lim\limits_{x \to 0} \dfrac{2x^2 + 5x + 7}{x}.$

13. $\lim\limits_{x \to 0} \dfrac{\sqrt{4 + x} - 2}{x}.$

14. $\lim\limits_{x \to 2} \dfrac{\sqrt{x + 7} - 3}{x - 2}.$

15. $\lim\limits_{x \to 1} \dfrac{\sqrt{x + 3} - 2}{x^2 - 1}.$

16. $\lim\limits_{x \to 0} \dfrac{\sqrt{9 + x} - 3}{x^2 + 2x}.$

17. $\lim\limits_{x \to 0} \dfrac{\sqrt{1 + x} - 1}{\sqrt{4 + x} - 2}.$

18. $\lim\limits_{x \to 1} \dfrac{\sqrt{2 - x} - 1}{2 - \sqrt{x + 3}}.$

Evaluate $\lim\limits_{x \to c} f(x)$ where $f(x)$ and c are given below.

19. $f(x) = \begin{cases} \dfrac{x^2 - 1}{x - 1} & \text{for } x \neq 1 \\ 3 & \text{for } x = 1 \end{cases}$, $c = 1$.

20. $f(x) = \begin{cases} \dfrac{x - 9}{\sqrt{x} - 3} & \text{for } x \neq 9 \\ 7 & \text{for } x = 9 \end{cases}$, $c = 9$.

The functions $f(x)$ and the values of a are given below. Evaluate the $\lim\limits_{h \to 0} [f(a + h) - f(a)]/h$ in each case.

21. $f(x) = 2x^2 + 3x + 1$, $a = 1$.

22. $f(x) = 3x^2 - 5x + 7$, $a = 2$.

23. $f(x) = x^2 - 1$, $x = 0$.

24. $f(x) = x^2 + x + 1$, $a = x$.

25. $f(x) = 2x^2 + 5x + 1$, $a = x$.

2.3 MORE ON LIMITS

(This section can be postponed or omitted if desired.)

In considering the limiting value of a function $f(x)$ as x approaches c, we must allow the possibility of x taking values that are both less than and greater than c. However it may happen in certain cases that the behavior of a given function is different for $x < c$ from its behavior when $x > c$. In such a case, we may wish to consider separately the possibilities that x might approach c from above or, alternatively, from below.

We say that *x approaches c from above*, and write $x \to c^+$, if x takes a sequence of values that get closer and closer to c but always remain greater than c. We say that *x approaches c from below*, and write $x \to c^-$, if x takes a sequence of values getting closer and closer to c but remaining less than c. If $f(x)$ approaches the limiting value L as $x \to c^+$, we write

$$\lim_{x \to c^+} f(x) = L.$$

If $f(x)$ approaches the limiting value M as $x \to c^-$, we write

$$\lim_{x \to c^-} f(x) = M.$$

Limits of this kind are called *one-sided limits*.

EXAMPLE   Investigate the limiting values of $f(x) = \sqrt{x - 1}$ as x approaches 1 from above and from below.

SOLUTION   As $x \to 1^+$, $x - 1$ approaches zero through positive values. Therefore

$$\lim_{x \to 1^+} \sqrt{x - 1} = 0.$$

On the other hand, as $x \to 1^-$, $x - 1$ still approaches zero but is always negative. Hence $\sqrt{x - 1}$ is not defined for $x < 1$, so $\lim_{x \to 1^-} \sqrt{x - 1}$ does not exist.

EXAMPLE   Investigate the limiting value of $f(x) = |x|/x$ as x approaches 0 from above and from below.

SOLUTION   For $x > 0$, $|x| = x$; therefore

$$f(x) = \frac{|x|}{x} = \frac{x}{x} = 1.$$

The given function has the value 1 for all $x > 0$, and therefore must have the limiting value 1 as x approaches 0 from above:

$$\lim_{x \to 0^+} \frac{|x|}{x} = 1.$$

For $x < 0$, $|x| = -x$, and

$$f(x) = \frac{|x|}{x} = \frac{-x}{x} = -1.$$

So $f(x)$ is identically equal to -1 for all $x < 0$, and therefore

$$\lim_{x \to 0^-} \frac{|x|}{x} = -1.$$

The graph of $y = f(x)$ is shown in Fig. 2.7. Note that $f(x)$ is not defined for $x = 0$ and that the graph makes a jump from -1 to $+1$ as x passes from below 0 to above 0.

The preceding examples illustrate two basic types of behavior. In the first case only one of the two limits from above and from below existed. In the second case, both limits existed but their values were different from one another. In both cases, the relevant two-sided limit, $\lim_{x \to c} f(x)$, does not exist. For a general $f(x)$, as illustrated in Fig. 2.8, if the graph of $f(x)$ makes a jump at $x = c$, then the two limits from above

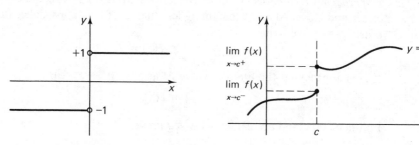

Figure 2.7   **Figure 2.8**

and from below are not equal to each other. Note that $\lim_{x \to c} f(x)$ exists if both $\lim_{x \to c^-} f(x)$ and $\lim_{x \to c^+} f(x)$ exist and are equal.

In deciding whether $\lim_{x \to c} f(x)$ exists we can be guided by the following rule. If $f(x)$ is defined by the *same* algebraic formula for values of x less than c, and for x greater than c, then we need not evaluate the two limits separately to check if they are equal. If $f(x)$ is defined by one formula for values of x less than c and by a different formula for values of x greater than c, then we should evaluate both the limits from above and below and should verify their equality.

EXAMPLE   Given $f(x) = 3x^2 + 5x - 9$, find $\lim_{x \to 1} f(x)$.

SOLUTION   Here $f(x)$ is defined for all values of x and, in particular, it is defined by the same formula for values of x less than 1 and for x greater than 1. So, we find the limit directly as

$$\lim_{x \to 1} f(x) = \lim_{x \to 1} (3x^2 + 5x - 9)$$

$$= 3(1)^2 + 5(1) - 9 = -1.$$

EXAMPLE   Given

$$f(x) = \begin{cases} \dfrac{x^2 - 4}{x - 2}, & x \neq 2, \\ 1, & x = 2, \end{cases}$$

evaluate $\lim_{x \to 2} f(x)$.

SOLUTION   Here again $f(x)$ is defined by the same formula for $x > 2$ and for $x < 2$. Therefore

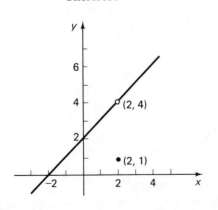

Figure 2.9

$$\lim_{x \to 2} f(x) = \lim_{x \to 2} \frac{x^2 - 4}{x - 2}$$

$$= \lim_{x \to 2} \frac{(x - 2)(x + 2)}{x - 2}$$

$$= \lim_{x \to 2} (x + 2)$$

$$= 2 + 2 = 4.$$

Note that in this example the graph of $f(x)$ (Fig. 2.9) is the straight line $y = x + 2$ with the point $(2, 4)$ missing, but with the separate point $(2, 1)$ included. This separate point has no effect on the value of the limit.

EXAMPLE   Given

$$f(x) = \begin{cases} 2x + 5 & \text{for } x > 3 \\ x^2 + 2 & \text{for } x \leq 3, \end{cases}$$

find $\lim\limits_{x \to 3} f(x)$.

SOLUTION   In this case, $f(x)$ is defined by two different formulas for values of $x < 3$ and for values of $x > 3$. So, we must find the limits separately from above and below. Since $f(x) = 2x + 5$ for $x > 3$, for the limit from above we find

$$\lim_{x \to 3^+} f(x) = \lim_{x \to 3} (2x + 5) = 2(3) + 5 = 11.$$

Similarly, for $x < 3$, we have $f(x) = x^2 + 2$, and therefore for the limit from below,

$$\lim_{x \to 3^-} f(x) = \lim_{x \to 3} (x^2 + 2) = 3^2 + 2 = 11.$$

Since

$$\lim_{x \to 3^+} f(x) = \lim_{x \to 3^-} f(x) = 11,$$

it follows that $\lim\limits_{x \to 3} f(x)$ exists and is equal to 11.

The graph of $f(x)$ in this case is shown in Fig. 2.10. Note that the graph changes type at $x = 3$, but does not make a jump at this point.

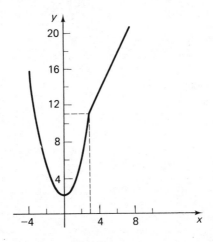

Figure 2.10

NOTE. The rule concerning "functions defined by the same algebraic formula" must be interpreted carefully when absolute values occur. For example, the function $f(x) = |x|/x$, which was discussed in an earlier example, is not defined by the same algebraic formula for both $x > 0$ and $x < 0$. This is because $|x|$ is equal respectively to x and to $-x$ in these two cases.

80

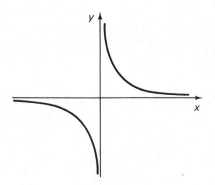

Figure 2.11

Let us consider the behavior of the function $f(x) = 1/x$ as x approaches 0 from above. Then x becomes smaller and smaller, always remaining positive, and its reciprocal $1/x$ gets larger and larger. In fact $1/x$ increases without bound as $x \to 0^+$. This is apparent from the graph of $y = 1/x$, shown in Fig. 2.11; the graph has a vertical asymptote at $x = 0$.

As a matter of convention, we write

$$\lim_{x \to 0^+} \frac{1}{x} = \infty.$$

This use of the limit notation does not imply that $\lim_{x \to 0^+} \frac{1}{x}$ exists in the ordinary sense of a limit. The infinity symbol ∞ simply indicates that $1/x$ becomes arbitrarily large and positive as x approaches zero from above.

Correspondingly, as x approaches zero from below, $1/x$ takes ever larger negative values. We indicate this by using the notation

$$\lim_{x \to 0^-} \frac{1}{x} = -\infty.$$

EXAMPLE   Evaluate the limits of $f(x) = 1/(x + 1)^2$ as x approaches -1 from above and from below.

SOLUTION   As $x \to -1^+$, $(x + 1)$ becomes smaller and smaller but is always positive. Therefore $1/(x + 1)^2$ takes larger and larger positive values. As $x \to -1^-$, $(x + 1)$ again approaches zero, but is negative. However, when it is squared, the result $(x + 1)^2$ is small and *positive*. Therefore $1/(x + 1)^2$ again takes large positive values. Therefore, we can write, according to the convention introduced above,

$$\lim_{x \to -1^+} \frac{1}{(x + 1)^2} = +\infty,$$

$$\lim_{x \to -1^-} \frac{1}{(x + 1)^2} = +\infty.$$

Let us conclude this section by giving very briefly the mathematically complete definitions of one- and two-sided limits. First of all, we consider limits from above: Let us define what is meant by the statement $\lim_{x \to c^+} f(x) = L$.

This means that $f(x)$ can be made as close as we like to L provided that x is restricted to lie close enough to c on the side $x > c$. For instance, $f(x)$ must lie between $L - 0.1$ and $L + 0.1$ for all values of x lying in some interval, say $c < x < c + h$; or $f(x)$ must lie between $L - 0.01$ and $L + 0.01$ for all values of x lying in some other interval, say $c < x < c + k$. In fact, we can choose any range of values about L that we like, no matter how small, and $f(x)$ must lie within this range provided we restrict x to a suitable interval to the right of c. That is, given any range, say between $L - \epsilon$ and $L + \epsilon$, where ϵ is an arbitrary positive number, then $f(x)$ lies in this range for all x in some interval, say $c < x < c + \delta$.

This leads us to the following formal definition.

DEFINITION   We say that $\lim_{x \to c^+} f(x) = L$ if, given any number $\epsilon > 0$, however small, there exists a positive number δ such that $L - \epsilon < f(x) < L + \epsilon$ for all x such that $c < x < c + \delta$.

The definition of limits from below is quite similar. The only change is that the interval of values of x must lie to the left of c, that is, $x < c$.

DEFINITION   We say that $\lim_{x \to c^-} f(x) = M$ if, given any $\epsilon > 0$, there exists a positive number δ such that $M - \epsilon < f(x) < M + \epsilon$ for all x such that $c - \delta < x < c$.

Finally, for two-sided limits, we must allow x to have values both above and below c.

DEFINITION   We say that $\lim_{x \to c} f(x) = L$ if, given any $\epsilon > 0$, there exists a positive number δ such that $L - \epsilon < f(x) < L + \epsilon$ for all x such that $c - \delta < x < c + \delta$ except possibly for $x = c$ itself.

EXERCISES 2.3

Evaluate the following one-sided limits.

1. $\lim_{x \to 1^+} \sqrt{x - 1}$.

2. $\lim_{x \to (1/2)^-} \sqrt{1 - 2x}$.

3. $\lim_{x \to (4/3)^+} \sqrt{4 - 3x}$.

4. $\lim_{x \to -1^-} \sqrt{x + 1}$.

5. $\lim_{x \to 1^+} \dfrac{|x - 1|}{x - 1}$.

6. $\lim_{x \to -1^-} \dfrac{x + 1}{|x + 1|}$.

7. $\lim_{x \to 3^-} \dfrac{|x - 3|}{9 - x^2}$.

8. $\lim_{x \to 2^-} \dfrac{x^2 - x - 2}{|x - 2|}$.

9. $\lim_{x \to 0^-} \dfrac{1}{x^2}$.

10. $\lim_{x \to -1^+} \dfrac{x - 1}{x + 1}$.

11. $\lim_{x \to 1^-} \dfrac{2 - x^2}{|x - 1|}$.

12. $\lim_{x \to 2^-} \dfrac{x^2 - 6}{(x - 2)^3}$.

Evaluate $\lim_{x \to c} f(x)$ where $f(x)$ and c are given below.

13. $f(x) = \begin{cases} 5x + 7 & \text{for } x > 2, \\ 2x + 3 & \text{for } x < 2, \end{cases} \quad c = 2.$

14. $f(x) = \begin{cases} 3x + 5 & \text{for } x < 1, \\ 1 - 2x & \text{for } x \geq 1, \end{cases}$ $c = 3.$

15. $f(x) = \begin{cases} 3x - 1 & \text{for } x < 2, \\ 2x + 3 & \text{for } x \geq 2, \end{cases}$ $c = 2.$

16. $f(x) = \begin{cases} \dfrac{x - 9}{\sqrt{x} - 3} & \text{for } x \neq 9, \\ 7 & \text{for } x = 9, \end{cases}$ $c = 9.$

2.4 CONTINUOUS FUNCTIONS

(This section can be postponed or omitted if desired. However it is recommended that the initial definitions be read.)

DEFINITION   A function $f(x)$ is said to be *continuous* at a point $x = c$ if the following three conditions are met:

   i. $f(x)$ is defined at $x = c$; that is, $f(c)$ is well-defined,

   ii. $\lim\limits_{x \to c} f(x)$ exists,

   iii. $\lim\limits_{x \to c} f(x) = f(c).$

If any one of these three conditions is not satisfied then the function is said to be *discontinuous* at $x = c$.

Roughly speaking, a function is continuous at $x = c$ if its graph passes through the value $x = c$ without a break. If the graph makes a jump or has a break as it passes through $x = c$, then the function is discontinuous.

EXAMPLE   The function $f(x) = |x|$ is continuous at $x = 0$.

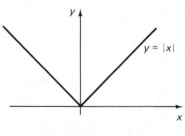

Figure 2.12

We note that $f(0) = |0| = 0$, so that condition (i) is satisfied. $\lim\limits_{x \to 0} f(x)$ also exists since as x approaches zero, $|x|$ approaches the limit zero. Finally, condition (iii) is met, since $\lim\limits_{x \to 0} f(x)$ and $f(0)$ are equal to each other, both being zero.

The graph of $y = |x|$ is shown in Fig. 2.12. The graph clearly passes through $x = 0$ without a break. It does have a corner (or change of slope) at $x = 0$, but this does not make it discontinuous.

EXAMPLE   Given

$$f(x) = \begin{cases} \dfrac{|x-2|}{x-2} & \text{when } x \neq 2 \\ 1 & \text{when } x = 2. \end{cases}$$

Discuss the continuity of $f(x)$ at $x = 2$.

SOLUTION   **i.**   The function is defined at $x = 2$ and $f(2) = 1$.

ii.   Even though $f(x)$ seems to be defined by the same algebraic expression for $x > 2$ and for $x < 2$, actually it is not. For $x > 2$ ($x - 2 > 0$),

$$f(x) = \frac{|x-2|}{x-2} = \frac{x-2}{x-2} = 1,$$

and for $x < 2$, ($x - 2 < 0$),

$$f(x) = \frac{|x-2|}{x-2} = \frac{-(x-2)}{x-2} = -1.$$

Thus,

$$\lim_{x \to 2^+} f(x) = \lim_{x \to 2}(1) = 1$$

and

$$\lim_{x \to 2^-} f(x) = \lim_{x \to 2}(-1) = -1.$$

Since these two limits from above and below are unequal, $\lim_{x \to 2} f(x)$ does not exist; therefore $f(x)$ is *not* continuous at $x = 2$. (In this case we need not test the third condition.)

The graph of $f(x)$ in this example (Fig. 2.13) makes a jump from -1 to $+1$ as x passes through the value 2.

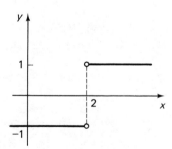

Figure 2.13

EXAMPLE   Given

$$f(x) = \begin{cases} \dfrac{x^2 - 9}{x - 3} & \text{if } x \neq 3 \\ 5 & \text{if } x = 3. \end{cases}$$

84   Is $f(x)$ continuous at $x = 3$?

i. Clearly $f(x)$ is defined at $x = 3$ and $f(3) = 5$.

ii. $\lim_{x \to 3} f(x) = \lim_{x \to 3} \dfrac{x^2 - 9}{x - 3} = \lim_{x \to 3} \dfrac{(x - 3)(x + 3)}{x - 3}$

$\qquad\qquad = \lim_{x \to 3} (x + 3) = 3 + 3 = 6.$

iii. $\lim_{x \to 3} f(x) = 6$ and $f(3) = 5$ are not equal.

In this case, the first two conditions are satisfied, but the third condition is *not* met, so the given function is discontinuous at $x = 3$. This is shown graphically in Fig. 2.14. The graph of $f(x)$ has a break at $x = 3$, and the isolated point $(3, 5)$ on the graph is not joined continuously to the rest of the graph.

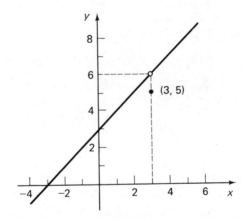

Figure 2.14

EXERCISES 2.4

Discuss the continuity of the following functions at $x = 0$ and sketch their graphs:

1. $f(x) = \dfrac{x^2}{x}.$

2. $g(x) = \sqrt{x^2}.$

3. $h(x) = \begin{cases} |x| & \text{for } x \neq 0 \\ 1 & \text{for } x = 0. \end{cases}$

4. $F(x) = \begin{cases} \dfrac{|x|}{x} & \text{if } x \neq 0 \\ 0 & \text{if } x = 0. \end{cases}$

5. $G(x) = \begin{cases} 0 & \text{if } x < 0 \\ 1 & \text{if } x > 0. \end{cases}$

6. $H(x) = \begin{cases} 0 & \text{if } x < 0 \\ x & \text{if } x > 0. \end{cases}$

Discuss the continuity of the following functions at the indicated points and sketch their graphs:

7. $f(x) = x^2 + 4x + 7$ at $x = 1$.

8. $g(x) = \dfrac{2x + 7}{x - 1}$ at $x = 1$.

9. $h(x) = \dfrac{3x + 1}{x + 1}$ at $x = 2$.

10. $f(x) = \begin{cases} \dfrac{|x - 3|}{x - 3} & \text{if } x \neq 3 \\ 0 & \text{if } x = 3 \end{cases}$ at $x = 3$.

11. $G(x) = \begin{cases} \dfrac{x^2 - 4}{x - 2} & \text{for } x \neq 2 \\ 4 & \text{for } x = 2 \end{cases}$ at $x = 2$.

12. $H(x) = \begin{cases} \dfrac{x^2 - 4}{x - 2} & \text{if } x \neq 2 \\ 5 & \text{if } x = 2 \end{cases}$ at $x = 1$.

Find the value of h in the following questions so that $f(x)$ is continuous at $x = 1$.

13. $f(x) = \begin{cases} x^2 - 3x + 4 & \text{if } x \neq 1 \\ h & \text{if } x = 1. \end{cases}$

14. $f(x) = \begin{cases} \dfrac{x^2 - 1}{x - 1} & \text{if } x \neq 1 \\ h & \text{if } x = 1. \end{cases}$

15. $f(x) = \begin{cases} \dfrac{x^2 - (h + 1)x + h}{x - 1} & \text{if } x \neq 1 \\ 2 & \text{if } x = 1. \end{cases}$

16. $f(x) = \begin{cases} hx + 3 & \text{if } x \geq 1 \\ 3 - hx & \text{if } x < 1. \end{cases}$

2.5 THE DERIVATIVE

We saw in Section 2.2 how the definition of the instantaneous velocity of a moving object leads naturally to a limiting process. The average velocity $\Delta s / \Delta t$ is first found for an interval of time from t to $t + \Delta t$; its limiting value is calculated as $\Delta t \to 0$.

$\Delta s/\Delta t$ might be described as the *average rate* of change of position s with respect to time, and its limit is the instantaneous rate of change of s with respect to t.

Now there are many examples of processes that develop in time and that can be described by one or more functions of t. In each case a corresponding defin,tion can be given of the rate of change of the appropriate quantity with respect to time. The following examples illustrate typical cases in point.

EXAMPLE   The size of a population of bacteria at time t (in minutes) is given by $w(t) = 2t^3$ mg. Find the instantaneous rate of growth of w at $t = 2$ min.

SOLUTION   The increment in w between $t = 2$ and $t = 2 + \Delta t$ is

$$\Delta w = w(2 + \Delta t) - w(2)$$
$$= 2(2 + \Delta t)^3 - 2(2)^3$$
$$= 2[8 + 12(\Delta t) + 6(\Delta t)^2 + (\Delta t)^3] - 16$$
$$= 24(\Delta t) + 12(\Delta t)^2 + 2(\Delta t)^3.$$

The instantaneous rate of growth of w is therefore given by

$$\lim_{\Delta t \to 0} \frac{\Delta w}{\Delta t} = \lim_{\Delta t \to 0} \frac{24(\Delta t) + 12(\Delta t)^2 + 2(\Delta t)^3}{\Delta t}$$
$$= \lim_{\Delta t \to 0} [24 + 12(\Delta t) + 2(\Delta t)^2]$$
$$= 24.$$

The population is therefore growing at the rate of 24 mg/min when $t = 2$.

EXAMPLE   In a chemical reaction in which the end-product is the substance BS, the amount of BS produced after t min is given by $y = f(t) = \sqrt{t}$ gm. Find the instantaneous rate of production of BS at $t = 2$ min.

SOLUTION   The instantaneous rate of production is, by definition:

$$\lim_{\Delta t \to 0} \frac{\Delta y}{\Delta t} = \lim_{\Delta t \to 0} \frac{f(2 + \Delta t) - f(2)}{\Delta t}$$
$$= \lim_{\Delta t \to 0} \frac{\sqrt{2 + \Delta t} - \sqrt{2}}{\Delta t}$$
$$= \lim_{\Delta t \to 0} \frac{(\sqrt{2 + \Delta t} - \sqrt{2})(\sqrt{2 + \Delta t} + \sqrt{2})}{\Delta t(\sqrt{2 + \Delta t} + \sqrt{2})}$$
$$= \lim_{\Delta t \to 0} \frac{(2 + \Delta t) - 2}{\Delta t(\sqrt{2 + \Delta t} + \sqrt{2})}$$
$$= \lim_{\Delta t \to 0} \frac{1}{\sqrt{2 + \Delta t} + \sqrt{2}} = \frac{1}{\sqrt{2 + 0} + \sqrt{2}}$$
$$= \frac{1}{2\sqrt{2}}.$$

Thus the instantaneous rate of generation of *BS* after 2 min. is equal $(1/2\sqrt{2})$ gm/min.

The instantaneous rate of change of a function such as in these examples is one case of what we call *derivatives* of functions. We shall now give a formal definition of the derivative of a function.

DEFINITION   Let $y = f(x)$ be a given function. Then the *derivative of y with respect to x*, denoted by dy/dx, is defined to be

$$\frac{dy}{dx} = \lim_{\Delta x \to 0} \frac{\Delta y}{\Delta x},$$

that is

$$\frac{dy}{dx} = \lim_{\Delta x \to 0} \frac{f(x + \Delta x) - f(x)}{\Delta x},$$

provided this limit exists.

The derivative is also given the name *differential coefficient*, and the operation of calculating the derivative of a function is called *differentiation*.

If the derivative of a function f exists at a particular point, then we say that f is *differentiable* at that point.

The derivative of $y = f(x)$ with respect to x is also denoted by any one of the following symbols:

$$\frac{d}{dx}(y), \quad \frac{df}{dx}, \quad \frac{d}{dx}(f), \quad y', \quad f'(x), \quad D_x y, \quad D_x f.$$

Every one of these notations means exactly the same thing as dy/dx.

NOTES.

1. dy/dx represents a single symbol and should not be interpreted as the ratio of two quantities dy and dx. To amplify the notation further, note that

   dy/dx denotes the derivative of y w. r. t. x if y is a function of the independent variable x;[1]

   du/dt denotes the derivative of u w. r. t. t if u is a function of the independent variable t;

   dx/du denotes the derivative of x w. r. t. u if x is a function of the independent variable u.

---

[1]*Note:* "w. r. t." means "with respect to."

From the definition,

$$\frac{dy}{dx} = \lim_{\Delta x \to 0} \frac{\Delta y}{\Delta x}, \quad \frac{du}{dt} = \lim_{\Delta t \to 0} \frac{\Delta u}{\Delta t}, \quad \text{and} \quad \frac{dx}{du} = \lim_{\Delta u \to 0} \frac{\Delta x}{\Delta u}, \text{ etc.}$$

2. Since $\Delta y = f(x + \Delta x) - f(x)$, as Δx approaches zero, Δy also approaches zero (provided that the function is continuous at x). However the limit of $\Delta y / \Delta x$ as $\Delta x \to 0$ does exists whenever $f(x)$ is differentiable.

3. If the derivative of a function exists at a point, then the function has to be continuous at that point. This does not mean that if the function is continuous at a point then the derivative will exist at that point: The function may be continuous but not differentiable. An example is given at the end of this section of a function that is continuous but not differentiable.

EXAMPLE   Find the derivative of $3x^2 + 2x$ with respect to x.

SOLUTION   Let

$$y = f(x) = 3x^2 + 2x. \tag{i}$$

Then

$$\begin{aligned}
y + \Delta y = f(x + \Delta x) &= 3(x + \Delta x)^2 + 2(x + \Delta x) \\
&= 3[x^2 + 2x \cdot \Delta x + (\Delta x)^2] + 2x + 2(\Delta x) \\
&= 3x^2 + 2x + \Delta x[6x + 2 + 3(\Delta x)].
\end{aligned} \tag{ii}$$

Subtracting (i) from (ii), we have

$$\Delta y = \Delta x[6x + 2 + 3(\Delta x)].$$

Dividing both sides by Δx,

$$\frac{\Delta y}{\Delta x} = 6x + 2 + 3(\Delta x).$$

Thus,

$$\frac{dy}{dx} = \lim_{\Delta x \to 0} \frac{\Delta y}{\Delta x} = \lim_{\Delta x \to 0} [6x + 2 + 3(\Delta x)]$$

$$= 6x + 2.$$

If $f(x) = 3x^2 + 2x$, then $f'(x) = 6x + 2$. In particular, $f'(1) = 6(1) + 2 = 8$.

EXAMPLE   Given $y = \sqrt{t + 1}$, find dy/dt.

SOLUTION   Let $y = f(t) = \sqrt{t + 1}$. Then

$$\begin{aligned}
\Delta y = f(t + \Delta t) - f(t) \\
= \sqrt{t + \Delta t + 1} - \sqrt{t + 1}
\end{aligned}$$

and

$$\frac{\Delta y}{\Delta t} = \frac{\sqrt{t + \Delta t + 1} - \sqrt{t + 1}}{\Delta t}.$$

Multiply the numerator and denominator by $\sqrt{t + \Delta t + 1} + \sqrt{t + 1}$.

$$\frac{\Delta y}{\Delta t} = \frac{\sqrt{t + \Delta t + 1} - \sqrt{t + 1}}{\Delta t} \cdot \frac{\sqrt{t + \Delta t + 1} + \sqrt{t + 1}}{\sqrt{t + \Delta t + 1} + \sqrt{t + 1}}$$

$$= \frac{(\sqrt{t + \Delta t + 1})^2 - (\sqrt{t + 1})^2}{\Delta t(\sqrt{t + \Delta t + 1} + \sqrt{t + 1})}$$

$$= \frac{(t + \Delta t + 1) - (t + 1)}{\Delta t(\sqrt{t + \Delta t + 1} + \sqrt{t + 1})}$$

$$= \frac{1}{\sqrt{t + \Delta t + 1} + \sqrt{t + 1}}.$$

Thus,

$$\frac{dy}{dt} = \lim_{\Delta t \to 0} \frac{\Delta y}{\Delta t} = \lim_{\Delta t \to 0} \frac{1}{\sqrt{t + \Delta t + 1} + \sqrt{t + 1}}$$

$$= \frac{1}{\sqrt{t + 1} + \sqrt{t + 1}} = \frac{1}{2\sqrt{t + 1}}.$$

Geometrical Interpretation

We have seen already that in the case where the independent variable in a function $y = f(t)$ represents "time," the derivative dy/dt gives the instantaneous rate of change of y. For example, if $s = f(t)$ represents the distance traveled by a moving object, then ds/dt provides the instantaneous velocity. Apart from this kind of application of derivatives, however, they also have a very great significance from the geometrical point of view.

If P and Q are the two points $(x, f(x))$ and $(x + \Delta x, f(x + \Delta x))$ on the graph of $y = f(x)$, then, as stated in Section 2.1,

$$\frac{\Delta y}{\Delta x} = \frac{f(x + \Delta x) - f(x)}{\Delta x}$$

represents the slope of the line segment PQ (Fig. 2.15). As Δx becomes smaller and

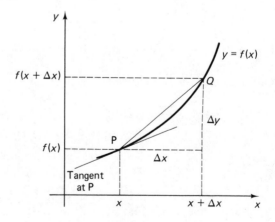

Figure 2.15

smaller, the point Q moves closer and closer to P, and the chord segment PQ becomes more and more nearly tangential. As $\Delta x \longrightarrow 0$, the slope of the chord PQ approaches the slope of the tangent line at P. Thus

$$\lim_{\Delta x \to 0} \frac{\Delta y}{\Delta x} = \frac{dy}{dx}$$

represents the slope of the tangent line to $y = f(x)$ at the point $P(x, f(x))$. As long as the curve $y = f(x)$ is "smooth" at P; that is, as long as we can draw a tangent, the limit will exist.

EXAMPLE   Find the slope of the tangent and the equation of the tangent line to the curve $y = x^2$ at the point $(2, 4)$.

SOLUTION   We have $y = f(x) = x^2$. Therefore

$$y + \Delta y = f(x + \Delta x)$$
$$= (x + \Delta x)^2$$
$$= x^2 + 2x \cdot \Delta x + (\Delta x)^2.$$

Subtracting,

$$\Delta y = 2x \cdot \Delta x + (\Delta x)^2,$$

and

$$\frac{\Delta y}{\Delta x} = 2x + \Delta x.$$

Therefore

$$\frac{dy}{dx} = \lim_{\Delta x \to 0} \frac{\Delta y}{\Delta x} = \lim_{\Delta x \to 0} (2x + \Delta x) = 2x.$$

Hence $f'(x) = 2x$. When $x = 2$, $f'(2) = 2(2) = 4$. Thus the slope of the tangent to $y = x^2$ at $(2, 4)$ is 4.

We can use the point–slope formula,

$$y - y_1 = m(x - x_1),$$

to obtain the equation of the tangent line. We must take (x_1, y_1) as the point $(2, 4)$, and the slope $m = 4$.

$$y - 4 = 4(x - 2)$$

or

$$y = 4x - 4.$$

EXAMPLE   Show that the function $f(x) = |x|$ is not differentiable at $x = 0$.

SOLUTION   We must take $x = 0$, so that $f(x) = f(0) = 0$ and $f(x + \Delta x) = f(0 + \Delta x) = |\Delta x|$. So

$$\Delta y = f(x + \Delta x) - f(x) = |\Delta x|.$$

Then

$$\frac{dy}{dx} = \lim_{\Delta x \to 0} \frac{\Delta y}{\Delta x} = \lim_{\Delta x \to 0} \frac{|\Delta x|}{\Delta x}.$$

But in Section 2.3, we discussed this limit as an example and we showed that the limit does not exist. In fact, the two limits from above and below exist but are unequal:

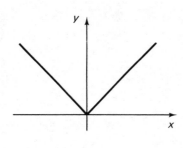

Figure 2.16

$$\lim_{\Delta x \to 0^+} \frac{|\Delta x|}{\Delta x} = 1,$$

$$\lim_{\Delta x \to 0^-} \frac{|\Delta x|}{\Delta x} = -1.$$

The graph of $y = |x|$ is shown in Fig. 2.16; for $x > 0$ the graph has a constant slope of 1, while for $x < 0$ it has a constant slope of -1. For $x = 0$ there is no slope since the graph has a corner at this value of x. This is why $|x|$ is not differentiable at $x = 0$.

EXERCISES 2.5

Find the derivatives of the following functions w. r. t. the independent variables involved:

1. $f(x) = 2x - 5$.

2. $g(u) = 3u^2 + 1$.

3. $f(t) = \dfrac{1}{t + 1}$.

4. $g(y) = \dfrac{1}{y}$.

5. Find dy/dx if:
   (a) $y = 3 - 2x^2$.   (b) $y = 3x + 7$.

6. Find du/dt if:

   (a) $u = 2t + 3$.   (b) $u = \dfrac{1}{2t + 1}$.

7. Find dx/dy if:

   (a) $x = \sqrt{y}$.   (b) $x = \dfrac{1}{y^2}$.

8. Find $f'(2)$ if $f(x) = 5 - 2x$.

9. Find $g'(4)$ if $g(x) = (x + 1)^2$.

10. Find $F'(3)$ if $F(t) = t^2 - 3t$.

11. Find $G'(1)$ if $G(u) = u^2 - u + 3$.

12. Find $h'(0)$ if $h(y) = y^2 + 7y$.

Find the slope of the tangent line to the graphs of the following functions at the indicated points. Determine also the equation of the tangent line in each case.

13. $y = \dfrac{x+1}{x}$ at $x = 1$.

14. $f(x) = \sqrt{x-1}$ at $x = 5$.

15. $f(x) = \dfrac{x+1}{x-1}$ at $x = 2$.

16. $g(t) = 5t^2 + 1$ at $t = -3$.

17. The size of a bacteria population at time t (measured in hours) is given by

$$p(t) = 10{,}000 + 1000t - 300t^2.$$

Determine the initial population $p(0)$ and the growth rate $p'(t)$ at time t.

18. Assume that a population grows according to the formula $p(t) = 30{,}000 + 60t^2$, where t is measured in days. Find the growth rate when (a) $t = 2$ days, (b) $t = 0$ days, and (c) $t = 5$ days.

19. The function $H(x)$ defined by

$$H(x) = \begin{cases} 0, & x < 0 \\ \frac{1}{2} & x = 0 \\ 1 & x > 0 \end{cases}$$

is called the *Heaviside step function*. Show that:
(a) $H(x)$ is neither continuous nor differentiable at $x = 0$.
(b) The function $f(x) = xH(x)$ is continuous but not differentiable at $x = 0$.
(c) The function $g(x) = x^2 H(x)$ is both continuous and differentiable at $x = 0$.

20. If $H(x)$ is as in exercise 19, a function of the type $F(x) = f(x)H(x - c)$ is often called a *threshold function*, since $F(x) = 0$ for $x < c$, and $F(x)$ "switches on" only for $x \geq c$. Find the conditions on $f(x)$ that guarantee that $F(x)$ is (a) continuous at $x = c$, and (b) differentiable at $x = c$.

2.6 DERIVATIVES OF POWER FUNCTIONS

It is clear from the previous section that finding derivatives of functions by directly using the definition of derivative is not always very easy and is generally time-consuming. This task can be appreciably lightened by the use of certain standard formulas. In this section, we shall develop formulas for finding the derivatives of power functions and combinations of power functions.

THEOREM 2.6.1

The derivative of a constant function is zero.

PROOF   Let

$$y = c. \tag{i}$$

be a constant function (where c is a constant). When x changes to $x + \Delta x$, y still has the same value c. Thus

$$y + \Delta y = c. \tag{ii}$$

Subtracting (i) from (ii), we have:

$$\Delta y = c - c = 0.$$

Therefore

$$\frac{\Delta y}{\Delta x} = \frac{0}{\Delta x} = 0$$

and

$$\frac{dy}{dx} = \lim_{\Delta x \to 0} \frac{\Delta y}{\Delta x} = \lim_{\Delta x \to 0} (0) = 0,$$

that is,

$$\frac{d}{dx}(c) = 0.$$

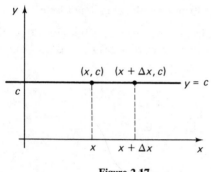

Figure 2.17

Geometrically the theorem asserts that the slope of the line $y = c$ is zero at every point on it. This is obviously true because the graph of $y = c$ (Fig. 2.17) is a horizontal line, and any horizontal line has zero slope.

EXAMPLES
$$\frac{d}{dx}(6) = 0; \quad \frac{d}{dt}\left(\frac{3}{2}\right) = 0.$$

The next theorem we shall prove concerns the derivative of the power function $f(x) = x^n$. We have the result that for any real value of n,

$$\boxed{\frac{d}{dx}(x^n) = nx^{n-1}.}$$

In words: *To find the derivative of any constant power of x, we decrease the power of x by 1 and multiply by the original exponent of x.*

EXAMPLES   **1.** $\dfrac{d}{dx}(x^7) = 7x^{7-1} = 7x^6.$

2. $\dfrac{d}{dy}(y^{3/2}) = \dfrac{3}{2}y^{(3/2)-1} = \dfrac{3}{2}y^{1/2}.$

3. $\dfrac{d}{dt}\left(\dfrac{1}{\sqrt{t}}\right) = \dfrac{d}{dt}(t^{-1/2}) = -\dfrac{1}{2}t^{(-1/2)-1} = -\dfrac{1}{2}t^{-3/2}.$

94

4. $\dfrac{d}{du}\left(\dfrac{1}{u^2}\right) = \dfrac{d}{du}(u^{-2}) = -2u^{-2-1} = -2u^{-3} = -\dfrac{2}{u^3}.$

5. $\dfrac{d}{dx}(x) = \dfrac{d}{dx}(x^1) = 1 \cdot x^{1-1} = x^0 = 1,$ because $x^0 = 1.$

In this section, we shall prove the formula for the derivative of x^n only for the case when n is a positive integer. The cases when n is a negative integer or a rational number will be proved in later sections. The proof of the present theorem makes use of the following standard result from algebra.

If n is a positive integer,

$$a^n - b^n = (a - b)(a^{n-1} + a^{n-2}b + a^{n-3}b^2 + \ldots + ab^{n-2} + b^{n-1}).$$

This result is easy to verify by multiplying out the two brackets on the right-hand side. It should be noted that the number of terms in the second bracket on the right is equal to n, the power of a or b on the left-hand side. For example, for

$$n = 2,\ a^2 - b^2 = (a - b)\underbrace{(a + b)}_{\text{2 terms}},$$

$$n = 3,\ a^3 - b^3 = (a - b)\underbrace{(a^2 + ab + b^2)}_{\text{3 terms}},$$

$$n = 4,\ a^4 - b^4 = (a - b)\underbrace{(a^3 + a^2b + ab^2 + b^3)}_{\text{4 terms}},\ \text{and so on.}$$

THEOREM 2.6.2

The derivative of x^n with respect to x is nx^{n-1} when n is a positive integer.

PROOF   Let

$$y = x^n. \tag{i}$$

When x changes to $x + \Delta x$, y changes to $y + \Delta y$ where

$$y + \Delta y = (x + \Delta x)^n. \tag{ii}$$

Subtracting (i) from (ii), we have:

$$\Delta y = (x + \Delta x)^n - x^n.$$

In order to simplify this expression for Δy we make use of the algebraic identity given above, setting $a = x + \Delta x$ and $b = x$. Then $a - b = (x + \Delta x) - x = \Delta x$, and

$$\Delta y = \Delta x\{(x + \Delta x)^{n-1} + (x + \Delta x)^{n-2} \cdot x + (x + \Delta x)^{n-3} \cdot x^2 \ldots$$
$$+ (x + \Delta x) \cdot x^{n-2} + x^{n-1}\}.$$

Dividing both sides by Δx and taking the limit as $\Delta x \longrightarrow 0$, we have:

$$\frac{dy}{dx} = \lim_{\Delta x \to 0} \frac{\Delta y}{\Delta x} = \lim_{\Delta x \to 0} \{(x + \Delta x)^{n-1} + (x + \Delta x)^{n-2} \cdot x$$
$$+ (x + \Delta x)^{n-3} \cdot x^2 \ldots + (x + \Delta x) \cdot x^{n-2} + x^{n-1}\}.$$

As $\Delta x \longrightarrow 0$, each of the terms in the bracket approaches the limit x^{n-1}. For example, the second term $(x + \Delta x)^{n-2} \cdot x \longrightarrow x^{n-2} \cdot x = x^{n-1}$ as $\Delta x \longrightarrow 0$. Furthermore there are n such terms added together, so

$$\frac{dy}{dx} = \underbrace{x^{n-1} + x^{n-1} + x^{n-1} + \ldots + x^{n-1} + x^{n-1}}_{n \text{ terms}}$$

$$= nx^{n-1},$$

as required.

We shall next prove some simple theorems on derivatives that will allow us to differentiate more complicated functions.

THEOREM 2.6.3

If $u(x)$ is a differentiable function of x and c is a constant, then

$$\frac{d}{dx}(cu) = c\frac{du}{dx};$$

that is, *the derivative of the product of a constant and a function of x is equal to the product of the constant and the derivative of the function.*

EXAMPLES   **1.** $\dfrac{d}{dx}(cx^n) = c\dfrac{d}{dx}(x^n) = c(nx^{n-1}) = ncx^{n-1}.$

2. $\dfrac{d}{dt}\left(\dfrac{4}{t}\right) = \dfrac{d}{dt}(4t^{-1}) = 4\dfrac{d}{dt}(t^{-1}) = 4(-1 \cdot t^{-2}) = -\dfrac{4}{t^2}.$

3. $\dfrac{d}{du}(2\sqrt{u}) = \dfrac{d}{du}(2u^{1/2}) = 2\dfrac{d}{du}(u^{1/2}) = 2 \cdot \dfrac{1}{2}u^{-1/2} = u^{-1/2}.$

THEOREM 2.6.4

If $u(x)$ and $v(x)$ are two differentiable functions of x, then

$$\frac{d}{dx}(u + v) = \frac{du}{dx} + \frac{dv}{dx};$$

that is, *the derivative of the sum of two functions is equal to the sum of the derivatives of the functions.*

This theorem can be readily extended to the sum of any number of functions and also to the differences between functions. For example,

$$\frac{d}{dx}(u - v) = \frac{du}{dx} - \frac{dv}{dx},$$

$$\frac{d}{dx}(u + v - w) = \frac{du}{dx} + \frac{dv}{dx} - \frac{dw}{dx}, \quad \text{and so on.}$$

EXAMPLE   Find the derivative of $x^3 + 7x^2 + 5$ w.r.t. x.

SOLUTION   Let $y = x^3 + 7x^2 + 5$. Then

$$\frac{dy}{dx} = \frac{d}{dx}(x^3 + 7x^2 + 5)$$

$$= \frac{d}{dx}(x^3) + \frac{d}{dx}(7x^2) + \frac{d}{dx}(5)$$

where we have used Theorem 2.6.4 to express the derivative of the sum $(x^3 + 7x^2 + 5)$ as the sum of the derivatives of x^3, $7x^2$ and 5. Evaluating these three derivatives we therefore obtain

$$\frac{dy}{dx} = 3x^2 + 7(2x) + 0$$

$$= 3x^2 + 14x.$$

EXAMPLE   Given $f(t) = (2t^2 + 1)(3t - 5)$, find $f'(t)$.

SOLUTION   Multiplying out the two factors on the right, we get

$$f(t) = (2t^2 + 1)(3t - 5)$$

$$= 6t^3 - 10t^2 + 3t - 5.$$

Then

$$f'(t) = \frac{d}{dt}(6t^3 - 10t^2 + 3t - 5)$$

$$= \frac{d}{dt}(6t^3) - \frac{d}{dt}(10t^2) + \frac{d}{dt}(3t) - \frac{d}{dt}(5)$$

$$= 6\frac{d}{dt}(t^3) - 10\frac{d}{dt}(t^2) + 3\frac{d}{dt}(t^1) - \frac{d}{dt}(5)$$

$$= 6(3t^2) - 10(2t^1) + 3(1 \cdot t^0) - 0$$

$$= 18t^2 - 20t + 3, \quad \text{because } t^0 = 1.$$

EXAMPLE   If $u = (6y^5 + 5y^2 - 3y + 2)/3y^2$, find du/dy.

SOLUTION   Here y is the independent variable and u the dependent variable. We note that u can be expressed as a sum of powers of y.

$$u = \frac{6y^5 + 5y^2 - 3y + 2}{3y^2}$$

$$= \frac{6y^5}{3y^2} + \frac{5y^2}{3y^2} - \frac{3y}{3y^2} + \frac{2}{3y^2}$$

$$= 2y^3 + \frac{5}{3} - \frac{1}{y} + \frac{2}{3y^2}$$

$$= 2y^3 + \tfrac{5}{3} - y^{-1} + \tfrac{2}{3}y^{-2}.$$

Therefore

$$\frac{du}{dy} = \frac{d}{dy}\left(2y^3 + \frac{5}{3} - y^{-1} + \frac{2}{3}y^{-2}\right)$$

$$= 2\frac{d}{dy}(y^3) + \frac{d}{dy}\left(\frac{5}{3}\right) - \frac{d}{dy}(y^{-1}) + \frac{2}{3}\frac{d}{dy}(y^{-2})$$

$$= 2(3y^2) + 0 - (-1 \cdot y^{-2}) + \tfrac{2}{3}(-2y^{-3})$$

$$= 6y^2 + y^{-2} - \tfrac{4}{3}y^{-3}$$

$$= 6y^2 + \frac{1}{y^2} - \frac{4}{3y^3}.$$

PROOF OF THEOREM 2.6.3

Let

$$y = cu(x). \tag{i}$$

Then if x is replaced by $x + \Delta x$, u becomes $u + \Delta u$ and y becomes $y + \Delta y$, so that

$$y + \Delta y = cu(x + \Delta x)$$
$$= c(u + \Delta u). \tag{ii}$$

Subtracting,

$$\Delta y = c\,\Delta u.$$

Dividing both sides by Δx we have

$$\frac{\Delta y}{\Delta x} = c\,\frac{\Delta u}{\Delta x}.$$

Taking the limit as $\Delta x \to 0$, we have

$$\lim_{\Delta x \to 0} \frac{\Delta y}{\Delta x} = \lim_{\Delta x \to 0}\left(c\,\frac{\Delta u}{\Delta x}\right) = c \lim_{\Delta x \to 0} \frac{\Delta u}{\Delta x};$$

that is,

$$\frac{dy}{dx} = c\,\frac{du}{dx},$$

as required. Alternatively we can write

$$\frac{d}{dx}[cf(x)] = c\,\frac{d}{dx}[f(x)].$$

PROOF OF THEOREM 2.6.4

Let

$$y = u(x) + v(x). \tag{i}$$

Let x be given an increment Δx. Since $y, u,$ and v are all functions of x, they change to $y + \Delta y, u + \Delta u$ and $v + \Delta v$, where

$$\begin{aligned} y + \Delta y &= u(x + \Delta x) + v(x + \Delta x) \\ &= (u + \Delta u) + (v + \Delta v) \end{aligned} \tag{ii}$$

Subtracting (i) from (ii),

$$\Delta y = \Delta u + \Delta v,$$

and dividing by Δx,

$$\frac{\Delta y}{\Delta x} = \frac{\Delta u}{\Delta x} + \frac{\Delta v}{\Delta x}.$$

If we now allow Δx to approach zero, we obtain

$$\lim_{\Delta x \to 0} \frac{\Delta y}{\Delta x} = \lim_{\Delta x \to 0} \frac{\Delta u}{\Delta x} + \lim_{\Delta x \to 0} \frac{\Delta v}{\Delta x},$$

that is,

$$\frac{dy}{dx} = \frac{du}{dx} + \frac{dv}{dx},$$

which proves the required result.

EXERCISES 2.6

Find the derivatives of the following functions w. r. t. the independent variable involved:

1. $f(x) = 4x^3 - 3x^2 + 7.$

2. $g(x) = 3x^4 - 7x^3 + 5x^2 + 8.$

3. $f(u) = 3u^2 + \dfrac{3}{u^2}.$

4. $h(t) = t - \dfrac{1}{t}.$

5. $F(y) = \sqrt{y} + \dfrac{1}{\sqrt{y}}.$

6. $G(x) = 3x^4 + (2x - 1)^2.$

7. $f(x) = (x - 7)(2x - 9)$.

8. $g(u) = (u + 1)(2u + 1)$.

9. $F(t) = 5 - 3t + 4t^2 - \dfrac{7}{t}$.

10. $G(y) = \dfrac{2y^2 + 3y - 7}{y}$.

11. Find $\dfrac{dy}{dx}$ if $y = x^3 + \dfrac{1}{x^3}$.

12. Find $\dfrac{du}{dx}$ if $u = x^2 - 7x + \dfrac{5}{x}$.

13. Find $\dfrac{dy}{du}$ if $y = u^3 - 5u^2 + \dfrac{7}{3u^2} + 6$.

14. Find $\dfrac{dx}{dt}$ if $x = \dfrac{t^3 - 5t^2 + 7t - 1}{t^2}$.

15. If $y = \sqrt{x}$, prove that $2y\dfrac{dy}{dx} = 1$.

16. If $u = \dfrac{1}{\sqrt{x}}$, prove that $2u^{-3}\dfrac{du}{dx} + 1 = 0$.

17. The distance traveled by a moving object at a time t is equal to $2t^3 - t^{1/2}$. Find the instantaneous velocity: (a) at time t; (b) at the time 4.

18. A ball is thrown vertically upwards with an initial velocity of 60 ft/sec. After t sec its height above the ground is given by $s = 60t - 16t^2$. Find its instantaneous velocity after t sec. What is special about $t = \frac{15}{8}$?

19. At the start of an experiment, a culture of bacteria is found to contain 10,000 individuals. The growth of the population was observed and was found that at any subsequent time t (hours) after the start of the experiment, the population size $p(t)$ could be expressed by the formula

$$p(t) = 2500(2 + t)^2.$$

Determine the formula for the rate of growth of the population at any time t, and in particular calculate the growth rate for $t = 15$ min and for $t = 2$ hr.

20. The proportion of seeds of a certain species of tree that scatter further than distance r from the base of the tree is given by

$$p(r) = \frac{3}{4}\left(\frac{r_0}{r}\right)^{1/2} + \frac{1}{4}\left(\frac{r_0}{r}\right),$$

where r_0 is a constant. Find the rate of change of the proportion with respect to distance and calculate $p'(2r_0)$.

21. During rapid (adiabatic) changes in pressure, the pressure p and density ρ of a gas vary according to the law $p\rho^{-\gamma} = c$, where γ and c are constants. Calculate $dp/d\rho$.

22. According to the Schütz–Borisoff law, the amount y of substrate transformed by an enzyme in a time interval t is given by $y = k\sqrt{cat}$, where c is the concentration of the enzyme, a is the initial concentration of the substrate, and k is a constant. What is the rate at which the substrate is being transformed?

2.7 DERIVATIVES OF PRODUCTS AND QUOTIENTS

In this section we shall prove and explain the use of two important theorems that provide useful techniques for differentiating complicated functions.

THEOREM 2.7.1

If $u(x)$ and $v(x)$ are any two differentiable functions of x, then

$$\boxed{\frac{d}{dx}(u \cdot v) = u\frac{dv}{dx} + v\frac{du}{dx}}$$

or

$$y' = uv' + vu';$$

that is, *the derivative of the product of two functions is equal to the first function times the derivative of the second plus the second function times the derivative of the first.* This result is called the *product rule*.

EXAMPLE   Find y' if $y = (5x^2 - 3x)(2x^3 + 8x + 7)$.

SOLUTION   The given function y can be written as a product $y = uv$ if we define

$$u = 5x^2 - 3x \quad \text{and} \quad v = 2x^3 + 8x + 7.$$

Then

$$u' = 10x - 3 \quad \text{and} \quad v' = 6x^2 + 8.$$

Therefore, by the product rule,

$$\begin{aligned} y' &= uv' + vu' \\ &= (5x^2 - 3x)(6x^2 + 8) + (2x^3 + 8x + 7)(10x - 3) \\ &= 50x^4 - 24x^3 + 120x^2 + 22x - 21, \end{aligned}$$

after simplification.

In this example we do not actually need the product rule in order to differentiate the given function. We could have calculated y' by multiplying out the brackets in y and expressing y as a sum of powers of x:

$$\begin{aligned} y &= (5x^2 - 3x)(2x^3 + 8x + 7) \\ &= 10x^5 - 6x^4 + 40x^3 + 11x^2 - 21x. \\ y' &= 10(5x^4) - 6(4x^3) + 40(3x^2) + 11(2x) - 21(1) \\ &= 50x^4 - 24x^3 + 120x^2 + 22x - 21. \end{aligned}$$

It will also be true of the other examples that we shall now give on the use of the product rule that they can be solved using the methods of the last section. However, later on we shall come across functions for which such an alternative method does not exist. For these later functions it will be essential to use the product rule in order to find their derivatives.

EXAMPLE   Given $f(t) = (2\sqrt{t} + 1)(t^2 + 3)$, find $f'(t)$.

SOLUTION   Using the product rule with $u = 2\sqrt{t} + 1$ and $v = t^2 + 3$, we have

$$f'(t) = (2\sqrt{t} + 1)\frac{d}{dt}(t^2 + 3) + (t^2 + 3)\frac{d}{dt}(2\sqrt{t} + 1)$$

$$= (2\sqrt{t} + 1)\frac{d}{dt}(t^2 + 3) + (t^2 + 3)\frac{d}{dt}(2t^{1/2} + 1)$$

$$= (2\sqrt{t} + 1)(2t) + (t^2 + 3)(2 \cdot \tfrac{1}{2}t^{-1/2})$$

$$= 4t^{3/2} + 2t + t^{3/2} + 3t^{-1/2}$$

$$= 5t^{3/2} + 2t + \frac{3}{\sqrt{t}}.$$

We can extend the rule to the product of three functions as is illustrated in the following example.

EXAMPLE   Find y' if $y = (x^2 + 1)(3x + 4)(5 - 3x^2)$.

SOLUTION   Let

$$u = x^2 + 1, \qquad v = 3x + 4, \qquad w = 5 - 3x^2, \quad \text{so that}$$

$$u' = 2x, \qquad v' = 3, \qquad w' = -6x, \quad \text{and} \quad y = uvw.$$

If we regard y as the product of (uv) times w, the product rule gives

$$y' = (uv)w' + w(uv)'$$

$$= (uv)w' + w(uv' + vu'),$$

where we have used the product rule a second time in evaluating $(uv)'$. Therefore

$$y' = uvw' + wuv' + wvu'$$

$$= (x^2 + 1)(3x + 4)(-6x) + (5 - 3x^2)(x^2 + 1)(3) + (5 - 3x^2)(3x + 4)(2x)$$

$$= -45x^4 - 48x^3 + 18x^2 + 16x + 15$$

after simplification.

PROOF OF THEOREM 2.7.1

Let $y = uv$. Then

$$y + \Delta y = (u + \Delta u) \cdot (v + \Delta v)$$

$$= uv + u \cdot \Delta v + v \cdot \Delta u + \Delta u \cdot \Delta v.$$

Subtracting,

$$\Delta y = u \cdot \Delta v + v \cdot \Delta u + \Delta u \cdot \Delta v,$$

or

$$\frac{\Delta y}{\Delta x} = u \frac{\Delta v}{\Delta x} + v \frac{\Delta u}{\Delta x} + \Delta u \cdot \frac{\Delta v}{\Delta x}.$$

Proceeding to limits as $\Delta x \longrightarrow 0$, we have

$$\lim_{\Delta x \to 0} \frac{\Delta y}{\Delta x} = u \cdot \lim_{\Delta x \to 0} \frac{\Delta v}{\Delta x} + v \cdot \lim_{\Delta x \to 0} \frac{\Delta u}{\Delta x} + \lim_{\Delta x \to 0} \Delta u \cdot \lim_{\Delta x \to 0} \frac{\Delta v}{\Delta x}.$$

(Note that the Limit Theorem 2.2.3(a) and (c) has been used.) In the last term on the right, $\Delta u \longrightarrow 0$ as $\Delta x \longrightarrow 0$, so we get

$$\frac{dy}{dx} = u \frac{dv}{dx} + v \frac{du}{dx}.$$

Thus

$$\text{if} \quad y = uv, \qquad y' = uv' + vu'.$$

THEOREM 2.7.2

If $u(x)$ and $v(x)$ are differentiable functions of x, then

$$\boxed{\frac{d}{dx}\left(\frac{u}{v}\right) = \frac{v \dfrac{du}{dx} - u \dfrac{dv}{dx}}{v^2}}$$

or

$$\left(\frac{u}{v}\right)' = \frac{vu' - uv'}{v^2}.$$

In words: *The derivative of a quotient of two functions is equal to the denominator times the derivative of the numerator minus the numerator times the derivative of the denominator, all divided by the denominator squared.* The result of this theorem is known as the *quotient rule*.

EXAMPLE   Find y' if

$$y = \frac{x^2 + 1}{x^3 + 4}.$$

SOLUTION   Using the quotient rule (with $u = x^2 + 1$ and $v = x^3 + 4$), we have

$$y' = \frac{(x^3 + 4) \dfrac{d}{dx}(x^2 + 1) - (x^2 + 1) \dfrac{d}{dx}(x^3 + 4)}{(x^3 + 4)^2}$$

$$= \frac{(x^3 + 4)(2x) - (x^2 + 1)(3x^2)}{(x^3 + 4)^2}$$

$$= \frac{2x^4 + 8x - (3x^4 + 3x^2)}{(x^3 + 4)^2}$$

$$= \frac{-x^4 - 3x^2 + 8x}{(x^3 + 4)^2}.$$

Let $y = u/v$. When x changes to $x + \Delta x$, then y changes to $y + \Delta y$, u to $u + \Delta u$ and v to $v + \Delta v$, so that

$$y + \Delta y = \frac{u + \Delta u}{v + \Delta v}.$$

Subtracting,

$$\Delta y = \frac{u + \Delta u}{v + \Delta v} - \frac{u}{v}$$

$$= \frac{v(u + \Delta u) - u(v + \Delta v)}{v(v + \Delta v)}$$

$$= \frac{v \, \Delta u - u \, \Delta v}{v(v + \Delta v)}.$$

Dividing by Δx we obtain

$$\frac{\Delta y}{\Delta x} = \frac{v \dfrac{\Delta u}{\Delta x} - u \dfrac{\Delta v}{\Delta x}}{v(v + \Delta v)}.$$

If we now proceed to the limit as $\Delta x \rightarrow 0$, so that

$$\frac{\Delta y}{\Delta x} \rightarrow \frac{dy}{dx}, \quad \frac{\Delta u}{\Delta x} \rightarrow \frac{du}{dx}, \quad \text{and} \quad \frac{\Delta v}{\Delta x} \rightarrow \frac{dv}{dx},$$

we have

$$\frac{dy}{dx} = \frac{v \dfrac{du}{dx} - u \dfrac{dv}{dx}}{v(v + 0)},$$

since the odd Δv in the denominator approaches zero. Thus we have proved the result stated in the theorem.

In the last section, we proved that $\dfrac{d}{dx}(x^n) = nx^{n-1}$ when n is a positive integer.

Using the quotient rule it is possible to verify this formula for the case when n is a negative integer. The proof is contained in the following theorem.

THEOREM 2.7.3

If n is any negative integer, the derivative of x^n w. r. t. x is nx^{n-1}.

PROOF

Define $m = -n$ so that m is a positive integer [for example, if $n = -7$, then $m = -(-7) = +7$]. Then

$$x^n = x^{-m} = \frac{1}{x^m}.$$

It follows therefore that we can write x^n as a quotient,

$$x^n = \frac{u(x)}{v(x)},$$

where $u(x) = 1$ and $v(x) = x^m$. From the quotient rule,

$$\frac{d}{dx}(x^n) = \frac{u'v - uv'}{v^2}.$$

Now since $u(x) = 1$, its derivative $u' = 0$. Furthermore, $v(x) = x^m$, and m is a positive integer, so we can use Theorem 2.6.2 to conclude that $v' = mx^{m-1}$. Therefore

$$\frac{d}{dx}(x^n) = \frac{(0)(x^m) - (1)(mx^{m-1})}{(x^m)^2}$$

$$= \frac{-mx^{m-1}}{x^{2m}} = -mx^{m-1-2m} = -mx^{-m-1}.$$

Recalling that $-m = n$, we conclude that

$$\frac{d}{dx}(x^n) = nx^{n-1},$$

as required.

EXERCISES 2.7

Using the product rule, find the derivatives of the following functions with respect to the variable involved:

1. $y = (x + 1)(x^3 + 3)$.

2. $y = (x^3 + 6x^2)(x^2 - 1)$.

3. $u = (7x + 1)(2 - 3x)$.

4. $u = (x^2 + 7x)(x^2 + 3x + 1)$.

5. $f(x) = (x^2 - 5x + 1)(2x + 3)$.

6. $g(t) = \left(t + \dfrac{1}{t}\right)\left(5t^2 - \dfrac{1}{t^2}\right)$.

7. $g(x) = (x^2 + 1)(3x - 1)(2x - 3)$.

8. $f(x) = (2x + 1)(3x^2 + 1)(x^3 + 7)$.

Use the quotient rule to find the derivatives of the following functions with respect to the independent variable involved:

9. $f(x) = \dfrac{x + 2}{x - 1}$.   **10.** $g(x) = \dfrac{3 - x}{x^2 - 3}$.

11. $y = \dfrac{t^2 - 7t}{t - 5}$.   **12.** $y = \dfrac{u^2 - u + 1}{u^2 + u + 1}$.

13. $x = \dfrac{\sqrt{u} + 1}{\sqrt{u} - 1}$.   **14.** $t = \dfrac{x^2 - 1}{x^2 + 1}$.

15. $y = \dfrac{1}{x^2 + 1}$.   **16.** $y = \dfrac{1}{(t + 1)^2}$.

17. Consider a sense organ which, when it receives a stimulus, produces a number of action potentials. Suppose at a particular time t sec from the start of stimulus, the total number of action potentials is given by

$$f(t) = 5t + \frac{3}{t^2 + 1} - 3.$$

Find the rate of change of the number of action potentials at any time t.

18. The weight of a certain stock of fish is given by $W = nw$, where n is the size of the stock and w the average weight of each fish. If n and w change with time t according to the formulas $n = (2t^2 + 3)$ and $w = (t^2 - t + 2)$, find the rate of change of W w.r.t. time.

19. The absolute temperature T of a gas is given by $T = cPV$, where P is its pressure, V its volume, and c is some constant depending on the mass of gas. If $P = (t^2 + 1)$ and $V = (2t + t^{-1})$ as functions of time t, find the rate of change of T w. r. t. t.

20. The density of algae in a water tank is equal to n/V, where n is the number of algae and V the volume of water in the tank. If n and V vary with time t according to the formulas $n = \sqrt{t}$ and $V = \sqrt{t} + 1$, calculate the rate of change of the density.

21. Let x be the size of a certain population of predators and y the size of the population of prey upon which they feed. As functions of time t, $x = t^2 + 4$ and $y = 2t^2 - 3t$. Let u be the number of prey to each predator. Find the rate of change of u.

2.8   DERIVATIVES OF COMPOSITE FUNCTIONS

Let $y = f(u)$ be a function of u, and $u = u(x)$ be a function of x. Then we can write

$$y = f[u(x)],$$

representing y as a function of x, called the composite function of f and u. It is denoted by $f \circ u(x)$ (Section 1.7).

The derivatives of composite functions can be found by the use of the following theorem.

THEOREM 2.8.1 (Chain Rule)

If y is a function of u and u is a function of x, then

$$\boxed{\frac{dy}{dx} = \frac{dy}{du} \cdot \frac{du}{dx}.}$$

The chain rule provides what is probably the most useful of all the aids to differentiation, as will soon become apparent. It is a tool that is seldom out of one's hands when working with the differential calculus, and the student should master its use as soon as possible. When using it to differentiate some complicated function, it is necessary at the start to spot how to break the given function into the composition of two simpler functions. The following examples provide some illustrations.

EXAMPLE Find dy/dx when $y = (x^2 + 1)^5$.

SOLUTION We could solve this problem by expanding $(x^2 + 1)^5$ as a polynomial in x. However it is much simpler to use the chain rule.

Observe that y can be written as a composite function in the following way:

$$y = u^5 \quad \text{where} \quad u = x^2 + 1.$$

Then

$$\frac{dy}{du} = 5u^4 \quad \text{and} \quad \frac{du}{dx} = 2x.$$

From the chain rule, therefore,

$$\frac{dy}{dx} = \frac{dy}{du} \cdot \frac{du}{dx}$$
$$= 5u^4 \cdot 2x$$
$$= 5(x^2 + 1)^4 \cdot 2x$$
$$= 10x(x^2 + 1)^4.$$

Another way of writing the chain rule is that if $y = f(u)$, then

$$\frac{dy}{dx} = f'(u)\frac{du}{dx}$$

(since $f'(u) = dy/du$). In particular, if $f(u) = u^n$, then $f'(u) = nu^{n-1}$. So we have the following special case of the chain rule:

$$\text{If} \quad y = [u(x)]^n \quad \text{then} \quad \frac{dy}{dx} = nu^{n-1}\frac{du}{dx}.$$

It is easier to remember these results in their verbal forms. Instead of using $u(x)$, let us use the word *inside* to stand for the function of x, which appears on the inside of the composition. Then

$$\text{If} \quad y = f(inside),$$

$$\text{then} \quad \frac{dy}{dx} = f'(inside) \cdot \text{derivative of } inside \text{ w. r. t. } x.$$

$$\text{If} \quad y = (inside)^n,$$

$$\text{then} \quad \frac{dy}{dx} = n(inside)^{n-1} \cdot \text{derivative of } inside \text{ w. r. t. } x.$$

Here *inside* stands for any differentiable function of x.

From the preceding example, $y = (x^2 + 1)^5$, we would understand *inside* to be $x^2 + 1$, and $y = f(inside) = (inside)^5$. Then immediately,

$$\frac{dy}{dx} = 5(inside)^4 \cdot \frac{d}{dx}(inside)$$

$$= 5(x^2 + 1)^4 \cdot \frac{d}{dx}(x^2 + 1)$$

$$= 5(x^2 + 1)^4 \cdot 2x$$

$$= 10x(x^2 + 1)^4,$$

giving the same answer as before.

EXAMPLE   Given $f(t) = 1/\sqrt{t^2 + 3}$, find $f'(t)$.

SOLUTION   Let $u = t^2 + 3$ so that

$$y = f(t) = \frac{1}{\sqrt{u}} = u^{-1/2}.$$

Then

$$\frac{du}{dt} = 2t \quad \text{and} \quad \frac{dy}{du} = -\frac{1}{2}u^{-3/2} = -\frac{1}{2}(t^2 + 3)^{-3/2}.$$

Thus by the chain rule,

$$\frac{dy}{dt} = \frac{dy}{du} \cdot \frac{du}{dt}$$

$$= -\tfrac{1}{2}(t^2 + 3)^{-3/2} \cdot 2t$$

$$= \frac{-t}{(t^2 + 3)^{3/2}}.$$

OR DIRECTLY

$$f(t) = \frac{1}{\sqrt{t^2 + 3}} = (t^2 + 3)^{-1/2}.$$

Here *inside* is $(t^2 + 3)$ and

$$f'(t) = -\frac{1}{2}(t^2 + 3)^{-(1/2)-1} \cdot \frac{d}{dt}(t^2 + 3)$$

$$= -\tfrac{1}{2}(t^2 + 3)^{-3/2} \cdot 2t$$

$$= -\frac{t}{(t^2 + 3)^{3/2}}.$$

EXAMPLE   Given $y = (x^2 + 5x + 1)(2 - x^2)^4$, find dy/dx.

SOLUTION   Using the product rule for derivatives, we get

$$\frac{dy}{dx} = (x^2 + 5x + 1)\frac{d}{dx}(2 - x^2)^4 + (2 - x^2)^4 \cdot \frac{d}{dx}(x^2 + 5x + 1).$$

In order to evaluate $(d/dx)(2 - x^2)^4$ we can use the chain rule with *inside* equal to $(2 - x^2)$. Thus we get

$$\frac{dy}{dx} = (x^2 + 5x + 1)\left[4(2 - x^2)^3 \cdot \frac{d}{dx}(2 - x^2)\right] + (2 - x^2)^4 \cdot (2x + 5)$$

$$= (x^2 + 5x + 1)[4(2 - x^2)^3(-2x)] + (2 - x^2)^4 \cdot (2x + 5)$$

$$= -8x(2 - x^2)^3(x^2 + 5x + 1) + (2x + 5)(2 - x^2)^4$$

$$= (2 - x^2)^3[-8x(x^2 + 5x + 1) + (2x + 5)(2 - x^2)]$$

$$= (2 - x^2)^3[10 - 4x - 45x^2 - 10x^3].$$

EXAMPLE   Given $y = \sqrt{x + 1}/\sqrt{3x + 4}$, find dy/dx.

SOLUTION   Let $u = (x + 1)/(3x + 4)$, so that $y = \sqrt{u} = u^{1/2}$. In order to use the chain rule we need du/dx, which can be found from the quotient rule:

$$\frac{du}{dx} = \frac{(3x + 4)\frac{d}{dx}(x + 1) - (x + 1)\frac{d}{dx}(3x + 4)}{(3x + 4)^2}$$

$$= \frac{(3x + 4) \cdot 1 - (x + 1) \cdot 3}{(3x + 4)^2} = \frac{1}{(3x + 4)^2}.$$

Also

$$\frac{dy}{du} = \frac{1}{2}u^{-1/2} = \frac{1}{2}\left(\frac{x + 1}{3x + 4}\right)^{-1/2} = \frac{1}{2}\left(\frac{3x + 4}{x + 1}\right)^{1/2}.$$

Therefore, by the chain rule,

$$\frac{dy}{dx} = \frac{dy}{dx} \cdot \frac{du}{dx}$$

$$= \frac{1}{2}\left(\frac{3x + 4}{x + 1}\right)^{1/2} \cdot \frac{1}{(3x + 4)^2}$$

$$= \frac{1}{2(x + 1)^{1/2}(3x + 4)^{3/2}}.$$

OR DIRECTLY

$$y = \left(\frac{x + 1}{3x + 4}\right)^{1/2}.$$

Here *inside* is $(x + 1)/(3x + 4)$. Therefore,

$$\frac{dy}{dx} = \frac{1}{2}\left(\frac{x + 1}{3x + 4}\right)^{-1/2} \cdot \frac{d}{dx}\left(\frac{x + 1}{3x + 4}\right)$$

$$= \frac{1}{2}\left(\frac{3x + 4}{x + 1}\right)^{1/2} \cdot \frac{(3x + 4) \cdot \frac{d}{dx}(x + 1) - (x + 1) \cdot \frac{d}{dx}(3x + 4)}{(3x + 4)^2}$$

$$= \frac{1}{2}\left(\frac{3x + 4}{x + 1}\right)^{1/2} \cdot \frac{(3x + 4)(1) - (x + 1)(3)}{(3x + 4)^2}$$

$$= \frac{1}{2(x + 1)^{1/2} \cdot (3x + 4)^{3/2}}.$$

The proof of the chain rule, if given in complete detail, would be a little more complicated than we wish to include. We shall therefore provide a proof which, although covering most cases that arise, does have certain restrictions on its range of applicability.

Let Δx be an increment in x. Since u and y are functions of x, they will change whenever x changes, so we denote their increments by Δu and Δy. As long as $\Delta u \neq 0$, we have

$$\frac{\Delta y}{\Delta x} = \frac{\Delta y}{\Delta u} \frac{\Delta u}{\Delta x}.$$

We now let $\Delta x \longrightarrow 0$. In this limit, we also have that $\Delta u \longrightarrow 0$ and $\Delta y \longrightarrow 0$, and so

$$\lim_{\Delta x \to 0} \frac{\Delta y}{\Delta x} = \lim_{\Delta x \to 0} \left(\frac{\Delta y}{\Delta u} \frac{\Delta u}{\Delta x} \right)$$

$$= \left(\lim_{\Delta x \to 0} \frac{\Delta y}{\Delta u} \right) \left(\lim_{\Delta x \to 0} \frac{\Delta u}{\Delta x} \right)$$

$$= \left(\lim_{\Delta u \to 0} \frac{\Delta y}{\Delta u} \right) \left(\frac{du}{dx} \right)$$

$$= \frac{dy}{du} \cdot \frac{du}{dx},$$

as required.

The reason why this proof is incomplete lies in the assumption that $\Delta u \neq 0$. For most functions $u(x)$, Δu will never vanish when Δx is sufficiently small (but $\Delta x \neq 0$). However, it is conceivable that the function $u(x)$ could be so peculiar in its behavior that Δu vanishes repeatedly as $\Delta x \longrightarrow 0$. For such an unusual function, the above proof would then break down. It is possible to modify the proof to cover such cases as this, but we shall not do so here.

Now that we know the chain rule, we are able to prove the following result.

THEOREM 2.8.2

If n is a rational number, then

$$\frac{d}{dx}(x^n) = nx^{n-1}.$$

PROOF

In this proof we shall assume that x^n is differentiable when n is a rational number. Using a different technique of proof entirely, it is possible to establish such differentiability, but we shall content ourselves with assuming that it is true.

Let $n = p/q$ where p and q are integers with $q > 0$. We observe that

$$(x^n)^q = x^{nq} = x^{(p/q) \cdot q} = x^p.$$

Therefore

$$\frac{d}{dx}[(x^n)^q] = \frac{d}{dx}(x^p) = px^{p-1}.$$

110

But by the chain rule (with *inside* equal to x^n)

$$\frac{d}{dx}[(x^n)^q] = q(x^n)^{q-1} \cdot \frac{d}{dx}(x^n)$$

$$= qx^{nq-n}\frac{d}{dx}(x^n)$$

$$= qx^{p-n}\frac{d}{dx}(x^n), \quad \text{because } nq = p.$$

Equating the two values of $\frac{d}{dx}[(x^n)^q]$, we have

$$qx^{p-n}\frac{d}{dx}(x^n) = px^{p-1}.$$

Therefore

$$\frac{d}{dx}(x^n) = \frac{px^{p-1}}{qx^{p-n}}$$

$$= \frac{p}{q}x^{(p-1)-(p-n)}$$

$$= \frac{p}{q}x^{n-1} = nx^{n-1} \quad \left(\text{because } \frac{p}{q} = n\right),$$

as required.

Thus we have proved so far that

$$\frac{d}{dx}(x^n) = nx^{n-1}$$

for *n* being any positive or negative integer or any positive or negative rational number. The result is also true in fact when *n* is any real number.

EXERCISES 2.8

Find the derivatives of the following functions with respect to the independent variable involved.

1. $y = (3x + 5)^7$.

2. $y = \sqrt{5 - 2t}$.

3. $u = (2x^2 + 1)^{3/2}$.

4. $x = (y^3 + 7)^6$.

5. $f(x) = \dfrac{1}{(x^2 + 1)^4}$.

6. $g(x) = \dfrac{1}{(x^2 + x + 1)^3}$.

7. $h(t) = \sqrt{t^2 + a^2}$.

8. $F(x) = \sqrt[3]{x^3 + 3x}$.

9. $G(u) = (u^2 + 1)^3(2u + 1)$.

10. $H(y) = (2y^2 + 3)^6(5y + 2)$.

11. $x = \dfrac{1}{\sqrt[3]{t^3 + 1}}$.

12. $y = \left(t + \dfrac{1}{t}\right)^{10}$.

13. $y = \dfrac{(x^2 + 1)^2}{x + 1}$.

14. $y = \sqrt{\dfrac{3x + 7}{5 + 2x}}$.

15. $x = \dfrac{t^2}{\sqrt{t^2 + 4}}$.

16. $Z = \dfrac{\sqrt{2x + 1}}{x + 2}$.

17. The rate R at which a chemical reaction progresses is equal to $\sqrt{T}$, where T is the temperature. If T varies with time t according to the formula $T = (3t + 1)/(t + 2)$, find the rate of change of R with respect to t.

18. The distance traveled by a moving object up to time t is given by $y = (3t + 1)\sqrt{t + 1}$. Find the instantaneous velocity at time t.

19. The proportion P of seeds that germinate depends on the soil temperature T. Suppose that under certain conditions, $P = T^7$, and that T varies with depth x below the surface as $T = (x^2 + 3)/(x + 3)$. Find the rate of change of P with respect to depth.

20. The size of a certain population at time t is $[(t^2 + 3t + 1)/(t + 1)]^6$. Find the rate of change of the population size.

2.9   HIGHER DERIVATIVES

If $y = f(t)$ is a function of time t, then as we have seen, the derivative $dy/dt = f'(t)$ represents the rate at which y changes. For example, if $s = f(t)$ is the distance traveled by a moving object, then $ds/dt = f'(t)$ gives the rate of change of distance or, in other words, the instantaneous *velocity* of the object. Let us denote this velocity by v. Then v is also a function of t, and as a rule can be differentiated to give the derivative dv/dt. This quantity represents the rate at which the velocity changes, that is, the *acceleration* of the moving object.

To calculate the acceleration, we must differentiate s and then differentiate the result once more. We have:

$$\text{Acceleration} = \frac{dv}{dt} = \frac{d}{dt}\left(\frac{ds}{dt}\right).$$

Acceleration is called the *second derivative* of s w. r. t. t and is usually denoted by $f''(t)$ or by d^2s/dt^2.

We can also differentiate the acceleration w.r.t. t since $f''(t)$ is, in general, a function of t that is differentiable. The result is denoted by $f'''(t)$ or by d^3s/dt^3 and is called the *third derivative* of s w.r.t. t. It represents the rate of change of the acceleration of the object. (There is no separate name like "velocity" or "acceleration" for this derivative.)

In problems having to do with moving objects, the second derivative, acceleration, is a quantity of prime importance. For example, the degree of safety of the braking system of an automobile depends on the maximum deceleration it can give (deceleration is just a negative acceleration). The medical effects of rocket launching on an astronaut depend on the level of acceleration to which he is subjected. The usefulness of the third derivative is much more restricted. (However one area in which it has found use is in the design of rapid transit systems, where the third derivative is used as a measure of the jolt received by a passenger when a subway train starts to move.)

We shall now examine higher-order derivatives in a more abstract context. Let $y = f(x)$ be a given function of x with derivative $dy/dx = f'(x)$. Technically, this is called the *first derivative* of y w.r.t. x. If $f'(x)$ is a differentiable function of x, its derivative is called the *second derivative* of y w.r.t x. If the second derivative is a differentiable function of x, its derivative is the *third derivative* of y, and so on.

The first and higher-order derivatives of y w.r.t. x are generally denoted by

$$\frac{dy}{dx}, \frac{d^2y}{dx^2}, \frac{d^3y}{dx^3}, \ldots, \frac{d^ny}{dx^n}$$

or

$$y', y'', y''', \ldots, y^{(n)}$$

or

$$f'(x), f''(x), f'''(x), \ldots, f^{(n)}(x).$$

From the definition of higher-order derivatives, it is clear that

$$\frac{d^2y}{dx^2} = \frac{d}{dx}\left(\frac{dy}{dx}\right), \quad \frac{d^3y}{dx^3} = \frac{d}{dx}\left(\frac{d^2y}{dx^2}\right), \quad \text{etc.}$$

EXAMPLE   Find the first and higher-order derivatives of $3x^4 - 5x^3 + 7x^2 - 1$.

SOLUTION   Let $y = 3x^4 - 5x^3 + 7x^2 - 1$. Then

$$\frac{dy}{dx} = \frac{d}{dx}(3x^4 - 5x^3 + 7x^2 - 1)$$

$$= 12x^3 - 15x^2 + 14x.$$

The second derivative of y is obtained by differentiating the first derivative.

$$\frac{d^2y}{dx^2} = \frac{d}{dx}\left(\frac{dy}{dx}\right)$$

$$= \frac{d}{dx}(12x^3 - 15x^2 + 14x)$$

$$= 36x^2 - 30x + 14.$$

Differentiating again, we obtain the third derivative.

$$\frac{d^3y}{dx^3} = \frac{d}{dx}\left(\frac{d^2y}{dx^2}\right)$$

$$= \frac{d}{dx}(36x^2 - 30x + 14)$$

$$= 72x - 30.$$

Continuing this process,

$$\frac{d^4y}{dx^4} = \frac{d}{dx}\left(\frac{d^3y}{dx^3}\right) = \frac{d}{dx}(72x - 30) = 72.$$

$$\frac{d^5y}{dx^5} = \frac{d}{dx}\left(\frac{d^4y}{dx^4}\right) = \frac{d}{dx}(72) = 0.$$

$$\frac{d^6y}{dx^6} = \frac{d}{dx}\left(\frac{d^5y}{dx^5}\right) = \frac{d}{dx}(0) = 0, \quad \text{and so on.}$$

In this particular example, all derivatives higher than the fourth are zero. This occurs because the fourth derivative is a constant.

EXAMPLE   Find the second derivative of $f(t) = (t^2 + 1)^5$.

SOLUTION   To find the first derivative, we use the chain rule. Thus

$$f'(t) = 5(t^2 + 1)^{5-1} \cdot \frac{d}{dt}(t^2 + 1)$$

$$= 5(t^2 + 1)^4 \cdot 2t$$

$$= 10t(t^2 + 1)^4.$$

Now $f'(t)$ is the product of two functions: $u = 10t$ and $v = (t^2 + 1)^4$. To find $f''(t)$, we shall use the product rule. Thus

$$f''(t) = (10t)\frac{d}{dt}(t^2 + 1)^4 + (t^2 + 1)^4 \cdot \frac{d}{dt}(10t)$$

$$= (10t)\left[4(t^2 + 1)^{4-1} \cdot \frac{d}{dt}(t^2 + 1)\right] + (t^2 + 1)^4 \cdot (10)$$

where we have used the chain rule to differentiate $v = (t^2 + 1)^4$. Therefore

$$f''(t) = (10t)[4(t^2 + 1)^3 \cdot 2t] + 10(t^2 + 1)^4$$

$$= 80t^2(t^2 + 1)^3 + 10(t^2 + 1)^4$$

$$= 10(t^2 + 1)^3[8t^2 + (t^2 + 1)]$$

$$= 10(t^2 + 1)^3(9t^2 + 1).$$

EXERCISES 2.9

Find the first and higher-order derivatives of the following functions w. r. t. the independent variable involved:

1. $y = 3x^5 + 7x^3 - 4x^2 + 12$.

2. $u = (t^2 + 1)^2$.

3. $f(x) = x^3 - 6x^2 + 9x + 16$.

4. $y(u) = (u^2 + 1)(3u - 2)$.

5. Find y'' if $y = \dfrac{x^2}{(x^2 + 1)}$.

6. Find $f'''(t)$ if $f(t) = \dfrac{(t - 1)}{(t + 1)}$.

7. Find $g^{(iv)}(u)$ if $g(u) = \dfrac{1}{(3u + 1)}$.

8. Find $\dfrac{d^2y}{dt^2}$ if $y = \sqrt{t^2 + 1}$.

9. Find $\dfrac{d^2u}{dx^2}$ if $u = \dfrac{1}{(x^2 + 1)}$.

10. Find $\dfrac{d^3y}{dx^3}$ if $y = \dfrac{(x^3 - 1)}{(x - 1)}$, $(x \neq 1)$.

11. Find the velocity and the acceleration of a moving object when the distance s traveled in time t is as given below.

(a) $s = 9t + 16t^2$.
(b) $s = 3t^3 + 7t^2 - 5t$.

12. If the distance s traveled in time t is given by $s = t(3 - t)$,

(a) at what times is the velocity zero?
(b) what is the value of acceleration when the velocity equals zero?

REVIEW EXERCISES FOR CHAPTER 2

1. State whether the following statements are true or false. If false, replace them with a correct statement.

(a) The derivative of the sum of two functions is equal to the sum of their derivatives.

(b) The derivative of the product of two functions is equal to the product of their derivatives.

(c) The derivative of the quotient of two functions is equal to the quotient of the derivatives of the two functions.

(d) $\dfrac{d}{dx}[f(x)]^n = n[f(x)]^{n-1}$.

(e) The derivative of y w.r.t. x represents the average rate of change of y w.r.t. x.

(f) The derivative of $f(y)$ w.r.t. u is $f'(y)$.

(g) If a function is continuous at a point, then it is differentiable at that point.

(h) If the derivative of a function does not exist at a point, then the function is not defined at that point.

(i) A function $f(x)$ is continuous at a point $x = c$ if and only if $\lim\limits_{x \to c} f(x)$ exists.

(j) A function must be defined at a point if the limit of the function exists at that point.

(k) If $\lim\limits_{x \to a} f(x)$ exists then it must be equal to $f(a)$.

(l) If the acceleration of a moving object is zero, then its velocity is also zero.

Evaluate the following limits.

2. $\lim\limits_{h \to 0} \dfrac{1}{h}\left(\dfrac{1}{x + h} - \dfrac{1}{x}\right)$.

3. $\lim\limits_{x \to a} \dfrac{\sqrt{x} - \sqrt{a}}{x - a}$.

4. $\lim\limits_{h \to 0} \dfrac{\sqrt{x + h} - \sqrt{x}}{h}$.

5. $\lim\limits_{x \to 0} \dfrac{x}{1 - \sqrt{1 - x}}$.

Determine which of the following functions are continuous at the indicated points.

6. $f(x) = \dfrac{x^3 + 3x}{x}$ at $x = 0$.

7. $f(x) = \begin{cases} \dfrac{|x^2 - 9|}{x - 3} & \text{if } x \neq 3 \\ 6 & \text{if } x = 3 \end{cases}$ at $x = 3$.

8. Determine the value of h if

$$f(x) = \begin{cases} x^2 + h & \text{for } x \neq 1 \\ 3 & \text{for } x = 1 \end{cases}$$

is continuous at $x = 1$.

9. Determine the value of the constant a if the function $f(x)$ defined below is continuous at $x = 2$.

$$f(x) = \begin{cases} ax^2 + 7x & \text{if } x \leq 2 \\ 3x^2 + 3a & \text{if } x > 2. \end{cases}$$

Find the derivatives of the following functions w.r.t. the independent variables involved:

10. $y = \dfrac{x}{\sqrt{x^2 + 1}}$.

11. $y = x\sqrt{x^2 + 4}$.

12. $f(x) = (x + 1)\sqrt{x + 3}$.

13. $f(x) = (2x + 1)^3(3x - 1)^4$.

14. $f(x) = \dfrac{(x^2 + 1)^3}{(x - 1)^4}$.

15. $f(x) = \dfrac{\sqrt{x + 1}}{\sqrt[3]{x + 1}}$.

16. $f(x) = \sqrt{\dfrac{x + \sqrt{x + 1}}{x^2 + 1}}$.

Find d^2y/dx^2 for the following functions.

17. $y = \sqrt[3]{x^3 + a^3}$.

18. $y = (3x - 7)^6(x + 1)^4$.

19. According to Michaelis and Menton, the initial rate of reaction v in an enzyme-catalyzed reaction depends on the concentration x of the substrate and is given by

$$v = \frac{Vx}{x + M},$$

where V and M are constants. Show that $v \to V$ as $x \to \infty$ and that for any finite value of x, v is always less than V. Find dv/dx, that is, the rate of change of the initial velocity of reaction w.r.t. the concentration of the substrate.

20. According to A. V. Hill, the relation between the load F acting on a muscle and the speed V of contraction or shortening of the muscle is given by

$$(F + a)V = (F_0 - F)b,$$

where a, b, F_0 are constants that depend on the particular species and type of muscle. Prove that the speed V approaches zero as $F \to F_0$ so that F_0 represents the maximum load under which the muscle can contract. Find dV/dF and dF/dV. Show that each of these derivatives is the reciprocal of the other.

3

Exponential and Logarithm Functions

3.1 EXPONENTIAL FUNCTIONS

We shall begin by reviewing certain basic properties of exponents. If c is any real number and p any natural number, then c^p (c raised to the power p) denotes the product of p c's all multiplied together. For example, $c^2 = c \cdot c$, $c^3 = c \cdot c \cdot c$ and so on. Further, $c^{1/p}$ is used to denote the pth root of c: $c^{1/p}$ is that number that, when raised to the power p, gives c. For example $c^{1/2}$ denotes the square root of c, $c^{1/3}$ denotes the cube root of c, and so on. In general we have the identity

$$(c^{1/p})^p = c.$$

If p and q are natural numbers, $c^{p/q}$ is used to denote the qth root of c^p. Alternatively it is equal to the qth root of c all raised to the power p:

$$c^{p/q} = (c^p)^{1/q} = (c^{1/q})^p.$$

Negative exponents are used to denote reciprocals. Thus c^{-2} is the reciprocal of c^2, $c^{-1/2}$ is the reciprocal of $c^{1/2}$, that is, of the square root of c. In general, for any positive rational number m,

$$c^{-m} = \frac{1}{c^m}.$$

We define $c^0 = 1$ for any nonzero real number c.

The following properties hold for real numbers a and b and rational numbers m and n (which may be positive or negative).

$$a^m \cdot a^n = a^{m+n}$$

$$\frac{a^m}{a^n} = a^{m-n}$$

$$(a^m)^n = a^{mn}$$

$$a^n \cdot b^n = (ab)^n.$$

In writing these properties certain restrictions must be borne in mind. For example, if m is a negative exponent, then a^m is not defined for $a = 0$. Similarly, a quantity a^m that involves an even root (e.g., $a^{1/2}$ or $a^{3/8}$) is only defined provided that $a \geq 0$. The following examples illustrate the use of these properties.

EXAMPLE Simplify

$$\frac{(\sqrt{3})^3 6^{2/3}}{\sqrt[6]{3}}.$$

SOLUTION

$$\frac{(\sqrt{3})^3 6^{2/3}}{\sqrt[6]{3}} = \frac{(3^{1/2})^3 (2 \cdot 3)^{2/3}}{3^{1/6}}$$

$$= \frac{3^{3/2} \cdot 2^{2/3} \cdot 3^{2/3}}{3^{1/6}}$$

$$= 3^{(3/2)+(2/3)-(1/6)} \cdot 2^{2/3}$$

$$= 3^2 (2^2)^{1/3} = 9\sqrt[3]{4}.$$

EXAMPLE Express the answers to the following calculations in scientific notation.

    i. $7.13 \times 10^{-4} + 6.21 \times 10^{-5}$.

    ii. $(7 \times 10^{-7}) \times (8 \times 10^{3})$.

    iii. $\dfrac{4 \times 10^{4}}{8 \times 10^{8}}$.

SOLUTION

    i. $\begin{aligned}7.13 \times 10^{-4} + 6.21 \times 10^{-5} &= 7.13 \times 10^{-4} + 0.621 \times 10^{-4}\\ &= (7.13 + 0.621) \times 10^{-4}\\ &= 7.751 \times 10^{-4}.\end{aligned}$

    ii. $\begin{aligned}(7 \times 10^{-7}) \cdot (8 \times 10^{3}) &= 56 \times 10^{-7} \times 10^{3}\\ &= 56 \times 10^{-4}\\ &= 5.6 \times 10^{-3}.\end{aligned}$

    iii. $\begin{aligned}\dfrac{4 \times 10^{4}}{8 \times 10^{8}} &= \dfrac{4}{8} \times 10^{4-8}\\ &= 0.5 \times 10^{-4}\\ &= 5 \times 10^{-5}.\end{aligned}$

EXAMPLE The weight of a certain microorganism is 5×10^{-8} gm. How many organisms are there in a population whose total weight is 0.25 gm?

SOLUTION

$$\begin{aligned}\text{The number of microorganisms} &= \frac{0.25}{5 \times 10^{-8}}\\ &= 0.05 \times 10^{8}\\ &= 5 \times 10^{6}.\end{aligned}$$

EXAMPLE The female of a certain species of insect produces 20 surviving female offspring per generation. There are four generations per year. What is the total number of female offspring that descend in 1 yr from a given insect?

SOLUTION In the first generation there are 20 female offspring. Each of these produces 20 further female offspring so that the total number in the second generation is $20 \times 20 = 20^{2}$. Again each of these produces 20 female offspring, so the third generation contains $20^{2} \times 20 = 20^{3}$ female insects. Similarly, the fourth generation contains 20^{4}. The total number of female descendants is therefore

$$20 + 20^{2} + 20^{3} + 20^{4} = 20 + 400 + 8000 + 160,000$$
$$= 168,420.$$

So far we have discussed the power a^{m} only in the case where m is a rational number. In such a case, a^{m} is defined in terms of powers and roots: $a^{p/q}$ is the qth root of a raised to the power p. When m is an irrational number, however, we cannot use this kind of definition. We shall discuss the problem of defining a^m for all real numbers m with reference to the following specific example.

Consider a culture of bacteria whose initial weight is 1 gm and that is growing in such a way that its weight doubles every hour. After 1 hr the weight will be 2 gm, after 2 hr it will have doubled again to 4 gm, after 3 hr it will be 8 gm, and so on. If we denote by $y(n)$ the weight of the culture after n hr, we see that

$$y(0) = 1, \qquad y(1) = 2, \qquad y(2) = 2^2 = 4, \qquad y(3) = 2^3 = 8, \quad \text{etc.}$$

In general, $y(n) = 2^n$.

We can also use this formula to calculate the weight of the culture at times that involve fractional parts of an hour rather than simply a whole number of hours. Thus, for instance, after $\frac{1}{2}$ hr the weight will be $y(\frac{1}{2}) = 2^{1/2} = \sqrt{2} = 1.414\ldots$ gm; after $\frac{3}{4}$ hr it will be $y(\frac{3}{4}) = 2^{3/4} = \sqrt[4]{8} = 1.682\ldots$ gm; after $1\frac{1}{2}$ hr it will be $y(\frac{3}{2}) = 2^{3/2} = \sqrt{8} = 2.828\ldots$ gm, and so on. In short, we can calculate the weight $y(x) = 2^x$ for any value of x that is a rational number.

The values so obtained can be plotted on a graph. In Fig. 3.1 the points (x, y) corresponding to $x = 0, \frac{1}{4}, \frac{1}{2}, \frac{3}{4}, 1, \frac{5}{4}, \frac{3}{2}, \frac{7}{4}$, and 2 are marked with circles. It is found that a smooth curve can be drawn through all of the points so obtained in such a way that $y = 2^x$ on the curve for any positive rational value of x. We can use this smooth curve to define the function $y = 2^x$ for all real values of x, both rational and irrational.

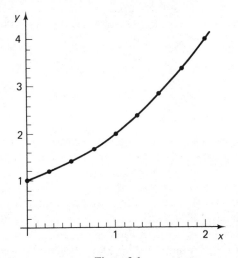

Figure 3.1

Let us suppose that we wish to calculate 2^x for some irrational value of x. We can do this by choosing a rational number x_1 that is very close to x, and calculating 2^{x_1}. Then by making x_1 close enough to x, we can make 2^{x_1} as close as we like to 2^x. For example, suppose that we want to calculate 2^π. We can take $x_1 = 3.14 = \frac{314}{100}$, which is quite close to π, and we can use $2^{(314/100)}$ as an approximation to 2^π. If this is not accurate enough, we can take $x_1 = 3.142 = \frac{3142}{1000}$ and use $2^{(3142/1000)}$ as an approximation to 2^π. Yet more accurately we can take $x_1 = 31,416/10,000$, and so on. By taking successively better approximations to π by rational numbers, we can get better and better approximations to the value of 2^π.

This process can be used for any irrational value of x to calculate the value of 2^x to any desired level of accuracy.

In a similar way we can define the function $y = a^x$ for any positive real number a. When $x = 0$, $y = a^0 = 1$; when $x = 2$, $y = a^2$; when $x = 3$, $y = a^3$, and so on for all positive integer values of x. When x is a positive rational number, a^x is defined in terms of an appropriate power and root: For example, when $x = \frac{7}{5}$, $a^x = \sqrt[5]{a^7}$. Finally, when x is an irrational number we can calculate a^x just as described above for 2^x: We choose a rational number x_1 sufficiently close to x and calculate a^{x_1}, which is approximately equal to a^x.

The function $y = a^x$ can also be defined for negative values of x in the usual way with reciprocals. For example, when $x = -1$, $y = a^{-1} = 1/a$; when $x = -\frac{1}{2}$, $y = a^{-1/2} = 1/\sqrt{a}$. In general, the value of $a^{-x} = 1/a^x$.

In Fig. 3.2 are the graphs of two functions $y = a^x$ and $y = b^x$ when $a > b > 1$. It is seen that for $x > 0$, these two functions grow at an ever-increasing rate as x increases. Since $a > b$, the graph of $y = a^x$ for positive values of x is situated above the graph of $y = b^x$ and increases more steeply.

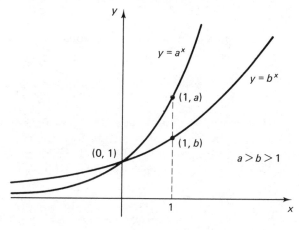

Figure 3.2

On the other hand, for $x < 0$ both functions decrease toward zero as x approaches $-\infty$. In this case the function a^x falls more steeply than b^x, and its graph is situated below the graph of $y = b^x$. The two graphs intersect when $x = 0$, since $a^0 = b^0 = 1$.

A function of the type $y = a^x$ is called an *exponential function*. When $a > 1$, the function is said to be a *growing exponential function*, whereas when $a < 1$ it is said to be a *decaying exponential function*. The graph of $y = a^x$ when $a < 1$ is illustrated in Fig. 3.3. When $a < 1$, a^x decreases as x gets larger and approaches zero as $x \rightarrow \infty$.

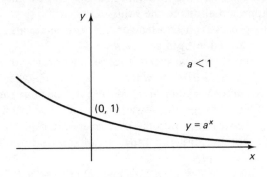

Figure 3.3

EXAMPLE   A culture of bacteria initially weighs 1 gm and is doubling in size every hour. How long will it take to reach a weight of 3 gm?

SOLUTION   We have seen above that after n hr, the weight is given by $y(n) = 2^n$. We need to calculate the value of n for which $y(n) = 3$:

$$2^n = 3.$$

If we take logs of both sides of this equation, we obtain

$$n \log 2 = \log 3$$
$$n = \frac{\log 3}{\log 2} = \frac{0.4771}{0.3010} = 1.58,$$

where the logs have been found from the table of common logarithms given in Appendix III. Thus it takes 1.58 hr for the culture to reach a weight of 3 gm. (We have assumed that the reader is familiar with the use of logarithms in carrying out arithmetical calculations. A brief review is given at the end of this section.)

EXAMPLE   The population of the earth at the beginning of 1976 was 4 billion and is growing at the rate of 2%/yr. What will be the population in the year 2000, assuming that the rate of growth remains unchanged?

SOLUTION   When a population grows by 2% a year, this means that its size at any time is 1.02 times what it was one year earlier. Thus at the beginning of 1977 the population will be $(1.02) \times 4$ billion. Furthermore, at the beginning of 1978 it will be 1.02 times the population at the beginning of 1977, that is, $(1.02)^2 \times 4$ billion. And so we continue with the population multiplying by a factor 1.02 for every year that passes. The population at the beginning of the year 2000, that is, after

24 years, will therefore be

$$(1.02)^{24} \times 4 \text{ billion} = 1.608 \times 4 \text{ billion}$$
$$= 6.43 \text{ billion}.$$

[Note that in order to work out, $(1.02)^{24}$ logs can be used in the following way:

From the table in Appendix III, we find that

$$\log (1.02) = 0.0086$$

Multiplying this by 24 we get 0.2064, and taking the antilog we find that

$$(1.02)^{24} = \text{antilog} (0.2064) = 1.608.]$$

It shou'd be pointed out that future projections over long periods of time such as that contained in this example (and also those in several of the following exercises) may not turn out to be very accurate. They indicate what will happen if present trends continue. A present trend can reasonably be expected to continue for a short time, so that short-range projections into the future are often quite reliable. Long-range projections, however, are less likely to be accurate.

Review of Logarithms

Logarithms will be defined and their basic properties established in the next section. However, the student will already have encountered their use in performing arithmetical computations, and the following review will serve as a reminder of the techniques involved. We assume that he or she is familiar with the use of tables for the evaluation of logarithms of numbers; more details of this can be found, if needed, in many elementary algebra texts.

In order to calculate the product of two numbers, we find their logarithms from the table in Appendix III, add these logs together, and then find the antilog of the result. We can write this operation in the form

$$\boxed{ab = \text{antilog} (\log a + \log b).}$$

EXAMPLE   Calculate 6.17×1.42.

SOLUTION
$$a = 6.17; \quad \log a = 0.7903$$
$$b = 1.42; \quad \log b = \underline{0.1523}$$
$$+ \underline{0.9426}$$
$$ab = \text{antilog} (0.9426) = 8.762.$$

Similarly, division can be performed by subtracting the log of the denominator from the log of the numerator, and then finding the antilog of the result:

$$\boxed{\frac{a}{b} = \text{antilog} (\log a - \log b).}$$

EXAMPLE   Calculate 6.17/1.42.

SOLUTION

$$a = 6.17; \quad \log a = 0.7903$$
$$b = 1.42; \quad \log b = \underline{0.1523}$$
$$- \; 0.6380$$

$$\frac{a}{b} = \text{antilog } (0.6380) = 4.345.$$

The calculation of powers of numbers follows the rule:

$$\boxed{a^n = \text{antilog } (n \log a).}$$

In order to calculate a^n, we find the log of a, multiply it by n, and find the antilog of the result.

EXAMPLE   Calculate $(6.17)^5$.

SOLUTION

$$a = 6.17; \quad \log a = 0.7903$$
$$n = 5; \qquad\qquad\qquad \underline{5}$$
$$\times \; 3.9515$$
$$a^n = \text{antilog } (3.9515) = 8945.$$

Another useful way of writing the rule for powers of numbers is as follows:

$$\boxed{\log (a^n) = n \log a.}$$

In this form the rule can be used to calculate the exponent n when a and a^n are both known.

EXAMPLE   For what exponent n is $2^n = 10$?

SOLUTION $\log (2^n) = n \log 2 = \log 10$. Therefore

$$n = \frac{\log 10}{\log 2} = \frac{1}{0.3010} = 3.32.$$

Finally, let us recall that numbers that are smaller than 1 have negative logarithms. For example,

$$\log (0.2) = \log \left(\tfrac{2}{10}\right) = \log 2 - \log 10$$
$$= 0.3010 - 1 = -0.6990.$$

However it is usually much more convenient, instead of using negative numbers for such logarithms, to use the bar notation instead. With this notation, for example, $\log (0.2)$ is denoted by $\bar{1}.3010$; this notation is an abbreviation for

$-1 + 0.3010$. Similarly,

$$\log (0.0002) = \log \frac{2}{10^4}$$

$$= \log 2 - \log (10^4)$$

$$= 0.3010 - 4.$$

Thus we use the notation

$$\log (0.0002) = \bar{4}.3010.$$

When calculating with such logarithms, it must be remembered that the barred part is a negative integer.

EXAMPLE   Calculate $0.617/14.2$.

SOLUTION

$$a = 0.617; \quad \log a = \bar{1}.7903$$

$$b = 14.2; \quad \log b = 1.1523$$

$$- \overline{\bar{2}.6380}$$

$$\frac{a}{b} = \text{antilog } (\bar{2}.6380) = 0.04345.$$

EXAMPLE   Calculate $(0.617)^5$.

SOLUTION

$$a = 0.617; \quad \log a = \bar{1}.7903$$

$$n = 5 \qquad\qquad\qquad 5$$

$$\overline{\bar{2}.9515}$$

$$a^n = \text{antilog } (\bar{2}.9515) = 0.08945.$$

EXERCISES 3.1

Simplify the following expressions.

1. $\sqrt{18}$.

2. $\sqrt{300}$.

3. $\dfrac{\sqrt{48}}{\sqrt{3}}$.

4. $\dfrac{\sqrt{4a^5}}{\sqrt[3]{8a^4}}$.

5. $\dfrac{\sqrt[3]{54}\,\sqrt[6]{2}}{\sqrt{6}}$.

6. $\dfrac{\sqrt[6]{5}}{\sqrt{15}\,\sqrt[3]{25}}$.

Express the answers to the following in scientific notation.

7. $0.0014 + 2.2 \times 10^{-4}$.

8. $0.013 \times 10^5 + 11{,}000$.

9. $(6 \times 10^6) \cdot (5 \times 10^{-5})$.

10. $(8 \times 10^{-7}) \cdot (2 \times 10^3)$.

11. $\dfrac{6 \times 10^6}{5 \times 10^{-5}}$.

12. $\dfrac{5 \times 10^{-5}}{8 \times 10^{-2}}$.

13. A certain female moth lays 120 eggs on average, and in 1 yr there may be as many as five generations. Assuming that one-third of the eggs survive and produce new moths, half of which are female, work out how many moths (male and female) descend from one female each year.

14. A certain breed of dog produces an average of two litters of pups per year, each litter containing an average of six pups, three of each sex. The female produces her first litter at the age of 2 yr. How many descendants would a given bitch have by the time she is 6 yr old?

15. A population of bacteria growing in a culture initially contains 10^4 organisms. If each cell divides, on the average, every 3 hr, how large will the population be after 12 hr and after 24 hr? How large will it be after 17 hr? How long will it take for the population size to reach 10^6?

16. A population of microorganisms is doubling in size every 75 min. If it initially weighs 0.1 gm, what will be its weight after 5 hrs and after 500 min? How long does it take to reach a weight of 0.5 gm?

17. The population of the earth at the beginning of 1976 was 4 billion. If the growth rate continues at 2%/yr, what will be the population in the year 2076? When will the population reach 10 billion?

18. With the data in exercise 17, what will be the population in the year 2026? When will the population reach 100 billion?

19. The population of China in 1970 was 750 million and was growing at 4%/yr. When would this population reach 2 billion, assuming that the same growth rate continued?

20. Using the data of the previous questions, calculate when the population of China would become equal to half the population of the earth.

21. The population of India in 1970 was 600 million, growing at an annual rate of 5%. When would the populations of India and China become equal, if these growth rates continued? Explain the apparent paradox between the answers to this exercise and the previous one.

22. A certain population of insects consists of two types: T_1 and T_2. Initially the population contains 90 T_1 insects and 10 T_2 insects. The T_1 population grows at 1%/day, and the T_2 population at 4%/ day, on the average. When will the population become equally divided between the two types?

3.2 INVERSE FUNCTIONS AND LOGARITHMS

Let $y = f(x)$ be a relation expressing the dependent variable y as a function of the independent variable x. Let us suppose that the equation $f(x) = y$ can be solved for x, thus allowing us to express x as a function of y. In this way, when x is expressed as a function of y, the roles of x and y are interchanged: y now becomes the independent variable and x the dependent variable. We write the relation expressing x as a function of y in the form $x = f^{-1}(y)$, and the function f^{-1} is called the *inverse* of the function f.

NOTE. $f^{-1}(y)$ is not to be confused with the inverse power $[f(y)]^{-1} = 1/f(y)$.

EXAMPLE   Find the inverse of the function $f(x) = 2x + 1$.

SOLUTION   Setting $y = f(x) = 2x + 1$, we must solve for x as a function of y.

$$2x = y - 1$$
$$x = \frac{y - 1}{2}.$$

Therefore the inverse function is given by

$$f^{-1}(y) = \frac{y - 1}{2}.$$

The graphs of $y = f(x)$ and $x = f^{-1}(y)$ are shown in Figs. 3.4 and 3.5, respec-

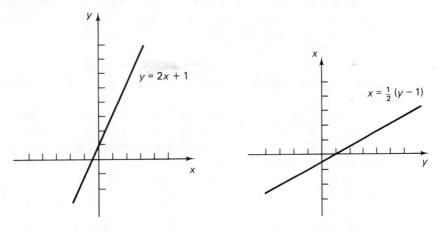

Figure 3.4   Figure 3.5

tively. Both graphs in this case are straight lines. Observe that when plotting the graph of $x = f^{-1}(y)$, the y-axis is taken as horizontal and the x-axis vertical because y is the independent variable.

EXAMPLE   Find the inverse of the function $f(x) = x^3$ and sketch its graph.

SOLUTION   Setting $y = f(x) = x^3$ we solve for x, obtaining

$$x = f^{-1}(y) = y^{1/3}.$$

The graphs of $y = f(x)$ and $x = f^{-1}(y)$ are shown in Figs. 3.6 and 3.7, respectively.

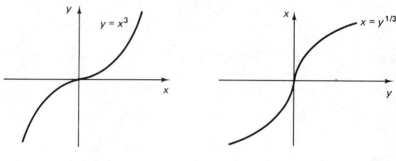

Figure 3.6   **Figure 3.7**

It can be seen from these two examples that the graphs of a function $y = f(x)$ and its inverse function $x = f^{-1}(y)$ are closely related. In fact the graph of the inverse function is obtained by flipping over the graph of the original function so that the coordinate axes become interchanged. For example, the student might try holding the graph of $y = x^3$ in front of a mirror in such a way that the y-axis is horizontal and the x-axis points vertically upwards. The reflection seen will be the graph of $x = y^{1/3}$, which is shown in Fig. 3.7.

The graphs of $y = f(x)$ and $x = f^{-1}(y)$ consist of precisely the same sets of points (x, y). The difference is only that the axes are drawn in different directions in the two cases.

Another way of seeing the relationship between the two graphs is to plot the functions $y = f(x)$ and $y = f^{-1}(x)$ on the same axes. Then the graph of either of these functions can be obtained by reflecting the other graph in the line $y = x$. Figure 3.8 shows the graphs of the function $f(x) = x^3$ and its inverse function drawn

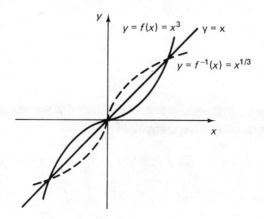

Figure 3.8

on the same axes. Clearly the two graphs are reflections of one another in the line $y = x$.

Not every function has an inverse. Consider, for example, the function $y = x^2$. Solving for x in terms of y, we obtain

$$x^2 = y,$$
$$x = \pm\sqrt{y}.$$

For any value of y in the region $y > 0$ there are two possible values of x; therefore we cannot say that x is a function of y. This is illustrated graphically in Figs. 3.9 and 3.10. Figure 3.9 shows the graph of $y = x^2$, which is a parabola opening upwards. Figure 3.10 shows the same graph but with the axes flipped over, the y-axis being horizontal and the x-axis vertical. For each $y > 0$ we have the two values $x = +\sqrt{y}$ and $x = -\sqrt{y}$; for example, when $y = 1$, x has the values $+1$ and -1, both satisfying the relation $y = x^2$.

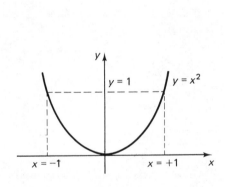

Figure 3.9

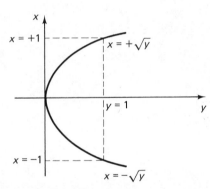

Figure 3.10

The graph in Fig. 3.10 corresponds to two functions rather than to one. The upper branch of the parabola is the graph of $x = +\sqrt{y}$, while the lower branch is the graph of $x = -\sqrt{y}$. Thus we can say that the function $y = x^2$ has two inverse functions: one given by $x = +\sqrt{y}$ and the other by $x = -\sqrt{y}$.

In a case such as this, it is possible to make the definition of f^{-1} unambiguous by restricting the values of x. For example, if x is restricted to the region $x \geq 0$, then $y = x^2$ has the unique inverse $x = +\sqrt{y}$. On the other hand, if x is restricted to the region $x \leq 0$, then the inverse is given by $x = -\sqrt{y}$. Placing a restriction on x in this way means restricting the domain of the original function f. We conclude therefore that in cases where a function $y = f(x)$ has more than one inverse function, the inverse can be made unique by placing a suitable restriction on the domain of f.

It is worth observing that a function $f(x)$ has a unique inverse whenever horizontal lines intersect its graph in only one point.

EXAMPLE   By placing suitable restrictions on its domain, find the inverse of the function $y = \sqrt{x^2 - a^2}$.

SOLUTION First solve the functional relation for x:

$$y^2 = x^2 - a^2$$
$$x^2 = y^2 + a^2$$
$$x = \pm\sqrt{y^2 + a^2}.$$

Again we see that there are two values of x corresponding to each value of y. This is illustrated in Fig. 3.11, which shows the graph of $y = \sqrt{x^2 - a^2}$. The domain of this function consists of the two regions $x \leq -a$ and $x \geq a$, and its range is $\{y \mid y \geq 0\}$. Figure 3.12 shows the same graph with the axes interchanged.

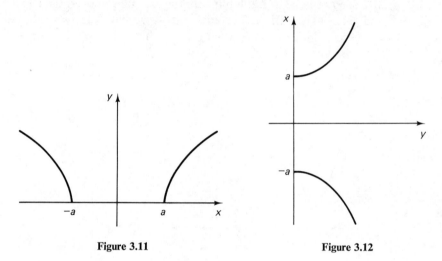

Figure 3.11   **Figure 3.12**

Clearly each vertical line in the region $y \geq 0$ intersects this second graph in two points.

In order to make the inverse unique, we can restrict x either to the region $x \geq a$ or to the region $x \leq -a$. In the first case the inverse is given by $x = +\sqrt{y^2 + a^2}$ and in the second case by $x = -\sqrt{y^2 + a^2}$. These two functions have as their respective graphs the upper and lower branches of the graph in Fig. 3.12.

We shall now consider the inverse of the exponential function. We have seen in the last section how the exponential function $y = a^x$ can be constructed. Let us now examine its inverse. It is not possible to express the inverse in terms of elementary functions (i.e., powers and radicals), and a new name must be invented for it. We write the inverse function of $y = a^x$ as $x = \log_a y$, with $\log_a y$ being termed the *logarithm of y with base a*. Thus we have the following definition of the logarithm:

$$x = \log_a y \quad \text{whenever} \quad y = a^x.$$

As with any other inverse function, the graph of the logarithm $x = \log_a y$ can be obtained from the graph of the exponential function $y = a^x$ by simply flipping the axes. The two graphs are illustrated in Figs. 3.13 and 3.14.

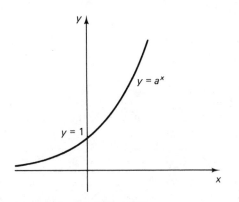

Figure 3.13   Figure 3.14

We note that $\log_a y$ is defined only for $y > 0$. In addition, $\log_a y$ is positive for $y > 1$ and negative for $y < 1$, and $\log_a 1 = 0$.

We shall now establish several basic properties of the logarithmic function. Let y_1 and y_2 be two positive real numbers and let

$$x_1 = \log_a y_1 \quad \text{and} \quad x_2 = \log_a y_2 .$$

Then

$$y_1 = a^{x_1} \quad \text{and} \quad y_2 = a^{x_2}.$$

It follows therefore that

$$y_1 y_2 = a^{x_1} \cdot a^{x_2} = a^{x_1 + x_2}$$

after using one of the fundamental properties of exponents. Consequently, from the definition of the logarithm, it follows that

$$x_1 + x_2 = \log_a y_1 y_2.$$

In other words, substituting for x_1 and x_2,

$$\boxed{\log_a y_1 y_2 = \log_a y_1 + \log_a y_2.}$$

A second result can be obtained by considering y_1/y_2:

$$\frac{y_1}{y_2} = \frac{a^{x_1}}{a^{x_2}} = a^{x_1} \cdot a^{-x_2} = a^{x_1 - x_2}.$$

Thus

$$x_1 - x_2 = \log_a \frac{y_1}{y_2},$$

or equivalently,

$$\boxed{\log_a \frac{y_1}{y_2} = \log_a y_1 - \log_a y_2.}$$

A special case of this result is obtained by setting $y_1 = 1$. When $y_1 = 1$, $\log_a y_1 = 0$, and we get

$$\log_a \frac{1}{y_2} = -\log_a y_2.$$

Fourthly, let $x = \log_a y$, so that $y = a^x$. Then $y^n = (a^x)^n = a^{xn}$. Thus $xn = \log_a y^n$, or

$$\log_a y^n = n \log_a y.$$

If in the definition of the logarithm we put $x = 1$, then $y = a^1 = a$. It follows from the inverse statement $\log_a y = x$ that

$$\log_a a = 1.$$

This statement is true for all $a > 0$.

These five properties form the basis for the use of logarithms in the performance of arithmetical operations involving multiplication, division, and the calculation of powers and roots. We have in fact already made use of them in working out some of the examples in Section 3.1.

EXAMPLE   Find the values of:

   (a) $\log_2 16$ (b) $\log_{1/3} 243$.

SOLUTION   (a)   Let $x = \log_2 16$. From the definition of logarithms it follows that $16 = 2^x$. But $16 = 2^4$, and so $x = 4$. Therefore $\log_2 16 = 4$.

   (b)   Let $x = \log_{1/3} 243$. From the definition, $243 = (\frac{1}{3})^x$. But $243 = 3^5 = (\frac{1}{3})^{-5}$. Therefore $x = -5$.

EXAMPLE   Evaluate $2^{\log_4 9}$.

SOLUTION   Let $x = \log_4 9$, so that $9 = 4^x$. The quantity we wish to calculate is therefore
$$2^{\log_4 9} = 2^x = (4^{1/2})^x = 4^{(1/2)x} = (4^x)^{1/2} = 9^{1/2} = 3.$$

The logarithms that are to be found in the usual tables are called *common logarithms* and are obtained by using the number 10 as base (that is, $a = 10$). Thus the common logarithm of a number y is $\log_{10} y$; however, to avoid cumbersome notation, the common logarithm is usually denoted by $\log y$, the base being omitted.[1] When the base is not written it should be understood to be 10.

The following two statements are equivalent,

$$x = \log y, \qquad y = 10^x.$$

---

[1] Note that in some books, the notation $\log y$ is used to mean the natural log of y (Section 3.4).

EXAMPLES   **(i)** Given $x = 1$. Then $y = 10^1 = 10$ and so $\log 10 = 1$.

(ii) Given $x = 2$. Then $y = 10^2 = 100$ and so $\log 100 = 2$.

(iii) Given $x = -1$. Then $y = 10^{-1} = 0.1$ and so $\log 0.1 = -1$.

(iv) To four figures we can find from the log table in Appendix III that $\log 3 = 0.4771$. This means that $3 = 10^{0.4771\cdots}$. Using the first property of logarithms proved above, it follows that

$$\log 30 = \log 3 + \log 10 = 1.4771$$

$$\log 300 = \log 3 + \log 100 = 2.4771$$

$$\log 0.3 = \log 3 + \log 0.1 = -1 + 0.4771 = -0.5229.$$

(Rather than set $\log 0.3 = -0.5229$, we usually write $\bar{1}.4771$ because this form makes the addition and subtraction of logarithms and the use of the log tables easier. $\bar{1}.4771$ simply stands for $-1 + .4771$.)

(v) Also from the tables we find that $\log 2 = 0.3010$. Thus

$$\log 6 = \log(3 \times 2) = \log 3 + \log 2 = 0.4771 + 0.3010 = 0.7781$$

$$\log 4 = \log 2^2 = 2 \log 2 = 0.6020.$$

EXAMPLE   Simplify the expression

$$E = \log 2 + 16 \log \tfrac{16}{15} + 12 \log \tfrac{25}{24} + 7 \log \tfrac{81}{80}.$$

SOLUTION

$$E = \log 2 + 16 \log \frac{2^4}{3 \cdot 5} + 12 \log \frac{5^2}{2^3 \cdot 3} + 7 \log \frac{3^4}{2^4 \cdot 5}$$

$$= \log 2 + 16(\log 2^4 - \log 3 - \log 5) + 12(\log 5^2 - \log 2^3 - \log 3)$$
$$+ 7(\log 3^4 - \log 2^4 - \log 5)$$

$$= \log 2 + 16(4 \log 2 - \log 3 - \log 5) + 12(2 \log 5 - 3 \log 2 - \log 3)$$
$$+ 7(4 \log 3 - 4 \log 2 - \log 5)$$

$$= \log 2(1 + 64 - 36 - 28) + \log 3(-16 - 12 + 28)$$
$$+ \log 5(-16 + 24 - 7)$$

$$= \log 2 + \log 5$$

$$= \log(2 \cdot 5)$$

$$= \log 10 = 1.$$

EXERCISES 3.2

Find the inverses of the following functions. Draw the graphs of the function and its inverse in each case.

133   **1.** $y = -3x - 4$.   **2.** $y = x - 1$.

3. $y = \sqrt{3x - 4}$.

4. $y = \sqrt{\frac{1}{4}x + 2}$.

5. $y = x^5$.

6. $y = x^{3/2}$.

By placing a suitable restriction on the domain of each of the following functions, find an inverse function.

7. $y = (x + 1)^2$.

8. $y = (3 - 2x)^2$.

9. $y = x^{2/3}$.

10. $y = \sqrt{x^2 + 1}$.

Find the values of the following expressions by using the definition of the logarithm.

11. $\log_2 512$.

12. $\log_{27} 243$.

13. $\log_{\sqrt{2}} 16$.

14. $\log_8 128$.

15. $\log_2 0.125$.

16. $\dfrac{\log_a 32}{\log_a 4}$.

17. $10^{\log 100}$.

18. $10^{\log 2}$.

19. $\log_4 (2^p)$.

20. $\log_2 (4^p)$.

21. $2^{\log_{1/2} 3}$.

22. $3^{\log_9 2}$.

In a chemical solution let $[H^+]$ denote the concentration of hydrogen ions measured in moles/ L (moles per liter). The pH of the solution is defined as pH $= -\log [H^+]$. Find the pH of solutions in which

23. $[H^+] = 3.1 \times 10^{-4}$.

24. $[H^+] = 0.21 \times 10^{-9}$.

Solve the following equations for x (do not use log tables):

25. $\log (10x + 5) - \log (x - 4) = \log 2$.

26. $\log_3 3 + \log_3 (x + 1) - \log_3 (2x - 7) = 4$.

27. A formula that is used to give the amount of heat generated by a human infant during a 24-hr period is $H = 0.0128 \, LW^{2/3}$, where L is the length of the infant in centimeters and W its weight in kilograms. How much heat is generated by an infant of length 60 cm and weight 3.8 kg?

3.3   LOG-LOG AND SEMI-LOG PLOTS

Let us suppose that in a certain experiment a quantity y is measured for a series of values of a variable x. Suppose also that there is reason to believe that x and y are related to each other by a linear relation of the form $y = ax + b$, where a and b are two constants that are to be determined from the experiment. For example, we may have some mathematical model of the experiment in question according to which x and y should be linearly related.

An equation of the form $y = ax + b$ has a graph that is a straight line when plotted with x and y as the usual Cartesian coordinates. This line has its slope equal to a and its y-intercept equal to b. Thus when we plot the experimental results on a graph, we should find that they all lie on, or at least close to, a straight line. (In practice they will not lie exactly on a straight line because of the presence of experimental errors.) By drawing the best straight line through the experimental points and then measuring its slope and intercept, we are able to determine the constants a and b.

EXAMPLE   It is believed that the length of a certain species of animal increases linearly with its age during the early period of its life. In an experiment the lengths of a number of the animals were measured at intervals during a period of 16 wk (weeks) from birth. The following results were obtained, in which x represents the number of weeks from birth and y the average length in centimeters of all of the measured animals.

x (weeks):	2	4	6	8	10	12	14	16
y (centimeters):	4.3	6.6	10.0	13.0	16.3	19.4	22.5	25.1

Assuming that $y = ax + b$, find constants a and b.

SOLUTION   After plotting x against y, the circled points in the Fig. 3.15 are obtained. The points lie close to a straight line, and the straight line that passes closest to the points has been included in the figure. The slope of this line is measured from the graph to be 1.5 cm/wk and the y-intercept is measured to be 1.2 cm. Therefore in the relation $y = ax + b$, $a = 1.5$ and $b = 1.2$. From these experimental data we conclude that

$$y = 1.5x + 1.2.$$

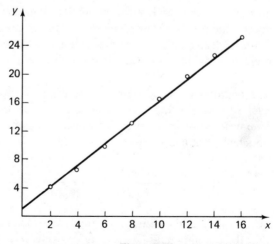

Figure 3.15

The preceding type of analysis of experimental data works as long as the two variables concerned are linearly related. But when y is a more complicated function of x it is not much use to plot the experimental data on a graph of y against x since the points cannot be expected to lie on a straight line. Suppose, for example, that we have reason to believe that x and y are related in the form $y = ax^b$, where again a and b are two constants to be determined. In this case a graph of y against x will not be a straight line (unless $b = 1$) but will have one of the forms shown in Fig. 3.16. When the experimental points are plotted it is very difficult to judge whether they do in fact lie on, or close to, a curve of the type $y = ax^b$, and it is difficult to determine a and b even if the curve can be drawn.

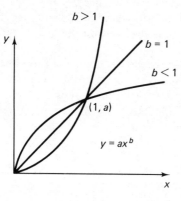

Figure 3.16

The problem becomes much easier if logarithms are used. For, taking logs of the equation $y = ax^b$, we obtain

$$\log y = \log (ax^b) = \log a + \log x^b = \log a + b \log x.$$

Let us introduce $Y = \log y$ and $X = \log x$ as new variables. Then this relation becomes

$$Y = bX + \log a.$$

Therefore if Y is plotted against X, the points should lie on a straight line whose slope is b and whose Y-intercept is $\log a$.

Thus by plotting the logarithms of the experimental values of x and y against one another, the points obtained should lie on (or close to) a straight line. If the best straight line is drawn through the points, then its slope determines the constant b and its y-intercept provides the logarithm of a. Such a graph is called a *log-log plot*.

EXAMPLE   The quantity of wood obtained from an average tree varies according to the age of the tree. The following table gives a set of experimental values of y, the volume of wood in hundreds of board feet, against x, the age of the tree in years obtained from measurements on fir trees.

x (age):	50	75	100	125	150	175	200
y (volume):	11	28	56	98	158	225	330

When x is plotted against y, the points shown in Fig. 3.17 are obtained. It appears that these points may lie on a curve given by a power function $y = ax^b$. Use a log-log plot to verify that this is so, and determine a and b.

136

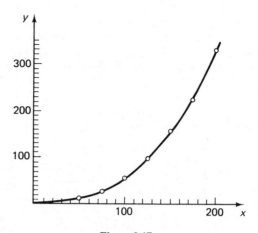

Figure 3.17

SOLUTION   The following table gives the values of $X = \log x$ and $Y = \log y$, obtained by taking logarithms of the experimental data.

x:	50	75	100	125	150	175	200
$X = \log x$:	1.70	1.88	2.00	2.10	2.18	2.24	2.30
$Y = \log y$:	1.04	1.45	1.75	1.99	2.20	2.35	2.52

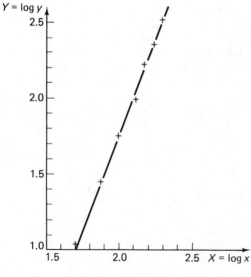

Figure 3.18

When Y is plotted against X, the points on Fig. 3.18 are obtained. It can be seen that they lie quite close to a straight line. The slope of the line is measured to be 2.5, which determines the value of the exponent b. In order to determine a we could measure the y-intercept of the straight line with the vertical Y-axis. However, since this point lies off the graph, we proceed as follows. We note that when $X = 2.0$, the value of Y on the straight line is equal to 1.75. Since these values must satisfy the equation $Y = bX + \log a$ with $b = 2.5$, we find that $\log a = 1.75 - (2.5)(2.0) = -3.25 = \bar{4}.75$. Therefore $a = 5.6 \times 10^{-4}$. Consequently, the relationship between the volume of lumber and the age of the tree from these results is

$$y = 5.6 \times 10^{-4} x^{5/2}.$$

When using log-log plots in practice, an alternative technique is to use special graph paper called logarithmic graph paper. An example of this type of paper is shown

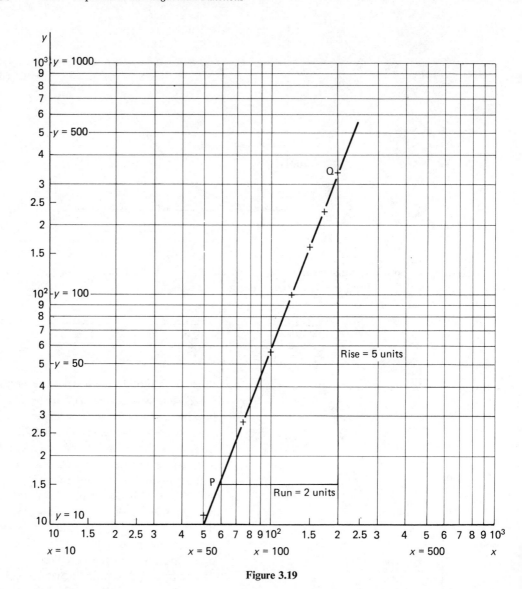

Figure 3.19

in Fig. 3.19. The spacing of the grid-lines on logarithmic graph paper is not uniform, but rather the lines are spaced according to the logarithms of the numbers to which they correspond. Because of this, it is not necessary to look up log x and log y before plotting on logarithmic paper; we simply plot the experimental values of x and y directly. In effect, the graph paper takes the logarithms for us.

    In Fig. 3.19 we have plotted the data from the preceding lumber example. When using logarithmic paper, the first thing to decide is how many cycles of 10 are involved in the data. For example, the values of x in our example range from 50 to 200, and overlap 2 cycles of 10, namely the cycle from 10 to 100 and the cycle from 100 to 1000. The values of y range from 11 to 330 and also overlap 2 cycles of 10. The cycles on the graph paper are labeled as required by the data. In our example we shall need to label only two cycles in each coordinate direction (Fig. 3.19).

The values of (x, y) when plotted from our example lie close to a straight line. In the original relation $y = ax^b$, the index b is equal to the slope of the line on the logarithmic graph paper. The slope of the line is obtained by measuring the rise and run between two points P and Q on the line and calculating slope = rise/run. It is important to measure rise and run using a ruler, *not* using the scales on the graph paper. The straight line in the figure turns out to have slope $b = 2.5$.

The constant a could be determined from the intercept with the vertical axis (which is the line $x = 1$). However in this example, this intercept is off the paper, so we must use some other point on the line in order to determine a, just as we did before. For example, we note that when $y = 10$, $x = 49.7$. Therefore $10 = a(49.7)^{2.5}$, so $a = 10(49.7)^{-2.5} = 5.6 \times 10^{-4}$, as before.

Allometric growth. An area in which logarithmic plots are frequently used is allometry, which is concerned with the comparative sizes of different parts of living organisms. Let x and y be the sizes of two different parts of the bodies of members of a certain species or of a certain group of species. For example, x could be the average total body weight and y the average brain weight of individuals of various ages belonging to the species *Homo sapiens*. As the age of the individual varies, the average quantities x and y vary, and it is commonly found that they are related by a power law of the form $y = ax^b$. Such a relation is known as Huxley's law of simple allometry. This law is also often used in comparing measurements among different species as well as between individuals in the same species.

In order to test whether a particular set of measurements follows simple allometry, one would use a log-log plot of x against y. Several examples are given in the exercises at the end of this section.

Semi-logarithmic Plots

Log-log plots are used when the variables are related through a power function, $y = ax^b$. We shall now consider another type of plot for use when y is an exponential function of x.

Let x and y be two variables that are related by an equation of the form $y = ba^x$, where a and b are two constants to be determined from a set of experimental data. Again a direct plot of x against y leads to a curved line, as illustrated by the figures in Section 3.1. When $a > 1$ the curve shows an exponential growth, while if $a < 1$ it shows an exponential decay. The extra constant b simply rescales the curve in the vertical direction when compared to the figures in Section 3.1.

Taking logs of the relation $y = ba^x$ yields

$$\log y = \log (ba^x) = \log b + \log a^x = \log b + x \log a.$$

Introducing $Y = \log y$, we find that

$$Y = (\log a)\, x + \log b.$$

Therefore, if Y is plotted against x, the experimental points should lie on, or close to, a straight line. The slope of the line is $\log a$ and the y-intercept is $\log b$. Thus by measuring the slope and intercept, the two constants a and b can be determined.

A graph of $Y = \log y$ against x is called a *semi-log plot*.

EXAMPLE   Shortly after consuming a substantial dose of whisky, the alcohol content of a certain individual's blood rises to a peak value of 0.22 mg/ml, and thereafter slowly decreases. If t is the time in hours after the maximum value is reached, and y the blood-alcohol level, the following table gives the experimentally measured values for this individual.

t:	0	0.5	0.75	1.0	1.5	2.0	2.5	3.0
y:	0.22	0.18	0.15	0.13	0.10	0.08	0.06	0.05

When y is plotted against t the experimental points correspond to the circles in Fig. 3.20. It appears that an exponential decay of the type $y = ba^t$ (with $a < 1$) may be sufficient to fit the data in this case. Use a semi-log plot to show that this is a reasonable hypothesis, and determine constants a and b.

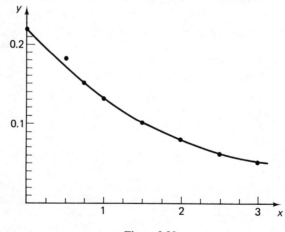

Figure 3.20

SOLUTION

t:	0	0.5	0.75	1.0	1.5	2.0	2.5	3.0
$Y = \log y$:	$\bar{1}.34$	$\bar{1}.26$	$\bar{1}.18$	$\bar{1}.11$	$\bar{1}.00$	$\bar{2}.90$	$\bar{2}.78$	$\bar{2}.79$

The values of $Y = \log y$ and t from the experimental data are given in the above table. The corresponding points are plotted in Fig. 3.21. (Note the way

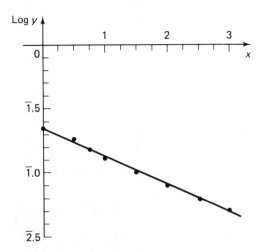

Figure 3.21

in which the negative logarithms are plotted. For example, in plotting $Y = \bar{1}.34$ we move down to the level $Y = -1$, then up a distance 0.34. This is the same as plotting $Y = -0.66$.) We see that on the graph of $Y = \log y$ versus t the experimental points lie close to a straight line. The slope of the line is -0.22, and since this must equal $\log a$, we find that $a = 0.60$. The Y-intercept turns out to be $\bar{1}.34$, which equals $\log b$, so $b = 0.22$. Thus we arrive at the formula $y = 0.22(0.60)^t$ for the alcohol content at time t.

In the next section we shall find a more convenient way of representing such formulas as this one involving exponential growth or decay.

Again, when making semi-log plots in practice, it is not necessary to calculate values of $\log y$ since special graph paper called semi-logarithmic graph paper can be used instead. An example of this paper is shown in Fig. 3.22. On semi-log paper, only one of the scales is logarithmic; the other scale is uniform.

The data for t and y from the preceding example are plotted in Fig. 3.22. We note the values of y range from 0.05 to 0.22 and hence overlap two cycles of 10: the cycle from 0.01 to 0.1 and the cycle from 0.1 to 1. These values are labeled accordingly along the y-axis. The t-axis is labeled in the usual way, and then the data values (t, y) are plotted directly. They lie close to a straight line.

In the equation $y = ba^t$, the value of b can be read directly as the y-intercept on the semi-log graph paper. It turns out to be 0.22 in our example. The value of a can be determined from the slope, but it is easier to work from some particular point on the straight line. For example, when $t = 2$, we find from the figure that $y = 0.079$ (the coordinates of the point P). Therefore $0.079 = ba^2 = 0.22a^2$, and

$$a = \left(\frac{0.079}{0.22}\right)^{1/2} = 0.60.$$

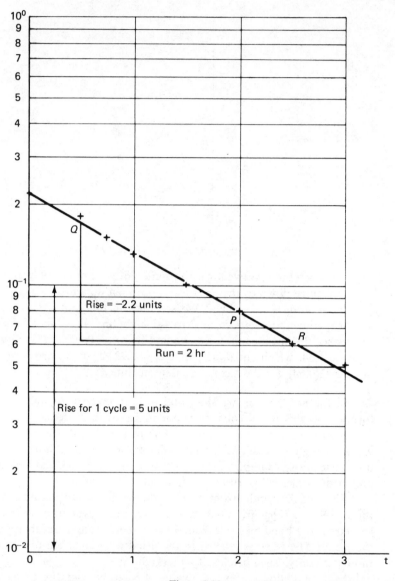

Figure 3.22

The use of the slope of a semi-log plot in order to determine the constant *a* is slightly tricky since one must remember to normalize the measured rise of the graph. This is illustrated in Fig. 3.22 where the rise and run between two points *Q* and *R* on the straight line are measured. The run is measured in units along the horizontal axis, and is in this case 2 hr. The rise is measured with a ruler and is found to be −2.2 units.

This must be normalized by dividing by the length of 1 cycle, which in this case is 5 units. Thus the normalized rise $= -2.2/5 = -0.44$. (Note that the rise is negative since the graph is falling.) Therefore we obtain the slope given by

$$\text{slope} = \frac{\text{normalized rise}}{\text{run}} = \frac{-0.44}{2} = -0.22.$$

Thus $\log a = -0.22 = \bar{1}.78$ and $a = 0.60$, as before.

Weber–Fechner law. One of the most common applications of semi-log plots arises in uses of the so-called Weber–Fechner law. This is a general law that is used to relate the magnitudes of stimulus and response in a wide variety of contexts. We let x represent the level of response to a certain type of stimulus and y represent the level of the stimulus. For example, y might be the amount of a certain drug injected into a subject and x might be some physiological variable that responds to the drug, such as pulse rate. Perhaps the most well-known example is the case where x is the response in terms of subjective sensation produced by a certain level y of sensory stimulus (for instance by levels of sound, in which case y would be the sound intensity). In cases such as these, the Weber–Fechner law states that x and y are related by an equation of the form

$$x = A \log y + B.$$

Alternatively, we can write this as $y = ba^x$, where $\log a = 1/A$ and $\log b = -B/A$.

The decibel scale of loudness of sounds is based on the logarithmic form of response implicit in the Weber–Fechner law, as also is the scale of brightness magnitudes, which is used to classify stars in astronomy.

In recent years the Weber–Fechner law has fallen into a degree of disrepute, and many writers recommend the use of a power law relationship between x and y as more accurately representing the empirical relationship between stimulus and response in many cases.

EXERCISES 3.3

1. In an experiment the weight W of the antlers of deer was measured for a number of deer of different ages. The results are given in the following table, W being in kilograms and the age A in months. Show that the data fit closely with the linear relation $W = mA + b$, and determine the constants m and b.

A:	20	22	30	34	42	43	46	54	56	68	70
W:	0.08	0.10	0.15	0.20	0.27	0.26	0.31	0.36	0.40	0.49	0.49

2. The average yield y in bushels per acre of corn grain in the United States varies from one year to another. The values for the period 1950–1971 are given in the following

table, in which t denotes the date starting with $t = 0$ in 1950 and increasing to $t = 21$ in 1971. Show that during this period a linear equation of the form $y = at + b$ fits these data reasonably well, and determine the values of a and b.

t:	0	1	2	3	4	5	6	7	8	9	10	11	12	13	14	15	16	17	18	19	20	21
y:	38	36	41	40	38	40	45	46	52	53	54	63	65	67	70	73	72	80	79	87	83	88

3. From one individual to another within the same species of mammal it is found that the brain volume V varies with the body weight W according to a power law $V = aW^b$. The following data are obtained from measurements on a number of adult chimpanzees (W in kg, V in cm^3). Show, using a log-log plot, that the power law fits these data quite well, and obtain the values of a and b.

W:	31	36	38	41	42	45	47	48	50	53	55	57
V:	365	380	382	395	397	410	410	415	420	427	437	440

4. The surface area S of human beings whose weights are W and heights H are given, on an average, by a formula of the type $S = aW^bH^c$, where a, b, and c are constants. The following table gives values of S in square meters obtained from measurements of a number of individuals of height 180 cm and different weights W (in kilograms).

W:	70	75	77	80	82	84	87	90	95	98
S:	2.10	2.12	2.15	2.20	2.22	2.23	2.26	2.30	2.33	2.37

Show that a power law $S = AW^b$ fits these data reasonably well, and determine A and b.

5. The following data are obtained for the mean cranium length in millimeters (x) and the mean face length in millimeters (y) for baboons of different ages. Show that these data follow the law $y = ax^b$, and find the values of a and b.

x:	78.5	100.25	108.9	114.7	118.25	122.0
y:	31.0	64.6	94.8	131.0	140.8	144.25

6. After an injection the concentration of a particular drug in the patient's blood rises rapidly to a level of 0.1 mg/cm^3, then slowly decreases. Measurements of the concentration at 1-hr intervals after the injection gave the following results:

t:	1	2	3	4	5	6	7	8
C:	0.086	0.063	0.045	0.032	0.022	0.016	0.011	0.008

Show that these measurements fit quite well an equation of the form $C = ak^t$ and determine the values of a and k.

7. A bacteriocide was added to a solution containing a population of 10^7 bacteria. At various times after this, the number of bacteria was measured and the following results were obtained.

t (time in minutes):	0	10	20	30	40	50	60	70	80
y (number of bacteria):	10^7	3.2×10^6	10^6	3.2×10^5	10^5	3.2×10^4	10^4	3.2×10^3	10^2

Use a semi-log plot to establish an equation relating y to t.

8. An experiment is conducted to test the effect of a certain drug on the adrenalin flow of dogs. If y is the number of milligrams of drug necessary to produce a flow of adrenalin x, it is believed that y is related to x according to the formula $y = ba^x$, where a and b are constants. (This is an example of the Weber–Fechner law.) The following values are measured:

x:	1	2	3	4	5	6
y:	3.2	6.1	11.4	23.0	49.0	95.0

Show that the hypothesized relationship between x and y is reasonably well-satisfied, and determine the constants a and b.

3.4 NATURAL LOGARITHMS AND EXPONENTIALS

Let us examine the function $y = a^x$ as illustrated in Fig. 3.2 for different values of a. When $x = 0$, $y = a^0 = 1$, so that the graph of $y = a^x$ passes through the point $(0, 1)$ for any positive value of the constant a. The slope of the graph as it passes through the point varies depending on a: The bigger the value of a, the greater the slope of the graph. If $a < 1$, the slope is negative at $(0, 1)$, while if $a > 1$ the slope is positive.

Let us select the particular value of a for which the slope of the graph at $x = 0$ is equal to 1. We shall denote this special value of a by the letter e. It turns out that e is an irrational number, having the value $e = 2.71828. \ldots$

Since the function $y = e^x$ has unit slope at $x = 0$,

$$y'(0) = \lim_{\Delta x \to 0} \frac{e^{0+\Delta x} - e^0}{\Delta x} = 1.$$

Therefore, since $e^0 = 1$,

$$\lim_{\Delta x \to 0} \frac{e^{\Delta x} - 1}{\Delta x} = 1. \tag{1}$$

This condition determines the value of e for us, although we shall not stop here to show how e is calculated.

Let us now evaluate the derivative of $y = e^x$ for a general x.

$$\frac{dy}{dx} = \lim_{\Delta x \to 0} \frac{e^{x+\Delta x} - e^x}{\Delta x}.$$

But using a basic property of exponents, $e^{x+\Delta x} = e^x \times e^{\Delta x}$, and so

$$\frac{dy}{dx} = \lim_{\Delta x \to 0} e^x \frac{e^{\Delta x} - 1}{\Delta x}$$

$$= e^x \lim_{\Delta x \to 0} \frac{e^{\Delta x} - 1}{\Delta x}$$

$$= e^x$$

after using the property (1) above.

Thus we have the important result that the derivative of the function e^x is identical with the function itself:

$$\boxed{\; y = e^x, \qquad \frac{dy}{dx} = e^x. \;}$$

The function e^x is also often denoted by exp x, and is called the *natural exponential function*. The reason why this particular exponential function is so important rests in the above property that its derivative is everywhere equal to the function itself. It is, apart from a constant factor, the only function that possesses this property. In fact, if y is a function of x that satisfies the condition that $dy/dx = y$ for all values of x, then it can be shown that y must be of the form $y = be^x$, where b is some constant. b can have any value. (We shall demonstrate this result in Section 10.2.) It is this fact that accounts for our interest in the number e and in exponentials and logarithms that have e as their base.

There are many examples in nature of functions that satisfy the condition that dy/dx is equal to y or, more generally, to a constant multiplied by y. Some of these examples will be discussed in Section 10.2. In such situations, the natural exponential function plays a central role.

It is possible to give alternative definitions of e and the function e^x other than the one above. The following definition provides further insight into the meaning of e. Consider some quantity, for example the size of some biological population, which is growing from year to year in such a way that the amount of growth in any year is x times the size at the beginning of the year. Then at the end of the year the population size is $(1 + x)$ times its value at the beginning of the year. [Another example that might be considered is the size of a savings deposit in a bank that gains interest at the rate of $100x \%$/yr. After 1 yr the deposit has increased to an amount $(1 + x)$ times its value at the beginning of the year.]

Now let us ask what would be the corresponding change in size if instead the population growth in each half-year were equal to $\frac{1}{2}x$ times the size at the beginning of that half-year. In such a case the population size would change by a factor $[1 + (x/2)]$ for each half-year, so that at the end of any full year, the size would equal $[1 + (x/2)]^2$ times its size at the beginning of that year.

We can continue in this way; for example, we can consider the possibility that the population growth in every four-month period is equal to $\frac{1}{3}x$ times the size at the beginning of that period. In general, we can divide the year into n equal periods of time and suppose that during each of these periods the population size changes by a factor $[1 + (x/n)]$. Since there are n such changes in a whole year, the population size changes during a year by a factor $[1 + (x/n)]^n$:

$$\frac{\text{size at end of year}}{\text{size at beginning of year}} = \left(1 + \frac{x}{n}\right)^n.$$

Now let us consider the possibility that n gets bigger and bigger; in fact let us allow $n \longrightarrow \infty$. Then it turns out that

$$\lim_{n\to\infty} \left(1 + \frac{x}{n}\right)^n = e^x.$$

If we put $x = 1$, we obtain from this equation that

$$\lim_{n\to\infty} \left(1 + \frac{1}{n}\right)^n = e.$$

We shall not prove these results here. However they do illustrate further the significance of the quantity e: If a quantity grows by a factor $[1 + (1/n)]$ during each period of time, where n is some large number, then during n such periods the quantity will grow by a factor that approximately equals e. The larger the value of n, the closer this factor will be to e.

Let us now return to some examples to illustrate the derivatives of e^x and related functions.

EXAMPLE   Find dy/dx when $y = be^{kx}$.

SOLUTION   The function $y = be^{kx}$ can be regarded as a composite function: $y = be^u$, where $u = kx$. Then

$$\frac{dy}{du} = b\frac{d}{du}(e^u) = be^u \quad \text{and} \quad \frac{du}{dx} = k.$$

So, from the chain rule,

$$\frac{dy}{dx} = \frac{dy}{du} \cdot \frac{du}{dx} = be^u \cdot k.$$

Note that the result of this example can be reexpressed in the form $dy/dx = ky$: The derivative of y is equal to k times the value of y itself.

EXAMPLE   Differentiate $xe^{\sqrt{x+1}}$ w.r.t. x.

SOLUTION   Setting $y = xe^{\sqrt{x+1}}$ we observe that y has the form of a product: $y = uv$ where $u = x$ and $v = e^{\sqrt{x+1}}$. Therefore, from the product rule,

$$\frac{dy}{dx} = \frac{d}{dx}(x)e^{\sqrt{x+1}} + x\frac{d}{dx}(e^{\sqrt{x+1}})$$

$$= e^{\sqrt{x+1}} + x\frac{d}{dx}(e^{\sqrt{x+1}}).$$

The remaining derivative here is found from the chain rule. Setting $z = \sqrt{x+1}$, we find

$$\frac{d}{dx}(e^{\sqrt{x+1}}) = \frac{d}{dx}(e^z) = \frac{d}{dz}(e^z) \cdot \frac{dz}{dx}$$

$$= e^z \cdot \frac{dz}{dx}$$

$$= e^z \cdot \tfrac{1}{2}(x+1)^{-1/2}.$$

Substituting, therefore, we get

$$\frac{dy}{dx} = e^{\sqrt{x+1}}\left[1 + \frac{1}{2}x(x+1)^{-1/2}\right].$$

Given a composite function of the form $y = e^{u(x)}$ where $u(x)$ is any differentiable function, the chain rule allows us to calculate its derivative:

$$\frac{dy}{dx} = \frac{dy}{du} \cdot \frac{du}{dx} = \frac{d}{du}(e^u) \cdot u'(x) = e^u u'(x).$$

Writing this in the verbal form we have used before, we can say that

$$\boxed{\frac{d}{dx}e^{inside} = e^{inside}\,\frac{d}{dx}(inside)}$$

where *inside* represents any differentiable function of x. For example, if we wanted to differentiate $e^{\sqrt{x+1}}$ as in the previous example, we would take *inside* $= \sqrt{x+1}$ and write

$$\frac{d}{dx}(e^{\sqrt{x+1}}) = e^{\sqrt{x+1}}\,\frac{d}{dx}(\sqrt{x+1})$$

$$= e^{\sqrt{x+1}} \cdot \tfrac{1}{2}(x+1)^{-1/2}.$$

We can also form logarithms with base e. These are called *natural logarithms* (or sometimes Napierian logarithms). They are denoted by the symbol ln:

$$y = e^x, \qquad x = \log_e y = \ln y$$

The natural logarithm has all of the properties proved earlier for logarithms with the base a. For example, $\ln(y_1 y_2) = \ln y_1 + \ln y_2$, $\ln(y^n) = n\ln y$, $\ln e = 1$, etc. We can, if we wish, use natural logarithms to perform arithmetical calculations in the same way as we use common logarithms, but to do so would require much more effort, since the value of ln 10 is not equal to 1, but is equal to 2.3026. Thus, for example, to find ln 30, we need to look up ln 3 in the tables and then add 2.3026 to it (since ln 30 = ln 3 + ln 10). This is much more lengthy than finding the common log of 30, for which we have to look up log 3 in the tables and then simply add 1 to it.

You might wonder why we need to bother at all with natural logarithms and exponentials. The basic reason lies in the above result concerning the derivative of the function e^x. Because of this result, it is somewhat simpler, as far as calculus is concerned, to handle the function e^x than to handle the more general exponential function a^x.

These two functions are related. Let $y = a^x$. Then, since $a = e^{\ln a}$, it follows that
$$y = (e^{\ln a})^x = e^{x \ln a}.$$

Thus any exponential function $y = a^x$ can be written in the equivalent form $y = e^{kx}$, where $k = \ln a$.

EXAMPLE   We found in the blood-alcohol example in Section 3.3 that at time t the content of alcohol was given by the formula
$$y = 0.22(0.60)^t.$$

We can write $(0.60)^t = e^{kt}$ where $k = \ln (0.60)$. The value of $\ln (0.60)$ can be found from standard tables of natural logarithms, and it turns out to be -0.51. (The method of obtaining this will be shown in an example below.) Thus
$$y = 0.22e^{-0.51t}.$$

This type of equation is the most common form in which to express exponential decay.

EXAMPLE   Find dy/dx when $y = a^x$.

SOLUTION   We can write $y = e^{kx}$ where $k = \ln a$. But from an earlier example we know that
$$\frac{d}{dx}(e^{kx}) = ke^{kx} = ka^x.$$

Therefore, $dy/dx = (\ln a) a^x$.

A function of the type $y = be^{kx}$ represents an exponential growth if $k > 0$ and an exponential decay if $k < 0$. (Note that $k = \ln a$ so $k > 0$ when $a > 1$ and $k < 0$ when $a < 1$. This corresponds to our discussion in Section 3.1 in which it was pointed out that $y = ba^x$ represents exponential growth if $a > 1$ and exponential decay if $a < 1$.)

Natural logarithms are related to common logarithms in the following way. Let $y = a^x$, so that $x = \log_a y$. We have seen above that we can also write $y = e^{x \ln a}$, so that $x \ln a = \ln y$. If we substitute $x = \log_a y$ into this equation, we see that

$$\boxed{\begin{aligned} \ln y &= (\log_a y)(\ln a) \\ \text{or} \quad \log_a y &= \frac{\ln y}{\ln a}. \end{aligned}}$$

This equation holds for any base a. If we take $a = 10$, $\log_a y$ becomes the common logarithm, $\log y$. Since $\ln (10) = 2.3026 \ldots$, it follows that
$$\ln y = (2.3026 \ldots) \log y,$$

EXAMPLE   Find: (a) $\ln (100)$;   (b) $\ln (0.60)$;   (c) $\ln (600)$.

SOLUTION   (a) $\ln (100) = \ln (10^2) = 2 \ln (10)$
$$= 2(2.3026 \ldots) = 4.6052.$$

(b)   There are two ways of proceeding to calculate ln (0.60). The first is to use common logarithms:

$$\log (0.60) = \bar{1}.7782 = -0.2218.$$

Therefore

$$\ln (0.60) = (2.3026)(-0.2218)$$
$$= -0.5107.$$

Alternatively, we can use tables of natural logarithms. A short table of these is given in Appendix III, from which we find that ln 6 = 1.7918. Therefore

$$\ln (0.6) = \ln \left(\tfrac{6}{10}\right)$$
$$= \ln 6 - \ln (10)$$
$$= 1.7918 - 2.3026$$
$$= -0.5108.$$

(The difference between these two answers is the result of round-off errors.)

(c)

$$\ln (600) = (2.3026) \log (600)$$
$$= (2.3026)(2.7782)$$
$$= 6.3971.$$

Alternatively,

$$\ln (600) = \ln 6 + \ln (100)$$
$$= 1.7918 + 4.6052$$
$$= 6.3970.$$

We have seen that $(d/dx)(e^x) = e^x$. We shall now evaluate the derivative of the function $y = \ln x$.

If $y = \ln x$, then $x = e^y$. Let us differentiate this second equation with respect to x:

$$\frac{d}{dx}(e^y) = \frac{d}{dx}(x) = 1.$$

But from the chain rule, we see that

$$\frac{d}{dx}(e^y) = \frac{d}{dy}(e^y) \cdot \frac{dy}{dx} = e^y \frac{dy}{dx}$$

since

$$\frac{d}{dy}(e^y) = e^y.$$

Therefore

$$e^y \frac{dy}{dx} = 1, \quad \text{and so} \quad \frac{dy}{dx} = \frac{1}{e^y} = \frac{1}{x}.$$

Thus if

$$y = \ln x, \quad \frac{dy}{dx} = \frac{1}{x}.$$

EXAMPLE   Find dy/dx if $y = \ln(x + c)$, where c is a constant.

SOLUTION y is a composite function, $y = \ln u$, where $u = x + c$. Therefore, from the chain rule,

$$\frac{dy}{dx} = \frac{dy}{du} \cdot \frac{du}{dx}$$

$$= \frac{d}{du}(\ln u) \cdot \frac{du}{dx}$$

$$= \left(\frac{1}{u}\right) \cdot (1)$$

$$= \frac{1}{x + c}.$$

In general, the chain rule allows us to differentiate any composite function of the form $y = \ln u(x)$ in the following way:

$$\frac{dy}{dx} = \frac{dy}{du} \cdot \frac{du}{dx} = \frac{d}{du}(\ln u) \cdot u'(x) = \frac{1}{u}u'(x).$$

Alternatively, in verbal form,

$$\frac{d}{dx}\ln(inside) = \frac{1}{inside}\frac{d}{dx}(inside),$$

where *inside* stands for any differentiable function of x.

EXAMPLE   Differentiate $\ln(x^2 + x - 2)$.

SOLUTION   Here we take $inside = (x^2 + x - 2)$. Then

$$\frac{d}{dx}\ln(x^2 + x - 2) = \frac{1}{(x^2 + x - 2)}\frac{d}{dx}(x^2 + x - 2)$$

$$= \frac{1}{(x^2 + x - 2)}(2x + 1)$$

$$= \frac{2x + 1}{x^2 + x - 2}.$$

When we require to differentiate the logarithm of a product or quotient of various expressions, it is often useful to simplify the given function first by making use of the properties of logarithms.

EXAMPLE   Find dy/dx when $y = \ln(e^x/\sqrt{x + 1})$.

SOLUTION   First simplifying y, we have

$$y = \ln\left(\frac{e^x}{\sqrt{x + 1}}\right) = \ln(e^x) - \ln(\sqrt{x + 1})$$

$$= x \ln e - \tfrac{1}{2}\ln(x + 1)$$

Therefore, immediately, (since $\ln e = 1$)

$$\frac{dy}{dx} = \ln e - \frac{1}{2}\frac{d}{dx}\ln(x+1) = 1 - \frac{1}{2(x+1)}.$$

EXAMPLE Find dy/dx if $y = \log x$.

SOLUTION In order to differentiate the common logarithm, we express it in terms of the natural logarithm,

$$y = \log(x) = \log_{10} x = \frac{\ln(x)}{\ln(10)}.$$

Therefore

$$\frac{dy}{dx} = \frac{1}{\ln(10)}\frac{d}{dx}\ln(x) = \frac{1}{\ln(10)} \cdot \frac{1}{x}$$

$$= \frac{0.4343\ldots}{x}$$

since

$$\frac{1}{\ln(10)} = \frac{1}{2.3026\ldots} = 0.4343\ldots.$$

EXERCISES 3.4

Find dy/dx for each of the following functions:

1. $y = xe^x$.

2. $y = \dfrac{e^x}{x}$.

3. $y = e^{x^2}$.

4. $y = e^{\sqrt{x}}$.

5. $y = \dfrac{e^{\sqrt{x}}}{e^x}$.

6. $y = e^{ax^3+bx^2+cx+d}$.

7. $y = \ln(x^2)$.

8. $y = \ln(ax^3 + bx^2 + cx + d)$.

9. $y = \dfrac{1}{\ln x}$.

10. $y = (\ln x)^2$.

11. $y = \log(e^x)$.

12. $y = \log(e^x - 1)$.

13. $y = x(\ln x - 1)$.

14. $y = x^2 \ln(x^2 + 1)$.

15. $y = \dfrac{\ln x}{x}$.

16. $y = \dfrac{x+1}{\ln(x+1)}$.

17. $y = \ln\dfrac{x+2}{\sqrt{x^2+1}}$.

18. $y = \ln\dfrac{e^x\sqrt{x-1}}{x^3}$.

19. $y = \sqrt{\ln x}$.

20. $y = \left(\dfrac{\ln x}{e^x}\right)^{1/3}$.

Given that ln 3 = 1.0986 and ln 4 = 1.3863 to four decimal places, evaluate the following:

21. ln 9.

22. ln 12.

23. ln 3000.

24. ln 40.

25. ln (0.04).

26. ln (0.003).

27. ln 6.

28. ln 5.4.

29. A certain population is growing according to the formula

$$y = p(1 - ce^{-kt})^3,$$

where p, c, and k are constants. Find the rate of growth at a general time t.

30. After injection, the concentration of a certain drug in the blood of a patient changes according to the formula $c = pt^2e^{-kt}$, where p and k are constants. Calculate the rate of increase of concentration at time t.

31. If w is the weight of an average animal of a certain species at age t, it is often found that

$$\ln w - \ln (A - w) = B(t - C)$$

where A, B, and C are certain constants. Express w as an explicit function of t. Show that

$$\frac{dw}{dt} = Bw\left(1 - \frac{w}{A}\right).$$

3.5   EXPONENTIAL GROWTH AND DECAY

There are many processes in nature that exhibit exponential growth or decay. One of the most important applications of exponential growth, which we have already mentioned in Section 3.1, is to the growth of biological populations. In general, population growth is a very complex phenomenon, being affected by environmental factors (such as the weather or the supply of food) and involving interactions between different populations (for example, one animal population might be the food source for another predator population or two species might compete for the same limited supply of food). In the case of human populations, social attitudes toward procreation also play a powerful role in determining the rate of growth. Thus any mathematical model of the growth of a population must usually take into account many factors of the type mentioned above.

In certain cases, however, it is possible to use the simple model of exponential growth. Consider, for example, a culture of bacteria growing in isolation in a laboratory, with a plentiful food supply available. For such a population it would usually be very satisfactory to assume that the population size increases as an exponential function of time, as we did in some of the examples and exercises in Section 3.1. Eventually it may happen that the culture begins to exhaust its food supply, in which case the exponential growth would slow down. But if there are no limiting factors such as this, exponential growth would continue indefinitely.

This example illustrates a common feature of the growth of biological populations in general. As long as the population size is small relative to the maximum population that could be supported by the given habitat, the growth will be exponential. Exponential growth breaks down when the environment begins to exert pressure on further growth.

In the case of human populations, an exponential growth model is quite good over short periods of time (for example, one or two decades). However over longer periods of time, predictions based on continued exponential growth at the present rate of growth will probably turn out to be quite misleading.

Consider a population that at time t is size y; suppose initially, at $t = 0$, the population is $y = b$ and that y increases by a factor a for each unit of time. For example, if the population increases by $4\%/\text{yr}$, then $a = 1.04$ and t must be measured in years; if the population is doubling every half-hour then $a = 4$ provided that t is measured in hours. Then at a general time t, the population size is given by

$$y = ba^t.$$

This is a typical formula for exponential growth.

This formula is normally written in a form using natural exponentials. We note that $a^t = e^{kt}$ where $k = \ln a$, and therefore

$$y = be^{kt}.$$

For exponential growth, k must be positive (i.e., $a > 1$).

Let us examine the derivative dy/dt. We have seen before that

$$\frac{dy}{dt} = \frac{d}{dt}(be^{kt}) = kbe^{kt}.$$

Therefore,

$$\frac{dy}{dt} = ky.$$

Consequently, the rate at which y is changing at any time is proportional to the value that y has at that time. This is a very important property that is characteristic of functions that grow exponentially in time. We shall discuss it further in Chapter 10 when we deal with differential equations.

It is possible to see from this property why exponential growth is relevant to populations. For example, in the case of microorganisms reproducing by cell division, all the organisms in the population continue to divide at approximately the same rate (as long as the food supply remains plentiful) and so the rate at which the population increases is directly proportional to the number of organisms currently in the population.

Let us now turn to processes involving exponential decay. An exponential decay process is one that is governed by a function of the type $y = be^{-kt}$, where b and k are positive constants and t is time. As t increases, y becomes steadily smaller and approaches zero as $t \rightarrow \infty$.

One very important example of exponential decay is the radioactive decay of certain elements whose atoms have unstable nuclei. For example, the isotope C^{14} of carbon is a radioactive element whose decay is used as a means of dating certain archaeological objects. In this case we take y to be the amount of radioactive material remaining at time t, and it turns out that

$$y = be^{-kt},$$

where b is the amount of material present initially (since $y = b$ when $t = 0$). The constant k is called the *decay constant*, and it varies from one radioactive element to another.

The rate of increase of y is negative (since y is in fact a decreasing function) and again turns out to be proportional to y itself:

$$\frac{dy}{dt} = \frac{d}{dt}(be^{-kt}) = -kbe^{-kt} = -ky.$$

This characteristic property is a natural consequence of the physical laws governing radioactivity, and explains why this phenomenon shows exponential decay: The number of nuclei that decay in a small time interval, and hence the rate of decay, is proportional to the number of radioactive nuclei remaining undecayed.

The time necessary for half of the radioactive material to decay is called the *half-life* of the material. We denote it by T. Then at $t = T$,

$$y = \tfrac{1}{2}b = be^{-kT}.$$

Therefore

$$e^{-kT} = \tfrac{1}{2}$$

$$kT = -\ln\left(\tfrac{1}{2}\right) = \ln 2$$

$$k = \frac{\ln 2}{T}.$$

Thus, when the half-life is specified, the constant k can be determined. For example, the half-life of C^{14} is 5570 years, so that for this material $k = \ln 2/5570 = 1.24 \times 10^{-4}$. (*Note:* The value of $\ln 2$ is found from the table in Appendix III to be equal to 0.6931.)

The principle of the carbon-dating method is as follows. The proportion of the radioactive isotope C^{14} to the regular isotope C^{12} of carbon occurring in the earth's atmosphere remains at a constant level, the amount of C^{14} that decays being exactly balanced by new C^{14}, which is formed by cosmic rays hitting the atmosphere. Living plants and animals absorb carbon from the atmosphere, and so contain a proportion of C^{14} that is more or less the same as that in the atmosphere. When the animal or vegetation dies, however, the absorption of new C^{14} ceases, and the proportion of C^{14} in the dead organic matter steadily decreases as a result of radioactive decay.

By measuring the proportion of C^{14} to C^{12} in organic matter obtained from an archaeological site, it is possible to calculate how long it is since the death of the animal or vegetable occurred. The following example illustrates the method of calculation.

EXAMPLE   The amount of radioactive carbon in a certain piece of human bone is measured to be 58% of the amount that occurs naturally in the atmosphere. How old is the bone?

SOLUTION   Let us take $t = 0$ to be the instant of time at which the owner of the bone passed away, and b to be the amount of C^{14} that was then present. Then t years later, the amount of C^{14} is given by

$$y = be^{-kt}, \qquad k = 1.24 \times 10^{-4}.$$

If we take t to be the present time, then we know that $y = 0.58b$, since the amount of C^{14} is measured to be 58% of the amount b that naturally occurs in living matter.

$$be^{-kt} = 0.58b$$

$$e^{-kt} = 0.58.$$

Therefore,

$$kt = -\ln(0.58) = 0.545$$

from the table of natural logarithms. It follows that

$$t = \frac{0.545}{k} = \frac{0.545}{1.24 \times 10^{-4}} = 4400.$$

Therefore the present time is 4400 yr later than the instant $t = 0$; in other words, the bone is 4400 yr old.

EXERCISES 3.5

Express the following functions in the form $y = ae^{kt}$:

1. $y = 2^t$.

2. $y = (1000)2^{t/3}$.

3. $y = 5(1.04)^t$.

4. $y = 6 \times 10^8(1.05)^t$.

5. The earth's population is at present 4 billion and is increasing by 2% each year. Express the population y at a time t years from now in the form $y = ae^{kt}$.

6. In a population of microorganisms, the organisms divide on the average every 15 min. If at time $t = 0$ the population size is 10^4, express the population size at time t (measured in hours) in the form ae^{kt}.

7. The population of Britain in 1600 is believed to have been about 5 million. Three hundred fifty years later it had increased to 50 million. What was the average percentage growth per year during that period? (Assume a uniform exponential growth.)

8. If a population increases from 5 million to 200 million over a period of 200 yr, what is the average percentage growth per year?

9. An organic specimen from an archaeological site is found to have a content of C^{14} that is 25% of the level naturally occurring in the atmosphere. How old is the specimen?

10. A piece of charcoal is found to have a C^{14} content equal to 77% of the natural level found in living matter. How old is the charcoal?

11. The half-life of radium is 1590 yr. If 10 gm of radium are left for 1000 yr, how much will remain?

12. If 20 gm of radium are left for 10,000 yr, how much will remain?

13. How many years does it take 10 gm of radium to decay so that only 8 gm of radium remain?

14. After how many years would 1 gm of radium decay so that 0.1 gm of radium remains?

15. A population growing exponentially would theoretically continue to grow forever. However, in practice, the growth of a biological population tends to level out eventually, for example, because there is a limited food supply available. One function that is commonly used to describe such restricted growth is the so-called logistic function,

$$y = \frac{p}{1 + ce^{-kt}},$$

where p, c, and k are constants. If $p = \frac{124}{3} \times 10^7$, $c = \frac{245}{3}$, and $e^{-100k} = \frac{4}{35}$, with t measured in years, what are the population sizes when $t = 0$, $t = 100$, and $t = 200$?

16. The weight of a culture of bacteria is given by

$$y = \frac{2}{1 + 3(2^{-t})}$$

when t is measured in hours. What are the weights when $t = 0, 1, 2,$ and 4?

17. For a population growing according to the logistic function given in exercise 15, show that at $t = 0$, $y = p/(1 + c)$, and that as $t \to \infty$, y approaches the limiting value p. Show also that at any value of t, $dy/dt = ky[1 - (y/p)]$.

18. Another function sometimes used to describe restricted growth of a population is the Gompertz function

$$y = pe^{-ce^{-kt}}$$

(again p, c, and k are constants). Show that at $t = 0$, $y = pe^{-c}$, and that as $t \to \infty$, $y \to p$. Show also that for any value of t,

$$\frac{dy}{dt} = ky \ln \frac{p}{y}.$$

REVIEW EXERCISES FOR CHAPTER 3

1. Are the following statements true or false? If false, replace them with a correct statement.
   (a) $a^m \cdot a^n = a^{mn}$ for all a.
   (b) $\dfrac{a^m}{a^n} = a^{m/n}$.

(c) $a^0 = 1$ for all $a \neq 0$.

(d) $\log m + \log n = \log (m + n)$ for $m, n > 0$.

(e) $\log m - \log n = \log (m - n)$ for $m, n > 0$.

(f) $\log (mn) = (\log m)(\log n)$ for $m, n > 0$.

(g) $\log \left(\dfrac{m}{n} \right) = \dfrac{(\log m)}{(\log n)}$ for $m > 0$ and $n > 1$.

(h) $\log (m^n) = (\log m)^n$.

(i) $\log_a 1 = 0$ for all $a \neq 0$.

(j) $\log_a a = 1$ for all a.

(k) The domain of e^{-x} is the set of all nonnegative real numbers.

(l) The domain of $f(x) = \ln x$ is $\{x \,|\, x \geq 0\}$.

(m) $\dfrac{d}{dx}[e^{f(x)}] = e^{f(x)}$.

(n) $\dfrac{d}{dx}(e^2) = e^2$.

(o) $\dfrac{d}{dx}[\ln f(x)] = \dfrac{1}{f(x)}$.

(p) $\dfrac{d}{dx}(\ln a) = \dfrac{1}{a}$.

(q) $\dfrac{d}{dx}(\ln |x|) = \dfrac{1}{x}$ for all $x \neq 0$.

Find the derivatives of the following functions w.r.t. the independent variable involved:

2. $\dfrac{e^x + 1}{e^x - 1}$.

3. $e^{2 \ln x}$.

4. e^{y^2}.

5. $\ln (2^x \sqrt{x - 1})$.

6. $e^u \ln (u^2)$.

7. $x^2 \ln (\sqrt{x})$.

8. $(e^t + \ln t)^5$.

9. $\ln (t^2 e^t)$.

10. $t^2 \ln (e^{3t})$.

11. $t \ln (2^t)$.

12. $\ln (\ln x)$.

13. e^{e^z}.

14. Find dy/dx if $t = \ln x$ and $y = e^t$.

15. During an epidemic, the number of infected individuals is given by $f(t) = at^p e^{-t}$ at time t (a and p are constants). Find the value of t for which $f'(t) = 0$. How would you describe the corresponding value of f?

16. The yield of grain y in bushels per acre is given by

$$y = y_0 + a(1 - e^{-kx}),$$

where x is the number of pounds per acre of fertilizer and y_0, a, and k are constants. What are the limiting values of y as $x \to 0$ and as $x \to \infty$? Find dy/dx.

17. An archaeological relic has a content of C^{14} equal to 1% of the value naturally occurring. How old is the relic?

18. If the population of the earth is 4 billion in 1976 and is increasing by 2% per year, when will the population be 6 billion?

19. A population of bacteria is doubling in size every 19 min. How long does it take to increase from 10^5 to 10^7 organisms?

20. When cancer cells are subjected to radiation treatment, the proportion of cells that survive the treatment is given by

$$P = e^{-kr},$$

where r is the radiation level and k a constant. It is found that 40% of the cancer cells survive when $r = 500$ Roentgens. What should the radiation level be in order to allow only 1% to survive?

21. In the case of a disease that affects virtually the whole of a given population, the following model can often be used:

$$\ln p - \ln(1 - p) = kt + C,$$

where p is the proportion of the population that have been infected at time t, and k and C are constants. Express p as a function of t, and show that

$$\frac{dp}{dt} = kp(1 - p).$$

What happens to p as $t \to \infty$?

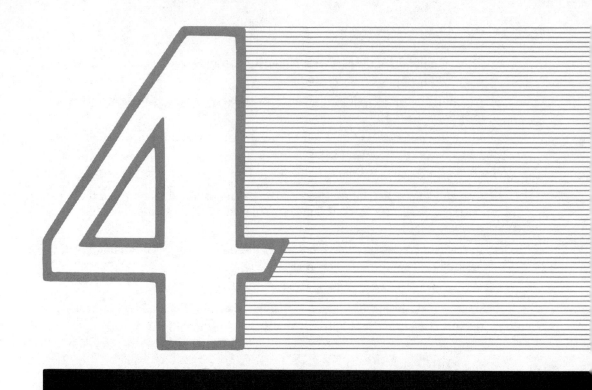

4

Applications
of
Derivatives

4.1 ANALYSIS OF CURVES

Let us suppose that we are given an explicit relationship, $y = f(x)$, between two variables, x and y. It often happens that we would like to get a rough qualitative picture of what the graph of this relation looks like. Of course we could do this by actually plotting a number of points (x, y), which we would calculate in the usual way from the given function $y = f(x)$. In fact, such a procedure would give us the precise graph of the relation. However, if all we need is a qualitative idea of the shape of the graph, it is often much easier to make use of certain properties of derivatives in order to sketch the graph. The first and second derivatives, $f'(x)$ and $f''(x)$, are effective tools in studying the nature of graphs, and here in Section 4.1 we shall discuss their use for this purpose. In Sections 4.2 and 4.3, we shall look at a related class of problems called optimization problems, in which one is interested in finding the largest or the smallest value of y satisfying a particular relation $y = f(x)$.

DEFINITION   A function $y = f(x)$ is said to be an *increasing function* over a certain interval of values of x if y increases with increase of x. That is, if x_1 and x_2 are any two values in the given interval with $x_2 > x_1$, then $f(x_2) > f(x_1)$.

A function $y = f(x)$ is said to be a *decreasing function* over an interval of its domain if y decreases with increase of x. That is, if $x_2 > x_1$ are two values of x in the given interval, then $f(x_2) < f(x_1)$.

Figures 4.1(a) and (b) illustrate respectively the graphs of an increasing and a decreasing function.

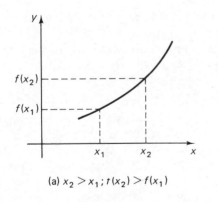

(a) $x_2 > x_1 ; f(x_2) > f(x_1)$ (b) $x_2 > x_1 ; f(x_2) < f(x_1)$

Figure 4.1

THEOREM 4.1.1

(a)   If $f(x)$ is an increasing function that is differentiable, then $f'(x) \geq 0$.

(b)   If $f(x)$ is a decreasing function that is differentiable, then $f'(x) \leq 0$.

PROOF   (a)   Let x and $x + \Delta x$ be two values of the independent variable, with $y = f(x)$ and $y + \Delta y = f(x + \Delta x)$ the corresponding values of the dependent variable.

161

Then

$$\Delta y = f(x + \Delta x) - f(x).$$

There are two cases to consider depending on whether $\Delta x > 0$ or $\Delta x < 0$. They are illustrated in Figs. 4.2 and 4.3, respectively.

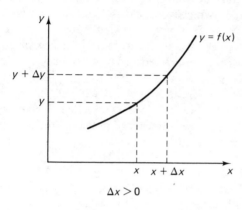

Figure 4.2   **Figure 4.3**

If $\Delta x > 0$, then $x + \Delta x > x$. Therefore, since $f(x)$ is an increasing function, $f(x + \Delta x) > f(x)$, so that $\Delta y > 0$. Consequently, both Δx and Δy are positive, so that $\Delta y/\Delta x > 0$. The second possibility is that $\Delta x < 0$. Then $x + \Delta x < x$, and $f(x + \Delta x) < f(x)$. Hence $\Delta y < 0$. In this case Δx and Δy are both negative, so that again $\Delta y/\Delta x > 0$.

In both cases, $\Delta y/\Delta x$ is positive. The derivative $f'(x)$ is the limit of $\Delta y/\Delta x$ as $\Delta x \to 0$, and since $\Delta y/\Delta x$ is always positive, it is clearly impossible for it to approach a negative number as a limiting value. Therefore the limit $f'(x) \geq 0$, as stated in the theorem.

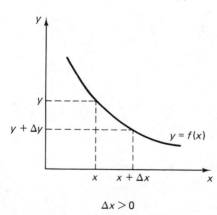

Figure 4.4

(b) The proof of part (b) of Theorem 4.1.1, when $f(x)$ is a decreasing function, is quite similar. The case when $\Delta x > 0$ is illustrated in Fig. 4.4. In this case $x + \Delta x > x$ and $f(x + \Delta x) < f(x)$ since f is decreasing. Consequently $\Delta y < 0$, and $\Delta y/\Delta x < 0$ (Δy is negative, Δx is positive). Taking the limit as $\Delta x \to 0$, therefore,

$$f'(x) = \lim_{\Delta x \to 0} \frac{\Delta y}{\Delta x} \leq 0.$$

Similarly, we see that when $\Delta x < 0$, for a decreasing function, $\Delta y > 0$, so that the ratio $\Delta y/\Delta x$ is still negative. Hence again, in the limit, $f'(x) \leq 0$.

This theorem has a converse that can be stated as follows.

THEOREM 4.1.2

(a) If $f'(x) > 0$ for all x in some interval, then $f(x)$ is an increasing function of x over that interval.

(b) If $f'(x) < 0$ for all x in some interval, then $f(x)$ is a decreasing function of x over that interval.

NOTE. Observe that in the second theorem the inequalities are strict.

The proof of this theorem will not be given. However it is intuitively an obvious result. In case (a), for example, the fact that $f'(x) > 0$ means, geometrically, that the tangent to the graph at any point has positive slope. If the graph of $f(x)$ always slopes upwards to the right, then y must increase as x increases. Correspondingly, in part (b), if $f'(x) < 0$ then the graph slopes downwards to the right and y decreases as x increases. A formal proof of this theorem can be given by making use of the *mean value theorem*, which is presented at the end of this section.

These theorems are used to determine the intervals where a function is increasing or decreasing: that is, where the graph is rising or falling.

EXAMPLE Find the values of x for which the function

$$f(x) = x^2 - 2x + 1$$

is increasing or decreasing.

SOLUTION
$$f(x) = x^2 - 2x + 1,$$
$$f'(x) = 2x - 2.$$

$f'(x) > 0$ implies that $2x - 2 > 0$, that is, $x > 1$. Thus $f(x)$ is increasing for all values of x in the interval $x > 1$. Similarly, $f'(x) < 0$ implies that $2x - 2 < 0$, i.e., $x < 1$. The function is decreasing for $x < 1$.

The graph of $y = f(x)$ is shown in Fig. 4.5. [Note that $f(1) = 0$, so the point $(1, 0)$ lies on the graph.] For $x < 1$ the graph slopes negatively and for

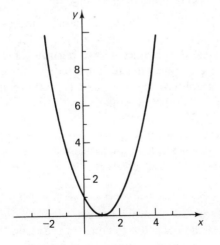

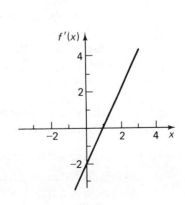

Figure 4.5

Figure 4.6

$x > 1$ it slopes positively. The graph of $f'(x)$ is shown in Fig. 4.6, and we observe that this graph lies below the x-axis for $x < 1$ and above the x-axis for $x > 1$.

EXAMPLE   Find the values of x for which the function

$$f(x) = x^3 - 3x$$

is increasing or decreasing. Sketch the graph of $y = f(x)$.

SOLUTION   We have

$$f'(x) = 3x^2 - 3$$
$$= 3(x^2 - 1)$$
$$= 3(x - 1)(x + 1).$$

To find the intervals of x in which $f'(x)$ is positive or negative we first find the values of x where $f'(x)$ is zero. These are the points at which $f'(x)$ can change from negative to positive values (or vice versa), and they represent the possible end-points of the intervals we are seeking.

In the present case, $f'(x) = 0$ when $(x - 1)(x + 1) = 0$; that is, when $x = -1$ or 1. Therefore the possible intervals in which we must examine the sign of $f'(x)$ are $x < -1$, $-1 < x < 1$, and $x > 1$.

i.   When $x > 1$, both $(x - 1)$ and $(x + 1)$ are positive. Therefore $f'(x) = 3(x - 1)(x + 1)$ is the product of positive numbers, and is positive. So in the region $x > 1$, $f'(x)$ is an increasing function.

ii.   When $-1 < x < 1$, $(x + 1)$ is still positive but $(x - 1)$ is negative. Therefore $f'(x)$ is the product of numbers one of which is negative, hence $f'(x) < 0$. So in the region $-1 < x < 1$, $f(x)$ is a decreasing function.

iii.   When $x < -1$, both $(x - 1)$ and $(x + 1)$ are negative, so $f'(x)$ is again positive. For $x < -1$, $f(x)$ is an increasing function.

We note that $f(-1) = 2$ and $f(1) = -2$, so the points $(-1, 2)$ and $(1, -2)$ lie on the graph of $y = f(x)$. The graph is sketched in Fig. 4.7, where use has been made of the information obtained above regarding the regions in which $f(x)$ is increasing or decreasing. One further piece of information has been used, namely that the graph crosses the x-axis ($y = 0$) when $x = 0$ and when $x = \pm\sqrt{3}$. [Setting $y = 0$ we get the equation $x^3 - 3x = 0$, or $x(x^2 - 3) = 0$, whose roots are 0 and $\pm\sqrt{3}$.]

The graph of $f'(x)$ is shown in Fig. 4.8. We note that this graph is above the x-axis in the regions where $f(x)$ is increasing and below the x-axis in the region $-1 < x < 1$ where $f(x)$ is decreasing.

We see from the above that the sign of the first derivative has a geometrical significance that is extremely useful when we need to obtain a qualitative sketch of the graph of a function. We shall now go on to consider the second derivative, which, as we shall see, also has an important geometrical interpretation.

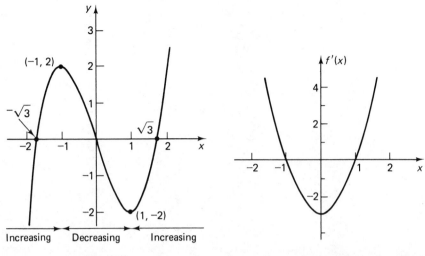

Figure 4.7	Figure 4.8

Consider a function $f(x)$ whose graph has the general shape indicated in Fig. 4.9. The slope of the graph is positive, $f'(x) > 0$, so y is an increasing function of x. Furthermore, the graph has the property that, as we move to the right (i.e., as x increases), the slope of the graph becomes steeper. That is, the derivative $f'(x)$ is also an increasing function of x. The graph of $f'(x)$ must have the form indicated qualitatively in Fig. 4.10.

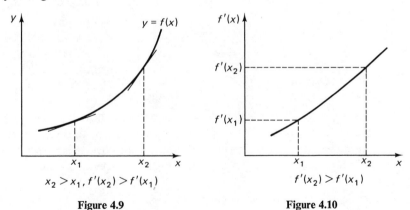

Figure 4.9	Figure 4.10

If $f'(x)$ is an increasing function of x, its derivative must be greater than or equal to zero by Theorem 4.1.1. That is, $f''(x) \geq 0$. Conversely, if $f'(x) > 0$ and $f''(x) > 0$, then the graph of $y = f(x)$ must slope upwards to the right and, furthermore, the graph must slope more and more steeply as x increases.

Now consider a function $f(x)$ whose graph has the form shown in Fig. 4.11. Here the graph slopes downwards to the right, $f'(x) < 0$, but the slope becomes less steep as x increases. Thus $f'(x)$ is increasing from large negative values toward zero, as indicated in Fig. 4.12. Again $f'(x)$ is an increasing function of x, so that $f''(x) \geq 0$.

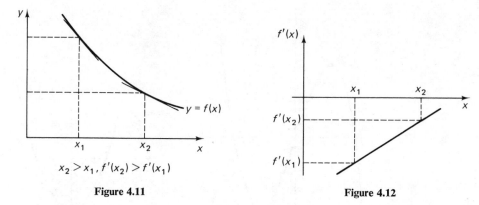

Figure 4.11

$$x_2 > x_1, f'(x_2) > f'(x_1)$$

Figure 4.11   **Figure 4.12**

Conversely, if $f'(x) < 0$ and $f''(x) > 0$ then the graph of $f(x)$ has the general form of this figure. That is, the graph slopes downwards to the right but becomes less steep as x increases.

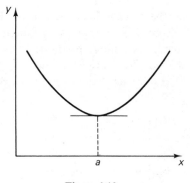

Figure 4.13

A third type of graph is shown in Fig. 4.13. This type consists essentially of an amalgamation of the two types discussed above. For $x < a, f'(x) < 0$, while for $x > a$, $f'(x) > 0$. Furthermore, $f'(x)$ increases from negative values to zero at $x = a$ and then to positive values for $x > a$. Thus $f'(x)$ is always increasing, so $f''(x) \geq 0$.

The geometrical property which characterizes the three types of graph discussed above is that they are all *concave upwards* (or convex downwards). We conclude therefore that if the graph of $y = f(x)$ is concave upwards then $f''(x) \geq 0$. Conversely, if $f''(x) > 0$ then the graph of $y = f(x)$ must be concave upwards.

Now let us consider the alternative possibility that the graph of $y = f(x)$ is *concave downwards*. The various cases corresponding to the three types discussed above are shown in Fig. 4.14. Figure 4.14(a) illustrates the case when $f'(x) > 0$ but the slope is becoming less steep as x increases. Figure 4.14(b) illustrates the case when $f'(x) < 0$

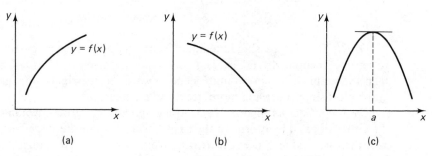

(a)   (b)   (c)

Figure 4.14

and the slope becomes steeper as x increases. Figure 4.14(c) shows the case when $f'(x) > 0$ for $x < a$ and $f'(x) < 0$ for $x > a$.

In each case, $f'(x)$ is a decreasing function of x. Therefore we conclude that when the graph of $y = f(x)$ is concave downwards, $f''(x) \leq 0$. Conversely, when $f''(x) < 0$, the graph of $y = f(x)$ is concave downwards.

EXAMPLE   Find the values of x for which the graph of

$$y = \tfrac{1}{6}x^4 - x^3 + 2x^2$$

is concave upwards or concave downwards.

SOLUTION
$$y = \tfrac{1}{6}x^4 - x^3 + 2x^2$$
$$y' = \tfrac{4}{6}x^3 - 3x^2 + 4x$$
$$y'' = 2x^2 - 6x + 4$$
$$= 2(x^2 - 3x + 2)$$
$$= 2(x - 1)(x - 2).$$

In order to find out whether the graph is concave upwards or downwards we must examine the sign of y''. We note that $y'' = 0$ when $x = 1$ and when $x = 2$, so we must consider the three regions $x < 1$, $1 < x < 2$, and $x > 2$. Note that when $x = 1$, $y = \tfrac{7}{6}$ and when $x = 2$, $y = \tfrac{8}{3}$. Thus the two points $(1, \tfrac{7}{6})$ and $(2, \tfrac{8}{3})$ lie in the graph.

When $x > 2$, both $(x-1)$ and $(x-2)$ are positive, so that $y'' = 2(x-1)(x-2)$ is the product of positive factors. Hence $y'' > 0$, and the graph is concave upwards for $x > 2$. When $1 < x < 2$, $(x - 1)$ is positive but $(x - 2)$ is negative. Hence $y'' = 2(x - 1)(x - 2)$ has one negative factor, so $y'' < 0$. So the graph is concave downwards in this region. When $x < 1$, both $(x - 1)$ and $(x - 2)$ are negative, so their product is positive. Hence $y'' > 0$ and the graph is concave upwards.

The graph of $y = f(x)$ in this example is shown in Fig. 4.15. Note the changes in convexity at the points $(1, \tfrac{7}{6})$ and $(2, \tfrac{8}{3})$ on the graph.

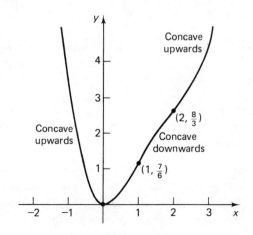

Figure 4.15

DEFINITION   A *point of inflection* on a curve is a point where the curve changes from concave upwards to concave downwards, or vice versa.

If $x = x_1$ is a point of inflection on the graph of $y = f(x)$, then on one side of x_1 the graph is concave upwards, i.e., $f''(x) > 0$, and on the other side of x_1 the graph is concave downwards, i.e., $f''(x) < 0$. Thus on passing from one side to the other of $x = x_1$, $f''(x)$ changes sign. At $x = x_1$ it is necessary either that $f''(x_1) = 0$ or that $f''(x_1)$ fails to exist [$f''(x)$ may become infinitely large as $x \longrightarrow x_1$].

In the preceding example, the graph of $y = \frac{1}{6}x^4 - x^3 + 2x^2$ has points of inflection at $x = 1$ and $x = 2$. For example, for $x < 1$ the graph is concave upwards, while for x just above 1, the graph is concave downwards. $x = 1$ is a point where the convexity changes, i.e., a point of inflection. The same is true for $x = 2$.

This example corresponds to a point of inflection at which $y'' = 0$. The following example illustrates the alternative possibility.

EXAMPLE   Find the points of inflection of $y = x^{1/3}$.

SOLUTION   We have

$$y' = \tfrac{1}{3}x^{-2/3},$$
$$y'' = \tfrac{1}{3}(-\tfrac{2}{3})x^{-5/3} = -\tfrac{2}{9}x^{-5/3}.$$

For $x > 0$, $x^{5/3}$ is positive, so $y'' < 0$. For $x < 0$, $x^{5/3}$ is negative, so $y'' > 0$. Thus the graph (Fig. 4.16) is concave upwards for $x < 0$ and concave downwards for $x > 0$. The value $x = 0$, at which

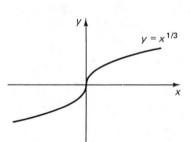

$y = x^{1/3}$

$y = 0$ also, is therefore a point of inflection. In this case y'' becomes indefinitely large as $x \longrightarrow 0$, so we have a point of inflection at which the second derivative fails to exist. (Note also that as $x \longrightarrow 0$, y' becomes infinite, so that the slope of the graph becomes vertical at the origin for this particular function.)

Figure 4.16

Observe that the tangent to a graph at a point of inflection always crosses the graph at that point. This is an unusual property for a tangent; as a rule the graph lies entirely on one side of the tangent line in the neighborhood of the point of tangency.

The use of the first and second derivatives of $f(x)$ as described in this section provides useful techniques for sketching the graph of $y = f(x)$. Let us summarize the steps involved in this procedure.

i. Find $f'(x)$. Find the coordinates of the points where $f'(x) = 0$: At these points the graph has horizontal tangents. Find the values of x for which $f'(x)$ is greater than or less than zero; these provide the regions in which the graph is increasing or decreasing.

ii. Find $f''(x)$. Find the regions for which $f''(x)$ is positive or negative; these provide the regions in which the graph is concave upwards or downwards. The points where f'' changes sign are the points of inflection.

Usually in addition to this we would make use of certain additional pieces of information in completing the sketch of the graph. These would, as a rule, fall into the following categories.

iii. A few explicit points on the graph are found. For example, the points of intersection with the coordinate axes are often useful. The intersection with the y-axis is obtained by putting $x = 0$, so that $y = f(0)$. The intersection with the x-axis is obtained by putting $y = 0$, which leads to the equation $f(x) = 0$, which must be solved for the corresponding values of x.

iv. The behavior of $f(x)$ as $x \to \pm\infty$ is examined. This determines the way in which the graph behaves as we move to the far right or the far left of the coordinate plane. For example, consider the graph of $y = \frac{1}{6}x^4 - x^3 + 2x^2$, which was discussed in a preceding example. When x becomes very large (positive or negative), y becomes approximately the same as the term with the largest power, $y \approx \frac{1}{6}x^4$. (This is true of any polynomial function: For sufficiently large values of x, a polynomial function behaves approximately like its term with the largest power.) As $x \to \pm\infty$, x^4 increases without bound and $y \to +\infty$. The graph of $y = \frac{1}{6}x^4 - x^3 + 2x^2$ becomes indefinitely higher, therefore, as x moves further and further towards either $+\infty$ or $-\infty$. As a second example, consider the graph of $y = x^3 - 3x$, which was also discussed earlier in this section. The highest power in this function is the cubic one, so as $x \to \pm\infty$, $y \approx x^3$. Again, as $x \to +\infty$, $y \to +\infty$, so the graph becomes higher and higher as x moves to the far right. However as $x \to -\infty$, $y \to -\infty$; therefore as x moves to the far left, the graph takes larger and larger negative values. These properties can be seen in the graph of this function (Fig. 4.7).

EXAMPLE Sketch the graph of $y = x^4 - 6x^2$.

SOLUTION $y = 0$ when $x^2(x^2 - 6) = 0$, that is, when $x = 0, \pm\sqrt{6}$. Thus the graph intersects the x-axis at three points: $x = 0$ and $\pm\sqrt{6}$. As x approaches positive or negative infinity, y becomes very large, i.e., $y \to +\infty$ (since $y \approx x^4$ as $x \to \pm\infty$).

$$y' = 4x^3 - 12x = 4x(x^2 - 3)$$
$$= 4x(x - \sqrt{3})(x + \sqrt{3}).$$

Thus $y' = 0$ when $x = 0$ or $\pm\sqrt{3}$. When $x = 0$, $y = 0$; when $x = \pm\sqrt{3}$, $y = (\sqrt{3})^4 - 6(\sqrt{3})^2 = 9 - 18 = -9$. Thus the graph has horizontal tangents at the points $(0, 0)$, $(\sqrt{3}, -9)$ and $(-\sqrt{3}, -9)$.

Examining the sign of y', we observe that when $x < -\sqrt{3}$, all three of the factors x, $(x - \sqrt{3})$, and $(x + \sqrt{3})$ are negative, so that $y' < 0$. Continuing with the other intervals, we obtain the results summarized in the table below:

INTERVAL	f'	f
$x < -\sqrt{3}$	< 0	decreasing
$-\sqrt{3} < x < 0$	> 0	increasing
$0 < x < \sqrt{3}$	< 0	decreasing
$x > \sqrt{3}$	> 0	increasing

Next we examine the second derivative.

$$y'' = 12x^2 - 12 = 12(x^2 - 1)$$
$$= 12(x + 1)(x - 1).$$

$y'' = 0$ when $x = \pm 1$, so we must consider the three intervals $x < -1, -1 < x < 1$, and $x > 1$. The signs of y'' in each of these intervals are summarized in the following table.

INTERVAL	f''	f
$x < -1$	> 0	concave upwards
$-1 < x < 1$	< 0	concave downwards
$x > 1$	> 0	concave upwards

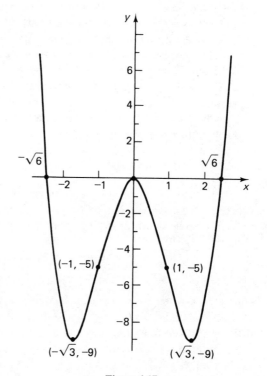

Figure 4.17

The values $x = \pm 1$ correspond to points of inflection since the convexity of the graph changes as x passes through these values. When $x = \pm 1, y = 1 - 6(1) = -5$. The points of inflection are therefore $(-1, -5)$ and $(1, -5)$. When $x = -1$, $y' = -4 + 12 = 8$. When $x = +1$, $y' = 4 - 12 = -8$. These provide the slopes of the graph at the two points of inflection. Combining all the above information we obtain the graph shown in Fig. 4.17.

The Mean Value Theorem

The mean value theorem plays a fundamental role in the proofs of many other theorems of calculus. For example, the enterprising student might try as an exercise to construct his own proof of Theorem 4.1.2 by making use of this theorem. We shall not give a proof of the mean value theorem here, but shall content ourselves with extablishing its plausibility. We shall begin by discussing a related theorem called Rolle's Theorem, which is actually a special case of the mean value theorem itself.

THEOREM 4.1.3 (Rolle's Theorem)

Let the function $f(x)$ be defined and continuous at all x in the closed interval $a \leq x \leq b$ and be differentiable at all x in the open interval $a < x < b$. Let $f(a) = f(b)$. There exists at least one number c satisfying $a < c < b$ at which $f'(c) = 0$.

Let us illustrate this theorem with some examples. Figures 4.18 and 4.19 show the graphs of some typical functions that satisfy the conditions of the theorem. The graphs are smooth (differentiable) between $x = a$ and $x = b$, and they occupy the same horizontal levels at $x = a$ and $x = b$. [This corresponds to the condition $f(a) = f(b)$.] In Fig. 4.18, the graph rises to a certain maximum height and then falls again to its original level as x approaches b. At the point $x = c$, at which the maximum height is attained, the tangent line is horizontal. Therefore $f'(c) = 0$ as required by the theorem.

In Fig. 4.19 there are two numbers, c and c', at which the tangent is horizontal; that is, $f'(c) = 0$ and $f'(c') = 0$. This of course does not conflict with the theorem, which states that there must exist *at least one* such number; in particular cases there may exist two or more points at which f' vanishes.

The mean value theorem can be deduced from Rolle's Theorem.

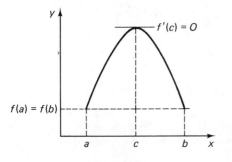

Figure 4.18

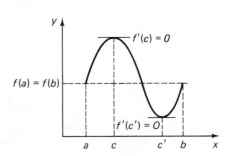

Figure 4.19

THEOREM 4.1.4 (Mean Value Theorem)

Let $f(x)$ be continuous on the closed interval $a \leq x \leq b$ and differentiable on the open interval $a < x < b$. Then there exists at least one number c satisfying $a < c < b$ such that

$$f'(c) = \frac{f(b) - f(a)}{b - a}.$$

PROOF

Define a function $g(x)$ as follows:

$$g(x) = f(x) - mx,$$

where m is to be chosen in such a way that $g(a) = g(b)$. $g(x)$ can be so-defined for all x such that $a \leq x \leq b$, and is continuous at each such x and differentiable for all x in $a < x < b$. Putting $x = a$ and $x = b$ we obtain

$$g(a) = f(a) - ma, \qquad g(b) = f(b) - mb.$$

Since we require $g(a) = g(b)$, we obtain the following equation, which leads to the value of m:

$$f(a) - ma = f(b) - mb,$$

and so

$$m = \frac{f(b) - f(a)}{b - a}.$$

The function $g(x)$ now satisfies all of the conditions of Rolle's Theorem. Hence there exists some number, c, such that $a < c < b$, at which $g'(c) = 0$. But

$$g'(x) = \frac{d}{dx}[f(x) - mx] = f'(x) - m,$$

and so

$$f'(c) - m = 0.$$

We conclude that there exists a number c satisfying $a < c < b$ such that $f'(c) = m$, and this proves the theorem.

The theorem has the following geometrical interpretation. On the graph of $y = f(x)$, the point A at which $x = a$ has coordinates $[a, f(a)]$, and the point B at which

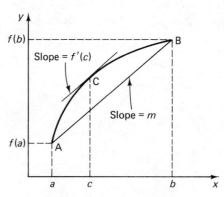

172

Figure 4.20

$x = b$ has coordinates $[b, f(b)]$ (Fig. 4.20). The slope of the chord AB is therefore given by

$$\text{slope of } AB = \frac{f(b) - f(a)}{b - a},$$

and this is the quantity denoted by m in the proof of the theorem. The theorem states, therefore, that there exists some point C on the graph whose x-coordinate lies between a and b, at which the tangent is parallel to the chord AB. At $x = c$, the tangent to the graph has slope $f'(c)$, and this slope is equal to m, the slope of AB.

EXERCISES 4.1

Find the values of x for which the following functions are: (a) increasing, (b) decreasing, (c) concave upwards, and (d) concave downwards. Also find the points of inflection, if any.

1. $y = x^2 - 6x + 7.$　　　　　　　　　**2.** $y = x^3 - 12x + 10.$

3. $f(x) = x^3 - 3x + 4.$　　　　　　　**4.** $f(x) = 2x^3 - 9x^2 - 24x + 20.$

5. $f(x) = x + \dfrac{1}{x}.$　　　　　　　　**6.** $f(x) = x^2 + \dfrac{1}{x^2}.$

7. $f(x) = \dfrac{x}{x + 1}.$　　　　　　　　**8.** $f(x) = \dfrac{x + 1}{x - 1}.$

9. $y = x + \ln x.$　　　　　　　　　　**10.** $y = x - e^x.$

11. $y = x \ln x.$　　　　　　　　　　**12.** $y = xe^{-x}.$

13. $y = x^5 - 5x^4 + 1.$　　　　　　　**14.** $y = x^7 - 7x^6.$

Sketch the graphs of the following functions:

15. $y = x^2 - 4x + 5.$　　　　　　　　**16.** $y = x^3 - 3x + 2.$

17. $y = 5x^6 - 6x^5 + 1.$　　　　　　**18.** $y = x^4 - 2x^2.$

19. $y = x^{2/3}.$　　　　　　　　　　　**20.** $y = x^{1/5}.$

21. $y = \ln x.$　　　　　　　　　　　**22.** $y = e^{-2x}.$

23. $xy = 2.$　　　　　　　　　　　　**24.** $xy = -1.$

4.2 MAXIMA AND MINIMA

Many of the important applications of derivatives involve finding the maximum or minimum values of a certain function. For example, a medical researcher might be interested in the amount of a drug, administered to a person, that gives the maximum reaction; an agriculturist might be interested in the amount of fertilizer that will yield a maximum return in the value of a crop; or a biologist might be concerned with the time at which the growth rate of a bacterial population is maximum, and so on.

Before we come to applications, however, let us discuss the theory of maxima and minima.

DEFINITION   **i.**   A function $f(x)$ is said to have a *local maximum* at $x = c$ if $f(c) > f(x)$ for all x sufficiently near c. Thus the points P and Q in Fig. 4.21 correspond to local maxima of the corresponding functions.

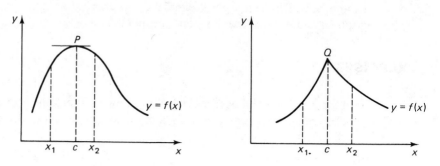

$$f(c) > f(x_1) \text{ and } f(c) > f(x_2)$$

Figure 4.21

ii.   A function $f(x)$ is said·to have a *local minimum* at $x = c$ if $f(c) < f(x)$ for all x sufficiently close to c. The points A and B in Fig. 4.22 correspond to local minima.

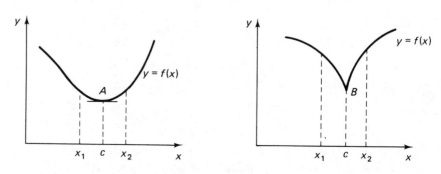

$$f(c) < f(x_1) \text{ and } f(c) < f(x_2)$$

Figure 4.22

A function may have more than one local maximum and more than one local minimum, as we see in Fig. 4.23. The points A, C, and E on the graph correspond to points where the function has local maxima, and the points B, D, and F correspond to points where the function has local minima.

A (*local*) *maximum or minimum value* of a function is the y-coordinate at the point at which the graph has a local maximum or minimum. A local minimum value of a function may be greater than a local maximum value.

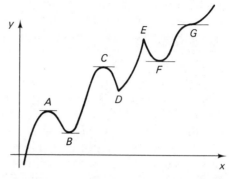

Figure 4.23

This can easily be seen from the graph in Fig. 4.23, where the ordinate of F is greater than the ordinate of A.

iii.   The term *extremum* is used to denote either a local maximum or a local minimum. Extrema is the plural of extremum.

iv.   The value $x = c$ is called a *critical value* for a continuous function f if

(a) either $f'(c) = 0$ or $f'(x)$ fails to exist at $x = c$ and if
(b) $f(c)$ is well-defined.

In the case where $f'(c) = 0$, the tangent to the graph of $y = f(x)$ is horizontal at $x = c$. This possibility is illustrated in Fig. 4.24(a). The second case, when

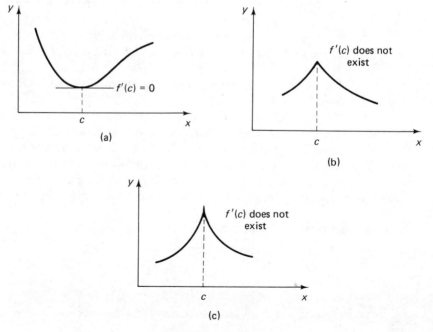

Figure 4.24

$f'(c)$ fails to exist, occurs with two possible types of behavior of the graph of $y = f(x)$: $f'(c)$ fails to exist when the graph has a corner at $x = c$; and $f'(c)$ fails to exist when the tangent to the graph becomes vertical at $x = c$ [so that $f'(x)$ becomes infinitely large as $x \to c$]. These two possibilities are illustrated in Fig. 4.24(b) and (c).

We emphasize the fact that for c to be a critical value, $f(c)$ must be well-defined. Consider, for example, $f(x) = x^{-1}$, whose derivative is $f'(x) = -x^{-2}$. Clearly $f'(x) \to -\infty$ as $x \to 0$. However $x = 0$ is not a critical value for this function since $f(0)$ does not exist.

It is clear from Fig. 4.24 that local extrema correspond to critical values of a function f. But not every critical value of a function corresponds to a local minimum or a local maximum The point P in Fig. 4.25(a), where the

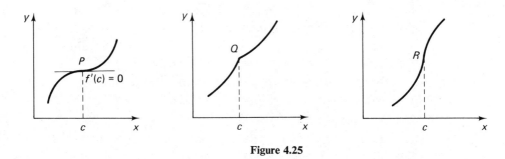

Figure 4.25

tangent is horizontal, corresponds to a critical value but is neither a local maximum nor a local minimum point. The points Q and R in Fig. 4.25(b) and (c) correspond to critical values at which $f'(c)$ fails to exist but that are not extrema of $f(x)$. In fact, P and R are points of inflection.

Let us now develop some tests for determining the maxima and minima of a function.

Consider first the case when the local extremum occurs at a critical value given by $f'(x) = 0$, that is, when the tangent line is horizontal at the point on the graph of f corresponding to the extremum. Then if the point is a local maximum the graph is concave downwards, and if the point is a local minimum the graph is concave upwards. But we know that whenever $f''(x) < 0$, the graph of f is concave downwards and whenever $f''(x) > 0$ the graph is concave upwards. Thus we have:

THEOREM 4.2.1 (Second Derivative Test)

Let $f(x)$ be twice differentiable at $x = c$. then

(i) $x = c$ is a local maximum of f whenever
$$f'(c) = 0 \quad \text{and} \quad f''(c) < 0;$$

(ii) $x = c$ is a local minimum of f whenever
$$f'(c) = 0 \quad \text{and} \quad f''(c) > 0.$$

EXAMPLE   Find the local maximum and minimum values of

$$x^3 + 2x^2 - 4x - 8.$$

SOLUTION   Let

$$f(x) = x^3 + 2x^2 - 4x - 8.$$
$$f'(x) = 3x^2 + 4x - 4.$$

To find the local maxima or minima we set $f'(x) = 0$, or

$$3x^2 + 4x - 4 = 0$$
$$(3x - 2)(x + 2) = 0,$$

which gives $x = \frac{2}{3}, -2$. Then

$$f''(x) = 6x + 4.$$

At $x = \frac{2}{3}$, $f''(\frac{2}{3}) = 6(\frac{2}{3}) + 4 = 8 > 0$. Hence, since $f''(x)$ is positive when $x = \frac{2}{3}$, $f(x)$ has a local minimum when $x = \frac{2}{3}$. The local minimum value is given by

$$f(\tfrac{2}{3}) = (\tfrac{2}{3})^3 + 2(\tfrac{2}{3})^2 - 4(\tfrac{2}{3}) - 8$$
$$= -\tfrac{256}{27}.$$

When $x = -2, f''(-2) = 6(-2) + 4 = -8 < 0$. Hence, since $f''(x)$ is negative when $x = -2$, $f(x)$ has a local maximum when $x = -2$. The local maximum value is given by

$$f(-2) = (-2)^3 + 2(-2)^2 - 4(-2) - 8 = 0.$$

Thus the only local maximum value of $f(x)$ is zero, which occurs when $x = -2$, and the only local minimum value is $-\frac{256}{27}$, which occurs when $x = \frac{2}{3}$.

Note that in the example, $f'(x)$ is well-defined for all values of x, so that the only critical values are those at which $f'(x)$ is zero.

EXAMPLE   Determine the local maxima and minima for $f(x) = (\ln x)/x$.

SOLUTION   Using the quotient rule, we have

$$f'(x) = \frac{x(\ln x)' - \ln x(x')}{x^2}$$

$$= \frac{x \cdot \dfrac{1}{x} - \ln x \cdot 1}{x^2}$$

$$= \frac{1 - \ln x}{x^2}.$$

For a local extremum, $f'(x) = 0$ or

$$\frac{1 - \ln x}{x^2} = 0.$$

$$1 - \ln x = 0.$$

$$\ln x = 1 = \ln e.$$

Therefore

$$x = e.$$

In this case we have only one critical point: $x = e$. [Note that $f'(x)$ becomes infinite as $x \to 0$. However $x = 0$ is not a critical point because $f(0)$ is not defined.]

Using the quotient rule again, we have:

$$f''(x) = \frac{x^2(1 - \ln x)' - (1 - \ln x) \cdot (x^2)'}{(x^2)^2}$$

$$= \frac{x^2\left(-\dfrac{1}{x}\right) - (1 - \ln x)(2x)}{x^4}$$

$$= \frac{2 \ln x - 3}{x^3}.$$

When $x = e$,

$$f''(e) = \frac{2 \ln e - 3}{e^3} = \frac{2 - 3}{e^3} = -\frac{1}{e^3} < 0,$$

where we have used the fact that $\ln e = 1$. Hence $f(x)$ has a local maximum when $x = 3$. In this case there are no local minima.

The second derivative test can be used for all local extrema at which $f'(c) = 0$ and $f''(c)$ does not equal zero. When $f''(x) = 0$ at a critical point $x = c$ or when $f''(c)$ fails to exist, then the second derivative test cannot be used to ascertain whether $x = c$ is a local maximum or minimum point. In such cases we must resort to a different test called the *first derivative test for local extrema*.

The first derivative test is also used for all critical points of the type at which $f'(c)$ fails to exist. It also happens sometimes even in cases where the second derivative test works, that the first derivative test can be simpler to use.

THEOREM 4.2.2 (First Derivative Test)

If $x = c$ is a critical value for $f(x)$, that is, either $f'(c) = 0$ or $f'(x)$ fails to exist as $x \to c$:

i. $x = c$ is a local *maximum* of f if $f'(x)$ changes sign from positive to negative

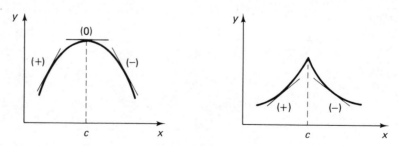

Local maximum at $x = c$

Figure 4.26

as x changes from just below c to just above c (Fig. 4.26). [The $(+)$, $(-)$, or (0) in brackets in the figure indicates the sign of the slope of the graph at given points.]

ii. $x = c$ is a local *minimum* of f if $f'(x)$ changes sign from negative to positive as x changes from just below c to just above c (Fig. 4.27).

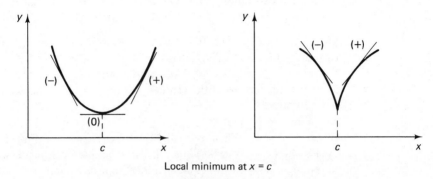

Local minimum at $x = c$

Figure 4.27

iii. $x = c$ is not a local extremum if $f'(x)$ does not change sign as x changes from just below $x = c$ to just above $x = c$ (Fig. 4.28). In such cases $x = c$ will be either a point of inflection or a corner on the graph of $f(x)$.

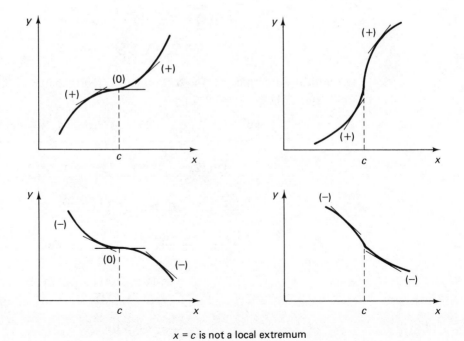

$x = c$ is not a local extremum

Figure 4.28

EXAMPLE　Find the local extrema for $f(x) = x^4 - 4x^3 + 7$.

SOLUTION　In this case

$$f'(x) = 4x^3 - 12x^2$$
$$= 4x^2 (x - 3).$$

For a local extremum, $f'(x) = 0$ or $4x^2(x - 3) = 0$, that is, $x = 0, 3$.

$$f''(x) = 12x^2 - 24x.$$

At $x = 3$, $f''(3) = 12(9) - 24(3) = 108 - 72 = 36 > 0$. Hence $x = 3$ is a local minimum point for $f(x)$.

At $x = 0$, $f''(0) = 0$. Thus we cannot use the second derivative test to determine the nature of the critical point $x = 0$. For this we must resort to the first derivative test.

When x is slightly negative, $x^2 > 0$ and $x - 3 < 0$ so that $f'(x) = 4x^2 (x - 3) < 0$. When x is slightly positive, $x^2 > 0$ and $x - 3 < 0$ so that $f'(x) = 4x^2(x - 3) < 0$. Thus $f'(x)$ is negative both just below and just above the point $x = 0$, and does not change sign as x increases through the value 0. Hence $x = 0$ is neither a maximum nor a minimum, but is in fact a point of inflection.

EXAMPLE　Investigate the local extrema for $f(x) = x^4/(x - 1)$.

SOLUTION　Using the quotient rule (with $u = x^4$, $v = x - 1$),

$$f'(x) = \frac{(x - 1) \cdot 4x^3 - x^4 \cdot 1}{(x - 1)^2}$$
$$= \frac{x^3(3x - 4)}{(x - 1)^2}.$$

For a local extremum, $f'(x) = 0$ or $x = 0, \frac{4}{3}$. [Note that $x = 1$ is not a critical value since $f(1)$ is not defined.]

$$f''(x) = \frac{(x - 1)^2 \frac{d}{dx}(3x^4 - 4x^3) - (3x^4 - 4x^3)\frac{d}{dx}(x - 1)^2}{(x - 1)^4}$$
$$= \frac{(x - 1)^2(12x^3 - 12x^2) - (3x^4 - 4x^3) \cdot 2(x - 1) \cdot 1}{(x - 1)^4}$$
$$= \frac{12x^2(x - 1)(x - 1) - 2x^3(3x - 4)}{(x - 1)^3}$$
$$= \frac{2x^2(3x^2 - 8x + 6)}{(x - 1)^3}.$$

At $x = \frac{4}{3}$,

$$f''(\tfrac{4}{3}) = \frac{2(\frac{4}{3})^2[3 \cdot \frac{16}{9} - 8 \cdot \frac{4}{3} + 6]}{(\frac{4}{3} - 1)^3}$$
$$= \frac{\frac{64}{27}}{\frac{1}{27}} = 64 > 0.$$

Hence $x = \frac{4}{3}$ is a local minimum point.

At $x = 0$, $f''(0) = 0$ and we must make use of the first derivative test in this case. When x is slightly negative, x^3 and $3x - 4$ are both negative and $(x - 1)^2$ is positive. Thus $f'(x) = x^3(3x - 4)/(x - 1)^2$ involves two negative quantities in the numerator and a positive quantity in the denominator; therefore it is positive. When x is slightly positive, x^3 and $(x - 1)^2$ are positive, whereas $3x - 4$ is negative, and $f'(x)$ is negative. Thus $f'(x)$ changes from positive to negative at $x = 0$, and by the first derivative test $x = 0$ is a local maximum point. Hence the given function has a local maximum at $x = 0$ and a local minimum at $x = \frac{4}{3}$.

Note that we could use the first derivative test for the point $x = \frac{4}{3}$ even though the second derivative test works there. Since the second differentiation is a bit messy, the first derivative test may even be preferable in this case.

EXAMPLE   Investigate the local extrema for $f(x) = (x - 1)^{2/3}$.

SOLUTION   We have

$$f'(x) = \frac{2}{3}(x - 1)^{-1/3} = \frac{2}{3\sqrt[3]{x - 1}}.$$

For no value of x does $f'(x)$ become zero. In this case the critical point occurs where $f'(x)$ becomes infinitely large, that is, $x = 1$. Whenever we are dealing with a critical point at which f' is infinite, we must always resort to the first derivative test to find out if it is a local maximum or a local minimum point.

When x is slightly less than one, $(x - 1) < 0$, and therefore $(x - 1)^{1/3} < 0$. So $f'(x) < 0$. When x is slightly greater than one, both $(x - 1)$ and $(x - 1)^{1/3}$ are positive, and $f'(x) > 0$. Thus $f'(x)$ changes sign from negative to positive at $x = 1$; hence $x = 1$ is a local minimum point. The graph of $f(x)$ is shown in Fig. 4.29.

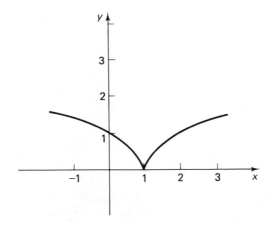

Figure 4.29

Absolute Maxima and Minima

The *absolute maximum* value of $f(x)$ over an interval $a \leq x \leq b$ of its domain is the largest value of $f(x)$ as x takes all the values from a to b. Similarly, the *absolute minimum* value of $f(x)$ is the smallest value of $f(x)$ as x increases from a to b. It is intuitively obvious that if $f(x)$ is continuous in $a \leq x \leq b$, the point at which $f(x)$ attains its absolute maximum must be either a local maximum of $f(x)$ or one of the endpoints a or b. A similar statement holds for the absolute minimum. Thus to find the absolute maximum and absolute minimum values of $f(x)$ over $a \leq x \leq b$, we simply select the largest and the smallest values from among the values of $f(x)$ at the critical points lying in $a \leq x \leq b$ and at the endpoints a and b. This is illustrated in the following examples.

EXAMPLE Determine the absolute maximum and minimum values of $f(x) = 1 + 12x - x^3$ over the interval $1 \leq x \leq 3$.

SOLUTION We have $f'(x) = 12 - 3x^2$. Since $f'(x)$ is defined for all x, the critical values of f are given by $f'(x) = 0$ or $x^2 = 4$, that is, $x = \pm 2$. But $x = -2$ is *not* within the given interval $1 \leq x \leq 3$. Thus we only consider the critical value $x = 2$, and the endpoints $x = 1$, $x = 3$. The values of $f(x)$ at these points are

$$f(1) = 1 + 12 - 1 = 12,$$
$$f(2) = 1 + 24 - 8 = 17, \quad \text{and}$$
$$f(3) = 1 + 36 - 27 = 10.$$

Thus the absolute maximum value of $f(x)$ is 17, which occurs at $x = 2$, and the absolute minimum is 10, which occurs at the endpoint $x = 3$. The graph of $y = 1 + 12x - x^3$ is shown in Fig. 4.30. Within the interval $1 \leq x \leq 3$, the graph has a single local maximum occurring at $x = 2$. The absolute minimum value occurs at the endpoint $x = 3$.

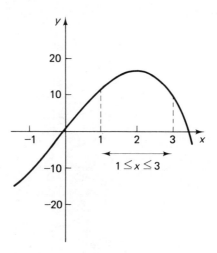

Figure 4.30

EXAMPLE   Find the absolute maximum and minimum values of $f(x) = (x - 2)^{1/3}/x$ in the interval $1 \leq x \leq 4$.

SOLUTION   Let us examine $f'(x)$ in order to find the critical points. From the quotient rule,

$$f'(x) = \frac{\frac{1}{3}(x - 2)^{-2/3}x - (x - 2)^{1/3}(1)}{x^2}$$

$$= \frac{\frac{1}{3}(x - 2)^{-2/3}}{x^2}[x - 3(x - 2)^{2/3}(x - 2)^{1/3}]$$

$$= \frac{(x - 2)^{-2/3}}{3x^2}[x - 3(x - 2)]$$

$$= \frac{1}{3x^2(x - 2)^{2/3}}(6 - 2x).$$

Then $f'(x) = 0$ when $6 - 2x = 0$, that is, when $x = 3$. Furthermore, $f'(x)$ fails to exist when $x = 0$ and $x = 2$. However $x = 0$ is not a critical point since $f(0)$ is not defined. Thus the critical points occur at $x = 2$ and $x = 3$, both of which lie within the given interval.

The values of $f(x)$ at the endpoints and critical points are as follows:

$$f(1) = \frac{(1 - 2)^{1/3}}{1} = -1,$$

$$f(2) = \frac{(2 - 2)^{1/3}}{2} = 0,$$

$$f(3) = \frac{(3 - 2)^{1/3}}{3} = \frac{1}{3} \approx 0.333,$$

$$f(4) = \frac{(4 - 2)^{1/3}}{4} = \frac{\sqrt[3]{2}}{4} \approx 0.315.$$

Of these values the largest is $\frac{1}{3}$, occurring when $x = 3$, and the smallest is -1, occurring when $x = 1$. These are respectively the absolute maximum and minimum values of $f(x)$ within the given interval.

When the interval with which we are concerned extends to $+\infty$ or $-\infty$, rather than being a closed interval with finite endpoints, we must modify our technique of finding absolute maxima and minima. Instead of the value of the function at the infinite endpoint we must use its limiting value as the variable approaches $+\infty$ or $-\infty$.

EXAMPLE   Find the absolute maximum and minimum values of

$$f(x) = \frac{x^2}{x^2 + 1}$$

in the interval $-1 \leq x < \infty$.

SOLUTION   First let us find the critical values of $f(x)$ as usual:

$$f'(x) = \frac{2x(x^2 + 1) - x^2(2x)}{(x^2 + 1)^2} = \frac{2x}{(x^2 + 1)^2},$$

where we have used the quotient rule. Therefore $f'(x) = 0$ when $x = 0$, and this is the only critical point. So for the given interval, f takes the following values at the left endpoint and at the critical point:

$$f(-1) = \frac{(-1)^2}{(-1)^2 + 1} = \frac{1}{2}, \quad \text{and}$$

$$f(0) = \frac{0^2}{0^2 + 1} = 0.$$

At the right endpoint ($x \to +\infty$), f has the limiting value as follows:

$$\lim_{x \to \infty} f(x) = \lim_{x \to \infty} \frac{x^2}{x^2 + 1} = \lim_{x \to \infty} \frac{1}{1 + \dfrac{1}{x^2}},$$

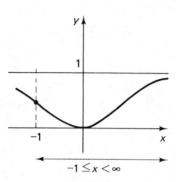

Figure 4.31

$-1 \leq x < \infty$

after dividing the numerator and denominator by x^2. As $x \to \infty$, $1/x^2 \to 0$, and

$$\lim_{x \to \infty} f(x) = 1.$$

Within the interval $-1 \leq x < \infty$, therefore, the absolute maximum and minimum values of $f(x)$ are respectively 1 and 0. This can be seen from the graph of $y = f(x)$ in Fig. 4.31.

The method of finding the maxima and minima of a function $f(x)$ can be summarized in the following steps:

Step 1. Find $f'(x)$ and solve the equation $f'(x) = 0$ for x.

Step 2. Test the values of x found in Step 1 for local maxima and minima.

(i) Find $f''(x)$ and evaluate this at each value found in Step 1. If c denotes any one of these values of x, then:

If $f''(c) < 0$, then $f(x)$ has a local maximum at $x = c$.

If $f''(c) > 0$, then $f(x)$ has a local minimum at $x = c$.

If $f''(c) = 0$, the test fails. In this case use the first derivative test as follows:

(ii) If $f'(x)$ changes sign from positive to negative at $x = c$, that is, if $f'(x) > 0$ for $x < c$, and $f'(x) < 0$ for $x > c$, then $f(x)$ has a local *maximum* at $x = c$. If $f'(x)$ changes sign from negative to positive at $x = c$, then $f(x)$ has a local *minimum* at $x = c$. If $f'(x)$ does not change sign at $x = c$, then $f(x)$ has a *point of inflection* at $x = c$.

Step 3. If $f'(x)$ fails to exist at some point for which $f(x)$ is defined, then examine this point for a possible local maximum or minimum by using the first derivative test as in Step 2(ii) above.

Step 4. In the event that we wish to find the absolute maximum or minimum value of $f(x)$ in some interval $a \leq x \leq b$, examine the values of $f(x)$ at all of the critical points that lie in the given interval and at the endpoints a and b. Select the largest or the smallest of these values to get the absolute maximum or absolute minimum, respectively.

EXERCISES 4.2

Find the values of x at the local maxima and minima of the following functions:

1. $f(x) = x^2 - 12x + 10$.

2. $f(x) = 1 + 2x - x^2$.

3. $f(x) = x^3 - 6x^2 + 7$.

4. $f(x) = x^3 - 3x + 4$.

5. $y = 2x^3 - 9x^2 + 12x + 6$.

6. $y = 4x^3 + 9x^2 - 12x + 5$.

7. $y = x^3 - 18x^2 + 96x$.

8. $y = x^3 - 3x^2 - 9x + 7$.

9. $y = x^5 - 5x^4 + 5x^3 - 10$.

10. $y = x^4 - 4x^3 + 3$.

11. $f(x) = x^3(x - 1)^2$.

12. $f(x) = x^4(x + 2)^2$.

13. $f(x) = x^{4/3}$.

14. $f(x) = x^{1/3}$.

15. $f(x) = x \ln x$.

16. $f(x) = xe^{-x}$.

Find the local maximum and minimum values of the following functions:

17. $f(x) = 2x^3 + 3x^2 - 12x - 15$.

18. $f(x) = \dfrac{x^3}{3} + ax^2 - 3xa^2$.

19. $f(x) = xe^x$.

20. $f(x) = xe^{-2x}$.

21. Show that $f(x) = x^3 - 3x^2 + 3x + 7$ has neither a local maximum nor minimum at $x = 1$.

22. Show that $f(x) = x + (1/x)$ has a local maximum and a local minimum value, but that the maximum value is less than the minimum value.

Find the absolute extrema of the following functions in the indicated intervals.

23. $f(x) = x^3 - 75x + 1$; $-1 \leq x \leq 6$.

24. $f(x) = \dfrac{x}{\sqrt{x - 1}}$; $\frac{5}{4} \leq x \leq 5$.

25. $f(x) = \dfrac{3x - 1}{x + 1}$; $1 \leq x < \infty$.

26. $f(x) = \dfrac{x^2 + x - 4}{x^2 - 1}$; $2 \leq x < \infty$.

4.3 APPLICATIONS OF MAXIMA AND MINIMA

Many practical situations arise when we want to maximize or minimize a certain quantity. The following example represents a typical case in point.

EXAMPLE   A conservationist is stocking a certain area of water with fish. The more fish he puts in, the more competition there will be for the available food supply, and so the fish will gain weight more slowly. In fact, he knows from previous experiments, that when there are n fish per unit area of water, the average amount of weight that each fish gains during one season is given by $w = (600 - 30n)$ gm. What is the value of n that leads to the maximum total production of weight of fish?

SOLUTION   The gain in weight of each fish is $w = 600 - 30n$. Since there are n fish per unit area, the total production per unit area, P, is equal to nw. Therefore,

$$P = n(600 - 30n)$$
$$= 600n - 30n^2.$$

The graph of P against n is shown in Fig. 4.32. P is zero when n is zero since there are then no fish to produce. As n increases, P increases to a maximum value and then decreases to zero again when $n = 20$. The reason why P decreases when n gets large is that for large values of n the fish put on very little weight, and even though there is a large number of them, the total production is small.

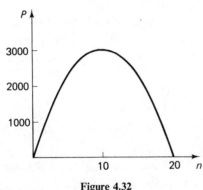

Figure 4.32

To find the value of n at which P is maximum, we differentiate and set the derivative $dP/dn = 0$.

$$\frac{dP}{dn} = 600 - 60n,$$

and $dP/dn = 0$ when $600 - 60n = 0$, that is, when $n = 10$. Thus the density of 10 fish per unit area gives the maximum total production. The maximum value of P is given by

$$P = 600(10) - 30(10)^2 = 3000 \text{ gm per unit area.}$$

It is obvious from the graph of P as a function of n that the value $n = 10$ corresponds to a maximum of P. However we can check this by using the second derivative:

$$\frac{d^2P}{dn^2} = -60.$$

The second derivative is negative (for all values of n), so the critical value $n = 10$ is a maximum of P.

Problems of this kind are called *maximization* (or *minimization*) problems. They are also often called *optimization* problems, since the aim in solving them is to find the optimum, or best, value that some quantity can have. Let us consider another example of a purely mathematical nature.

EXAMPLE   Find two numbers whose sum is 16 and whose product is the largest possible.

SOLUTION   Let the two numbers be x and y, so that $x + y = 16$. If $P = xy$ denotes the product of the two numbers, then we are required to find the values of x and y that make P a maximum.

We cannot differentiate P immediately, since it is a function of two variables, x and y. These two variables are not independent, however, but are related through the condition $x + y = 16$. We must use this condition to eliminate one of the variables from P, thus leaving P as a function of a single variable. We have $y = 16 - x$; therefore

$$P = xy = x(16 - x) = 16x - x^2.$$

We must find the value of x that makes P a maximum.

$$\frac{dP}{dx} = 16 - 2x$$

and

$$\frac{dP}{dx} = 0 \quad \text{when} \quad 16 - 2x = 0, \text{ i.e., } x = 8.$$

The second derivative $d^2P/dx^2 = -2 < 0$, and so $x = 8$ corresponds to a maximum of P. When $x = 8$, $y = 8$ also, and the maximum value of P is then equal to 64.

The solution of optimization problems of the type given above is often found to be one of the most difficult areas of the differential calculus. The main obstacle arises at the level of translating the given "word problem" into the necessary equations. Once the equations have been constructed, it is usually much more straightforward to complete the solution by using the appropriate bit of calculus. This task of phrasing word problems in terms of mathematical equations is one that occurs repeatedly in all branches of applied mathematics, and it is something that the applied student should master if he wishes his calculus courses ever to be more than a decorative feature on his university transcript.

Unfortunately it is not possible to lay down hard and fast rules by means of which any word problem can be translated into equations. However there are a few guiding principles that are useful to bear in mind:

1. Identify all of the variables involved in the problem and denote each by a symbol.

In the first example the variables were n, the number of fish per unit area; w, the average gain in weight per fish; and P, the total production of fish weight per unit area. In the second example the variables were x and y, the two numbers, and P, their product.

2. Identify the variable that is to be maximized or minimized and express it in terms of the other variables in the problem.

In the first example, the total production P is maximized, and we wrote $P = nw$, expressing P in terms of n and w. In the second example, the product P of x and y is maximized, and of course $P = xy$.

3. Identify all of the relationships between the variables. Express these relationships mathematically.

In the first example the relationship $w = 600 - 30n$ was given. In the second example the relationship between x and y was that their sum was equal to 16, so we wrote the mathematical equation $x + y = 16$.

4. Express the quantity to be maximized or minimized in terms of one of the other variables. In order to do this, use is made of the relationships obtained in Step 3 above in order to eliminate all but one of the variables.

In the first example we have $P = nw$ and $w = 600 - 30n$, so, eliminating w, we obtain P in terms of n: $P = n(600 - 30n)$. In the second example we have $P = xy$ and $x + y = 16$, so, eliminating y, we obtain $P = x(16 - x)$.

5. Having expressed the required quantity as a function of one variable, calculate its critical points and test each for local maximum or minimum.

Let us follow through these steps in another example.

EXAMPLE   A tank is to be constructed with a square horizontal base and rectangular vertical sides. There is no top. The tank must hold 4 m³ of water. The material of which the tank is to be constructed costs $10/m². What dimensions for the tank minimize the cost of material?

SOLUTION   *Step 1.* The variables in the problem are the dimensions of the tank and the total area of its sides, which determines the amount of material used in its construction. We let x denote the length of one side of the base and y denote the height of the tank (Fig. 4.33). The quantity to be minimized is the total cost of materials, which we denote by C.

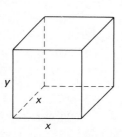

Figure 4.33

Step 2. C is equal to the area of the tank multiplied by $10, which is the cost per unit area. The base is a square with side x and has its area equal to x^2. Each of the four sides is a rectangle with sides x and y, and therefore has area equal to xy. The total area of base plus four sides is therefore $x^2 + 4xy$. Consequently, we can write

$$C = 10(x^2 + 4xy).$$

Step 3. We observe that the quantity to be minimized (or maximized) is expressed as a function of two variables and we need a relation between x and y so that one of these variables can be eliminated. This relation is obtained from the requirement stated in the problem that the volume of the tank must be 4 m³. Now the volume equals the area of the base times the height, that is, x^2y, and so we have the condition

$$x^2y = 4.$$

Step 4. Consequently, $y = 4/x^2$, and so

$$C = 10\left[x^2 + 4x\left(\frac{4}{x^2}\right)\right] = 10\left[x^2 + \frac{16}{x}\right].$$

Step 5. We can at last differentiate to find the critical points of C:

$$\frac{dC}{dx} = 10\left(2x - \frac{16}{x^2}\right) = 20\left(x - \frac{8}{x^2}\right) = 0,$$

so, $x - 8/x^2 = 0$. Therefore, $x^3 = 8$ or $x = 2$. The base of the tank should therefore have a side 2 m in length. The height of the tank is then given by

$$y = \frac{4}{x^2} = \frac{4}{2^2} = 1 \text{ m.}$$

It is easily verified that $d^2C/dx^2 > 0$ when $x = 2$, so that this value of x provides a local minimum of C.

The following example is along similar lines, but is a little more difficult.

EXAMPLE   A grain silo is to be built in the form of a vertical cylinder with a hemispherical roof (Fig. 4.34). The silo is to be capable of storing 10,000 ft³ of grain. (Assume

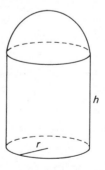

Figure 4.34

the grain will be stored only in the cylindrical part, not in the roof.) The hemispherical roof costs twice as much per unit area to manufacture as the cylindrical sides cost. What dimensions should the silo have in order to minimize its total cost?

Let the radius of the silo be r and the height of the straight sides be h, both measured in feet. Then the volume of the cylindrical part is equal to the area of its base times its height, that is $(\pi r^2) \cdot h$. Since this must equal 10,000 ft³, we have the equation

$$\pi r^2 h = 10,000. \tag{1}$$

This is the condition that relates the two variables r and h that we have introduced.

The quantity to be minimized is the total cost, which we denote by C. This is equal to the cost of the cylinder plus the cost of the hemispherical roof.

The cost of the cylinder equals the area of its sides multiplied by the cost per unit area. The area equals the height times the perimeter, that is $h \cdot (2\pi r) = 2\pi rh$. But the cost per unit area was not stated in the problem, so we introduce a symbol for it, say c. Then the cost of constructing the cylindrical side is equal to $2\pi rhc$.

The cost of constructing the hemispherical roof equals its area multiplied by $2c$, since the roof costs twice as much per unit area as the sides. The area of a hemisphere is $2\pi r^2$ (half the area of a complete sphere), and the cost of the roof is equal to $(2\pi r^2) \cdot 2c = 4\pi r^2 c$. Thus the total cost is

$$C_, = 2\pi rhc + 4\pi r^2 c = 2\pi c(rh + 2r^2).$$

As in the previous example, the function to be minimized involves two variables, and one of them must be eliminated before differentiating. Since from Eq. (1) above, $h = 10,000/\pi r^2$, it follows that

$$C = 2\pi c\left(2r^2 + \frac{10,000}{\pi r}\right).$$

We wish to find the value of r which makes C a minimum; therefore we set $dC/dr = 0$.

$$\frac{dC}{dr} = 2\pi c\left(4r - \frac{10,000}{\pi r^2}\right)$$

and

$$\frac{dC}{dr} = 0 \quad \text{when} \quad 4r - \frac{10,000}{\pi r^2} = 0,$$

or

$$r^3 = \frac{10,000}{4\pi} = 795.8.$$

Therefore,

$$r = \sqrt[3]{795.8} = 9.27.$$

The corresponding value of h is given by

$$h = \frac{10,000}{\pi r^2} = \frac{10,000}{\pi(9.27)^2} = 37.07.$$

Thus the optimum dimensions of the silo are radius 9.27 ft and height of side 37.07 ft.

To complete the problem we should verify that the value of r we have found does indeed minimize C. To do this, we calculate the second derivative,

$$\frac{d^2C}{dr^2} = 2\pi c\left(4 + \frac{10,000}{\pi} \cdot \frac{2}{r^3}\right)$$

and substitute for the value of r, namely $r = 9.27$. It is clear that d^2C/dr^2 is positive, and therefore C is a minimum as required.

Optimization problems arise frequently in the field of medicine.

EXAMPLE   When a person coughs, the radius of the main air passage in his lungs decreases. It is desired to calculate the value of this radius which gives the maximum velocity of air through the passage during the cough.

During the inhalation that occurs before the cough, a pressure P is built up in the lungs, which causes the radius of the air passage to contract. The simplest assumption we can make is that the contraction is linearly proportional to the pressure, and we write $(R_0 - R) = kP$, where R is the radius of the passage under the pressure P, R_0 is the radius before the pressure is built up, and k is the constant of proportionality.

When the pressure is released, it forces the air through the contracted airway. It is a result of fluid dynamics that when a fluid flows through a pipe of radius R under the action of a pressure P, the average velocity that it achieves is given by $V = hPR^2$, where h is a certain constant. Thus we have

$$V = \left(\frac{h}{k}\right)R^2(R_0 - R) = aR^2(R_0 - R),$$

where a is used to denote (h/k) for the sake of brevity. In order to find the value of R that makes V a maximum, we set $dV/dR = 0$:

$$\frac{dV}{dR} = a(2R_0R - 3R^2) = 0.$$

Therefore,

$$R(2R_0 - 3R) = 0,$$

and so

$$R = 0 \quad \text{or} \quad R = \frac{2R_0}{3}.$$

To find out which of these values of R corresponds to a maximum of V, we differentiate again and apply the second derivative test:

$$V'' = 2aR_0 - 6aR.$$

When $R = 0$, $V'' = 2aR_0 > 0$, so V is a minimum. When $R = 2R_0/3$, $V'' = 2aR_0 - 6a(\frac{2}{3}R_0) = -2aR_0 < 0$, and V is a maximum. Therefore, the radius that gives the maximum velocity of expelled air is equal to $2R_0/3$, that is, two-thirds the radius of the air passage at atmospheric pressure.

Maximization problems also occur in the study of epidemics. We can represent the progress of an epidemic by means of two functions: $f(t)$, which denotes the number of people infected at time t, and $g(t)$, which denotes the total number of people who have been or still are infected up to time t. The difference $g(t) - f(t)$ is equal to the number of people who have been infected in the past but have either recovered or died prior to time t.

Typical graphs of $f(t)$ and $g(t)$ are shown in Fig. 4.35. The graph of $f(t)$ rises to a maximum, and then eventually falls to zero as the number of infected individuals declines. The graph of $g(t)$ rises steadily, and for large values of t, $g(t)$ approaches a limiting value G, which equals the total number of individuals who were at some stage infected by the epidemic.

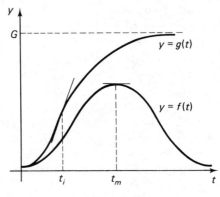

Figure 4.35

During the course of an epidemic there are two points of major importance insofar as they affect the demand for medical services. One is the point t_m, at which $f(t)$ attains its maximum, since this is the time at which the greatest number of individuals are sick at the same time. The value $f(t_m)$ determines the maximum demand that will be placed on medical resources.

The derivative $g'(t)$ gives the *rate* at which new individuals become infected by the epidemic. In the early stages this rate typically starts off small, increases as the epidemic spreads more and more rapidly, and then eventually decreases again as the epidemic dies down. The point $t = t_i$, at which $g'(t)$ attains its maximum value, is significant in that it represents the time at which the infection is spreading most rapidly. Once t_i is passed, the rate at which new cases occur is a decreasing function of time; when this occurs, the epidemic is said to be "under control." Geometrically, the point $t = t_i$ corresponds to the point at which the graph of $g(t)$ is steepest or, equivalently, to the point at which that graph has a point of inflection.

EXAMPLE   During the course of an epidemic, the number of people infected at time t is given by

$$N = f(t) = At^{5/2}e^{-t},$$

where t is measured in weeks from the start of the epidemic, and A is a constant. Find the value of t at which the number of infected people is a maximum, and find also the maximum value of N.

SOLUTION   Using the product rule, we have

$$\frac{dN}{dt} = At^{5/2} \frac{d}{dt}(e^{-t}) + e^{-t}\frac{d}{dt}(At^{5/2})$$

$$= -At^{5/2}e^{-t} + \tfrac{5}{2}At^{3/2}e^{-t}$$

$$= At^{3/2}e^{-t}(-t + \tfrac{5}{2}).$$

Thus $dN/dt = 0$ when $t = 0$ and when $t = \tfrac{5}{2}$ (the factor e^{-t} never vanishes for any finite value of t). It is readily shown using either the first or second derivative test that $t = \tfrac{5}{2}$ corresponds to a local maximum of N. The maximum value of N is therefore

$$N_{\max} = f(\tfrac{5}{2}) = A(\tfrac{5}{2})^{5/2}e^{-5/2} = 0.811A.$$

EXERCISES 4.3

1. Find two numbers whose sum is 10 and whose product is maximum.

2. Find two numbers whose sum is 8 and the sum of their squares is a minimum.

3. Find two positive numbers whose sum is 75 and the product of one times the square of the other is a maximum.

4. Find two positive numbers whose sum is 12 and the sum of their cubes is a minimum.

5. Show that among all the rectangles of area 100 cm², the one with smallest perimeter is the square of side 10 cm.

6. What is the area of the largest rectangle that can be drawn inside a circle of radius a?

7. What is the area of the largest rectangle that can be drawn inside a semicircle of radius a?

8. A farmer wishes to enclose a rectangular paddock using only 100 yd of fencing. What is the largest area he can enclose?

9. Repeat exercise 8 for the case in which one side of the paddock makes use of an existing fence, and only three new sides need to be constructed using the 100 yd of available fencing.

10. A handbill is to contain 48 in.² of printed matter with 3-in. margins at the top and bottom and 1-in. margins on each side. What dimensions for the handbill will consume the least amount of paper?

11. A cistern is to be constructed to hold 324 ft³ of water. The cistern has a square base and four vertical sides, all made of concrete, and a square top made of steel. If the steel costs twice as much per unit area as the concrete, determine the dimensions of the cistern that minimize the total cost of construction.

12. Repeat exercise 11 in the case when the shape of the cistern is a cylinder with a circular base and top.

13. The yield of fruit from each tree of an apple orchard decreases as the density at which the trees are planted increases. When there are n trees per acre, the average number of apples per tree is known to be equal to $(900 - 10n)$ for a particular variety of apple (when n lies between 30 and 60). What value of n gives the maximum total yield of apples per acre?

14. A forest company plans to log a certain area of fir trees after a given number of years. The average number of board feet obtained per tree over the given period is known to be equal to $(50 - 0.5x)$, where x is the number of trees per acre, and when x lies between 35 and 80. What density of trees should be maintained in order to maximize the amount of timber per acre?

15. The growth rate of a certain population is given by

$$\frac{dy}{dt} = ky(p - y),$$

where k and p are constants. Find the value of y for which the growth rate is maximum.

16. The growth rate y of a population whose size is x is given by

$$y = \tfrac{1}{5}x - \tfrac{1}{2000}x^2,$$

when the time is measured in days. For what population size x will the growth rate be maximum?

17. The function f that determines the concentration in the blood of a certain drug at time t after injection is known to be

$$f(t) = \frac{A}{r - s}(e^{-st} - e^{-rt}),$$

where A, r, and s are constants, with $r > s$. Find the value of t at which the concentration reaches a maximum.

18. The reaction as a function of time (measured in hours) to two drugs is given by
$$R_1(t) = te^{-t}, \qquad R_2(t) = te^{-2t^2}.$$

Which drug has the larger maximum reaction?

19. The reaction of a drug at time t after it has been administered is given by $R(t) = t^2 e^{-t}$. At what time is the reaction maximum?

20. During the first few months of its life, a child increases in weight approximately according to the formula

$$W = \frac{100\pi e^{c(t-5/3)}}{1 + e^{c(t-5/3)}},$$

where t is age measured in months and c is a constant. At what age is the child's weight increasing most rapidly?

21. The size of a population of bacteria that is introduced to a nutrient grows according to the formula

$$N(t) = 5000 + \frac{30,000t}{100 + t^2},$$

where the time t is measured in hours. Determine the maximum size of the population.

22. During the course of an epidemic, the proportion of the population infected after a time t is equal to

$$\frac{t^2}{5(1 + t^2)^2}$$

(t is measured in months, and the epidemic starts at $t = 0$). Find the maximum proportion of population that becomes infected. Find also the time at which the proportion of infected individuals is increasing most rapidly.

23. An individual suffering from a certain disease is administered an amount x of a suitable drug. His probability of being cured is then

$$\frac{\sqrt{x}}{3(1 + x)}.$$

Find the value of x that gives him the maximum probability of being cured.

24. The velocity V of a certain enzyme reaction obeys the relationship

$$V = a(b - x)(c + x),$$

where a, b, and c are positive constants and x is the amount of substrate decomposed. Prove that the maximum velocity occurs when the amount of substrate decomposed is $\frac{1}{2}(b - c)$.

25. The yield y (bushels per acre) of a certain crop of wheat is given by $y = a(1 - e^{-kx}) + b$, where a, b, and k are constants, and x is the number of pounds per acre of fertilizer. The profit from the sale of the wheat is given by $P = py - c_o - cx$, where p is the profit per bushel, c is the cost per pound of fertilizer, and c_o is an overhead cost. Determine how much fertilizer must be used in order to maximize the profit P.

26. The size of a certain bacterial population at time t (in hours) is given by $y = a[1 + \frac{1}{2}e^t]^{-1}$, where a is a constant. A biologist plans to observe the population over the 2-hr period from $t = 0$ to $t = 2$. What will be the largest and the smallest growth rates that he observes?

27. When a substance, for example, a drug or a tracer, is injected into a blood vessel, it is carried along by the blood stream and also spreads out along the blood vessel by diffusion. At any point along the blood vessel, the concentration of the injected substance increases to a maximum and then falls off. As a function of time t, the concentration c may often be taken to be

$$c = at^{-1/2}e^{-b(t+kt^{-1})},$$

where a, b, and k are constants. For what value of t $(t > 0)$ does c achieve its maximum?

28. When a dose d of a certain drug is administered to a patient his blood pressure drops by an amount p given by $p = \frac{1}{3}d^2(c - d)$, where c is a constant. Find the dosage that provides the greatest drop in blood pressure.

29. Organic waste, when deposited in a lake, decreases the oxygen content of the water. If t denotes the time in days after the waste is deposited, then it is found experimentally that the oxygen content is given in a particular instance by

$$y = t^3 - 30t^2 + 6000$$

for $0 \leq t \leq 25$. Find the maximum and minimum values of y during the first 25 days following the depositing of the waste.

30. If an intensity x of light falls on a plant, the rate of photosynthesis y, measured in appropriate units, is found experimentally to be given by $y = 150x - 25x^2$ as long as $0 \leq x \leq 5$. Find the maximum and minimum values of y when x lies in the interval $1 \leq x \leq 5$.

4.4 NEWTON'S METHOD

A value of x that satisfies the equation $f(x) = 0$ is called a *root* of the equation. For example, $x = 3$ is a root of the equation $x^2 - 5x + 6 = 0$, because on substituting $x = 3$ into the equation we have

$$3^2 - 5(3) + 6 = 0 \quad \text{or} \quad 9 - 15 + 6 = 0,$$

which is true. So the equation is satisfied when $x = 3$.

Geometrically, a real root of the equation $f(x) = 0$ is a value of x where the graph of $y = f(x)$ meets the x-axis.

If the equation $f(x) = 0$ is linear or quadratic in x, we know how to solve it using standard formulas. For example, if $f(x) = Ax^2 + Bx + C$, a quadratic function, then we know that there are two roots given by the formula

$$x = \frac{1}{2A}(-B \pm \sqrt{B^2 - 4AC}).$$

For third- and fourth-order equations, corresponding formulas do exist, though they are complicated and not very familiar. For polynomial equations of order higher than the fourth or for nonpolynomial equations such as $x \ln x + 3x = 1$, there are, generally speaking, no algebraic formulas that give the roots.

Newton's method provides a means of finding the roots of an equation by making use of the first derivative, and it has these advantages: It can be used to solve $f(x) = 0$ when $f(x)$ is a polynomial of any degree or a nonpolynomial function. Furthermore, we always get a numerical answer in decimal form for the root. The main disadvantage of Newton's method is that the arithmetic can become involved unless an electronic calculator is used.

Basically Newton's method is a technique for approximating the root by repetitive calculations. Suppose that $f(x) = 0$ has a root at $x = r$, and assume the shape of the graph of $y = f(x)$ to be as shown in Fig. 4.36, crossing the x-axis at the point $(r, 0)$. Let x_1 be any value of x close to r and let us construct the tangent line to $y = f(x)$ at the point $[x_1, f(x_1)]$. Let this line meet the x-axis at $x = x_2$. It is clear from the figure that x_2 lies considerably closer to r than x_1 does.

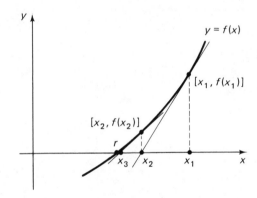

Figure 4.36

Now let us repeat the process: Construct the tangent line to the curve at $[x_2, f(x_2)]$ and let it meet the x-axis at $x = x_3$. Again, x_3 lies even closer to r than x_2. We can repeat the process again, getting a value x_4 yet closer to r; then again getting x_5, then x_6, and so on. So we obtain a sequence of numbers $x_1, x_2, x_3, x_4, \ldots$ that approach closer and closer to the root r. Eventually one of these numbers will be sufficiently close to r that we can take it as giving an approximate value of the exact root r itself. In practice it is usually necessary to calculate only a few numbers in the sequence before a sufficiently accurate value of the root is obtained.

Before deriving the general formulas for use with Newton's method, let us consider its application to the particular case when $f(x) = x^2 - 2$. The root of the equation $f(x) = 0$ occurs in this case when $x^2 = 2$, that is, when $x = \sqrt{2}$. The answer we shall get from Newton's method will therefore be equal to the square root of 2, expressed in decimal form.

The graph of $y = x^2 - 2$ is shown in Fig. 4.37. We know that the square root of 2 lies somewhere between 1 and 2, so let us take $x_1 = 2$ as our starting value. What

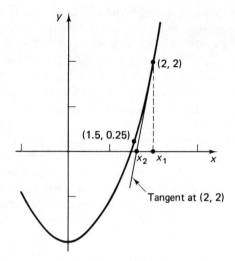

Figure 4.37

we must do is construct the tangent to the graph at the point $x = 2$, $y = 2$, and then find the point x_2 where this tangent meets the x-axis.

The slope of $y = x^2 - 2$ is given by

$$\frac{dy}{dx} = 2x.$$

When $x = 2$, this slope is equal to 4. The tangent line is therefore the straight line that has slope 4 and that passes through the point $(2, 2)$. From the point–slope formula its equation is

$$y - 2 = 4(x - 2)$$

or

$$y = 4x - 6.$$

This meets the x-axis where $y = 0$ and $x = x_2$. Therefore,

$$0 = 4x_2 - 6 \quad \text{or} \quad x_2 = 1.5.$$

We must now repeat the process, using the tangent line at the point where $x = 1.5$. The corresponding value of y is:

$$y = x_2^2 - 2 = (1.5)^2 - 2 = 0.25.$$

The slope of the tangent line is equal to dy/dx evaluated at $x = x_2$, that is, $dy/dx = 2x_2 = 2(1.5) = 3$.

From the point–slope formula, the equation of the tangent is therefore

$$y - 0.25 = 3(x - 1.5)$$

or

$$y = 3x - 4.25.$$

This meets the x-axis where $y = 0$ and $x = x_3$, and so

$$0 = 3x_3 - 4.25$$

or

$$x_3 = \frac{4.25}{3} = 1.41667.$$

The value of $\sqrt{2}$ to four decimal places is 1.4142, so we are already pretty close to the answer. Let us repeat the calculation once more and find x_4. When $x = 1.41667$, the corresponding value of y is $y = (1.41667)^2 - 2 = 0.006944$. The slope of the tangent is equal to $2x_3 = 2(1.41667) = 2.83334$. Therefore, the equation of the tangent is

$$y - 0.006944 = (2.83334)(x - 1.41667).$$

Setting $y = 0$ in this equation, we obtain the point $x = x_4$ where the line meets the x-axis:

$$(2.83334)(x_4 - 1.41667) = -0.006944$$

$$x_4 - 1.41667 = \frac{-0.006944}{2.83334} = -0.00245$$

$$x_4 = 1.41667 - 0.00245 = 1.41422.$$

We can see how the successive values $x_1 = 2$, $x_2 = 1.5$, $x_3 = 1.4167$, and $x_4 = 1.4142$ get progressively closer to the root of the given equation, $\sqrt{2}$. To four decimal places, x_4 agrees precisely with the value of $\sqrt{2}$.

Now let us develop some general formulas that will make the application of Newton's method more direct. We shall consider the root of some general function $f(x)$. The slope of the tangent to $y = f(x)$ at $x = x_1$ is $f'(x_1)$. Thus using the point–slope formula the equation of the tangent, passing through the point $[x_1, f(x_1)]$ and having the slope $f'(x_1)$, is

$$y - f(x_1) = f'(x_1)(x - x_1).$$

This tangent meets the x-axis ($y = 0$) at the point $x = x_2$, so

$$0 - f(x_1) = f'(x_1)(x_2 - x_1),$$

or,

$$x_2 - x_1 = \frac{-f(x_1)}{f'(x_1)};$$

therefore

$$x_2 = x_1 - \frac{f(x_1)}{f'(x_1)}.$$

This formula allows us to calculate x_2 once x_1 has been chosen.

Now the step from x_2 to x_3 is exactly analogous to the step from x_1 to x_2, and so x_3 will be given by a similar formula to the one above, but with x_1 replaced by x_2 on the right-hand side:

$$x_3 = x_2 - \frac{f(x_2)}{f'(x_2)}.$$

In fact we can use this same formula for the general step in the sequence. In going from x_n to x_{n+1}, we have

$$x_{n+1} = x_n - \frac{f(x_n)}{f'(x_n)}.$$

The question remains as to how the initial value x_1 is chosen. This is usually selected by trial and error. If $f(x)$ is a continuous function in the interval $a \leq x \leq b$ and if $f(a)$ and $f(b)$ are of opposite signs, then the equation $f(x) = 0$ must have at least one root lying between a and b. This is because the graph of $y = f(x)$ has to cross the x-axis somewhere between a and b. Thus what we do is to try a number of values of x, calculating the values of $f(x)$ at each one, in order to find intervals of the x-axis in which $f(x)$ changes sign. We usually start out by evaluating $f(x)$ at a number of integer values of x.

Having found an interval $a \leq x \leq b$ in which a root lies, we choose x_1 to be a point inside the interval. As a rule it does not matter which point we choose, and it is quite common to choose one of the endpoints, $x_1 = a$ or $x_1 = b$.

EXAMPLE   Use Newton's method to find the positive real root of $x^4 + x - 3 = 0$.

SOLUTION   Let $f(x) = x^4 + x - 3$. By trial we have

$$f(0) = 0 + 0 - 3 = -3,$$
$$f(1) = 1 + 1 - 3 = -1, \quad \text{and}$$
$$f(2) = 16 + 2 - 3 = 15.$$

Thus $f(x)$ changes sign in $1 < x < 2$, and hence a real root of $f(x) = 0$ lies between 1 and 2.

Let us take our first approximation as $x_1 = 1$. [We note that $f(1)$ is closer to zero than $f(2)$, so the root is probably closer to 1 than to 2.] Then

$$x_2 = x_1 - \frac{f(x_1)}{f'(x_1)}$$

$$= x_1 - \frac{x_1^4 + x_1 - 3}{4x_1^3 + 1}$$

$$= \frac{x_1(4x_1^3 + 1) - (x_1^4 + x_1 - 3)}{4x_1^3 + 1}$$

$$= \frac{3x_1^4 + 3}{4x_1^3 + 1}. \tag{i}$$

Setting $x_1 = 1$,

$$x_2 = \frac{3 + 3}{4 + 1} = \frac{6}{5} = 1.2.$$

Applying Newton's method again [changing x_1 to x_2 and x_2 to x_3 in (i)] we have

$$x_3 = \frac{3x_2^4 + 3}{4x_2^3 + 1}$$

$$= \frac{3(1.2)^4 + 3}{4(1.2)^3 + 1} \quad \text{when} \quad x_2 = 1.2$$

$$= \frac{9.2208}{7.9120} = 1.165.$$

Applying Newton's method once again, we have

$$x_4 = \frac{3x_3^4 + 3}{4x_3^3 + 1}$$

$$= \frac{3(1.165)^4 + 3}{4(1.165)^3 + 1}$$

$$= \frac{8.5262}{7.3247} = 1.16403.$$

Comparing x_3 and x_4, which are quite close together, we can conclude that, correct to two decimal places, the real root is 1.16. Most likely, x_4 is correct to three places (1.164), but in order to be sure of this we need to calculate x_5. It turns out that $x_5 = 1.16404$, so that x_4 and x_5 agree to four places of decimals, and the exact root must be 1.1640 to four places.

Newton's method can also be used to find the decimal estimate of some irrational numbers in the way that in our opening example we found the value of $\sqrt{2}$ to four decimal places. For example, suppose we wish to use Newton's method to evaluate the cube root of 2, that is $\sqrt[3]{2}$. Then if we let $x = \sqrt[3]{2}$ and take cubes of both sides of this equation, we obtain that $x^3 = 2$, or $x^3 - 2 = 0$. Therefore the cube root of 2 can be calculated by finding the root of the equation $x^3 - 2 = 0$. Newton's method can be used to do this.

Newton's method can also be used to find the roots of non-algebraic equations.

EXAMPLE   Use Newton's method to find the positive root of $e^x - x - 6 = 0$ correct to two decimal places.

SOLUTION   Let $f(x) = e^x - x - 6$. Evaluating $f(x)$ at $x = 0, 1, 2, 3, \ldots$ we have

$$f(0) = e^0 - 0 - 6 = 1 - 6 = -5 = -\text{ve.}$$
$$f(1) = e - 1 - 6 = 2.7183 - 1 - 6 = -\text{ve.}$$
$$f(2) = e^2 - 2 - 6 = 7.3891 - 2 - 6 = -\text{ve.}$$
$$f(3) = e^3 - 3 - 6 = 20.0855 - 3 - 6 = +\text{ve.}$$
$$f(4) = e^4 - 4 - 6 = 54.5981 - 10 = +\text{ve.}$$

It is clear from the above table that $f(x)$ changes sign in the interval $2 < x < 3$. Also for $x > 3$, e^x grows much faster than $x + 6$ and therefore $f(x)$ is always positive. Thus there is only one positive root of $f(x) = 0$, and this lies between 2 and 3. By Newton's method,

$$x_2 = x_1 - \frac{f(x_1)}{f'(x_1)}$$

$$= x_1 - \frac{e^{x_1} - x_1 - 6}{e^{x_1} - 1}$$

$$= \frac{(x_1 - 1)e^{x_1} + 6}{e^{x_1} - 1}. \tag{i}$$

Let us take the first approximation as $x_1 = 2$. Then from (i), the second approximation is given by

$$x_2 = \frac{e^2 + 6}{e^2 - 1} = \frac{13.3891}{6.3891} = 2.0956.$$

Applying Newton's method again, we have from (i),

$$x_3 = \frac{(x_2 - 1)e^{x_2} + 6}{e^{x_2} - 1}$$

$$= \frac{(1.0956)e^{2.0956} + 6}{e^{2.0956} - 1}$$

$$= \frac{14.9076}{7.1303} = 2.0907.$$

Clearly the result does not change between the second and third approximations when calculated up to two decimal places. Thus, correct up to 2 decimal places, the real root of the given equation is 2.09.

Newton's method is an example of a class of numerical techniques called iteration methods. An *iteration method* is generally any method of calculation that involves repeating the same basic step over and over again getting closer and closer approximations to the required answer at each repetition.

It is not always the case that Newton's method will work. For example if we are unlucky enough to choose as our starting value x_1 a value at which $f'(x_1) = 0$, then x_2 will not be defined (Fig. 4.38). There also exist certain specific types of curves for which Newton's method fails. Figure 4.39 indicates a situation in which the iterations

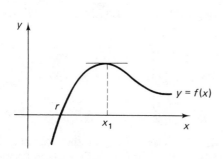

Figure 4.38

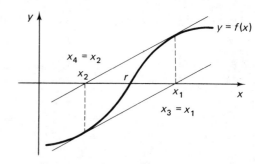

Figure 4.39

x_3, x_5, and so on are equal to the starting value x_1, while $x_2 = x_4 = x_6$, etc. In addition, even for functions for which the method works, it may happen that if x_1 is chosen too far away from the root, the iterations x_2, x_3, etc. can fail to approach a limiting value.

In practice, it is usually easy enough to spot when something like this is going wrong just by observing whether the successive approximations are getting closer together.

It is an advantage to use a starting value x_1 that is as close as possible to the root itself. This reduces the likelihood of the method failing and also means that fewer iterations will be needed, as a rule, to get a sufficiently accurate value of the root. It will be observed that in the examples considered above, having found an interval $a \leq x \leq b$ such that $f(a)$ and $f(b)$ are of opposite signs, we chose as x_1 the value a or b for which the function was closest to zero. For example, in the first example $[f(x) = x^4 + x - 3]$ we found that $f(1) = -1$ and $f(2) = 15$, so we chose $x_1 = 1$. However a better starting value would have been obtained by choosing x_1 to be one-sixteenth of the distance from 1 to 2 because f changes from -1 to 15 as x changes from 1 to 2 (Fig. 4.40). That is, $x_1 = 1.06$ would be a better starting value than $x_1 = 1$.

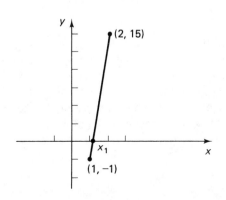

Figure 4.40

This type of interpolation between a and b to find x_1 will usually (but not always!) give a better starting value than choosing a or b themselves.

In locating the approximate positions of the roots of a complicated equation it is often useful to use a graphical technique. We shall illustrate the type of consideration involved using the third example above: $e^x - x - 6 = 0$. We can write this equation in the form

$$e^x = x + 6.$$

Thus the roots of this equation are actually the points of intersection of the two graphs, $y = e^x$ and $y = x + 6$.

We see both of these graphs in Fig. 4.41. It is quite clear that for positive x the two curves intersect in only one point P, and the corresponding value $x = r$ is the

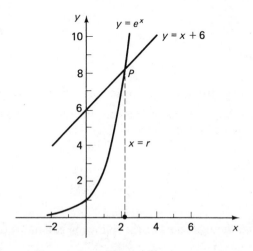

Figure 4.41

required root. Without drawing the graphs accurately we can estimate by eye that r is somewhere between 2 and 3.

EXAMPLE   Find the approximate locations of the roots of the equation

$$\ln x - 2x + 5 = 0.$$

SOLUTION   The roots are the values of x at the points of intersection of the two graphs $y = \ln x$ and $y = 2x - 5$. These are shown in Fig. 4.42. Clearly there are two points of intersection, P and Q. At P, x is very close to zero and $y \simeq -5$. Thus $\ln x \simeq -5$, that is, x is approximately equal to $e^{-5} = 0.0067$. At Q it is clear from the graph that x lies somewhere between 3 and 4. It is readily verified that the function $\ln x - 2x + 5$ indeed changes sign between 3 and 4.

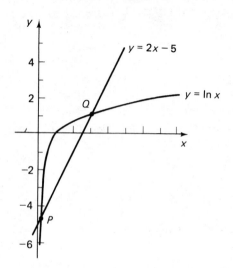

Figure 4.42

EXERCISES 4.4

[Use an electronic calculator to solve the following problems.]

Use two applications of Newton's method to find the approximate values of the following irrational numbers:

1. $\sqrt{2}$.   2. $\sqrt{5}$.

3. $\sqrt[3]{6}$.   4. $\sqrt[3]{28}$.

5. $\sqrt[4]{3}$.   6. $\dfrac{1}{\sqrt{3}}$.

Find all the real roots of the following equations correct to 2 decimal places.

7. $x^2 - 3x + 1 = 0$.   8. $2x^2 + 5x - 6 = 0$.

9. $x^3 - x - 8 = 0$.   10. $x^3 + x^2 + 3x - 10 = 0$.

11. $x^3 - 4x^2 + x + 3 = 0.$

12. $x^4 - 5x^3 + x + 6 = 0.$

13. $e^x - x - 3 = 0.$

14. $e^{2x} - x^2 - 10 = 0.$

15. $\ln x + x - 2 = 0.$

16. $x \ln x + x - 7 = 0.$

17. After an injection the concentration $C(t)$ of a certain drug in a patient's blood is given by

$$C(t) = te^{-t/2}$$

in milligrams per cubic centimeter, where t is the time measured in hours. The next injection is to be given when $C(t)$ falls to 0.05 mg/cm³. Calculate how many hours later it must be given.

18. The number of board-feet per tree, B, for a certain type of fir tree is given by $5 \times 10^{-2}x^{5/2}$, where x is the age of the tree in years. The sale value of the lumber is 10¢ per board-foot and the cost of cutting it is $(10^4 + B)$ cents per tree (the cost rises with the size of the tree). If the forest company wishes to make an average profit of $3 per tree per year, how long should it let the trees grow before cutting?

4.5 THE DIFFERENTIALS *dx* AND *dy*

Let $y = f(x)$ be a differentiable function of the independent variable x. Up to now we have used dy/dx to denote the derivative of y with respect to x and treated dy/dx as a single symbol, not as a ratio of dy and dx. Now we shall define the new concept of a *differential* so that dx and dy will have meaning separately; this will permit us to think of dy/dx either as the symbol for the derivative of y with respect to x or as the ratio of dy and dx.

DEFINITION   Let $y = f(x)$ be a differentiable function of x. Then

i. dx called the *differential of the independent variable x*, is simply an arbitrary increment of x, that is

$$dx = \Delta x.$$

ii. dy, the *differential of the dependent variable y*, is a function of x and dx given by

$$dy = f'(x)\, dx.$$

The differential dy is also denoted by df.

It is obvious from the above definitions of differentials dx and dy that:

a. if $dx = 0$, then $dy = 0$;

b. if $dx \neq 0$, then the ratio of dy divided by dx is given by

$$\frac{dy}{dx} = \frac{f'(x)\, dx}{dx} = f'(x),$$

and therefore is equal to the derivative of y with respect to x. There is nothing strange about the last result because we deliberately defined dy as the product of $f'(x)$ and dx in order that the result in (b) above would turn out to be true.

EXAMPLE If $y = x^3 + 5x + 7$, find dy.

SOLUTION Let $y = f(x)$ so that $f(x) = x^3 + 5x + 7$. Then
$$f'(x) = 3x^2 + 5,$$
and, by definition,
$$dy = f'(x)\,dx$$
$$= (3x^2 + 5)\,dx.$$

It should be noted that $dx\,(= \Delta x)$ is another independent variable, and the value of dy depends on the *two* independent variables x and dx.

Geometrically, we know that $f'(x)$ represents the slope of the tangent to the curve $y = f(x)$ at the point x (Fig. 4.43). Thus if P is the point on the curve whose abscissa is x, and PT is the tangent at P, then $f'(x)$ is the slope of the tangent line PT.

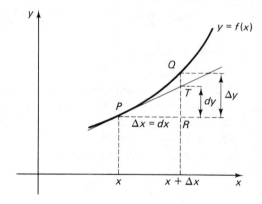

Figure 4.43

The slope is the ratio of rise and run. Let us consider a run of PR from the value x of the abscissa to $x + \Delta x$. Then $PR = \Delta x = dx$. In this case the tangent PT has the rise TR, so that
$$\frac{\text{rise}}{\text{run}} = \frac{TR}{PR} = \text{slope of tangent line } PT = f'(x).$$

In other words, $TR = f'(x) \cdot PR = f'(x)\,dx = dy$. Thus dy is the length of the line segment TR.

We note also that Δy, the increment in y, is the length of the line segment QR. It is clear that, in general, even though $dx = \Delta x$, the differential dy is not the same as the increment in y, that is, $dy \neq \Delta y$.

EXAMPLE If $y = x^3 + 3x$, find dy and Δy when $x = 2$ and $\Delta x = 0.01$.

SOLUTION If $y = f(x) = x^3 + 3x$, then
$$f'(x) = 3x^2 + 3.$$

Therefore

$$dy = f'(x)\, dx$$
$$= (3x^2 + 3)\, dx.$$

When $x = 2$, $\Delta x = 0.01$,

$$dy = (12 + 3)(0.01)$$
$$= 0.15.$$

By definition,

$$\Delta y = f(x + \Delta x) - f(x)$$
$$= f(2 + 0.01) - f(2)$$
$$= f(2.01) - f(2)$$
$$= [(2.01)^3 + 3(2.01)] - [2^3 + 3(2)]$$
$$= [8.120601 + 6.03] - 14$$
$$= 14.150601 - 14$$
$$= 0.150601.$$

Thus, $dy = 0.15$ and $\Delta y = 0.150601$, showing that the differential and increment of y are not quite equal to each other.

It is clear from Fig. 4.43 that if $dx = \Delta x$ is very small, then dy can be taken as a good estimate of Δy even though the two are not equal. The fact that dy is a good approximation to Δy (when Δx is small) is useful because it is usually easier to compute $f'(x)\, dx$ than $\Delta y = f(x + \Delta x) - f(x)$. Thus

$$\Delta y \simeq dy = f'(x)\, dx = f'(x)\, \Delta x$$

or

$$f(x + \Delta x) = y + \Delta y$$
$$\simeq f(x) + f'(x)\, \Delta x.$$

EXAMPLE   Find an approximation to the value of $\sqrt{16.1}$.

SOLUTION   In this case we are interested in the value of $f(x) = \sqrt{x}$ when $x = 16.1$. Since we do not know the square root of 16.1, we take 16.1 as $x + \Delta x$ and select x, a number near 16.1 whose square root is known. We should clearly select $x = 16$, and $\Delta x = 16.1 - 16 = 0.1$.

$$f(x) = \sqrt{x} = x^{1/2} \quad \text{gives}$$
$$f'(x) = \frac{1}{2} x^{-1/2} = \frac{1}{2\sqrt{x}}.$$
$$f(x + \Delta x) \simeq f(x) + f'(x) \cdot \Delta x$$

Substituting $x = 16$ and $\Delta x = 0.1$, we have

$$f(16.1) \simeq f(16) + f'(16)(0.1)$$

or

$$\sqrt{16.1} \simeq \sqrt{16} + \frac{1}{2\sqrt{16}}(0.1)$$

$$\simeq 4 + \tfrac{1}{8}(0.1)$$

$$\simeq 4.0125.$$

(The true value of $\sqrt{16.1}$ is 4.012481 to six decimal places, so our approximation is pretty good.)

In the approximate formula $f(x + \Delta x) \simeq f(x) + f'(x)\,\Delta x$, let us change the notation a bit and write $x = a$:

$$f(a + \Delta x) \simeq f(a) + f'(a)\,\Delta x.$$

Now let us replace $a + \Delta x$ by x. Then $\Delta x = x - a$, and we end up with the following formula:

$$f(x) \simeq f(a) + (x - a)f'(a).$$

But the right-hand side here is a linear function of x. If we define $m = f'(a)$ and $c = f(a) - af'(a)$, then the approximation becomes simply $f(x) \simeq mx + c$.

Thus we have established the very important result that, over a short interval of x, any differentiable function $f(x)$ can be approximated by a linear function.

Use is often made of this result in devising mathematical models of complex phenomena. Suppose that x and y are two biological variables that are related in some complex and not very well understood way. Then regardless of the degree of complexity of the relationship, as long as it is smooth we can approximate it by a linear model $y = mx + c$ for certain constants m and c provided that the range of variation of x is sufficiently restricted. Linear models of this kind are frequently used in the biological sciences as a starting point in the analysis of complex phenomena.

Differentials are useful in estimating the effects of experimental errors in measured quantities. Let x be an empirical variable that is measured with a certain amount of error, and let $y = f(x)$ be some other variable that is calculated from the measured value of x. If the value of x that is used in calculating y is in error, then of course the calculated value of y will also be incorrect.

Let x be the true value of the measured variable and $x + dx$ the measured value, so that dx is now the error in this variable. The true value of the calculated variable is $y = f(x)$, but the calculated value is $f(x + dx)$. Thus the error in y is equal to $f(x + dx) - f(x)$. If dx is small, which can usually be presumed to be the case, we can approximate this error by the differential dy. Thus we arrive at the result that the error in y is given approximately by $dy = f'(x)\,dx$.

EXAMPLE   The Fick method of measuring cardiac output y (that is, the number of cubic centimeters per minute of bloodflow) is based on the formula

$$y = \frac{f}{x - z},$$

where f is the number of cubic centimeters per minute of CO_2 that is exhaled, x is the concentration of CO_2 in the mixed venous blood entering the lung, and

z is the concentration of CO_2 in the arterial blood leaving the lung. The quantities f and z can be measured reasonably accurately, but x is difficult to measure. Assume that the measured values are $f = 400$ cm^3/min, $x = 0.55$, and $z = 0.48$ (the units of x and z are cubic centimeters of dissolved CO_2 per cubic centimeter of blood). Assume that the measured values of f and z are accurate, but that the percentage error in x may be as high as 2%. Find the corresponding percentage error in y.

SOLUTION   The percentage error in x is given by

$$100 \frac{dx}{x} = 2.$$

Therefore $dx = 0.02\, x = (0.02)(0.55) = 0.011$. The calculated value of y is

$$y = \frac{400}{0.55 - 0.48} = \frac{400}{0.07} = 5714 \text{ cm}^3/\text{in}.$$

The error in y is given approximately by

$$dy = f'(x)\, dx = \frac{-f\, dx}{(x - z)^2} = -\frac{400}{(0.07)^2}(0.011) = -898.$$

Ignoring the minus sign, we therefore obtain the percentage error in y to be

$$100 \frac{dy}{y} = 100 \frac{898}{5714} = 16\%.$$

We see that a small (2%) error in x has been magnified to a substantial (16%) error in y.

We close this section by summarizing a number of formulas for differentials. These are easily obtained from corresponding formulas for the derivatives:

DERIVATIVES	DIFFERENTIALS
1. $\dfrac{d}{dx}(c) = 0.$	$d(c) = 0.$
2. $\dfrac{d}{dx}(cu) = c\dfrac{du}{dx}.$	$d(cu) = c\, du.$
3. $\dfrac{d}{dx}(u + v) = \dfrac{du}{dx} + \dfrac{dv}{dx}.$	$d(u + v) = du + dv.$
4. $\dfrac{d}{dx}(uv) = u\dfrac{dv}{dx} + v\dfrac{du}{dx}.$	$d(uv) = u\, dv + v\, du.$
5. $\dfrac{d}{dx}\left(\dfrac{u}{v}\right) = \dfrac{v\dfrac{du}{dx} - u\dfrac{dv}{dx}}{v^2}.$	$d\left(\dfrac{u}{v}\right) = \dfrac{v\, du - u\, dv}{v^2}.$
6. $\dfrac{d}{dx}(u^n) = nu^{n-1}\dfrac{du}{dx}.$	$d(u^n) = nu^{n-1}\, du.$
7. $\dfrac{d}{dx}(x^n) = nx^{n-1}.$	$d(x^n) = nx^{n-1}\, dx.$
8. $\dfrac{d}{dx}(e^x) = e^x.$	$d(e^x) = e^x\, dx.$
9. $\dfrac{d}{dx}(\ln x) = \dfrac{1}{x}.$	$d(\ln x) = \dfrac{1}{x}\, dx.$

EXERCISES 4.5

Find dy for the following functions:

1. $y = x^2 + 7x + 1.$ **2.** $y = (t^2 + 1)^4.$

3. $y = t \ln t.$ **4.** $y = ue^{-u}.$

5. $y = \ln (z^2 + 1).$ **6.** $y = \dfrac{x + 1}{x^2 + 1}.$

7. $xy + x^2 = 3.$ **8.** $yz^2 + z^2 - 4yz = 1.$

9. $\ln (yz) = 1 + z.$ **10.** $xe^y + e^x = 1.$

11. Find dy for $y = x^3$ when $x = 2$, $dx = 0.01$.

12. Find du for $u = t^2 + 3t + 1$ when $t = -1$, $dt = 0.02$.

13. Find dx for $x = y \ln y$ when $y = 1$, $dy = 0.003$.

14. Find df for $f(x) = xe^x$ when $x = 0$, $\Delta x = -0.01$.

Find dy and Δy for the following functions:

15. $y = 3x^2 + 5$ when $x = 2$, $\Delta x = 0.01$.

16. $y = \sqrt{t}$ when $t = 4$, $dt = 0.41$.

17. $y = \ln u$ when $u = 3$, $du = 0.06$.

18. $y = \sqrt{x + 2}$ when $x = 2$, $dx = 0.84$.

19. Use differentials to approximate the cube root of 9.

20. Use differentials to approximate the fourth root of 17.

21. Use differentials to approximate the fifth root of 31.

22. Use differentials to approximate the value of $(4.01)^3 + \sqrt{4.01}$.

23. The radius of a sphere is equal to 8 cm with a possible error of ± 0.002 cm. The volume is calculated assuming that the radius is exactly 8 cm. Use differentials to estimate the maximum error in the calculated volume.

24. Use differentials to determine the approximate increase in the surface area of a soap bubble when its radius increases from 2 cm to 2.001 cm.

25. If the volume of a sphere is to be determined to within a percentage error that does not exceed 2%, what is the maximum percentage error that can be allowed in the measured value of the radius?

26. The acceleration due to gravity, g, is determined by measuring the period of swing of a pendulum. If the length of the pendulum is l and the measured period is T, then g is given by the formula

$$g = \frac{4\pi^2 l}{T^2}.$$

Find the percentage error in g if:

(a) l is measured accurately but T has an error of 1%, and

(b) T is measured accurately but l has an error of 2%.

4.6   IMPLICIT DIFFERENTIATION

When y is a given function of x, with $y = f(x)$, then we often say that y is an *explicit function* of the independent variable x. Examples of explicit functions are:

$$y = 3x^2 - 7x + 5, \qquad y = 5x + \frac{1}{x-1}, \quad \text{etc.}$$

Sometimes the fact that y is a function of x is implied by some functional relation of the type $F(x, y) = 0$, in which both x and y appear as arguments of the function F on the left-hand side. Then we say that y is an *implicit function* of x.

EXAMPLE   Consider $xy + 3y - 7 = 0$. In this relation we have a function on the left involving both x and y, which means that y is an implicit function of x. In this case we can solve for y:

$$y(x + 3) = 7,$$

$$y = \frac{7}{x+3}.$$

Thus we can express y as an explicit function. In this example, the given implicit function is equivalent to a certain explicit function. This is not always the case, as shown in the following two examples.

EXAMPLE   If $x^2 + y^2 = 4$, then y is an implicit function of x. In this case we can again solve for y and write

$$y^2 = 4 - x^2$$

$$y = +\sqrt{4 - x^2} \quad \text{or} \quad y = -\sqrt{4 - x^2}.$$

These last two functions are explicit functions. Thus the implicit relation $x^2 + y^2 = 4$ is equivalent to the two explicit functions

$$y = +\sqrt{4 - x^2} \quad \text{and} \quad y = -\sqrt{4 - x^2}.$$

EXAMPLE $y^5 + x^3 - 3xy = 0$. Here y is an implicit function of x. In this case we cannot solve for y in terms of x; that is, we cannot express y as an explicit function of x by means of any algebraic formula.

When y is an implicit function of x we can find dy/dx without actually solving the relation for y in terms of x. When we differentiate an expression in y w.r.t. x, we make use of the chain rule. For example,

$$\frac{d}{dx}(y^3) = \frac{d}{dy}(y^3) \cdot \frac{dy}{dx} = 3y^2 \frac{dy}{dx}.$$

In general,

$$\frac{d}{dx}(f(y)) = f'(y)\frac{dy}{dx}.$$

EXAMPLE   Find dy/dx if $x^2 + y^2 = 4$.

SOLUTION   Differentiate each term w.r.t. x.

$$\frac{d}{dx}(x^2) = 2x, \qquad \frac{d}{dx}(y^2) = 2y\frac{dy}{dx}, \qquad \frac{d}{dx}(4) = 0.$$

Therefore after differentiating the given relation we obtain

$$2x + 2y\frac{dy}{dx} = 0$$

$$2y\frac{dy}{dx} = -2x$$

$$\frac{dy}{dx} = -\frac{2x}{2y} = -\frac{x}{y}.$$

CHECK   Let us check this result using an equivalent explicit function.

$$y = \sqrt{4 - x^2} = (4 - x^2)^{1/2}.$$

Using the chain rule,

$$\frac{dy}{dx} = \frac{1}{2}(4 - x^2)^{(1/2)-1} \cdot \frac{d}{dx}(4 - x^2)$$

$$= \frac{1}{2}(4 - x^2)^{-1/2} \cdot (-2x)$$

$$= \frac{-x}{\sqrt{4 - x^2}} = -\frac{x}{y},$$

which is the same result as found above.

NOTE.   Had we taken the other explicit function, $y = -\sqrt{4 - x^2}$, we still would have found that $dy/dx = -x/y$.

When we evaluate dy/dx from an implicit relation $F(x, y) = 0$ we are assuming that x is the independent variable and y the dependent variable. However, given the implicit relation $F(x, y) = 0$, we could instead regard y as being the independent variable with x being a function of y. In that case we should evaluate the derivative dx/dy.

EXAMPLE Find dx/dt if $x^3 + t^3 = 3xt$.

SOLUTION Here x is an implicit function of t. Differentiating both sides w.r.t. t we have

$$\frac{d}{dt}(x^3) + \frac{d}{dt}(t^3) = 3\frac{d}{dt}(xt)$$

$$3x^2\frac{dx}{dt} + 3t^2 = 3\left(x \cdot 1 + t\frac{dx}{dt}\right),$$

where we have used the product rule for the right-hand side. Collecting the terms involving dx/dt onto the left-hand side, we obtain

$$\frac{dx}{dt}(x^2 - t) = x - t^2$$

$$\frac{dx}{dt} = \frac{x - t^2}{x^2 - t}.$$

EXAMPLE Given $x^2 + y^2 = 4xy$, find dx/dy.

SOLUTION Here x is an implicit function of y. Differentiating both sides w.r.t. y, we have

$$\frac{d}{dy}(x^2) + \frac{d}{dy}(y^2) = 4\frac{d}{dy}(xy)$$

$$2x\frac{dx}{dy} + 2y = 4\left(x \cdot 1 + y\frac{dx}{dy}\right)$$

$$2\frac{dx}{dy}(x - 2y) = 2(2x - y)$$

$$\frac{dx}{dy} = \frac{2x - y}{x - 2y}.$$

Higher-order derivatives can also be calculated from an implicit relation. The method consists first of finding the first derivative in the manner outlined above, and then differentiating the resulting expression with respect to the independent variable.

EXAMPLE Find d^2y/dx^2 when $x^3 + y^3 = 3x + 3y$.

SOLUTION Here x is the independent variable since we are required to find derivatives w.r.t. x. So, differentiating implicitly w.r.t. x, we obtain

$$3x^2 + \frac{d}{dx}(y^3) = 3 + 3\frac{dy}{dx},$$

i.e.,

$$3x^2 + 3y^2\frac{dy}{dx} = 3 + 3\frac{dy}{dx}.$$

Therefore

$$(1 - y^2)\frac{dy}{dx} = x^2 - 1,$$

213

and

$$\frac{dy}{dx} = \frac{x^2 - 1}{1 - y^2}.$$

Differentiating again w.r.t. x and using the quotient rule we get

$$\frac{d^2y}{dx^2} = \frac{d}{dx}\left(\frac{x^2 - 1}{1 - y^2}\right)$$

$$= \frac{(1 - y^2)\frac{d}{dx}(x^2 - 1) - (x^2 - 1)\frac{d}{dx}(1 - y^2)}{(1 - y^2)^2}$$

$$= \frac{2x(1 - y^2) + 2y(x^2 - 1)\frac{dy}{dx}}{(1 - y^2)^2}.$$

At this stage we observe that the expression for the second derivative still involves the first derivative. Hence to complete the solution we must substitute $dy/dx = (x^2 - 1)/(1 - y^2)$:

$$\frac{d^2y}{dx^2} = \frac{2x(1 - y^2) + 2y(x^2 - 1)\left(\frac{x^2 - 1}{1 - y^2}\right)}{(1 - y^2)^2}$$

$$= \frac{2x(1 - y^2)^2 + 2y(x^2 - 1)^2}{(1 - y^2)^3}.$$

In the last step we have multiplied through in the numerator and denominator of the fraction by $(1 - y^2)$.

EXERCISES 4.6

Find the explicit function or functions corresponding to the following implicit relations:

1. $xy + x - y = 0$.

2. $x^2 - y^2 + x + y = 0$.

3. $x^2 + y^2 + 2xy = 4$.

4. $\sqrt{x} + \sqrt{y} = 1$.

5. $xy^2 + yx^2 = 6$.

6. $4x^2 + 9y^2 = 36$.

Find dy/dx when

7. $x^2 + y^2 + 2y = 15$.

8. $\sqrt{x} + \sqrt{y} = 1$.

9. $x^3 + y^3 = a^3$ (a is constant).

10. $x^2 - xy + y^2 = 3$.

11. $(y - x)(y + 2x) - 12 = 0$.

12. $x^4 + y^4 = 2x^2y^2 + 3$.

13. $xy^2 + yx^2 = 6$.

14. $x^2y^2 + x^2 + y^2 = 3$.

15. $x^5 + y^5 = 5xy$.

16. $\frac{x^2}{a^2} - \frac{y^2}{b^2} = 1$ (a and b are constants)

17. $xy + e^y = 1$.

18. $\frac{x}{y} + \ln\left(\frac{y}{x}\right) = 6$.

19. $xy + \ln(xy) = -1$.

20. $x^2 + y^2 = 4e^{x+y}$.

21. Find $\dfrac{dx}{dt}$ if $3x^2 + 5t^2 = 15$.

22. Find $\dfrac{du}{dy}$ if $u^2 + y^2 + u - y = 1$.

23. Find $\dfrac{dx}{dy}$ if $x^3 + y^3 = xy$.

24. Find $\dfrac{dt}{dx}$ if $x^3 + t^3 + x^3t^3 = 9$.

Find the equation of the tangent to the following curves at the given points:

25. $x^3 + y^3 - 3xy = 3$: $(1, 2)$.

26. $x^2 + y^2 = 2x + y + 15$: $(-3, 1)$.

27. Find $\dfrac{d^2y}{dx^2}$ if $x^2 + y^2 = 4xy$.

28. Find $\dfrac{d^2u}{dt^2}$ if $u^5 + t^5 = 5ut$.

29. Find $\dfrac{d^2x}{dy^2}$ if $x^3 + y^3 - 3xy = 1$.

30. Find $\dfrac{d^2y}{dx^2}$ if $(x + 2)(y + 3) = 7$.

31. Let x and y be the sizes of two populations, one of which preys on the other. At any time x and y satisfy the implicit relation

$$\frac{(x + ty - h)^2}{a^2} + \frac{(y - tx - k)^2}{b^2} = 1,$$

where a, b, h, k, and t are certain constants. Calculate dy/dx.

32. By writing $y = x^{p/q}$ in the form $y^q - x^p = 0$, use implicit differentiation to prove that $(d/dx)(x^n) = nx^{n-1}$ when n is a rational number p/q.

4.7   PARAMETRIC EQUATIONS

It is often useful to study functions and their graphs in so-called parametric form. The following example illustrates this approach.

Suppose that a ball is thrown horizontally from the top of a cliff, with a velocity of 10 ft/sec. If there were no force of gravity, the ball would continue to move horizontally with a constant velocity, and after a time interval of t sec would have traveled a distance of $10t$ ft. But because of gravity, the ball accelerates downwards and develops a vertical velocity. If we take the x-axis horizontal and y-axis vertical, then the path of the ball is a curve in the xy plane (Fig. 4.44).

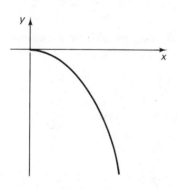

Figure 4.44

The path can be described by an equation expressing y as a function of x, as we did before. However this type of representation does not give complete information about the motion of the ball. For example, if we are simply given the equation of the path in the form $y = F(x)$, we cannot calculate where the ball will be after 2 sec or what its velocity will be after 2 sec. To study such questions we must know the values of x and y in terms of

t; that is, x and y must be expressed in the forms

$$x = f(t), \qquad y = g(t)$$

as functions of t.

In the present example the horizontal distance x is unaffected by the force of gravity and therefore is given by $x = 10t$. Similarly, the vertical distance traveled by the ball in t sec is the same as if the ball were just released from rest and falling vertically. Thus it is given by $y = -16t^2$ (the negative sign because the y-axis is taken to point upwards). The motion of the ball is described by the equations

$$x = 10t, \qquad y = -16t^2.$$

These are known as the *parametric equations* of the path, and t is called the *parameter*.

From the parametric equations we can calculate the position of the ball at any time after it is thrown. For example, 2 sec after the start of its flight ($t = 2$) the coordinates of the ball are

$$x = 10(2) = 20 \text{ ft}$$
$$y = -16(2)^2 = -64 \text{ ft}.$$

So the position of the ball is the point $(20, -64)$.

We now give a formal definition of the parametric equations of a curve in the xy-plane.

DEFINITION   Let f and g be two functions with a common domain D; that is, $f(t)$ and $g(t)$ are well-defined real numbers for all $t \in D$. Then

$$x = f(t), \qquad y = g(t)$$

define the *parametric equations* of a curve, and t is called the *parameter*. Usually we insist that the two functions f and g be continuous functions and also, as a rule, differentiable.

If we eliminate the parameter t from the two parametric equations, we get the Cartesian equation of the path or curve in x and y. In the example of the thrown ball,

$$x = 10t; \qquad t = \frac{x}{10},$$

and using this value of t in the second equation, we have

$$y = -16\left(\frac{x}{10}\right)^2 = -\frac{4}{25}x^2.$$

This equation represents a parabola in the xy plane.

To draw the graph of a curve given by parametric equations we give different values to the parameter and calculate the corresponding values of x and y. Then we plot these points (x, y) and join them by a smooth curve. In drawing the graph we do not plot the values of the parameter. We shall illustrate the drawing of the graph of parametric equations by the following example.

EXAMPLE   Draw the graph of the curve given by the parametric equations

$$x = t^2, \qquad y = 2t.$$

Find also the direct relation between x and y.

SOLUTION   The domains of the functions that define x and y are clearly the set of all real numbers. Thus t can be any real number. The values of x, y for various values of t are given in the following table:

t:	...	-3	-2	-1	0	1	2	3	4	...
x:	...	9	4	1	0	1	4	9	16	...
y:	...	-6	-4	-2	0	2	4	6	8	...

After plotting the points (x, y) given in the table and joining them by a smooth curve, the graph is as shown in Fig. 4.45.

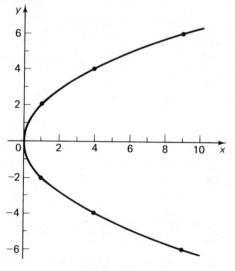

Figure 4.45

To find the direct relation between x and y, we eliminate t from the two equations. From $y = 2t$, we have $t = y/2$, and using this value of t in x,

$$x = \left(\frac{y}{2}\right)^2 \quad \text{or} \quad y^2 = 4x.$$

NOTE.   The graph of a pair of parametric equations need not represent y as a function of x, as is clear from the above example. But it is in most cases possible to restrict the values of the parameter in such a way that y is a well-defined function of x. For example, in the above example, if we take only $t \geq 0$, then the graph of the parametric equations is the upper part of the parabola, and as such y is a function of x.

We are now ready to find the formula for dy/dx for curves given by parametric equations. Let

$$x = f(t), \quad \text{and} \quad y = g(t)$$

be the parametric equations of a curve with domain D such that y is also a function of x. y is a function of x and x is a function of t. Thus y can be viewed as a composite function of t, and by the chain rule we can write

$$\frac{dy}{dt} = \frac{dy}{dx} \cdot \frac{dx}{dt}$$

If $dx/dt \neq 0$, then from this equation we have

$$\boxed{\frac{dy}{dx} = \frac{dy/dt}{dx/dt}.}$$

EXAMPLE   Given $x = 1 + t^2$, $y = 1/t$; find dy/dx.

SOLUTION $x = 1 + t^2$ implies $dx/dt = 2t$ and $y = 1/t = t^{-1}$ implies $dy/dt = (-1)t^{-2} = -1/t^2$. Therefore

$$\frac{dy}{dx} = \frac{dy/dt}{dx/dt} = \frac{-1/t^2}{2t} = -\frac{1}{2t^3}.$$

EXAMPLE   Find the equation of the tangent to the graph of

$$x = 1 + t, \quad y = t^2$$

at the point $(2, 1)$.

SOLUTION   We must first find the value of t that corresponds to the given point $(2, 1)$. We have

$$x = 2 = 1 + t \quad \text{and} \quad y = 1 = t^2,$$

which give the value $t = 1$.
$dx/dt = 1$ and $dy/dt = 2t$. Therefore

$$\frac{dy}{dx} = \frac{dy/dt}{dx/dt} = \frac{2t}{1} = 2t.$$

When $t = 1$, $dy/dx = 2$. Thus the slope of the tangent at $(2, 1)$ is $m = 2$.
Using the point–slope formula,

$$y - y_1 = m(x - x_1),$$

the equation of the tangent line at $(2, 1)$ is

$$y - 1 = 2(x - 2),$$

or

$$y = 2x - 3.$$

The calculation of higher derivatives for functions given in parametric form is illustrated in the following example.

EXAMPLE   Find d^2y/dx^2 when x and y are given parametrically by $x = 4at^3 + 1, y = 2at^2$.

SOLUTION   We have $dx/dt = 12at^2$ and $dy/dt = 4at$. Therefore the first derivative is:

$$\frac{dy}{dx} = \frac{dy/dt}{dx/dt} = \frac{4at}{12at^2} = \frac{1}{3t}.$$

In order to calculate the second derivative, let u denote dy/dx. Then as a function of the parameter t, $u = \frac{1}{3}t^{-1}$. The required second derivative is

$$\frac{d^2y}{dx^2} = \frac{d}{dx}\left(\frac{dy}{dx}\right) = \frac{du}{dx}.$$

Therefore

$$\frac{du}{dx} = \frac{du/dt}{dx/dt} = \frac{-\frac{1}{3}t^{-2}}{12at^2} = -\frac{1}{36at^4}.$$

NOTE.   In the case of second derivatives, it is a common mistake to calculate them for equations in parametric form by setting

$$\frac{d^2y}{dx^2} = \frac{d^2y/dt^2}{d^2x/dt^2}.$$

This is *incorrect*.

EXERCISES 4.7

Find the relation between x and y corresponding to the following parametric equations. In each case give the domain of the parametric representation.

1. $x = \dfrac{1}{t}, \quad y = t^2.$ **2.** $x = \sqrt{t - 4}, \quad y = t + 1.$

3. $x = t^2, \quad y = t^3.$ **4.** $x = t^2 + t, \quad y = \sqrt{t^2 - t}.$

Sketch the graph of the following parametric equations and also find the direct relation between x and y.

5. $x = t + 1, \quad y = 3t - 2.$ **6.** $x = 2t, \quad y = t^2 + 1.$

Find dy/dx if

7. $x = 2t + 3, \quad y = 2 - t.$ **8.** $x = t^2 + t, \quad y = t^2.$

9. $x = at^2, \quad y = bt^3.$ **10.** $x = \dfrac{1 - t^2}{1 + t^2}, \quad y = \dfrac{2t}{1 + t^2}.$

11. $x = \sqrt{t} + 1, \quad y = \sqrt[3]{t}.$ **12.** $x = \theta^2 + 3\theta, \quad y = \theta + 1.$

13. $x = u^2 + 1, \quad y = u + 3.$

14. $x = at^2 + bt + c, \quad y = pt^2 + qt + r$
       where $a, b, c, p, q,$ and r are constants.

15. If $r = e^t$ and $\theta = 2t$, find $dr/d\theta$.

16. If $x = a(1 - e^{-t})$ and $y = b(1 - e^{-t}) - ct$ find dy/dx.

Find the equation of the tangent to the following curves at the given points.

17. $x = \dfrac{1}{t}$, $y = t^2$; $(1, 1)$.

18. $x = \sqrt{t - 4}$, $y = t + 1$; $(1, 6)$.

19. $x = t^2 + t$, $y = 2t - 1$; $(2, -5)$.

20. $x = \dfrac{\theta}{1 + \theta}$, $y = \theta^2$; $\left(\dfrac{2}{3}, 4\right)$.

Find the equation of the normal to the following curves at the given points. (The normal is the straight line perpendicular to the tangent, through the given point.)

21. $x = t^2$, $y = t^3$; $(1, 1)$.

22. $x = t^2 + t$, $y = t^2$; $(0, 1)$.

23. Find $\dfrac{d^2 y}{dx^2}$ if $x = at^2$, $y = 2at$ $(a = \text{constant})$.

24. Find $\dfrac{d^3 y}{dx^3}$ if $x = 2t + 1$, $y = t^3 + 2$.

25. Find $\dfrac{d^2 y}{dx^2}$ if $x = \dfrac{1 - t^2}{1 + t^2}$, $y = \dfrac{2t}{1 + t^2}$.

26. Find $\dfrac{d^2 y}{dx^2}$ if $x = \dfrac{3t^2}{1 + t^3}$, $y = \dfrac{1 - t^3}{1 + t^3}$.

27. If $x = \ln \theta$ and $y = \theta^2 - 1$, evaluate dy/dx and $d^2 y/dx^2$.

28. If $x = \ln \theta - \theta$ and $y = \ln \theta + \theta$, evaluate dy/dx and $d^2 y/dx^2$.

29. A squirrel wishes to jump from one branch to another, which is 4 ft lower and also 5 ft away in the horizontal direction. Assuming that he jumps horizontally, with what velocity must he leap? When he arrives at the second branch, in what direction will he be traveling?

30. A ball is thrown into the air with a velocity of 40 ft/sec at an angle of 45° to the horizontal. If the x-axis is taken horizontal and the y-axis vertical, the origin being at the initial point of the ball's flight, then the position of the ball at time t is given by $x = 20\sqrt{2}\,t$, $y = 20\sqrt{2}\,t - 16t^2$. Calculate the slope of the path t sec after the ball is thrown. At what value of t does the ball reach its maximum height?

31. Repeat exercise 30 for the case when the ball is thrown at an angle of 60° to the horizontal. In this case, $x = 20t$ and $y = 20\sqrt{3}\,t - 16t^2$.

32. In a certain chemical reaction, when an amount u of a catalyst is added, the amount of the end-product BS generated is $x = u/(u + 1)$ gm. The increase in temperature of the reacting mixture is $T = 10\sqrt{u}/(\sqrt{u} + 1)$°C. Initially the amount of catalyst is sufficient to generate 0.5 gm of BS. If u is changed in such a way as to produce an increment $dx = 0.05$ gm, use differentials to calculate the corresponding increment dT in temperature.

REVIEW EXERCISES FOR CHAPTER 4

1. State whether the following statements are true or false. Justify your answers.

   (a) If a function f has a maximum or a minimum at $x = c$, then $f'(c)$ must be zero.

   (b) If $f'(c) = 0$, then the function f has either a maximum or a minimum at $x = c$.

   (c) The tangent to the graph of a function at a point where it is maximum or minimum is either horizontal or vertical.

   (d) The tangent to the graph of a function at a point of inflection is always horizontal.

   (e) If $f(x)$ is concave upwards in an interval of its domain, then both $f'(x)$ and $f''(x)$ must be positive.

   (f) A local maximum value of a function is always greater than a local minimum value of the same function.

   (g) The differential coefficient of $f(x)$ w.r.t. x is $f'(x)$.

   (h) The differential of x^3 is $3x^2$.

   (i) If $f(x)$ is continuous at $x = a$ then $f(x) \approx f(a) + (x - a)f'(a)$ for x sufficiently close to a.

   (j) For a linear function f, $df = \Delta f$.

   (k) For a nonlinear function $y = f(x)$, $dy \neq \Delta y$.

   (l) To find the differential of a dependent variable, the function need not be differentiable w.r.t. the independent variable involved.

   (m) The function $f(x)$ is increasing or decreasing in $a \leq x \leq b$ as $f(x) > 0$ or $f(x) < 0$ respectively.

   (n) In an implicit relation of the form $f(x, y) = 0$, x and y are both independent variables.

   (o) If $x = f(t)$ and $y = g(t)$ are parametric equations of a curve, then the curve always represents the graph of a function.

Find the values of x for which the following functions are (a) increasing; (b) decreasing; (c) concave upwards; (d) concave downwards. Also find the points of inflection if any.

2. $f(x) = \dfrac{1}{x^2 + 1}.$ **3.** $g(x) = \dfrac{x}{x^2 + 1}.$

4. $y = e^{x^2}.$ **5.** $y = xe^{x^2}.$

6. Find the maxima and minima of $f(t) = 2t^3 - 3t^2 + 1.$

7. Find the absolute extrema of $g(x) = \sqrt{x^2 - 4}$ in $2 \leq x \leq 3.$

8. Determine two values of the constant c so that $f(x) = x + (c/x)$ can have

   (a) a local maximum at $x = -1.$

   (b) a local minimum at $x = 2.$

9. Determine the constant k in such a way that the function $f(x) = x^3 + (k/x^2)$ can have

(a) a local minimum at $x = 1$.

(b) a local maximum at $x = -2$.

(c) a point of inflection at $x = 2$.

10. Determine the constants A and B so that the function $f(x) = x^3 + Ax^2 + Bx + C$ can have

(a) a maximum at $x = -2$ and a minimum at $x = 1$.

(b) a maximum at $x = -3$ and a point of inflection at $x = -2$.

11. Find the restrictions on the constants A, B, and C in order that $f(x) = Ax^2 + Bx + C$ can have a local minimum.

12. During a certain influenza epidemic, the proportion of the population in the lower mainland who are infected is denoted by $y(t)$, where t is the time in weeks since the start of the epidemic. It is found that

$$y(t) = \frac{t}{4 + t^2}.$$

(a) What is the physical interpretation of dy/dt?

(b) For what value of t is y maximum?

(c) For what values of t is y increasing and decreasing?

13. In many mammals the blood flow would become turbulent if the Reynolds number relating to the flow exceeded the value

$$R = A \ln r - Br,$$

where r is the radius of the aorta and A and B are positive constants. Find the radius r that makes this critical Reynolds number a maximum.

14. The value of a certain fruit crop (in dollars) is given by

$$V = A(1 - e^{-KI}),$$

where A and K are constants and I is the number of pounds per acre of insecticide with which the crop is sprayed. If the cost of spraying is given by $C = BI$, where B is a constant, find the value of I that makes $V - C$ a maximum. What is the interpretation of your result when $AK < B$?

15. In a certain location the concentration of water in the soil is given in terms of depth x by the formula

$$c = 1 - e^{-x^2}.$$

Find the depth at which c is increasing most rapidly. Use Newton's method to find the depth at which $c = 0.5$.

16. Use Newton's method to find the approximate value of $\sqrt{7}$ correct to two decimal places.

Use differentials to approximate the values of the following expressions:

17. $(2.1)^4$.   **18.** $(4.1)^3 + (4.1)^{-1/2}$.

19. Find $\dfrac{dy}{dx}$ if $x^{2/3} + y^{2/3} = a^{2/3}$.   **20.** Find $\dfrac{dy}{dx}$ if $\sqrt[5]{x} - \sqrt[5]{y} = \sqrt[5]{a}$.

21. Find $\dfrac{dy}{dx}$ if $x = \dfrac{t}{(t+1)}$, $\quad y = \dfrac{t^2}{(t^2+1)}$.

22. Find $\dfrac{dy}{dt}$ if $y = u^3$, $\quad t = \sqrt{u} - 1$.

Find d^2y/dx^2 for the following relations:

23. $x^2 + y^2 - 3xy = 0$.   **24.** $x^2 + y^2 - 4xy = 1$.

25. $x = \dfrac{y}{y^2+1}$.   **26.** $x = \dfrac{y^2+1}{y+1}$.

27. $x = (t^3 + 1)$, $y = (t^2 + 2)$.

28. $x = f(t)$, $\quad y = g(t)$ where f and g are certain twice-differentiable functions of t.

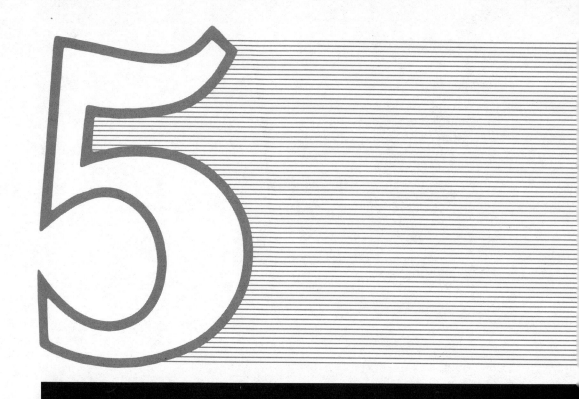

5

Trigonometric Functions

5.1 DEFINITIONS AND ELEMENTARY USES

There are two sets of units that are commonly used to measure angles: degrees and radians. The angle corresponding to one complete revolution is given the value 360° (degrees) or 2π radians. Thus 1 radian is equal to $(360/2\pi)$ degrees, or approximately 57.3°. The values of some of the more commonly used angles are given in terms of radians as follows:

$$180° = \pi \text{ radians}; \qquad 90° = \frac{\pi}{2} \text{ radians}; \qquad 45° = \frac{\pi}{4} \text{ radians};$$

$$60° = \frac{\pi}{3} \text{ radians}; \qquad 30° = \frac{\pi}{6} \text{ radians}; \qquad 120° = \frac{2\pi}{3} \text{ radians}.$$

For applications of trigonometry involving calculus it is most convenient to express angles in terms of radian measure. From now on we shall assume that, unless otherwise stated, *all angles are given in radians*.

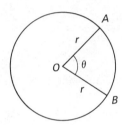

Figure 5.1

The reason why radian measure is preferred to degree measure lies in the fact that the length of an arc of a circle and the area of a sector of a circle are given by simple formulas when the angle subtended at the center by the arc or sector is expressed in radians. Consider the arc AB in Fig. 5.1, subtending an angle θ at the center O of the circle to which it belongs. The angle AOB is a fraction $(\theta/2\pi)$ of one complete revolution. Hence the arc AB has a length equal to this same fraction $(\theta/2\pi)$ of the total circumference of the circle. Thus

$AB = (\theta/2\pi)(2\pi r) = \theta r.$

Similarly the sector AOB has an area that is equal to a fraction $(\theta/2\pi)$ of the total area of the circle. Thus

$$\text{area } AOB = \left(\frac{\theta}{2\pi}\right)(\pi r^2) = \frac{1}{2}\theta r^2.$$

EXAMPLE What is the area of the sector of a circle of radius 2 cm that subtends an angle of 50° at the center?

SOLUTION We must first convert the angle to radians:

$$50° = 50\left(\frac{2\pi}{360}\right) \text{ radians} = 0.873 \text{ radians}.$$

Then the area of the sector is equal to

$$\tfrac{1}{2}(0.873)(2^2) = 1.746 \text{ cm}^2.$$

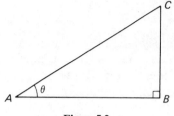

Figure 5.2

The trigonometric functions are defined in terms of a right-angled triangle (Fig. 5.2). Let ABC be such a triangle, right-angled at B, and let θ denote the angle at A. Then the sine and cosin of θ are defined by $\sin \theta = BC/AC$, and $\cos \theta = AB/AC$. In words, the sine of an angle is the ratio of the opposite side to the hypotenuse, and the cosine is the ratio of the adjacent side to the hypotenuse.

In terms of the sine and cosine, four other trigonometric functions, the tangent, cotangent, secant, and cosecant, are defined in the following way:

$$\tan \theta = \frac{\sin \theta}{\cos \theta} = \frac{BC}{AB}$$

$$\cot \theta = \frac{\cos \theta}{\sin \theta} = \frac{1}{\tan \theta} = \frac{AB}{BC}$$

$$\sec \theta = \frac{1}{\cos \theta} = \frac{AC}{AB}$$

$$\operatorname{cosec} \theta = \frac{1}{\sin \theta} = \frac{AC}{BC}.$$

The sine and cosine of any angle never exceed 1 in magnitude. On the other hand, the secant and cosecant are always greater than or equal to 1 in magnitude. The tangent and cotangent can have values that are less than, equal to, or greater than 1 in magnitude.

There are a few special angles the values of whose sines and cosines are worth remembering. These particular values are set out in the following table in a form easy to remember.

θ (degrees)	0°	30°	45°	60°	90°
θ (radians)	0	$\dfrac{\pi}{6}$	$\dfrac{\pi}{4}$	$\dfrac{\pi}{3}$	$\dfrac{\pi}{2}$
$\sin \theta$	$\dfrac{\sqrt{0}}{2}$	$\dfrac{\sqrt{1}}{2}$	$\dfrac{\sqrt{2}}{2}$	$\dfrac{\sqrt{3}}{2}$	$\dfrac{\sqrt{4}}{2}$
$\cos \theta$	$\dfrac{\sqrt{4}}{2}$	$\dfrac{\sqrt{3}}{2}$	$\dfrac{\sqrt{2}}{2}$	$\dfrac{\sqrt{1}}{2}$	$\dfrac{\sqrt{0}}{2}$

Several of the values in the table can of course be written more simply: for example,

$$\frac{\sqrt{0}}{2} = 0, \qquad \frac{\sqrt{4}}{2} = 1.$$

Most of the values in the table can be seen immediately from the definition of the sine and cosine by examining special cases of the triangle ABC in Fig. 5.2. For example, when $\theta = 0$, BC must be of zero length and $AC = AB$. Then $\sin \theta = BC/AC = 0$, and $\cos \theta = AB/AC = 1$. Similarly, when $\theta = \pi/2 = 90°$, AB is zero, and then $\sin \theta = BC/AC = 1$, and $\cos \theta = AB/AC = 0$.

The values of $\sin \pi/4$ and $\cos \pi/4$ can also be calculated. When $\theta = \pi/4 = 45°$, the angle at C in the triangle is also equal to $\pi/4$ and the triangle is therefore isosceles; $BC = AB$. From Pythagoras's theorem, the square of the hypotenuse is given by $AC^2 = BC^2 + AB^2 = 2BC^2$. Taking square roots we obtain $AC = \sqrt{2}\,BC$. Therefore $\sin \theta = \sin(\pi/4) = BC/AC = 1/\sqrt{2} = \sqrt{2}/2$, as given in the preceding table. Similarly, $\cos \theta = AB/AC = BC/AC = 1/\sqrt{2}$.

The other trigonometric functions of these special angles can readily be calculated. For example,

$$\sec \frac{\pi}{6} = \frac{1}{\cos \dfrac{\pi}{6}} = \frac{2}{\sqrt{3}},$$

$$\tan \frac{\pi}{3} = \frac{\sin \dfrac{\pi}{3}}{\cos \dfrac{\pi}{3}} = \frac{\dfrac{\sqrt{3}}{2}}{\dfrac{1}{2}} = \frac{\sqrt{3}}{2} \cdot \frac{2}{1} = \sqrt{3}.$$

The values of the trigonometric functions for general angles less than $\pi/2$ are found in various standard collections of tables.

The trigonometric functions have many applications, one of which is illustrated by the following examples.

EXAMPLE   A tree stands on level ground. An observer situated 350 ft from the base of the tree observes that its top subtends an angle of 22° above the horizontal. How tall is the tree?

SOLUTION   In Fig. 5.3, A denotes the position of the observer. We see that $BC/AB = \tan 22°$ $= 0.4040$, where the tangent of 22° has been found from the standard tables. Therefore the height of the tree is given by

$$BC = (350)(0.4040) = 141.4 \text{ ft.}$$

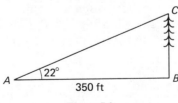

Figure 5.3

NOTE.   In examples such as this one, which require the use of trigonometric tables, it is more appropriate to give angles in degrees rather than radians since the tables must be entered with angles in degrees and minutes.

EXAMPLE   An observer on board a ship sailing within sight of land measures by means of a sextant the angle of elevation of a certain mountain peak above the horizontal. The measured angle is 4°. According to the chart the given peak has an altitude of 7000 ft. How far is the ship away from the base of the mountain?

SOLUTION   From Fig. 5.4, $BC/AB = \tan 4° = 0.06993$. Therefore the distance of the ship is given by

$$AB = \frac{(7000)}{(0.06993)} = 100,100 \text{ ft}$$

$$= 19.0 \text{ statute miles.}$$

So far we have defined the trigonometric functions only for angles less than 90°. They are defined for larger angles, and also for negative angles, in the following way. Let O be the origin of a system of Cartesian coordinates whose axes are Ox and Oy, and let there be drawn a circle whose radius is unity and whose center is O (Fig. 5.5). Then for any angle θ we can find a point A on the circle such that OA makes an angle

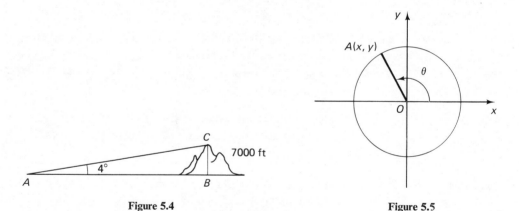

Figure 5.4   Figure 5.5

θ with the axis Ox. Positive angles are measured anticlockwise from Ox, and negative angles are measured clockwise. If the angle concerned exceeds 2π, then we would turn through more than one revolution in rotating from the direction Ox to the direction OA. In the case illustrated in Fig. 5.5, θ is about $2\pi/3$ (or 120°).

If A has coordinates (x, y), then we define

$$\sin \theta = y, \qquad \cos \theta = x.$$

If θ is less than $\pi/2$, this definition agrees with the earlier one in terms of a right-angled triangle.

When θ lies between 0 and $\pi/2$, so that A is in the first quadrant, both x and y are positive, and $\sin \theta$ and $\cos \theta$ are both positive. The same is true for $\tan \theta$ (and all the other trigonometric functions for that matter). When θ lies between $\pi/2$ and π,

228

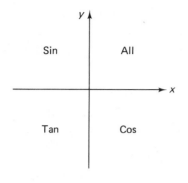

Figure 5.6

x is negative. Thus $\cos\theta$ and $\tan\theta$ are negative but $\sin\theta$ is positive in the second quadrant. In the third quadrant (θ between π and $3\pi/2$), both *x* and *y* are negative. Therefore $\sin\theta$ and $\cos\theta$ are both negative but $\tan\theta$ is positive in this quadrant. Finally, in the fourth quadrant (θ between $3\pi/2$ and 2π), $\cos\theta$ is positive and $\sin\theta$ and $\tan\theta$ are both negative. The functions that are positive in each quadrant are indicated in Fig. 5.6.

The graphs of $\sin\theta$ and $\cos\theta$ for values of θ between about $-360°$ and $+720°$ (i.e., between -2π and $+4\pi$ radians) are shown in Fig. 5.7. The function $\sin\theta$ starts out at zero when $\theta = 0$, increases to 1 as θ increases to $\pi/2$, and then decreases to zero as θ increases to π. As θ increases from π to 2π, $\sin\theta$ becomes negative, decreasing to -1 when $\theta = 3\pi/2$, and then increasing back to zero when $\theta = 2\pi$. This basic cycle of behavior of the function then repeats itself: The graph of $\sin\theta$ for θ between 2π and 4π is an exact copy of the graph for θ between 0 and 2π, simply shifted along the θ-axis. In fact the basic cycle continues to repeat indefinitely in both the positive and negative θ-directions: The graph of $\sin\theta$ for θ between 58π and 60π, for example, is a replica of its graph between 0 and 2π.

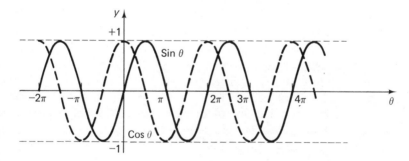

Figure 5.7

The domain of the function $y = \sin x$ is the set of all real numbers, while the range is the set $\{y \,|\, -1 \le y \le 1\}$.

The graph of $\cos\theta$ is identical with the graph of $\sin\theta$ except that it is displaced back along the θ axis through a distance $\pi/2$. As θ increases from 0 to π, $\cos\theta$ decreases from 1 to -1; then as θ increases further from π to 2π, $\cos\theta$ increases from -1 back to 1. Again this basic pattern repeats indefinitely.

Note from the graphs of $\sin\theta$ and $\cos\theta$ the following special values:

$$\sin\pi = \sin 2\pi = 0, \qquad \cos\pi = -1, \qquad \cos 2\pi = 1.$$

For angles greater than $\pi/2$, the other four trigonometric functions are defined in terms of $\sin \theta$ and $\cos \theta$ in the same way as for acute angles:

$$\tan \theta = \frac{\sin \theta}{\cos \theta}, \qquad \cot \theta = \frac{\cos \theta}{\sin \theta},$$

$$\sec \theta = \frac{1}{\cos \theta}, \qquad \operatorname{cosec} \theta = \frac{1}{\sin \theta}.$$

The graph of $\tan \theta$ is shown in Fig. 5.8. $\tan \theta$ becomes infinite at values of θ for which $\cos \theta = 0$, that is, for $\theta = \pm(\pi/2), \pm(3\pi/2)$, etc. It can be seen that the graph has vertical asymptotes at these values of θ. Between each adjacent pair of asymptotes, $\tan \theta$ increases from $-\infty$ to $+\infty$.

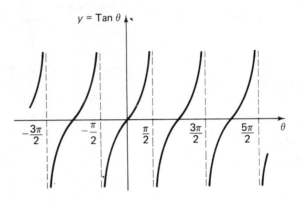

Figure 5.8

The domain of the function $y = \tan x$ is the set $\{x \mid x \neq \pm(\pi/2), \pm(3\pi/2), \ldots\}$, that is, the set of all real numbers except $\pm(\pi/2), \pm(3\pi/2)$, etc. The range of this function is the set of all real numbers.

Returning to the definition of $\sin \theta$ and $\cos \theta$, we observe that any point $A(x, y)$ on the unit circle satisfies the relation $x^2 + y^2 = 1$. But $x = \cos \theta$ and $y = \sin \theta$ by definition, so that for any angle θ we have the trigonometric identity (often called the Pythagorean identity)

$$\sin^2 \theta + \cos^2 \theta = 1.$$

Dividing through this equation in turn by $\cos^2 \theta$ and $\sin^2 \theta$ we obtain two further identities:

$$\frac{\sin^2 \theta}{\cos^2 \theta} + 1 = \frac{1}{\cos^2 \theta}, \quad \text{or} \quad \boxed{\tan^2 \theta + 1 = \sec^2 \theta};$$

$$1 + \frac{\cos^2 \theta}{\sin^2 \theta} = \frac{1}{\sin^2 \theta}, \quad \text{or} \quad \boxed{1 + \cot^2 \theta = \operatorname{cosec}^2 \theta.}$$

It is often useful to be able to express the sines and cosines of the sum or the difference of two angles in terms of the sines and cosines of the two angles themselves.

There are four formulas, as follows:

$$\cos(\theta_1 + \theta_2) = \cos\theta_1 \cos\theta_2 - \sin\theta_1 \sin\theta_2$$
$$\cos(\theta_1 - \theta_2) = \cos\theta_1 \cos\theta_2 + \sin\theta_1 \sin\theta_2$$
$$\sin(\theta_1 + \theta_2) = \sin\theta_1 \cos\theta_2 + \cos\theta_1 \sin\theta_2$$
$$\sin(\theta_1 - \theta_2) = \sin\theta_1 \cos\theta_2 - \cos\theta_1 \sin\theta_2.$$

The corresponding formulas for the tangent function are:

$$\tan(\theta_1 + \theta_2) = \frac{\tan\theta_1 + \tan\theta_2}{1 - \tan\theta_1 \tan\theta_2},$$

$$\tan(\theta_1 - \theta_2) = \frac{\tan\theta_1 - \tan\theta_2}{1 + \tan\theta_1 \tan\theta_2}.$$

If, in the above formulas for the sums of angles, we set $\theta_1 = \theta_2 = \theta$, we obtain the so-called double-angle formulas:

$$\cos 2\theta = \cos^2\theta - \sin^2\theta,$$
$$\sin 2\theta = 2\sin\theta\cos\theta,$$
$$\tan 2\theta = \frac{2\tan\theta}{1 - \tan^2\theta}$$

We can make use of these identities in order to obtain expressions for the sines and cosines of angles that lie outside the range between 0 and $\pi/2$. For example, if in the sum formula for the cosine we set $\theta_1 = \pi/2$ and $\theta_2 = \theta$, we obtain

$$\cos\left(\frac{\pi}{2} + \theta\right) = \cos\left(\frac{\pi}{2}\right)\cos\theta - \sin\left(\frac{\pi}{2}\right)\sin\theta.$$

Since $\cos(\pi/2) = 0$ and $\sin(\pi/2) = 1$, this becomes

$$\boxed{\cos\left(\frac{\pi}{2} + \theta\right) = -\sin\theta.}$$

Similarly, from the sum formula for the sine, we obtain

$$\sin\left(\frac{\pi}{2} + \theta\right) = \sin\left(\frac{\pi}{2}\right)\cos\theta + \cos\left(\frac{\pi}{2}\right)\sin\theta,$$

i.e.,

$$\boxed{\sin\left(\frac{\pi}{2} + \theta\right) = \cos\theta.}$$

A variety of relations of this type can be derived. Setting $\theta_1 = \pi$ and $\theta_2 = \theta$ in each of the four sum and difference formulas, we obtain that

$$\boxed{\begin{array}{ll} \sin(\pi - \theta) = \sin\theta, & \sin(\pi + \theta) = -\sin\theta \\ \cos(\pi - \theta) = -\cos\theta, & \cos(\pi + \theta) = -\cos\theta. \end{array}}$$

And setting $\theta_1 = 2\pi$ and $\theta_2 = \theta$ in the difference formulas, we find that

$$\boxed{\begin{aligned} \sin(2\pi - \theta) &= \sin(-\theta) = -\sin\theta, \\ \cos(2\pi - \theta) &= \cos(-\theta) = \cos\theta. \end{aligned}}$$

Note that in these formulas we have also used the fact that $(2\pi - \theta)$ represents the same angle as $(-\theta)$.

These identities are useful in calculating the sines and cosines of angles that lie outside the range given in the usual sets of trigonometric tables.

EXAMPLE   Find **(a)** $\sin 154°$; **(b)** $\cos 221°$; **(c)** $\cos(19\pi/10)$.

SOLUTION   **(a)** $\sin 154° = \sin(180° - 26°) = \sin 26° = 0.4384$.
(b) $\cos 221° = \cos(180° + 41°) = -\cos 41° = -0.7547$.
(c)   In order to use the tables the angle must be converted from radians to degrees:

$$\frac{19\pi}{10} = \frac{19\pi}{10}\left(\frac{360}{2\pi}\right)^{\circ} = 342°.$$

Then

$$\cos\left(\frac{19\pi}{10}\right) = \cos(342°) = \cos(360° - 18°) = \cos 18° = 0.9511.$$

Note that in each of these examples, the quantity required is first expressed in terms of the sine or cosine of an angle lying between 0 and 90° by use of one of the foregoing relations. This sine or cosine is then obtained from the standard tables.

EXERCISES 5.1

Convert the following angles to radian measure:

1. $15°$.

2. $270°$.

3. $150°$.

4. $330°$.

5. $396°$.

Convert the following angles to degrees:

6. $\dfrac{\pi}{10}$.

7. $\dfrac{3\pi}{4}$.

8. 5π.

9. $\dfrac{11\pi}{6}$.

10. $\dfrac{7\pi}{2}$.

Calculate the length of the arc and the area of the sector of a circle with the given radius and subtending the given angle at the center in each of the following exercises:

11. Radius = 5 cm, angle = 31°.

12. Radius = 3 in., angle = 1.5 radians.

13. Radius = 3 in., angle = 2 radians.

14. Radius = 4 cm, angle = 120°.

15. If $\sin \theta = \frac{1}{4}$, use the Pythagorean identities to evaluate $\cos \theta$, $\tan \theta$, and $\sec \theta$.

16. If $\tan \theta = \frac{3}{4}$, find $\sin \theta$ and $\cos \theta$.

17. If $\sec \theta = 5$, find $\tan \theta$ and $\sin \theta$.

18. If $\cos \theta = \frac{1}{5}$, find $\tan \theta$ and $\operatorname{cosec} \theta$.

Express the following in terms of $\sin \theta$, $\cos \theta$, $\tan \theta$, or $\cot \theta$, whichever is appropriate:

19. $\sin (\pi + \theta)$.

20. $\cos (90° - \theta)$.

21. $\cos (270° - \theta)$.

22. $\sin \left(\dfrac{\pi}{2} + \theta \right)$.

23. $\cos (2\pi - \theta)$.

24. $\tan (\pi - \theta)$.

25. $\tan \left(\dfrac{\pi}{2} + \theta \right)$.

26. $\tan (270° - \theta)$.

27. Using standard tables, find:
  (a) $\sin 183°$.
  (b) $\cos (11\pi/18)$.
  (c) $\tan 281°$.

28. Using standard tables, find:
  (a) $\sin 160°$.
  (b) $\cos 252°$.
  (c) the tangent of 5 radians.

29. Prove the answers to exercises 19, 21, 23, and 25 when θ is acute by using the definitions of sine and cosine given in Fig. 5-5.

30. Prove the answers to exercises 20, 22, 24, and 26 using the formulas for sums and differences.

31. A kite string is 200 ft long and makes an angle of 37° with the horizontal. How high off the ground is the kite (assuming the string is straight)?

32. An individual standing 100 ft from the base of a tall building observes that the elevation of the top of the building is 80° above the horizontal. How high is the building?

33. An observer measures the angle of elevation of the top of a tree to be 80° above the horizontal. He then moves a distance of 20 ft further away from the tree (on level ground) and observes that the angle of elevation has decreased to 60°. How tall is the tree?

34. Two points *A* and *B* lie on a north–south line, with *A* lying 100 yards north of *B*. A third point *C* is observed by a surveyor to have a direction from *A* of 60° east of north and a direction from *B* of 30° east of north. How far is *C* from *A* and from *B*?

35. The navigator of a ship observes that a certain light lies in a direction 25° east of north from his vessel. After the ship has traveled a distance of 1 mile along its course, which is due north, he observes that the same light now lies in a direction 45° east of north. How close to the light will the ship pass?

36. An airplane is traveling at a constant height of 1000 ft heading directly over the head of an anti-aircraft gun. At the instant that the gun starts to fire, the direction of the aircraft makes an angle of 45° above the horizontal. If the velocity of the bullets is five times that of the airplane, how far ahead of the plane must the gunner aim in order to hit it?

37. Prove that $\cos 3\theta = 4 \cos^3 \theta - 3 \cos \theta$.

38. Derive a corresponding expression for $\sin 3\theta$ in terms of $\sin \theta$.

5.2 LIMITS OF TRIGONOMETRIC FUNCTIONS

Our immediate aim is to obtain expressions for the derivatives of the functions $\sin x$ and $\cos x$. Before we can do so, however, we must establish certain important results concerning the limits of various trigonometric functions.

We are primarily interested in the behavior of functions such as $\sin x$ and $\cos x$ as their argument x approaches zero. First of all, we observe that as $x \longrightarrow 0$, the function $\sin x$ approaches closer and closer to zero while the function $\cos x$ approaches closer and closer to the value 1. Thus in terms of limits, we can write

$$\lim_{x \to 0} \sin x = 0, \qquad \lim_{x \to 0} \cos x = 1.$$

It is possible to say much more than this about the limiting behavior of these two functions as $x \longrightarrow 0$. Consider the following table in which we have set out the values of $\sin x$ and x itself (in radians) for a sequence of angles decreasing from 10° to 1°.

x (degrees)	10°	5°	2°	1°
x (radians)	0.17453	0.08727	0.03491	0.017453
sin x	0.17365	0.08716	0.03490	0.017452

It is apparent from this table that for small angles x the value of $\sin x$ is almost equal to the value of x in radians, and that the smaller x becomes, the more closely do $\sin x$ and x approach one another in value; for example, for an angle of 10°, $\sin x$ and x differ by about 1%, but for an angle of 2° they differ by only 0.03%.

Clearly then, for small x,

$$\sin x \simeq x.$$

This result is known as the *small-angle approximation* to $\sin x$.

When x is small, the value of $\cos x$ is approximately equal to 1: $\cos x \simeq 1$. From this we can obtain a corresponding small-angle approximation for the function $\tan x$. Since $\sin x \simeq x$ and $\cos x \simeq 1$ for small x, we see that

$$\tan x = \frac{\sin x}{\cos x} \simeq \frac{x}{1} = x.$$

Thus, for small angles, $\tan x \simeq x$ also.

Since $\cos x$ approaches 1 as $x \to 0$, the quantity $(1 - \cos x)$ tends to the limiting value zero. In the following table the values of $1 - \cos x$ are given for a series of small angles, and for comparison the values of $\frac{1}{2}x^2$ are also given (x being measured in radians).

x (degrees)	$10°$	$5°$	$2°$
$1 - \cos x$	0.015192	0.003805	0.00060917
$\frac{1}{2}x^2$	0.015231	0.003808	0.00060924

It can be seen that these two quantities, $(1 - \cos x)$ and $\frac{1}{2}x^2$, are almost equal to one another, and, moreover, that the difference between them decreases as x decreases. They differ by about 0.3% for an angle of 10° and by about 0.01% for an angle of 2°. Thus we have the further approximation for small angles that

$$1 - \cos x \simeq \tfrac{1}{2}x^2,$$

or

$$\cos x \simeq 1 - \tfrac{1}{2}x^2.$$

These approximations are often useful in simplifying expressions involving trigonometric functions.

EXAMPLE   An observer on board a ship measures the angle of elevation of a certain mountain peak whose height is 7000 ft. The measured angle is 4° (Fig. 5.9). How far is the ship from the base of the mountain?

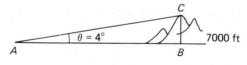

Figure 5.9

SOLUTION

$$AB = \frac{BC}{\tan \theta} = \frac{7000}{\tan \theta}.$$

Since θ is small (only 4°), we can approximate $\tan \theta$ by the radian measure of θ:

$$\tan \theta \simeq 4\left(\frac{2\pi}{360}\right) = \frac{\pi}{45} = 0.06981.$$

Therefore

$$AB \simeq \frac{7000}{0.06981} = 100{,}300 \text{ ft.}$$

We note that this approximate answer is almost the same as the precise result obtained when this example was considered in Section 5.1.

The small-angle approximations can be re-expressed in terms of limits. The first of these is the most basic, and is given in the following theorem.

THEOREM 5.2.1

$$\lim_{\theta \to 0} \frac{\sin \theta}{\theta} = 1.$$

PROOF   Let θ be any angle between 0 and $\pi/2$, and let AOB be a sector of a circle of radius r subtending an angle θ at the center (Fig. 5.10). Also in the figure, AD is the tangent to the circle at A, and AC is the perpendicular from A to OB.

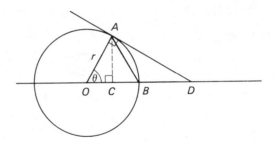

Figure 5.10

The area of the sector OAB is equal to $\frac{1}{2}r^2\theta$ (θ must be measured in radians). The area of the triangle OAB is equal to $\frac{1}{2}(OB)(AC)$ (half the base times the height). Since $OB = r$ and $AC = OA \sin \theta = r \sin \theta$, the area of the triangle is $\frac{1}{2}r^2 \sin \theta$.

The area of the triangle OAB is clearly smaller than the area of the sector OAB, and so

$$\tfrac{1}{2}r^2 \sin \theta < \tfrac{1}{2}r^2\theta.$$

Dividing through by $\frac{1}{2}r^2\theta$ we obtain the result that

$$\frac{\sin \theta}{\theta} < 1.$$

The inequality is true for any angle θ lying between 0 and $\pi/2$.

Now consider the triangle OAD. Because this triangle has a right angle at A, its area is $\frac{1}{2}(OA)(AD)$, which equals $\frac{1}{2}r^2 \tan \theta$ (since $OA = r$ and $AD = OA$

$\tan\theta$). But clearly the area of triangle OAD is greater than the area of the sector OAB; therefore

$$\tfrac{1}{2}r^2\tan\theta > \tfrac{1}{2}r^2\theta.$$

Dividing through by $\tfrac{1}{2}r^2\theta$ we obtain that

$$\frac{\tan\theta}{\theta} > 1.$$

Multiplying both sides by $\cos\theta$ we get

$$\frac{\sin\theta}{\theta} > \cos\theta.$$

If we combine the two inequalities that we have just obtained, the result is that

$$\cos\theta < \frac{\sin\theta}{\theta} < 1.$$

Now let θ approach zero. As it does so, $\cos\theta$ becomes closer and closer to 1. Therefore $(\sin\theta)/\theta$ is squeezed between 1 and $\cos\theta$, with $\cos\theta$ approaching 1 as $\theta \to 0$. It follows that $(\sin\theta)/\theta$ must also approach 1 as $\theta \to 0$, and so we have proved that

$$\lim_{\theta\to 0+}\frac{\sin\theta}{\theta} = 1.$$

In order to complete the proof of the theorem, we must consider negative values of θ and also show that

$$\lim_{\theta\to 0-}\frac{\sin\theta}{\theta} = 1.$$

The proof of this second part is quite similar to that given above and we leave it as an exercise for the student.

Two further important limits are evaluated in the next two examples.

EXAMPLE   Show that $\displaystyle\lim_{\theta\to 0}\frac{1-\cos\theta}{\theta^2} = \frac{1}{2}.$

SOLUTION   Multiplying the numerator and denominator by $(1 + \cos\theta)$, we obtain

$$\lim_{\theta\to 0}\frac{1-\cos\theta}{\theta^2} = \lim_{\theta\to 0}\frac{(1-\cos\theta)(1+\cos\theta)}{\theta^2(1+\cos\theta)}$$

$$= \lim_{\theta\to 0}\frac{1-\cos^2\theta}{\theta^2(1+\cos\theta)}$$

$$= \lim_{\theta\to 0}\frac{\sin^2\theta}{\theta^2(1+\cos\theta)}$$

$$= \lim_{\theta\to 0}\left[\left(\frac{\sin\theta}{\theta}\right)^2 \cdot \frac{1}{1+\cos\theta}\right].$$

Now as $\theta \to 0$, $(\sin \theta)/\theta$ approaches the limiting value 1 and $\cos \theta$ also approaches 1. Therefore

$$\lim_{\theta \to 0} \frac{1 - \cos \theta}{\theta^2} = (1)^2 \cdot \frac{1}{1+1} = \frac{1}{2}.$$

EXAMPLE   Show that $\lim_{\theta \to 0} \dfrac{1 - \cos \theta}{\theta} = 0$.

(Note that this limit differs from the previous one in having θ in the denominator instead of θ^2.)

SOLUTION $\lim_{\theta \to 0} \dfrac{1 - \cos \theta}{\theta^2} = \lim_{\theta \to 0} \dfrac{1 - \cos \theta}{\theta^2} \theta$,

$$= (\tfrac{1}{2})(0) = 0,$$

since the first factor, $(1 - \cos \theta)/\theta^2$, approaches the limiting value $\frac{1}{2}$ while the second factor, θ, approaches zero.

The basic limit established in Theorem 5.2.1 can be used to evaluate more complex limits involving the trigonometric functions. The following examples illustrate some of the procedures involved.

EXAMPLE   Show that $\lim_{x \to 0} \left(\dfrac{\sin px}{qx} \right) = \dfrac{p}{q}$.

SOLUTION   Let us denote px by θ. Then

$$\frac{\sin px}{qx} = \frac{\sin \theta}{q \left(\dfrac{\theta}{p} \right)} = \frac{p}{q} \cdot \frac{\sin \theta}{\theta}.$$

As $x \to 0$, $\theta \to 0$. Therefore

$$\lim_{x \to 0} \frac{\sin px}{qx} = \lim_{\theta \to 0} \frac{p}{q} \cdot \frac{\sin \theta}{\theta}$$

$$= \frac{p}{q} \lim_{\theta \to 0} \frac{\sin \theta}{\theta} = \frac{p}{q}(1) = \frac{p}{q}.$$

EXAMPLE   Find $\lim_{x \to 0} \dfrac{\sin x}{1 + \cos x}$.

SOLUTION   The limit in this example can be evaluated simply by substituting $x = 0$ into the function involved, since upon substitution we do not obtain an indefinite form for the answer. Thus

$$\lim_{x \to 0} \frac{\sin x}{1 + \cos x} = \frac{\sin (0)}{1 + \cos(0)} = \frac{0}{1+1} = 0.$$

EXERCISES 5.2

1. $\lim\limits_{\theta \to \pi/2} \dfrac{\sin \theta}{\theta}$.

2. $\lim\limits_{\theta \to \pi/3} \left(\dfrac{1 - \cos \theta}{\theta^2}\right)$.

3. $\lim\limits_{x \to \pi/2} \dfrac{x}{\cos x}$.

4. $\lim\limits_{\theta \to 0} \dfrac{1 + 2 \sin \theta + \sin^2 \theta}{\sqrt{4 + \cos^2 \theta}}$.

5. $\lim\limits_{\theta \to 2\pi} \operatorname{cosec} \theta$.

6. $\lim\limits_{x \to 0} \dfrac{\sin 4x}{3x}$.

7. $\lim\limits_{x \to 0} \dfrac{x}{\tan x}$.

8. $\lim\limits_{x \to 0} \dfrac{1 - \cos 2x}{x}$.

9. $\lim\limits_{x \to 0} \dfrac{1 - \cos 2x}{x^2}$.

10. $\lim\limits_{x \to 0} \dfrac{1 - \cos x}{\sin^2 x}$.

11. $\lim\limits_{x \to 0} \dfrac{(1 - \cos x)^2}{\tan^2 x}$.

12. $\lim\limits_{x \to 0} \left(\dfrac{\tan^2 x}{1 - \cos x}\right)$.

13. $\lim\limits_{x \to 0} \dfrac{\sin x}{1 - \cos x}$.

14. $\lim\limits_{x \to 0} \dfrac{\sec x - \tan 2x}{\sec x - \sin 2x}$.

15. A person stands on the edge of a cliff 200 ft high, looking out to sea. How far can he see? (radius of the earth $= 2.09 \times 10^7$ ft)

16. A ship's funnels are 75 ft above the waterline. How far out to sea is it when it disappears over the horizon of an observer standing on the shore?

5.3 DERIVATIVES OF TRIGONOMETRIC FUNCTIONS

In this section we shall obtain formulas for the derivatives of the six trigonometric functions. We shall begin by proving the following very important results for the derivatives of the sine and cosine.

THEOREM 5.3.1

$$\frac{d}{dx}(\sin x) = \cos x$$

$$\frac{d}{dx}(\cos x) = -\sin x.$$

PROOF

In order to prove the first result set $y = \sin x$. Then by definition

$$\frac{dy}{dx} = \lim_{\Delta x \to 0} \frac{\Delta y}{\Delta x} = \lim_{\Delta x \to 0} \frac{\sin (x + \Delta x) - \sin x}{\Delta x}.$$

239

Using the formula for the sine of a sum of two quantities,

$$\sin(x + \Delta x) = \sin x \cos \Delta x + \cos x \sin \Delta x,$$

$$\frac{dy}{dx} = \lim_{\Delta x \to 0} \frac{\sin x \cos \Delta x + \cos x \sin \Delta x - \sin x}{\Delta x}$$

$$= \lim_{\Delta x \to 0} \left[\frac{\cos x \sin \Delta x}{\Delta x} + \frac{\sin x (\cos \Delta x - 1)}{\Delta x} \right]$$

$$= \cos x \lim_{\Delta x \to 0} \frac{\sin \Delta x}{\Delta x} + \sin x \lim_{\Delta x \to 0} \frac{\cos \Delta x - 1}{\Delta x}.$$

The two limits in this expression correspond to results established in the preceding section:

$$\lim_{\Delta x \to 0} \frac{\sin \Delta x}{\Delta x} = 1 \quad \text{and} \quad \lim_{\Delta x \to 0} \left[\frac{1 - \cos \Delta x}{\Delta x} \right] = 0.$$

Therefore

$$\frac{dy}{dx} = \cos x(1) + \sin x(0) = \cos x,$$

as required.

The proof of the second result follows a similar line of reasoning. In this case set $y = \cos x$. Then

$$\frac{dy}{dx} = \lim_{\Delta x \to 0} \frac{\Delta y}{\Delta x} = \lim_{\Delta x \to 0} \frac{\cos(x + \Delta x) - \cos x}{\Delta x}$$

$$= \lim_{\Delta x \to 0} \frac{\cos x \cos \Delta x - \sin x \sin \Delta x - \cos x}{\Delta x}$$

$$= \lim_{\Delta x \to 0} \left[\cos x \frac{\cos \Delta x - 1}{\Delta x} - \sin x \frac{\sin \Delta x}{\Delta x} \right]$$

$$= \cos x(0) - \sin x(1)$$

$$= -\sin x.$$

Again the two standard limits from the last section have been used.

EXAMPLE   What is the slope of the graph of $y = \sin x$ when

(a) $x = 0$;   (b) $x = \pi/3$;

(c) $x = \pi/2$;   (d) $x = \pi$?

SOLUTION   For a general x, the slope of $y = \sin x$ is given by y', where $y' = \cos x$ in this case.

(a)   For $x = 0$, slope $= \cos(0) = 1$.

(b)   For $x = \pi/3$, slope $= \cos(\pi/3) = \frac{1}{2}$.

(c)   For $x = \pi/2$, slope $= \cos(\pi/2) = 0$.

(d)   For $x = \pi$, slope $= \cos(\pi) = -1$.

EXAMPLE   Find dy/dx when $y = \cos 3x$.

SOLUTION   We can regard y as a composite function by writing $y = \cos u$ and $u = 3x$. Then

$$\frac{dy}{du} = \frac{d}{du}(\cos u) = -\sin u.$$

$$\frac{du}{dx} = \frac{d}{dx}(3x) = 3.$$

Then by the chain rule,

$$\frac{dy}{dx} = \frac{dy}{du} \cdot \frac{du}{dx} = (-\sin u)(3) = -3 \sin 3x.$$

In general, let *inside* stand for any differentiable function of x. Then from the chain rule we can obtain the derivatives of composite functions of the form $\cos$ (*inside*) and $\sin$ (*inside*):

If $y = \sin$ (*inside*)   then $\dfrac{dy}{dx} = \cos$ (*inside*) $\cdot \dfrac{d}{dx}$ (*inside*).

If $y = \cos$ (*inside*)   then $\dfrac{dy}{dx} = -\sin$ (*inside*) $\cdot \dfrac{d}{dx}$ (*inside*).

EXAMPLE   Find dy/dx when $y = \sin(x^2 + 1)$.

SOLUTION   Here *inside* $= (x^2 + 1)$.

$$\frac{dy}{dx} = \cos(x^2 + 1)\frac{d}{dx}(x^2 + 1)$$

$$= 2x \cos(x^2 + 1).$$

EXAMPLE   Find dy/dx when $y = x^2 \sin x$.

SOLUTION   We can write y as a product; $y = uv$ where $u = x^2$ and $v = \sin x$. From the product rule, therefore,

$$\frac{dy}{dx} = \frac{d}{dx}(x^2) \cdot \sin x + x^2 \frac{d}{dx}(\sin x)$$

$$= 2x \sin x + x^2 \cos x.$$

EXAMPLE   Find dy/dx when $y = (\cos 3x)/(\sin x)$.

SOLUTION   From the quotient rule

$$\frac{dy}{dx} = \frac{\sin x \dfrac{d}{dx}(\cos 3x) - \cos 3x \dfrac{d}{dx}(\sin x)}{(\sin x)^2}$$

$$= \frac{\sin x (-3 \sin 3x) - \cos 3x (\cos x)}{\sin^2 x}$$

$$= \frac{-3 \sin x \sin 3x - \cos x \cos 3x}{\sin^2 x}.$$

241

The derivatives of the other four trigonometric functions can be obtained by expressing these functions in terms of sine and cosine. The results are contained in the following theorem.

THEOREM 5.3.2

$$\frac{d}{dx}(\tan x) = \sec^2 x.$$

$$\frac{d}{dx}(\cot x) = -\operatorname{cosec}^2 x.$$

$$\frac{d}{dx}(\sec x) = \sec x \tan x.$$

$$\frac{d}{dx}(\operatorname{cosec} x) = -\cot x \operatorname{cosec} x.$$

The first of these results is very useful, while the other three are of less common use.

PROOF

Setting $y = \tan x$, we can express y as a quotient, $y = u/v$, where $u = \sin x$, $v = \cos x$. From the quotient formula, therefore,

$$\frac{dy}{dx} = \frac{\cos x \dfrac{d}{dx}(\sin x) - \sin x \dfrac{d}{dx}(\cos x)}{(\cos x)^2}$$

$$= \frac{(\cos x)(\cos x) - (\sin x)(-\sin x)}{(\cos x)^2}$$

$$= \frac{\cos^2 x + \sin^2 x}{\cos^2 x}$$

$$= \frac{1}{\cos^2 x} = \sec^2 x.$$

In this derivation we have used the fact that, for any x, $\cos^2 x + \sin^2 x = 1$.

The proof of the formula for the derivative of $\cot x$ follows in a quite similar way, and we leave it as an exercise for the student.

In the case of the secant function, $y = \sec x = 1/(\cos x)$, we can again express $y = u/v$, where in this case $u = 1$ and $v = \cos x$. So, from the quotient formula,

$$y' = \frac{\cos x \dfrac{d}{dx}(1) - (1)\dfrac{d}{dx}(\cos x)}{(\cos x)^2}$$

$$= \frac{(0)(\cos x) - (1)(-\sin x)}{(\cos x)^2}$$

$$= \frac{\sin x}{\cos^2 x}$$

$$= \frac{\sin x}{\cos x} \cdot \frac{1}{\cos x}$$

$$= \tan x \sec x.$$

The formula for the derivative of cosec x is established in a similar way, and again we shall leave it as an exercise.

EXAMPLE   Find dy/dx when $y = \tan \sqrt{x}$.

SOLUTION   We can write y as a composite function, $y = \tan u$, where $u = x^{1/2}$. Then

$$\frac{dy}{du} = \sec^2 u, \qquad \frac{du}{dx} = \frac{1}{2} x^{-1/2},$$

and by the chain rule,

$$\frac{dy}{dx} = \frac{dy}{du} \cdot \frac{du}{dx} = (\sec^2 u)\left(\frac{1}{2} x^{-1/2}\right)$$

$$= \frac{1}{2\sqrt{x}} \sec^2 \sqrt{x}.$$

Many problems of optimization involve the differentiation of trigonometric functions. This is often true in cases where the problem contains some element of a geometrical nature. We shall give an example of this type.

EXAMPLE   It is desired to construct a telephone link between two points A and B, which are situated on opposite banks of a river whose width is 1 km. A lies 2 km upstream from B. It costs twice as much to lay a length of cable underwater as it does to lay the same length overland. At what angle should the cable cross the river in order to minimize the total cost of laying it?

SOLUTION   Let the telephone line cross the river from A to a point C making an angle x with the direct route across (Fig. 5.11). Then from the trigonometry of the triangle ACD, noting that $AD = 1$ km, we see that $AC = AD \sec x = \sec x$ and $CD = AD \tan x = \tan x$. Therefore $BC = BD - CD = 2 - \tan x$.

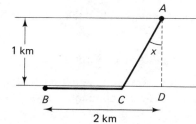

Figure 5.11

If the segment BC costs c dollars per km, then AC costs $2c$ dollars per km (twice as much for the underwater section). So the total cost, which we denote by "$", is given by

$$\$ = c(BC) + 2c(AC)$$

$$= c(2 - \tan x) + 2c (\sec x)$$

$$= c[2 - \tan x + 2 \sec x].$$

We wish to calculate the value of x that makes the total cost, $, a minimum. Thus we must set $d\$/dx = 0$:

$$\frac{d\$}{dx} = c[-\sec^2 x + 2 \sec x \tan x] = 0,$$

i.e.,

$$2 \sec x \tan x = \sec^2 x.$$

Multiplying this equation by $\cos^2 x$ (noting that $\sec x \times \cos x = 1$ and $\tan x \times \cos x = \sin x$), we obtain

$$2 \sin x = 1$$
$$\sin x = \tfrac{1}{2}.$$

Thus $x = \pi/6$ (i.e., $30°$). The minimum cost is achieved by laying the line across the river at an angle of $30°$ to the direct route across.

Let us now examine a number of problems concerned with the motion of leaping animals, for example, a motion such as the jump of a frog or grasshopper, the spring of a cat, the hop of a kangaroo, the leap of a fish, and so on (but not, for example, the motion of a flying squirrel for which aerodynamic effects are important).

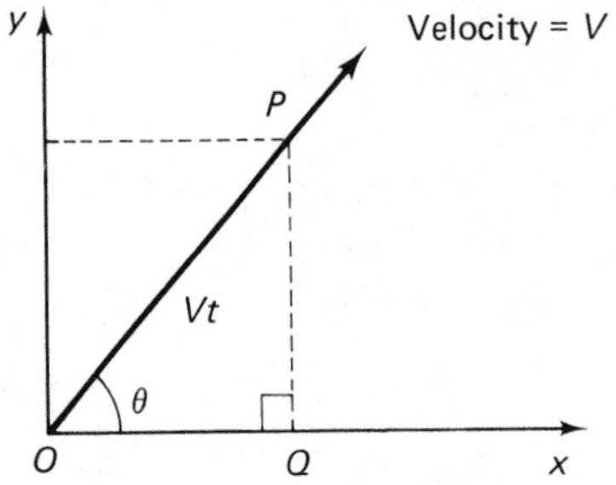

Figure 5.12

We take the x-axis to be horizontal and the y-axis to point vertically upwards with the origin at the point from which the animal leaps. We suppose that it launches itself with a velocity **V** ft/sec in a direction that makes an angle θ with the horizontal x-axis (Fig. 5.12).

If there were no force of gravity, the animal would continue to move in the same direction with velocity **V**. After t seconds it would reach a point P such that the distance $OP = \mathbf{V}t$. In the triangle OPQ, $OQ = OP \cos \theta = \mathbf{V}t \cos \theta$ and $PQ = OP \sin \theta = \mathbf{V}t \sin \theta$. Therefore, if there were no gravity, the coordinates of the animal at time t sec would be

$$x = \mathbf{V}t \cos \theta, \qquad y = \mathbf{V}t \sin \theta.$$

We have seen earlier that the effect of gravity is to make a body fall through a vertical distance of $s = 16t^2$ ft in t sec. This amount must be subtracted from the above value of y in order to include the effect of gravity. Thus the actual position of the animal at time t is given by

$$x = \mathbf{V}t \cos \theta, \qquad y = \mathbf{V}t \sin \theta - 16t^2.$$

We observe that these two equations are parametric equations of the path of the animal, with the time t being the parameter. We can eliminate t and obtain the equation of the path in terms of x and y only. For, from the first equation,

$$t = \frac{x}{V \cos \theta},$$

and substituting this value of t into the equation for y we obtain

$$y = \mathbf{V} \sin \theta \left(\frac{x}{V \cos \theta} \right) - 16 \left(\frac{x}{V \cos \theta} \right)^2$$

$$= x \tan \theta - \frac{16x^2}{V^2 \cos^2 \theta}.$$

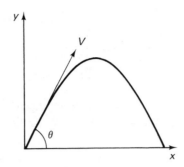

Figure 5.13

The graph of this function is actually a parabola, and has the form shown in Fig. 5.13.

EXAMPLE A porpoise leaps from the water with a velocity of 20 ft/sec in a direction making an angle of 60° with the horizontal. How high above the surface of the water does his path reach?

SOLUTION Here $\mathbf{V} = 20$ and $\theta = 60°$, so that $\cos \theta = \frac{1}{2}$ and $\tan \theta = \sqrt{3}$. Therefore the equation of the path is

$$y = \sqrt{3}\,x - \frac{16x^2}{20^2(\frac{1}{2})^2} = \sqrt{3}\,x - \frac{4}{25}x^2.$$

At the highest point on the path $dy/dx = 0$. Therefore,

$$\frac{dy}{dx} = \sqrt{3} - \frac{4}{25}(2x) = 0;$$

that is,

$$x = \frac{25\sqrt{3}}{8}.$$

To obtain the height we must calculate the value of y when x takes this value:

$$y = \sqrt{3}\left(\frac{25\sqrt{3}}{8}\right) - \frac{4}{25}\left(\frac{25\sqrt{3}}{8}\right)^2$$

$$= \frac{75}{8} - \frac{75}{16} = \frac{75}{16} = 4.7.$$

The porpoise therefore leaps 4.7 ft out of the water.

EXAMPLE A certain grasshopper can jump with maximum velocity of 10 ft/sec. What is the farthest horizontal distance he can jump?

SOLUTION Here $\mathbf{V} = 10$, and we must find the value of θ such that the horizontal distance is maximum.

When the grasshopper lands, we have $y = 0$, so that

$$y = x \tan \theta - \frac{16x^2}{\mathbf{V}^2 \cos^2 \theta} = 0.$$

This equation has two roots. One of them is $x = 0$; this corresponds to the point at which the grasshopper launches itself. The second root is the value of x at the point where it lands, and is

$$x = \frac{V^2 \cos^2 \theta}{16} \cdot \tan \theta = \frac{V^2}{16} \sin \theta \cos \theta.$$

This equation gives the horizontal distance traveled, and we must maximize it, that is, we must set $dx/d\theta = 0$. Using the product rule,

$$\frac{dx}{d\theta} = \frac{V^2}{16} \left[\cos \theta \frac{d}{d\theta} (\sin \theta) + \sin \theta \frac{d}{d\theta} (\cos \theta) \right]$$

$$= \frac{V^2}{16} [\cos \theta \cdot \cos \theta + \sin \theta (-\sin \theta)]$$

$$= \frac{V^2}{16} (\cos^2 \theta - \sin^2 \theta).$$

Therefore $dx/d\theta = 0$ when $\cos^2 \theta - \sin^2 \theta = 0$. Dividing through by $\cos^2 \theta$, we obtain $1 - \tan^2 \theta = 0$, and so $\tan \theta = \pm 1$. Since θ must lie between 0 and 90°, we must take $\tan \theta = +1$, or $\theta = 45°$. We see therefore that the maximum distance is traveled when the grasshopper launches itself at an angle of 45°.

When $\theta = 45°$, $\sin \theta = \cos \theta = 1/\sqrt{2}$, and so the distance traveled is

$$x = \frac{V^2}{16} \left(\frac{1}{\sqrt{2}} \right) \left(\frac{1}{\sqrt{2}} \right) = \frac{V^2}{32}.$$

Substituting $V = 10$, we get $x = 10^2/32 = 3.1$. So the farthest the grasshopper can jump is 3.1 ft.

We leave it as an exercise for the student to verify that the critical number $\theta = 45°$ does indeed correspond to a maximum value of x.

EXERCISES 5.3

Find dy/dx for the following functions:

1. $y = \sin 4x.$

2. $y = \tan 2x.$

3. $y = x \cos x.$

4. $y = x^{-1} \sec x.$

5. $y = \sin x \cos x.$

6. $y = \tan x \cot x.$

7. $y = \sin^3 x|\cos^5 x.$

8. $y = \sin^2 x \tan^4 x.$

9. $y = \dfrac{\sin x}{1 + \cos x}.$

10. $y = \dfrac{\sec x}{1 + \sec x}.$

—11. $y = \dfrac{\sqrt{x}}{1 - \tan x}.$

12. $y = \dfrac{\cot x}{2x + 3}.$

13. $y = \operatorname{cosec}^2 x.$

14. $y = \sqrt{\tan x}.$

15. $y = e^{a \sin x}$.

16. $y = xe^{\sin x}$.

17. $y = \cos (\ln x)$

18. $y = \sin (\cos x)$.

19. $y = \ln (\cos x)$.

20. $y = \ln (\sin^2 x)$.

21. $y = \ln (x \sin x)$.

22. $y = \ln \left(\dfrac{\sin x}{1 + \cos x} \right)$.

Find $d^2 y/dx^2$ for the following functions:

23. $y = 3 \cos x$.

24. $y = \tan x$.

25. $y = e^{\sin x}$.

26. $y = \ln (\sin x)$.

Find dy/dx and $d^2 y/dx^2$ for the following pairs of parametric equations:

27. $x = \sin 2\theta$, $y = \cos \theta$.

28. $x = a(t - \sin t)$, $y = a(1 - \cos t)$.

29. If $y = A \cos kx + B \sin kx$, where A, B, and k are constants, show that

$$\frac{d^2 y}{dx^2} + k^2 y = 0.$$

30. If $y = e^{-pt}(A \cos kt + B \sin kt)$, where A, B, p, and k are constants, show that

$$\frac{d^2 y}{dt^2} + 2p \frac{dy}{dt} + (p^2 + k^2)y = 0.$$

31. A population of bacteria is growing in such a way that its weight after a time t hr is given by $w = e^t(2 - \cos t)$. Find the rate of growth of the population at $t = 0$ and at $t = \pi/2$.

32. During the course of a year the average proportion p of the population suffering from influenza is given by

$$p = A(\sin 2\pi t + a \cos 4\pi t) + B,$$

where A, B, and a are constants ($A > 0$) and the time variable t increases from 0 to 1 during 1 yr. Show that if $|a| < \frac{1}{4}$, p has only one maximum and one minimum, and find the values of t at which they occur. Show that the maximum value of p is $A(1 - a) + B$, and find the minimum value.

33. A frog can leap with a velocity of 8 ft/sec. What is the maximum height it can reach?

34. A fish can leap from the water with a velocity of 10 ft/sec; however the angle at which it can leap cannot exceed 60°. What is the maximum height it can reach?

35. A cat can spring with a velocity of 14 ft/sec. What is the maximum horizontal distance it can travel in one bound?

36. The velocity with which a kangaroo can jump depends on the angle at which he leaves the ground. Assume that $V = a(1 + \sin \theta)$, where a is a constant. If the kangaroo travels by continuous hopping, show that his maximum speed over the ground is achieved if he hops each time at an angle of 30° to the horizontal. Calculate the maximum speed he can achieve.

5.4 THE INVERSE TRIGONOMETRIC FUNCTIONS

In Section 3.2 we introduced the notion of the inverse of a function. Our aim in the present section is to apply this idea to the three main trigonometric functions, sine, cosine, and tangent. Before doing so, however, we shall briefly review the definition of the inverse of a function.

If f is any function, its inverse f^{-1} is defined in the following way. Let y be any value in the range of f and let x be the value in the domain of f such that $y = f(x)$. Then we write $x = f^{-1}(y)$. In other words, if the equation $y = f(x)$ is solved for x, and x is expressed as a function of y, then we obtain the inverse function f^{-1}.

The graph of the function $x = f^{-1}(y)$ is the same as the graph of the function $y = f(x)$. However for the function $x = f^{-1}(y)$, the y-axis is taken in the horizontal direction and the x-axis is taken to be vertical, since y is now regarded as the independent variable and x is regarded as a function of y.

As an illustration of these remarks, consider the function $y = f(x) = 2 - \frac{1}{4}x^3$, whose graph is shown in Fig. 5.14. Solving for x in terms of y, we get

$$\frac{1}{4}x^3 = 2 - y$$
$$x = (8 - 4y)^{1/3} = f^{-1}(y).$$

The graph of $x = f^{-1}(y)$ is shown in Fig. 5.15, with the y-axis drawn horizontally and

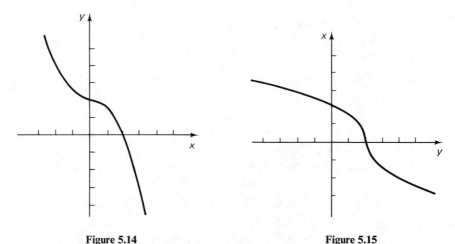

Figure 5.14   Figure 5.15

the x-axis vertically. The set of points (x, y) is the same for the two graphs, but the axes are drawn differently.

It may happen that, corresponding to a particular value of y in the range of f, there are more than one value of x in the domain such that $y = f(x)$. In such a case, the inverse function is not well-defined. The simplest example of this is the function $f(x) = x^2$. Here, for any $y > 0$, there are two values of x such that $y = f(x)$, given by $x = \pm\sqrt{y}$. Therefore the inverse function f^{-1} is not defined in this case since we do not know whether $f^{-1}(y)$ would mean the positive square root or the negative square root of y.

248

We can get around this difficulty by placing a suitable restriction on the values of x. For example, we could restrict the allowed values of x to the nonnegative real numbers. If we do so, then there is only one value of x such that $y = x^2$, namely $x = \sqrt{y}$; the positive square root. Note that by restricting the values of x we are in fact placing a restriction on the domain of the original function f. So by restricting the domain of the function $f(x) = x^2$ to the nonnegative real numbers, we ensure that an inverse function is well-defined.

This is in fact an example of a general procedure in cases where there are more than one value of x such that $y = f(x)$. A restriction is placed on the domain of f in such a way that only one of these values of x lies within the restricted domain. This allows us to define a unique inverse function, $x = f^{-1}(y)$.

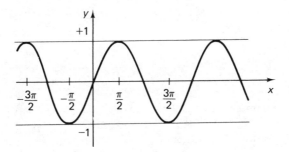

Figure 5.16

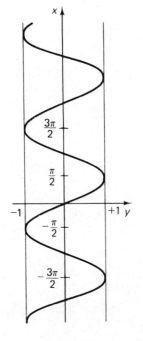

Figure 5.17

Now let us consider the function $f(x) = \sin x$, whose graph is shown in Fig. 5.16. The range of $f(x)$ is the interval $-1 \le y \le 1$, and the domain is the set of all real numbers. In Fig. 5.17, the same graph has been plotted with the y-axis horizontal and the x-axis vertical. This graph clearly does not define a function, since any vertical line intersects the graph in more than one point (in an infinite number of points in fact).

In other words, for any value of y ($-1 \le y \le 1$) there are infinitely many values of x such that $y = \sin x$. For example, if we take $y = 0$, x can have any of the values $\{\ldots, -2\pi, -\pi, 0, \pi, 2\pi, \ldots\}$.

If we want to define a unique inverse function for $y = \sin x$, we must place a restriction on the domain. What we do is to restrict x to the interval $-\pi/2 \le x \le \pi/2$. Within this range, there is one and only one value of x such that $y = \sin x$. This value of x is written $x = \text{Sin}^{-1} y$ (note the capital S) and is called the *principal value of the inverse sine* of y. The graph of $x = \text{Sin}^{-1} y$ is shown in

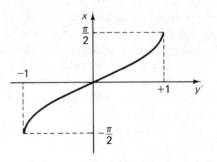

Figure 5.18

Fig. 5.18. The domain of the function Sin⁻¹ is the set $\{y \mid -1 \leq y \leq 1\}$, and the range is the set $\{x \mid -\pi/2 \leq x \leq \pi/2\}$.

EXAMPLE   Evaluate $\text{Sin}^{-1}\left(\frac{1}{2}\right)$.

SOLUTION   Let $x = \text{Sin}^{-1}\left(\frac{1}{2}\right)$. This means that $\sin x = \frac{1}{2}$ or, in other words, x is the angle whose sine is equal to $\frac{1}{2}$. There are many angles whose sine is equal to $\frac{1}{2}$, for example, $\pi/6$ (30°), $5\pi/6$ (150°), $-7\pi/6$ ($-210°$), $-11\pi/6$ ($-330°$), and so on. In order to find the principal inverse sine we must select the one of these angles that lies between $-\pi/2$ and $+\pi/2$. Therefore

$$\text{Sin}^{-1}\left(\frac{1}{2}\right) = \frac{\pi}{6}.$$

EXAMPLE   Evaluate $\text{Sin}^{-1}(-1)$.

SOLUTION   We require the angle whose sine is equal to -1. Since the principal inverse sine must lie in the interval $-\pi/2 \leq \text{Sin}^{-1} y \leq \pi/2$, we take the result

$$\text{Sin}^{-1}(-1) = -\frac{\pi}{2}.$$

EXAMPLE   If $y = \sin\left(\frac{3\pi}{4}\right)$, what is $\text{Sin}^{-1} y$?

SOLUTION   It is a temptation to say $\text{Sin}^{-1} y = 3\pi/4$. However this would not be correct since $3\pi/4$ does not lie between $-\pi/2$ and $+\pi/2$. Since $y = \sin(3\pi/4) = 1/\sqrt{2}$, the quantity required is $\text{Sin}^{-1}(1/\sqrt{2})$. The answer is $\pi/4$.

The inverse cosine function is quite similar to the inverse sine. Figure 5.19 shows the graph of $y = \cos x$ and Fig. 5.20 shows the same graph with the directions of the x- and y-axes interchanged. As with the sine function, the second graph does not define a function, since there are many values of x corresponding to each y in the interval $-1 \leq y \leq 1$.

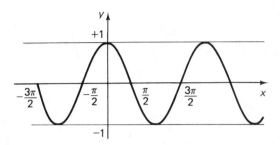

Figure 5.19

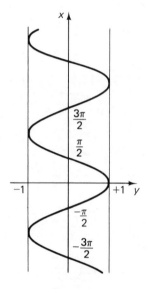

Figure 5.20

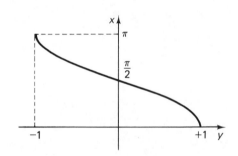

Figure 5.21

Thus in order to obtain a well-defined inverse cosine we must place a restriction on the domain of the original cosine function. We restrict x to the interval $0 \le x \le \pi$ so that the graph of the inverse function consists of the segment drawn in Fig. 5.21. With this restriction we obtain the so-called *principal value of the inverse cosine*, which is denoted by $x = \text{Cos}^{-1} y$.

The domain of the function Cos^{-1} is the set $\{y \,|\, -1 \le y \le 1\}$, and the range is the set $\{x \,|\, 0 \le x \le \pi\}$.

EXAMPLE   Evaluate $\text{Cos}^{-1}(0)$ and $\text{Cos}^{-1}(-1)$.

SOLUTION   The values of Cos^{-1} must always lie between 0 and π inclusive. Hence

$$\text{Cos}^{-1}(0) = \frac{\pi}{2} \quad \text{and} \quad \text{Cos}^{-1}(-1) = \pi (\text{not } -\pi).$$

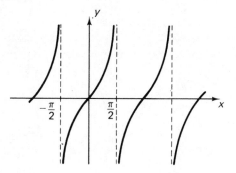

Figure 5.22

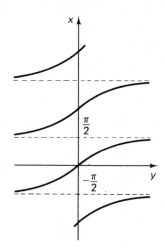

Figure 5.23

For the tangent function we obtain the graphs shown in Figs. 5.22 and 5.23. Figure 5.22 shows the graph of $y = \tan x$ with the axes drawn in the usual way, while Fig. 5.23 shows the same graph with the y-axis horizontal and the x-axis vertical. Again we have the same problem in Fig. 5.23, that a vertical line inter

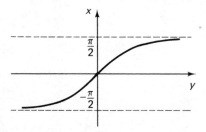

Figure 5.24

sects the graph in an infinite number of points, and so the graph does not define a function.

The restriction placed on x is $-\pi/2 < x < \pi/2$. With this restriction, we obtain the principal value of the inverse tangent, which is written $x = \text{Tan}^{-1} y$. Its graph is shown in Fig. 5.24.

The domain of the function Tan^{-1} is the set of all real numbers and the range is the set $\{x \,|\, -\pi/2 < x < \pi/2\}$.

EXAMPLE   Evaluate $\text{Tan}^{-1}(-1)$ and $\text{Tan}^{-1}(\tan 2\pi/3)$.

$$\mathrm{Tan}^{-1}\,(-1) = -\frac{\pi}{4}$$

$$\tan\left(\frac{2\pi}{3}\right) = -\sqrt{3}, \quad \text{so that}$$

$$\mathrm{Tan}^{-1}\left(\tan\frac{2\pi}{3}\right) = \mathrm{Tan}^{-1}\,(-\sqrt{3}) = -\frac{\pi}{3}.$$

In the remainder of this section we shall find expressions of the derivatives of the three inverse trigonometric functions discussed above.

First of all, consider $y = \mathrm{Sin}^{-1}\,x$. (Note that x is again being used to denote the independent variable and y to denote the dependent one.) Its graph is shown in Fig. 5.25.

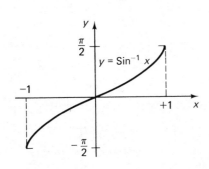

If $y = \mathrm{Sin}^{-1}\,x$, then $x = \sin y$. We can use implicit differentiation to differentiate this relation with respect to x. The result is

$$\frac{d}{dx}(x) = \frac{d}{dx}(\sin y), \quad \text{i.e.,} \quad 1 = \cos y \cdot \frac{dy}{dx},$$

and so

Figure 5.25

$$\frac{dy}{dx} = \frac{1}{\cos y}.$$

We should like dy/dx to be expressed as a function of x rather than y. From the trigonometric identity $\cos^2 y + \sin^2 y = 1$, it follows that

$$\cos y = \pm\sqrt{1 - \sin^2 y} = \pm\sqrt{1 - x^2},$$

since $x = \sin y$. Hence

$$\frac{dy}{dx} = \pm\frac{1}{\sqrt{1 - x^2}}.$$

Which sign should we take? It is clear from the above graph that the principal inverse sine function always has positive slope (except at $x = \pm 1$, where the slope is undefined). Hence we must take the positive sign:

$$\frac{dy}{dx} = \frac{1}{\sqrt{1 - x^2}}$$

or

$$\boxed{\frac{d}{dx}(\mathrm{Sin}^{-1}\,x) = \frac{1}{\sqrt{1 - x^2}}.}$$

Another way in which we can see that the positive sign is the right one is as follows. Since $y = \mathrm{Sin}^{-1}\,x$ is the principal inverse sine, y must lie between $-\pi/2$ and $+\pi/2$. Therefore $\cos y > 0$. Consequently $\cos y = +\sqrt{1 - x^2}$, the positive square root being taken.

For the function $y = \text{Cos}^{-1} x$, $x = \cos y$. Hence, by implicit differentiation,

$$1 = -\sin y \frac{dy}{dx}, \qquad \frac{dy}{dx} = -\frac{1}{\sin y}.$$

But

$$\sin y = \pm\sqrt{1 - \cos^2 y} = \pm\sqrt{1 - x^2},$$

so that

$$\frac{dy}{dx} = \pm\frac{1}{\sqrt{1 - x^2}}.$$

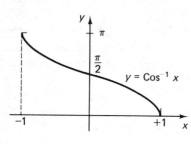

In this case the graph of $y = \text{Cos}^{-1} x$ (Fig. 5.26) has negative slope, and so the negative sign must be for the derivative:

$$\frac{dy}{dx} = \frac{-1}{\sqrt{1 - x^2}}$$

or

Figure 5.26

$$\frac{d}{dx}(\text{Cos}^{-1} x) = \frac{-1}{\sqrt{1 - x^2}}.$$

(Note also that since $y = \text{Cos}^{-1} x$, y lies between 0 and π. Hence $\sin y \geq 0$, so we must take the positive square root when we write $\sin y = \sqrt{1 - x^2}$.)

Finally consider $y = \text{Tan}^{-1} x$. Then $x = \tan y$, and by implicit differentiation

$$1 = \sec^2 y \frac{dy}{dx}, \qquad \frac{dy}{dx} = \frac{1}{\sec^2 y}.$$

But $\sec^2 y = 1 + \tan^2 y = 1 + x^2$ since $\tan y = x$. Therefore

$$\frac{dy}{dx} = \frac{1}{1 + x^2}$$

or

$$\frac{d}{dx}(\text{Tan}^{-1} x) = \frac{1}{1 + x^2}.$$

EXAMPLE   Evaluate dy/dx when $y = \text{Sin}^{-1}(3x)$.

SOLUTION   Write y as a composite function, $y = \text{Sin}^{-1}(u)$, where $u = 3x$. Then

$$\frac{dy}{du} = \frac{1}{\sqrt{1 - u^2}}, \qquad \frac{du}{dx} = 3.$$

By the chain rule,

$$\frac{dy}{dx} = \frac{dy}{du} \cdot \frac{du}{dx} = \frac{1}{\sqrt{1 - u^2}} \, (3)$$

$$= \frac{3}{\sqrt{1 - 9x^2}}.$$

In the last step we substituted $u = 3x$ under the square root.

EXAMPLE   Find $(d/dx)[x \operatorname{Tan}^{-1} \sqrt{x}]$.

SOLUTION   Setting $y = x \operatorname{Tan}^{-1} \sqrt{x}$ we first observe that y can be written in the form of a product, $y = u(x)v(x)$, where $u(x) = x$ and $v(x) = \operatorname{Tan}^{-1} \sqrt{x}$. In order to make use of the product formula we need to evaluate u' and v'. The first of these is easy: $u'(x) = 1$. We note that $v(x)$ is a composite function: $v = \operatorname{Tan}^{-1} w$ where $w = \sqrt{x}$. Then

$$\frac{dv}{dw} = \frac{1}{1 + w^2}, \qquad \frac{dw}{dx} = \frac{1}{2\sqrt{x}}.$$

So by the chain rule,

$$\frac{dv}{dx} = \frac{dv}{dw} \cdot \frac{dw}{dx} = \frac{1}{1 + w^2} \cdot \frac{1}{2\sqrt{x}}.$$

Substituting $w = \sqrt{x}$, so that $1 + w^2 = 1 + x$, we arrive at

$$\frac{dv}{dx} = \frac{1}{2\sqrt{x}\,(1 + x)}.$$

Now, using the product formula,

$$\frac{dy}{dx} = u\frac{dv}{dx} + \frac{du}{dx}v$$

$$= (x)\frac{1}{2\sqrt{x}\,(1 + x)} + (1) \operatorname{Tan}^{-1} \sqrt{x}$$

$$= \frac{\sqrt{x}}{2(1 + x)} + \operatorname{Tan}^{-1}\sqrt{x}.$$

In this section we have dealt with the inverses of the sine, cosine, and tangent functions and their derivatives. It is also possible to construct inverses of the secant, cosecant, and cotangent functions in a way quite similar to that used here, and to obtain corresponding expressions for their derivatives.

EXERCISES 5.4

Evaluate the following:

1. $\operatorname{Sin}^{-1}(1)$.

2. $\operatorname{Sin}^{-1}\left(\dfrac{-1}{\sqrt{2}}\right)$.

3. $\operatorname{Sin}^{-1}\left[\sin\left(\dfrac{-3\pi}{2}\right)\right]$.

4. $\operatorname{Sin}^{-1}\left[\sin\left(\dfrac{5\pi}{6}\right)\right]$.

5. $\operatorname{Sin}^{-1}[\sin(-2)]$.

6. $\operatorname{Sin}^{-1}\left[\cos\left(\dfrac{3\pi}{4}\right)\right]$.

7. $\operatorname{Cos}^{-1}(1)$.

8. $\operatorname{Cos}^{-1}\left(\dfrac{-\sqrt{3}}{2}\right)$.

9. $\text{Cos}^{-1}\left(\dfrac{-1}{\sqrt{2}}\right).$

10. $\text{Cos}^{-1}\left[\cos\left(\dfrac{-\pi}{6}\right)\right].$

11. $\text{Cos}^{-1}\left[\cos\left(\dfrac{4\pi}{3}\right)\right].$

12. $\text{Cos}^{-1}\left[\sin\left(\dfrac{5\pi}{6}\right)\right].$

13. $\text{Tan}^{-1}(-1).$

14. $\text{Tan}^{-1}(\sqrt{3}).$

15. $\text{Tan}^{-1}[\tan(\pi)].$

16. $\text{Tan}^{-1}\left[\tan\left(\dfrac{-5\pi}{6}\right)\right].$

Differentiate the following functions:

17. $\text{Sin}^{-1}(7x).$

18. $\text{Cos}^{-1}(x^2).$

19. $\text{Tan}^{-1}(x+\sqrt{x}).$

20. $x\,\text{Sin}^{-1}(2x).$

21. $x^2\,\text{Tan}^{-1}(\tfrac{1}{2}x).$

22. $[\text{Sin}^{-1}x]^2.$

23. $\text{Sin}^{-1}\sqrt{x}.$

24. $\text{Tan}^{-1}(\tan x).$

25. $\sqrt{\text{Cos}^{-1}x}.$

26. $e^{\text{Tan}^{-1}x}.$

Calculate dy/dx and d^2y/dx^2 for the following functions given in parametric form:

27. $x = \text{Tan}^{-1}t, \quad y = \text{Sin}^{-1}t.$

28. $x = \text{Cos}^{-1}t, \quad y = \text{Sin}^{-1}t.$

29. During the course of an epidemic, the total number of people who have been infected up to time t is given by

$$N = A[\text{Tan}^{-1}k(t-t_0) + \text{Tan}^{-1}kt_0].$$

What is the limiting value of N as $t \to \infty$? The epidemic is said to be under control when N passes the point of inflection and dN/dt begins to decrease. At what value of t does this occur, and what is the corresponding value of N?

5.5   PERIODIC FUNCTIONS

It was pointed out in Section 5.1 that the graphs of the functions $y = \sin x$ and $y = \cos x$ show a cyclic pattern of behavior. The segment of the graph of either of the two functions lying between $x = 0$ and $x = 2\pi$ repeats itself continually as we move along the x-axis. The graph of $y = \sin x$, for example, for $2\pi \le x \le 4\pi$, is a precise copy of its graph for $0 \le x \le 2\pi$; the graph for $4\pi \le x \le 6\pi$ is a further copy of this same segment, and so on.

Functions that have this property of repeating themselves indefinitely are called *periodic functions*. A somewhat different example of a periodic function is shown in Fig. 5.27. Starting at $x = 0$, the function in this graph increases linearly with x until it reaches the value 1 when $x = 1$. Then it decreases linearly, reaching the value zero again when $x = 2$. This basic triangular pattern is then repeated indefinitely as x further increases. The function in this example is called a sawtooth function.

The *period* of a periodic function is the distance along the x-axis occupied by the basic segment of its graph, which determines the cyclic behavior. In the case of the

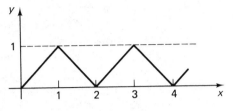

Figure 5.27

functions sin x and cos x, the segment between $x = 0$ and $x = 2\pi$ repeats indefinitely, and so these functions have a period equal to 2π. The sawtooth function in Fig. 5.27 has a period equal to 2. The function $y = \tan x$ is periodic with period equal to π, since the basic segment in the interval $-\pi/2 \leq x \leq \pi/2$ repeats cyclically.

EXAMPLE   What are the periods of the functions $y = \cos 3x$ and $y = 2 \cos 3x$?

SOLUTION   The graph of $y = \cos 3x$ is shown in Fig. 5.28. It is very similar to the graph of $y = \cos x$ except that it is squashed in the x-direction. When $x = \pi/3$, $y = \cos 3(\pi/3) = \cos \pi = -1$; when $x = 2\pi/3$, $y = \cos 3(2\pi/3) = \cos(2\pi) = 1$. Thus the function $y = \cos 3x$ makes one complete cycle from 1 to -1 and back to 1 again as x increases from 0 to $2\pi/3$. The period of this function is therefore $2\pi/3$.

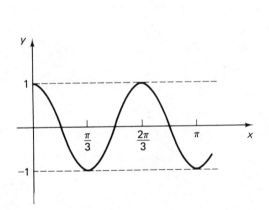

Figure 5.28

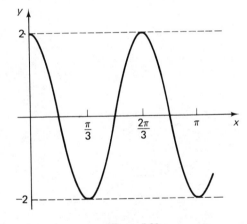

Figure 5.29

The function $y = 2 \cos 3x$ (Fig. 5.29) is similar to the preceding function except that instead of oscillating between $+1$ and -1, it oscillates between $+2$ and -2. Its period is still $2\pi/3$.

Applications of periodic function to cases of practical interest usually arise in situations where the independent variable x represents "time," and the function concerned relates to some cyclic or periodic process unfolding in time. In the biological sciences, there are many phenomena that show such cyclic behavior, generally called biological rhythms. For example, the reading of an electrocardiogram shows a periodic pattern, the period being the length of time between heartbeats; the hormone level in women follows the regular menstrual cycle, whose period is about one month; many animals show seasonal variations whose period is one year. In addition to examples such as these occurring within one biological organism there are examples of "social" phenomena that exhibit periodicity, for example, certain biological populations vary in a periodic manner, and the incidences of certain diseases show cyclic seasonal variations.

Processes in the physical sciences that show periodic behavior usually do so with a high degree of precision. For example, in the case of the pendulum of a clock, each swing duplicates the previous swing more or less exactly. In the case of biological phenomena, however, periodic behavior is usually not exact. Although each cycle may be closely similar to the preceding cycles, generally small variations occur from one cycle to the next that make the process only approximately periodic. Nevertheless it is often useful to idealize the process and to pretend that it is exactly periodic; by doing so we allow the use of certain powerful mathematical tools that are available for analyzing periodic functions.

One of the most useful of these tools is the so-called *Fourier series*, which is based on the trigonometric functions discussed in this chapter. Our aim in the remainder of this section is to provide a short introduction to this topic.

Consider the function $y = a \cos \omega t$, where a and ω are constants and the variable t is regarded as time. If we introduce a new variable $x = \omega t$, this function becomes $y = a \cos x$. On the xy-plane this function has the usual cosine form (Fig. 5.30), except

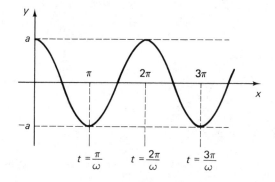

Figure 5.30

that it oscillates between $\pm a$ instead of between ± 1. When $x = 0$, $y = a$; when $x = \pi$, $y = -a$; when $x = 2\pi$, $y = a$ and so on. We call a the *amplitude* of y.

Now the original variable t is simply proportional to x: $t = x/\omega$. Thus when $x = 0$, $t = 0$ also; when $x = \pi$, $t = \pi/\omega$; and when $x = 2\pi$, $t = 2\pi/\omega$. But $x = 2\pi$ corresponds to exactly one basic cycle of the function, that is, to one period. Therefore, in terms of the t-variable, the period of the function $y = a \cos \omega t$ is $2\pi/\omega$.

We shall use T to denote the period: $T = 2\pi/\omega$. The constant ω is called the *angular frequency*.

A closely related function is $y = a \cos \omega(t - t_0)$, where t_0 is a third constant. The graph of this function (Fig. 5.31) is the same as that of $y = a \cos \omega t$ except that it is shifted along the t-axis by an amount t_0. So $y = a$ when $t = t_0$, $y = -a$ when $t = t_0 + (\pi/\omega,)$ $y = a$ when $t = t_0 + (2\pi/\omega,)$ and so on. The amplitude of this function is still a, and its period is still $T = 2\pi/\omega$.

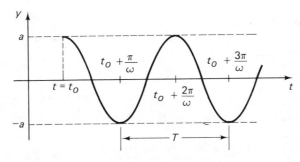

Figure 5.31

EXAMPLE

Let $t_0 = \pi/2\omega$. Then $\omega t_0 = \pi/2$, and the function becomes

$$y = a \cos(\omega t - \omega t_0)$$

$$= a \cos\left(\omega t - \frac{\pi}{2}\right)$$

$$= a \sin \omega t.$$

The graph of this function is shown in Fig. 5.32. The first maximum occurs at $t = t_0 = \pi/2\omega$, and the next maximum occurs exactly one period later, at $t = t_0 + T = (\pi/2\omega) + (2\pi/\omega) = (5\pi/2\omega)$.

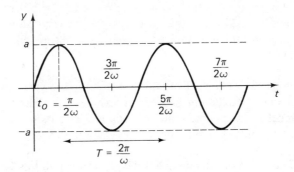

Figure 5.32

Using the difference formula for cosines, we can write the function $y = a \cos \omega(t - t_0)$ in the form

$$y = a \cos(\omega t - \omega t_0)$$

$$= a \cos \omega t \cos \omega t_0 + a \sin \omega t \sin \omega t_0.$$

Let us denote $a \cos \omega t_0 = A$ and $a \sin \omega t_0 = B$; both of these quantities are constants. Then y has the form

$$y = A \cos \omega t + B \sin \omega t.$$

This is an equivalent way of writing the function $y = a \cos \omega (t - t_0)$, and is sometimes more convenient.

EXAMPLE   When $t_0 = \pi/2\omega$,

$$A = a \cos \omega t_0 = a \cos (\pi/2) = 0,$$

$$B = a \sin \omega t_0 = a \sin (\pi/2) = a.$$

So

$$y = A \cos \omega t + B \sin \omega t = a \sin \omega t,$$

which is the same result as obtained in the preceding example.

EXAMPLE   What is the amplitude of the function $y = \cos \omega t + 2 \sin \omega t$?

SOLUTION   We note that

$$A^2 + B^2 = a^2 \cos^2 \omega t_0 + a^2 \sin^2 \omega t_0$$

$$= a^2 (\cos^2 \omega t_0 + \sin^2 \omega t_0)$$

$$= a^2(1) = a^2.$$

The amplitude is therefore given in terms of A and B by $a = \sqrt{A^2 + B^2}$. In this example, $A = 1$ and $B = 2$. Therefore

$$a = \sqrt{1^2 + 2^2} = \sqrt{5}.$$

The function $y = a \cos \omega (t - t_0)$ or, equivalently, $y = A \cos \omega t + B \sin \omega t$ is often termed a *sinusoidal* or a *simple harmonic* function with period $T = 2\pi/\omega$. Many processes that occur in practice can be represented by a function of this type by making an appropriate choice of the constants a, ω, and t_0. In particular, the constant ω would be chosen to give the right period for the process in question.

EXAMPLE   A certain biological variable y is observed to vary approximately in a sinusoidal manner, oscillating between the values $y = 1$ and $y = 2$ on a 24-hr cycle. It reaches its maximum at 3 o'clock P.M. and its minimum at 3 o'clock A.M. every day. Find a formula for y as a function of t.

SOLUTION   The graph of y against t has the form shown in Fig. 5.33. Here time t is measured in hours starting with $t = 0$ at midnight and increasing to $t = 24$ at the following midnight. The maximum of y then occurs at $t = 15$.

The first thing to note is that y does not oscillate between two values $\pm a$ but between 1 and 2. Therefore we should subtract the average 1.5 and consider the variable $(y - 1.5)$. This function (Fig. 5.34) oscillates between the values ± 0.5. Consequently we can take the amplitude $a = 0.5$.

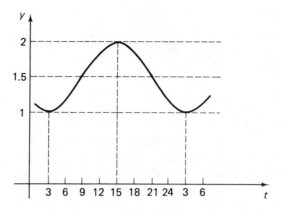

Figure 5.33

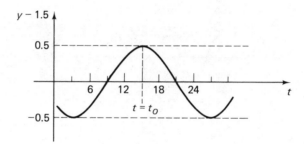

Figure 5.34

The period $T = 2\pi/\omega = 24$ hr. Therefore we must take $\omega = 2\pi/24 = \pi/12$. Finally, $t_0 = 15$, since t_0 equals the time at which the process achieves its maximum. Therefore

$$y - 1.5 = a \cos \omega (t - t_0)$$

$$= 0.5 \cos \frac{\pi}{12}(t - 15),$$

$$y = 1.5 + 0.5 \cos \frac{\pi}{12}(t - 15).$$

Not all periodic processes are sinusoidal. For example, the readout from an electrocardiogram is a much more complex periodic function. Figure 5.35 shows an

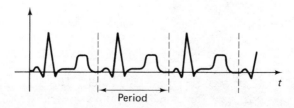

Period

Figure 5.35

idealized version of such a readout, which is clearly not sinusoidal, but nevertheless periodic. The sawtooth function discussed at the beginning of this section is another example of a periodic function that is not sinusoidal.

More complex periodic functions of this kind can be obtained by combining a number of different sinusoidal functions with angular frequencies that are multiples of the basic angular frequency ω. Let us begin by considering the example

$$y = \cos x + \tfrac{1}{2} \cos 2x.$$

The function $\cos x$ oscillates between ± 1 with a period 2π, while the function $\tfrac{1}{2} \cos 2x$ oscillates between $\pm\tfrac{1}{2}$ with a period π. The graphs of these two functions are shown in Fig. 5.36. If we consider the sum of these two functions, we obtain a function whose

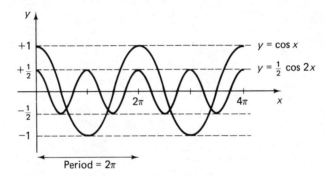

Figure 5.36

period is 2π and that has the graph shown in Fig. 5.37. So by combining two sinusoidal functions, $\cos x$ and $\tfrac{1}{2} \cos 2x$, we have obtained a new function that is still periodic but definitely not sinusoidal.

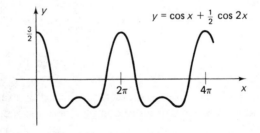

Figure 5.37

The reason why the new function has the period 2π is that $\tfrac{1}{2} \cos 2x$ can be regarded as having a period 2π instead of π since the segment of this function between 0 and 2π can be seen to repeat indefinitely. This segment, in fact, consists of two "basic" periods of the function. Thus both $\cos x$ and $\tfrac{1}{2} \cos 2x$ repeat periodically with period 2π; hence their sum must do the same.

As a second example consider the function

$$y = \cos x + \tfrac{1}{2} \sin 2x.$$

Figure 5.38 shows the graphs of $y = \cos x$ and $y = \tfrac{1}{2} \sin 2x$, while Fig. 5.39 shows the graph of the sum $y = \cos x + \tfrac{1}{2} \sin 2x$. Again both $\cos x$ and $\tfrac{1}{2} \sin 2x$ can be regarded as having a period 2π, even though $\tfrac{1}{2} \sin 2x$ has the basic period π. The result is that the sum function is periodic with period 2π.

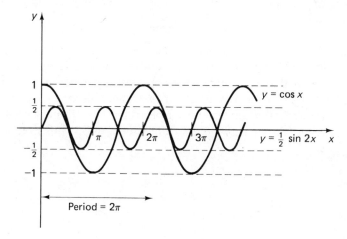

Figure 5.38

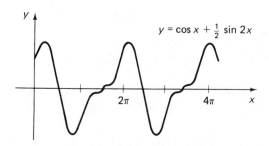

Figure 5.39

Again we see that, by taking a combination of trigonometric functions, a periodic function is obtained that is much more complex than a simple sinusoidal type of function.

Now let us generalize this idea. Let $a_0, a_1, \ldots, a_n$ and $b_1, b_2, \ldots, b_n$ be two given sets of real numbers, and consider the function

$$y = p(x) = a_0 + a_1 \cos x + b_1 \sin x + a_2 \cos 2x + b_2 \sin 2x + \ldots$$

$$+ a_n \cos nx + b_n \sin nx.$$

A function of this type is called a *trigonometric polynomial of degree n*.

No matter what the values of the coefficients $\{a_k\}$ and $\{b_k\}$, a trigonometric polynomial is always periodic with period 2π. This arises because each of the terms in the polynomial has period 2π. For example, the term $a_k \cos kx$ is periodic with a basic period equal to $2\pi/k$, but it can be regarded as also having a period of 2π. The interval of 2π contains exactly k basic periods of this function.

By making various choices for the coefficients $\{a_k\}$ and $\{b_k\}$ and for the degree n, trigonometric polynomials that have very complex graphs can be generated. Figure 5.40, for example, shows the graph of the trigonometric polynomial

$$y = \sin x + \tfrac{1}{3} \sin 3x + \tfrac{1}{5} \sin 5x + \tfrac{1}{7} \sin 7x,$$

whose degree is 7. In this polynomial the coefficients are as follows: $b_1 = 1, b_3 = \tfrac{1}{3}$, $b_5 = \tfrac{1}{5}, b_7 = \tfrac{1}{7}, b_2 = b_4 = b_6 = 0$, and all the coefficients $a_0, a_1, a_2, \ldots$ are zero. It can be seen that as x increases from zero, this function rises rapidly and makes a number of small oscillations about the value $\pi/4$, then as x passes through π it falls rapidly to negative values and makes a number of small oscillations about $-\pi/4$. As x passes through 2π the function rises again to positive values and the pattern repeats.

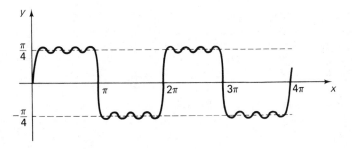

Figure 5.40

There is a fundamental theorem concerned with trigonometric polynomials that states the following: Let $f(x)$ be any function that is continuous, that is periodic with period 2π, and such that its derivative $f'(x)$ exists for all except a finite set of values of x (that is, the graph of $f(x)$ has a finite number of corners). Then it is possible to find a trigonometric polynomial $p(x)$ such that the difference $|f(x) - p(x)|$ is as small as desired for all x. That is, there exists a trigonometric polynomial $p(x)$ that is arbitrarily close to any continuous periodic function $f(x)$ whose graph is continuous and is smooth except at a finite number of points.

How do we find the coefficients $\{a_k\}$ and $\{b_k\}$ in this trigonometric polynomial $p(x)$ that approximates $f(x)$? The coefficients that must be used are called the *Fourier coefficients* of $f(x)$, and there are formulas that allow them to be calculated. However these formulas involve the use of integration, which we have not yet covered. They can be found in more advanced textbooks that deal with Fourier series in greater detail.

What are the periods of the following functions?

1. $y = 4 \sin 5x$.

2. $y = -3 \cos 2x$.

3. $y = \tan (3x - 1)$.

4. $y = \dfrac{\cos 7x}{2 + \cos 7x}$.

What are the amplitudes of the following functions?

5. $y = 3 \sin x$.

6. $y = 4 \sin (2t - 1)$.

7. $y = -\sin t + 3 \cos t$.

8. $y = 3 \sin 4t + 4 \cos 4t$.

9. $y = \sin 2t + \cos \left(2t - \dfrac{\pi}{4}\right)$.

10. $y = A \sin \left(t + \dfrac{\pi}{6}\right) + B \cos \left(t - \dfrac{\pi}{3}\right)$.

11. A variable y varies sinusoidally between the values -1 and $+3$ with a period of 2 sec. Taking $t = 0$ to coincide with an instant at which $y = 3$, obtain a formula for y as a function of t.

12. A biological variable y varies roughly sinusoidally with a period of 28 days, attaining its maximum at $t = 12$ (t being time measured in days). If the maximum and minimum values of y are 0.42 and 0.28, obtain a formula for y as a function of time.

13. During flight the wings of a bird oscillate up and down in a manner that is approximately sinusoidal. Let y be the angle that the wings make at time t with the upward vertical. Then, assuming that y varies sinusoidally with a period of 0.2 sec, the upper and lower positions being 70° respectively above and below the horizontal position, write down an expression for y as a function of t. (Take $t = 0$ to correspond to the horizontal position of the wings.)

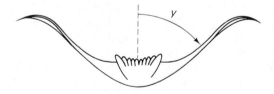

14. The respiratory cycle of a resting human being has a duration of approximately 5 sec. Assuming that the rate of flow y of air into the lungs varies sinusoidally as a function of time with a maximum amplitude of 0.5 liters/sec, express y as a function of t. Take $t = 0$ first to correspond to the instant when the lungs are empty at the end of expiration, and then repeat the question with $t = 0$ corresponding to the instant when the lungs are full at the end of inspiration.

What are the degrees and the shortest periods of the following trigonometric polynomials?

15. $y = 4 \cos x + \cos 3x$.

16. $y = \frac{1}{2} \sin x + \sin 3x - 2 \sin 5x$.

17. $y = \cos \pi t + \frac{1}{4} \cos 4\pi t$.

18. $y = 1 + \sin 2t + \cos 8t$.

19. Two biological oscillators produce sinusoidal outputs with the same amplitudes and angular frequencies, a and ω, but that are out of phase with one another. The outputs are respectively $y_1 = a \cos \omega t$ and $y_2 = a \cos (\omega t - \alpha)$, where α is a constant that specifies the degree to which the two outputs are out of phase. The two outputs are combined, giving a total input to the rest of the organism of $y = y_1 + y_2$. Show that y is still sinusoidal with angular frequency ω but with an amplitude given by $a\{2(1 + \cos \alpha)\}^{1/2} = 2a \cos (\alpha/2)$. Hence if $\alpha = 0$, the amplitude of the combined output is $2a$, but as α increases to π, this amplitude falls to zero.

REVIEW EXERCISES FOR CHAPTER 5

1. Are the following statements true or false? If false, give the corresponding correct statements.

   (a) The sine and cosine of any angle cannot be less than -1 or greater than $+1$ in value.

   (b) The sine, cosine, and tangent can be defined for all angles.

   (c) The relation $\sin^2 \theta + \cos^2 \theta = 1$ can only be proved for angles θ that lie between 0 and $\pi/2$.

   (d) $\sin (180° - x) = \sin x$ for all angles x.

   (e) $\cos \left(\dfrac{\pi}{2} + x \right) = \sin x$ for all angles x.

   (f) The function $y = \sin x$ is a periodic function with period 4π. Its shortest period is equal to 2π.

   (g) $\dfrac{d}{dx} \cos f(x) = -\sin f(x)$.

   (h) The values of $\text{Sin}^{-1} x$, $\text{Cos}^{-1} x$, and $\text{Tan}^{-1} x$ must lie between $-\pi/2$ and $+\pi/2$ (inclusive).

   (i) For any value of x in the interval $-1 < x < 1$, $\text{Cos}^{-1} x + \text{Sin}^{-1} x = \pi/2$.

   (j) $\text{Tan}^{-1} x = \text{Sin}^{-1} x / \text{Cos}^{-1} x$.

Calculate the following limits:

2. $\lim\limits_{x \to 0} \dfrac{\sin 2x}{\tan 3x}$.

3. $\lim\limits_{x \to 0} \dfrac{1 - \cos x}{x \sin x}$.

4. $\lim\limits_{x \to 0} \cot x \tan x$.

5. $\lim\limits_{x \to 0} x^2 \operatorname{cosec}^2 2x$.

6. $\lim\limits_{x \to \pi/2} \left(\dfrac{\pi}{2} - x \right) \tan x.$

7. $\lim\limits_{x \to \pi/3} \dfrac{\sin 3x}{\sin 9x}.$

Find dy/dx for the following functions:

8. $y = \sin x \sqrt{\cos x}.$

9. $y = \sin (\text{Sin}^{-1} x^2).$

10. $y = \text{Sin}^{-1} \sqrt{1 - x^2}.$

11. $y = \ln (\tan x).$

12. $y = \sqrt{1 + \tan^2 x}.$

13. $y = x \cot x.$

14. A frog can leap with a velocity of 8 ft/sec. What is the greatest horizontal distance it can leap?

15. For the kangaroo in exercise 36 of Section 5.3, at what angle must he jump in order to travel the greatest horizontal distance in a single bound?

Calculate the values of the following:

16. $\text{Sin}^{-1} \left[\sin \dfrac{5\pi}{2} \right].$

17. $\text{Cos}^{-1} \left[\cos \dfrac{-7\pi}{6} \right].$

18. $\text{Tan}^{-1} \left[\cot \dfrac{3\pi}{2} \right].$

19. A biological variable z has a period of 24 hr and varies sinusoidally between the values 5 and 9. Taking $t = 0$ to coincide with midnight, and measuring t in hours, find a formula expressing z as a function of t, given that $z = 5$ at 4 o'clock P.M.

20. If n is the average number of attacks of measles per 10,000 population in North America, then the approximate variation of $\log n$ throughout the year is given by the trigonometric polynomial

$$\log n = 1.5 + 0.7 \cos \left(\frac{\pi t}{6} \right) + 0.15 \sin \left(\frac{\pi t}{3} \right),$$

where t is measured in months and $t = 0$ is taken at the beginning of April. What are the period and degree of this polynomial? Sketch the graph of $\log n$ against t.

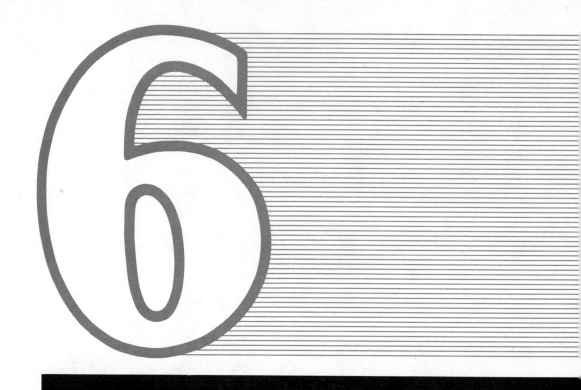

6

Integration

6.1 ANTIDERIVATIVES

So far in our study of calculus we have been concerned with the process of differentiation, that is, the calculation and use of the derivatives of functions. This aspect of the subject is called *differential calculus*. We shall now turn to the second area of study within the general area of calculus, called *integral calculus*, in which we are concerned with the opposite process to differentiation.

We saw before that, if $s(t)$ is the distance traveled in time t by a moving object, then the instantaneous velocity is $v(t) = s'(t)$, the derivative of $s(t)$. In order to calculate v, we simply differentiate $s(t)$. However it may happen that we already know the velocity function $v(t)$ and wish to calculate the distance s traveled. In such a situation, we know the derivative $s'(t)$ and need to find the function $s(t)$, a step opposite to that of differentiation. As another example, we may be concerned with a chemical reaction in which the rate at which a certain product is formed is a known function of time, and we may wish to calculate the amount formed up to a general time. Or we might know the rate $p'(t)$ at which a certain population is growing, and wish to calculate the population size at time t, $p(t)$.

The process of finding the function when its derivative is given is called *integration*, and the function is called the *integral* or *antiderivative* of the given derivative. Let $f(x)$ be the derivative of $F(x)$, so that $dF/dx = f(x)$. Then $F(x)$ is an integral or antiderivative of $f(x)$. We write this statement in the form

$$\int f(x)\, dx = F(x),$$

which is read as "the integral of $f(x)$, dx, is equal to $F(x)$." The function $f(x)$ to be integrated is called the *integrand*, and the symbol $\int$ is the "integral sign." The symbol

$$\int \ldots dx$$

stands for "integral, with respect to x, of . . . ." It is the inverse of the symbol

$$\frac{d}{dx} \cdots$$

which means "derivative, with respect to x, of . . . ." The integral sign and dx go together; the integral sign means the operation of integration and the dx specifies that the "variable of integration" is x. The integrand is always put between the integral sign and the differential of the variable of integration.

To evaluate $\int f(x)\, dx$, we just have to think of a function $F(x)$ whose derivative is $f(x)$. For example, to evaluate $\int (1/x)\, dx$, we think of a function whose derivative is $1/x$. Since

$$\frac{d}{dx}(\ln x) = \frac{1}{x},$$

we conclude that

$$\int \frac{1}{x}\, dx = \ln x.$$

However, it should be observed that this answer is not unique because the functions $(\ln x + 1)$, $(\ln x - 3)$, and $(\ln x + \frac{1}{2})$ all have derivative $1/x$. In fact, for any constant C, $\ln x + C$ is an integral of $1/x$. We write

$$\int \frac{1}{x}\, dx = \ln x + C.$$

The constant C, which can have any arbitrary value, is called the *constant of integration*.

This aspect is common to all antiderivatives, namely that they are not unique; any constant can be added to them without destroying their property of being the antiderivative of the given function. However this is the only ambiguity that there is: If $F(x)$ is any antiderivative of $f(x)$, then any other antiderivative of $f(x)$ differs from $F(x)$ only by a constant. (This result will be proved at the end of this section.) Therefore we can say that if $F'(x) = f(x)$, then the general antiderivative of $f(x)$ is given by

$$\int f(x)\, dx = F(x) + C,$$

where C is an arbitrary constant.

Since the constant of integration is arbitrary—that is, it can be any real number—the integral so obtained is given the more complete name, *indefinite integral*. Sometimes different methods of evaluating an integral can give different forms for the answer, but the student will soon realize that the two answers will always differ by a constant.

From the definition of the integral, it is clear that

$$\frac{d}{dx}\left[\int f(x)\, dx \right] = f(x);$$

the processes of differentiation and integration neutralize each other.

We shall establish a number of simple and standard formulas for integration. The first of these is known as the *power formula* and gives us the rule to integrate any power of x except the inverse of x. It is:

$$\int x^n\, dx = \frac{x^{n+1}}{n+1} + C \quad (n \neq -1).$$

Thus to integrate any power of x except the inverse first power, we must increase the power by 1, then divide by the new exponent, and finally add the arbitrary constant of integration.

This formula is obtained by reversing the corresponding formula for differentiation. We observe that

$$\frac{d}{dx}\left(\frac{x^{n+1}}{n+1} \right) = \frac{1}{n+1} \frac{d}{dx}(x^{n+1})$$

$$= \frac{1}{n+1}(n+1)x^n = x^n.$$

Therefore, since the derivative of $x^{n+1}/(n+1)$ is x^n, an antiderivative of x^n must be $x^{n+1}/(n+1)$. The general antiderivative is then obtained by adding the constant of integration.

EXAMPLES   (a) $\displaystyle \int x^3\,dx = \frac{x^{3+1}}{3+1} + C = \frac{x^4}{4} + C \quad (n = 3).$

(b) $\displaystyle \int \frac{1}{x^2}\,dx = \int x^{-2}\,dx = \frac{x^{-2+1}}{-2+1} + C = \frac{x^{-1}}{-1} + C$

$$= -\frac{1}{x} + C \quad (n = -2).$$

(c) $\displaystyle \int \frac{1}{\sqrt{t}}\,dt = \int t^{-1/2}\,dt = \frac{t^{-(1/2)+1}}{(-\frac{1}{2}+1)} + C$

$$= 2\sqrt{t} + C \quad (n = -\tfrac{1}{2}).$$

A number of formulas giving antiderivatives of simple functions are given in the following table of standard integrals. Each formula is stated twice, once using x as a variable and once using u. The purpose of this will become clear in section 6.2 when we introduce the method of substitutions.

TABLES OF STANDARD ELEMENTARY INTEGRALS

1. (a) $\displaystyle \int x^n\,dx = \frac{x^{n+1}}{n+1} + C \quad (n \neq -1).$ 1. (a) $\displaystyle \int u^n\,du = \frac{u^{n+1}}{n+1} + C \quad (n = -1).$

   (b) $\displaystyle \int \frac{1}{x}\,dx = \ln|x| + C.$ (b) $\displaystyle \int \frac{1}{u}\,du = \ln|u| + C.$

2. $\displaystyle \int e^x\,dx = e^x + C.$ 2. $\displaystyle \int e^u\,du = e^u + C.$

3. $\displaystyle \int \sin x\,dx = -\cos x + C.$ 3. $\displaystyle \int \sin u\,du = -\cos u + C.$

4. $\displaystyle \int \cos x\,dx = \sin x + C.$ 4. $\displaystyle \int \cos u\,du = \sin u + C.$

5. $\displaystyle \int \sec^2 x\,dx = \tan x + C.$ 5. $\displaystyle \int \sec^2 u\,du = \tan u + C.$

6. $\displaystyle \int \mathrm{cosec}^2 x\,dx = -\cot x + C.$ 6. $\displaystyle \int \mathrm{cosec}^2 u\,du = -\cot u + C.$

7. $\displaystyle \int \sec x \tan x\,dx = \sec x + C.$ 7. $\displaystyle \int \sec u \tan u\,du = \sec u + C.$

8. $\displaystyle \int \mathrm{cosec}\,x \cot x\,dx = -\mathrm{cosec}\,x + C.$ 8. $\displaystyle \int \mathrm{cosec}\,u \cot u\,du = -\mathrm{cosec}\,u + C.$

9. $\displaystyle \int \frac{1}{\sqrt{1-x^2}}\,dx = \mathrm{Sin}^{-1} x + C.$ 9. $\displaystyle \int \frac{1}{\sqrt{1-u^2}}\,du = \mathrm{Sin}^{-1} u + C.$

10. $\displaystyle \int \frac{1}{1+x^2}\,dx = \mathrm{Tan}^{-1} x + C.$ 10. $\displaystyle \int \frac{1}{1+u^2}\,du = \mathrm{Tan}^{-1} u + C.$

All of these results are obtained by simply reversing corresponding results for derivatives. Formula 1(b) requires some comment. For $x > 0$ this formula is straightforward, since then $|x| = x$, and we know that

$$\frac{d}{dx} \ln x = \frac{1}{x}.$$

Since $1/x$ is the derivative of $\ln x$, it follows that the antiderivative of $1/x$ must be $\ln x$ plus the constant of integration.

When $x < 0$, we have that $|x| = -x$. Therefore

$$\frac{d}{dx} \ln |x| = \frac{d}{dx} \ln (-x) = \frac{1}{(-x)}(-1) = \frac{1}{x},$$

where the chain rule has been used in carrying out the differentiation. Thus $1/x$ is the derivative of $\ln |x|$ for $x < 0$, as well as for $x > 0$. Therefore the antiderivative of $1/x$ must be $\ln |x| + C$, as given in the table.

Now we shall prove two theorems that will simplify the algebra of integration.

THEOREM 6.1.1

The integral of the product of a constant and a function of x is equal to the constant times the integral of the function. That is, if c is a constant,

$$\int c\, f(x)\, dx = c \int f(x)\, dx.$$

EXAMPLES   **(a)** $\displaystyle\int 3x^2\, dx = 3 \int x^2\, dx = 3\left(\frac{x^3}{3}\right) + C = x^3 + C.$

(b) $\displaystyle\int 2e^x\, dx = 2 \int e^x\, dx = 2e^x + C.$

(c) $\displaystyle\int 5\, dx = 5 \int 1\, dx = 5x + C.$

PROOF OF THEOREM 6.1.1

We have

$$\frac{d}{dx}\left[c \int f(x)\, dx \right] = c \frac{d}{dx}\left[\int f(x)\, dx \right] = c\, f(x).$$

Therefore $c\, f(x)$ is the derivative of $c \int f(x)\, dx$, and so from the definition of the antiderivative, it follows that $c \int f(x)\, dx$ must be the antiderivative of $c\, f(x)$. In other words,

$$\int c\, f(x)\, dx = c \int f(x)\, dx,$$

which proves the result.

From this theorem it follows that we can move any multiplicative constant across the integral sign. (*Caution:* Variables must not be moved across the integral sign.)

THEOREM 6.1.2

The integral of the sum of two functions is equal to the sum of their integrals; that is,

$$\int [f(x) + g(x)]\, dx = \int f(x)\, dx + \int g(x)\, dx.$$

NOTE. This result can be extended to the difference of two functions or an algebraic sum of any finite number of functions.

EXAMPLE Find the integral of $[x - (3/x)]^2$.

SOLUTION We want to evaluate $\int [x - (3/x)]^2\, dx$.

First we expand $[x - (3/x)]^2$ to express the integrand as a sum of power functions. Thus

$$\int \left(x - \frac{3}{x} \right)^2 dx = \int \left(x^2 - 6 + \frac{9}{x^2} \right) dx$$

$$= \int x^2\, dx - \int 6\, dx + \int 9x^{-2}\, dx$$

$$= \int x^2\, dx - 6 \int 1\, dx + 9 \int x^{-2}\, dx$$

$$= \frac{x^{2+1}}{2+1} - 6x + 9 \frac{x^{-2+1}}{-2+1} + C$$

$$= \frac{x^3}{3} - 6x - \frac{9}{x} + C.$$

EXAMPLE Find the antiderivative of $\dfrac{3 - 5t + 7t^2 + t^3}{t^2}$.

SOLUTION In this case we want to evaluate $\displaystyle\int \frac{3 - 5t + 7t^2 + t^3}{t^2}\, dt$.

$$\int \frac{3 - 5t + 7t^2 + t^3}{t^2}\, dt = \int \left(\frac{3}{t^2} - \frac{5}{t} + 7 + t \right) dt$$

$$= 3 \int t^{-2}\, dt - 5 \int \frac{1}{t}\, dt + 7 \int 1\, dt + \int t\, dt$$

$$= 3 \frac{t^{-2+1}}{-1} - 5 \ln|t| + 7t + \frac{t^{1+1}}{2} + C$$

$$= -\frac{3}{t} - 5 \ln|t| + 7t + \frac{t^2}{2} + C.$$

PROOF OF THEOREM 6.1.2

$$\frac{d}{dx} \left[\int f(x)\, dx + \int g(x)\, dx \right] = \frac{d}{dx} \left[\int f(x)\, dx \right] + \frac{d}{dx} \left[\int g(x)\, dx \right]$$

$$= f(x) + g(x).$$

273

Therefore $f(x) + g(x)$ is the derivative of $\int f(x)\,dx + \int g(x)\,dx$, and so by the definition of the antiderivative,

$$\int [f(x) + g(x)]\,dx = \int f(x)\,dx + \int g(x)\,dx.$$

EXAMPLE The rate of growth of a colony of fruit flies at time $t\,(t \geq 1)$ is equal to $10(t + 2)/t$. When $t = 1$ there are 20 flies in the colony. Calculate the number of flies at a general value of $t\,(t > 1)$.

SOLUTION Let the size of the colony at time t be denoted by $p(t)$. We are given that $p'(t) = 10(t + 2)/t$. But $p(t)$ is the antiderivative of $p'(t)$, and so

$$p(t) = \int \frac{10(t + 2)}{t}\,dt$$

$$= 10 \int \left(1 + \frac{2}{t}\right) dt \qquad = 10\left[\int 1\,dt + \int \frac{2}{t}\,dt\right.$$

$$= 10[t + 2 \ln |t| + C],$$

where C is the constant of integration. We are also given that when $t = 1$, $p = 20$. Therefore, setting $t = 1$, we obtain:

$$p(1) = 20 = 10[1 + 2 \ln 1 + C]$$

$$= 10(1 + C).$$

Therefore $1 + C = 2$, or $C = 1$. Consequently we can substitute this value of C into the expression for $p(t)$, obtaining the result

$$p(t) = 10[t + 2 \ln |t| + 1]$$

for the population size at time t.

EXAMPLE During daylight hours the velocity of a migrating goose is given by $v = 20 - t/3$ (mph), where t is time measured in hours starting with $t = 0$ at dawn. How many miles has the goose traveled up to time t? How far does the goose fly in a 12-hr day?

SOLUTION Let $s(t)$ be the distance traveled between dawn $(t = 0)$ and the time t. Then, of course, $s(0) = 0$. Also, the derivative $s'(t)$ equals the velocity so that

$$s'(t) = 20 - \frac{t}{3}.$$

Integrating, we find $s(t)$:

$$s(t) = \int (20 - \tfrac{1}{3}t)\,dt \qquad = 20\left(\int 1\,dt - \frac{1}{3}\int t\,dt\right.$$

$$= 20t - \tfrac{1}{3}(\tfrac{1}{2}t^2) + C.$$

In order to determine the value of C we set $t = 0$, since we know that $s(0) = 0$.

We find that

$$s(0) = 0 = 20(0) - \tfrac{1}{6}(0)^2 + C,$$

from which it follows that $C = 0$. Therefore

$$s(t) = 20t - \tfrac{1}{6}t^2,$$

which gives the required distance traveled up to time t.

To find the distance flown in a full day we set $t = 12$. We get

$$s(12) = 20(12) - \tfrac{1}{6}(12)^2 = 240 - 24 = 216.$$

The goose flies 216 miles in the 12-hr day.

It was stated earlier in this section that the only ambiguity in the antiderivative of a function consists of the arbitrary constant of integration. This result, which underlies much of the work in the above examples, is proved in the following theorem.

THEOREM 6.1.3

If $F(x)$ and $G(x)$ are antiderivatives of the same function $f(x)$, both defined on some interval $a < x < b$, then F and G differ by a constant on that interval.

PROOF   Since F and G are antiderivatives of f,

$$F'(x) = f(x) \quad \text{and} \quad G'(x) = g(x).$$

Let $H(x) = F(x) - G(x)$ be the difference between these two functions. Then

$$H'(x) = F'(x) - G'(x) = f(x) - f(x) = 0.$$

So the function H has a derivative that is identically zero. We shall prove from this that H must be a constant function, and this will complete the proof that F and G differ by a constant.

Let x_1 and x_2 be two arbitrary points in the given interval $a < x < b$. By the mean value theorem (Theorem 4.1.4), there exists a number c lying between x_1 and x_2 such that

$$H'(c) = \frac{H(x_1) - H(x_2)}{x_1 - x_2}.$$

But $H'(c) = 0$ since H' is identically zero in $a < x < b$. Consequently, $H(x_1) - H(x_2) = 0$, or $H(x_1) = H(x_2)$. Since x_1 and x_2 can be any two points in $a < x < b$, it follows that $H(x)$ has a constant value for all x in $a < x < b$.

EXERCISES 6.1

Write down the integrals of the following:

1. (a) x^7. $= \tfrac{1}{8}x^8 + c$

   (c) $\dfrac{1}{\sqrt{x}}$.

   (b) $\sqrt{x}$. $= \int x^{1/2} = \dfrac{x^{1.5}}{1.5} + C$

   (d) 7.

$\int \dfrac{1}{x^{1/2}} dx = \int x^{-1/2} dx = \dfrac{x^{1/2}}{1/2} + c$

$\int 7\, dx = 7\int 1\, dx = 7x$

2. $7x^2 - 3x + 8 - \dfrac{1}{\sqrt{x}} + \dfrac{1}{x} + \dfrac{1}{x^2}.$

3. $x^7 + 7x + \dfrac{x}{7} + \dfrac{7}{x}.$

4. $e^x + x^e + e + x.$

5. $x^2\left(x + \dfrac{2}{\sqrt{x}}\right).$

6. $\left(\sqrt{x} + \dfrac{1}{\sqrt{x}}\right)\left(x - \dfrac{2}{x}\right).$

Find the antiderivatives of the following functions with respect to the independent variable involved:

7. $4x^3 + 3x^2 + 2x + 1 + \dfrac{1}{x} + \dfrac{1}{x^3}.$

8. $3e^t - 5t^3 + 7 + \dfrac{3}{t}.$

9. $\sqrt{u}\,(u^2 + 3u + 7).$

10. $\dfrac{2y^3 + 7y^2 - 6y + 9}{3y}.$

11. $\sqrt{x}\,(x + 1)(2x - 1).$

12. $\dfrac{(t - t^2)^2}{t\sqrt{t}}.$

Evaluate the following integrals:

13. $\displaystyle\int \dfrac{1 + 3x + 7x^2 - 2x^3}{x^2}\,dx.$

14. $\displaystyle\int \dfrac{(2t + 1)^2}{3t}\,dt.$

15. $\displaystyle\int \left(3\theta^2 - 6\theta + \dfrac{9}{\theta} + 4e^\theta\right)d\theta.$

16. $\displaystyle\int (\sqrt{2}\,y + 1)^2\,dy.$

17. $\displaystyle\int \tan^2 x\,dx$

18. $\displaystyle\int \cot^2\theta\,d\theta.$

[Hint for exercises 17 and 18: Use $\tan^2\theta = \sec^2\theta - 1$, $\cot^2\theta = \operatorname{cosec}^2\theta - 1$.]

19. If the velocity of motion at time t is $v = (t + \sqrt{t}\,)^3$, find the distance traveled between the times $t = 0$ and $t = 4$.

20. During a chemical reaction the rate at which the end-product BS is produced at time $t(t \geq 1)$ is equal to $(\sqrt{t} + 1)/t^2$ mg/min. At the time $t = 1$ the reaction is started. How much BS is produced up to time t?

$t \geq 1$

6.2 METHOD OF SUBSTITUTION

Not all integrals, as they are given, can be evaluated directly by the use of standard integrals discussed in the previous section. Often the given integral can be reduced to a standard integral already known by a change of the variable of integration. Such a method is called the *method of substitution* and corresponds to the chain rule in differentiation.

First of all, we shall discuss linear substitutions in which we set a linear function of the original variable equal to a new variable. This is explained in the following theorem.

$$\text{If} \quad \int f(x)\, dx = F(x) + C, \quad \text{then}$$

$$\int f(ax + b)\, dx = \frac{1}{a} F(ax + b) + C,$$

$\frac{1}{a} \int x$

where a and b are any two constants ($a \neq 0$). In other words, in order to integrate $f(ax + b)$, we treat $(ax + b)$ as if it were a single variable, and then divide the resulting integral by a, the coefficient of x.

Before proving this theorem, let us illustrate it with some examples. First of all, we have seen in the last section that the rule for integrating a power of x is

$$\int x^n\, dx = \frac{x^{n+1}}{n + 1} + C \quad (n \neq -1).$$

This corresponds to setting $f(x) = x^n$ and $F(x) = x^{n+1}/(n + 1)$ in the statement of the theorem. Then, according to the theorem, we can replace the argument x by $ax + b$:

$$f(ax + b) = (ax + b)^n, \qquad F(ax + b) = \frac{(ax + b)^{n+1}}{(n + 1)}.$$

The theorem then gives the following result:

$$\int (ax + b)^n\, dx = \frac{1}{a} \frac{(ax + b)^{n+1}}{n + 1} + C.$$

For example, when $a = 2$, $b = 3$, and $n = 4$ we obtain the formula

$$\int (2x + 3)^4\, dx = \frac{1}{2} \frac{(2x + 3)^{4+1}}{4 + 1} + C$$

$$= \frac{(2x + 3)^5}{10} + C. \tag{i}$$

When $f(x) = 1/x$, $F(x) = \ln |x|$, since we have the standard formula

$$\int \frac{1}{x}\, dx = \ln |x| + C.$$

If x is replaced by $ax + b$, the two functions become

$$f(ax + b) = \frac{1}{ax + b}, \qquad F(ax + b) = \ln |ax + b|.$$

Then the theorem provides the following result:

$$\int f(ax + b)\, dx = \frac{1}{a} F(ax + b) + C,$$

i.e.,

$$\int \frac{1}{ax + b}\, dx = \frac{1}{a} \ln |ax + b| + C. \tag{ii}$$

For example, when $a = 3$ and $b = -2$, this formula becomes

$$\int \frac{1}{3x - 2} \, dx = \frac{1}{3} \ln|3x - 2| + C.$$

It is easy enough to verify that the results contained in these examples are correct. For example, in order to prove that the result (i) above is correct we must show that the derivative of the function $[(2x + 3)^5/10 + C]$, which appears on the right-hand side, is the given integrand. But from the chain rule,

$$\frac{d}{dx}\left[\frac{1}{10}(2x + 3)^5 + C\right] = \frac{1}{10} \cdot 5(2x + 3)^4 \frac{d}{dx}(2x + 3) + \frac{d}{dx}(C)$$

$$= \tfrac{1}{2}(2x + 3)^4(2) + 0 = (2x + 3)^4.$$

So we see that the required integrand is obtained; hence (i) is correct.

In the same way we can verify that the result (ii) is correct by showing that the derivative of the right-hand side is the integrand $1/(ax + b)$. Setting $u = ax + b$ and using the chain rule, we find that

$$\frac{d}{dx} \ln|ax + b| = \frac{d}{du} \ln|u| \cdot \frac{du}{dx} = \frac{1}{u}(a) = \frac{a}{ax + b}.$$

Therefore

$$\frac{d}{dx}\left[\frac{1}{a} \ln|ax + b| + C\right] = \frac{1}{ax + b},$$

from which it follows that

$$\int \frac{1}{ax + b} \, dx = \frac{1}{a} \ln|ax + b| + C.$$

The proof of the general theorem follows the same pattern as our proofs of the particular formulas in these examples.

PROOF OF THEOREM 6.2.1

Since $\int f(x) \, dx = F(x) + C$, it is necessary that

$$f(x) = \frac{dF}{dx}.$$

Therefore

$$\frac{d}{dx}\left[\frac{1}{a} F(ax + b)\right] = \frac{1}{a} \frac{d}{dx} F(ax + b)$$

$$= \frac{1}{a} \cdot \frac{d}{du} F(u) \cdot \frac{du}{dx},$$

where we have introduced the new variable $u = ax + b$ and made use of the chain rule. Since

$$\frac{d}{du} F(u) = f(u) = f(ax + b) \quad \text{and} \quad \frac{du}{dx} = a,$$

we have that

$$\frac{d}{dx}\left[\frac{1}{a}F(ax + b)\right] = f(ax + b).$$

Then, from the definition of the integral, it follows that $(1/a) F(ax + b)$ is the antiderivative of $f(ax + b)$, that is,

$$\int f(ax + b) \, dx = \frac{1}{a} F(ax + b) + C,$$

as required.

Theorem 6.2.1 is a powerful tool and can be used to generalize each integral in the table given in Section 6.1 by replacing x by $ax + b$ $(a \neq 0)$. Some of these integrals are given below, with the constants of integration omitted.

1.	$\int x^n \, dx = \frac{x^{n+1}}{n+1}$ $(n \neq -1)$	1.	$\int (ax + b)^n \, dx = \frac{1}{a} \cdot \frac{(ax + b)^{n+1}}{n+1}$ $(n \neq -1)$
	$= \ln\|x\|$ $(n = -1)$.		$= \frac{1}{a} \cdot \ln\|ax + b\|$ $(n = -1)$.
2.	$\int e^x \, dx = e^x$.	2.	$\int e^{ax+b} \, dx = \frac{e^{ax+b}}{a}$.
3.	$\int \sin x \, dx = -\cos x$.	3.	$\int \sin(ax + b) \, dx = \frac{-\cos(ax + b)}{a}$.
4.	$\int \sec^2 x \, dx = \tan x$.	4.	$\int \sec^2(ax + b) \, dx = \frac{\tan(ax + b)}{a}$.
5.	$\int \frac{1}{\sqrt{1 - x^2}} \, dx = \mathrm{Sin}^{-1} x$	5.	$\int \frac{1}{\sqrt{1 - (ax + b)^2}} \, dx = \frac{1}{a} \mathrm{Sin}^{-1}(ax + b)$.

It can be seen from this table that the integrals of $f(x)$ and $f(ax + b)$ are essentially similar in form.

EXAMPLE   Evaluate

$$\int (3x - 7)^5 \, dx.$$

SOLUTION   From the general result (1) in the above table,

$$\int (ax + b)^n \, dx = \frac{(ax + b)^{n+1}}{a(n + 1)} + C.$$

We must set $a = 3$, $b = -7$, and $n = 5$ in this general formula in order to evaluate the required integral:

$$\int (3x - 7)^5 \, dx = \frac{(3x - 7)^{5+1}}{(3)(5 + 1)} + C$$
$$= \tfrac{1}{18}(3x - 7)^6 + C.$$

EXAMPLE   Evaluate

$$\int \sin(5 - 3x) \, dx.$$

Setting $a = -3$ and $b = 5$ in formula (3) of the table we obtain

$$\int \sin (5 - 3x)\, dx = \frac{-\cos (5 - 3x)}{(-3)} + C$$

$$= \tfrac{1}{3} \cos (5 - 3x) + C.$$

We can look at this type of linear substitution from a different point of view. Let us write $u = ax + b$ in the integral $\int f(ax + b)\, dx$, and let us pretend that the dx that appears at the end of the integral can be treated as a differential. Then $du = a\, dx$, or $dx = (1/a)\, du$, and we get that

$$\int f(ax + b)\, dx = \int f(u) \cdot \frac{1}{a}\, du$$

$$= \frac{1}{a} \int f(u)\, du$$

$$= \frac{1}{a} F(u) + C.$$

$$= \frac{1}{a} F(ax + b) + C,$$

which is the correct answer. So in making a linear substitution, we get the right answer by treating the dx in the integral as if it were the differential of x. (We shall see later that this is also true for more general substitutions.)

Let us repeat the last two examples working from this new viewpoint.

EXAMPLE Evaluate

$$\int (3x - 7)^5\, dx.$$

SOLUTION Put $3x - 7 = u$, so that

$$3\, dx = du \quad \text{or} \quad dx = \tfrac{1}{3}\, du.$$

Then

$$\int (3x - 7)^5\, dx = \int u^5 \cdot \tfrac{1}{3}\, du$$

$$= \tfrac{1}{3} \int u^5\, du$$

$$= \frac{1}{3} \cdot \frac{u^6}{6} + C$$

$$= \tfrac{1}{18}(3x - 7)^6 + C$$

because $u = 3x - 7$.

EXAMPLE Evaluate

$$\int \sin (5 - 3x)\, dx.$$

SOLUTION   Put $5 - 3x = u$, so that

$$-3\, dx = du \quad \text{or} \quad dx = -\tfrac{1}{3}\, du.$$

Then

$$\int \sin (5 - 3x)\, dx = \int \sin u \cdot (-\tfrac{1}{3}\, du)$$

$$= -\tfrac{1}{3} \int \sin u\, du$$

$$= \tfrac{1}{3} \cos u + C$$

$$= \tfrac{1}{3} \cos (5 - 3x) + C$$

because $u = 5 - 3x$.

The general substitution method is stated in the following theorem.

THEOREM 6.2.2

If $F'(x) = f(x)$, then

$$\int f[g(x)]g'(x)\, dx = F[g(x)] + C$$

for any differentiable function $g(x)$ that is not a constant function.

Let us illustrate this theorem with a couple of examples before we prove it. Again let us start with the power formula,

$$\int x^n\, dx = \frac{x^{n+1}}{n+1} + C, \quad (n \neq -1),$$

which corresponds to setting $f(x) = x^n$ and $F(x) = x^{n+1}/(n+1)$. Then according to the theorem we must replace the argument x in these two functions by the function $g(x)$:

$$f[g(x)] = [g(x)]^n, \qquad F[g(x)] = \frac{[g(x)]^{n+1}}{(n+1)}.$$

The theorem then states in this particular case that

$$\int [g(x)]^n g'(x)\, dx = \frac{[g(x)]^{n+1}}{n+1} + C, \qquad (n \neq -1).$$

In this result, $g(x)$ can be any differentiabe function that is not constant. For example, let us take $g(x) = x^2 + 1$ and $n = 4$. Then $g'(x) = 2x$, and we obtain that

$$\int (x^2 + 1)^4 2x\, dx = \frac{(x^2 + 1)^{4+1}}{4+1} + C.$$

After division by 2, this becomes

$$\int (x^2 + 1)^4 x\, dx = \frac{(x^2 + 1)^5}{10} + C',$$

where $C' = C/2$. (Note that C' can still be any arbitrary constant, since dividing by 2 does not remove the arbitrariness.)

As a further example let us take $g(x) = \ln x$ and $n = 2$. Since then $g'(x) = 1/x$, we get the result

$$\int \frac{(\ln x)^2}{x}\, dx = \frac{(\ln x)^3}{3} + C.$$

It is clear that by choosing different functions $f(x)$ and $g(x)$, a great variety of different integrals can be evaluated. In practice, when using this substitution method to evaluate a given integral, it is necessary to spot how to choose these functions in such a way that the given integrand is expressed in the form $f[g(x)]g'(x)$, with f being a sufficiently simple function. We shall elaborate on this later, but first let us pause to prove the theorem.

PROOF OF THEOREM 6.2.2

Set $u = g(x)$. Then, from the chain rule,

$$\frac{d}{dx}F[g(x)] = \frac{d}{dx}F(u) = \frac{d}{du}F(u) \cdot \frac{du}{dx}$$
$$= f(u)\, g'(x)$$
$$= f[g(x)]g'(x).$$

Therefore, from the definition of the antiderivative, it follows that

$$\int f[g(x)]g'(x)\, dx = F[g(x)] + C,$$

as required.

Note that, as with the linear substitution, we would get the right answer if we simply substituted $u = g(x)$ into the given integral, pretending that dx can be treated as a differential. For $du = g'(x)\, dx$,

$$\int f[g(x)]g'(x)\, dx = \int f(u)\, du = F(u) + C,$$
$$= F[g(x)] + C.$$

When using the substitution method in practice, this is usually the easiest way in which to look upon it.

EXAMPLE   Evaluate

$$\int (x^2 + 3x - 7)^5(2x + 3)\, dx.$$

SOLUTION   We observe that the differential of $(x^2 + 3x - 7)$ is equal to $(2x + 3)\, dx$, which appears in the integral. Therefore we set $x^2 + 3x - 7 = u$. Then $(2x + 3)\, dx =$

du. Using this substitution the given integral reduces to

$$\int (x^2 + 3x - 7)^5(2x + 3)\, dx = \int u^5\, du$$

$$= \frac{u^6}{6} + C$$

$$= \tfrac{1}{6}(x^2 + 3x - 7)^6 + C.$$

where we have substituted back the value of *u*.

EXAMPLE   Evaluate

$$\int \frac{1}{x \ln x}\, dx.$$

SOLUTION   The given integral is

$$\int \frac{1}{x \ln x}\, dx = \int \frac{1}{\ln x} \cdot \frac{1}{x}\, dx.$$

[handwritten: u = ln x, du = 1/x dx]

Note that we have separated the integrand in such a way that the combination $(1/x)\, dx$ occurs as a distinct factor. This combination is the differential of ln *x*, and, moreover, the rest of the integrand is also a simple function of ln *x*. So we put ln *x* = *u*. Then $(1/x)\, dx = du$.

The given integral now reduces to

$$\int \frac{1}{x \ln x}\, dx = \int \frac{1}{\ln x} \cdot \frac{1}{x}\, dx$$

$$= \int \frac{1}{u} \cdot du$$

$$= \ln |u| + C$$

$$= \ln |\ln x| + C$$

after substituting *u* = ln *x*.

We observe from these examples that the appropriate technique in using the substitution method is to *look for a function* $u = g(x)$ *whose differential* $g'(x)\, dx$ *occurs in the original integral.* The choice of substitution is by nature ambiguous, but the student will soon learn from experience to spot the right one to make.

EXAMPLE   Evaluate

$$\int \sin^6 t \cos t\, dt.$$

SOLUTION   Clearly cos *t dt*, the differential of sin *t*, appears in the integral, and so we put sin *t* = *u*. Then cos *t dt* = *du*.

[handwritten: du = ... d/dt sin t × dt]

We observe that the remaining factor in the integrand, $\sin^6 t$, is also a simple function of u. Therefore

$$\int \sin^6 t \cos t \, dt = \int u^6 \, du$$

$$= \frac{u^7}{7} + C$$

$$= \tfrac{1}{7} \sin^7 t + C$$

because $u = \sin t$.

Sometimes the appropriate exact differential itself may not appear in the integral, but the function that does appear must be multiplied or divided by a certain constant. This is illustrated by the following examples.

EXAMPLE   Evaluate

$$\int (x + 1)\sqrt{x^2 + 2x + 7} \, dx.$$

SOLUTION   We have

$$\int \sqrt{x^2 + 2x + 7} \cdot (x + 1) \, dx.$$

Now $(x + 1) \, dx$ is not the exact differential of $x^2 + 2x + 7$. But if we multiply $(x + 1) \, dx$ by 2, we get $(2x + 2) \, dx$, which is now the exact differential of $x^2 + 2x + 7$. Therefore we multiply and divide the integrand by 2, obtaining

$$\int (x + 1)\sqrt{x^2 + 2x + 7} \, dx = \tfrac{1}{2} \int \sqrt{x^2 + 2x + 7} \cdot (2x + 2) \, dx,$$

and now put

$$x^2 + 2x + 7 = y, \quad \text{so that} \quad (2x + 2) \, dx = dy.$$

Then

$$\int \sqrt{x^2 + 2x + 7} \cdot (x + 1) \, dx = \tfrac{1}{2} \int (x^2 + 2x + 7)^{1/2} \cdot (2x + 2) \, dx$$

$$= \tfrac{1}{2} \int y^{1/2} \, dy$$

$$= \tfrac{1}{2} \frac{y^{(1/2+1)}}{(\tfrac{1}{2} + 1)} + C$$

$$= \tfrac{1}{3} y^{3/2} + C$$

$$= \tfrac{1}{3}(x^2 + 2x + 7)^{3/2} + C$$

because $y = x^2 + 2x + 7$.

EXAMPLE   Evaluate

$$\int e^{3 + 2 \tan y} \sec^2 y \, dy.$$

SOLUTION $\sec^2 y \, dy$ is the differential of $3 + 2 \tan y$ when adjusted by a suitable constant. Thus the differential of $3 + 2 \tan y$ appears in the integral and so we put $3 + 2 \tan y = u$. Then

$$2 \sec^2 y \, dy = du \quad \text{or} \quad \sec^2 y \, dy = \tfrac{1}{2} \, du.$$

$u = 3 + 2\tan y$

Therefore

$du = 2\sec^2 y \, dy$

$$\int e^{3+2\tan y} \sec^2 y \, dy = \int e^u \cdot \tfrac{1}{2} \, du$$

$$= \tfrac{1}{2} \int e^u \, du$$

$$= \tfrac{1}{2} e^u + C$$

$$= \tfrac{1}{2} e^{3+2\tan y} + C$$

because $u = 3 + 2 \tan y$.

EXERCISES 6.2

Make use of a linear substitution or direct use of Theorem 6.2.1 to evaluate the following integrals:

1. $\int (2x + 1)^7 \, dx.$

2. $\int \frac{1}{(3x - 1)^2} \, dx.$

3. $\int \frac{1}{2y - 1} \, dy.$

4. $\int \frac{1}{1 - 3t} \, dt.$

5. $\int e^{3x+2} \, dx.$

6. $\int e^{2-5x} \, dx.$

7. $\int \sin (2x + 1) \, dx.$

8. $\int \csc^2 (3 - 2u) \, du.$

9. $\int \frac{dt}{\sqrt{t + 1}}.$

10. $\int \frac{dx}{1 + (x - 1)^2}.$

Use an appropriate substitution to evaluate the following antiderivatives:

11. $\int (x^2 + 7x + 3)^4 (2x + 7) \, dx.$

12. $\int (x + 2)(x^2 + 4x + 2)^{10} \, dx.$

13. $\int \frac{2x + 3}{(x^2 + 3x + 1)^3} \, dx.$

14. $\int \frac{4x - 1}{(2x^2 - x + 1)} \, dx.$

15. $\int \frac{2x + 3}{x^2 + 3x + 1} \, dx.$

16. $\int \frac{x + 1}{x^2 + 2x - 1} \, dx.$

17. $\int t e^{t^2} \, dt.$

18. $\int \cos \theta \, e^{\sin \theta} \, d\theta.$

19. $\int \frac{e^{\sin^{-1} x}}{\sqrt{1 - x^2}} \, dx.$

20. $\int \frac{e^{\sqrt{t}}}{\sqrt{t}} \, dt.$

21. $\int \frac{1}{x} (\ln x)^3 \, dx.$

22. $\int \frac{\ln (x + 1)}{x + 1} \, dx.$

23. $\int \frac{1}{x(1 + \ln x)} \, dx.$

24. $\int \frac{1}{(x + 2) \ln (x + 2)} \, dx.$

25. $\int \frac{1}{x} \sin (\ln x) \, dx.$

26. $\int y \sin (y^2) \, dy.$

27. $\int x^2 \sec^2 (x^3) \, dx.$

28. $\int x^{n-1} \cos (x^n) \, dx.$

29. $\int \frac{e^x}{(1 + e^x)^2} \, dx.$

30. $\int (x + 2)\sqrt{x^2 + 4x + 1} \, dx.$

31. $\int \frac{3t^2 + 1}{t(t^2 + 1)} \, dt.$

32. $\int \frac{y}{\sqrt{1 + y^2}} \, dy.$

33. $\int \sin^3 2t \cos 2t \, dt.$

34. $\int \cos^3 \theta \sin \theta \, d\theta.$

35. $\int \tan^5 \theta \sec^2 \theta \, d\theta.$

36. $\int \cot^4 x \operatorname{cosec}^2 x \, dx.$

37. $\int \sec^3 x \tan x \, dx.$

38. $\int \operatorname{cosec}^5 x \cot x \, dx.$

39. $\int \frac{\sec^2 x \, dx}{1 + \tan x}.$

40. $\int \frac{\cos x}{\sin x} \, dx.$

41. $\int \tan x \, dx.$

42. $\int \frac{\cos x}{(1 + \sin x)^2} \, dx.$

6.3   TABLES OF INTEGRALS

In the previous section we introduced the method of substitution by means of which certain complex integrals can be reduced to one or other of the standard integrals listed in Section 6.1. Besides the substitutions mentioned in this section, there are others that are in common use; in particular, the trigonometric substitutions that we shall discuss in Section 6.5 are especially important. Besides the substitution method, there are other techniques that are useful when it comes to evaluating integrals, and certain of these will be discussed in later sections of this chapter.

In general, the evaluation of integrals requires considerable skill and often ingenuity. The variety of techniques available for the purpose is an indication of this fact. Moreover, it is not possible to give hard and fast rules as to which method or substitution will work in a given situation, but it is necessary to develop through experience an intuition for which method is likely to work best.

In the face of these difficulties, by far the most convenient way for the student to evaluate integrals is by use of a table of integrals. A table of integrals is merely a list of a large number of integrals together with their values. In order to evaluate a given integral, it is possible to extract the answer from the table, substituting the values of any constants as necessary. A number of such tables exist, some more complete than others; in Appendix II will be found a fairly short table of integrals, which is nevertheless sufficient to allow the evaluation of all integrals appearing in our examples and exercises. The integrals in this table are classified under certain headings to facilitate the use of the table. For example, all integrands involving a factor of the type $\sqrt{ax + b}$ are listed together, integrands involving trigonometric functions are listed together, as are those involving exponential functions, and so on.

EXAMPLE   Evaluate

$$\int \frac{1}{(4 - x^2)^{3/2}} \, dx.$$

SOLUTION   We must look through the table of integrals until we find an integral of the same form as that given above. The section entitled "Integrals Containing $\sqrt{a^2 - x^2}$" is the appropriate place to look, and in formula 37 we find

$$\int \frac{1}{(a^2 - x^2)^{3/2}} \, dx = \frac{1}{a^2} \frac{x}{\sqrt{a^2 - x^2}}.$$

This holds for any value of the constant a, so if we set $a = 2$ we obtain the required integral:

$$\int \frac{1}{(4 - x^2)^{3/2}} \, dx = \frac{1}{4} \frac{x}{\sqrt{4 - x^2}} + C$$

after adding in the constant of integration.

EXAMPLE   Evaluate

$$\int \frac{1}{2x^2 - 3x + 4} \, dx.$$

SOLUTION   If we compare this with the standard integral

$$\int \frac{1}{ax^2 + bx + c} \, dx,$$

which appears in the table of integrals, we have $a = 2$, $b = -3$, and $c = 4$. Therefore

$$b^2 - 4ac = (-3)^2 - 4(2)(4) = 9 - 32 = -23 < 0.$$

When $b^2 - 4ac < 0$ we have from the tables (formula 77) that

$$\int \frac{1}{ax^2 + bx + c} \, dx = \frac{2}{\sqrt{4ac - b^2}} \operatorname{Tan}^{-1} \left(\frac{2ax + b}{\sqrt{4ac - b^2}} \right).$$

Substituting the values of a, b, and c, therefore,

$$\int \frac{1}{2x^2 - 3x + 4} = \frac{2}{\sqrt{23}} \operatorname{Tan}^{-1} \left(\frac{4x - 3}{\sqrt{23}} \right) + C,$$

where C is the constant of integration, which should always be included.

Sometimes the use of tables is not quite so straightforward as this, and it may be necessary to use the tables two or more times in evaluating certain integrals. The following examples illustrate this point.

EXAMPLE   Evaluate

$$\int \frac{1}{x^2\sqrt{2-3x}}\, dx.$$

SOLUTION   If we look up the integrals involving $\sqrt{ax+b}$ in the table, we find that formula 23 states that

$$\int \frac{1}{x^n\sqrt{ax+b}}\, dx = -\frac{\sqrt{ax+b}}{(n-1)bx^{n-1}}$$
$$-\frac{(2n-3)a}{(2n-2)b}\int \frac{1}{x^{n-1}\sqrt{ax+b}}\, dx \quad (n\neq 1).$$

In our example, $n=2$, $a=-3$, and $b=2$. Therefore

$$\int \frac{1}{x^2\sqrt{2-3x}}\, dx = -\frac{\sqrt{2-3x}}{2x} + \frac{3}{4}\int \frac{1}{x\sqrt{2-3x}}\, dx. \tag{i}$$

To evaluate the integral on the right in (i), we again look up the table of integrals that involve $\sqrt{ax+b}$, and formula 22 gives

$$\int \frac{1}{x\sqrt{ax+b}}\, dx = \frac{1}{\sqrt{b}}\ln\left|\frac{\sqrt{ax+b}-\sqrt{b}}{\sqrt{ax+b}+\sqrt{b}}\right|, \quad \text{if } b>0.$$

Putting $a=-3$, $b=2$ in this we have

$$\int \frac{1}{x\sqrt{2-3x}}\, dx = \frac{1}{\sqrt{2}}\ln\left|\frac{\sqrt{2-3x}-\sqrt{2}}{\sqrt{2-3x}+\sqrt{2}}\right|.$$

Using this value on the right-hand side in (i),

$$\int \frac{1}{x^2\sqrt{2-3x}}\, dx = -\frac{\sqrt{2-3x}}{2x} + \frac{3}{4\sqrt{2}}\ln\left|\frac{\sqrt{2-3x}-\sqrt{2}}{\sqrt{2-3x}+\sqrt{2}}\right| + C,$$

where we have again added the constant of integration C.

EXAMPLE   Evaluate

$$\int \sin^5 x\, dx.$$

SOLUTION   From the table of integrals involving trigonometric functions, formula 87 gives:

$$\int \sin^n x\, dx = -\frac{1}{n}\sin^{n-1} x \cos x + \frac{n-1}{n}\int \sin^{n-2} x\, dx \quad (n\geq 2). \tag{i}$$

In our case we have $n=5$, and (i) becomes

$$\int \sin^5 x\, dx = -\tfrac{1}{5}\sin^4 x \cos x + \tfrac{4}{5}\int \sin^3 x\, dx. \tag{ii}$$

To evaluate $\int \sin^3 x \, dx$ we put $n = 3$ in (i):

$$\int \sin^3 x \, dx = -\tfrac{1}{3} \sin^2 x \cos x + \tfrac{2}{3} \int \sin x \, dx \qquad \text{(iii)}$$

and formula 79 gives

$$\int \sin x \, dx = -\cos x.$$

Using this in (iii), we get

$$\int \sin^3 x \, dx = -\tfrac{1}{3} \sin^2 x \cos x - \tfrac{2}{3} \cos x,$$

and substituting this into (ii) (and adding the constant of integration) we have

$$\int \sin^5 x \, dx = -\tfrac{1}{4} \sin^4 x \cos x + \tfrac{4}{5}(-\tfrac{1}{3} \sin^2 x \cos x - \tfrac{2}{3} \cos x) + C$$

$$= -\tfrac{1}{4} \sin^4 x \cos x - \tfrac{4}{15} \sin^2 x \cos x - \tfrac{8}{15} \cos x + C.$$

EXAMPLE   Evaluate

$$\int x^2(1 - x^2)^{3/2} \, dx.$$

SOLUTION   In formula 50 of the Table of Integrals we find the following integral:

$$\int x^n(a^2 - x^2)^{3/2} \, dx = \frac{1}{n+1} x^{n+1}(a^2 - x^2)^{3/2} + \frac{3}{n+1} \int x^{n+2}\sqrt{a^2 - x^2} \, dx.$$

Setting $n = 2$ and $a = 1$, therefore, we obtain that

$$\int x^2(1 - x^2)^{3/2} = \tfrac{1}{3}x^3(1 - x^2)^{3/2} + \int x^4\sqrt{1 - x^2} \, dx.$$

In order to evaluate the integral on the right-hand side of this equation we must make use of formula 51. Setting $a = 1$ and $n = 4$ in this formula, we get

$$\int x^4\sqrt{1 - x^2} \, dx = -\tfrac{1}{6}x^3(1 - x^2)^{3/2} + \tfrac{3}{6} \int x^2\sqrt{1 - x^2} \, dx.$$

Therefore

$$\int x^2(1 - x^2)^{3/2} \, dx = \tfrac{1}{3}x^3(1 - x^2)^{3/2} - \tfrac{1}{6}x^3(1 - x^2)^{3/2} + \tfrac{1}{2} \int x^2\sqrt{1 - x^2} \, dx$$

$$= \tfrac{1}{6}x^3(1 - x^2)^{3/2} + \tfrac{1}{2} \int x^2\sqrt{1 - x^2} \, dx.$$

Finally, making use of formula 44 with $a = 1$, we have that

$$\int x^2\sqrt{1 - x^2} \, dx = -\frac{x(1 - x^2)^{3/2}}{4} + \frac{x\sqrt{1 - x^2}}{8} + \frac{1}{8} \operatorname{Sin}^{-1} x,$$

and so the required integral is found to be

$$\int x^2(1 - x^2)^{3/2} = (\tfrac{1}{6}x^3 - \tfrac{1}{8}x)(1 - x^2)^{3/2} + \tfrac{1}{16}x\sqrt{1 - x^2} + \tfrac{1}{16} \operatorname{Sin}^{-1} x + C,$$

where we have added the constant of integration C.

Sometimes before the Table of Integrals can be used, it is necessary to make a change of variable by means of substitution in order to reduce the given integral to one that appears in the table.

EXAMPLE   Evaluate

$$\int \frac{e^x}{(e^x + 2)(3 - e^x)} \, dx.$$

SOLUTION   In this case we do not find the integral in the tables. Let us change the variable of integration first. Clearly $e^x \, dx$, the differential of e^x, appears in the integral, so we put $e^x = y$. Then $e^x \, dx = dy$ and the given integral now becomes

$$\int \frac{e^x}{(e^x + 2)(3 - e^x)} \, dx = \int \frac{1}{(y + 2)(3 - y)} \, dy. \qquad \text{(i)}$$

A general integral of this form is given in the table in formula 15;

$$\int \frac{1}{(ax + b)(cx + d)} \, dx = \frac{1}{bc - ad} \ln \left| \frac{cx + d}{ax + b} \right| \quad (bc - ad \neq 0).$$

In our example, $a = 1$, $b = 2$, $c = -1$, $d = 3$, and, of course, $x = y$. Thus

$$\int \frac{1}{(y + 2)(3 - y)} \, dy = \frac{1}{(2)(-1) - (1)(3)} \ln \left| \frac{-y + 3}{y + 2} \right| + C$$

$$= -\frac{1}{5} \ln \left| \frac{3 - y}{y + 2} \right| + C,$$

where C is the constant of integration.

Substituting back the value $y = e^x$, therefore,

$$\int \frac{e^x}{(e^x + 2)(3 - e^x)} \, dx = -\frac{1}{5} \ln \left| \frac{3 - e^x}{e^x + 2} \right| + C.$$

Whenever an integral is evaluated using a table of integrals we can readily verify that the answer obtained is correct by differentiating it: The result of differentiation should be the original integrand. For instance, it is readily verified by standard methods of differentiation that

$$\frac{d}{dx} \left(-\frac{1}{5} \ln \left| \frac{3 - e^x}{e^x + 2} \right| \right) = \frac{e^x}{(e^x + 2)(3 - e^x)},$$

and this result provides a check on the answer to the preceding worked example.

The student may wonder how tables of integrals are constructed in the first place. There are in fact a number of techniques, apart from the general substitution method, that are useful in evaluating integrals and that are used in constructing tables of the type given in the Appendix. In the following three sections of this chapter we shall provide an introduction to the three most important of these techniques.

Provided that the student has developed sufficient skill at using tables of integrals, the techniques given in the following sections will not be used frequently. However they will still be of some use, since an integral will sometimes be encountered that is not listed in the student's table. In such a case, one or other of these additional techniques may be useful in transforming the given integral into one that is listed.

EXERCISES 6.3

Make use of integral tables to evaluate the following integrals:

1. $\displaystyle\int \frac{1}{x^2 - 3x + 1}\, dx.$

2. $\displaystyle\int \frac{1}{x^2 - 2x + 3}\, dx.$

3. $\displaystyle\int \frac{x}{(2x - 3)^2}\, dx.$

4. $\displaystyle\int \frac{y}{(3y + 7)^5}\, dy.$

5. $\displaystyle\int \frac{\sqrt{3x + 1}}{x}\, dx.$

6. $\displaystyle\int \frac{t}{(2t + 3)^{5/2}}\, dt.$

7. $\displaystyle\int \frac{1}{t\sqrt{16 + t^2}}\, dt.$

8. $\displaystyle\int \frac{x^2}{\sqrt{x^2 + 25}}\, dx.$

9. $\displaystyle\int \frac{x^2}{\sqrt{x^2 - 9}}\, dx.$

10. $\displaystyle\int \frac{\sqrt{x^2 - 9}}{x}\, dx.$

11. $\displaystyle\int (t^2 - 4)^{3/2}\, dt.$

12. $\displaystyle\int \frac{u^2}{(25 - u^2)^{3/2}}\, du.$

13. $\displaystyle\int \sin^4 7x\, dx.$

14. $\displaystyle\int \cos^3 2x\, dx.$

15. $\displaystyle\int \sin^2 x \cos^3 x\, dx.$

16. $\displaystyle\int \cos x \cos 2x\, dx.*$

17. $\displaystyle\int x^2 \sin x\, dx.$

18. $\displaystyle\int y^3 \cos 2y\, dy.$

19. $\displaystyle\int \tan^5 \theta\, d\theta.$

20. $\displaystyle\int \cot^3 2\theta\, d\theta.$

21. $\displaystyle\int x^3 e^{2x}\, dx.$

22. $\displaystyle\int y^2 e^{3-y}\, dy.$

23. $\displaystyle\int \theta \operatorname{Tan}^{-1} \theta\, d\theta.$

24. $\displaystyle\int x \operatorname{Sin}^{-1} 2x\, dx.$

25. $\displaystyle\int \frac{1}{x^2\sqrt{x - 1}}\, dx.$

26. $\displaystyle\int \frac{1}{y^2\sqrt{y^2 + 3}}\, dy.$

27. $\displaystyle\int e^{2x} \sin 3x\, dx.$

28. $\displaystyle\int \frac{\cos 2x}{e^{3x}}\, dx.$

29. $\displaystyle\int \frac{e^x}{(1 - e^x)(2 - 3e^x)}\, dx.$

30. $\displaystyle\int \frac{\cos t}{(2 + 3\sin t)(1 - 2\sin t)}\, dt.$

---

*Hint for exercise 16: write $\cos 2x = 2\cos^2 x - 1$.

6.4 METHOD OF PARTIAL FRACTIONS

The method of partial fractions can often be used to simplify integrals whose integrands contain a rational algebraic function, the numerator and denominator of which are polynomials. Let us begin by illustrating the method with a simple example.

EXAMPLE Evaluate

$$\int \frac{1}{(x-1)(x+2)}\, dx.$$

SOLUTION Consider the following identities:

$$\frac{1}{x-1} - \frac{1}{x+2} = \frac{(x+2)-(x-1)}{(x-1)(x+2)}$$

$$= \frac{3}{(x-1)(x+2)}.$$

In the first step, we have simply combined the two fractions with their common denominator $(x-1)(x+2)$. It follows therefore that the integrand in the given integral can be written as

$$\frac{1}{(x-1)(x+2)} = \frac{1}{3}\left(\frac{1}{x-1} - \frac{1}{x+2}\right).$$

In this form we say that the given integrand has been split into *partial fractions*. Note that each fraction on the right-hand side has a denominator that contains only one linear factor of the form $(ax+b)$. Therefore

$$\int \frac{1}{(x-1)(x+2)}\, dx = \frac{1}{3}\int \frac{1}{x-1}\, dx - \frac{1}{3}\int \frac{1}{x+2}\, dx$$

$$= \tfrac{1}{3}\ln|x-1| - \tfrac{1}{3}\ln|x+2| + C$$

$$= \frac{1}{3}\ln\left|\frac{x-1}{x+2}\right| + C,$$

after using formula 6 in Appendix II.

This result could, of course, also be obtained by using formula 15 in the table.

In general, when the integrand is a rational algebraic function of the type $f(x)/g(x)$, the degree of $f(x)$ being *less* than the degree of $g(x)$ and $g(x)$ being a product of linear factors, then we first express $f(x)/g(x)$ as a sum of fractions with linear denominators, called *partial fractions*. After we have resolved $f(x)/g(x)$ into partial fractions, each partial fraction can be easily integrated with the help of standard formulas.

The method of partial fractions for the case when the denominator $g(x)$ consists of nonrepeated linear factors proceeds as follows. To each linear nonrepeated factor in the denominator we write a fraction with the same linear factor in the denominator and a constant in the numerator. Then these constants in the numerators are evaluated by giving certain values to x on both sides of the equation. This is illustrated in the following example.

EXAMPLE   Evaluate

$$\int \frac{(x+1)}{(x-1)(x+2)} \, dx.$$

SOLUTION   First we resolve the integrand into partial fractions. Let

$$\frac{x+1}{(x-1)(x+2)} = \frac{A}{x-1} + \frac{B}{x+2}, \tag{i}$$

where A and B are certain constants to be determined. Multiplying both sides of (i) by $(x-1)(x+2)$ we get:

$$x + 1 = A(x + 2) + B(x - 1).$$

This equation is an identity that must hold for all values of x.

Put $x = 1$: $1 + 1 = A(1 + 2) + B(0)$ or $A = \frac{2}{3}$,

Put $x = -2$: $-2 + 1 = A(0) + B(-2 - 1)$ or $B = \frac{1}{3}$.

Then from (i),

$$\frac{x+1}{(x-1)(x+2)} = \frac{2}{3(x-1)} + \frac{1}{3(x+2)}.$$

Therefore

$$\int \frac{x+1}{(x-1)(x+2)} \, dx = \int \left[\frac{2}{3(x-1)} + \frac{1}{3(x+2)} \right] dx$$

$$= \frac{2}{3} \int \frac{1}{x-1} \, dx + \frac{1}{3} \int \frac{1}{x+2} \, dx$$

$$= \frac{2}{3} \ln|x - 1| + \frac{1}{3} \ln|x + 2| + C,$$

where we have used formula 6 in Appendix II for the last step.

Note that in evaluating the constants A and B, we substituted the two values of x that are the roots of the denominator. This is usually the simplest way of finding the unknown constants.

When the numerator in the given integrand is not of degree lower than the denominator, we first divide the denominator into the numerator until a remainder is obtained whose degree is less than that of the denominator. This is illustrated in the following example.

EXAMPLE   Evaluate

$$\int \frac{x^3}{x^2 - x - 2} \, dx.$$

The integrand is $x^3/(x^2 - x - 2)$. Here the degree of the numerator is *not* less than the degree of the denominator, so we first divide the numerator by the denominator using the method of long division:

$$x^2 - x - 2\,\big|\,\overline{x^3 \qquad\qquad}\,\big|\,x + 1$$
$$\underline{x^3 - x^2 - 2x}$$
$$x^2 + 2x$$
$$\underline{x^2 - x - 2}$$
$$3x + 2$$

Therefore

$$\frac{x^3}{x^2 - x - 2} = x + 1 + \frac{3x + 2}{x^2 - x - 2}. \qquad\qquad \text{(i)}$$

Now we resolve $(3x + 2)/(x^2 - x - 2)$ into partial fractions. First resolve $x^2 - x - 2$ into linear factors:

$$x^2 - x - 2 = (x - 2)(x + 1).$$

Let the partial fractions be as follows:

$$\frac{3x + 2}{x^2 - x - 2} = \frac{3x + 2}{(x - 2)(x + 1)} = \frac{A}{x - 2} + \frac{B}{x + 1}. \qquad\qquad \text{(ii)}$$

Multiplying through by $(x - 2)(x + 1)$,

$$3x + 2 = A(x + 1) + B(x - 2).$$

To find A we put $x = 2$:

$$3(2) + 2 = A(2 + 1) + B(0), \quad A = \tfrac{8}{3};$$

to find B we put $x = -1$:

$$3(-1) + 2 = A(0) + B(-1 - 2), \quad B = \tfrac{1}{3}.$$

Therefore, from (ii),

$$\frac{3x + 2}{x^2 - x - 2} = \frac{8}{3(x - 2)} + \frac{1}{3(x + 1)},$$

and from (i),

$$\frac{x^3}{x^2 - x - 2} = x + 1 + \frac{8}{3(x - 2)} + \frac{1}{3(x + 1)}.$$

Thus

$$\int \frac{x^3}{x^2 - x - 2}\,dx = \int \left(x + 1 + \frac{8}{3} \cdot \frac{1}{x - 2} + \frac{1}{3} \cdot \frac{1}{x + 1} \right) dx$$

$$= \frac{x^2}{2} + x + \frac{8}{3}\ln|x - 2| + \frac{1}{3}\ln|x + 1| + C.$$

The method of partial fractions can be used when the denominator of the integrand involves more than two linear factors.

EXAMPLE   Evaluate

$$\int \frac{1}{(x^2 - 1)(x - 2)} \, dx.$$

SOLUTION   The integrand is split into partial fractions as follows:

$$\frac{1}{(x^2 - 1)(x - 2)} = \frac{1}{(x - 1)(x + 1)(x - 2)} = \frac{A}{x - 1} + \frac{B}{x + 1} + \frac{C}{x - 2}.$$

Note that one fraction occurs on the right for each of the linear factors in the denominator. Multiplying both sides of this equation by $(x - 1)(x + 1)(x - 2)$ we obtain

$$1 = A(x + 1)(x - 2) + B(x - 1)(x - 2) + C(x - 1)(x + 1).$$

To determine A, set $x = 1$, since then the coefficients of B and C become zero:

$$1 = A(1 + 1)(1 - 2) = A(2)(-1), \quad A = -\tfrac{1}{2}.$$

To determine B, set $x = -1$:

$$1 = B(-1 - 1)(-1 - 2) = 6B, \quad B = \tfrac{1}{6}.$$

To determine C, set $x = 2$:

$$1 = C(2 - 1)(2 + 1) = 3C, \quad C = \tfrac{1}{3}.$$

Therefore

$$\int \frac{1}{(x^2 - 1)(x - 2)} \, dx = \int \left[-\frac{1}{2(x - 1)} + \frac{1}{6(x + 1)} + \frac{1}{3(x - 2)} \right] dx$$

$$= -\tfrac{1}{2} \ln |x - 1| + \tfrac{1}{6} \ln |x + 1| + \tfrac{1}{3} \ln |x - 2| + C.$$

In some cases the integrand of a given integral can be reduced to a rational algebraic function by means of a substitution.

EXAMPLE   Evaluate

$$\int \frac{1}{x(1 + \ln x)(1 + 2 \ln x)} \, dx.$$

SOLUTION   The given integral can be written as

$$\int \frac{1}{(1 + \ln x)(1 + 2 \ln x)} \cdot \frac{1}{x} \, dx.$$

Since $(1/x) \, dx$ is the differential of $\ln x$ we put

$$\ln x = y \quad \text{and} \quad \frac{1}{x} \, dx = dy.$$

Then

$$\int \frac{1}{x(1 + \ln x)(1 + 2 \ln x)} \, dx = \int \frac{1}{(1 + y)(1 + 2y)} \, dy$$

$$= \int \left(\frac{-1}{1 + y} + \frac{2}{1 + 2y} \right) dy,$$

(after splitting the integrand into partial fractions).

$$= -\int \frac{1}{1 + y} \, dy + 2 \int \frac{1}{1 + 2y} \, dy$$

$$= -\frac{\ln |1 + y|}{1} + 2 \frac{\ln |1 + 2y|}{2} + C$$

$$= \ln |1 + 2y| - \ln |1 + y| + C$$

$$= \ln \left| \frac{1 + 2y}{1 + y} \right| + C$$

$$= \ln \left| \frac{1 + 2 \ln x}{1 + \ln x} \right| + C$$

because $y = \ln x$.

It should be mentioned that the method of partial fractions is not restricted to the case when the denominator of the integrand consists of a product of simple linear factors. It can be used when the denominator contains repeated factors, for example, the integral

$$\int \frac{1}{(x - 1)(x - 2)^2(x - 3)^3} \, dx,$$

in which the factors $(x - 2)$ and $(x - 3)$ are repeated, can be evaluated by partial fractions. In this case we would write the integrand as

$$\frac{1}{(x - 1)(x - 2)^2(x - 3)^3} = \frac{A}{x - 1} + \frac{B}{x - 2} + \frac{C}{(x - 2)^2}$$

$$+ \frac{D}{x - 3} + \frac{E}{(x - 3)^2} + \frac{F}{(x - 3)^3},$$

where $A, B, \ldots, F$ are six constants that have to be determined.

The method can also be used when the denominator contains factors that are polynomials of degree two or higher. For example, the integral

$$\int \frac{1}{(x^2 + 1)(x - 1)} \, dx$$

can be evaluated by writing the integrand as

$$\frac{1}{(x^2 + 1)(x - 1)} = \frac{Ax + B}{x^2 + 1} + \frac{C}{x - 1}.$$

The three constants A, B, and C again have to be determined.

We do not propose to go into detail here regarding these extensions of the method of partial fractions.

EXERCISES 6.4

Evaluate the following integrals:

1. $\displaystyle\int \frac{1}{x(x+1)}\,dx.$

2. $\displaystyle\int \frac{x-1}{(x-2)(x-3)}\,dx.$

3. $\displaystyle\int \frac{y+3}{y^2-1}\,dy.$

4. $\displaystyle\int \frac{1}{t^2-a^2}\,dt.$

5. $\displaystyle\int \frac{3z-4}{z^2-3z+2}\,dz.$

6. $\displaystyle\int \frac{4}{x(x^2-1)}\,dx.$

7. $\displaystyle\int \frac{y^2+1}{(y^2-1)(y+2)}\,dy.$

8. $\displaystyle\int \frac{1}{(t-1)(t-2)(t-3)}\,dt.$

9. $\displaystyle\int \frac{1}{t-t^3}\,dt.$

10. $\displaystyle\int \frac{x^2+1}{x^2-1}\,dx.$

11. $\displaystyle\int \frac{x^2-x-7}{x^2-3x-4}\,dx.$

12. $\displaystyle\int \frac{u^3}{u^2-1}\,du.$

13. $\displaystyle\int \frac{\cos x\,dx}{(1+\sin x)(2-\sin x)}.$

14. $\displaystyle\int \frac{e^x}{(e^x+1)(3+e^x)}\,dx.$

15. $\displaystyle\int \frac{\sec^2 x}{(3-\tan x)(2+\tan x)}\,dx.$

16. $\displaystyle\int \frac{\sin t\,dt}{(1-\cos t)(2+\cos t)}.$

6.5 TRIGONOMETRIC SUBSTITUTIONS

The trigonometric substitutions are particular examples of the general substitution method. They are often of use when the integrand contains factors of the type (a^2-x^2), (a^2+x^2), or (x^2-a^2), and particularly when such a factor appears underneath a radical sign. Their use is based on the trigonometric square identities (or Pythagorean identities), which are as follows:

$$1 - \sin^2 \theta = \cos^2 \theta$$
$$1 + \tan^2 \theta = \sec^2 \theta \quad \text{or} \quad \sec^2 \theta - 1 = \tan^2 \theta.$$

The substitutions themselves consist of writing either $x = a \sin \theta$ or $x = a \tan \theta$ or $x = a \sec \theta$, where θ is the new variable of integration and a is a certain constant determined by the given integrand. The choice of substitution depends on which of the three factors appears in the integrand, and is summarized in the following table.

If the Integral Involves	The Appropriate Substitution Is
(i) $a^2 - x^2$	$x = a \sin \theta$
(ii) $a^2 + x^2$	$x = a \tan \theta$
(iii) $x^2 - a^2$	$x = a \sec \theta$

The purpose of these substitutions is to change the factor $a^2 - x^2$, $a^2 + x^2$, or $x^2 - a^2$ that appears in the integrand to a single squared trigonometric function. For example, if we put $x = a \sin \theta$ in $a^2 - x^2$, we have:

$$a^2 - x^2 = a^2 - a^2 \sin^2 \theta = a^2(1 - \sin^2 \theta) = a^2 \cos^2 \theta.$$

Thus the factor $(a^2 - x^2)$ becomes a simple function of the new variable θ. If this factor appears under a square root sign an even simpler function of θ is obtained since $\sqrt{a^2 - x^2} = a \cos \theta$.

EXAMPLE   Evaluate

$$\int \frac{1}{\sqrt{a^2 - x^2}} \, dx, \quad a > 0.$$

SOLUTION   Here the substitution $a^2 - x^2 = u$ cannot work because its differential $-2x \, dx$ does not appear in the integral nor can it be obtained by multiplying by any constant. Instead we put

$$x = a \sin \theta, \quad \text{so that}$$

$$dx = a \cos \theta \, d\theta, \quad \theta = \text{Sin}^{-1} \frac{x}{a}$$

Then

$$\int \frac{1}{\sqrt{a^2 - x^2}} \, dx = \int \frac{1}{\sqrt{a^2 - a^2 \sin^2 \theta}} \cdot a \cos \theta \, d\theta$$

$$= \int \frac{1}{a\sqrt{1 - \sin^2 \theta}} \cdot a \cos \theta \, d\theta$$

$$= \int \frac{1}{a \cos \theta} \cdot a \cos \theta \, d\theta$$

$$= \int 1 \, d\theta = \theta + C$$

$$= \text{Sin}^{-1} \frac{x}{a} + C.$$

NOTE.   When we take $\sqrt{1 - \sin^2 \theta} = \cos \theta$, we are being consistent with the principal value of $\theta = \text{Sin}^{-1}(x/a)$ for which $-\pi/2 < \theta < \pi/2$ since $\cos \theta$ is positive for $-\pi/2 < \theta < \pi/2$.

EXAMPLE   Evaluate

$$\int \frac{x}{\sqrt{x^2 - 9}} \, dx.$$

SOLUTION   In this example we have a choice of methods. First of all we can write

$$\int \frac{x}{\sqrt{x^2 - 9}} \, dx = \frac{1}{2} \int \frac{2x \, dx}{\sqrt{x^2 - 9}}.$$

Since in this case the differential of $x^2 - 9$, which is $2x \, dx$, appears, we put

$$x^2 - 9 = y \quad \text{and then} \quad 2x \, dx = dy.$$

Therefore

$$\int \frac{x}{\sqrt{x^2 - 9}} \, dx = \frac{1}{2} \int \frac{2x \, dx}{\sqrt{x^2 - 9}}$$

$$= \frac{1}{2} \int \frac{dy}{\sqrt{y}}$$

$$= \tfrac{1}{2} \int y^{-1/2} \, dy$$

$$= \frac{1}{2} \cdot \frac{y^{-1/2+1}}{(-\frac{1}{2} + 1)}$$

$$= \sqrt{y} + C$$

$$= \sqrt{x^2 - 9} + C$$

because $y = x^2 - 9$.

ALTERNATIVELY   Because the integrand involves $\sqrt{x^2 - a^2} = \sqrt{x^2 - 9}$ with $a = 3$, we can make use of the trigonometric substitution

$$x = a \sec \theta \quad \text{or} \quad x = 3 \sec \theta.$$

Then

$$dx = 3 \sec \theta \tan \theta \, d\theta.$$

Therefore

$$\int \frac{x}{\sqrt{x^2 - 9}} \, dx = \int \frac{3 \sec \theta}{\sqrt{9 \sec^2 \theta - 9}} \cdot 3 \sec \theta \tan \theta \, d\theta$$

$$= \int \frac{9 \sec^2 \theta \tan \theta}{3\sqrt{\sec^2 \theta - 1}} \, d\theta$$

$$= 3 \int \frac{\sec^2 \theta \tan \theta}{\tan \theta} \, d\theta$$

$$= 3 \int \sec^2 \theta \, d\theta$$

$$= 3 \tan \theta + C.$$

The final step is to express this result in terms of the original variable x. We have

$$\tan \theta = \sqrt{\sec^2 \theta - 1} = \sqrt{(x/3)^2 - 1} = \tfrac{1}{3}\sqrt{x^2 - 9}.$$

Therefore the given integral becomes

$$\int \frac{x}{\sqrt{x^2 - 9}} \, dx = \sqrt{x^2 - 9} + C,$$

the same answer as found by the earlier method.

In the preceding example we had to find the value of one trigonometric function (tan θ) when the value of a second trigonometric function (sec θ) was already known.

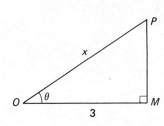

Figure 6.1

Problems of this type commonly arise when trigonometric substitutions are employed, and they can always be resolved by using one or more of the square identities (Pythagorean identities). Perhaps an easier method of solving such problems is to construct a right-angled triangle in which the given angle is one of angles. For instance, in the preceding example, sec $\theta = x/3$, and the appropriate triangle is shown in Fig. 6.1. Theta is the angle at O, and since the secant is the hypothenuse divided by the adjacent side, we can label $OP = x$ and $OM = 3$. Then from Pythagoras's theorem, $PM = \sqrt{OP^2 - OM^2} = \sqrt{x^2 - 9}$. Finally,

$$\tan \theta = \frac{PM}{OM} = \frac{\sqrt{x^2 - 9}}{3} = \frac{1}{3}\sqrt{x^2 - 9},$$

as before.

EXAMPLE   Evaluate

$$\int \frac{1}{(a^2 + x^2)^{3/2}} \, dx.$$

SOLUTION   Since the integrand involves $(a^2 + x^2)$ we put $x = a \tan \theta$. Then $dx = a \sec^2 \theta \, d\theta$ and so

$$\int \frac{1}{(a^2 + x^2)^{3/2}} \, dx = \int \frac{1}{(a^2 + a^2 \tan^2 \theta)^{3/2}} \cdot a \sec^2 \theta \, d\theta$$

$$= \int \frac{1}{(a^2 \sec^2 \theta)^{3/2}} \cdot a \sec^2 \theta \, d\theta$$

$$= \int \frac{1}{a^3 \sec^3 \theta} \cdot a \sec^2 \theta \, d\theta$$

$$= \frac{1}{a^2} \int \frac{1}{\sec \theta} \, d\theta$$

$$= \frac{1}{a^2} \int \cos \theta \, d\theta$$

$$= \frac{1}{a^2} \sin \theta + C.$$

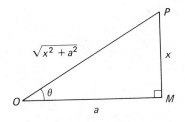

Figure 6.2

We must express $\sin \theta$ in terms of x when we have $\tan \theta = x/a$. From the right-angled triangle OMP (Fig. 6.2), in which θ is the angle at O, we have:

$$\tan \theta = \frac{MP}{OM} = \frac{x}{a} \quad \text{(given)}.$$

Therefore we can take $MP = x$ and $OM = a$. Then

$$OP^2 = MP^2 + OM^2 = x^2 + a^2 \quad \text{or} \quad OP = \sqrt{a^2 + x^2}.$$

Thus

$$\sin \theta = \frac{MP}{OP} = \frac{x}{\sqrt{a^2 + x^2}},$$

and so

$$\int \frac{1}{(a^2 + x^2)^{3/2}} \, dx = \frac{1}{a^2} \cdot \frac{x}{\sqrt{a^2 + x^2}} + C.$$

EXERCISES 6.5

Make use of trigonometric substitutions to evaluate the following integrals:

1. $\displaystyle\int \frac{1}{a^2 + x^2} \, dx.$

2. $\displaystyle\int \frac{dt}{4 + t^2}$

3. $\displaystyle\int \frac{dy}{\sqrt{9 - y^2}}.$

4. $\displaystyle\int \frac{d\theta}{16 - \theta^2}.$

5. $\displaystyle\int \frac{dx}{x\sqrt{x^2 - 16}}, \quad (x > 4).$

6. $\displaystyle\int \frac{dy}{y\sqrt{y^2 - 25}}, \quad (y > 5).$

7. $\displaystyle\int \frac{dt}{(t^2 + 9)^{3/2}}.$

8. $\displaystyle\int \frac{dx}{(25 - x^2)^{3/2}}.$

9. $\displaystyle\int \frac{dx}{(x^2 - a^2)^{3/2}}.$

10. $\displaystyle\int \frac{x}{\sqrt{9 - x^2}} \, dx.$

11. $\displaystyle\int \frac{x}{(a^2 - x^2)^{3/2}} \, dx.$

12. $\displaystyle\int \frac{\theta \, d\theta}{\theta^2 + a^2}.$

13. $\displaystyle\int \frac{\theta \, d\theta}{\theta^2 + 3}.$

14. $\displaystyle\int \frac{x \, dx}{x^2 + 5}.$

6.6 INTEGRATION BY PARTS

The method of integration by parts can often be used to evaluate an integral whose integrand consists of a product of two functions. It is analogous to the product formula of differential calculus, and is in fact derived from it.

From differential calculus we know that

$$\frac{d}{dx}[u(x)v(x)] = u'(x)v(x) + u(x)v'(x)$$

or

$$u(x)v'(x) = \frac{d}{dx}[u(x)v(x)] - u'(x)v(x).$$

Integrating both sides with respect to x, we get

$$\int u(x)v'(x)\, dx = u(x)v(x) - \int u'(x)v(x)\, dx.$$

Now let us set $u(x) = f(x)$ and $v'(x) = g(x)$. Then we can write $v(x) = G(x)$, where $G(x)$ denotes the integral of $g(x)$, and the above equation becomes

$$\int f(x)g(x)\, dx = f(x)G(x) - \int f'(x)G(x)\, dx.$$

This formula expresses the integral of the product $f(x)g(x)$ in terms of the integral of the product $f'(x)G(x)$. It is useful because in many cases the integral of $f'(x)G(x)$ is easier to evaluate than the integral of the original product $f(x)g(x)$. The following example illustrates this point.

EXAMPLE Evaluate

$$\int x \sin x\, dx.$$

SOLUTION Choose $f(x) = x$ and $g(x) = \sin x$ so that the given integral is equal to $\int f(x)g(x)\, dx$. Then $f'(x) = 1$ and $G(x)$, the integral of $g(x)$, is given by $G(x) = -\cos x + C_1$, C_1 being a constant of integration. Substituting these values into the formula for integration by parts we obtain that

$$\int f(x)g(x)\, dx = f(x)G(x) - \int f'(x)G(x)\, dx,$$

$$\int x \sin x\, dx = x(-\cos x + C_1) - \int (1)(-\cos x + C_1)\, dx$$

$$= -x \cos x + C_1 x + \int (\cos x - C_1)\, dx$$

$$= -x \cos x + C_1 x + \sin x - C_1 x + C$$

$$= -x \cos x + \sin x + C,$$

where C again is a constant of integration.

The integral in this example could also be found using formula 92 in Appendix II. The student should verify that the answer obtained is the same as that above.

NOTE. It should be observed that the first constant of integration C_1 in the above example, which arises in integrating $g(x)$ to obtain $G(x)$, cancels from the final answer. This is always the case when integrating by parts. Therefore, in practice, we never bother to include a constant of integration in $G(x)$, but simply take $G(x)$ to be any particular antiderivative of $g(x)$.

When using this method it is important to make the right selection of $f(x)$ and $g(x)$ in expressing the original integrand as a product. Otherwise the integral of $f'(x)G(x)$ may turn out to be no easier to evaluate than the integral of $f(x)g(x)$. One obvious criterion in choosing f and g is that we must be able to integrate $g(x)$ in order to write down $G(x)$. Usually we should choose $g(x)$ in such a way that its antiderivative $G(x)$ is a fairly simple function.

The following guidelines will be helpful in deciding the choice of f and g.

a. If the integrand is the product of a positive integral power of x (x, x^2, x^3, etc.) and an exponential or trigonometric function, it is often useful to take $f(x)$ as the given power of x. The preceding example illustrates this type of choice.

b. If the integrand contains as a factor either a logarithmic or inverse trigonometric function it is often useful to choose this function as $f(x)$. If the integrand consists simply of a logarithmic or inverse trigonometric function we can take $g(x) = 1$. The following examples illustrate this.

EXAMPLE   Evaluate

$$\int x \ln |x + 1| \, dx.$$

SOLUTION   Choose $f(x) = \ln |x + 1|$ and $g(x) = x$. Then

$$f'(x) = \frac{1}{x + 1}, \qquad G(x) = \frac{1}{2} x^2.$$

Substituting into the formula for integration by parts we obtain

$$\int f(x)g(x) \, dx = f(x)G(x) - \int f'(x)G(x) \, dx,$$

$$\int \ln |x + 1| \cdot x \, dx = \ln |x + 1| \cdot \frac{1}{2} x^2 - \int \frac{1}{x + 1} \cdot \frac{1}{2} x^2 \, dx$$

$$= \frac{1}{2} x^2 \ln |x + 1| - \frac{1}{2} \int \left(x - 1 + \frac{1}{x + 1} \right) dx.$$

In the last step we have divided out the fraction $x^2/(x + 1)$ in the manner that was discussed in Section 6.4. It follows therefore that

$$\int x \ln |x + 1| \, dx = \tfrac{1}{2} x^2 \ln |x + 1| - \tfrac{1}{2} [\tfrac{1}{2} x^2 - x + \ln |x + 1|] + C$$

$$= \tfrac{1}{2} (x^2 - 1) \ln |x + 1| - \tfrac{1}{4} x^2 + \tfrac{1}{2} x + C.$$

EXAMPLE   Evaluate

$$\int \text{Sin}^{-1} y \, dy.$$

SOLUTION   In this case we can express the integrand as a product by writing $f(y) = \text{Sin}^{-1} y$ and $g(y) = 1$. Then

$$f'(y) = \frac{1}{\sqrt{1 - y^2}} \quad \text{and} \quad G(y) = y.$$

Integrating by parts, therefore, we obtain

$$\int \text{Sin}^{-1} y \, dy = y \, \text{Sin}^{-1} y - \int \frac{1}{\sqrt{1 - y^2}} y \, dy.$$

In order to evaluate the integral on the right-hand side we make the substitution $1 - y^2 = u$, so that $-2y \, dy = du$. Then

$$\int \frac{y}{\sqrt{1 - y^2}} \, dy = \int \frac{1}{u^{1/2}} \left(-\frac{1}{2} \, du \right) = -\frac{1}{2} \int u^{-1/2} \, du$$

$$= -u^{1/2} + \text{constant}$$

$$= -\sqrt{1 - y^2} + \text{constant}.$$

Therefore

$$\int \text{Sin}^{-1} y \, dy = y \, \text{Sin}^{-1} y + \sqrt{1 - y^2} + C.$$

[Note that instead of substituting $1 - y^2 = u$ we could evaluate the second integral by substituting $y = \sin x$.]

The method of integration by parts can be summarized as follows. The given integrand is split into the product of two functions $f(x)$ and $g(x)$ in such a way that the following criteria are met:

$f(x)$ should have a relatively simple derivative $f'(x)$;
$g(x)$ should have a simple antiderivative $G(x)$;
the product $f'(x)G(x)$ should form a simpler
integrand than the given integrand $f(x)g(x)$.

The following table lists a number of basic situations in which integration by parts can be used. In each case the relevant functions f and g are given, together with the new integrand $f'(x)G(x)$, which remains to be integrated. It will be observed that this new integrand is in each case simpler than the original integrand.

In the first three examples the simplification occurs in the new integrand in that the power of x is reduced from n to $(n - 1)$. In order to evaluate the resulting integrals it is necessary to integrate by parts again in order to reduce the power of n still further. This process of integration by parts is continued until the power of n is reduced to zero and we are left with an integrand that simply consists of a trigonometric or exponential function.

304   function.

ORIGINAL INTEGRAND	$f(x)$	$g(x)$	NEW INTEGRAND $f'(x)G(x)$
$x^n \sin x$ $(n \geq 1)$	x^n	$\sin x$	$-nx^{n-1} \cos x$
$x^n \cos x$ $(n \geq 1)$	x^n	$\cos x$	$nx^{n-1} \sin x$
$x^n e^{ax}$ $(n \geq 1)$	x^n	e^{ax}	$(n/a)x^{n-1}e^{ax}$
$x^n \ln x$	$\ln x$	x^n	$(n+1)^{-1} x^n$
$x^n \mathrm{Sin}^{-1} x$	$\mathrm{Sin}^{-1} x$	x^n	$(n+1)^{-1}x^{n+1}(1-x^2)^{-1/2}$
$x^n \mathrm{Cos}^{-1} x$	$\mathrm{Cos}^{-1} x$	x^n	$-(n+1)^{-1}x^{n+1}(1-x^2)^{-1/2}$

EXAMPLE Evaluate

$$\int x^2 \sin x \, dx.$$

SOLUTION Using integration by parts with $f(x) = x^2$ and $g(x) = \sin x$, we get

$$\int x^2 \sin x \, dx = -x^2 \cos x + \int 2x \cos x \, dx.$$

For the remaining integral we integrate by parts again, this time taking $f(x) = x$ and $g(x) = \cos x$:

$$\int x \cos x \, dx = x \sin x - \int \sin x \, dx$$

$$= x \sin x + \cos x.$$

Therefore

$$\int x^2 \sin x \, dx = -x^2 \cos x + 2(x \sin x + \cos x) + C,$$

where we have finally added the constant of integration.

EXERCISES 6.6

Evaluate the following integrals:

1. $\int x \ln x \, dx.$

2. $\int x^3 \ln x \, dx.$

3. $\int x^n \ln x \, dx.$

4. $\int \ln (x + 1) \, dx.$

5. $\int \ln x \, dx.$

6. $\int \ln (x^2) \, dx.$

7. $\int x e^x \, dx.$

8. $\int x e^{-x} \, dx.$

9. $\int xe^{mx}\,dx.$

10. $\int \frac{x}{e^{2x}}\,dx.$

11. $\int \mathrm{Cos}^{-1} x\,dx.$

12. $\int \mathrm{Tan}^{-1} x\,dx.$

13. $\int \ln(x^x)\,dx.$

14. $\int x\cos x\,dx.$

15. $\int x\sin 2x\,dx.$

16. $\int (2x+3)\cos x\,dx.$

17. $\int x^2 \cos 2x\,dx.$

18. $\int x^2 e^x\,dx.$

19. $\int (x+1)^2 e^{2x}.$

20. $\int (x^3 - 5x + 7)\sin x\,dx.$

REVIEW EXERCISES FOR CHAPTER 6

1. State whether the following statements are true or false. If false, replace them by the corresponding correct statement.
   (a) The antiderivative of an integrable function is unique.
   (b) The integral of the sum of two functions is equal to the sum of their integrals.
   (c) The integral of the product of two functions is equal to the product of their integrals.

   (d) $\int \frac{d}{dx}[f(x)]\,dx = f(x).$

   (e) $\frac{d}{dx}\left[\int f(t)\,dt\right] = f(t).$

   (f) If $f'(x) = g'(x)$, then $f(x) = g(x)$.

   (g) $\int \ln x\,dx = \frac{1}{x} + C.$

   (h) $\int e^x\,du = e^x + C.$

   (i) $\int \tan x\,dx = \sec^2 x + C.$

   (j) $\int [f(x)]^n\,dx = \frac{[f(x)]^{n+1}}{n+1} + C, \quad n \neq -1.$

   (k) $\int x^n\,dx = \frac{x^{n+1}}{n+1} + C$ for all n.

   (l) $\int xf(x)\,dx = x\int f(x)\,dx.$

   (m) $\int \frac{1}{x^2}\,dx = \ln x^2 + C.$

   (n) $\int e^{x^2}\,dx = e^{x^3/3} + C.$

   (o) $\int e^t\,dt = \frac{e^{t+1}}{t+1} + C.$

   Make use of an appropriate substitution to evaluate the following integrals:

2. $\int e^x \cos(e^x + 1)\,dx.$

3. $\int \frac{1}{x(1 + \ln x)}\,dx.$

4. $\int \frac{e^x}{e^x + 1}\,dx.$

5. $\int \sin^5 t\,dt.$

6. $\int \cos^3 t \, dt$.

7. $\int \sec^4 u \tan u \, du$.

8. $\int \frac{\operatorname{Sin}^{-1} x}{\sqrt{1 - x^2}} \, dx$.

9. $\int \frac{(\operatorname{Tan}^{-1} u)^2}{1 + u^2} \, du$.

Making use of tables, or otherwise, evaluate the following integrals:

10. $\int \sin^3 x \, dx$.

11. $\int \frac{1}{x^2 \sqrt{2x + 1}} \, dx$.

12. $\int \frac{3x + 2}{(x - 1)(2x + 1)} \, dx$.

13. $\int \frac{3x - 1}{(x - 1)(x + 2)(x - 3)} \, dx$.

14. $\int \sqrt{4x^2 - 9} \, dx$.

15. $\int \sqrt{9x^2 + 36} \, dx$.

16. $\int x^2 \sqrt{9x^2 + 16} \, dx$.

17. $\int (25t^2 - 9)^{3/2} \, dt$.

18. $\int x \ln |x + 1| \, dx$.

19. $\int x^5 \ln x \, dx$.

20. $\int \frac{e^x}{(1 + e^x)(2 + e^x)} \, dx$.

21. $\int \frac{1}{t(1 + \ln t)(2 - \ln t)} \, dt$.

22. $\int x^3 \sin x \, dx$.

23. $\int \tan^6 x \, dx$.

24. $\int e^{2x} \sin 3x \, dx$.

25. $\int \frac{\cos 2x}{e^{3x}} \, dx$.

7

The
Definite
Integral

7.1 SUMMATION NOTATION

The sigma notation is a convenient way of expressing sums that involve large numbers of terms. It is very commonly used in statistics and many other branches of mathematics. We shall make use of this notation when defining definite integrals in the next section.

According to the sigma notation (or summation notation) the sum

$$x_1 + x_2 + x_3 + \ldots + x_n$$

is abbreviated by the expression

$$\sum_{i=1}^{n} x_i.$$

This is read as "the sum of x_i as i goes from 1 to n." The Greek letter Σ (capital sigma) corresponds to our S in the English alphabet and suggests the word *sum*. Thus

$$x_1 + x_2 + x_3 + \ldots + x_n = \sum_{i=1}^{n} x_i. \qquad (1)$$

The subscript i used in the sigma notation on the right-hand side of (1) above is called the *summation index*. It can be replaced by any other letter, say j or k, that is not already being used to stand for something else, and the value of the sum will not be changed. Thus

$$\sum_{i=1}^{n} x_i = \sum_{j=1}^{n} x_j = \sum_{k=1}^{n} x_k, \quad \text{etc.}$$

In the sum $\sum_{i=1}^{n} x_i$, the index of summation i (which is indicated below Σ) takes the values $1, 2, 3, \ldots, n$. The starting value (1 in this case) is indicated below Σ, and the last value (n in this case) is indicated above Σ. Thus to expand a sum given in the Σ-notation we give all possible integer values to the index of summation in the expression that follows Σ and then add all the terms. For example, in order to expand $\sum_{k=3}^{7} f(x_k)$, we note that the index k takes the values 3, 4, 5, 6, 7 (the starting value 3 is indicated below Σ and the last value 7 is indicated above Σ). Therefore in the expression that follows Σ, that is, in $f(x_k)$, we replace k by 3, 4, 5, 6, 7 and then add all the terms so obtained. Thus

$$\sum_{k=3}^{7} f(x_k) = f(x_3) + f(x_4) + f(x_5) + f(x_6) + f(x_7).$$

The only thing that changes from one term to the next is the numeral in the place indicated by the index of summation (k in this case).

Following are a few other examples of sums given by the Σ-notation.

i. $\sum_{k=1}^{7} k^3 = 1^3 + 2^3 + 3^3 + 4^3 + 5^3 + 6^3 + 7^3.$

ii. $\sum_{k=1}^{5} a^k = a^1 + a^2 + a^3 + a^4 + a^5.$

iii. $\sum\limits_{i=2}^{5} \dfrac{3i}{i-1} = \dfrac{3(2)}{2-1} + \dfrac{3(3)}{3-1} + \dfrac{3(4)}{4-1} + \dfrac{3(5)}{5-1}$.

iv. $\sum\limits_{j=1}^{4} \dfrac{j+1}{j^2+1} = \dfrac{1+1}{1^2+1} + \dfrac{2+1}{2^2+1} + \dfrac{3+1}{3^2+1} + \dfrac{4+1}{4^2+1}$.

v. $\sum\limits_{p=1}^{100} \ln p = \ln 1 + \ln 2 + \ln 3 + \ldots + \ln 100$.

It should be observed that the number of terms in the expansion of $\sum\limits_{k=m}^{n}$ is equal to $(n-m+1)$. Thus $\sum\limits_{k=3}^{7} x_k$ contains $7-3+1 = 5$ terms, $\sum\limits_{i=1}^{10} x^i$ contains $10-1+1 = 10$ terms, and so on.

THEOREM 7.1.1

If m and n are integers, with $n \geq m$, then

(a) $\sum\limits_{k=m}^{n} c = (n-m+1)c$, where c is a constant.

(b) $\sum\limits_{k=m}^{n} (x_k \pm y_k) = \sum\limits_{k=m}^{n} x_k \pm \sum\limits_{k=m}^{n} y_k$.

(c) $\sum\limits_{k=m}^{n} cx_k = c \sum\limits_{k=m}^{n} x_k$, where c is a constant.

(d) $\sum\limits_{k=m}^{n} (x_k - x_{k-1}) = x_n - x_{m-1}$.

PROOF

(a) The number of terms in $\sum\limits_{k=m}^{n} c$ is $(n-m+1)$, and each term in the expansion is equal to c because the expression that follows Σ, that is c, does not involve the index of summation k. Thus

$$\sum_{k=m}^{n} c = \underbrace{c + c + c + \ldots + c}_{(n-m+1) \text{ terms}}$$

$$= (n-m+1)c.$$

(b) $\sum\limits_{k=m}^{n} (x_k \pm y_k) = (x_m \pm y_m) + (x_{m+1} \pm y_{m+1}) + \ldots + (x_n \pm y_n)$

$$= (x_m + x_{m+1} + \ldots + x_n) \pm (y_m + y_{m+1} + \ldots + y_n)$$

$$= \sum_{k=m}^{n} x_k \pm \sum_{k=m}^{n} y_k.$$

(c) $\sum\limits_{k=m}^{n} cx_k = cx_m + cx_{m+1} + \ldots + cx_n$

$$= c(x_m + x_{m+1} + \ldots + x_n)$$

$$= c \sum_{k=m}^{n} x_k.$$

(d) $\sum_{k=m}^{n} (x_k - x_{k-1}) = (x_m - x_{m-1}) + (x_{m+1} - x_m) + (x_{m+2} - x_{m+1})$
$$+ \ldots + (x_{n-1} - x_{n-2}) + (x_n - x_{n-1})$$
$$= x_n - x_{m-1}$$

because all other terms cancel with each other. (Some of these cancellations are indicated by slashes in the preceding expression.)

COROLLARY   In particular, when $m = 1$, the above results become

(a) $\sum_{k=1}^{n} c = nc.$

(b) $\sum_{k=1}^{n} (x_k \pm y_k) = \sum_{k=1}^{n} x_k \pm \sum_{k=1}^{n} y_k.$

(c) $\sum_{k=1}^{n} cx_k = c \sum_{k=1}^{n} x_k.$

(d) $\sum_{k=1}^{n} (x_k - x_{k-1}) = x_n - x_0.$

EXAMPLE   Expand the following sums:

(a) $\sum_{k=1}^{7} (2).$ (b) $\sum_{p=3}^{7} (5).$

(c) $\sum_{k=-4}^{5} (3).$ (d) $\sum_{k=1}^{n} (3).$

SOLUTION   (a) $\sum_{k=1}^{7} (2)$ contains $(7 - 1 + 1) = 7$ terms, and each of them is equal to 2. Thus

$$\sum_{k=1}^{7} (2) = 7(2) = 14.$$

(b) $\sum_{p=3}^{7} (5)$ contains $7 - 3 + 1 = 5$ terms, and each of them is equal to 5. Thus

$$\sum_{p=3}^{7} (5) = 5(5) = 25.$$

(c) $\sum_{k=-4}^{5} (3)$ contains $5 - (-4) + 1 = 10$ terms, and each of them is equal to 3. Thus

$$\sum_{k=-4}^{5} (3) = 10(3) = 30.$$

(d) $\sum_{k=1}^{n} 3$ contains $n - 1 + 1 = n$ terms. Thus

$$\sum_{k=1}^{n} 3 = n(3) = 3n.$$

THEOREM 7.1.2

$$\text{(a)} \quad \sum_{k=1}^{n} k = 1 + 2 + 3 + \ldots + n = \frac{n(n+1)}{2}.$$

$$\text{(b)} \quad \sum_{k=1}^{n} k^2 = 1^2 + 2^2 + 3^2 + \ldots + n^2 = \frac{n(n+1)(2n+1)}{6}.$$

$$\text{(c)} \quad \sum_{k=1}^{n} k^3 = 1^3 + 2^3 + 3^3 + \ldots + n^3 = \left[\frac{n(n+1)}{2}\right]^2.$$

PROOF

(a) $\sum_{k=1}^{n} k = 1 + 2 + 3 + \ldots + (n-1) + n.$

The terms in this sum form an arithmetic progression. In order to evaluate this sum let us first write down the sum again with the terms on the right in the opposite order:

$$\sum_{k=1}^{n} k = n + (n-1) + (n-2) + \ldots + 2 + 1.$$

Now let us add these two sums together term by term:

$$2\sum_{k=1}^{n} k = [1 + n] + [2 + (n-1)] + [3 + (n-2)] + \ldots$$
$$+ [(n-1) + 2] + [n + 1]$$
$$= \underbrace{(n+1) + (n+1) + (n+1) + \ldots + (n+1) + (n+1)}_{n \text{ terms}}$$
$$= n(n+1).$$

Therefore

$$\sum_{k=1}^{n} k = \frac{n(n+1)}{2}.$$

(b) To prove (b) we shall make use of the following result:
$$k^3 - (k-1)^3 = k^3 - (k^3 - 3k^2 + 3k - 1)$$
or
$$k^3 - (k-1)^3 = 3k^2 - 3k + 1.$$

This is an identity that is true for all values of k. Putting $k = 1, 2, 3, \ldots, n$, we obtain the following sequence of equations:

$$1^3 - 0^3 = 3 \cdot 1^2 - 3 \cdot 1 + 1$$
$$2^3 - 1^3 = 3 \cdot 2^2 - 3 \cdot 2 + 1$$
$$3^3 - 2^3 = 3 \cdot 3^2 - 3 \cdot 3 + 1$$
$$- - - - - - - - - -$$
$$- - - - - - - - - -$$
$$n^3 - (n-1)^3 = 3 \cdot n^2 - 3 \cdot n + 1.$$

If these equations are all added vertically we observe that most of the terms on the left-hand side cancel out and we are left with

$$n^3 - 0^3 = 3(1^2 + 2^2 + 3^2 + \ldots + n^2) - 3(1 + 2 + 3 + \ldots + n)$$
$$+ \underbrace{(1 + 1 + \ldots + 1)}_{n \text{ terms}}$$

$$n^3 = 3 \sum_{k=1}^{n} k^2 - 3\frac{n(n + 1)}{2} + n,$$

where we have made use of Theorems 7.1.1(a) and 7.1.2(a). Thus

$$3 \sum_{k=1}^{n} k^2 = n^3 - n + \frac{3n(n + 1)}{2}$$
$$= n(n + 1)(n - 1) + \tfrac{3}{2}n(n + 1)$$
$$= n(n + 1)[n - 1 + \tfrac{3}{2}]$$
$$= n(n + 1)\left(\frac{2n + 1}{2}\right).$$

Hence

$$\sum_{k=1}^{n} k^2 = \frac{n(n + 1)(2n + 1)}{6}.$$

which proves the result.

(c)   The proof of this part is left as an exercise for the reader. (*Hint:* Make use of the identity $k^4 - (k - 1)^4 = 4k^3 - 6k^2 + 4k - 1$.)

NOTES.

1. The above results can also be proved by the method of mathematical induction.
2. The results of this theorem will be used as standard formulas in the following section.

EXAMPLE   Evaluate the sum of the squares of the first 100 natural numbers.

SOLUTION

$$1^2 + 2^2 + 3^2 + \ldots + 100^2 = \sum_{k=1}^{100} k^2$$
$$= \frac{100(100 + 1)(2 \cdot 100 + 1)}{6}$$
$$= \frac{100(101)(201)}{6}$$
$$= 338{,}350,$$

where we have made use of Theorem 7.1.2(b) for $n = 100$.

EXAMPLE Evaluate the following sum:

$$7^3 + 8^3 + 9^3 + \ldots + 30^3.$$

SOLUTION The given sum can be written as

$$7^3 + 8^3 + 9^3 + \ldots + 30^3 = (1^3 + 2^3 + 3^3 + \ldots + 30^3)$$
$$- (1^3 + 2^3 + 3^3 + \ldots + 6^3)$$
$$= \sum_{k=1}^{30} k^3 - \sum_{k=1}^{6} k^3$$
$$= \left[\frac{30(30+1)}{2}\right]^2 - \left[\frac{6(6+1)}{2}\right]^2$$
$$= (465)^2 - (21)^2$$
$$= 216{,}225 - 441$$
$$= 215{,}784.$$

EXAMPLE Evaluate the following sums:

(a) $\displaystyle\sum_{i=1}^{n} (2i + 3)$.

(b) $\displaystyle\sum_{k=1}^{50} (3k^2 + 2k + 1)$.

(c) $\displaystyle\sum_{p=1}^{4} (2p^3 + 7p + 3)$.

SOLUTION (a) Making use of Theorem 7.1.1 we can write

$$\sum_{i=1}^{n} (2i + 3) = \sum_{i=1}^{n} 2i + \sum_{i=1}^{n} 3$$
$$= 2\sum_{i=1}^{n} i + \sum_{i=1}^{n} 3$$
$$= 2\frac{n(n+1)}{2} + 3n$$
$$= n(n+4),$$

where we have used Theorem 7.1.2(a) for the first summation and Theorem 7.1.1(a) for the second summation.

(b) First let us find $\displaystyle\sum_{k=1}^{n} (3k^2 + 2k + 1)$. Using Theorems 7.1.1 and 7.1.2 we have

$$\sum_{k=1}^{n} (3k^2 + 2k + 1) = 3\sum_{k=1}^{n} k^2 + 2\sum_{k=1}^{n} k + \sum_{k=1}^{n} 1$$
$$= 3\left[\frac{n(n+1)(2n+1)}{6}\right] + 2\left[\frac{n(n+1)}{2}\right] + n(1)$$
$$= \frac{n(n+1)(2n+1)}{2} + n(n+1) + n.$$

$2 \sin x \cos x$

If we now replace n by 50 on both sides we obtain

$$\sum_{k=1}^{50} (3k^2 + 2k + 1) = \frac{50(51)(101)}{2} + 50(51) + 50$$

$$= 128{,}775 + 2550 + 50$$

$$= 131{,}375.$$

(c)   In this case we could first compute the sum for any n and then put $n = 4$. However the number of terms in this case is 4, which is small, so that it is in actual fact simpler to expand the sum and compute it directly:

$$\sum_{k=1}^{4} (2k^3 + 7k + 3) = (2 \cdot 1^3 + 7 \cdot 1 + 3) + (2 \cdot 2^3 + 7 \cdot 2 + 3)$$

$$+ (2 \cdot 3^3 + 7 \cdot 3 + 3) + (2 \cdot 4^3 + 7 \cdot 4 + 3)$$

$$= (2 + 7 + 3) + (16 + 14 + 3) + (54 + 21 + 3)$$

$$+ (128 + 28 + 3)$$

$$= 12 + 33 + 78 + 159$$

$$= 282.$$

EXAMPLE   Given $x_1 = 3$, $x_2 = 5$, $x_3 = -1$, and $x_4 = 2$, find

(a) $\displaystyle\sum_{k=1}^{3} x_k^2$.

(b) $\displaystyle\sum_{k=1}^{4} (x_k - 2)^2$.

(c) $\displaystyle\left[\sum_{k=1}^{4} (x_k - 2) \right]^2$.

SOLUTION   (a) $\displaystyle\sum_{k=1}^{3} x_k^2 = x_1^2 + x_2^2 + x_3^2$

$$= 3^2 + 5^2 + (-1)^2$$

$$= 9 + 25 + 1$$

$$= 35.$$

(b) $\displaystyle\sum_{k=1}^{4} (x_k - 2)^2 = (x_1 - 2)^2 + (x_2 - 2)^2 + (x_3 - 2)^2 + (x_4 - 2)^2$

$$= (3 - 2)^2 + (5 - 2)^2 + (-1 - 2)^2 + (2 - 2)^2$$

$$= 1 + 9 + 9 + 0$$

$$= 19.$$

(c) $\displaystyle\sum_{k=1}^{4} (x_k - 2) = (x_1 - 2) + (x_2 - 2) + (x_3 - 2) + (x_4 - 2)$

$$= (3 - 2) + (5 - 2) + (-1 - 2) + (2 - 2)$$

$$= 1 + 3 - 3 + 0$$

$$= 1.$$

Thus $\displaystyle\left[\sum_{k=1}^{4} (x_k - 2) \right]^2 = 1^2 = 1.$

EXAMPLE Given that $\sum_{i=1}^{5} x_i = 13$ and $\sum_{i=1}^{5} x_i^2 = 49$, find

(a) $\sum_{i=1}^{5} (3x_i - 2)$.

(b) $\sum_{i=1}^{5} (2x_i + 3)^2$.

SOLUTION (a) $\sum_{i=1}^{5} (3x_i - 2) = \sum_{i=1}^{5} 3x_i - \sum_{i=1}^{5} 2$

$$= 3 \sum_{i=1}^{5} x_i - 5(2)$$
$$= 3(13) - 10$$
$$= 29.$$

(b) $\sum_{i=1}^{5} (2x_i + 3)^2 = \sum_{i=1}^{5} (4x_i^2 + 12x_i + 9)$

$$= 4 \sum_{i=1}^{5} x_i^2 + 12 \sum_{i=1}^{5} x_i + 5(9)$$
$$= 4(49) + 12(13) + 45$$
$$= 196 + 156 + 45$$
$$= 397.$$

EXERCISES 7.1

Evaluate the following sums:

1. $\sum_{k=1}^{4} (2k - 3)$.

2. $\sum_{k=0}^{3} (k^2 + 7)$.

3. $\sum_{p=2}^{5} (p^2 + p - 1)$.

4. $\sum_{i=-3}^{3} (i^2 - i + 2)$.

5. $\sum_{i=2}^{4} \frac{i}{i-1}$.

6. $\sum_{q=1}^{4} \left(\frac{q^2 + 1}{q} \right)$.

7. $\sum_{n=1}^{3} \frac{1}{n(n+1)}$.

8. $\sum_{k=0}^{5} \left(\frac{1}{k+1} - \frac{1}{k+2} \right)$.

9. $\sum_{k=1}^{n} (2k - 1)$.

10. $\sum_{k=1}^{n} (3k + 2)$.

11. $\sum_{j=1}^{n} (j^2 + j + 1)$.

12. $\sum_{j=1}^{n} (2j^2 - j + 3)$.

13. $\sum_{k=1}^{n} (k + 1)(2k - 1)$.

14. $\sum_{k=1}^{n} (k - 1)(k + 1)$.

15. $\sum_{k=1}^{n} (k^3 + 7k - 1)$.

16. $\sum_{p=1}^{n} (p - 1)(p^2 + p + 1)$.

316

17. $\sum\limits_{p=1}^{20} (p^2 + 7p - 6)$.

18. $\sum\limits_{r=1}^{30} (r^3 + 1)$.

19. $\sum\limits_{k=1}^{25} (k + 1)(k + 3)$.

20. $\sum\limits_{k=1}^{20} (k + 1)(k^2 + 1)$.

21. $\sum\limits_{p=11}^{50} k^2$.

22. $\sum\limits_{k=6}^{20} (2k^2 + 5k - 3)$.

23. Make use of the identity $(k + 1)^2 - k^2 = 2k + 1$ to evaluate the sum $\sum\limits_{k=1}^{n} k$.

24. Make use of the identity $k^2 - (k - 1)^2 = 2k - 1$ to evaluate the sum $\sum\limits_{k=1}^{n} k$.

25. Given $x_1 = 1$, $x_2 = -2$, $x_3 = 3$, $x_4 = 7$, and $x_5 = 4$, evaluate

(a) $\sum\limits_{p=1}^{5} (2x_p - 3)$.

(b) $\sum\limits_{p=1}^{5} (x_p + 2)^2$.

26. Given $x_1 = 1$, $x_2 = 2$, $x_3 = 3$, $x_4 = 4$, $x_5 = 5$, $y_1 = 3$, $y_2 = -1$, $y_3 = 7$, $y_4 = -2$, and $y_5 = -1$, find:

(a) $\sum\limits_{p=1}^{5} x_p y_p$.

(b) $\sum\limits_{p=1}^{5} x_p^2 y_p$.

(c) $\sum\limits_{k=1}^{5} (x_k - y_k)^2$.

27. Given $\sum\limits_{i=1}^{7} x_i = 13$ and $\sum\limits_{i=1}^{7} x_i^2 = 63$, find:

(a) $\sum\limits_{i=1}^{7} (5 - 2x_i)$.

(b) $\sum\limits_{p=1}^{7} (3x_p - 1)^2$.

28. Given $\sum\limits_{i=1}^{10} x_i^2 = 15$, $\sum\limits_{i=1}^{10} (x_i + y_i)^2 = 73$, and $\sum\limits_{i=1}^{10} y_i^2 = 26$, find $\sum\limits_{p=1}^{10} x_p y_p$.

7.2 AREAS UNDER CURVES

The calculation of the areas of rectangles and triangles is very simple: The area of a rectangle is obtained by multiplying its base times its height. The area of a triangle is given by half the base times the height. The area of any other plane figure that is bounded by *straight line segments* can also easily be calculated by subdividing the figure into a number of triangles and rectangles. This is illustrated in Fig. 7.1. The area is then given by the sum of the areas of the triangles and rectangles into which the given figure has been subdivided. In particular, the area of any polygon is obtained by dividing the polygon into a number of triangles.

When the plane figure is not bounded by straight lines, then the area can be calculated by the method of successive approximations. The Greek mathematicians were the first to use this method to calculate the area of a circle. First they approximated the area of the circle by inscribing a square. Then they improved the approximation by inscribing an octagon, then a 16-sided polygon and so on (Fig. 7.2). Obviously

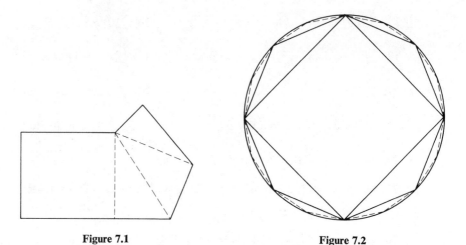

Figure 7.1

Figure 7.2

each new polygon with more sides provides a better approximation to the area of the circle than the previous one. The areas of the inscribed polygons are always smaller than the area of the circle, but as the number of sides becomes larger, the area approaches that of the circle.

The Greeks also made use of circumscribed polygons that, as the number of sides increases, approach the area of a circle through larger values.

We shall use a similar technique to define and calculate the area A, which is bounded on one side by the graph of a certain function $y = f(x)$ and on the other sides by the vertical lines $x = a$, $x = b$, and the x-axis (Fig. 7.3). For simplicity, we shall suppose that $f(x) \geq 0$ for $a \leq x \leq b$. Let $n > 1$ be a positive integer, and let us divide the interval $a \leq x \leq b$ into n equal subintervals each of length h. Then $h = (b - a)/n$. Let the dividing points be

$$x_1, x_2, x_3, \ldots, x_{n-1}$$

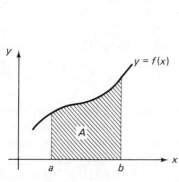

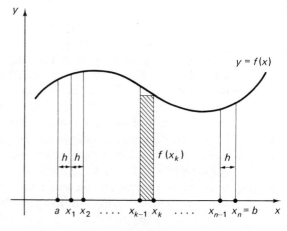

Figure 7.3

Figure 7.4

with $b = x_n$ (Fig. 7.4). Then $x_1 = a + h$, $x_2 = a + 2h$, $x_3 = a + 3h$, ..., and so on. In general, the kth dividing point is $x_k = a + kh$, and the last one is

$$x_n = a + nh = a + (b - a) = b.$$

On the kth subinterval, $x_{k-1} \leq x \leq x_k$, we erect a rectangle of height equal to the value of $f(x)$ at the right-hand endpoint, that is $f(x_k)$. The area of this rectangle is equal to $f(x_k) \times h$. A similar rectangle is erected on each of the n intervals, and we take the sum of the areas of the n rectangles as an approximation to the true area A under the curve. Thus, denoting the sum of the areas of the rectangles by A_n, we have

$$A_n = \sum_{k=1}^{n} f(x_k) \cdot h = \sum_{k=1}^{n} f(a + kh) \cdot h.$$

Figures 7.5 and 7.6 illustrate the approximation for $n = 8$ and $n = 16$. In general,

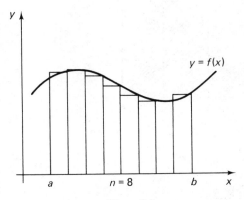

Figure 7.5

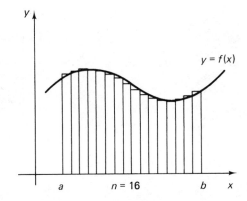

Figure 7.6

it will readily be accepted that as n gets larger and larger the sum A_n of areas of the rectangles approximates the true area A more and more closely. In fact by taking n sufficiently large, we can make A_n as close as we desire to A; thus we can write the area A as the limit of A_n as $n \to \infty$ (or $h \to 0$), that is,

$$A = \lim_{n \to \infty} \sum_{k=1}^{n} f(x_k) \cdot h$$

or

$$A = \lim_{n \to \infty} \sum_{k=1}^{n} f(a + kh) \cdot h,$$

where

$$h = \frac{b - a}{n}.$$

NOTE. From the purely logical point of view, these equations actually form the *definition* of the area under the curve $y = f(x)$.

EXAMPLE   Approximate the area under the curve $y = x^2$ from $x = 0$ to $x = 4$ by dividing the area into four rectangles.

SOLUTION   Here $a = 0$, $b = 4$, $n = 4$, and $f(x) = x^2$ (Fig. 7.7). Thus

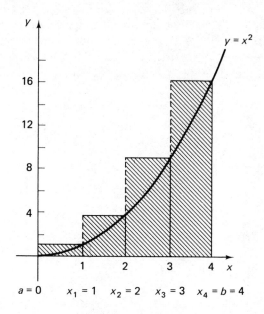

Figure 7.7

$$h = \frac{b - a}{4} = \frac{4 - 0}{4} = 1,$$

$$x_1 = a + h = 1,$$

$$x_2 = a + 2h = 2,$$

and, similarly, $x_3 = 3$, $x_4 = 4$. Further,

$$f(x_1) = x_1^2 = 1^2 = 1,$$

$$f(x_2) = x_2^2 = 2^2 = 4,$$

$$f(x_3) = 3^2 = 9,$$

$$f(x_4) = 4^2 = 16.$$

Thus

$$A_4 = \sum_{k=1}^{4} f(x_k) \cdot h$$

$$= h[f(x_1) + f(x_2) + f(x_3) + f(x_4)]$$

$$= 1[1 + 4 + 9 + 16] = 30 \text{ square units.}$$

320   *NOTE.*   As is clear from Fig. 7.7, the true area A is less than this value.

EXAMPLE Evaluate the area under the curve $y = 3x^2 + 2x + 5$ between the lines $x = 1$ and $x = 3$.

SOLUTION Here we want the exact area A. Let us first evaluate A_n. In this case $a = 1$, $b = 3$, and $f(x) = 3x^2 + 2x + 5$. Thus, for a general number n of intervals in the subdivision,

$$h = \frac{b - a}{n} = \frac{3 - 1}{n} = \frac{2}{n}.$$

Consequently,

$$x_k = a + kh = 1 + \frac{2k}{n}.$$

Then

$$f(x_k) = 3x_k^2 + 2x_k + 5$$

$$= 3\left(1 + \frac{2k}{n}\right)^2 + 2\left(1 + \frac{2k}{n}\right) + 5$$

$$= 3\left(1 + \frac{4k}{n} + \frac{4k^2}{n^2}\right) + 2 + \frac{4k}{n} + 5$$

$$= \frac{12}{n^2}k^2 + \frac{16}{n}k + 10$$

By definition,

$$A_n = \sum_{k=1}^{n} f(x_k) \cdot h$$

$$= \sum_{k=1}^{n} \left(\frac{12}{n^2}k^2 + \frac{16}{n}k + 10\right) \cdot \frac{2}{n}$$

$$= \frac{24}{n^3} \sum_{k=1}^{n} k^2 + \frac{32}{n^2} \sum_{k=1}^{n} k + \frac{20}{n} \sum_{k=1}^{n} 1$$

$$= \frac{24}{n^3} \cdot \frac{n(n + 1)(2n + 1)}{6} + \frac{32}{n^2} \cdot \frac{n(n + 1)}{2} + \frac{20}{n} \cdot n,$$

where we have used the standard sum formulas of Theorem 7.1.2. Then, after simplification,

$$A_n = 4\frac{(n + 1)(2n + 1)}{n^2} + 16\frac{n + 1}{n} + 20.$$

Thus

$$A = \lim_{n \to \infty} A_n$$

$$= \lim_{n \to \infty} \left[4\frac{(n + 1)(2n + 1)}{n^2} + 16\frac{n + 1}{n} + 20\right]$$

$$= \lim_{n \to \infty} \left[4\left(1 + \frac{1}{n}\right)\left(2 + \frac{1}{n}\right) + 16\left(1 + \frac{1}{n}\right) + 20\right]$$

$$= 4(1 + 0)(2 + 0) + 16(1 + 0) + 20$$

$$= 8 + 16 + 20 = 44 \text{ square units.}$$

EXERCISES 7.2

In each of the following exercises find the approximation A_n to the area bounded by the given curve $y = f(x)$, the x-axis and the lines $x = a$ and $x = b$ by dividing the area into n rectangles.

1. $y = 2x + 3$, $x = 1$, $x = 5$, and $n = 4$.

2. $y = 5 - 2x$, $x = 0$, $x = 2$, and $n = 6$.

3. $y = x^2$, $x = 0$, $x = 4$, and $n = 8$.

4. $y = 16 - x^2$, $x = 0$, $x = 3$, and $n = 6$.

5. $y = x^2 + 2$, $x = 0$, $x = 2$, and $n = 4$.

6. $y = 2x^2 - 3x + 1$, $x = -1$, $x = 3$, and $n = 4$.

7. $y = \sin x$, $x = 0$, $x = \pi$, and $n = 4$.

8. $y = \cos x$, $x = 0$, $x = \dfrac{\pi}{2}$, and $n = 5$.

9. $y = \ln\left(1 + \dfrac{1}{x}\right)$, $x = 1$, $x = 6$, and $n = 5$.

10. $y = 2^{-x}$, $x = 0$, $x = 3$, and $n = 3$.

Using the summation formulas of the previous section, evaluate the exact areas under the following curves bounded by the x-axis and the given vertical lines by taking the limit of A_n as $n \to \infty$.

11. $y = 3x + 2$, $x = 1$, and $x = 3$.

12. $y = 5x^2$, $x = 0$, and $x = 2$.

13. $y = 4 - x^2$, $x = 0$, and $x = 2$.

14. $y = 2x^2 + 3x - 1$, $x = 1$, and $x = 4$.

15. $y = x^3$, $x = 0$, and $x = 3$.

16. $y = 1 + x^3$, $x = 0$, and $x = 2$.

7.3 DEFINITE INTEGRALS

Let $f(x)$ be a continuous function defined on the closed interval $a \le x \le b$. Then the *definite integral* of $f(x)$ from $x = a$ to $x = b$, denoted by $\int_a^b f(x)\,dx$, is defined as

$$\int_a^b f(x)\,dx = \lim_{n \to \infty} \sum_{k=1}^{n} f(a + kh) \cdot h,$$

where $h = (b - a)/n$. The real numbers a and b are known as the *limits of integration*.

From the above definition it is clear that if $f(x) \geq 0$ in $a \leq x \leq b$, then the definite integral $\int_a^b f(x)\, dx$ represents the *area* bounded by the curve $y = f(x)$, the x-axis, and the vertical lines $x = a$, $x = b$.

In the last example of Section 7.2 we evaluated one such definite integral by dividing the corresponding area into rectangles and calculating the limit of the total area of the rectangles as their number tended to infinity. Clearly such a method of calculation is lengthy and cumbersome, and it is highly desirable to find an alternative means of calculating areas that requires less effort. Such an alternative method does exist and is based on a theorem called the *fundamental theorem of calculus*, perhaps the most remarkable theorem in calculus. It establishes a simple and elegant relationship between the definite integral of a function $f(x)$ and the antiderivative of $f(x)$.

THEOREM 7.3.1 (Fundamental Theorem of Calculus)

If $f(x)$ is a continuous function of x in $a \leq x \leq b$, and $F(x)$ is any antiderivative of $f(x)$, then

$$\int_a^b f(x)\, dx = F(b) - F(a).$$

We shall prove this theorem later, after doing some examples to show its use. But first note the following. When using the fundamental theorem it is usual as a matter of convenience to use square brackets with super- and subscripts on the right-hand side in the following way:

$$\int_a^b f(x)\, dx = \left[F(x) \right]_a^b = F(b) - F(a).$$

We read this as "The definite integral of $f(x)$ from $x = a$ to $x = b$ is $F(x)$ at b minus $F(x)$ at a." The bracket notation in the middle means that the function inside the bracket must be evaluated at the two values of the argument that are indicated after the bracket. The difference between these two values of the function is then taken in this order: value at the top argument minus value at the bottom argument.

In evaluating definite integrals we drop the constant of integration from the antiderivative of $f(x)$ because this constant of integration cancels in the final answer. Let $F(x) + C$ be any antiderivative of $f(x)$, where C is a constant of integration. Then, by the above theorem,

$$\int_a^b f(x)\, dx = \left[F(x) + C \right]_a^b$$
$$= [F(b) + C] - [F(a) + C]$$
$$= F(b) - F(a),$$

and C has disappeared from this answer.

EXAMPLE Evaluate the following definite integrals:

(a) $\int_a^b x^4 \, dx$.

(b) $\int_1^3 \frac{1}{t} \, dt$.

(c) $\int_0^\pi \sin u \, du$.

SOLUTION (a) $\int x^4 \, dx = \frac{x^5}{5}$. Thus

$$\int_a^b x^4 \, dx = \left[\frac{x^5}{5}\right]_a^b = \frac{b^5}{5} - \frac{a^5}{5} = \frac{1}{5}(b^5 - a^5).$$

(b) $\int_1^3 \frac{1}{t} \, dt = \left[\ln |t|\right]_1^3 = \ln |3| - \ln |1|$

$= \ln 3 - \ln 1 = \ln 3$

because $\ln 1 = 0$.

(c) $\int_0^\pi \sin u \, du = \left[-\cos u\right]_0^\pi = (-\cos \pi) - (-\cos 0)$

$= [-(-1)] - (-1) = 1 + 1 = 2.$

When evaluating definite integrals where the antiderivative is found by the method of substitution, it is important to note that the limits of integration also change when the variable of integration changes. This is illustrated in the following example.

EXAMPLE Evaluate

$$\int_1^2 xe^{x^2} \, dx.$$

SOLUTION Let

$$I = \int_1^2 xe^{x^2} \, dx.$$

To find an antiderivative of xe^{x^2} we can make use of the substitution method. We write the given integral as

$$I = \frac{1}{2} \int_1^2 e^{x^2} \cdot 2x \, dx.$$

Since $2x \, dx$, the differential of x^2, occurs in the integral, we put $x^2 = u$ so that $2x \, dx = du$. When $x = 1$, $u = 1^2 = 1$, and when $x = 2$, $u = 2^2 = 4$. Therefore

$$I = \frac{1}{2} \int_1^4 e^u \, du.$$

Note that in terms of the new variable u the limits of integration are 1 and 4. Then

$$I = \frac{1}{2}\left[e^u \right]_1^4 = \frac{1}{2}(e^4 - e^1)$$
$$= \frac{1}{2}e(e^3 - 1).$$

We can use this method of antiderivatives to find the area bounded by the curve $y = f(x)$, the x-axis, and the vertical lines $x = a$, $x = b$. This area is given by $\int_a^b f(x)\,dx$, provided that $f(x) \geq 0$ in $a \leq x \leq b$.

EXAMPLE   Evaluate the area bounded by the curve $y = 3x^2 + 2x + 5$, the x-axis, and the lines $x = 1$, $x = 3$.

SOLUTION   Clearly $f(x) = 3x^2 + 2x + 5$ is nonnegative for values of x in $1 \leq x \leq 3$. Thus the required area is given by

$$\int_1^3 (3x^2 + 2x + 5)\,dx = \left[x^3 + x^2 + 5x \right]_1^3$$
$$= [3^3 + 3^2 + 5(3)] - [1^3 + 1^2 + 5(1)]$$
$$= (27 + 9 + 15) - (1 + 1 + 5)$$
$$= 51 - 7 = 44 \text{ square units.}$$

This result agrees with the answer found in the last solved example of Section 7.2. In this earlier example the same area was calculated as the limit of the sum of the areas of an appropriate set of rectangles. In the present example the fundamental theorem has been used.

We shall now prove the fundamental theorem of calculus.

PROOF OF THEOREM 7.3.1

We shall prove the theorem for the particular case when $f(x)$ is a nonnegative increasing function in $a < x < b$, although the proof can readily be extended to all continuous functions.

When $f(x) \geq 0$ we know that the definite integral represents the area under the curve $y = f(x)$. Let us define the area function $A(x)$, which represents the area under the curve $y = f(x)$ from the value a to the value x of the abscissa, where $a \leq x \leq b$. $A(x)$ is the shaded area in Fig. 7.8. Thus $A(a) = 0$ because the area under the curve bounded by the vertical lines at a and again at a is zero. Further, $A(b)$ is clearly the area under the curve from a to b, that is,

$$A(b) = \int_a^b f(x)\,dx.$$

When x is changed to $x + \Delta x$ ($\Delta x > 0$), the area $A(x)$ also increases to $A + \Delta A$, where $A + \Delta A$ is the area under the curve between the values a and

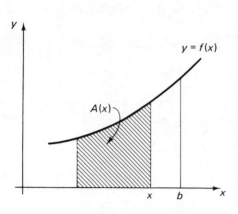

Figure 7.8   **Figure 7.9**

$x + \Delta x$ of the abscissa (Fig. 7.9). It is reasonable to expect that ΔA is equal to the area of the shaded strip in Fig. 7.9. This is a plausible assumption that, however, we shall not prove rigorously.

The area ΔA is greater than the area of the inscribed rectangle whose height is $f(x)$ and width Δx; the area ΔA is less than the area of the circumscribed rectangle whose height is $f(x + \Delta x)$ and width is Δx. Thus

$$f(x) \cdot \Delta x < \Delta A < f(x + \Delta x) \cdot \Delta x.$$

Dividing throughout by Δx, we get

$$f(x) < \frac{\Delta A}{\Delta x} < f(x + \Delta x).$$

Since $f(x + \Delta x) \to f(x)$ as $\Delta x \to 0$, upon taking the limits of the above inequalities as $\Delta x \to 0$, we obtain

$$\lim_{\Delta x \to 0} \frac{\Delta A}{\Delta x} = f(x)$$

or

$$A'(x) = f(x).$$

Because $F(x)$ is an antiderivative of $f(x)$, it follows also that $F'(x) = f(x)$. Thus $F(x)$ and $A(x)$ are both antiderivatives of $f(x)$, and hence they can differ from each other by at most a constant; that is,

$$A(x) = F(x) + C, \tag{i}$$

where C is some constant. Putting $x = a$ and remembering that $A(a) = 0$, we have

$$A(a) = F(a) + C = 0,$$

and so,

$$C = -F(a).$$

Replacing C by $-F(a)$ in (i) above we get

$$A(x) = F(x) - F(a).$$

Finally, setting $x = b$, we arrive at the result that

$$A(b) = F(b) - F(a);$$

or, after replacing $A(b)$ by the definite integral,

$$\int_a^b f(x)\, dx = F(b) - F(a),$$

which proves the result.

NOTE. The fundamental theorem of calculus (Theorem 7.3.1) was stated and proved only for continuous functions $f(x)$. In actual fact it is true for a much wider class of functions, and it is possible to use this theorem to calculate areas beneath graphs that have discontinuities.

THEOREM 7.3.2

If $f(t)$ is continuous in $a \leq t \leq x$, then

$$\frac{d}{dx}\left(\int_a^x f(t)\, dt\right) = f(x).$$

PROOF   Let $F(t)$ be an antiderivative of $f(t)$; then, by Theorem 7.3.1,

$$\int_a^x f(t)\, dt = \left[F(t)\right]_a^x = F(x) - F(a).$$

This is a function of x and hence can be differentiated with respect to x. Thus

$$\frac{d}{dx}\left[\int_a^x f(t)\, dt\right] = \frac{d}{dx}[F(x) - F(a)] = F'(x).$$

But $F'(t) = f(t)$ since $F(t)$ is an antiderivative of $f(t)$, and so

$$\frac{d}{dx}\left[\int_a^x f(t)\, dt\right] = f(x).$$

EXAMPLE   Evaluate

$$\frac{d}{dx}\left[\int_1^x \frac{t}{\ln(t+1)}\, dt\right].$$

SOLUTION   By Theorem 7.3.2 we have

$$\frac{d}{dx}\left[\int_1^x \frac{t}{\ln(t+1)}\, dt\right] = \frac{x}{\ln(x+1)}.$$

We need not first evaluate the integral and then differentiate.

EXAMPLE   Evaluate:

i. $\dfrac{d}{dx}\left[\displaystyle\int_1^3 u \sin^7 u \, du\right].$

ii. $\displaystyle\int_0^1 \dfrac{d}{dx}(x^3 \, \text{Sin}^{-1} x) \, dx.$

SOLUTION   **i.**   In this case it is important to note that the definite integral $\displaystyle\int_1^3 u \sin^7 u \, du$ has some constant value and is *not* a function of x. Thus

$$\frac{d}{dx}\left(\int_1^3 u \sin^7 u \, du\right) = 0.$$

ii.   From the definition of antiderivatives, if $F'(x) = f(x)$,

$$\int f(x) \, dx = \int F'(x) \, dx = F(x) + C.$$

Thus

$$\int \frac{d}{dx}(x^3 \, \text{Sin}^{-1} x) \, dx = x^3 \, \text{Sin}^{-1} x + C,$$

and so

$$\int_0^1 \frac{d}{dx}(x^3 \, \text{Sin}^{-1} x) \, dx = \left[x^3 \, \text{Sin}^{-1} x \right]_0^1 \quad \text{(we can forget about } C)$$
$$= 1^3 \, \text{Sin}^{-1} 1 - 0 \cdot \text{Sin}^{-1} 0$$
$$= \frac{\pi}{2} - 0 = \frac{\pi}{2}.$$

NOTE.   It is worthwhile to notice the difference between the two questions (i) and (ii). The positions of the integral sign and the differentiation operator d/dx are reversed.

We close this section by giving some simple properties of definite integrals.

THEOREM 7.3.2

(i) $\displaystyle\int_a^a f(x) \, dx = 0.$

(ii) $\displaystyle\int_a^b f(x) \, dx = -\int_b^a f(x) \, dx.$

(iii) $\displaystyle\int_a^b f(x) \, dx = \int_a^c f(x) \, dx + \int_c^b f(x) \, dx$

where c is any other number.

PROOF   Let $F(x)$ be any antiderivative of $f(x)$. Then by using the fundamental theorem of calculus we have:

i. $\displaystyle\int_a^a f(x)\,dx = \Big[\,F(x)\,\Big]_a^a = F(a) - F(a) = 0.$

ii. $\displaystyle\int_a^b f(x)\,dx = \Big[\,F(x)\,\Big]_a^b = F(b) - F(a)$

and

$$\int_b^a f(x)\,dx = \Big[\,F(x)\,\Big]_b^a = F(a) - F(b)$$

so that

$$\int_a^b f(x)\,dx = -\int_b^a f(x)\,dx.$$

iii.   The proof of this part is left as an exercise for the reader.

EXERCISES 7.3

Evaluate the following definite integrals.

1. $\displaystyle\int_0^1 x^2\,dx.$

2. $\displaystyle\int_{-1}^3 x^3\,dx.$

3. $\displaystyle\int_1^2 (3t^2 - 5t + 7)\,dt.$

4. $\displaystyle\int_0^5 (u^2 + u + 1)\,du.$

5. $\displaystyle\int_0^{\pi/2} \sin t\,dt.$

6. $\displaystyle\int_{-1}^3 (4x - 2e^x)\,dx.$

7. $\displaystyle\int_e^{e^2} \frac{\ln t}{t}\,dt.$

8. $\displaystyle\int_0^2 y^2 \cos(y^3)\,dy.$

9. $\displaystyle\int_0^1 \frac{e^{\mathrm{Tan}^{-1} x}}{1 + x^2}\,dx.$

10. $\displaystyle\int_0^1 e^x \sin(e^x)\,dx.$

11. $\displaystyle\int_2^2 \frac{x^2 + 2x - 8}{e^x + \ln x}\,dx.$

12. $\displaystyle\int_3^3 e^{x^2}\,dx.$

13. $\displaystyle\int_1^1 e^{\mathrm{Sin}^{-1} u}\,du.$

14. $\displaystyle\int_5^5 \frac{\sin^2 x - (\ln x)^5}{7x + 2}\,dx.$

15. $\displaystyle\int_0^1 \frac{d}{dt}\!\left[\frac{e^t + 2t - 1}{3 + \ln(1 + t)}\right]dt.$

16. $\displaystyle\int_0^1 \frac{d}{dx}(\mathrm{Tan}^{-1} x)^2\,dx.$

17–22. Evaluate the areas in questions (11–16) of the previous Exercises 7.2 by using the definite integral.

Evaluate the following derivatives:

23. $\displaystyle\frac{d}{dx}\left(\int_2^x \frac{e^t \ln t}{1 + t^2}\,dt\right).$

24. $\displaystyle\frac{d}{dt}\left(\int_1^t x^3\,\mathrm{Sin}^{-1} x\,dx\right).$

25. $\dfrac{d}{dx}\left(\displaystyle\int_{x}^{3}\dfrac{\sin u}{1+u}\,du\right)$.

26. $\dfrac{d}{du}\left(\displaystyle\int_{u}^{1}(\mathrm{Tan}^{-1}\,t)^4\,dt\right)$.

27. $\dfrac{d}{dt}\left(\displaystyle\int_{1}^{2}\dfrac{\mathrm{Cos}^{-1}\,x}{1+x^2}\,dx\right)$.

28. $\dfrac{d}{dx}\left(\displaystyle\int_{1}^{3}(\ln x)^7\,dx\right)$.

7.4 MORE ON AREAS

In the last section we proved that the area under the curve $y = f(x)$ bounded by the lines $x = a$, $x = b$, and $y = 0$ (x-axis) is given by the definite integral $\displaystyle\int_{a}^{b} f(x)\,dx$ in the case when $f(x) \ge 0$ in $a \le x \le b$.

Consider now the corresponding area bounded by the curve $y = f(x)$, the lines $x = a$, $x = b$, and the x-axis in the case when $f(x) \le 0$ for $a \le x \le b$. The area in question clearly lies below the x-axis, as shown in Fig. 7.10.

Let us define $g(x) = -f(x)$ so that $g(x) \ge 0$ for $a \le x \le b$. The area bounded by $y = g(x)$ [or $y = -f(x)$], the lines $x = a$, $x = b$, and the x-axis lies above the x-axis (Fig. 7.11). This area, as in the last section, is given by the definite integral

$$\int_{a}^{b} g(x)\,dx = -\int_{a}^{b} f(x)\,dx.$$

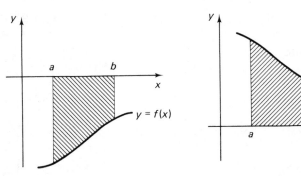

Figure 7.10   Figure 7.11

Comparing Figs. 7.10 and 7.11, it is clear that the two shaded areas are equal in magnitude; this is the case since one area can be obtained simply by reflecting the other in the x-axis. Thus the area below the x-axis, bounded by the curve $y = f(x)$ and the lines $x = a$ and $x = b$, is given by the definite integral

$$-\int_{a}^{b} f(x)\,dx.$$

EXAMPLE   Find the area bounded by $y = x^2 - 9$, $x = 0$, $x = 2$, and the x-axis.

SOLUTION   The graph of $y = x^2 - 9$ lies below the x-axis for $0 \le x \le 2$. The required area (shown shaded in Fig. 7.12) is given by

$$-\int_0^2 (x^2 - 9)\,dx = \int_0^2 (9 - x^2)\,dx$$

$$= \left[9x - \frac{x^3}{3} \right]_0^2$$

$$= \left[9(2) - \frac{2^3}{3} \right] - \left[9(0) - \frac{0^3}{3} \right]$$

$$= 18 - \tfrac{8}{3} - 0$$

$$= 15\tfrac{1}{3} \text{ square units.}$$

Let us now consider the area bounded by the curve $y = f(x)$, the lines $x = a$, $x = b$, and the x-axis, in the case when $f(x)$ is sometimes positive and sometimes negative in the interval $a \le x \le b$ (Fig. 7.13). Such an area consists of certain parts

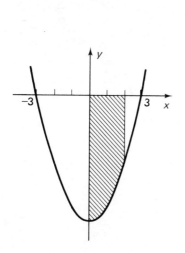

Figure 7.12

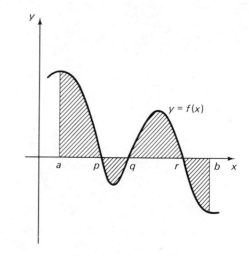

Figure 7.13

below the x-axis and certain parts above the x-axis. We shall assume that we can find the points where the graph of $y = f(x)$ crosses the x-axis, that is, the values of x for which $f(x) = 0$. In the figure we have illustrated the case when there are three such points, denoted by $x = p$, q, and r. In the case shown in Fig. 7.13,

$$f(x) \ge 0 \quad \text{for} \quad a \le x \le p,$$
$$f(x) \le 0 \quad \text{for} \quad p \le x \le q,$$
$$f(x) \ge 0 \quad \text{for} \quad q \le x \le r,$$

and

$$f(x) \le 0 \quad \text{for} \quad r \le x \le b.$$

331

In a problem of this type we calculate the area of each subinterval separately and then determine the required area from the sum of all these areas. In Fig. 7.13 the areas between $x = a$ and $x = p$ and between $x = q$ and $x = r$ lie above the x-axis, while the areas between $x = p$ and $x = q$ and between $x = r$ and $x = b$ lie below the x-axis. Therefore the required area is equal to

$$\int_a^p f(x)\,dx + \left[-\int_p^q f(x)\,dx\right] + \int_q^r f(x)\,dx + \left[-\int_r^b f(x)\,dx\right].$$

EXAMPLE   Find the area bounded by the x-axis, the curve $y = (x - 1)(x - 2)(x - 3)$, and the lines $x = 0$, $x = 4$.

SOLUTION   The graph of $y = (x - 1)(x - 2)(x - 3)$ crosses the x-axis where $y = 0$, that is, $(x - 1)(x - 2)(x - 3) = 0$, which gives $x = 1, 2, 3$ (Fig. 7.14). Thus we

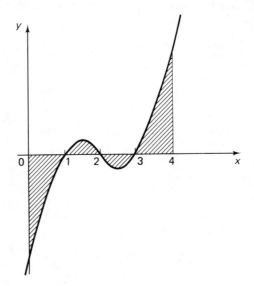

Figure 7.14

divide the given interval $0 \leq x \leq 4$ into the four subintervals: $0 \leq x \leq 1$, $1 \leq x \leq 2$, $2 \leq x \leq 3$, and $3 \leq x \leq 4$. To find whether the curve $y = (x - 1)(x - 2)(x - 3)$ lies above or below the x-axis in each of these subintervals, we need only to observe the sign of y, which is easily determined from the following table:

$x \longrightarrow$	$0 < x < 1$	$1 < x < 2$	$2 < x < 3$	$3 < x < 4$
$(x - 1)$	$-$	$+$	$+$	$+$
$(x - 2)$	$-$	$-$	$+$	$+$
$(x - 3)$	$-$	$-$	$-$	$+$
$y = (x - 2)(x - 2)(x - 3)$	$-$	$+$	$-$	$+$

Thus the graph of $y = (x - 1)(x - 2)(x - 3)$ lies above the x-axis for $1 < x < 2$ and $3 < x < 4$ and below the x-axis for $0 < x < 1$ and $2 < x < 3$. Hence the required area is

$$A = \left[-\int_0^1 (x - 1)(x - 2)(x - 3)\, dx \right] + \int_1^2 (x - 1)(x - 2)(x - 3)\, dx$$
$$+ \left[-\int_2^3 (x - 1)(x - 2)(x - 3)\, dx \right] + \int_3^4 (x - 1)(x - 2)(x - 3)\, dx.$$

Let us denote by $F(x)$ the indefinite integral

$$F(x) = \int (x - 1)(x - 2)(x - 3)\, dx$$
$$= \int (x^3 - 6x^2 + 11x - 6)\, dx$$
$$= \tfrac{1}{4}x^4 - 2x^3 + \tfrac{11}{2}x^2 - 6x.$$

Then

$$A = -[F(1) - F(0)] + [F(2) - F(1)] - [F(3) - F(2)] + [F(4) - F(3)]$$
$$= F(0) - 2F(1) + 2F(2) - 2F(3) + F(4)$$
$$= 0 - 2[\tfrac{1}{4} - 2 + \tfrac{11}{2} - 6] + 2[\tfrac{1}{4} \cdot 2^4 - 2 \cdot 2^3 + \tfrac{11}{2} \cdot 2^2 - 6 \cdot 2]$$
$$- 2[\tfrac{1}{4} \cdot 3^4 - 2 \cdot 3^3 + \tfrac{11}{2} \cdot 3^2 - 6 \cdot 3]$$
$$+ [\tfrac{1}{4} \cdot 4^4 - 2 \cdot 4^3 + \tfrac{11}{2} \cdot 4^2 - 6 \cdot 4]$$
$$= \tfrac{9}{2} - 4 + \tfrac{9}{2} + 0$$
$$= 5 \text{ square units.}$$

Area Between Two Curves

Let us now consider the area bounded by the two curves $y = f(x)$, $y = g(x)$, and the lines $x = a$, $x = b$. We shall suppose that $f(x) \geq g(x) \geq 0$ in $a \leq x \leq b$ so that both curves lie above the x-axis and the curve $y = f(x)$ lies above the curve $y = g(x)$. This area is shaded in Fig. 7.15. Clearly this area is the difference between the area

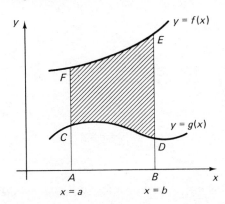

Figure 7.15

bounded by $y = f(x)$ and the x-axis and the area bounded by $y = g(x)$ and the x-axis; that is, the area $CDEF$ between the two curves is equal to the area $ABEF$ minus the area $ABDC$.

Thus the required area is given by

$$\int_a^b f(x)\, dx - \int_a^b g(x)\, dx = \int_a^b [f(x) - g(x)]\, dx.$$

EXAMPLE   Find the area between the curves $y = x^2 + 5$, $y = x^3$ and the lines $x = 1$, $x = 2$.

SOLUTION   The graph of $y = x^2 + 5$ lies above the curve $y = x^3$ in the interval $1 < x < 2$. Thus the required area A (shown shaded in Fig. 7.16) is given by

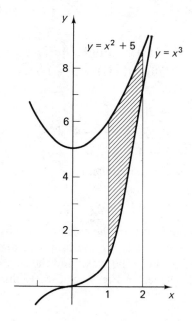

Figure 7.16

$$A = \int_1^2 [(x^2 + 5) - x^3]\, dx$$

$$= \left[\frac{x^3}{3} + 5x - \frac{x^4}{4}\right]_1^2$$

$$= (\tfrac{8}{3} + 10 - 4) - (\tfrac{1}{3} + 5 - \tfrac{1}{4})$$

$$= 3\tfrac{7}{12} \text{ square units.}$$

EXAMPLE   Find the area of the region that is enclosed by the curves $y = -x^2$ and $y = x^2 - 8$.

SOLUTION   In this case we are not given the limits of integration. The first step is to sketch the graphs of the two curves in order to determine the required area they enclose and the limits of integration. Sketches of the two curves are shown in Fig. 7.17, in which the enclosed area is shaded.

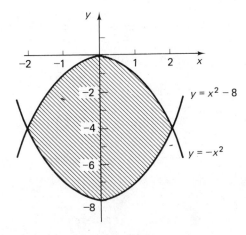

Figure 7.17

To find the points of intersection of two curves we must treat the two equations of the curves as simultaneous equations and solve them for x and y. In this particular example, equating the two values of y, we have:

$$y = -x^2 = x^2 - 8 \quad \text{or} \quad 2x^2 - 8 = 0.$$

Therefore

$$x = \pm 2.$$

Thus, for the area that is shown shaded in Fig. 7.17, x varies from -2 to $+2$. Hence

$$\text{Area} = \int_{-2}^{2} [(-x^2) - (x^2 - 8)]\, dx$$

$$= \int_{-2}^{2} (8 - 2x^2)\, dx$$

$$= \left[8x - \tfrac{2}{3}x^3 \right]_{-2}^{2}$$

$$= (16 - \tfrac{16}{3}) - (-16 + \tfrac{16}{3})$$

$$= \tfrac{64}{3} \text{ square units.}$$

EXAMPLE   Find the area bounded by the curves $y = \sin x$ and $y = \cos x$ between $x = 0$ and $x = \pi/2$.

SOLUTION The two curves $y = \sin x$ and $y = \cos x$ intersect where $\sin x = \cos x$ or $\tan x = 1$ (Fig. 7.18). In the interval $0 \le x \le \pi/2$, there is one value of x at which $\tan x = 1$, namely $x = \pi/4$. In this case we divide the problem into two parts: for $0 \le x \le \pi/4$, $\cos x \ge \sin x$; for $\pi/4 \le x \le \pi/2$, $\sin x \ge \cos x$. Thus the required area is given by

$$A = \int_0^{\pi/4} (\cos x - \sin x)\, dx + \int_{\pi/4}^{\pi/2} (\sin x - \cos x)\, dx$$

$$= \Big[\sin x + \cos x \Big]_0^{\pi/4} + \Big[-\cos x - \sin x \Big]_{\pi/4}^{\pi/2}$$

$$= \left(\sin \frac{\pi}{4} + \cos \frac{\pi}{4} \right) - (\sin 0 + \cos 0)$$

$$+ \left(-\cos \frac{\pi}{2} - \sin \frac{\pi}{2} \right) - \left(-\cos \frac{\pi}{4} - \sin \frac{\pi}{4} \right)$$

$$= \left(\frac{1}{\sqrt{2}} + \frac{1}{\sqrt{2}} \right) - (0 + 1) + (0 - 1) - \left(-\frac{1}{\sqrt{2}} - \frac{1}{\sqrt{2}} \right)$$

$$= \frac{4}{\sqrt{2}} - 2$$

$$= (2\sqrt{2} - 2) \text{ square units.}$$

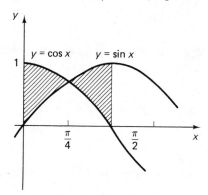

Figure 7.18

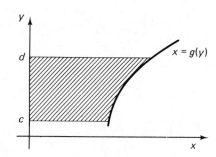

Figure 7.19

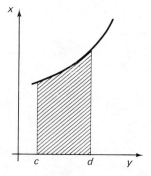

Figure 7.20

We close this section by giving the expression for the area bounded by the curve $x = g(y)$, the y-axis, and the horizontal lines $y = c$, $y = d$. This area (shown shaded in Fig. 7.19) is given by

$$\int_c^d g(y)\, dy,$$

where $d \ge c \ge 0$. We can see this if we redraw the figure with the y-axis horizontal and the x-axis vertical, as shown in Fig. 7.20. The area in question then becomes the area between the curve and the horizontal axis, and is given by the appropriate definite integral. The names of the variables x and y are simply interchanged.

EXAMPLE Find the area bounded by the parabola $y^2 = 4x$, the y-axis, and the horizontal lines $y = 1$, $y = 3$.

SOLUTION The required area is shown in Fig. 7.21. Here $x = y^2/4$, so that $g(y) = y^2/4$. Thus the required area is

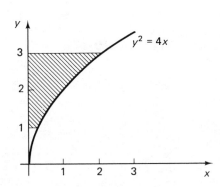

Figure 7.21

$$\int_1^3 \frac{y^2}{4}\, dy = \left[\frac{1}{4} \cdot \frac{y^3}{3}\right]_{y=1}^3$$
$$= \tfrac{1}{12}(3^3 - 1^3)$$
$$= \tfrac{1}{12}(27 - 1)$$
$$= \tfrac{13}{6} \text{ square units.}$$

EXERCISES 7.4

In each of the following exercises, find the area bounded by the curve $y = f(x)$, the x-axis, and the lines $x = a$ and $x = b$.

1. $y = -x^2$; $x = 0$, $x = 3$.
2. $y = 1 - \sqrt{x}$; $x = 1$, $x = 9$.
3. $y = -e^x$; $x = \ln 2$, $x = \ln 5$.
4. $y = x^3$; $x = -1$, $x = 1$.
5. $y = x^2 - 4$; $x = 0$, $x = 3$.
6. $y = x^2 - 3x + 2$; $x = 0$, $x = 3$.
7. $y = \sin x$; $x = 0$, $x = 3\pi$.
8. $y = \cos x$; $x = 0$, $x = 2\pi$.

Find the area that lies between the following pairs of curves and between the two given vertical lines.

9. $y = x^2$, $y = 3x$; $x = 1$, $x = 2$.
10. $y = x^2$, $y = 2x - 1$; $x = 0$, $x = 2$.
11. $y = \sqrt{x}$, $y = x^2$; $x = 0$, $x = 1$.
12. $y = x^2$, $y = x^3$; $x = 0$, $x = 2$.
13. $y = e^x$, $y = x^2$; $x = 0$, $x = 1$.
14. $y = x^3$, $y = 3x - 2$; $x = 0$, $x = 2$.

Determine the area of the region enclosed between the following pairs of curves:

15. $y = x^2$; $y = 2 - x^2$.   **16.** $y = x^2$; $y = \sqrt{x}$.

17. $y = x^3$; $y = x^2$.   **18.** $y = x^2$; $y = 2x$.

Find the area bounded by the following curves and lines:

19. $y = x^2$, $y = 0$, $y = 4$, and $x = 0$ (y-axis).

20. $y^2 = x$, $y = 0$, $y = 2$, and $x = 0$.

7.5 VOLUMES OF REVOLUTION

Let R be a plane region that lies entirely on one side of a fixed straight line L in its plane (Fig. 7.22). When the plane region R is revolved about the line L, a solid results that is called a *solid of revolution*. The fixed line L is called the *axis* of the solid of revolution.

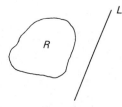

Figure 7.22

A number of examples of solids of revolution are illustrated in Fig. 7.23. For example, if a semicircle is revolved about its bounding diameter, the solid of revolution generated is a sphere. When a plane rectangle is revolved about one of its sides, the solid of revolution is a cylinder. When a right triangle is revolved about one of its edges other than the hypotenuse, it generates a cone. If a circle revolves about a line that does not intersect it, the solid of revolution generated looks like a doughnut and is called a *torus*.

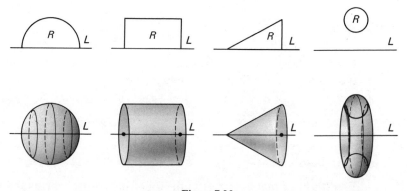

Figure 7.23

Consider the plane section of a solid of revolution by a plane that is perpendicular to the axis of revolution. Clearly such a plane cuts the solid in a region that is either a circle or is bounded by two (or possibly more) concentric circles. For example, a section through the cone perpendicular to the axis gives a circle, while a section through the torus gives a region lying between two concentric circles.

In Section 7.2 the determination of the plane area under a curve rested on two ideas, namely the formula for the area of the rectangle and a method for approximating any plane region by a large number of rectangular slices. In the same way, the determination of the volumes of solids of revolution also involves two basic steps. The first makes use of the fact that the volume V of a right circular cylinder of radius r and height h is $V = \pi r^2 h$ (Fig. 7.24). The second step is to approximate any solid of revolution by a large number of right circular cylinders obtained by making slices perpendicular to the axis of revolution.

Let R be the plane region bounded by $y = f(x)$, the x-axis, and the lines $x = a$, $x = b$, where $f(x) \geq 0$ in $a \leq x \leq b$. Furthermore, let the interval $a \leq x \leq b$ be subdivided into n intervals with the dividing points $x_k = a + kh$ ($k = 1, 2, \ldots$), where $h = (b - a)/n$ and $x_0 = a$ and $x_n = b$ (Fig. 7.25). The approximating rectangle

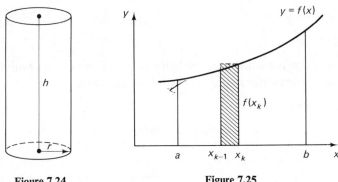

Figure 7.24   **Figure 7.25**

for the area under the curve in the interval $x_{k-1} \leq x \leq x_k$ has an altitude $f(x_k)$ and a width $x_k - x_{k-1} = h$.

When the region R is revolved about the x-axis to generate a solid of revolution, each of the n rectangles generates a thin circular cylinder. The cylinder generated by a rectangle in the kth subinterval has a radius $f(x_k)$ and height (or width) h, so that its volume is

$$\pi[f(x_k)]^2 \cdot h$$

(Fig. 7.26). The volume V of the solid of revolution can be approximated by the sum of the volumes of the n circular cylinders, that is,

$$V \simeq \sum_{k=1}^{n} \pi[f(x_k)]^2 \cdot h.$$

As $h \to 0$ or $n \to \infty$, this approximating volume gives the true volume of the solid of revolution. Thus

$$V = \lim_{n \to \infty} \sum_{k=1}^{n} \pi[f(x_k)]^2 \cdot h = \int_a^b \pi[f(x)]^2 \, dx,$$

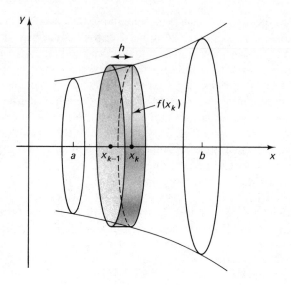

Figure 7.26

where the definition of the definite integral given at the beginning Section 7.3 has been used. Thus we have proved the following result.

THEOREM 7.5.1

Let $f(x)$ be a continuous nonnegative function in $a \leq x \leq b$. Then the volume of the solid of revolution that is formed by revolving the area bounded by $y = f(x)$, the x-axis, and the lines $x = a$, $x = b$ about the x-axis is given by

$$V = \pi \int_a^b [f(x)]^2 \, dx.$$

EXAMPLE   Find the volume of the solid generated by revolving about the x-axis the area bounded by the curve $y = \sqrt{x}$, the x-axis, and the vertical lines $x = 0$, $x = 4$.

SOLUTION   Figure 7.27 shows the region to be revolved and the solid of revolution formed whose volume is required. This solid of revolution is called a *paraboloid*. According to the formula in Theorem 7.5.1, the required volume is

$$V = \pi \int_0^4 (\sqrt{x})^2 \, dx$$

$$= \pi \int_0^4 x \, dx$$

$$= \pi \left[\frac{x^2}{2} \right]_0^4$$

$$= \pi \left[\frac{4^2}{2} - \frac{0^2}{2} \right]$$

$$= 8\pi \text{ cubic units.}$$

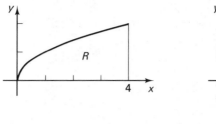

Figure 7.27

EXAMPLE   The triangular region bounded by the lines $y = (r/h)x$, $x = h$, and the x-axis is revolved about the x-axis. Find the volume of the solid generated.

SOLUTION   The solid generated is a right circular cone as illustrated in Fig. 7.28. The radius of the base of the cone is equal to r and the height is h. The volume is

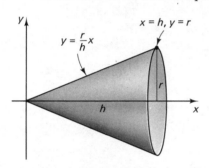

Figure 7.28

$$V = \pi \int_0^h \left(\frac{r}{h}x\right)^2 dx$$

$$= \pi \frac{r^2}{h^2} \int_0^h x^2 \, dx$$

$$= \pi \frac{r^2}{h^2} \left[\frac{x^3}{3}\right]_0^h$$

$$= \pi \frac{r^2}{h^2} \left[\frac{h^3}{3} - 0\right]$$

$$= \frac{\pi}{3} r^2 h \text{ cubic units.}$$

We shall now consider the case when the region R that is revolved in forming the solid of revolution is not bounded by the x-axis as in the previous example, but instead is bounded by the two curves $y = f(x)$ and $y = g(x)$ (as well as the lines $x = a$, $x = b$). We suppose that $f(x) \geq g(x) \geq 0$ for $a \leq x \leq b$, as illustrated in Fig. 7.29. Then the volume of the solid generated by revolving the region R about the x-axis is the difference between the volume generated by revolving the region between $y = f(x)$ and the x-axis and the volume generated by revolving the region between $y = g(x)$ and the x-axis (in both cases with x in the interval $a \leq x \leq b$); thus

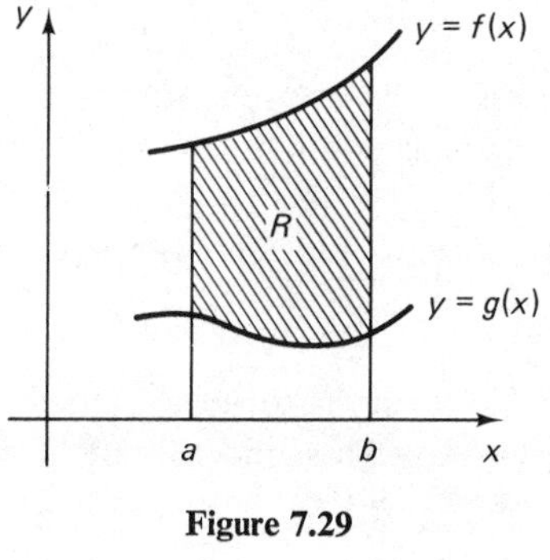

$$V = \pi \int_a^b [f(x)]^2 \, dx - \pi \int_a^b [g(x)]^2 \, dx$$

$$= \pi \int_a^b \{[f(x)]^2 - [g(x)]^2\} \, dx.$$

Figure 7.29

EXAMPLE   Find the volume of the solid generated by revolving about the x-axis, the region enclosed between the two parabolas $y = x^2$ and $y = \sqrt{x}$.

SOLUTION   The region R to be revolved is shown shaded in Fig. 7.30. To find the limits of integration we find the points of intersection of the two parabolas $y = x^2$ and $y = \sqrt{x}$. Equating the two values of y we get

$$x^2 = \sqrt{x} \quad \text{or} \quad \sqrt{x}(x^{3/2} - 1) = 0,$$

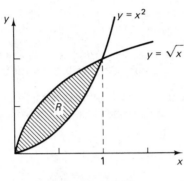

which gives $x = 0$ or $x = 1$. In the interval $0 \le x \le 1$, the curve $y = \sqrt{x}$ lies above the curve $y = x^2$. Thus the required volume is

$$V = \pi \int_0^1 [(\sqrt{x})^2 - (x^2)^2] \, dx$$

$$= \pi \int_0^1 (x - x^4) \, dx$$

$$= \pi \left[\frac{x^2}{2} - \frac{x^5}{5} \right]_0^1$$

$$= \pi [(\tfrac{1}{2} - \tfrac{1}{5}) - 0]$$

$$= \frac{3\pi}{10} \text{ cubic units.}$$

Figure 7.30

EXAMPLE   A circle of radius r, with its center at the point $(0, R)$ (with $R > r$), is revolved around the x-axis to form a solid of revolution. Find the volume of the resulting torus.

SOLUTION   The equation of the circle (Fig. 7.31), from the general formula, is

$$(x - 0)^2 + (y - R)^2 = r^2.$$

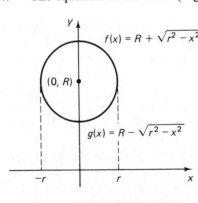

Figure 7.31

Therefore

$$(y - R)^2 = r^2 - x^2$$

and so, after solving for y,

$$y = R \pm \sqrt{r^2 - x^2}.$$

Thus we have two functions: The upper semicircle corresponds to the positive square root $y = R + \sqrt{r^2 - x^2} = f(x)$, and the lower semicircle to the negative square root $y = R - \sqrt{r^2 - x^2} = g(x)$. In both cases x varies from $-r$ to r.

To form the torus, the area between these two semicircles is revolved about the x-axis. Thus the volume of revolution is given by

$$V = \pi \int_{-r}^{r} \{[f(x)]^2 - [g(x)]^2\}\, dx$$

$$= \pi \int_{-r}^{r} [(R + \sqrt{r^2 - x^2})^2 - (R - \sqrt{r^2 - x^2})^2]\, dx$$

$$= \pi \int_{-r}^{r} 4R\sqrt{r^2 - x^2}\, dx$$

$$= 4\pi R \left[\frac{1}{2} r^2 \operatorname{Sin}^{-1} \frac{x}{r} + \frac{1}{2} x\sqrt{r^2 - x^2}\right]_{-r}^{r}$$

after using formula 42 in Appendix II. Therefore

$$V = 2\pi R[r^2 \operatorname{Sin}^{-1}(1) - r^2 \operatorname{Sin}^{-1}(-1) + 0 - 0]$$

$$= 2\pi R r^2 \left[\frac{\pi}{2} - \left(-\frac{\pi}{2}\right)\right]$$

$$= 2\pi^2 R r^2.$$

NOTE. If a curve whose equation is written in the form $x = g(y)$ is revolved about the y-axis, then the resulting solid of revolution that lies between $y = c$ and $y = d$ has a volume given by

$$\pi \int_{c}^{d} [g(y)]^2\, dy.$$

EXERCISES 7.5

The region R given in each of the exercises below is revolved about the x-axis. Find the volume of solid formed.

1. R is bounded by $y = x^2$, $x = 0$, $x = 2$, and the x-axis ($y = 0$).
2. R is bounded by $y = x + 1$, $x = 0$, $x = 2$, and $y = 0$.
3. R is bounded by $y = \sin x$, $x = 0$, $x = \pi$, and $y = 0$.
4. R is bounded by $y = \cos x$, $x = 0$, $x = \pi/2$, and $y = 0$.
5. R is bounded by $y = \sqrt{4 - x^2}$, $x = -2$, $x = 2$, and $y = 0$.
6. R is bounded by $y = 1/x$, $x = 1$, $x = 3$, and $y = 0$.
7. R is bounded by $y = x^{1/3}$, $x = 0$, $x = 8$, and $y = 0$.
8. R is enclosed between $y = x^2$ and $y = x^3$.
9. R is enclosed between $y = \sqrt{4 - x^2}$ and $x - \sqrt{3}\,y + 2 = 0$.
10. R is enclosed between $y = \sqrt{4 - x^2}$, $x - \sqrt{3}\,y + 2 = 0$, and $y = 0$.
11. R is enclosed between $y = x^2$ and $y = x$.
12. R is enclosed between $y = x^2$ and $y = x + 2$.

13. Using the methods of integration, find the volume of a sphere of radius r.

14. The region bounded by the x-axis and the upper half of the ellipse

$$\frac{x^2}{a^2} + \frac{y^2}{b^2} = 1$$

is revolved about the x-axis. Find the volume of the solid generated (ellipsoid of revolution).

15. The trunk of a fir tree is 50 ft high, with diameters of 2 ft at the bottom and 1 ft at the top. What volume of wood does it contain? (Assume that the sides are straight.)

7.6 CENTERS OF MASS

Consider a finite number n of particles of masses $m_1, m_2, \ldots, m_n$ placed along the x-axis at distances $x_1, x_2, \ldots, x_n$ respectively from the origin (Fig. 7.32). Then

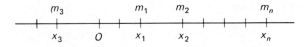

Figure 7.32

the *moment* of this system of n masses about the origin is defined to be

$$m_1 x_1 + m_2 x_2 + \ldots + m_n x_n = \sum_{k=1}^{n} m_k x_k.$$

The total mass of the system is

$$m = m_1 + m_2 + \ldots + m_n = \sum_{k=1}^{n} m_k.$$

The center of mass of the system of particles is defined to be the point $\bar{x}$, which has the property that if the total mass were concentrated at $\bar{x}$ then the moment about the origin would be unchanged. Now if the total mass $m = \sum_{k=1}^{n} m_k$ were concentrated at $\bar{x}$, the moment would simply be $m\bar{x}$. Therefore

$$m\bar{x} = \sum_{k=1}^{n} m_k x_k,$$

that is,

$$\bar{x} = \frac{\displaystyle\sum_{k=1}^{n} m_k x_k}{m} = \frac{\displaystyle\sum_{k=1}^{n} m_k x_k}{\displaystyle\sum_{k=1}^{n} m_k},$$

which determines the center of mass of the system.

The center of mass has the property that the system will balance if supported by a knife-edge placed at the position $\bar{x}$ (the x-axis line itself is considered weightless). For this reason the center of mass is also often called *the center of gravity* of the system. We note that the position $\bar{x}$ of the center of mass does not depend on the choice of the origin.

EXAMPLE   The principle of moments underlies the simple seesaw. Suppose that one boy weighs 30 kg and sits 2 meters (m) from the pivot of a seesaw (Fig. 7.33). If a

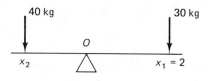

Figure 7.33

second boy weighs 40 kg, how far from the pivot must he sit in order for the seesaw to balance?

SOLUTION   Let us take the origin to be at the pivot of the seesaw and let the coordinates of the two boys be $x_1 = 2$ m and x_2, which is unknown. The two masses are $m_1 = 30$ and $m_2 = 40$. The seesaw will balance then if the center of mass of the two boys occurs at the pivot, that is, $\bar{x} = 0$. Therefore

$$m_1 x_1 + m_2 x_2 = 30(2) + 40 x_2 = (m_1 + m_2)\bar{x} = 0.$$

That is,

$$60 + 40 x_2 = 0,$$

i.e.,

$$x_2 = -\tfrac{60}{40} = -1.5.$$

So the second boy must sit 1.5 m from the pivot.

EXAMPLE   Four masses of 5, 7, 3, and 10 units are placed along the x-axis at the points $(3, 0)$, $(-2, 0)$, $(-5, 0)$, and $(7, 0)$ respectively. Find the center of mass of the system.

SOLUTION   Here $m_1 = 5$, $m_2 = 7$, $m_3 = 3$, and $m_4 = 10$. Also $x_1 = 3$, $x_2 = -2$, $x_3 = -5$, and $x_4 = 7$. Then the center of mass is given by

$$\bar{x} = \frac{\sum\limits_{k=1}^{4} m_k x_k}{\sum\limits_{k=1}^{4} m_k} = \frac{m_1 x_1 + m_2 x_2 + m_3 x_3 + m_4 x_4}{m_1 + m_2 + m_3 + m_4}$$

$$= \frac{5(3) + 7(-2) + 3(-5) + 10(7)}{5 + 7 + 3 + 10}$$

$$= \frac{15 - 14 - 15 + 70}{25} = \frac{56}{25}.$$

Now consider a system of n particles of masses $m_1, m_2, \ldots, m_n$ located at the points $(x_1, y_1), (x_2, y_2), \ldots, (x_n, y_n)$ respectively in the xy-plane. Then the *moment* of the system *about the y-axis* is defined to be

$$M_y = m_1 x_1 + m_2 x_2 + \ldots + m_n x_n = \sum_{k=1}^{n} m_k x_k,$$

and the *moment about the x-axis* is defined to be

$$M_x = m_1 y_1 + m_2 y_2 + \ldots + m_n y_n = \sum_{k=1}^{n} m_k y_k.$$

The total mass of the system is

$$m = m_1 + m_2 + \ldots + m_n = \sum_{k=1}^{n} m_k.$$

If we imagine the masses to be supported by a weightless tray and assume that each mass occupies exactly one point, then the *center of mass* is the point at which the tray can be supported by a single pinpoint support in such a way as to balance perfectly in a horizontal position (Fig. 7.34). Mathematically, the center of mass (or center-of gravity) is the point $(\bar{x}, \bar{y})$ such that

$$m\bar{x} = M_y, \qquad m\bar{y} = M_x.$$

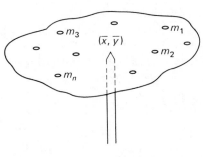

Figure 7.34

Thus the center of mass $(\bar{x}, \bar{y})$ of the system of n masses is given by

$$\bar{x} = \frac{m_1 x_1 + m_2 x_2 + \ldots + m_n x_n}{m_1 + m_2 + \ldots + m_n} = \frac{\sum_{k=1}^{n} m_k x_k}{\sum_{k=1}^{n} m_k}.$$

$$\bar{y} = \frac{m_1 y_1 + m_2 y_2 + \ldots + m_n y_n}{m_1 + m_2 + \ldots + m_n} = \frac{\sum_{k=1}^{n} m_k y_k}{\sum_{k=1}^{n} m_k}.$$

EXAMPLE   Find the center of mass of the system of masses of 2, 3, and 4 located at the points $(1, 2)$, $(3, -7)$, and $(5, 1)$ respectively.

SOLUTION   Here $m_1 = 2$, $m_2 = 3$, and $m_3 = 4$; $(x_1, y_1) = (1, 2)$, $(x_2, y_2) = (3, -7)$, and $(x_3, y_3) = (5, 1)$. Then

$$\bar{x} = \frac{m_1 x_1 + m_2 x_2 + m_3 x_3}{m_1 + m_2 + m_3} = \frac{2(1) + 3(3) + 4(5)}{2 + 3 + 4} = \frac{31}{9}.$$

$$\bar{y} = \frac{m_1 y_1 + m_2 y_2 + m_3 y_3}{m_1 + m_2 + m_3} = \frac{2(2) + 3(-7) + 4(1)}{2 + 3 + 4} = -\frac{13}{9}.$$

The center of mass is therefore the point $(\frac{31}{9}, -\frac{13}{9})$.

Let us now consider *homogeneous laminas*. (The word lamina is used to mean a thin sheet or layer of material, for example, a piece of plywood or of sheet-steel.) We say that a lamina is homogeneous if two pieces of it have equal weights whenever their areas are equal. For a homogeneous lamina, we define the density to be the mass per unit area. Therefore the mass of a homogeneous lamina of density D and area A is given by

$$m = DA.$$

We now wish to define the center of mass of a homogeneous lamina in a way that is to be consistent with our experience with systems of particles. First of all we observe from our common experience that a rectangular sheet of uniform thickness can be balanced at its geometric center, so it is natural for us to define the center of mass of a homogeneous rectangular lamina to be its geometric center (Fig. 7.35). In the same

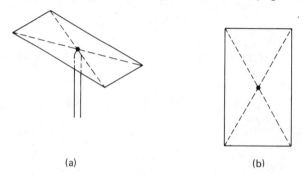

(a)   (b)

Figure 7.35

manner, we define the center of mass of a uniform circular region to be its geometric center.

With these definitions of the centers of mass of uniform rectangular and circular laminas, it is possible to find the center of mass of regions that are combinations of any number of rectangles and circles. In doing so, we treat each rectangle or circle as if all of its mass were concentrated at its center. This is illustrated by the following example.

EXAMPLE   A region is made up of a combination of rectangles of uniform density D. The shape and dimensions of the region are illustrated in Fig. 7.36. Find the center of mass.

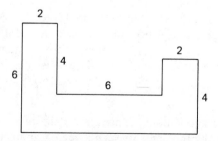

Figure 7.36

SOLUTION   First of all we select the axes of coordinates as shown in the Fig. 7.37. Then the centers of mass of the three rectangles are the points P_1, $(-4, 3)$; P_2, $(0, 1)$; P_3, $(4, 2)$, and their total masses are $12D$, $12D$, and $8D$ respectively. (These are obtained by multiplying the area of each rectangle by the density D.) We may treat the region as a system of three point masses located at the centers of mass, that is, a mass $12D$ at P_1, a mass $12D$ at P_2, and a mass $8D$ at P_3. Then, as before, we have

$$\bar{x} = \frac{12D(-4) + 12D(0) + 8D(4)}{12D + 12D + 8D} = -\frac{16D}{32D} = -\frac{1}{2}$$

and

$$\bar{y} = \frac{12D(3) + 12D(1) + 8D(2)}{12D + 12D + 8D} = \frac{64D}{32D} = 2.$$

The center of mass is therefore the point $(-\frac{1}{2}, 2)$, marked G in the figure.

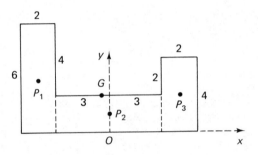

Figure 7.37

REMARKS. In the above example, the numerators for $\bar{x}$ and $\bar{y}$ are the sums of the moments about the y-axis and about the x-axis respectively. The denominator in each case is the total mass of the whole region.

As long as the plane region is of uniform density, that is, is homogeneous (which we shall assume to be the case), the actual value of density plays no part in determining the center of mass. It is clear from the above example that D cancels in the computation of $\bar{x}$ and $\bar{y}$. Therefore, from now on, we shall always *assume that the density has the value 1* so that the *mass of any region is equal to its area.*

We shall now establish formulas for the coordinates of the center of mass of a plane region that is bounded by the graph of a function $y = f(x)$, the x-axis, and the lines $x = a$, $x = b$. As indicated before, we assume that the density of the region is 1, so that the total mass of the region is its area, given by

$$m = \int_a^b f(x)\, dx.$$

As when finding the area under the curve, we use the approximation method for defining the moments of such a region about the coordinate axes. Let us divide the interval $a \le x \le b$ into n equal parts each of length $h = (b - a)/n$, and let x_k

348

($k = 0, 1, 2, \ldots, n$) be the endpoints of these subintervals. We construct a rectangle on each subinterval and approximate the area under the curve by the area of the corresponding rectangle (Fig. 7.38). In particular, the k-th rectangle erected on the interval $x_{k-1} \leq x \leq x_k$ has height $f(x_k)$ and width $x_k - x_{k-1} = h$. Its center of mass is its geometric center, which has the coordinates $[x_k - (h/2), \frac{1}{2} f(x_k)]$.

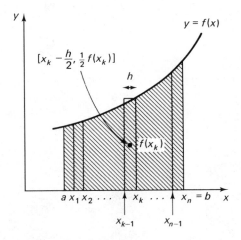

Figure 7.38

The mass of this rectangle is equal to its area, which is $h \times f(x_k)$. We can imagine the mass of each of the n rectangles to be concentrated at the center of the rectangle, and then the moment about the y-axis of these n rectangles is given by

$$M_y = \sum_{k=1}^{n} \left[f(x_k) \cdot h \cdot \left(x_k - \frac{h}{2} \right) \right].$$

Similarly the moment about the x-axis of the n rectangles is given by

$$M_x = \sum_{k=1}^{n} [f(x_k) \cdot h \cdot \tfrac{1}{2} f(x_k)].$$

As $n \longrightarrow \infty$, the sum of the areas of the rectangles approaches the true area under the curve, and in the same way the moments about the y-axis and x-axis of the rectangles approach the true moments of the area under the curve. We note that as $n \longrightarrow \infty$, $h \longrightarrow 0$ and $x_k - (h/2) \longrightarrow x_k$. Thus for a plane region bounded by $y = f(x)$, the x-axis, and the lines $x = a$, $x = b$, the moments about the y-axis and x-axis are given by

$$M_y = \lim_{n \to \infty} \sum_{k=1}^{n} \left[\left(x_k - \frac{h}{2} \right) \cdot f(x_k) \cdot h \right] = \int_a^b x f(x) \, dx.$$

$$M_x = \lim_{n \to \infty} \sum_{k=1}^{n} \{ \tfrac{1}{2} [f(x_k)]^2 \cdot h \} = \int_a^b \tfrac{1}{2} [f(x)]^2 \, dx.$$

The center of mass $(\bar{x}, \bar{y})$ is then defined as before by the equations

$$M_y = m\bar{x}, \qquad M_x = m\bar{y},$$

so that

$$\bar{x} = \frac{\displaystyle\int_a^b x f(x)\, dx}{\displaystyle\int_a^b f(x)\, dx}, \qquad \bar{y} = \frac{\frac{1}{2}\displaystyle\int_a^b [f(x)]^2\, dx}{\displaystyle\int_a^b f(x)\, dx}$$

because

$$m = \int_a^b f(x)\, dx.$$

EXAMPLE   Find the center of mass of the plane region bounded by $y = \sqrt{x}$, the x-axis, and the lines $x = 1$, $x = 4$.

SOLUTION   Here $f(x) = \sqrt{x}$, $a = 1$, $b = 4$ (Fig. 7.39). Thus

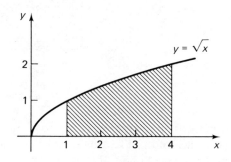

Figure 7.39

$$\int_a^b x f(x)\, dx = \int_1^4 x \cdot \sqrt{x}\, dx$$

$$= \int_1^4 x^{3/2}\, dx$$

$$= \left[\frac{x^{5/2}}{\frac{5}{2}}\right]_1^4$$

$$= \tfrac{2}{5}[4^{5/2} - 1^{5/2}]$$

$$= \tfrac{2}{5}[2^5 - 1]$$

$$= \tfrac{62}{5},$$

$$\tfrac{1}{2}\int_a^b [f(x)]^2\, dx = \tfrac{1}{2}\int_1^4 (\sqrt{x})^2\, dx$$

$$= \tfrac{1}{2}\int_1^4 x\, dx$$

$$= \frac{1}{2}\left[\frac{x^2}{2}\right]_1^4$$

$$= \tfrac{1}{4}[4^2 - 1^2] = \tfrac{15}{4},$$

and

$$\int_a^b f(x)\,dx = \int_1^4 \sqrt{x}\,dx = \int_1^4 x^{1/2}\,dx$$

$$= \left[\frac{x^{3/2}}{\frac{3}{2}}\right]_1^4 = \frac{2}{3}[4^{3/2} - 1^{3/2}]$$

$$= \frac{2}{3}[2^3 - 1] = \frac{14}{3}.$$

Thus the center of mass $(\bar{x}, \bar{y})$ is given by

$$\bar{x} = \frac{\displaystyle\int_a^b x f(x)\,dx}{\displaystyle\int_a^b f(x)\,dx} = \frac{\frac{62}{5}}{\frac{14}{3}} = \frac{62}{5}\cdot\frac{3}{14} = \frac{93}{35}.$$

$$\bar{y} = \frac{\frac{1}{2}\displaystyle\int_a^b [f(x)]^2\,dx}{\displaystyle\int_a^b f(x)\,dx} = \frac{\frac{15}{4}}{\frac{14}{3}} = \frac{15}{4}\cdot\frac{3}{14} = \frac{45}{56}.$$

That is, the center of mass is the point $(\frac{93}{35}, \frac{45}{56})$.

EXERCISES 7.6

In each of the following exercises, the masses and the coordinates of the system of particles in the xy-plane are given. Find the moments of the system about the coordinate axes and also the coordinates of the center of mass.

1. 2, $(3, 1)$; 5, $(-1, 2)$; 3, $(2, -1)$.

2. 5, $(1, -2)$; 4, $(2, 5)$; 6, $(-2, 3)$; 2, $(3, 1)$.

3. 1, $(2, 0)$; 3, $(5, -1)$; 2, $(1, -2)$; 4, $(3, 1)$.

4. 2, $(0, 0)$; 8, $(-1, 1)$; 4, $(3, 0)$; 6, $(0, 5)$.

Find the center of mass of the following plane regions of uniform density.

5.

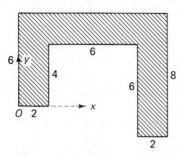

6.

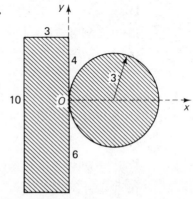

7.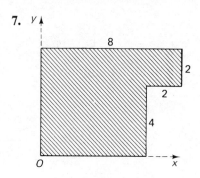

8.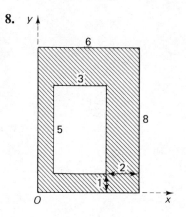

9.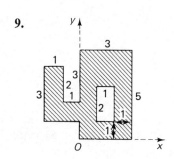

10.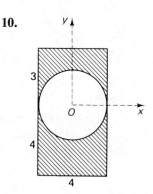

In each of the following exercises find the center of mass of the plane region R. The region R is bounded by:

11. $y = x^2$, $x = 0$, $x = 3$, and $y = 0$ (x-axis).

12. $y = x^2 + 3x + 1$, $x = 1$, $x = 2$, and $y = 0$.

13. $y = x^3$, $x = 0$, $x = 1$, and $y = 0$.

14. $y = \dfrac{1}{x}$, $x = 1$, $x = 4$, and $y = 0$.

15. $y = \sqrt{a^2 - x^2}$, $x = 0$, $x = a$, and $y = 0$ (quarter-circle).

16. $y = \sqrt{a^2 - x^2}$, $x = -a$, $x = a$, and $y = 0$ (semicircular region).

17. $y = \sin x$, $x = 0$, $x = \pi$, and $y = 0$.

18. $y = \cos x$, $x = 0$, $x = \dfrac{\pi}{2}$, and $y = 0$.

19. $y = \ln x$, $x = 1$, $x = 2$, and $y = 0$.

20. $y = \dfrac{b}{a}x$, $x = 0$, $x = a$, and $y = 0$ (triangular region).

7.7 THE NATURAL LOGARITHM

In Chapter 3 the logarithm function was defined as the inverse of the exponential function. The exponential function itself, $f(x) = e^x$, can immediately be given a definite meaning when x is an integer or a rational number. But when x is an irrational number, we define the value of e^x in Chapter 3 by an approximation technique: The value of x was approximated by a rational number and it was claimed that the resulting value of e^x was approximately equal to the true value. This definition is unsatisfactory—or at least incomplete—and in this section we propose to provide a more rigorous definition of these two functions. The method of approach is first to use integral calculus to define $\ln x$ for all positive real numbers x: then to define the exponential function as the inverse of the logarithm.

DEFINITION   For $x > 0$, the *natural logarithm* of x is defined by the following definite integral;

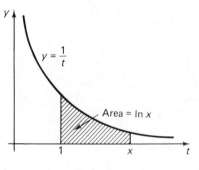

Figure 7.40

$$\ln x = \int_1^x \frac{1}{t}\, dt.$$

In other words, when $x > 1$, $\ln x$ is defined to be the area bounded by the curve $y = 1/t$, the t-axis, and the vertical lines $t = 1$ and $t = x$ (Fig. 7.40).

For $x < 1$, $\ln x$ is the negative of the area beneath the curve $y = 1/t$ and between the lines $t = x$ and $t = 1$, since the limits in the above definite integral must be interchanged to give the area.

All the properties of logarithms that were proved in Chapter 3 by using the properties of exponential functions can also be proved by using this definition of $\ln x$.

THEOREM 7.7.1

If x and y are any positive real numbers, then

(a) $\ln (xy) = \ln x + \ln y$.

(b) $\ln \left(\dfrac{x}{y}\right) = \ln x - \ln y$.

(c) $\ln 1 = 0$.

(d) $\ln x^n = n \ln x$ if n is any rational number.

(a)   By the definition of natural logarithm, we have

$$\ln (xy) = \int_1^{xy} \frac{1}{t}\, dt$$

$$= \int_1^x \frac{1}{t}\, dt + \int_x^{xy} \frac{1}{t}\, dt. \tag{1}$$

In the second integral on the right, let us make a change of variable by writing $t = xu$ so that $dt = x\, du$. When $t = x$, $u = 1$; when $t = xy$, $u = y$. Thus

$$\int_x^{xy} \frac{1}{t}\, dt = \int_1^y \frac{1}{xu} \cdot x\, du = \int_1^y \frac{1}{u}\, du = \int_1^y \frac{1}{t}\, dt.$$

Therefore, from (1) above,

$$\ln (xy) = \int_1^x \frac{1}{t}\, dt + \int_1^y \frac{1}{t}\, dt$$

$$= \ln x + \ln y$$

by definition. This proves part (a) of the theorem.

(b)   To prove part (b), we note that

$$\ln x = \ln \left(y \cdot \frac{x}{y} \right)$$

$$= \ln y + \ln \left(\frac{x}{y} \right),$$

where part (a) has been used to express the logarithm of the product $y(x/y)$ as the sum of the logarithms of y and (x/y). Therefore

$$\ln \left(\frac{x}{y} \right) = \ln x - \ln y.$$

(c)   In part (b) above, setting $y = x$, we have

$$\ln \left(\frac{x}{x} \right) = \ln x - \ln x$$

or

$$\ln 1 = 0.$$

This result also follows directly from the definition of $\ln x$ and the properties of definite integrals since

$$\ln 1 = \int_1^1 \frac{1}{t}\, dt = 0.$$

(d)   To prove this part, we proceed in stages. First of all, let us consider the case when n is a positive integer. For $y = x$, part (a) implies that

$$\ln x^2 = \ln (x \cdot x) = \ln x + \ln x = 2 \ln x,$$
$$\ln x^3 = \ln (x \cdot x^2) = \ln x + \ln x^2 = \ln x + 2 \ln x$$
$$= 3 \ln x.$$

Proceeding in this way, step by step, we prove that

$$\ln x^n = n \ln x,$$

if n is a positive integer.

If n is a negative integer, we write $n = -m$ and $x^n = x^{-m} = 1/x^m$, where m is a positive integer. Thus

$$\ln x^n = \ln \left(\frac{1}{x^m}\right) = \ln 1 - \ln x^m.$$

But $\ln 1 = 0$ by (c) and $\ln x^m = m \ln x$ (since m is a positive integer), and therefore

$$\ln x^n = 0 - m \ln x$$

or

$$\ln x^n = n \ln x$$

after setting $m = -n$.

When n is a rational number of the type p/q, where p is any integer positive or negative and q is a positive integer, we define $z = x^{1/q}$ so that $z^q = x$. Then $\ln x = q \ln z$ since q is a positive integer. Therefore

$$\ln x^n = \ln x^{p/q} = \ln (x^{1/q})^p$$
$$= \ln z^p = p \ln z$$
$$= \frac{p}{q} \cdot q \ln z = \frac{p}{q} \ln x$$
$$= n \ln x.$$

THEOREM 7.7.2

If $f(x) = \ln x$ then

(a) $f'(x) = 1/x$, $x > 0$.

(b) $f(x)$ is an increasing function of x.

(c) $\ln x \longrightarrow +\infty$ as $x \longrightarrow +\infty$.

(d) $\ln x \longrightarrow -\infty$ as $x \longrightarrow 0+$.

(e) the range of f is the set of all real numbers.

PROOF   (a) We have

$$f(x) = \ln x = \int_1^x \frac{1}{t} \, dt, \qquad (x > 0).$$

Then by the fundamental theorem of calculus (Theorem 7.3.2), we have

$$f'(x) = \frac{d}{dx} \left(\int_1^x \frac{1}{t} \, dt \right) = \frac{1}{x}, \qquad x > 0.$$

(b)   Since $f'(x) > 0$ for all $x > 0$, it follows that the function $f(x) = \ln x$ is increasing in its domain $x > 0$.

(c)   To prove this part we must show that $\ln x$ increases without bound as x increases indefinitely. Let n be a positive integer and let $x > 2^n$. Since $\ln x$ is an increasing function of x as proved above in part (b), then $x > 2^n$ implies that

$$\ln x > \ln (2^n) = n \ln 2.$$

Now $\ln 2$ is a positive constant (as it represents the area under the curve $y = 1/t$ between $t = 1$, $t = 2$). Therefore

$$n \ln 2 \longrightarrow +\infty \quad \text{as} \quad n \longrightarrow \infty.$$

But since $x > 2^n$, $x \longrightarrow \infty$ as $n \longrightarrow \infty$, and therefore

$$\ln x \longrightarrow +\infty \quad \text{as} \quad x \longrightarrow \infty.$$

(d)   In this case, if n is a positive integer, we note that $(1/2^n) \longrightarrow 0$ as $n \longrightarrow \infty$. Thus, if $0 < x < (1/2^n)$, then by the fact that $\ln x$ is an increasing function of x, we have

$$\ln x < \ln \left(\frac{1}{2^n}\right) = -n \ln 2.$$

As $n \longrightarrow \infty$, $-n \ln 2 \longrightarrow -\infty$ and $x \longrightarrow 0$ through positive values because $0 < x < (1/2^n)$. Thus

$$\ln x \longrightarrow -\infty \quad \text{as} \quad x \longrightarrow 0+.$$

(e)   From parts (b), (c), and (d) above, it follows that the range of $f(x) = \ln x$ is the set of all real numbers, since $\ln x$ is an increasing function that increases from $-\infty$ to $+\infty$ as x increases from 0 to ∞. [Note that $f(x)$ must be continuous since it is differentiable; hence its range cannot have any gaps in it.]

The domain of $f(x) = \ln x$ is the set of all positive real numbers and the range is the set of all real numbers. Since $f(x)$ is increasing in its domain (as proved in the preceding theorem), it has a unique inverse, which is defined below.

DEFINITION   The inverse of the natural logarithmic function is called the *exponential function* and is denoted by e^x if x is the independent variable. The defining equations for e^x are as follows:

$$y = e^x \quad \text{whenever} \quad x = \ln y.$$

The domain of e^x is the set of all real numbers and the range is the set of all positive real numbers. The basic properties of the exponential function are given in the following theorem.

THEOREM 7.7.3

If $g(x) = e^x$ then

(a) $g(x)$ is increasing for all values of x.

(b) $e^x \cdot e^y = e^{x+y}$.

(c) $e^x/e^y = e^{x-y}$.

(d) $(e^x)^n = e^{nx}$ if n is rational.

(e) $e^x \rightarrow +\infty$ as $x \rightarrow +\infty$.

(f) $e^x \rightarrow 0$ as $x \rightarrow -\infty$.

PROOF

(a) Let $P = e^x$ and $Q = e^y$ with $y > x$. We wish to show that $Q > P$. From the definition of inverse functions we know that

$$P = e^x \quad \text{implies} \quad x = \ln P$$

and

$$Q = e^y \quad \text{implies} \quad y = \ln Q.$$

Since $\ln t$ is an increasing function in its domain the inequality $y > x$ will hold true only if $Q > P$. Thus $y > x$ implies $Q > P$, that is, $g(x) = e^x$ is an increasing function of x.

(b) As in part (a) above, we let $P = e^x$ and $Q = e^y$. Then

$$x = \ln P \quad \text{and} \quad y = \ln Q.$$

Therefore

$$x + y = \ln P + \ln Q$$
$$= \ln (PQ).$$

Thus

$$e^{x+y} = PQ = e^x \cdot e^y,$$

which proves the result.

The proof of other parts is left as an exercise for the student.

REVIEW EXERCISES FOR CHAPTER 7

1. State whether the following statements are true or false. If false, replace them by the corresponding correct statements.

   (a) $\left(\sum\limits_{k=1}^{n} x_k \right)^2 = \sum\limits_{k=1}^{n} x_k^2.$

   (b) $\sum\limits_{k=1}^{n} (x_k - c) = \sum\limits_{k=1}^{n} x_k - C$, where C is a constant.

   (c) If $f(x)$ is continuous in $a \le x \le b$, then $\int_a^b f(x)\,dx$ represents the area bounded by the curve $y = f(x)$, the x-axis, and the lines $x = a$, $x = b$.

   (d) If $\int_a^b f(x)\,dx = \int_b^a f(x)\,dx$ then $\int_a^b f(x)\,dx = 0$.

   (e) $\dfrac{d}{dx}\left[\int_a^x f(t)\,dt \right] = f'(x).$

   (f) $\dfrac{d}{dx}\left[\int_a^b f(x)\,dx \right] = \int_a^b \dfrac{d}{dx}[f(x)]\,dx.$

   (g) If $F(x)$ is an antiderivative of $f(x)$ then

   $$\frac{d}{dx}\left[\int_a^x f(t)\,dt \right] = F'(x).$$

(h) $\int_a^b f(t)\, dt$ is always some real number.

(i) $\int_a^b f(x)\, dx$ and $\int_a^b f(t)\, dt$ are different from one another.

(j) The sum of the moments of a system of n masses situated along the x-axis about their center of mass is always zero.

(k) If the center of mass $(\bar{x}, \bar{y})$ of any plane region is taken as the origin, then the moments about the x-axis and y-axis of the region are each zero.

(l) By definition, ln 2 is the area under the curve $y = 1/x$, bounded by the x-axis and the lines $x = 0$, $x = 2$.

(m) $\ln x \rightarrow 0$ as $x \rightarrow 1$.

(n) $\ln x \rightarrow \infty$ as $x \rightarrow 0+$.

(o) $\ln (1 + 2 + 3) = \ln 1 + \ln 2 + \ln 3$.

2. If

$$\bar{x} = \frac{1}{n} \sum_{k=1}^{n} x_k,$$

prove that

$$\sum_{k=1}^{n} (x_k - \bar{x}) = 0$$

and

$$\sum_{k=1}^{n} (x_k - \bar{x})^2 = \sum_{k=1}^{n} x_k^2 - \frac{1}{n} \left(\sum_{k=1}^{n} x_k \right)^2.$$

3. Prove that $\displaystyle\int_0^{\pi/2} \sin x\, dx = \int_0^{\pi/2} \cos x\, dx$.

4. Prove that $\displaystyle\int_a^{ab} \frac{1}{t}\, dt = \int_1^b \frac{1}{t}\, dt$.

Find the area bounded by $y = f(x)$, the x-axis, and the lines $x = a$, $x = b$, where $f(x)$, a, and b have the following values:

5. $f(x) = \sin^2 x$, $a = 0$, $b = 3\pi$.

6. $f(x) = \cos x$, $a = 0$, $b = \pi$.

7. $f(x) = \ln x$, $a = 1$, $b = 2$.

8. $f(x) = e^x$, $a = -2$, $b = 2$.

9. Find the volume of the solid of revolution formed by revolving about the x-axis the portion of the circle $x^2 + y^2 = 1$, which lies above the two lines $y = x$ and $y = -x$.

Find the center of mass of the following plane regions. The region R is bounded by

10. $y = e^x$, $x = 0$, $x = 2$, and $y = 0$.

11. $y = \dfrac{1}{x^2}$, $x = 1$, $x = 3$, and $y = 0$.

12. $y = x^{2/3}$, $x = 0$, $x = 1$, and $y = 0$.

Probability

8.1 SAMPLE SPACES AND EVENTS

Historically, the theory of probability originated in investigations conducted by Pascal and Fermat in the middle of the seventeeth century at the instigation of certain figures in the gambling world of the time. Today, besides its applications to games of chance, probability theory has become an important tool in such diverse fields as engineering, meteorology, insurance and actuarial work, business operations, and of course the biomedical areas. In the latter field perhaps the best-known application is the theory of genetic inheritance, originated over a hundred years ago by Mendel.

There exist many situations in which the precise outcome of an observation cannot be predicted, even though the set of all possible outcomes can be listed. For example, if a coin is tossed, we cannot predict whether it will land heads or tails; however we do know that it will land on one or the other, so that the set of possible outcomes is known. Similarly, when a sperm and ovum unite, we cannot predict (at the present time) whether the resulting offspring will be male or female, although we can be reasonably certain that it will be one or the other. In order to develop the theory of such random processes we introduce the following definitions.

DEFINITION   The set of all possible outcomes of an observation is called the *sample space* and is denoted by S. Each element of this set (that is, each outcome) is called a *sample point*. The sample space is said to be *finite* if the number of outcomes is finite. In the first few sections of this chapter we shall deal only with finite sample spaces.

EXAMPLES   (a)   The observation consists of tossing a coin. In this case there are two outcomes, namely heads (H) or tails (T) (assuming that the coin cannot land on its edge!) so that the sample space consists of just two elements. Using set notation we can write

$$S = \{H, T\}.$$

(b)   An experiment is conducted to measure the efficacy of a certain rat poison. Ten rats are administered a measured dose of the poison and the number of them that die is observed. The outcome of this experiment can be any number from 0 to 10, and we can denote the sample space by

$$S = \{0, 1, 2, \ldots, 10\}.$$

Tree diagrams are useful for listing the sample spaces of more complicated observations or experiments.

EXAMPLE   Consider the flipping of two coins. Figure 8.1 is the beginning of the tree diagram. It contains a starting point and branches to H and T, the possible outcomes from flipping the first coin.

360

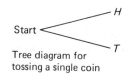

Tree diagram for
tossing a single coin

Figure 8.1

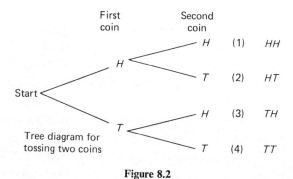

Tree diagram for
tossing two coins

Figure 8.2

Figure 8.2 completes the tree diagram by showing the possible outcomes from flipping the second coin; if the first was H, the second could be either H or T; if the first was T, the second could again be either H or T. The elements of the sample space are now obtained by reading from the *start* to the ends of branches (1), (2), (3), and (4). Thus the sample space is

$$S = \{(H, H), (H, T), (T, H), (T, T)\}.$$

The ordered n-tuple notation is used for sample spaces, but it can be dropped if no ambiguity arises. The above sample space is often written as

$$S = \{HH, HT, TH, TT\}.$$

We shall, in the field of probability, be concerned with what are called *events.* Let us consider as an example the tossing of two coins, for which we have already seen that the sample space is {HH, HT, TH, TT}. Then an example of the type of event with which we might be concerned is the event that "both coins fall alike." Looking at the list of points in the sample space we see that two of the four points satisfy the requirement of both coins falling alike, namely the outcomes HH and TT. We can say, therefore, that the event that both coins fall alike *consists of* the set of outcomes {HH, TT}.

As a second example, the event that "the first coin falls heads" occurs if and only if the outcome is either HH or HT. Thus this event can be identified with the set of outcomes {HH, HT}.

The event that "at least one coin falls heads" can be identified with the set {HH, HT, TH} of outcomes.

We see then that, in each of these examples, an event is associated with a certain set of outcomes of the observation in question. This leads us to make the following formal defintion.

DEFINITION   Any subset of a sample space S for a particular observation is called an *event*. Usually the letter E is used to denote events.

EXAMPLE   Consider the experiment of rolling a die whose six faces have 1, 2, 3, 4, 5, and 6 spots respectively. The sample space is $S = \{1, 2, 3, 4, 5, 6\}$ if we observe the number of spots on the face that lands uppermost.

i. $E_1 = \{2, 4, 6\}$ represents the event that the die will show an even number of spots.

ii. $E_2 = \{4, 5, 6\}$ represents the event that the die will show a number of spots greater than three.

iii. Suppose we are interested in the event that the die will show up *seven* spots. Clearly this will never happen, because no face on the die has seven spots. Thus this subset is empty, that is $E_3 = \varnothing$. Such an event is termed an *impossible event*.

iv. The event that either an odd or an even number of spots shows up is given by the subset $E_4 = \{1, 2, 3, 4, 5, 6\}$. Clearly this is all of the sample space and this event is certain to happen. Thus if an event consists of all the sample points of the sample space, then this event is termed a *certain event*.

It is often useful to represent sample spaces and events by means of what is called a *Venn diagram*. The sample space S itself is represented by a number of points enclosed within a rectangular boundary, each of the points representing one of the outcomes of the observation. Any event would be represented by a closed region within the rectangle contains all of the points corresponding to the outcomes in that event. For example, the sample space for rolling a die is shown in Fig. 8.3(a), while in Fig. 8.3(b) are shown the events E_1 and E_2 from the previous example. These events are specified simply by encircling the corresponding outcomes.

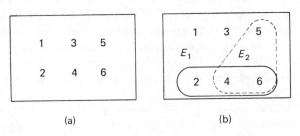

(a)   (b)

Figure 8.3

There exist many situations in which a given event can be related to two or more other events. For example, consider the tossing of two coins for which the sample space consists of the four outcomes $S = \{HH, HT, TH, TT\}$. Then $E_1 = \{HT, TH\}$ represents the event that both coins fall differently and $E_2 = \{TT, TH\}$ is the event that the first coin falls tails. Suppose we are interested in the event that either the first coin falls tails or else both coins fall differently. This event will occur whenever E_1 happens or E_2 happens or both E_1 and E_2 happen together. That is, this event has the possibilities $\{HT, TH, TT\}$, which are the sample points contained in either of the two events E_1 or E_2. The Venn diagram for this experiment is shown in Fig. 8.4. The event that either E_1 or E_2 or both occur consists of all three of the encircled points. More generally, we have the following definition.

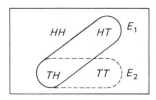

Figure 8.4

DEFINITION   Let E_1 and E_2 be two events of a sample space S. Then the *union* of E_1 and E_2, denoted by $E_1 \cup E_2$, is the set of all the sample points that are either in E_1 or E_2 or both. Thus

$$E_1 \cup E_2 = \{x \,|\, x \in E_1 \text{ or } x \in E_2 \text{ or both}\}.$$

$E_1 \cup E_2$ is read as "E_1 union E_2." In terms of Venn diagrams, $E_1 \cup E_2$ represents the entire shaded region shown in Fig. 8.5.

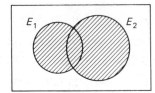

Figure 8.5

EXAMPLE   Given a sample space $S = \{1, 2, 3, 4, 5, 6\}$ for the rolling of a die, let E_1 be the event that an even number turns up, $E_1 = \{2, 4, 6\}$; E_2 be the event that an odd number turns up, $E_2 = \{1, 3, 5\}$; and E_3 be the event that the number turned up is less than 4, $E_3 = \{1, 2, 3\}$. Then

(a) $E_1 \cup E_3 = \{1, 2, 3, 4, 6\}$ is the event that either the number turned up is even *or* the number turned up is less than 4, or both.

(b) $E_1 \cup E_2 = \{1, 2, 3, 4, 5, 6\}$ is the event that either the number turned up is even *or* it is odd. Clearly in this case, $E_1 \cup E_2 = S$, the whole sample space.

Let us once again consider the sample space $S = \{HH, HT, TH, TT\}$ for the flipping of two coins. Let $E_1 = \{HT, TH, TT\}$ be the event that at least one coin falls tails and $E_2 = \{HH, HT, TH\}$ be the event that at least one coin falls heads. Consider the event that at least one coin falls tails *and* at least one coin falls heads. This event has the possibilities $\{HT, TH\}$, which are the sample points common to the events E_1 and E_2. This leads us to the following definition.

DEFINITION   Let E_1 and E_2 be two events of a sample space S. Then the *intersection* of these events, denoted by $E_1 \cap E_2$, is the set of all the sample points that belong both to E_1 and E_2. Thus

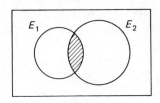

$$E_1 \cap E_2 = \{x \mid x \in E_1 \text{ and } x \in E_2\}.$$

We read $E_1 \cap E_2$ as "E_1 intersection E_2." The event $E_1 \cap E_2$ is shown shaded in Fig. 8.6. It consists of the sample points

Figure 8.6   that are common to E_1 and E_2.

EXAMPLE   Consider a sample space $S = \{0, 1, 2, 3, 4, 5, 6\}$. Let $E_1 = \{2, 3, 5\}$, $E_2 = \{1, 3, 5\}$, and $E_3 = \{0, 1, 2, 4\}$. Then

(a) $E_1 \cap E_2 = \{3, 5\}$.

(b) $E_3 \cap E_2 = \{1\}$.

(c) $E_1 \cap E_3 = \{2\}$.

EXAMPLE   Consider the sample space $S = \{HH, HT, TH, TT\}$ for the flipping of two coins. Let

$E_1 = \{HH, TT\}$ be the event that both coins fall alike;

$E_2 = \{HT, TH\}$ be the event that both coins fall differently;

$E_3 = \{HH, HT, TH\}$ be the event that at least one coin falls heads;

$E_4 = \{TT, HT\}$ be the event that the second coin falls tails.

Then

$E_1 \cap E_4 = \{TT\}$ is the event that both coins fall alike *and* the second coin falls tails.

$E_2 \cap E_3 = \{HT, TH\}$ is the event that at least one coin falls heads *and* both coins fall differently.

$E_1 \cap E_2 = \varnothing$ is the event that both coins fall alike *and* both coins fall differently.

Note that $E_1 \cap E_2$ is an empty set because the sets E_1 and E_2 have nothing in common. In other words, $E_1 \cap E_2$ is an *impossible event*.

DEFINITION   Two events E_1 and E_2 of a sample space are said to be *mutually exclusive* if there is no sample point that is contained in both E_1 and E_2, that is, if $E_1 \cap E_2 = \varnothing$. In other words E_1 and E_2 are mutually exclusive if they cannot occur at the same time. Figure 8.7 represents two mutually exclusive events. The regions representing E_1 and E_2 have no sample points in

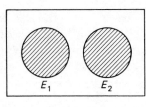

Figure 8.7   common.

EXAMPLE Consider the sample space $\{1, 2, 3, 4, 5, 6\}$. Let $E_1 = \{2, 4, 6\}$, $E_2 = \{1, 5\}$, and $E_3 = \{2, 3\}$. Then E_1 and E_2 are mutually exclusive because $E_1 \cap E_2 = \emptyset$; E_1 and E_3 are *not* mutually exclusive because $E_1 \cap E_3 = \{2\}$ is not an empty set. E_2 and E_3 are mutually exclusive.

EXAMPLE Consider the sample space for drawing a single card from a deck of 52 cards. Then the events "drawing an 8" and "drawing a 10" are mutually exclusive because the drawn card cannot be both an 8 and a 10 at the same time. On the other hand, the events "drawing an ace" and "drawing a heart" are *not* mutually exclusive because it is possible to draw a card that is an ace as well as a heart.

DEFINITION Let E be an event in a sample space S. Then the complement E' of the event E with respect to the sample space S is the set of all outcomes in S that are not in E. Thus

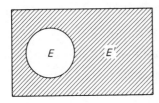

Figure 8.8

$$E' = \{x \,|\, x \in S \text{ and } x \notin E\}.$$

In terms of Venn diagrams, E' is the region inside the rectangle and outside the region that represents the event E (Fig. 8.8) It follows from the definition of E' that the events E and E' are mutually exclusive.

EXAMPLE Consider the sample space $S = \{1, 2, 3, 4, 5, 6\}$ for the roll of a die. Let $E_1 = \{1, 2, 3\}$, $E_2 = \{4, 6\}$. Then

$$E_1' = \text{set of all outcomes in } S \text{ that are not in } E_1$$
$$= \{4, 5, 6\}.$$

$$E_2' = \text{set of all outcomes in } S \text{ that are not in } E_2$$
$$= \{1, 2, 3, 5\}.$$

EXERCISES 8.1

Write down the sample spaces for the following experiments:

1. Five persons are tested for cancer and the number of persons who have this disease is observed.

2. Three coins are tossed, each of them falling heads and tails with no other possibility.

3. A coin is tossed and at the same time a die is rolled. (The six faces of the die are marked with numbers $1, 2, \ldots, 6$.)

4. Two dice are rolled and the numbers turned up on the dice are observed.

5. The order in which a mouse, an elk, and a rabbit arrive at a lake is observed.

6. Three rats are chosen from a cage that has three brown and two white rats. The colors of the selected rats are observed.

7. A card is selected from among the four aces in a deck of playing cards.

8. A spade card is selected from a deck of cards.

9. A red card is selected from a deck of cards.

10. A box contains three white balls and two black balls that are exactly alike except for color. Two balls are drawn from the box and their colors are observed.

For exercise 4 above, write down the events that the sum of the numbers turned up on the two dice is equal to the following values:

11. 10.   12. 5.

13. 7.   14. 13.

A card is drawn from a pack of 52 cards. Let the events be defined as follows:

E_1: the card drawn is a heart.

E_2: the card drawn is a black card.

E_3: the card drawn has a denomination less than 7 (ace counts low).

E_4: the card drawn is an ace.

Then express the following events in terms of sets as well as in words:

15. $E_1 \cup E_2$.   16. $E_1 \cap E_2$.

17. $E_1 \cup E_3$.   18. $E_1 \cap E_3$.

19. $E_3 \cap E_4$.   20. $E_1 \cup E_4$.

21. E_2'.   22. E_3'.

23. Two coins are tossed. Which of the following events are mutually exclusive?

$E_1 =$ coins fall alike.

$E_2 =$ coins fall at least one head.

$E_3 =$ coins fall two heads.

$E_4 =$ coins fall differently.

$E_5 =$ coins fall at least one tail.

$E_6 =$ coins fall three heads.

A card is drawn from a deck of 52 cards. Which of the following pairs of events are mutually exclusive?

24. "The drawn card is a 7" and "the card is an ace."

25. "The card is a 4" and "the card is a spade."

26. "The card is a face card" and "the card is a heart."

27. "The card is a 5" and "the card is a face card."

8.2 PROBABILITY

When an ordinary coin is tossed the probability that it will land heads is equal to one half. Most people would accept this statement as correct, but let us ask ourselves precisely what it means. In fact, the word probability is widely used in everyday speech with several different shades of meaning, and our opening statement has two or three interpretations. Perhaps the most straightforward interpretation is based on the so-called *frequency definition of probability*.

Let the coin be tossed N times and let it come down heads K times. Then the ratio K/N represents the proportion of the tosses that fall heads. Of course, K/N may not be equal to $\frac{1}{2}$; we all know, for example, that if a coin is tossed twice it does not have to land once heads and once tails, but it can quite easily land heads both times, or tails both times. However, if the coin is tossed a very large number of times, it can be expected to fall heads about half of these times. That is, if N is very large, then the proportion K/N of heads will almost always be close to $\frac{1}{2}$. If we envisage N approaching infinity, then the proportion K/N can be expected to approach $\frac{1}{2}$ as its limiting value.

This type of approach allows us to define the term *probability* in a more general context. Let us consider an experiment that has n possible outcomes. The sample space consists of n points, which we can label $O_1, O_2, \ldots, O_n$. Let the experiment be repeated N times and let the outcome O_1 occur K_1 times, the outcome O_2 occur K_2 times, and so on. (Of course, $K_1 + K_2 + \ldots + K_n = N$ since each of the N repetitions of the experiment must lead to one of the outcomes $O_1, \ldots, O_n$.) The ratios $K_1/N, K_2/N, \ldots, K_n/N$ give the proportions of the repetitions that lead respectively to the outcomes $O_1, O_2, \ldots, O_n$. The sum of these proportions is equal to 1:

$$\frac{K_1}{N} + \frac{K_2}{N} + \ldots + \frac{K_n}{N} = \frac{K_1 + K_2 + \ldots + K_n}{N} = \frac{N}{N} = 1.$$

We define the *probability* of each outcome to be the limiting value of the corresponding proportion as $N \longrightarrow \infty$:

$$p_1 = \lim_{N \to \infty} \frac{K_1}{N}, p_2 = \lim_{N \to \infty} \frac{K_2}{N}, \ldots, p_n = \lim_{N \to \infty} \frac{K_n}{N},$$

where p_1 is the probability of outcome O_1, p_2 the probability of outcome O_2, and so on.

EXAMPLE If a die is rolled there are six possible outcomes, which we can label simply by the number of spots on the top face. If the die is rolled a large number of times we can expect (at least if the die is evenly balanced) that the proportion of each of these outcomes will be close to one-sixth. Thus in this case, the probabilities of all six outcomes are equal:

$$p_1 = p_2 = \ldots = p_6 = \tfrac{1}{6}.$$

EXAMPLE The probability that an individual chosen at random from the Canadian population has blue eyes is 0.1 and that he or she has brown eyes is 0.9. Here the experiment consists of selecting an individual from the population of Canada.

If this experiment is repeated a very large number of times then the proportion of the individuals chosen with blue eyes will be close to 10% and the proportion with brown eyes will be close to 90%.

An event E was defined in Section 8.1 as consisting of a subset of the outcomes in the sample space for a particular observation or experiment. The event occurs whenever one of the outcomes in this subset occurs. In N repetitions of the experiment, the proportion of times that E occurs is equal to the sum of the proportions for all of the outcomes that belong to the event E.

EXAMPLE   Let the experiment consist of rolling a die. In N rolls, let the outcome be 1 on K_1 of the rolls, let it be 2 on K_2 of the rolls, and so on. Then, of course, since the sample space is {1, 2, 3, 4, 5, 6}, we must have $K_1 + K_2 + \ldots + K_6 = N$. Let E be the event that the number rolled is odd. Then $E = \{1, 3, 5\}$. The number of the N repetitions on which E occurs must be $K_1 + K_3 + K_5$. The proportion of the rolls on which E occurs is then

$$\frac{K_1 + K_3 + K_5}{N} = \frac{K_1}{N} + \frac{K_3}{N} + \frac{K_5}{N}.$$

Thus we see that the proportion of times when E occurs is equal to the sum of the proportions when the three outcomes 1, 3, and 5 occur.

When N becomes very large, the proportion of times when an event E occurs becomes closer and closer to the sum of the probabilities of the outcomes that belong to E. It is therefore natural to define the *probability of an event E* to be equal to the sum of the probabilities of all of the outcomes that belong to E. It is denoted by $P(E)$.

REMARKS.

1. The proportion of times when an event occurs cannot be negative and cannot exceed 1. Therefore we always must have

$$0 \leq P(E) \leq 1.$$

2. If $E = \varnothing$, that is the event is impossible, then $P(E) = 0$.
3. If $E = S$, that is the event E is certain to happen since it consists of the whole set of possible outcomes, then $P(E) = P(S) = 1$.

Very often we find ourselves dealing with observations for which all of the outcomes in the sample space are equally likely to occur. For example, if a coin is tossed the two outcomes "heads" or "tails" can usually be presumed to be equally likely. Or if a well-balanced die is rolled, the outcomes 1, 2, 3, 4, 5, or 6 should be equally likely to occur. If the die is rolled a large number of times N, each of the six numbers should come up as readily as the rest; that is, each number should appear about one-sixth of the time as long as N is large enough.

The outcomes of an experiment are said to be *equally likely* if their probabilities are all equal to one another. If the sample space contains n sample points then the probability of each outcome must be $1/n$ if they are equally likely (since the sum of the probabilities of all the outcomes must equal $P(S)$, which is 1).

It follows that if an event E contains a number k of sample points in an experiment in which the outcomes are equally likely, then the probability $P(E) = k/n$:

$$P(E) = \frac{\text{number of sample points in } E}{\text{number of sample points in } S} = \frac{k}{n}.$$

EXAMPLE What is the probability of throwing a number greater than 4 with a standard die?

SOLUTION When the die is rolled the outcome can be any one of the six numbers 1, 2, 3, 4, 5, or 6. That is, $S = \{1, 2, 3, 4, 5, 6\}$. The event E of throwing a number greater than 4 will consist of the outcomes 5 or 6. Thus $E = \{5, 6\}$. In this case

$$k = \text{number of sample points in } E = 2$$

$$n = \text{number of sample points in } S = 6.$$

Thus

$$P(E) = \frac{k}{n} = \frac{2}{6} = \frac{1}{3}.$$

EXAMPLE What is the probability of throwing a 7 with a standard die?

SOLUTION As before, the sample space is $S = \{1, 2, 3, 4, 5, 6\}$. The event of throwing a 7 is clearly an impossible event because $7 \notin S$, that is, $E = \varnothing$. Thus we have $k = 0$ because the empty set has no element, and $n = 6$. Therefore

$$P(E) = \frac{k}{n} = \frac{0}{6} = 0.$$

EXAMPLE Find the probability of throwing at least two heads by tossing three fair coins.

SOLUTION The sample space for tossing three coins is easily obtained from the tree diagram in Fig. 8.9. Here

$$S = \{HHH, HHT, HTH, HTT, THH, THT, TTH, TTT\}.$$

The event of throwing at least two heads is

$$E = \{HHH, HHT, HTH, THH\}.$$

In this case

$$k = \text{number of elements in } E = 4;$$

$$n = \text{number of elements in } S = 8.$$

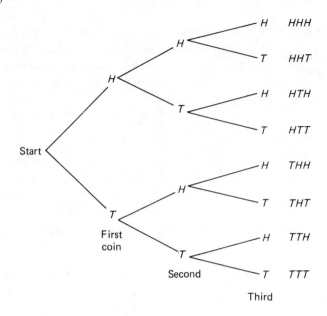

Figure 8.9

Thus

$$P(E) = P(\text{at least two heads})$$

$$= \frac{k}{n} = \frac{4}{8} = \frac{1}{2}.$$

EXAMPLE   Find the probability of throwing a sum of 9 with the roll of two dice.

SOLUTION   The sample space can easily be obtained by a tree diagram, and is listed below in ordered pair notation.

$$\begin{array}{cccccc}
(1, 1) & (1, 2) & (1, 3) & (1, 4) & (1, 5) & (1, 6) \\
(2, 1) & (2, 2) & (2, 3) & (2, 4) & (2, 5) & (2, 6) \\
(3, 1) & (3, 2) & (3, 3) & (3, 4) & (3, 5) & (3, 6) \\
(4, 1) & (4, 2) & (4, 3) & (4, 4) & (4, 5) & (4, 6) \\
(5, 1) & (5, 2) & (5, 3) & (5, 4) & (5, 5) & (5, 6) \\
(6, 1) & (6, 2) & (6, 3) & (6, 4) & (6, 5) & (6, 6)
\end{array}$$

In the ordered pair (x, y), x denotes the number on the first die and y denotes the number on the second die. There are 36 ordered pairs in this sample space, that is, $n = 36$. The sum of 9 on the two dice would result if any one of the four pairs $(3, 6)$, $(4, 5)$, $(5, 4)$, or $(6, 3)$ were rolled. Thus there are four sample points in the event "a sum of 9 is rolled," that is, $k = 4$. Hence

$$P(\text{sum of 9 rolled}) = \frac{k}{n} = \frac{4}{36} = \frac{1}{9}.$$

EXAMPLE There are two children in a family. What is the probability that both are boys? (Assume that each child is equally likely to be either boy or girl.)

SOLUTION If B denotes a boy and G a girl, then the sample space can be obtained from the tree diagram in Fig. 8.10. Thus

$$S = \{BB, BG, GB, GG\}$$

and the event that both are boys is $E = \{BB\}$. Clearly $n = 4$ and $k = 1$. Therefore

$$P(\text{both boys}) = \frac{k}{n} = \frac{1}{4}.$$

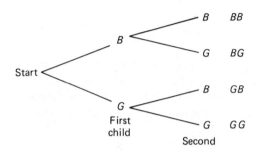

Figure 8.10

EXAMPLE A card is drawn from a deck of cards. What is the probability that it is a 5 or a 7?

SOLUTION There are 52 cards in a deck and any one of them can be drawn. Thus the sample space will consist of $n = 52$ sample points. The event of drawing a 5 or a 7 consists of $k = 8$ sample points, because the card drawn can be any one of the four cards marked 5 or any one of the four cards marked 7. Thus

$$P(5 \text{ or } 7) = \tfrac{8}{52} = \tfrac{2}{13}.$$

Probability Formulas

The preceding few examples have illustrated the calculation of probabilities working from the basic definition. In more complicated examples, such a method of calculation can become exceedingly lengthy, and it is often possible to shortcut a great deal of computation by using certain so-called probability formulas. In the remainder of this section we shall discuss some of these formulas and their uses.

Let us begin by considering the following example: Two dice are rolled and we require the probability that the sum of the two scores is different from 9.

In a preceding example we calculated the probability that the sum of the two scores is *equal* to 9. This event corresponds to the four outcomes (3, 6), (4, 5), (5, 4), and (6, 3) and so has probability $\frac{4}{36}$ or $\frac{1}{9}$. The event that the sum of the two scores is different from 9 corresponds to the set of all the outcomes except these four. Since there are 32 other outcomes, the probability of getting a sum different from 9 is equal to $\frac{32}{36}$ or $\frac{8}{9}$.

Observe that in this example we are dealing with two events that are complementary to one another: The event that the sum is equal to 9 and the event that it is different from 9. These two events have probabilities of $\frac{1}{9}$ and $\frac{8}{9}$ respectively, these two probabilities adding to 1. This property of complementary events generalizes as follows.

FORMULA 1   If E' is the complementary event to E then

$$\boxed{P(E') = 1 - P(E).}$$

PROOF $P(E)$ is equal to the sum of probabilities of all of the outcomes belonging to E. E' consists of all of the outcomes that do not belong to E and so $P(E')$ is the sum of probabilities of these outcomes. Therefore the sum $P(E) + P(E')$ is equal to the sum of probabilities of all the possible outcomes in S, hence must equal $P(S)$ or 1:

$$P(E) + P(E') = 1.$$

The stated formula follows immediately.

EXAMPLE   In a certain community the probability of a 70-year-old individual living to be 80 is 0.64. What is the probability of an individual who is 70 and a member of the community dying sometime in the next 10 years?

SOLUTION   If E denotes the event that a 70-year-old individual will live for the next 10 years, then the complementary event E' will be that a 70-year-old person will die within the next 10 years. We are given that $P(E) = 0.64$ and we want $P(E')$. Now

$$P(E') = 1 - P(E) = 1 - 0.64 = 0.36.$$

Thus the probability of an individual who is 70 years old now dying within the next 10 years is 0.36.

To illustrate the second probability formula, let us consider the following example. Ten rats are administered a certain drug and the number of rats that die within 36 hr is observed. The probability of 7 or more rats dying is $\frac{4}{5}$ and the probability of 7 or fewer rats dying is $\frac{3}{5}$. What is the probability that exactly 7 rats die?

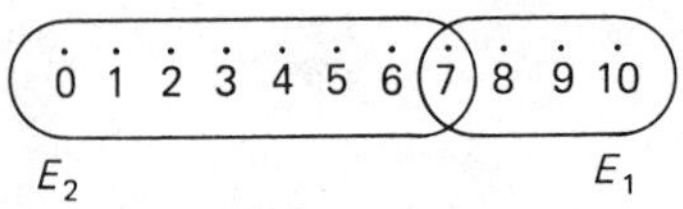

The sample space for this observation consists of the eleven outcomes listed in Fig. 8.11. E_1, the event that 7 or more rats die, contains the four outcomes $\{7, 8, 9, 10\}$, which are labeled E_1 in the

Figure 8.11

figure. Similarly the event E_2 that 7 or fewer rats die contains the eight outcomes $\{0, 1, 2, 3, 4, 5, 6, 7\}$. The event that exactly 7 rats die is the intersection $E_1 \cap E_2$, and we require the probability of this event, $P(E_1 \cap E_2)$.

We observe that the union $E_1 \cup E_2$ comprises the whole of the sample space, and hence its probability must be 1:

$$P(E_1 \cup E_2) = 1.$$

But this probability is equal to the sum of the probabilities of all of the outcomes in E_1 and E_2, and we can write

$$P(E_1 \cup E_2) = P(\{0, 1, 2, 3, 4, 5, 6, 7\}) + P(\{8, 9, 10\})$$
$$= P(E_2) + P(\{8, 9, 10\}).$$

Now the set of outcomes $\{8, 9, 10\}$ form the event E_1 except that the outcome 7 is missing. Therefore we can write

$$P(E_1) = P(\{8, 9, 10\}) + P(\{7\})$$

or,

$$P(\{8, 9, 10\}) = P(E_1) - P(\{7\}).$$

So, substituting into the preceding equation, we obtain

$$P(E_1 \cup E_2) = P(E_2) + P(E_1) - P(\{7\}). \tag{1}$$

We know that

$$P(E_1 \cup E_2) = 1, \qquad P(E_1) = \tfrac{4}{5} \quad \text{and} \quad P(E_2) = \tfrac{3}{5}.$$

Substituting these values, we obtain

$$1 = \tfrac{3}{5} + \tfrac{4}{5} - P(\{7\}) = \tfrac{7}{5} - P(\{7\})$$

and so

$$P(\{7\}) = \tfrac{7}{5} - 1 = \tfrac{2}{5}.$$

The probability of precisely 7 rats dying is $\tfrac{2}{5}$.

In this example the event that 7 rats die is the intersection $E_1 \cap E_2$. Equation (1) is therefore an example of the following general formula.

FORMULA 2   Let E_1 and E_2 be two events that are subsets of the same sample space S. Then

$$\boxed{P(E_1 \cup E_2) = P(E_1) + P(E_2) - P(E_1 \cap E_2).}$$

PROOF   Figure 8.12 illustrates the Venn diagram for the experiment in question. The sets of sample points corresponding to E_1 and E_2 are shown. The region indicated by I consists of sample points that are in E_1 but are outside E_2. Similarly, II consists of sample points that are in E_2 but are not in E_1. Section III denotes the sample points that are in both E_1 and E_2, that is, the intersection $E_1 \cap E_2$. Now

$$P(E_1) = \text{sum of probabilities of the outcomes in I and III,}$$

$$P(E_2) = \text{sum of probabilities of the outcomes in II and III.}$$

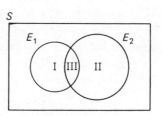

Figure 8.12

Therefore if $P(E_1)$ and $P(E_2)$ are added, we obtain the sum of probabilities of the outcomes in all three regions I, II, and III, except that those in III are counted twice. But the sum of probabilities in III is equal to $P(E_1 \cap E_2)$, and so

$$P(E_1) + P(E_2) - P(E_1 \cap E_2)$$

$$= \text{sum of probabilities of the outcomes in I, II, and III}$$

$$= P(E_1 \cup E_2),$$

as required.

In particular, if the events E_1 and E_2 are mutually exclusive, $E_1 \cap E_2 = \emptyset$, so that $P(E_1 \cap E_2) = 0$. The above formula reduces to

$$P(E_1 \cup E_2) = P(E_1) + P(E_2).$$

EXAMPLE   Find the probability of drawing an ace or a spade from a deck of 52 cards in a single draw.

SOLUTION   A single card can be drawn from a deck of 52 cards in 52 different ways, so that the sample space consists of 52 elements. The event E_1 of "drawing an ace" can be achieved in 4 ways because there are 4 aces. Thus $P(E_1) = \frac{4}{52}$. The event E_2 of "drawing a spade" can be achieved in 13 ways because there are 13 spade cards in the deck. Thus $P(E_2) = \frac{13}{52}$. There is only one card that is both an ace and a spade, that is, the event $E_1 \cap E_2$ can happen in only one way; therefore $P(E_1 \cap E_2) = \frac{1}{52}$. We are interested in the probability of drawing an ace *or* a spade, that is, $P(E_1 \cup E_2)$. Using formula 2 we have

$$P(E_1 \cup E_2) = P(E_1) + P(E_2) - P(E_1 \cap E_2)$$

$$= \tfrac{4}{52} + \tfrac{13}{52} - \tfrac{1}{52}$$

$$= \tfrac{16}{52} = \tfrac{4}{13}.$$

EXAMPLE   Find the probability of throwing a sum of 7 or a sum of 9 with the roll of 2 dice.

SOLUTION   The sample space for the roll of 2 dice consists of $6 \times 6 = 36$ ordered pairs of sample points. Let E_1 be the event that a sum of 7 is obtained and E_2 be the event that a sum of 9 is obtained. Then we have

$$E_1 = \{(6, 1), (5, 2), (4, 3), (3, 4), (2, 5), (1, 6)\}$$

and

$$E_2 = \{(6, 3), (5, 4), (4, 5), (3, 6)\}.$$

Clearly $E_1 \cap E_2 = \varnothing$, that is, the events E_1 and E_2 are mutually exclusive. The event of throwing a sum of 7 or 9 is $E_1 \cup E_2$. Thus

$$P(E_1 \cup E_2) = P(E_1) + P(E_2)$$
$$= \tfrac{6}{36} + \tfrac{4}{36}$$
$$= \tfrac{10}{36} = \tfrac{5}{18}.$$

We now introduce the idea of independence of events, which can loosely be defined as follows. Two events E_1 and E_2 are said to be *independent* if the occurrence of one is not affected by the occurrence or nonoccurrence of the other.

For example, consider the flip of a coin and the roll of a die. There is no reason to believe that the result of the flip of the coin will affect the result of the roll of the die. If E_1 is, say, the event that the coin falls heads and E_2 is the event that the die falls 5 or 6, then these two events are independent. Similarly, when two coins are tossed, the result for the second coin is in no way affected by what the first coin turns up. In other words, the results of throws of two coins are independent events.

FORMULA 3   If E_1 and E_2 are two independent events then the probability of occurrence of E_1 and E_2 together is given by

$$\boxed{P(E_1 \cap E_2) = P(E_1) \cdot P(E_2).}$$

The above formula can easily be extended to a finite number of independent events. If $E_1, E_2, \ldots, E_m$ are m independent events, then the probability of occurrence of these m events together is given by

$$P(E_1 \cap E_2 \cap \ldots \cap E_m) = P(E_1) \cdot P(E_2) \ldots P(E_m).$$

This formula is often called the *multiplication rule* because the right-hand side is the product of the probabilities of each of the events.

EXAMPLE   Find the probability of flipping a head with a coin and then drawing an ace from a deck of 52 cards.

SOLUTION   The probability of getting a head (H) with the flip of a coin is $P(H) = \tfrac{1}{2}$. The probability of drawing an ace (A) from a pack of 52 cards is $P(A) = \tfrac{4}{52}$. The two events "flipping a head" and "drawing an ace" are independent events. Therefore

$$P(A \text{ and } H) = P(A) \cdot P(H)$$
$$= \tfrac{4}{52} \cdot \tfrac{1}{2} = \tfrac{1}{26}.$$

EXAMPLE   A card is drawn from a deck of 52 cards and then it is replaced, the cards are shuffled, and a second card is drawn. Find the probability that the first card is a 7 and the second card is a spade.

SOLUTION   The result of the first drawing and the result of the second drawing are independent events because the result of the second drawing is not affected by the first drawing. This is so since the first drawn card is replaced in the deck before the second drawing. Now we can draw a card out of 52 cards in 52 ways. The card 7 can be drawn in 4 ways and hence the probability of drawing a 7 is

$$P(7) = \tfrac{4}{52} = \tfrac{1}{13}.$$

Similarly, the probability of drawing a spade in the second drawing is

$$P(\text{spade}) = \tfrac{13}{52} = \tfrac{1}{4}.$$

Using formula 3, the probability of drawing a 7 in the first drawing and a spade in the second drawing is

$$P(7 \text{ and spade}) = P(7) \cdot P(\text{spade})$$
$$= \tfrac{1}{13} \cdot \tfrac{1}{4} = \tfrac{1}{52}.$$

Note that if in this example the first card is not replaced before the second is drawn then the two drawings cannot be treated as independent. In such a case the result of the second draw *is* influenced by the result of the first.

When the probabilities of two events are known and we also know the probability of their joint occurrence, then formula 3 can be used to test whether the two events are independent or not. This we illustrate in the following example.

EXAMPLE   The probability that a person over 60 years old in a certain community drinks alcohol is $\tfrac{2}{3}$ and the probability that a person over 60 years old has heart disease is $\tfrac{2}{15}$. The probability that a person over 60 years old both drinks alcohol and has heart disease is $\tfrac{1}{16}$. Are "drinking alcohol" and "heart disease" independent events?

SOLUTION   Let E_1 denote the event that a person over 60 years old drinks alcohol and E_2 be the event that a person over 60 years of age has heart disease. Then we are given that

$$P(E_1) = \tfrac{2}{3}, \qquad P(E_2) = \tfrac{2}{15}.$$

The event that a person over 60 years of age drinks alcohol and has heart disease as well is $E_1 \cap E_2$, and we are given that

$$P(E_1 \cap E_2) = \tfrac{1}{16}.$$

Now

$$P(E_1) \cdot P(E_2) = \tfrac{2}{3} \cdot \tfrac{2}{15}$$
$$= \tfrac{4}{75} \neq \tfrac{1}{16} = P(E_1 \cap E_2).$$

Therefore the two events E_1 and E_2, that is, drinking and heart disease, are *not* independent.

EXERCISES 8.2

A card is drawn from a well-shuffled pack of 52 cards. Find the probability of drawing:

1. A heart.

2. A 7 or a 9.

3. Not a spade.

4. A 6 or a diamond.

5. A 4 or a 5 or a heart.

6. Neither a king nor a queen.

A family has three children. Find the probability that:

7. There are two boys.

8. There are 3 girls.

9. There is at least 1 boy.

10. There is at least 1 boy and 1 girl.

11. There is at least 1 boy or 1 girl.

Two dice are rolled. Find the probability of obtaining:

12. A sum of 5.

13. A sum of 8 or a sum of 6.

14. Not a sum of 9.

15. Ten rats are injected with a certain drug and the number of rats that die within 2 hr is recorded. Suppose that the probability that exactly 7 rats die is $\frac{2}{5}$ and the probability that 8, 9, or 10 rats die is $\frac{3}{10}$. Find the probability that:
(a) 7 or more rats die.
(b) 6 or less rats die.

16. In a large population of fruit flies, 30% of the flies have a wing mutation, 40% have an eye mutation, and 15% have both eye and wing mutations. A fly is chosen from this population. What is the probability that it has at least one of the mutations?

17. Each of five rats is injected with a certain poison and the number of rats dying within 24 hr is observed. Suppose that the probability of one or more rats dying is $\frac{7}{10}$ and the probability of 4 or less rats dying is $\frac{3}{10}$. Find the probability that:
(a) 5 rats die.
(b) No rats die.
(c) Exactly 3 rats die.
(d) 1, 2, 3, or 4 rats die.

18. Seven rats are administered a certain drug and the number of rats dying within 3 days is observed. Suppose that the probability of 5 or more rats dying is $\frac{7}{10}$ and the probability of 5 or less rats dying is $\frac{2}{3}$. Find the probability that:
(a) 6 or 7 rats die.
(b) 4 or less rats die.
(c) Exactly 5 rats die.
(d) All the rats die.

19. The probability that a person over 50 years of age is a smoker in a certain community is $\frac{3}{5}$ and the probability that a person over 50 years of age has a cancer is $\frac{1}{20}$. The probability that a person over 50 years of age will be a smoker and have a cancer is $\frac{1}{25}$. Are smoking and cancer disorders independent events?

20. Two cards are drawn from a pack of 52 cards one by one with replacement, that is, the first card is replaced before the second card is drawn. Find the probability that:

(a)   The first card is an ace and the second is a heart.

(b)   The first card is a 5 and the second is a 10.

21. A card is chosen at random from a deck of 52 cards. Let E_1, E_2, E_3, and E_4 be the events defined as follow: E_1 = the card is at least as high as an 8, E_2 = the card is an ace, E_3 = the card is a spade, and E_4 = the card is a 5. Determine $P(E_1)$, $P(E_2)$, $P(E_3)$, and $P(E_4)$. Use formula 3 to determine which pairs of events are independent.

8.3 COUNTING METHODS

It will be clear from the examples of the last section that we very often find our-selves dealing with observations for which the outcomes are all equally likely. In this situation, in order to compute the probability of any event, all we need to know is the number of outcomes that lead to the event in question and also the total number of outcomes of the observation. One way of evaluating these two numbers, which was essentially the method used in Section 8.2, is to make a complete list of all of the out-comes and simply to count up those that belong to the event, and also to count the total number of them. But for some observations the number of outcomes may be exceedingly large, and listing them all may be a long and difficult task. The following examples illustrate this point.

EXAMPLES   (a)   What is the probability of rolling at least 2 sixes in 6 rolls of a die? The total number of outcomes when a die is rolled 6 times is 46,656, and it would clearly be ridiculous to attempt to list them all.

(b)   What is the probability of drawing a full house in a poker hand of 5 cards? What is the probability of drawing a royal flush? And so on? The total number of different poker hands, which is the number of points in the sample space, is about $2\frac{1}{2}$ million.

(c)   Among a social group of a certain species of animal containing m males and f females, how many different mating patterns are possible? The answer to this depends of course on the mating behavior of the species in question. But to illustrate the large number of patterns that are possible, consider a group con-sisting of 10 males and 10 females in which complete promiscuity is the modus vivendi. The number of possible mating patterns for the group as a whole is then about 10^{30}.

(d)   If a blond-haired, blue-eyed woman marries a brown-haired, brown-eyed man, what is the probability that their first child will be a blond-haired, brown-eyed boy?

Although in this particular example the number of combinations of off-spring is not too large, in general problems concerned with genetic inheritance can involve quite large numbers of possiblities. For example, the number of possible types of offspring that can be produced by a given man and woman is about 7×10^{13}.

What is needed in situations like this (which in fact arise very often) is a method of counting the numbers of outcomes of various events without going to the length of listing them. In this section we shall briefly outline several techniques that are useful to this end.

The following fundamental principle underlies many counting methods. Consider an experiment that consists of several (i.e., p) independent operations. Let the first operation have n_1 possible outcomes, the second operation n_2 possible outcomes, and so on, the pth operation having n_p outcomes. Then the total number of outcomes of the whole experiment is given by the product

$$n_1 \cdot n_2 \cdot \ldots \cdot n_p.$$

EXAMPLES (a) Let the experiment consist of flipping 3 coins. We can break this experiment up into 3 independent operations, each operation being the flip of one of the coins. Each of these has two outcomes, heads or tails. The total number of outcomes of the experiment is therefore given by the product $2 \times 2 \times 2 = 8$.

(b) Let the experiment consist of rolling 2 dice. Each of these rolls has 6 possible outcomes. Therefore the complete experiment consisting of the 2 rolls has a number of outcomes given by the product $6 \times 6 = 36$.

(c) Suppose that we have available 10 rats for experimental purposes and wish to select 3 of the rats for three different experiments. In how many ways can the selection be made?

A rat can be selected for the first experiment in 10 different ways, since any one of the available rats can be chosen. Having selected the first rat, however, there remain only 9 rats available for the second experiment, so that the second choice can be made in 9 ways. Similarly the third rat can be chosen in only 8 ways, once the first two rats have been removed. Consequently, the number of ways in which the three rats can be chosen is given by the product $10 \times 9 \times 8 = 720$.

When counting numbers of outcomes the following notation is exremely useful: The product of the first n natural numbers is called *factorial n* and is denoted by $n!$ Thus

$$n! = 1 \cdot 2 \cdot 3 \cdot \ldots \cdot (n-1) \cdot n.$$

For example,

$$1! = 1 \qquad\qquad , \qquad 2! = 1 \cdot 2 = 2$$
$$3! = 1 \cdot 2 \cdot 3 = 6 \qquad , \qquad 4! = 1 \cdot 2 \cdot 3 \cdot 4 = 24$$
$$5! = 1 \cdot 2 \cdot 3 \cdot 4 \cdot 5 = 120, \qquad 6! = 1 \cdot 2 \cdot 3 \cdot 4 \cdot 5 \cdot 6 = 720.$$

We can also write $n!$ as follows:

$$n! = [1 \cdot 2 \cdot 3 \cdot \ldots \cdot (n-1)] \cdot n$$
$$= [(n-1)!]n$$

or

$$\boxed{n! = n(n-1)!.}$$

This equation is called the reduction formula for factorial n. Thus

$$7! = 7(6!), \qquad 25! = 25 \cdot (24!).$$

We can apply this formula repeatedly to obtain:

$$n! = n(n-1)!$$
$$= n(n-1)(n-2)!$$
$$= n(n-1)(n-2)(n-3)!, \quad \text{and so on.}$$

If we put $n = 1$ in the reduction formula we obtain the result that

$$1! = 1(0!) \quad \text{or} \quad 1 = 0!.$$

Thus, we must define

$$\boxed{0! = 1}$$

in order to have the reduction formula valid for all natural numbers including $n = 1$.

Now let us return to the question of counting outcomes. Consider a set consisting of n different objects from which r objects are to be selected, and let us ask in how many ways the selection can be made. The answer to this question depends on the way in which the selection is made. First let us consider the case in which each of the objects selected is replaced in the original set before the next selection is made (so that any given one of the objects may be selected two or more times). This method of selection is called *selection with replacement*.

When selecting with replacement, the first object can be chosen in n different ways, since there are n objects in the set from which to choose, and any one of them may be chosen. Since the selected object is returned to the set before making the second choice, the second object can also be selected in n ways. Similarly, each of the other r objects can be selected in this same number of ways. Hence the r objects together can be selected in a number of ways equal to the product

$$\underbrace{n \cdot n \cdot \ldots \cdot n}_{r \text{ factors}} = n^r.$$

Next let us consider the case in which the r objects are selected one-by-one without replacement; that is, the first object is not replaced before the second is selected, the first two are not replaced before the third is selected, and so on. A selection made in this way is called a *permutation* or an *ordered selection* of r objects from n.

As before, the first object may be chosen in n ways. However since the first object is not replaced, there are only $(n-1)$ objects from which the second object can be chosen. Hence the second object can be selected in $(n-1)$ ways. Similarly the third object can be selected in $(n-2)$ ways once the first two have been chosen, and so on. The last (rth) object can be chosen in $(n-r+1)$ ways. Thus the number of ways in which the whole permutation can be selected is given by the product

$$n(n-1)(n-2) \cdot \ldots \cdot (n-r+1).$$

This product is denoted by the symbol $_nP_r$ and is the number of permutations of r objects from among n possible choices. The number $_nP_r$ is given by the product of r consecutive integers, starting with n and decreasing to $(n-r+1)$. For example,

$$_8P_3 = 8(8-1)(8-2) = 8 \cdot 7 \cdot 6 = 336$$

$$_{15}P_2 = 15 \cdot 14 = 210.$$

We can also express $_nP_r$ in terms of factorials as follows:

$$_nP_r = n(n-1) \cdot \ldots \cdot (n-r+1)$$

$$= \frac{n(n-1) \cdot \ldots \cdot (n-r+1)(n-r)(n-r-1) \cdot \ldots \cdot 2 \cdot 1}{(n-r)(n-r-1) \cdot \ldots \cdot 2 \cdot 1}.$$

In this step we have introduced the factors $(n-r), (n-r-1), \ldots, 2, 1$ into both numerator and denominator, and so have not changed anything. But now the numerator is equal to $n!$ and the denominator is equal to $(n-r)!$ so that

$$\boxed{_nP_r = \frac{n!}{(n-r)!}.}$$

NOTE.   Putting $r = n$ we obtain the result that

$$\boxed{_nP_n = \frac{n!}{0!} = n!.}$$

EXAMPLES   **(a)**

$$_8P_3 = 8 \cdot 7 \cdot 6$$

$$= \frac{8 \cdot 7 \cdot 6 \cdot 5 \cdot 4 \cdot 3 \cdot 2 \cdot 1}{5 \cdot 4 \cdot 3 \cdot 2 \cdot 1} = \frac{8!}{5!}.$$

(b)

$$_{15}P_2 = 15 \cdot 14$$

$$= \frac{15 \cdot 14 \cdot 13 \cdot 12 \cdot \ldots \cdot 2 \cdot 1}{13 \cdot 12 \cdot \ldots \cdot 2 \cdot 1} = \frac{15!}{13!}.$$

EXAMPLE   From a group of 8 people, it is required to select individuals to participate in 5 different tests. In how many ways can the selection be made?

SOLUTION   Since the tests are all *different* the order in which the 5 are chosen is significant. The number of ways in which the choice can be made is therefore the number of permutations,

$$_8P_5 = 8 \cdot 7 \cdot 6 \cdot 5 \cdot 4 = 6720.$$

Finally let us consider the case when r objects are selected from n without replacement, but when the order in which the objects are chosen is of no significance. For instance, in the last example, we could suppose that the 5 individuals are chosen from the available 8 in order to participate in the *same* test. Then the order in which the 5 are chosen would be immaterial, and all we would need to know would be which 5 formed the chosen group.

In order to simplify things let us suppose that 3 individuals are to be chosen from among 4 and let us label the available 4 by the letters a, b, c, d. If the order in which the choice is made is important then the number of choices is equal to $_4P_3 = 4 \times 3 \times 2 = 24$. We can list these 24 choices as follows:

$$
\begin{array}{cccccc}
\text{bcd,} & \text{bdc,} & \text{cbd,} & \text{cdb,} & \text{dcb,} & \text{dbc} \\
\text{acd,} & \text{adc,} & \text{cad,} & \text{cda,} & \text{dac,} & \text{dca} \\
\text{abd,} & \text{adb,} & \text{bad,} & \text{bda,} & \text{dab,} & \text{dba} \\
\text{abc,} & \text{acb,} & \text{bac,} & \text{bca,} & \text{cab,} & \text{cba.}
\end{array}
$$

Now we observe from this list that each group of 3 individuals appears 6 times, corresponding to the 6 different ways in which the 3 can be ordered. If the order in which the 3 are selected is of no importance, all of these 6 permutations of each group are equivalent to one another; for example, bcd is equivalent to bdc, or to cbd, and so on. When order is immaterial the number of different choices is only equal to 4, corresponding to the 4 rows listed above.

Now consider the general problem in which r objects are chosen from among n, the order of selection being of no significance. Each such choice is called a *combination of r objects from among n*, and the number of combinations is denoted by the symbol $\binom{n}{r}$.

Any permutation of r objects from among n can be formed by first deciding on which r objects are to be chosen and then arranging these r objects in an appropriate order. The number of permutations $_nP_r$ is therefore equal to the number of ways of choosing particular combinations of r objects from among n multiplied by the number of ways in which each combination can be arranged in an order. That is,

$$_nP_r = \binom{n}{r} \cdot N(r),$$

where $N(r)$ is the number of ordered arrangements of the chosen r objects. But $N(r)$ must be equal to $_rP_r = r!$, the number of permutations of r objects from among r.

Therefore

$$\,_nP_r = \binom{n}{r} r!$$

and so

$$\binom{n}{r} = \frac{\,_nP_r}{r!} = \frac{n!}{(n-r)!r!}.$$

We can also write the number of combinations in the form

$$\binom{n}{r} = \frac{n(n-1) \cdot \ldots \cdot (n-r+1)}{r(r-1) \cdot \ldots \cdot 2 \cdot 1}.$$

Note that both the numerator and denominator in this fraction contain r consecutive integers as factors.

EXAMPLES   (a) $$\binom{8}{3} = \frac{8(8-1)(8-2)}{3(3-1)(3-2)} = \frac{8 \cdot 7 \cdot 6}{3 \cdot 2 \cdot 1} = 56.$$

(b) $$\binom{7}{5} = \frac{7 \cdot 6 \cdot 5 \cdot 4 \cdot 3}{5 \cdot 4 \cdot 3 \cdot 2 \cdot 1} = \frac{7 \cdot 6}{2 \cdot 1} = 21.$$

In the first case 3 factors occur in numerator and denominator, while in the second, 5 factors occur.

REMARKS.

$$\binom{n}{n} = \binom{n}{0} = 1.$$

$$\binom{n}{r} = \binom{n}{n-r} \quad \text{for all } r.$$

This second property can be seen as follows. When any group of r objects is chosen from among n, the number of objects left behind, unchosen, is equal to $(n-r)$. The number of ways of choosing the r must therefore be equal to the number of ways of deciding on the $(n-r)$ objects that are not to be chosen. But we can choose the set of $(n-r)$ objects in $\binom{n}{n-r}$ ways, and so this must be equal to $\binom{n}{r}$. (We can also prove this result directly from the formula expressing $\binom{n}{r}$ in terms of factorials.)

EXAMPLES   (a) $$\binom{7}{5} = \binom{7}{7-5} = \binom{7}{2} = \frac{7 \cdot 6}{2 \cdot 1} = 21.$$

(b) $$\binom{50}{48} = \binom{50}{50-48} = \binom{50}{2} = \frac{50 \cdot 49}{2 \cdot 1} = 25 \cdot 49 = 1225.$$

Let us summarize all the above results in the following theorem.

THEOREM 8.3.1

When r objects are selected from a given set of n different objects, the number of

(a) selections with replacement is n^r

(b) selections without replacement is

$$
\begin{aligned}
_nP_r &= n(n-1)(n-2) \cdot \ldots \cdot (r \text{ factors}) \\
&= n(n-1)(n-2) \cdot \ldots \cdot (n-r+1) \\
&= \frac{n!}{(n-r)!}
\end{aligned}
$$

(c) combinations is

$$
\binom{n}{r} = \frac{n!}{r!(n-r)!}.
$$

EXAMPLE Three cards are selected one-by-one with replacement from a pack of 52 cards. What is the probability of selecting a 6, a 7, and a spade, in that order?

SOLUTION Using Theorem 8.3.1(a), 3 cards can be selected one-by-one with replacement in $52 \times 52 \times 52 = 52^3$ ways. A 6-card can be selected out of the 4 available 6-cards in 4 ways. Similarly a 7 and a spade can be selected in 4 and 13 ways respectively. Therefore a 6, a 7, and a spade together can be selected in $4 \times 4 \times 13 = 208$ ways. Thus the sample space contains 52^3 points and the event contains 208 points. Therefore the required probability is

$$
P(E) = \frac{208}{52^3} = \frac{1}{676}.
$$

EXAMPLE What is the probability of a poker hand (5 cards) containing 4 of a kind?

SOLUTION First of all let us work out the total number of poker hands that are possible; this will give us the number of points n in the sample space. A poker hand consists of a set of 5 cards, so what we need is the number of ways of choosing 5 cards from among the 52 cards in the deck. This is

$$
n = \binom{52}{5} = \frac{52 \cdot 51 \cdot 50 \cdot 49 \cdot 48}{1 \cdot 2 \cdot 3 \cdot 4 \cdot 5}.
$$

The event whose probability we wish to calculate is that the poker hand contains 4 cards of the same denomination (i.e., 4 aces, or 4 sevens, etc.). If there are k different ways in which such a hand can be chosen, then the required probability is equal to k/n.

Consider first the possibility that the hand contains 4 aces. The number of hands that have this property is equal simply to the number of ways in which the fifth card can be selected, namely 48 (since there are 48 cards remaining after the 4 aces have gone). Similarly there are 48 poker hands containing 4 twos, 48 hands containing 4 threes, and so on. Since there are 13 possible denominations the number of ways of choosing a hand with 4 of a kind, regardless of denomination is given by

$$k = 13 \cdot 48.$$

The required probability is therefore

$$P = \frac{k}{n} = \frac{13 \cdot 48 \cdot 1 \cdot 2 \cdot 3 \cdot 4 \cdot 5}{52 \cdot 51 \cdot 50 \cdot 49 \cdot 48} = \frac{1}{4165}.$$

EXAMPLE   Four rats are selected for an experiment from a group of 6 white and 4 brown rats. In how many ways can the selection be made so that the selected group contains

(a)   two brown rats or

(b)   at least two brown rats?

SOLUTION   **(a)**   We have to select, in all, 4 rats for the experiment. Since the selection has to contain 2 brown rats, the other 2 rats must be white. Now we can select 2 brown rats out of 4 given brown rats in $\binom{4}{2} = 6$ ways, and we can select 2 white rats out of 6 given white rats in $\binom{6}{2} = \frac{6 \cdot 5}{2 \cdot 1} = 15$ ways. Using the fundamental principle of counting we see that we can select 2 brown and 2 white rats in $6 \times 15 = 90$ ways.

(b)   In this case, since the selection has to contain *at least* 2 brown rats, the selection can have either 2 brown (and 2 white) or 3 brown (and 1 white) or 4 brown (and no white) rats. The first type of selection containing 2 brown and 2 white rats can be made in 90 ways [see part (a) above]. For the second type of selection 3 brown rats can be selected from among the 4 given brown rats in $\binom{4}{3} = 4$ ways, and 1 white rat can be selected out of 6 white rats in $\binom{6}{1} = 6$ ways. Thus 3 brown and 1 white rats together can be selected in $4 \times 6 = 24$ ways. For the third alternative, there is just one way of choosing all 4 brown rats and no white ones. Hence for the selection to have at least 2 brown rats it can be made in $90 + 24 + 1 = 115$ ways.

A number of interesting counting problems arise in the analysis of courtship and mating arrangements in social groups of animals. Let us consider courtship in a group of animals of a certain species whose behavior is such that during any period of time each male may court any or all of the females in the group and each female may similarly court any number of the males. If the group is observed over a certain period, we should like to know how many different courtship patterns may possibly be found.

Consider first a single male–female pair, denoted by M and F. Then M may or may not court F during the given time interval and F may or may not court M. So the number of courting patterns between M and F is equal to the product 2×2, or 4. We can represent these patterns diagrammatically as follows:

$$M \rightleftarrows F, \quad M \rightarrow F, \quad M \leftarrow F, \quad M \quad F.$$

For example, the second diagram represents the possibility that M courts F but F does not court M; the fourth diagram corresponds to the case of no courtship in either direction.

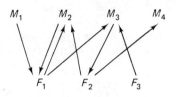

Figure 8.13

Consider now a group consisting of 4 males and 3 females. Many courtship patterns are of course possible within such a group, one pattern being illustrated in Fig. 8.13. In this particular pattern the males M_1 and M_2 both court the female F_1, while F_1 herself courts M_2 and M_3. M_3 courts female F_2, who in turn courts the males M_2 and M_4. Finally female F_3 courts M_3.

The number of male–female pairs in this group is equal to the number of ways of choosing the male, namely 4, multiplied by the number of ways of choosing the female, namely 3. Hence there are 12 such pairs. Between each male–female pair there are, as we saw above, 4 possible courtship arrangements. Therefore, by the fundamental principle of counting, the number of possible courting patterns for the group as a whole is equal to the product

$$\underbrace{4 \cdot 4 \cdot \ldots \cdot 4}_{\text{12 factors}} = 4^{12}.$$

(This is quite a large number, about 17 million.)

Some further examples on courtship patterns are included in the exercises.

EXERCISES 8.3

Evaluate the following:

1. $_{10}P_2$.

2. $_6P_4$.

3. $_5P_5$.

4. $\binom{10}{2}$.

5. $\binom{30}{28}$.

6. $\binom{20}{17}$.

7. $\dfrac{20!}{18!}$.

8. $\dfrac{10!}{7!}$.

9. $\dfrac{8!}{7! + 6!}$.

10. Given $(n + 1)! = 20n!$, find the value of n.

11. Three cards are selected one-by-one with replacement from a deck of 52 cards. What is the probability of selecting an ace, a king, and a queen, in that order?

12. Three rats are selected from a group of 5 white and 4 brown rats for a certain experiment. What is the probability that:
    (a) all the selected rats are white?
    (b) the selected rats are 2 white and 1 brown?
    (c) all the selected rats are brown?

13. Two flower seeds are randomly selected from a bag containing 10 seeds for red flowers and 5 seeds for white flowers. What is the probability that:
    (a) both result in white flowers?
    (b) one of each color is selected?

14. Three flower seeds are selected from a package that contains 5 seeds for white, 6 seeds for red, and 4 for yellow flowers. What is the probability that:
    (a) all 3 seeds produce red flowers?
    (b) the 3 seeds produce flowers of different colors?
    (c) the 3 seeds produce flowers of the same color?

15. A sample of 6 individuals is selected for a certain test from a group containing 20 smokers and 10 nonsmokers. What is the probability that the sample contains 4 smokers?

16. A preparation in a test tube contains 30 pine pollen grains and 10 oak pollen grains. A random sample of 5 grains is chosen. What is the probability that:
    (a) all 5 grains are oak?
    (b) all 5 grains are pine?
    (c) there are at least 3 oak grains in the sample?

17. A group of animals consists of m males and f females. How many different court-ship patterns are possible within the group? (Assume that each female may or may not court each male and vice versa.)

18. Five married couples spend an evening together. Each man may or may not flirt with each woman, and vice versa, except that no one flirts with his/her wife or husband. How many flirtation patterns are possible?

19. Repeat exercise 17 in the case when s of the males are indisposed and do not court any female ($s \leq m$). Assume that the males which are out of action may still themselves be courted by females.

20. Repeat exercise 17 in the case of a species of animal whose courtship behavior is as follows. Each female may or may not court any male in the group. The males do not initiate courtship themselves, but may or may not respond to the courtship advances of any female who indicates an interest in them.

8.4 APPLICATIONS IN GENETICS

The various counting methods, and probability theory in general, find many applications in the field of genetics, the science of heredity. In this section we shall touch upon a few of these applications.

The information that determines the structure of each living organism is stored in elements called *chromosomes*, which are present in each cell of the organism. With the exception of the reproductive germ cells (that is, the sperm cells of the male and the egg cells of the female), each cell contains an identical set of chromosomes. For example, in the human being, each cell contains 46 chromosomes. Normally, when any cell divides into two new cells, each of the new cells contains an exact copy of the 46 chromosomes that were present in the parent cell.

The exception to this occurs with the formation of the germ cells. The chromosomes in the normal cells of the organism occur in pairs, and when a cell divides to produce two germ cells, only one of each pair of chromosomes enters each of the resulting germ cells. Thus, for example, the sperm or egg cells in humans contain 23 chromosomes each.

When fertilization occurs a sperm and egg cell unite, and the resulting fertilized egg once again contains the full complement of 46 chromosomes. As this cell divides and grows into an offspring this set of 46 chromosomes is duplicated in each cell and determines the inherited characteristics of the new organism. Since 23 of these chromosomes have been derived from each parent the offspring will have a combination of characteristics, some inherited from one parent and some from the other.

Consider an organism in which there are n chromosome pairs in each cell ($n = 23$ for humans), and let us calculate how many different germ cells can be formed using the chromosomes of any individual. For any pair of chromosomes there are two choices of the particular chromosome that is to enter the germ cell. Since there are n pairs, the number of ways of choosing the n chromosomes in the germ cell is given by the product

$$\underbrace{2 \cdot 2 \cdot \ldots \cdot 2}_{n \text{ terms}} = 2^n.$$

For example, each human being has the capability of making 2^{23} different types of sperm or ovum (egg cell).

The male organism can generate 2^n different chromosome arrangements in his sperm cells and the female can generate 2^n different arrangements in her egg cells. The number of different chromosome arrangements in a fertilized ovum from any male–female union is therefore equal to $2^n \times 2^n = 2^{2n}$. For example, any man and woman are capable of producing 2^{46} different chromosome arrangements in their offspring. This number is quite large (about 7×10^{13}). It is this large variety in the possible chromosome type of the offspring that gives bisexual reproduction its evolutionary advantage over asexual reproduction.

The genetic information contained in the chromosomes is stored in long molecules of DNA (deoxyribonucleic acid). The DNA molecule consists of a long backbone whose shape is the famous helix of Crick and Watson, along which are distributed strings of chemical units called *nucleotides*. There are four commonly occurring nucleotides: adenine, cytosine, guanine, and thymine, denoted respectively by the letters A, C, G, and T. The order in which these nucleotides occur along the DNA molecule determines the information stored by the molecule.

Consider the DNA molecules that contain only three nucleotides, and let us ask how many different molecules of this kind exist. Examples of such molecules might be GCT or CAA and so on, where the three letters are used to denote the order in which the nucleotides occur along the backbone. Then there are 4 ways in which the first nucleotide can be chosen, 4 ways in which the second can be chosen and 4 ways in which the third can be chosen. Hence the number of ways of choosing the whole string of three is given by the product $4 \times 4 \times 4 = 4^3 = 64$. So there are 64 different DNA molecules containing three nucleotides.

In actual fact, DNA molecules are much longer than this, often containing upwards of a thousand nucleotides. For any such molecule there are four ways of choosing each nucleotide in the string; hence the total number of different DNA molecules containing n nucleotides is given by the product

$$\underbrace{4 \cdot 4 \cdot \ldots \cdot 4}_{n \text{ factors}} = 4^n.$$

For example, there are 4^{1000} different DNA molecules that contain 1000 nucleotides. This number is very large, being about 10^{602}, a number which, if written out, would consist of a one followed by six hundred and two zeros. This huge number of different DNA molecules gives some indication of the variety of information that can be stored by these molecules.

The way in which the DNA molecules in the chromosomes influence the development of the organism is by controlling the production of various proteins within the cells of the organism. The presence or absence of certain of these proteins then determines various properties of the organism; for example, one type of protein may determine whether a person's eye color is blue or brown, or another type of protein may determine whether a rabbit's coat is long or short hair. Each DNA molecule in a given chromosome is divided into segments called *genes*, each gene controlling the production of one type of protein. One chromosome will in general be responsible for the production of many different proteins.

The way in which the proteins are generated is briefly as follows. The nucleotides along a particular gene segment of a DNA molecule are divided into groups of three, and each group of three "codes for" the production of a certain amino acid. For example, the triple ACG of adjacent nucleotides codes for one amino acid, the triple CCC codes for a second amino acid, and so on. A protein molecule is a long molecule that consists of a string of amino acids, and when a protein is being formed by a particular gene, the order in which the amino acids are added to the protein is determined by the order in which the associated triplets of nucleotides occur along the gene. There are about 20 different amino acids, some of which are coded for by more than one triplet of nucleotides.

Any particular gene that controls the manufacture of a certain protein usually occurs throughout the species in question in several different forms. These forms, or *alleles*, differ from one another to some extent in the arrangement of nucleotides along the gene. The simplest case to consider, and one that occurs frequently in practice, is that in which the gene in question has two alleles, which we denote by A and a. The most significant difference between alleles is usually their degree of activity, one allele being capable of producing more of the protein than the other.

As noted earlier, the chromosomes are divided into pairs, one of each pair for any given organism coming from each of its parents. Thus each gene occurs twice within any organism. In the case of a gene with two different alleles there are therefore three different possibilities for the gene pair: AA, Aa, or aa. These three are referred to as *genotypes*. (Note that there is no difference between Aa and aA.)

It can happen that the three genotypes exhibit three different physical characteristics. For example, for the gene that controls the color of sweet pea flowers, those plants of genotype AA have red flowers, those of genotype aa have white flowers, while those of genotype Aa have pink flowers. (Mendel's original experiments concerned this particular gene.) However it commonly happens that the two genotypes AA and Aa are physically indistinguishable from one another, and in this case we say that the allele A is *dominant* while the allele a is *recessive*. For example, in the case of the gene controlling eye color in humans, the genotypes AA and Aa both have brown eyes while the genotype aa has blue eyes.

The genotypes AA and aa are called *homozygous* or *pure* while Aa is called *heterozygous* or *hybrid*.

We wish to address ourselves to the general question: Given parents of certain genotypes, what will be the probabilities of the offspring being of the three possible genotypes? For example, if both parents are hybrid (Aa) genotypes, the offspring may be of any of the three types AA, aa, or Aa, and we would like to calculate the proportions of offspring that belong to each of these categories.

First of all, consider the case of a parent organism of pure genotype, say AA. When the chromosomes are divided into two groups during formation of a germ cell, no matter how the division occurs, the resulting germ cell must contain an A-type gene, since no a-type gene is present. If two such parents mate, both sperm and ovum contain A-type genes, so that the resulting fertilized egg will be of pure AA genotype. Thus if both parents are AA genotypes the offspring are also all AA genotypes.

Similarly, if both parents are aa genotypes the offspring are all aa genotypes like the parents. If one parent is AA and the other parent is aa then all the offspring are hybrid genotypes (Aa).

Now let us consider the case when one parent is hybrid while the other is pure genotype. When the chromosomes divide in the hybrid parent half of the resulting germ cells contain an A-type gene and the other half contain an a-type gene. On the other hand, all of the germ cells in the pure genotype parent contain the same type of gene; for example, in an AA parent, all germ cells contain an A-type gene. In such a case, half of the fertilized eggs will be of AA genotype and half will be of Aa genotype.

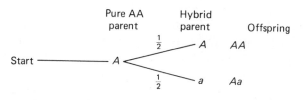

Figure 8.14

The formation of these two possible genotypes is illustrated in Fig. 8.14. In the first stage of the diagram the A-type gene is received from the parent of AA genotype; then in the second stage the two possibilities of A- or a-type genes are received from the hybrid parent, resulting respectively in AA and Aa offspring. Each of these genotypes occurs with probability $\frac{1}{2}$.

Similarly, if parents of genotypes Aa and aa mate, half the offspring are of type Aa and half are of type aa.

Finally let us consider the case when two individuals of genotype Aa mate—the so-called *hybrid cross*. The tree diagram for this cross is shown in Fig. 8.15. Since the

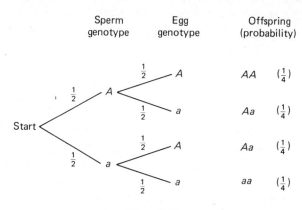

Figure 8.15

male parent is hybrid the genotype of the sperm can be A or a with equal probability. The egg is also equally likely to be of genotype A or a. The offspring can therefore be AA with probability $\frac{1}{4}$, aa with probability $\frac{1}{4}$, and Aa with probability $\frac{1}{2}$.

EXAMPLE (a) If a sweet pea with red flowers (genotype AA) is crossed with a sweet pea with white flowers (type aa) all the offspring have pink flowers (type Aa).

(b) If a red sweet pea is crossed with a pink sweet pea, half the offspring have red flowers and half have pink flowers.

(c) If two pink sweet peas are crossed, one quarter of the offspring have red flowers, one quarter have white flowers and one half have pink flowers.

The above results can be applied to cases when the three genotypes AA, Aa, and aa have physically distinguishable characteristics. However if the A gene is dominant and a is recessive, there are only two physical types that are distinguishable from one another: Genotypes AA and Aa both exhibit the dominant physical characteristic and can be grouped together and indicated by the letter D, while genotype aa exhibits the recessive physical characteristic and can be indicated by the letter R. The results of various types of matings as far as the physical characteristics of the offspring are concerned is then summarized in the following table.

PARENT 1	PARENT 2		
	AA	Aa	aa
AA	All D	All D	All D
Aa	All D	$\frac{3}{4}$ D; $\frac{1}{4}$ R	$\frac{1}{2}$ D; $\frac{1}{2}$ R
aa	All D	$\frac{1}{2}$ D; $\frac{1}{2}$ R	All R

For example, in the case of a hybrid cross, we have seen that $\frac{1}{4}$ of the offspring are AA, $\frac{1}{4}$ are aa, and $\frac{1}{2}$ are Aa. But if the A gene is dominant the offspring of types AA and Aa show the same dominant physical characteristic. These two types together constitute $\frac{3}{4}$ of the offspring. The remaining $\frac{1}{4}$ are of genotype aa and hence are of recessive physical type.

EXAMPLE   Consider eye color in humans, for which the brown-eye gene (A) is dominant over the blue-eye gene (a).

(a)   If one parent is of type Aa (hence has brown eyes) and the other has blue eyes (hence must be of genotype aa) then there are equal probabilities that the children will have brown eyes or blue eyes.

(b)   If both parents are of genotype Aa (with brown eyes), the probability that the offspring will have brown eyes is $\frac{3}{4}$ and the probability that it will have blue eyes is $\frac{1}{4}$.

NOTE.   The phenomenon of eye color is in actual fact more complicated than the above simple model of a single gene with two alleles.

Let us now turn to the question of the distribution of genotypes throughout a population, and in particular to the way in which this distribution changes from one generation to the next. We shall consider a large population of N organisms in which, at some instant, a proportion f are of genotype AA, g are of genotype Aa, and h are of genotype aa. (Hence $f + g + h = 1$. The actual number of organisms of type AA is fN, and so on.)

Among the population there are $2N$ genes of the type under consideration. The fN individuals of type AA have $2fN$ genes of type A; the gN individuals of type Aa have gN genes of type A and also gN of type a; the hN individuals of type aa have $2hN$ genes of type a. Thus within the population there are $(2fN + gN)$ genes of type

A and $(2hN + gN)$ genes of type a. We are therefore led to define the so-called *gene frequencies p and q* for the types A and a respectively as follows:

$$p = \frac{2fN + gN}{2N} = f + \frac{1}{2}g, \qquad q = \frac{2hN + gN}{2N} = h + \frac{1}{2}g.$$

p is the proportion of all the genes in the population that are of type A and q is the proportion that are of type a. Of course $p + q = 1$.

We assume that within the population mating is random, that is, independent of genotype. Then the formation of the gene pair within a fertilized egg is illustrated in Fig. 8.16. The sperm and egg can each have genes A or a with respective probabilities

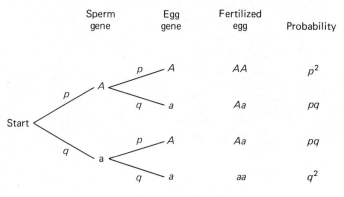

Figure 8.16

p and q. The resulting fertilized egg is genotype AA with probability p^2, Aa with probability $2pq$, and aa with probability q^2.

In the first generation of offspring let $f_1, g_1,$ and h_1 be the proportions respectively of genotypes AA, Aa, and aa. Then the result we have established is that

$$f_1 = p^2, \qquad g_1 = 2pq, \qquad h_1 = q^2.$$

(Here we have equated proportions with probabilities. This is valid as long as the number of offspring in each generation is large.)

The gene frequencies in this first generation of offspring are the same, namely p and q, as in the parent population. For example, the frequency of A-type genes is given by

$$f_1 + \tfrac{1}{2}g_1 = p^2 + \tfrac{1}{2}(2pq) = p^2 + pq = p(p + q) = p$$

since $p + q = 1$. Similarly the frequency of a-type genes is equal to q.

It follows from this result that when the offspring join the adult population and begin mating, the gene frequencies are unaltered from their initial values p and q. Hence the proportions of the different genotypes in subsequent generations of offspring remain the same as the proportions in the first generation, namely

$$\text{AA}: p^2; \qquad \text{Aa}: 2pq; \qquad \text{aa}: q^2.$$

Eventually the older members of the population die off, leaving these values as the proportions of the three genotypes in the whole population.

This result is known as the *Hardy–Weinberg law* in population genetics: In a large population in which mating is independent of genotype, the proportions of the population belonging to the three genotypes eventually stabilize at the values p^2, $2pq$, and q^2 for types AA, Aa, and aa respectively, where p and q are the gene frequencies for types A and a.

EXAMPLE   A rabbit colony is established with 20% white rabbits, 40% black rabbits, and 40% black-and-white rabbits. Assuming that fur color is controlled by a single gene with 2 alleles, AA being black, aa white, and Aa black-and-white, find the eventual proportions of the three genotypes in the population.

SOLUTION   The initial proportions of the three genotypes are given as follows:

$$AA: f = 0.4; \qquad Aa: g = 0.4; \qquad aa: h = 0.2.$$

Therefore the gene frequencies are given by

$$p = f + \tfrac{1}{2}g = 0.6, \qquad q = h + \tfrac{1}{2}g = 0.4.$$

The eventual proportions, according to the Hardy–Weinberg law, are therefore as follows:

$$\text{Proportion of AA} = p^2 = 0.36.$$
$$\text{Proportion of Aa} = 2pq = 2(0.6)(0.4) = 0.48.$$
$$\text{Proportion of aa} = q^2 = 0.16.$$

The population will eventually comprise 36% black, 16% white, and 48% black-and-white rabbits.

Let us finally consider one further aspect of genetics, that of sex-linked characteristics. A male has two chromosomes, the X and Y chromosomes, which are not paired; a female has a pair of X chromosomes and no Y chromosomes. When a male cell divides to produce two sperm cells the X chromosome goes to one sperm cell and the Y chromosome goes to the other. Thus half of the sperm cells contain X chromosomes and half contain Y chromosomes. The sex of the offspring is determined by which of these two types of sperm manages to fertilize the female ovum; if an X sperm succeeds the fertilized ovum will contain two X chromosomes and hence will be female, while if a Y sperm succeeds the fertilized ovum will have one X and one Y chromosome and therefore will be male.

It follows that any male offspring definitely gets his father's Y chromosome. Hence the characteristic of maleness, which is determined by the Y chromosome, will definitely be inherited by any son from his father and will definitely not be inherited by a daughter from her father.

Consider now the more interesting case of a characteristic determined by a gene that occurs on the X chromosome. We shall again consider the case of a gene with two alleles, A and a. Then a female having two X chromosomes can be classified as one of the three genotypes AA, Aa, or aa. However a male, having only one X chromosome, must be classified as either A or a according to the allele of his single gene.

A male offspring inherits one X chromosome from his mother, none from his father. Hence if the mother is AA, her son will always be A; if she is aa, her son will always be a; if she is Aa, her son can be either type A or a, with equal probabilities.

A female offspring inherits her father's X chromosome and one of her mother's two X chromosomes. Hence the genotypes of female offspring of various types of matings are as given in the table.

FATHER	MOTHER		
	AA	Aa	aa
A	All AA	Half AA, half Aa	All Aa
a	All Aa	Half Aa, half aa	All aa

Now let us examine the distribution of sex-linked genes throughout a population. Let the females in the population consist of proportions f, g, and h of the three genotypes AA, Aa, and aa respectively, and let the males consist of proportions P and Q of the two genotypes A and a. The gene frequencies are therefore

$$p = f + \tfrac{1}{2}g, \qquad q = h + \tfrac{1}{2}g$$

for A and a genes respectively among the females and P and Q for A and a genes among the males.

Let f_1, g_1, and h_1 be the proportions of the three genotypes among the first generation of female offspring and P_1 and Q_1 the proportions of A and a genotypes among the first generation of male offspring. Then, since each male offspring inherits his X chromosome from his mother, it follows that $P_1 = p$ and $Q_1 = q$. Furthermore it is readily seen that $f_1 = Pp$, $g_1 = Pq + Qp$, and $h_1 = Qq$ (see the tree diagram in Fig. 8.17). The gene frequencies in this first generation of female offspring are

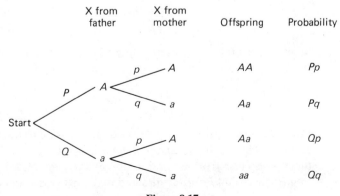

Figure 8.17

therefore:

$$p_1 = f_1 + \tfrac{1}{2}g_1 = Pp + \tfrac{1}{2}(Pq + Qp)$$
$$= \tfrac{1}{2}(Pp + Pq) + \tfrac{1}{2}(Pp + Qp)$$
$$= \tfrac{1}{2}P(p + q) + \tfrac{1}{2}p(P + Q)$$
$$= \tfrac{1}{2}(P + p)$$

since $p + q = P + Q = 1$. Similarly,

$$q_1 = h_1 + \tfrac{1}{2}g_1 = \tfrac{1}{2}(Q + q).$$

We see from these results that the gene frequency P_1 in any generation of male offspring is equal to the frequency p in the previous generation of females, and the frequency p_1 in any generation of female offspring is equal to the average of the frequencies p and P in the males and females of the previous generation.

EXAMPLE   In a certain population the distribution of a sex-linked gene with two alleles is as follows. Among the females 10% are AA, 40% are Aa, and 50% are aa; among the males 80% are A and 20% are a. Determine how the gene frequencies P and p for the A allele vary over the next eight generations.

SOLUTION   The initial proportions of the three genotypes among the females in the population are given as follows:

$$AA: f = 0.1; \qquad Aa: g = 0.4; \qquad aa: h = 0.5.$$

Therefore the initial gene frequencies are

$$p = f + \tfrac{1}{2}g = 0.3, \qquad q = h + \tfrac{1}{2}g = 0.7.$$

Among the males the initial gene frequencies are given to be

$$P = 0.8, \qquad Q = 0.2.$$

For the first generation of offspring the gene frequencies for males and females, P_1 and p_1, are given respectively by

$$P_1 = p = 0.3$$
$$p_1 = \tfrac{1}{2}(P + p) = \tfrac{1}{2}(0.8 + 0.3) = 0.55.$$

Similarly, for the second generation of offspring, the corresponding frequencies P_2 and p_2 are given by

$$P_2 = p_1 = 0.55$$
$$p_2 = \tfrac{1}{2}(P_1 + p_1) = \tfrac{1}{2}(0.3 + 0.55) = 0.425.$$

We can continue in this way to calculate the gene frequencies in successive generations. The results for the first eight generations of offspring are given in the table. It is clear from this table that as the generations advance, the two gene frequencies p and P for females and males get closer and closer together.

p	P
0.3	0.8
0.55	0.3
0.425	0.55
0.4875	0.425
0.4563	0.4875
0.4719	0.4563
0.4641	0.4719
0.4680	0.4641
0.4660	0.4680

This result is in fact always true: Eventually the gene frequencies for a sex-linked gene among the males and females of a population stabilize at values that are equal to one another. Furthermore, the frequencies of the three genotypes among the females eventually settle down at values equal to the Hardy–Weinberg proportions. This follows from the above results that from one generation to the next, $P_1 = p$ and $p_1 = \frac{1}{2}(P + p)$.

The way in which P and p change from generation to generation is therefore as shown in Fig. 8.18. It is clear from this that p and P approach closer and closer to one

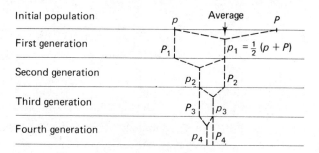

Figure 8.18

another as the generations advance, showing that eventually the gene frequencies for male and female groups in the population become effectively equal to one another.

When $P = p$ and $Q = q$ the proportions of the three genotypes among females become as follows:

$$AA: Pp = p^2; \qquad Aa: Pq + Qp = 2pq; \qquad aa: Qq = q^2.$$

Thus we see that the Hardy–Weinberg law also applies to the distribution of genotypes among females for sex-linked characteristics, as for other characteristics. However there is one difference: For nonsex-linked characteristics the Hardy–Weinberg proportions are reached exactly in the first generation; for sex-linked characteristics it takes several generations before the Hardy–Weinberg proportions are approximately attained.

EXERCISES 8.4

1. Each cell of a certain organism contains three chromosome pairs. How many different types of germ cells can any individual form? How many different offspring can a given male–female union produce?

2. Repeat exercise 1 for an organism whose cells contain eight pairs of chromosomes.

3. Some living organisms are capable of self-fertilization, wherein the male and female germ cells derive from the same individual organism. If the cells contain only a single pair of chromosomes, how many different chromosome arrangements in an offspring are possible? Enumerate the possibilities.

4. Repeat exercise 3 for an organism whose cells contain three pairs of chromosomes.

5. For an organism with n chromosome pairs, how many different types of offspring can be produced by one organism by means of self-fertilization?

6. How many different DNA molecules can be formed with four nucleotides?

7. How many different DNA molecules can be formed with five nucleotides but in which the nucleotide thymine does not occur?

8. How many different DNA molecules can be formed with five nucleotides in which no two adjacent nucleotides are the same?

9. If amino acids were coded for by pairs of nucleotides rather than triplets, what is the greatest number of different amino acids that could be produced by elements of a DNA molecule?

Explain the following observations by constructing a suitable genetic model. In each case assume that the relevant characteristic is controlled by a single gene.

10. Two black guinea pigs are mated. Three-quarters of the offspring turn out to be black and one-quarter turn out to be white.

11. A brown-haired man mates with a blonde woman. Half the children have brown hair and half have blonde hair.

12. A tall bean plant is crossed with a short bean plant. All the offspring are of medium height.

13. A tall plant of another species is crossed with a short plant; all the offspring are tall.

14. Two rabbits have coats of medium length. Of their progeny, one-quarter have short coats, one-quarter have long coats, and one-half have medium-length coats.

15. Two tall pea plants are crossed and the seeds are planted. The resulting crop of plants are allowed to fertilize themselves at random. It is observed that one in sixteen of the resulting second generation of plants is short, the other $\frac{15}{16}$ being tall.

16. Two pink sweet peas are crossed and the seeds are planted. The resulting crop of plants are allowed to fertilize themselves at random. What proportions of the

resulting second generation will be red, pink, and white? What will be the color proportions in the third generation?

17. A field is planted with sweet peas, 10% being red, 60% pink, and 30% white. If the resulting offspring are allowed to fertilize themselves, what will be the proportions of red, pink, and white flowers in the following generation of plants?

18. Repeat exercise 17 given that the initial planting consists of 40% red and 60% pink sweet peas.

19. The height of a species of animal is controlled by a single gene with two alleles; the genotypes AA and Aa are both tall, and aa is short. If the gene frequencies are p and q for A and a respectively, what will be the eventual proportions of tall and short animals in the population?

20. A population of the animals in exercise 19 exists in a steady state vis-à-vis the Hardy–Weinberg law with 84% being tall and 16% being short. A second population of the same size as the first also exists in steady state with 36% being tall and 64% being short. If the two populations are mixed, what will be the eventual proportions of tall and short animals in the combined population?

21. A gene exists with three alleles. How many pure and how many hybrid genotypes are there?

22. Repeat exercise 21 for the case of a gene with n alleles.

23. In a population, the proportions of genotypes AA, Aa, and aa among the males are f, g, and h respectively and among the females are f', g', and h'. If mating is random, show that after two generations the Hardy–Weinberg proportions (AA: p^2; Aa: $2pq$; aa: q^2) are attained throughout the population. Find the values of p and q.

8.5 THE BINOMIAL DISTRIBUTION

In this section we shall deal with the particular type of random process in which there are exactly two mutually exclusive outcomes. Such events are known as *Bernoulli trials*, named after the 18th-century Swiss mathematician Jacob Bernoulli. For example, guessing the sex of an unborn child is a Bernoulli trial, because there are only two possible outcomes, male or female, and these two outcomes are mutually exclusive. Similarly, the flip of a single coin is a Bernoulli trial, since it has only two mutually exclusive outcomes, head or tail.

The study of Bernoulli trials has applications in many areas. For example, in medicine, a certain drug may or may not cure a specific disease, or in genetics, certain genes may be present in one or another of two possible forms.

It is customary to denote the two outcomes of a Bernoulli trial by the terms success (S) and failure (F). If p denotes the probability of success in a Bernoulli trial, then the probability of failure is $q = 1 - p$ because the two outcomes (success and failure) are mutually exclusive and the two probabilities p and q must add up to 1. We assume that the probability of success in each Bernoulli trial remains unchanged from one trial to the next.

For example, let us consider patients suffering from a certain disease that is commonly fatal. In such a case we can define "success" to occur if the patient recovers and "failure" to occur if the patient dies. Then p is the probability of recovery and q the probability of death for any such patient. (It would in principle be quite possible, though perhaps macabre, to define success and failure the other way around. In such a case the values of p and q would be interchanged.)

Now let us consider a group of three such patients, and let us calculate the probabilities that all three recover, that two recover, that only one recovers, and that all three die. The set of outcomes can be listed as follows:

$$\{SSS, SSF, SFS, FSS, SFF, FSF, FFS, FFF\}.$$

Here, for example, SSF means that the first two patients recover (successes) and the third dies (failure).

The outcome SSS, in which each patient recovers, has a probability $p \times p \times p = p^3$ since each Bernoulli trial has a probability of p of resulting in a success.

Similarly the outcome SSF has probability $p \times p \times q = p^2q$ since the two successes each have probability p and the failure has probability q. In the same way the other two outcomes SFS and FSS, in which there are two successes and one failure, also have probabilities p^2q. Therefore the total probability of two patients surviving and one dying is equal to the sum of the probabilities of the three outcomes SSF, SFS, and FSS, and is given by $3p^2q$.

Continuing in this way we see that the probability of one patient surviving and two dying is equal to $3pq^2$, and the probability of all three dying is equal to q^3.

EXAMPLE   Let "success" be rolling a six with a die. Then $p = \frac{1}{6}$ and $q = \frac{5}{6}$. The probability of getting 2 sixes when rolling 3 dice is equal to

$$3p^2q = 3(\tfrac{1}{6})^2(\tfrac{5}{6}) = \tfrac{5}{72}.$$

These results are special cases of a general expression for the probability of obtaining r successes in a sequence of n Bernoulli trials ($0 \le r \le n$). The general result is given in the following theorem.

THEOREM 8.5.1

If p is the probability of success and q the probability of failure in a single Bernoulli trial, then the probability of exactly r successes in a sequence of n independent trials is

$$P(r) = \binom{n}{r} p^r q^{n-r}.$$

EXAMPLE A fair coin is tossed five times. What is the probability of getting

(a) three heads?

(b) at least three heads?

SOLUTION The tossing of a coin is a Bernoulli trial because there are two mutually exclusive outcomes: Head (success) and tail (failure). The probability of getting a head in a single trial is $p = \frac{1}{2}$. Thus $q = 1 - p = \frac{1}{2}$. Since the coin is tossed 5 times we have $n = 5$ trials.

(a) In this case $r = 3$ because we want 3 heads (successes). Using Theorem 8.5.1 the required probability of 3 heads (i.e., 3 successes) is

$$P(3) = \binom{n}{r} p^r q^{n-r}$$

$$= \binom{5}{3} \left(\frac{1}{2}\right)^3 \left(\frac{1}{2}\right)^{5-3}$$

$$= \frac{5 \cdot 4 \cdot 3}{3 \cdot 2 \cdot 1} \cdot \left(\frac{1}{2}\right)^5$$

$$= 10 \cdot \frac{1}{32} = \frac{5}{16}.$$

(b) In this case r, the number of successes, can equal 3, 4, or 5. Since the events of getting 3 heads, 4 heads, and 5 heads are all mutually exclusive, the probability of getting 3, 4, or 5 heads (that is, at least 3 heads) is equal to the *sum* of the probabilities of these events. In other words, the required probability is

$P(r \geq 3) = P(3) + P(4) + P(5)$

$$= \binom{5}{3} \left(\frac{1}{2}\right)^3 \left(\frac{1}{2}\right)^{5-3} + \binom{5}{4} \left(\frac{1}{2}\right)^4 \left(\frac{1}{2}\right)^{5-4} + \binom{5}{5} \left(\frac{1}{2}\right)^5 \left(\frac{1}{2}\right)^{5-5}$$

$$= \left(\frac{1}{2}\right)^5 \left[\binom{5}{3} + \binom{5}{4} + \binom{5}{5} \right]$$

$$= \frac{1}{32} \left[\frac{5 \cdot 4 \cdot 3}{3 \cdot 2 \cdot 1} + \frac{5 \cdot 4 \cdot 3 \cdot 2}{4 \cdot 3 \cdot 2 \cdot 1} + \frac{5 \cdot 4 \cdot 3 \cdot 2 \cdot 1}{5 \cdot 4 \cdot 3 \cdot 2 \cdot 1} \right]$$

$$= \frac{1}{32}[10 + 5 + 1] = \frac{16}{32} = \frac{1}{2}.$$

EXAMPLE In Vancouver, 30% of the population have a certain blood type. What is the probability that exactly four out of a randomly selected group of 10 Vancouverites will have that blood type?

SOLUTION p, the probability that an individual has the given blood type, is 30% or 0.3. Then

$$q = 1 - p = 1 - 0.3 = 0.7.$$

The examination of each individual for blood type constitutes a Bernoulli trial because the 2 possibilities that each person has or does not have the given blood type are mutually exclusive and are the only 2 possibilities. We require the probability of four successes (i.e., 4 persons having the given blood type) in 10 trials (10 persons). Thus $n = 10$, $r = 4$. Therefore the required probability is

$$P(4) = \binom{n}{r} p^r q^{n-r}$$

$$= \binom{10}{4} (0.3)^4 (0.7)^{10-4}$$

$$= (210)(0.3)^4 (0.7)^6 \approx 0.2.$$

EXAMPLE   By noting the family history of a certain couple it is established that the probability that any child of theirs will have a certain birth defect is $\frac{1}{10}$. If the couple has 4 children, what is the probability that at least 1 child will have the defect?

SOLUTION   The probability that each child has the defect (success) is $p = \frac{1}{10} = 0.1$ and then $q = 1 - p = 0.9$. In this case $n = 4$ since there are 4 trials (i.e., 4 children). Thus the probability of having at least 1 child with the defect is equal to

$$P(r \geq 1) = P(1) + P(2) + P(3) + P(4),$$

where

$$P(r) = \binom{n}{r} p^r q^{n-r} = \binom{4}{r} (0.1)^r (0.9)^{4-r}.$$

Thus the required probability is

$$P(r \geq 1) = \binom{4}{1}(0.1)^1(0.9)^3 + \binom{4}{2}(0.1)^2(0.9)^2 + \binom{4}{3}(0.1)^3(0.9)^1$$

$$+ \binom{4}{4}(0.1)^4(0.9)^0$$

$$= 4(0.0729) + 6(0.0081) + 4(0.0009) + 0.0001$$

$$= 0.2916 + 0.0486 + 0.0036 + 0.0001$$

$$= 0.3439.$$

ALTERNATIVELY.   The event E' of having no defective child is the complement of the event E of having at least 1 defective child. Now

$$P(E') = P(r = 0) = \binom{n}{r} p^r q^{n-r}$$

$$= \binom{4}{0}(0.1)^0(0.9)^4 = 0.6561.$$

Therefore

$$P(E) = 1 - P(E')$$

$$= 1 - 0.6561 = 0.3439.$$

Since the n trials contain r successes there must be $(n - r)$ failures. Consider any sequence of r successes and $(n - r)$ failures. Since all the n trials [r successes and $(n - r)$ failures] are independent, the probability of this sequence of r successes and $(n - r)$ failures is the product of r p's and $(n - r)$ q's, that is, it is $p^r q^{n-r}$.

The r successes can occur in n trials in different ways. Any particular way is obtained by selecting the r trials that are to result in success. The number of ways in which n trials can result in r successes is the same as the number of ways in which r objects can be selected out of n different objects. This can be done in

$$\binom{n}{r}$$ ways.

Thus n trials can result in $\binom{n}{r}$ sequences of r successes and $(n - r)$ failures.

Each of these sequences has the probability $p^r q^{n-r}$. Therefore the probability of r successes in n trials is

$$P(r) = \binom{n}{r} p^r q^{n-r}.$$

The quantity $(q + p)^n$ can be expanded by means of the binomial theorem in the following way:

$$(q + p)^n = \binom{n}{0}q^n + \binom{n}{1}q^{n-1}p + \binom{n}{2}q^{n-2}p^2 + \cdots$$

$$+ \binom{n}{r}q^{n-r}p^r + \cdots + \binom{n}{n}p^r.$$

If p represents the probability of success and $q = 1 - p$ the probability of failure, then the first, second, third, $\ldots$, $(n + 1)$th terms in the binomial expansion of $(q + p)^n$ represent the probabilities of $0, 1, 2, \ldots, n$ successes in a sequence of n independent Bernoulli trials. So if n is a positive integer, the various terms in the binomial expansion of $(q + p)^n$ give a complete picture of the various probabilities of different numbers of successes. It is for this reason that the probabilities of Theorem 8.5.1, namely $P(r)$ $= \binom{n}{r}p^r q^{n-r}$, $r = 0, 1, 2, \ldots, n$, are called *binomial probabilities*.

Since $p + q = 1$ the left-hand side in the above expansion is equal to 1. This means that the probabilities of $0, 1, 2, \ldots, n$ successes add up to 1. It is intuitively clear that this must be the case since the sample space for n Bernoulli trials consists of $\{0, 1, 2, \ldots, n \text{ successes}\}$.

If x is a variable that takes the values $0, 1, 2, \ldots, r, \ldots, n$ with probabilities $\binom{n}{0}q^n, \binom{n}{1}q^{n-1}p, \binom{n}{2}q^{n-2}p^2, \ldots, \binom{n}{r}q^{n-r}p^r, \ldots, \binom{n}{n}p^n$ respectively, then x is said to be a *binomial variate*, and the values of x and their corresponding probabilities form the *binomial distribution*.

A fair coin is tossed four times. What is the probability of getting:

1. one head? **2.** at least one head?

A fair die is rolled five times. Find the probability of rolling four spots

3. exactly two times. **4.** at least two times.

The probability that a certain couple will have a left-handed child is $\frac{1}{5}$.

5. If the couple has five children, what is the probability that exactly two are left-handed?

6. If the couple has four children, what is the probability that exactly two are right-handed?

7. If the couple has six children, what is the probability that at least one is left-handed?

8. 20% of a certain breed of rabbits are born with long hair. What is the probability that in a litter of six rabbits, exactly two will have long hair?

9. In a certain population of fruit flies, 25% have a certain eye mutation. If four flies are chosen at random, what is the probability that exactly two have the eye mutation?

10. Suppose that 40% of the patients diagnosed as having a certain disease die from it. What is the probability that exactly one will die from a group of four who have this disease?

11. Six rats are administered a certain dose of poison and the number of rats dying within 24 hr is observed. Suppose that each rat has a probability $\frac{1}{4}$ of dying and that the survival of each rat is independent of the survival of the other rats. What is the probability that:
   (a) four rats die? (b) all the rats die?

12. 85% of the trees planted by a certain landscaping firm survive. What is the probability that 8 or more out of a group of 10 trees planted will survive?

13. It is known that 60% of the offspring of a certain species of dog are black-eyed. The eye-color of one offspring is not related to that of another. What is the probability that there are at least one-third black-eyed pups in a litter of nine?

14. A rabbit is trained to touch one of two levers upon command. Suppose that the probability of his touching the correct lever on hearing the command is 0.75. If his responses to the commands given on different occasions are independent, what is the probability that out of five tries, he touches the correct lever three or four times?

15. In a certain population the probability of an individual having color blindness is 0.02. What is the probability that exactly 2 individuals will be color blind out of a group of 10 persons selected from this population?

16. Assuming the birth of boys and girls to be equally likely, what proportion of families with exactly four children should be expected to have two boys and two girls?

17. The examination of fossilized pollen grains found in the various layers of lake sediment is used to provide information on the type of vegetation that surrounded the lake at the time when the particular layer of sediment was formed. The proportion of pollen grains in the sediment that derive from fir trees of one species or another is 0.6. If 10 grains are examined, what are the probabilities that: (a) 6, (b) 7, (c) 5 of the grains turn out to be fir pollen?

18. In exercise 17 suppose that 100 grains are examined. If $P(r)$ is the probability of finding r fir pollen grains, find the value of $P(60)/P(59)$ and $P(60)/P(61)$. What conclusion can you draw regarding the outcome $r = 60$?

8.6 CONTINUOUS PROBABILITY DISTRIBUTIONS

So far we have considered only phenomena for which the outcomes form a discrete set. For example, the number of heads in 10 throws of a coin must take one of the discrete set of values $\{0, 1, 2, \ldots, 10\}$, the number of males in 20 births must be some integer between 0 and 20, or the number of atomic decays occurring in a piece of radioactive material during a given interval of time must be some integer between 0 and ∞. However there are many experiments in which the variable measured may take any of a continuous range of values, not simply a discrete set of values. For example, if the experiment consists of selecting a tree of a certain species at random from a given forest area and measuring its height, the result may turn out to be any real number between zero and some maximum value depending on the particular species of tree. Or the experiment may consist of measuring the time taken by a rat to thread a maze; again the result can turn out to be any real number lying in a certain interval—say between 4 seconds (sec), which might represent the fastest time feasible for the distance involved through the maze, and 30 sec, which might represent the time in which even the dumbest rat can find his way through.

In order to deal with situations of this type we must extend the idea of probability to cases in which the variable X, which represents the outcome of the experiment, can take any real value in a certain interval, say $a \leq X \leq b$. Such a variable is called a *continuous random variable*. For a continuous random variable the sample space consists not of a discrete set of points but of the continuous interval of values from a to b.

DEFINITION   Let X be a continuous random variable that can take values in the interval $a \leq X \leq b$. Then the *probability distribution function* of X, denoted by $F(x)$, is defined for any x satisfying $a \leq x \leq b$ as the probability that X turns out to be less than or equal to x:

$$F(x) = P(a \leq X \leq x).$$

As an example, let the variable X be the height of a randomly chosen 20-year-old girl. Then X is, of course, a continuous random variable, and we can suppose that it takes values in the range 4.0 ft to 6.5 ft. (Although S can take values outside this range, these occur with very low probability, and we ignore them.) The distribution function $F(x)$ is shown in Fig. 8.19. From the figure we see, for example, that the probability that the height is less that 5 ft is 0.1. The probability that X is less than 5.9 ft is 0.95. The probability that X is less that 5.5 ft is 0.6, and so on.

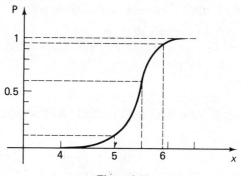

Figure 8.19

As a second example, let the random variable X be the lifespan of a randomly chosen individual born in North America between 1850 and 1870. In this case X can take values between 0 years and, let us say, 100 years, although again a few cases do occur outside this range. The distribution function for X is shown in Fig. 8.20. Note in this case, for example, that

$$P(0 \leq X \leq 60) = 0.25$$
$$P(0 \leq X \leq 68) = 0.5$$
$$P(0 \leq X \leq 14) = 0.15.$$

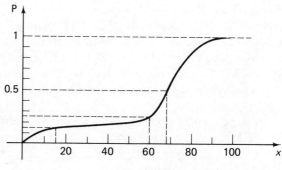

Figure 8.20

REMARKS.

i. $F(a)$ represents the probability that $X = a$. We shall assume that this is equal to zero; most cases of interest satisfy this condition: $F(a) = 0$.

ii. $F(b)$ represents the probability that X lies between a and b. This is certain to occur; hence $F(b) = 1$.

iii. Let $x_1 \leq x_2$. Then the event that $a \leq X \leq x_2$ is the union of the two events $a \leq X \leq x_1$ and $x_1 < X \leq x_2$. Since these two events are disjoint (i.e., they cannot occur simultaneously) we have

$$F(x_2) = P(a \leq X \leq x_2) = \underbrace{P(a \leq X \leq x_1)}_{F(x_1)} + P(x_1 < X \leq x_2).$$

Therefore

$$F(x_2) - F(x_1) = P(x_1 < X \leq x_2). \tag{a}$$

In other words, the probability that X turns out to lie between x_1 and x_2 is given by the difference between the values of $F(x)$ at x_2 and x_1.

iv. Since the probability on the right-hand side of equation (a) above cannot be negative, it follows that $F(x_2) - F(x_1) \geq 0$. That is,

$$F(x_2) \geq F(x_1) \quad \text{whenever} \quad x_2 \geq x_1.$$

It follows therefore that $F(x)$ is an increasing function of x, and so it increases from 0 when $x = a$ up to 1 when $x = b$.

EXAMPLE   A certain continuous random variable X takes values between 0 and 1. Its probability distribution function is

$$F(x) = \tfrac{4}{3}(x - \tfrac{1}{4}x^4).$$

Find the probabilities $P(X \leq \tfrac{1}{2})$, $P(\tfrac{1}{3} < X \leq \tfrac{1}{2})$, $P(X > \tfrac{1}{3})$.

SOLUTION   Observe that $F(0) = 0$ and $F(1) = 1$. The probability that X is less than or equal to $\tfrac{1}{2}$ is

$$P(X \leq \tfrac{1}{2}) = F(\tfrac{1}{2}) = \tfrac{4}{3}[\tfrac{1}{2} - \tfrac{1}{4}(\tfrac{1}{2})^4]$$
$$= \tfrac{4}{3}(\tfrac{1}{2} - \tfrac{1}{64}) = \tfrac{31}{48}.$$

Also,

$$P(X \leq \tfrac{1}{3}) = F(\tfrac{1}{3}) = \tfrac{4}{3}[\tfrac{1}{3} - \tfrac{1}{4}(\tfrac{1}{3})^4]$$
$$= \tfrac{4}{3}(\tfrac{1}{3} - \tfrac{1}{324}) = \tfrac{107}{243}.$$

It follows therefore that

$$P(\tfrac{1}{3} < X \leq \tfrac{1}{2}) = P(X \leq \tfrac{1}{2}) - P(X \leq \tfrac{1}{3})$$
$$= \tfrac{31}{48} - \tfrac{107}{243} = \tfrac{799}{3888}.$$

Finally, the event that $X > \tfrac{1}{3}$ is the complementary event to $\{X \leq \tfrac{1}{3}\}$. Therefore

$$P(X > \tfrac{1}{3}) = 1 - P(X \leq \tfrac{1}{3})$$
$$= 1 - \tfrac{107}{243} = \tfrac{136}{243}.$$

DEFINITION In most (but not all) cases of interest, the probability distribution function $F(x)$ increases smoothly from 0 to 1 as x increases from a to b. In such a case we define the *density function $f(x)$* as the derivative of $F(x)$:

$$f(x) = F'(x).$$

Since $F(a) = 0$, it follows that

$$F(x) = \int_a^x f(t)\, dt.$$

[This must hold since $F(x)$ is an antiderivative of $f(x)$ and the constant of integration must be chosen in such a way that $F(a)$ vanishes.]

Figures 8.21 and 8.22 show examples of distribution functions together with the corresponding density functions. We observe that $f(x)$ is equal to the slope of $F(x)$ so that the density function is greatest at values of x, where the probability distribution function is rising most rapidly. In the second example, $F(x)$ rises rapidly when x is just greater than 0, then levels off, and finally rises rapidly again as x approaches its upper limit b. As a result the corresponding density function is large for x near 0 and is large over a range of values of x close to b, but is almost zero in between, where the graph of $F(x)$ is almost horizontal.

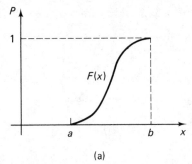

(a)

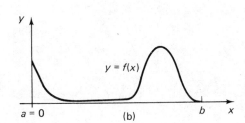

(b)

Figure 8.21

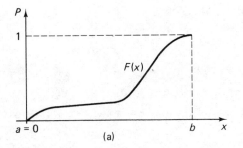

(a)

(b)

Figure 8.22

EXAMPLE In the preceding worked example $F(x) = \frac{4}{3}(x - \frac{1}{4}x^4)$ for $0 \le x \le 1$. The density function is therefore given by

$$f(x) = \frac{d}{dx}\left[\frac{4}{3}\left(x - \frac{1}{4}x^4\right)\right]$$

$$= \frac{4}{3}(1 - x^3).$$

The graphs of $F(x)$ and $f(x)$ in this example are shown in Fig. 8.23. Observe that

$$\int_0^x f(t)\, dt = \int_0^x \frac{4}{3}(1 - t^3)\, dt = \frac{4}{3}(x - \frac{1}{4}x^4) = F(x).$$

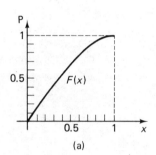
(a)

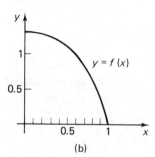
(b)

Figure 8.23

REMARKS.

i. Since $F(x)$ is an increasing function its derivative is never negative. Hence any density function $f(x)$ is nonnegative:

$$f(x) \ge 0 \quad \text{for} \quad a \le x \le b.$$

ii. If $c \le d$ are any two real numbers between a and b, the probability that $c < X \le d$ is given by

$$P(c < X \le d) = F(d) - F(c)$$

$$= \int_a^d f(t)\, dt - \int_a^c f(t)\, dt$$

$$= \int_c^d f(t)\, dt.$$

iii. By allowing c to approach d in the preceding equation we obtain the probability that X is actually equal to d:

$$P(X = d) = \lim_{c \to d-} P(c < X \le d) = \int_d^d f(t)\, dt = 0.$$

The event that $X = d$ is, of course, not impossible, yet we see that the probability of its occurrence is equal to zero. This is a general feature of continuous random variables that have a density function, namely that the

409

probability that they take any one specific value is equal to zero. In order to get a nonzero probability we must consider an interval of values for the variable.

iv. The integral of $f(x)$ from a to b is equal to 1:

$$\int_a^b f(x)\, dx = F(b) = 1.$$

v. In view of the relationship between definite integrals and areas under curves, we see that

$$P(c \le X \le d) = \int_c^d f(t)\, dt$$

is equal to the area underneath the graph of $y = f(x)$ lying between the vertical lines $x = c$ and $x = d$ (Fig. 8.24). It is this association of probabilities with areas under the graph of $f(x)$ that gives the density function its intuitive appeal.

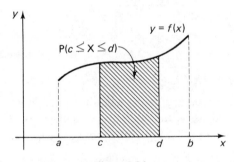

Figure 8.24

Observe that in the above equation we have used the fact that $P(c \le X \le d) = P(c < X \le d)$. This is so because $P(X = c)$ is equal to zero.

vi. Another interpretation of the density function that is often useful is that $f(x)\, dx$ can be regarded as the probability that X lies in the infinitesimally small interval between the values x and $x + dx$.

vii. We often need to consider continuous random variables whose values lie not on a finite interval $a \le X \le b$ but on a semi-infinite interval of the type $a \le X < \infty$ or on the whole infinite interval $-\infty < X < \infty$. In such cases we must allow $b \longrightarrow \infty$ and (in the second case) $a \longrightarrow -\infty$. We use notation like the following in order to express probabilities in such cases.

$$\int_a^\infty f(x)\, dx = \lim_{b \to \infty} \int_a^b f(x)\, dx$$

$$\int_{-\infty}^\infty f(x)\, dx = \lim_{b \to \infty} \lim_{a \to -\infty} \int_a^b f(x)\, dx.$$

For example, if X takes values in $-\infty < X < \infty$, then the probability that $X \leq x$ is given by

$$P(X \leq x) = \int_{-\infty}^{x} f(t)\, dt = \lim_{a \to -\infty} \int_{a}^{x} f(t)\, dt.$$

EXAMPLE X is a continuous random variable on the interval $-\infty < X < \infty$ with density function given by

$$f(x) = \frac{2}{\pi}(x^2 + 4)^{-1}.$$

Find $P(X \leq x)$.

SOLUTION   Consider first $P(a \leq X \leq x)$:

$$P(a \leq X \leq x) = \int_{a}^{x} f(t)\, dt = \frac{2}{\pi} \int_{a}^{x} \frac{1}{t^2 + 4}\, dt$$

$$= \frac{2}{\pi} \left[\frac{1}{2} \operatorname{Tan}^{-1} \frac{t}{2} \right]_{a}^{x}$$

$$= \frac{1}{\pi} \left[\operatorname{Tan}^{-1} \left(\frac{x}{2} \right) - \operatorname{Tan}^{-1} \left(\frac{a}{2} \right) \right]$$

where we have used formula 26 in the table of integrals in Appendix II. If we now allow $a \to -\infty$ and use the fact that

$$\lim_{a \to -\infty} \operatorname{Tan}^{-1} \left(\frac{a}{2} \right) = -\frac{\pi}{2},$$

we obtain that

$$P(X \leq x) = \frac{1}{\pi} \left[\operatorname{Tan}^{-1} \left(\frac{x}{2} \right) + \frac{\pi}{2} \right].$$

[Observe that in the limit as $x \to \infty$, the right-hand side tends to 1, as it must, since certainly $P(-\infty < X < \infty) = 1$.]

EXAMPLE   The probability that a certain plant will die within x hours in a certain environment is estimated to be $[1 - (1 + x^2)^{-1}]$. Determine the probabilities that the plant will die within 2 hr and that it will survive more than 3 hr. Find the corresponding density function.

SOLUTION   The random variable X is in this case the observed lifetime of the plant. X can take values in the interval $0 \leq X < \infty$, and the probability distribution function is given to be

$$F(x) = P(X \leq x) = 1 - \frac{1}{1 + x^2}.$$

Note that $F(0) = 0$ and that $F(x) \to 1$ as $x \to \infty$. The probability of the plant dying within 2 hr is

$$P(X \leq 2) = F(2) = 1 - \frac{1}{1 + 4} = \frac{4}{5}.$$

The probability of the plant dying within 3 hr is

$$P(X \leq 3) = F(3) = 1 - \frac{1}{1+9} = \frac{9}{10}.$$

Therefore the probability of its surviving longer than 3 hr is

$$P(X > 3) = 1 - P(X \leq 3) = 1 - \tfrac{9}{10} = \tfrac{1}{10}.$$

Finally, the density function is given by

$$f(x) = F'(x) = \frac{d}{dx}[1 - (1 + x^2)^{-1}]$$

$$= \frac{2x}{(1 + x^2)^2}.$$

Our aim in the remainder of this section will be to introduce certain parameters that are helpful in characterizing the distribution of a continuous random variable. The first of these is called the *mean* of the random variable and is denoted by the letter μ (mu). Let $f(x)$ be the density function of the variable in question, taking values on the interval $a \leq x \leq b$. Then μ is defined by

$$\boxed{\mu = \int_a^b xf(x)\, dx.}$$

If a very large number of measurements of X are made and from these measurements the average value of X is calculated, then this average will become closer and closer to μ as the number of measurements increases toward infinity. In this sense, the mean can be regarded as providing an *average value* of the random variable X. Another term that is used for μ is *expected value* or *expectation*; these terms derive from the application of probability theory to games of chance, where "expectation" is to be equated with "average winnings."

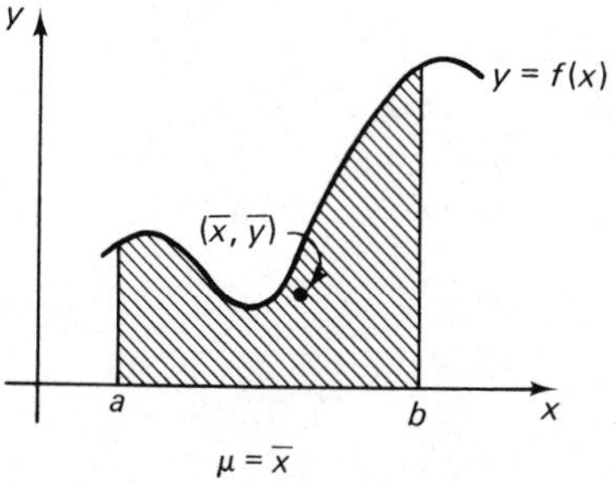

The mean μ is closely related to the centroid (i.e., center of gravity) of the geometrical region beneath the graph of the density function (Fig. 8.25). The x-coordinate of the centroid of the plane area lying between $y = f(x)$, the x-axis, and the lines $x = a$, $x = b$ is given by

$\mu = \bar{x}$

Figure 8.25

$$\bar{x} = \frac{\int_a^b xf(x)\, dx}{\int_a^b f(x)\, dx}.$$

If $f(x)$ represents a density function on the interval $a \leq x \leq b$, then $\int_a^b f(x)\, dx = 1$ and therefore

$$\bar{x} = \int_a^b xf(x)\,dx.$$

Thus geometrically, the mean μ can be taken as the x-coordinate of the centroid of the area bounded by the density curve, the x-axis, and the lines $x = a$, $x = b$.

The second parameter we shall introduce is called the *variance* of the random variable and is denoted by the symbol σ^2 (sigma squared). It is defined by

$$\sigma^2 = \int_a^b (x - \mu)^2 f(x)\,dx.$$

σ, the positive square root of the variance, is called the *standard deviation*.

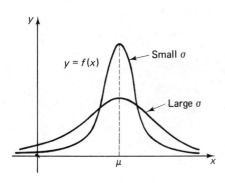

Figure 8.26

The mean of a random variable provides a measure of the center of the probability distribution. The standard deviation σ, on the other hand, provides a measure of the width of the distribution. If X is highly likely to be close to the mean μ then σ is small. If X has substantial probabilities of lying some distance from μ, the value of σ will be larger. Figure 8.26 illustrates two different probability densities with the same μ but with different values of σ. The graph with small σ is concentrated close to $x = \mu$, whereas the graph with large σ is spread out much more widely.

The formula for σ^2 can be simplified further as below:

$$\sigma^2 = \int_a^b (x - \mu)^2 f(x)\,dx$$

$$= \int_a^b (x^2 - 2\mu x + \mu^2) f(x)\,dx$$

$$= \int_a^b x^2 f(x)\,dx - 2\mu \int_a^b xf(x)\,dx + \mu^2 \int_a^b f(x)\,dx$$

$$= \int_a^b x^2 f(x)\,dx - 2\mu \cdot \mu + \mu^2 \cdot 1$$

or

$$\sigma^2 = \int_a^b x^2 f(x)\,dx - \mu^2.$$

EXAMPLE   (Uniform Distribution)

Consider the function $f(x) = 1/(b - a)$ on $a \le x \le b$. Then $f(x)$ is constant throughout the interval $a \le x \le b$. Clearly $f(x)$ is greater than zero and continuous on $a \leq x \leq b$. Also,

$$\int_a^b f(x)\,dx = \int_a^b \frac{1}{b-a}\,dx$$

$$= \frac{1}{b-a}\int_a^b 1\,dx$$

$$= \frac{1}{b-a}\Big[x\Big]_a^b$$

$$= \frac{1}{b-a}(b-a) = 1.$$

Thus $f(x) = 1/(b-a)$ defines a density function on $a \leq x \leq b$.
The distribution function $F(x)$ is given by

$$F(x) = \int_a^x f(t)\,dt = \int_a^x \frac{1}{b-a}\,dt$$

$$= \frac{1}{b-a}\Big[t\Big]_a^x = \frac{x-a}{b-a} \quad \text{on } a \leq x \leq b.$$

The mean μ of this distribution is

$$\mu = \int_a^b xf(x)\,dx = \int_a^b x \cdot \frac{1}{b-a}\,dx$$

$$= \frac{1}{b-a}\Big[\frac{x^2}{2}\Big]_a^b = \frac{1}{b-a} \cdot \frac{b^2-a^2}{2}$$

$$= \frac{b+a}{2}.$$

The standard deviation σ of this distribution is given by

$$\sigma^2 = \int_a^b x^2 f(x)\,dx - \mu^2$$

$$= \int_a^b x^2 \cdot \frac{1}{b-a}\,dx - \mu^2$$

$$= \frac{1}{b-a}\Big[\frac{x^3}{3}\Big]_a^b - \Big(\frac{a+b}{2}\Big)^2$$

$$= \frac{1}{b-a} \cdot \frac{b^3-a^3}{3} - \frac{(b+a)^2}{4}$$

$$= \tfrac{1}{3}(b^2+ba+a^2) - \tfrac{1}{4}(b^2+2ab+a^2)$$

$$= \tfrac{1}{12}(a^2-2ab+b^2) = \frac{(b-a)^2}{12}.$$

Therefore

$$\sigma = \frac{1}{2\sqrt{3}}(b-a).$$

The uniform distribution corresponds to the situation in which all measurements of X within the interval $a \leq X \leq b$ are equally likely.

Another particular distribution that is often of use in the biological sciences is the exponential distribution. This has density function given by

$$f(x) = ce^{-cx} \quad (0 \leq x < \infty).$$

It can be shown that the mean for this density function is given by $\mu = 1/c$, and the standard deviation is also $\sigma = 1/c$.

The exponential distribution is used in survival problems, in which we are concerned with the probability of a particular organism surviving for a time x in a given environment. This probability can often be assumed to be given by the exponential density function. Another application arises in situations concerning the occurrence of random events in time, in which the random variable X is the time until the next event occurs. For example, we may be concerned with the occurrence of cases of leukemia in a large city, and the variable of interest is the time it will take for the next case to occur. Or we may be interested in the time before the next nuclear decay in a piece of radioactive material. In each case these random times can be taken to have an exponential distribution.

EXERCISES 8.6

In each of the following exercises, determine the value of the constant c so that $f(x)$ is a density function on the given interval. Determine also the distribution function in each case.

1. $f(x) = c \sin x$ on $0 \leq x \leq \pi$.

2. $f(x) = cx(4 - x)$ on $0 \leq x \leq 4$.

3. $f(x) = \dfrac{c}{(1 + x)^3}$ on $0 \leq x < \infty$.

4. $f(x) = \dfrac{1}{2}e^{-cx}$ on $0 \leq x < \infty$.

5. $f(x) = \dfrac{c}{1 + x^2}$ on $-\infty < x < \infty$.

6. $f(x) = \begin{cases} c(3x - x^2) & \text{on } 0 \leq x \leq 3 \\ 0 & \text{elsewhere.} \end{cases}$

7. $f(x) = cx^2 e^{-x^3}$ on $0 \leq x < \infty$.

8. $f(x) = c \cos x$ on $-\pi/2 \leq x \leq \pi/2$.

9. The lifespan (measured in days) of a certain species of plant in a given environment is a continuous random variable with density function $f(x) = \dfrac{1}{100}e^{-x/100}$. Determine:
   (a) the distribution function.
   (b) the average or expected lifespan of the plants.
   (c) the probability that a given plant will die within 50 days.

10. The density function of a certain continuous response is $f(x) = \frac{1}{2}\sin x$ on $0 \leq x \leq \pi$. If X is a continuous random variable representing this response, determine $P(X \leq \pi/2)$, $P(\pi/6 \leq X < \pi/2)$, and $P(X < 5\pi/6)$.

11. The distribution function of a certain continuous response is $F(x) = 1 - 1/(1 + x)^2$. If X denotes this continuous response determine the probabilities:
    (a) $P(X < 1)$.   (b) $P(X = 2)$.   (c) $P(1 < X < 3)$.

12. The probability that a certain species of animal born at $t = 0$ will be killed before time t (measured in weeks) is $1 - e^{-t/20}$.
    (a) If T denotes the continuous random variable, which represents the lifetime of any given animal, determine the density function of T.
    (b) What is the average number of weeks before the animal is killed?

13. Let T denote the digestion time (measured in hours) of a unit of food. Then T is a continuous random variable. Suppose that its density function is $f(x) = 9x \times e^{-3x}$ on $0 \leq x < \infty$. Determine the distribution function and use it to determine the probability that the food is completely digested within 2 hr. What is the probability that food takes longer than 3 hr to be digested? What is the average time required for the digestion of food?

8.7   THE NORMAL DISTRIBUTION

In this section we discuss the most important and most commonly used continuous probability distribution, the *normal distribution*.

The probability density function for the normal distribution is

$$f(x) = \frac{1}{\sigma\sqrt{2\pi}}e^{-(x-\mu)^2/2\sigma^2} \quad \text{on} - \infty < x < \infty.$$

The mean and variance of the distribution are the quantities μ and σ^2, respectively, which occur in the above formula. A continuous random variable X having the density function $f(x)$ defined above is said to be a normal random variable with mean μ and standard deviation σ. The graph of $f(x)$ is a bell-shaped curve that is symmetrical about $x = \mu$ and extends indefinitely in both directions from the central line $x = \mu$. The graph has a maximum at $x = \mu$ and approaches closer and closer to the x-axis as x moves away from μ in either direction. The graph of $f(x)$ shown in Fig. 8.27 is called a *normal curve*.

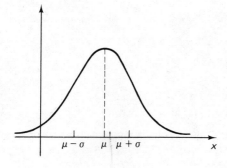

Figure 8.27

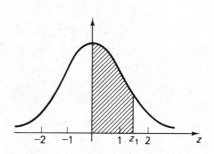

Figure 8.28

A normal random variable for which the mean μ is 0 and the standard deviation σ is equal to 1 is called a *standard normal variable* and is usually denoted by the letter Z. Its density function is given by

$$f(z) = \frac{1}{\sqrt{2\pi}} e^{-z^2/2}.$$

Again the graph of $f(z)$ is a bell-shaped curve, but centered about $z = 0$ (Fig. 8.28).

If z_1 is any real positive number then the probability $P(0 \leq Z \leq z_1)$ that the standard normal variable Z lies between 0 and z_1 is given by the area shaded in Fig. 8.28. In other words,

$$P(0 \leq Z \leq z_1) = \int_0^{z_1} \frac{1}{\sqrt{2\pi}} e^{-z^2/2} \, dz.$$

The values of the integral on the right for different values of z_1 are given in a table in Appendix III.

EXAMPLE   From the table we find that

$$P(0 \leq Z \leq 1.00) = 0.3413.$$

In other words, there is a probability of 34.13% that a standard normal random variable lies between 0 and 1.

The areas under the standard normal curve are not given in the table for negative values of z_1. In fact we do not need them since they can be calculated by using the symmetry of the normal curve. The area under the standard normal curve between $Z = -a$ and $Z = 0$ is equal to the area between $Z = 0$ and $Z = a$, i.e.,

$$\boxed{P(-a \leq Z \leq 0) = P(0 \leq Z \leq a).}$$

EXAMPLE   Given that Z is a standard normal variable, use tables to evaluate the following probabilities:

(a) $P(1 \leq Z \leq 2)$.   (b) $P(-1.5 \leq Z \leq 0.5)$.
(c) $P(-1.42 \leq Z \leq 1.42)$.   (d) $P(Z \geq 2.41)$.
(e) $P(Z \geq -2)$.   (f) $P(Z \leq -1.78)$.

SOLUTION   (a) The area desired is shown in Fig. 8.29.

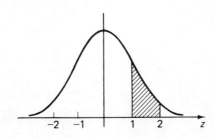

Figure 8.29

$$P(1 \leq Z \leq 2) = P(0 \leq Z \leq 2) - P(0 \leq Z \leq 1)$$
$$= 0.4772 - 0.3413$$
$$= 0.1359 \quad \text{(after using tables).}$$

(b) As is clear from Fig. 8.30,

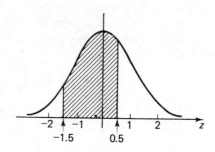

Figure 8.30

$$P(-1.5 \leq Z \leq 0.5) = P(-1.5 \leq Z \leq 0) + P(0 \leq Z \leq 0.5)$$
$$= P(0 \leq Z \leq 1.5) + P(0 \leq Z \leq 0.5) \quad \text{(where we}$$
have used the symmetry property of the normal curve)
$$= 0.4332 + 0.1915 \quad \text{(after using tables)}$$
$$= 0.6247.$$

(c) $P(-1.42 \leq Z \leq 1.42) = P(-1.42 \leq Z \leq 0) + P(0 \leq Z \leq 1.42)$
$$= P(0 \leq Z \leq 1.42) + P(0 \leq Z \leq 1.42)$$
$$= 2P(0 \leq Z \leq 1.42)$$
$$= 2(0.4222) = 0.8444.$$

(d) $P(Z \geq 2.41) = P(Z \geq 0) - P(0 \leq Z \leq 2.41) \quad$ (Fig. 8.31)
$$= 0.5 - 0.4920$$
$$= 0.0080.$$

Note that we used the fact that $P(Z \geq 0)$ is equal to half the total area under the normal curve, that is, to 0.5.

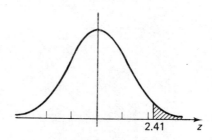

Figure 8.31

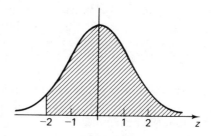

Figure 8.32

(e) $P(Z \geq -2) = P(-2 \leq Z \leq 0) + P(Z \geq 0)$ (Fig. 8.32)

$\qquad = P(0 \leq Z \leq 2) + P(Z \geq 0)$

$\qquad = 0.4772 + 0.5$

$\qquad = 0.9772.$

(f) $P(Z \leq -1.78) = P(Z \leq 0) - P(-1.78 \leq Z \leq 0)$

$\qquad = P(Z \leq 0) - P(0 \leq Z \leq 1.78)$

$\qquad = 0.5 - 0.4625$

$\qquad = 0.0375.$

Now let us return to the case of a general normal variable X whose mean is μ and standard deviation σ. Any such variable can be converted to a standard normal variable Z by means of the transformation

$$Z = \frac{X - \mu}{\sigma}.$$

If X is a normal random variable with mean μ and standard deviation σ, then Z as defined by this formula is a standard normal variable (i.e., has mean 0 and standard deviation 1). The relationship between X and Z is illustrated in Fig. 8.33, which shows

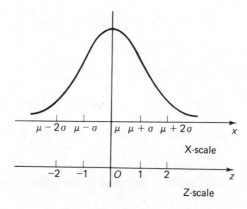

Figure 8.33

that the change from X to Z represents a shift of origin together with a re-scaling. When $X = \mu, Z = O$; when $X = \mu + \sigma, Z = 1$; when $X = \mu + 2\sigma, Z = 2$, and so on. Thus the point $X = \mu$ is shifted to the origin $Z = O$, and the scale is changed in such a way that a distance of σ on the X-axis corresponds to a distance of 1 on the Z-axis.

This transformation, together with the tables of the standard normal probabilities, can be used to calculate probabilities for any normal random variable. This is illustrated in the following examples.

EXAMPLE   Let X be a normal random variable with mean 10 and standard deviation 4. Determine the probabilities:

(a) $P(12 \leq X \leq 15)$.

(b) $P(X \geq 7)$.

SOLUTION   The first step is to transform X to the standard normal variable Z by means of the formula $Z = (X - \mu)/\sigma$. In this case $\mu = 10$ and $\sigma = 4$ so that

$$Z = \frac{X - 10}{4}.$$

(a)   When $X = 12$, $Z = (12 - 10)/4 = 0.5$; when $X = 15$, $Z = (15 - 10)/4 = 1.25$. Therefore

$$P(12 \leq X \leq 15) = P(0.5 \leq Z \leq 1.25)$$
$$= P(0 \leq Z \leq 1.25) - P(0 \leq Z \leq 0.5)$$
$$= 0.3944 - 0.1915$$
$$= 0.2029.$$

(b)   When $X = 7$, $Z = (7 - 10)/4 = -0.75$. Therefore

$$P(X \geq 7) = P(Z \geq -0.75)$$
$$= P(-0.75 \leq Z < 0) + P(Z \geq 0)$$
$$= P(0 \leq Z \leq 0.75) + P(Z \geq 0)$$
$$= 0.2734 + 0.5$$
$$= 0.7734.$$

EXAMPLE   The heights of plants of a certain species are normally distributed, the mean (average) height being 30 cm and the standard deviation being 5 cm. What proportion of the plants are greater than 40 cm in height?

SOLUTION   If X denotes the height of any plant of the given species, then X is a normal random variable with mean $\mu = 30$ and standard deviation $\sigma = 5$. The corresponding standard normal variable is $Z = (X - \mu)/\sigma = (X - 30)/5$. When $X = 40$, $Z = (40 - 30)/5 = 2$. Therefore the probability that X exceeds 40 is given by:

$$P(X \geq 40) = P(Z \geq 2)$$
$$= P(Z \geq 0) - P(0 \leq Z \leq 2)$$
$$= 0.5 - 0.4772 = 0.0228.$$

Therefore 2.28 % of the plants have heights in excess of 40 cm.

EXAMPLE   The lengths (at adulthood) of a certain species of fish are normally distributed with a mean of 60 cm. If 1.22 % of the fish are of length less than 51 cm, determine the standard deviation.

SOLUTION   Denoting the length of a randomly chosen fish by X, we are given that X is a normal random variable with mean $\mu = 60$ and an unknown standard deviation σ. The corresponding standard normal variable is therefore $Z = (X - \mu)/\sigma = (X - 60)/\sigma$. When $X = 51$, $Z = (51 - 60)/\sigma = -9/\sigma$. The probability that X is less than 51 is given to be 1.22 %, or 0.0122. Therefore $P(Z \leq -9/\sigma) = 0.0122$. From the symmetry of the normal curve (Fig. 8.34) it follows that

$$P\left(Z \geq \frac{9}{\sigma}\right) = 0.0122$$

and so

$$P\left(0 \leq Z \leq \frac{9}{\sigma}\right) = 0.5 - 0.0122$$

$$= 0.4878.$$

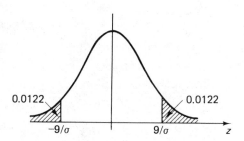

0.0122   0.0122

$-9/\sigma$ $9/\sigma$ z

Figure 8.34

But from the standard normal table given in Appendix III we find that $P(0 \leq Z \leq 2.25) = 0.4878$. It follows therefore that

$$\frac{9}{\sigma} = 2.25,$$

i.e.,

$$\sigma = \frac{9}{2.25} = 4.$$

The standard deviation in the length of the fish is therefore 4 cm.

The normal probability distribution is commonly used when the random variable in question represents a physical dimension of some biological organism, for instance, the weight of a randomly chosen 30-year-old human male, or the volume of lumber in a 100-year-old Douglas fir tree. Examples such as the yield per acre of corn in a certain region or the marks obtained by a class of students in an examination would also usually be taken as normally distributed.

An important and somewhat different use of the normal distribution arises as an approximation to the binomial distribution when the number of trials is large. Let the number of Bernoulli trials be n, with p being the probability of success and $q = 1 - p$ the probability of failure. Then, as we saw in Section 8.5, the probability of r successes is given by

$$P(r) = \binom{n}{r} p^r q^{n-r}.$$

When n is large it can become very tedious to calculate $P(r)$ from this formula, but instead it is possible to make use of normal probabilities in order to find an approximate value of $P(r)$.

It turns out that the probability $P(r)$ can be obtained from the normal probability curve when n is large by setting the mean $\mu = np$ and standard deviation $\sigma = \sqrt{npq}$. Then $P(r)$ is approximately equal to the area under the normal curve lying between the values $r - \frac{1}{2}$ and $r + \frac{1}{2}$ of the independent variable.

EXAMPLE   A coin is flipped 16 times. The probability of exactly 5 heads occurring is given exactly by the above binomial formula with $n = 16$, $p = q = \frac{1}{2}$, and $r = 5$:

$$P(5) = \binom{16}{5}\left(\frac{1}{2}\right)^5\left(\frac{1}{2}\right)^{16-5}$$

$$= \frac{16 \cdot 15 \cdot 14 \cdot 13 \cdot 12}{1 \cdot 2 \cdot 3 \cdot 4 \cdot 5}\left(\frac{1}{2}\right)^{16}$$

$$= 4368\left(\frac{1}{2}\right)^{16}$$

$$= 0.0667.$$

Instead we can use the normal probabilities with

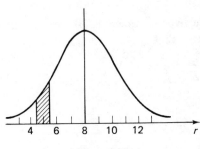

Figure 8.35

$$\mu = np = 16(\tfrac{1}{2}) = 8,$$
$$\sigma = \sqrt{npq} = \sqrt{16(\tfrac{1}{2})(\tfrac{1}{2})} = 2.$$

$P(5)$ is then given by the area under the normal curve (Fig. 8.35), with $\mu = 8$ and $\sigma = 2$ lying between the values 4.5 and 5.5 of the independent variable. The corresponding standard normal variable is

$$Z = \frac{r - \mu}{\sigma} = \frac{r - 8}{2}.$$

When $r = 4.5$, $Z = (4.5 - 8)/2 = -1.75$; when $r = 5.5$, $Z = (5.5 - 8)/2 = -1.25$. Thus

$$P(5) \simeq P(-1.75 < Z < -1.25)$$
$$= P(1.25 < Z < 1.75)$$
$$= P(0 < Z < 1.75) - P(0 < Z < 1.25)$$
$$= 0.4599 - 0.3944$$
$$= 0.0655.$$

Clearly, we have obtained an answer that is quite close to the exact value of $P(5)$.

This normal approximation to binomial probabilities can be used provided that n is sufficiently large. A good rule of thumb is that the two quantities np and nq should both be greater than about 5. The larger these two quantities are, the better the approximation will be.

The approximation of a binomial distribution by a normal distribution is particularly useful when we have to calculate the probability of less than or more than a certain number of successes, which involves using the binomial probability formula repeatedly to evaluate the many terms that occur. Suppose that the probability of a patient staying more than 24 hr in an emergency ward of a hospital is 0.32 and that we are interested in the probability that out of 60 patients admitted on a certain day, at most 10 will stay more than 24 hr. The problem can be treated as a binomial distribution because each patient who is admitted to the emergency ward has two mutually exclusive possibilities in that he will either stay more than 24 hr or he will not. The probability p of success (a stay of more than 24 hr) is 0.32. We have $n = 60$. Then, using the binomial probability formula, the required probability of at most 10 patients staying more than 24 hr is

$$P(k \leq 10) = P(0) + P(1) + P(2) + P(3) + P(4) + P(5)$$
$$+ P(6) + P(7) + P(8) + P(9) + P(10),$$

where

$$P(k) = \binom{n}{k} p^k q^{n-k} = \binom{60}{k}(0.32)^k(0.68)^{60-k}$$

Evidently, calculating these terms involves a tremendous amount of work. On the other hand, if we use the normal approximation, the calculation becomes very simple. According to this approximation we introduce a normal random variable X with mean μ and standard deviation σ where

$$\mu = np = (60)(0.32) = 19.2$$

and

$$\sigma = \sqrt{np(1 - p)} = \sqrt{60(0.32)(0.68)} = 3.6.$$

Then, for example, the probability of 10 successes is approximately equal to the area

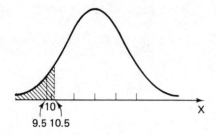

Figure 8.36

under the normal curve between $X = 9.5$ and 10.5 (Fig. 8.36). The probability of 9 successes is approximately equal to the area between $X = 8.5$ and $X = 9.5$. And so we continue. It is clear that the probability of 10 or fewer successes is approximately equal to the whole shaded area in the figure. That is,

$$P(k \leq 10) \approx P(X \leq 10.5).$$

The corresponding standard normal variable is

$$Z = \frac{X - \mu}{\sigma} = \frac{X - 19.2}{3.6}.$$

When $X = 10.5$, $Z = (10.5 - 19.2)/3.6 = -2.42$. Therefore

$$
\begin{aligned}
P(X \leq 10.5) &= P(Z \leq -2.42) \\
&= P(Z \leq 0) - P(-2.42 \leq Z \leq 0) \\
&= 0.5 - P(0 \leq Z \leq 2.42) \\
&= 0.5 - 0.4922 \\
&= 0.0078.
\end{aligned}
$$

EXERCISES 8.7

Given that Z is a standard normal variable, determine the following probabilities:

1. $P(Z \geq 1.78)$.

2. $P(Z \leq 1.25)$.

3. $P(Z \geq -1.2)$.

4. $P(Z \leq -2.58)$.

5. $P(1.29 \leq Z \leq 2.15)$.

6. $P(-2.74 \leq Z \leq -1.4)$.

7. $P(-1.3 \leq Z \leq 1.3)$.

8. $P(-1.45 \leq Z \leq 2.01)$.

Given that X is a normal random variable with mean 15.8 and standard deviation 2.1, determine the following probabilities:

9. $P(X \geq 18.74)$.

10. $P(X \leq 18.635)$.

11. $P(14.708 \leq X \leq 22.184)$.

12. $P(18.32 \leq X \leq 20.63)$.

13. The time T required for digestion of one unit of a certain food is normally distributed with mean 35 minutes (min) and standard deviation 4 min. What is the probability that:
   (a) a unit of food is digested in less than 40 min?
   (b) a unit of food is digested in less than 28 min?

14. The weights of live chickens on Mr. Jack's farm are normally distributed with mean 5.8 lb and standard deviation 1.6 lb. Find:

(a) the probability that a chicken selected at random will weigh at least 7 lb.

(b) the percentage of chickens that will weigh less than 5 lb.

15. A normal distribution has the mean $\mu = 75.0$. Find its standard distribution if 11.9% of the area under the curve lies to the right of 79.13.

16. A random variable has a normal distribution with standard deviation 1.4. Determine the mean of the distribution if the probability that the random variable will take a value less than 59 is 0.9772.

17. The annual rainfall in a certain region is a normally distributed variable with standard deviation 2.5 in. Determine the average annual rainfall in a given year if the probability of more than 32 in. of rain in that year is 0.0548.

18. The acidity of human blood measured on the pH scale is a normal random variable with mean 7.2. Determine the standard deviation if the probability that the pH level is greater than 7.47 is 0.0359.

19. A fair coin is tossed 100 times. Use the normal approximation to determine the probability of:
    (a) getting at most 60 heads?
    (b) getting at least 40 heads but not more than 65 heads?

20. A die is rolled 60 times. Use the normal approximation to estimate the probability of getting 6 spots at the most 10 times.

21. In a large population of smokers 20% have a lung disorder. A sample of 400 smokers is taken from this population and tested for lung disease. Use the normal approximation to estimate the probability that the sample contains:
    (a) at least 100 persons with lung disease.
    (b) at least 70 persons and not more than 95 persons with lung disease.
    (c) at most 75 persons with lung disease.

22. In a large population of fruit flies 30% have an eye mutation. A random sample of 200 flies is taken from this population. What is the probability that at least 55 and no more than 70 flies in the sample have the eye mutation?

23. It is known that 12% of the people of a certain large community are left-handed. If 500 persons are selected from this community, what is the probability that there are at most 45 people who are left-handed?

24. A medical treatment for a certain disease is effective in 90% of the cases where it is used. The treatment is given to 2000 people having the disease. What is the probability that at least 1780 persons are cured?

REVIEW EXERCISES FOR CHAPTER 8

1. Are the following statements true or false? If false, replace them by the corresponding correct statements.
    (a) If two events are mutually exclusive then they are independent.
    (b) The probability of any event is a nonnegative real number.

(c)   If E_1 and E_2 are two independent events then

$$P(E_1 \cap E_2) = P(E_1) + P(E_2).$$

(d)   If E_1 and E_2 are two mutually exclusive events then

$$P(E_1 \cap E_2) = 0.$$

(e)   In a binomial probability the Bernoulli trials are independent.

(f)   The area under the normal curve is equal to 1.

(g)   The mean and standard deviation of a normal distribution are equal to 0 and 1 respectively.

(h)   If Z is a standard normal variable and $X = 12 + 3Z$, then X is a normal variable with mean 3 and standard deviation 12.

(i)   If E_1 and E_2 are two independent events then

$$P(E_1 \cup E_2) = P(E_1) + P(E_2) - P(E_1)P(E_2).$$

(j)   Two independent events cannot occur simultaneously.

(k)   If a die is rolled 180 times, the probability of getting 33 or more sixes is approximately equal to 0.81.

2.   Of 10 girls in a class 4 have blue eyes. If 2 of the girls are chosen at random, what is the probability that
(a)   both have blue eyes?   (b)   neither has blue eyes?
(c)   at least one has blue eyes?

3.   Two people are selected at random from a group of 10 married couples. What is the probability that
(a)   they are husband and wife?
(b)   one is male and one is female?

4.   A biologist has 10 rabbits and 15 monkeys available for a certain experiment. Half of the rabbits are white and one-third of the monkeys are white. If he selects one of these animals, what is the probability that it is either a rabbit or a white animal?

5.   Of the patients examined at the local clinic 30% have high blood pressure, 35% have excessive weight and 15% have both. What is the probability that a patient selected at random will have at least one of these characteristics? Are the events "excessive weight" and "high blood pressure" independent? Explain.

6.   In a certain community 40% of the people smoke, 32% of the people drink, and 60% either smoke or drink or do both. What percentage of the people smoke as well as drink?

7.   The probability that a man will live 10 more years is $\frac{1}{5}$ and the probability that his wife will live 10 more years is $\frac{1}{4}$. What is the probability that
(a)   both will live 10 more years?
(b)   at least one of them will live for 10 more years?
(c)   only the wife will live for 10 more years?

8. Let E_1 be the event that a family has children of both sexes and E_2 be the event that a family has at most one boy. Show that the events E_1 and E_2 are
   (a)   independent if the family has 3 children.
   (b)   are not independent if the family has 2 children.

9. If the events E_1 and E_2 are independent, show that E_1 and E_2' are also independent. (*Hint:* Make a Venn diagram for $E_1 \cap E_2'$.)

10. A family has 8 children. Assuming that the probability of any child being a boy is $\frac{1}{2}$, find the probability that the family will have fewer girls than boys.

11. Let X be a continuous random variable with density function $f(x) = \frac{1}{20}x + k$ on $0 \leq x \leq 4$. Evaluate k and find $P(2 \leq X \leq 3)$.

12. Suppose the maximum daily temperature during May is normally distributed with mean $60°$ and standard deviation $5°$. Find the probability that the temperature will be at least $63°$.

13. Assuming the heights H of 1000 students are normally distributed with mean 1.6 meters (m) and standard deviation 8 cm, find the number N of students with heights
   (a)   between 1.72 and 1.82 m.
   (b)   greater than or equal to 1.8 m.

14. A fair coin is tossed 300 times. Use the normal approximation to determine the probability of getting at least 175 heads.

15. A well-known birth control medicine is effective in 99% of the cases. If this medicine is used by 10,000 women, what is the probability that at most 75 women get pregnant?

16. The proportion of persons having high blood pressure in a large population is 4%. If 500 people are selected at random from this population and tested for high blood pressure, what is the probability that at least 15 and no more than 25 of them have high blood pressure?

17. It is known that in a large population, 40% of the people are overweight. If 200 persons are selected from this population, what is the probability that at most 60 of them will be overweight?

18. In a large maternity hospital, on the average, 2% of the children are born with some defect. In a certain week there are 450 births registered. What is the probability that:
   (a)   no child is born with a defect in this week?
   (b)   at most 5 children are born with a defect?

9

Functions of Several Variables

9.1 FUNCTIONS AND DOMAINS

So far we have restricted our attention to cases in which the dependent variable is a function of a single independent variable, $y = f(x)$. However in reality, in many, perhaps most, applications, we come across situations in which one quantity depends not on just one other variable but on several variables. For instance, we have used before the example where the average yield of a certain crop depends on the amount of fertilizer used. But in actual fact, the yield depends not only on the amount of fertilizer, but also on a number of other factors, such as the amount of rainfall during the growing season, the average temperature during that period, and the intrinsic quality of the soil. To take another example, the weight of a particular harvest of fish may depend on the amount of food available to them, the number of predators, and also on the number of fish left alive from the previous harvest.

EXAMPLE   The volume V occupied by a gas is proportional to its absolute temperature T and inversely proportional to its pressure P. Express V as a function of P and T.

SOLUTION   We are given that $V \propto T$. Another way of saying this is that the ratio (V/T) does not depend on temperature. However this ratio does depend on the pressure P since we are told that $V \propto 1/P$. Therefore we can write

$$\frac{V}{T} = \frac{k}{P},$$

where k is a certain constant. It follows that

$$V = \frac{kT}{P},$$

This final equation expresses V as a function of the two independent variables T and P.

EXAMPLE   A certain predator animal spends its time hunting in two regions, denoted by 1 and 2. It first hunts in region 1, the amount of food w_1 that it catches being proportional to the time t_1 that it spends in this region. Then it moves to region 2, spending a time t_2 there. The amount of food w_2 that it catches in region 2 is proportional to t_2, but the "constant" of proportionality is multiplied by a factor $e^{-\alpha w_1}$. Express the total amount of food as a function of the two times t_1 and t_2.

SOLUTION   Since $w_1 \propto t_1$, we can write $w_1 = at_1$, a being a certain constant. The meaning of this constant is that it is the average rate at which the predator catches food in region 1 (for example the number of grams per hour). The amount w_2 is proportional to t_2, but with a factor $e^{-\alpha w_1}$ occurring in the proportionality. Therefore

$$w_2 = be^{-\alpha w_1}t_2,$$

where b is a second constant.

We note that when $w_1 = 0$ we get $w_2 = bt_2$. Therefore the meaning of the constant b is that it equals the average rate at which the predator catches food in region 2 when he has not eaten at all in region 1 ($w_1 = 0$). When $w_1 > 0$, the rate at which he catches food in region 2 is equal to $be^{-\alpha w_1}$, which is smaller than b. We can write

$$w_2 = be^{-\alpha a t_1} t_2.$$

Therefore the total amount of food eaten, $w = w_1 + w_2$, is given as a function of t_1 and t_2 by

$$w = at_1 + bt_2 e^{-\alpha a t_1}.$$

In cases such as these we need to introduce several independent variables. For most of this chapter we shall consider the case of two independent variables, and we shall usually use x and y to stand for them. Toward the end of each section we shall consider the generalization to three or more independent variables, which in most respects is quite straightforward. The dependent variable will usually be denoted by z, and we use the notation $z = f(x, y)$ to indicate that z is a function of both x and y.

We first give a formal definition of a function of two variables.

DEFINITION   Let D be a set of pairs of real numbers (x, y), and let f be a rule that specifies a unique real number for each pair (x, y) in D. Then we say that f is a *function of the two variables* x and y and the set D is the *domain* of f. The value of f at the pair (x, y) is denoted by $f(x, y)$ and the set of all of these values is called the *range* of f.

EXAMPLE   Let $f(x, y) = 2x + y$. Calculate the values of f at the pairs $(0, 0)$, $(1, 2)$, and $(2, -4)$. Find the domain and range of f.

SOLUTION   The values of a function of two variables are obtained simply by substituting the given values of x and y:

$$f(0, 0) = 2(0) + 0 = 0,$$
$$f(1, 2) = 2(1) + 2 = 4,$$
$$f(2, -4) = 2(2) + (-4) = 0.$$

In this case the value of f is a well-defined real number for all real values of x and y, and the domain is the set of all pairs (x, y) of real numbers.

The range of f is the set of all real numbers. For, if we are given any real number z, we can always find x and y such that $z = f(x, y) = 2x + y$. In fact we can let x be arbitrary and take $y = z - 2x$.

We can associate any pair (x, y) of real numbers with a point in the Cartesian plane by using x and y as the Cartesian coordinates of the point. Thus the domain D of a function of two variables can be viewed as a subset of points in the xy-plane. In the preceding example all pairs of real numbers (x, y) belong to the domain, so that from the geometrical viewpoint we can say that the domain consists of the whole xy-plane.

The range of a function of two variables is a subset of the real numbers, just as it is for a function of one variable.

EXAMPLE Given $f(x, y) = \sqrt{4 - x^2 - y^2}$, calculate $f(0, 0), f(0, 2), f(1, -1)$, and $f(1, 2)$. Find the domain of f and represent it graphically.

SOLUTION Substituting the given values of x and y we obtain

$$f(0, 0) = \sqrt{4 - 0^2 - 0^2} = \sqrt{4} = 2,$$
$$f(0, 2) = \sqrt{4 - 0^2 - 2^2} = \sqrt{0} = 0,$$
$$f(1, -1) = \sqrt{4 - 1^2 - (-1)^2} = \sqrt{2},$$
$$f(1, 2) = \sqrt{4 - 1^2 - 2^2} = \sqrt{-1} \quad \text{(not defined)}.$$

The pair $(1, 2)$ therefore does not belong to the domain of f.

In order for $f(x, y)$ to be a well-defined real number the quantity under the radical sign must be nonnegative. Thus

$$4 - x^2 - y^2 \geq 0$$
$$x^2 + y^2 \leq 4.$$

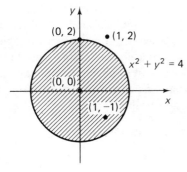

Figure 9.1

Thus the domain of f consists of those points (x, y) such that $x^2 + y^2 \leq 4$. Geometrically, $x^2 + y^2 = 4$ is the equation of a circle centered at the origin and with a radius of 2, and the inequality $x^2 + y^2 \leq 4$ holds at points inside and on this circle. These points form the domain D (Fig. 9.1). The point $(1, 2)$ lies outside the circle, consistent with our earlier finding that $f(1, 2)$ does not exist.

We leave it as an exercise to the student to show that the range of the function $z = \sqrt{4 - x^2 - y^2}$ is the set $\{z \mid 0 \leq z \leq 2\}$.

EXAMPLE Represent graphically the domain of the function $f(x, y) = \ln(x + y)$.

SOLUTION The logarithm function is only defined when its argument $(x + y)$ is strictly positive. So the domain consists of those pairs (x, y) that satisfy the condition $x + y > 0$. The graph of this linear inequality is shown in Fig. 9.2.

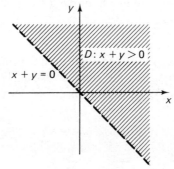

Figure 9.2

EXAMPLE   Represent graphically the domain of the function

$$f(x, y) = \ln (x + y)/(x - 1).$$

SOLUTION   As in the preceding example, the logarithm is only defined provided that $x + y$ > 0. However, in this example, there is a second factor, $(x - 1)^{-1}$, which is not defined for $x = 1$. Therefore the domain is the same as in the preceding example except that all points on the line $x = 1$ are excluded (Fig. 9.3).

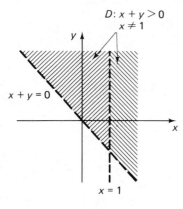

$D: x + y > 0$
$x \neq 1$

$x + y = 0$

$x = 1$

Figure 9.3

As in the case of functions of one variable, a function of two variables can be defined implicitly by a relation involving all three of the variables x, y, and z. The explicit form of such a function is obtained by solving the given implicit relation for z. However such explicit solutions may not be possible in terms of simple algebraic functions. Moreover, it is possible that an implicit relation may correspond to more than one explicit function, just as for functions of one variable.

EXAMPLE   Write the relation $x^2 + y^2 + z^2 = 4$ as an explicit function.

SOLUTION   We must solve for z:

$$z^2 = 4 - x^2 - y^2$$

$$z = \pm\sqrt{4 - x^2 - y^2}.$$

Thus we see that the given implicit relation corresponds to two explicit functions, $z = \sqrt{4 - x^2 - y^2}$ and $z = -\sqrt{4 - x^2 - y^2}$.

EXAMPLE   The implicit relation

$$e^{xyz} = x + y + z$$

cannot be solved in explicit form. However it does determine z as a function of x and y. (Actually, for certain values of x and y it corresponds to two functions as for the preceding example, but we shall not prove this fact.)

Now let us consider the generalization of the preceding ideas to functions of more than two variables. When there are three independent variables it is usual to denote them by x, y, and z, and to denote the dependent variable by w. The function consists of a rule that assigns a real number to each set of three values (x, y, z) of the independent variables. We write the value as $w = f(x, y, z)$.

EXAMPLE   If

$$f(x, y, z) = \frac{\sqrt{9 - x^2 - y^2}}{x + z},$$

432   evaluate $f(0, 0, 1)$, $f(1, -1, 4)$, and $f(-1, 2, 1)$. Find the domain of f.

SOLUTION   The value of f at any given (x, y, z) is obtained by substituting the given values of x, y, and z into the algebraic expression that defines f:

$$f(0, 0, 1) = \frac{\sqrt{9 - 0^2 - 0^2}}{0 + 1} = \frac{\sqrt{9}}{1} = 3,$$

$$f(1, -1, 4) = \frac{\sqrt{9 - 1^2 - (-1)^2}}{1 + 4} = \frac{\sqrt{7}}{5},$$

$$f(-1, 2, 1) = \frac{\sqrt{9 - (-1)^2 - 2^2}}{-1 + 1} = \frac{\sqrt{4}}{0}, \text{ which is undefined.}$$

In the last case the given point $(-1, 2, 1)$ does not belong to the domain of f.

In order for $f(x, y, z)$ to be well-defined it is necessary that the quantity under the radical sign be nonnegative and also that the denominator of f be nonzero. Thus we have the conditions

$$9 - x^2 - y^2 \geq 0, \quad \text{or} \quad x^2 + y^2 \leq 9,$$

and $x + z \neq 0$. Using set notation we can write the domain as

$$D = \{(x, y, z) \mid x^2 + y^2 \leq 9, \quad x + z \neq 0\}.$$

EXAMPLE   If $f(x, y, z) = (x + y + z)^{-1} \sin \pi(x + y + z)$, evaluate $f(1, 0, \frac{1}{2})$, $f(-1, 3, 2)$, and $f(1, 4, -5)$. Find the domain of f.

SOLUTION
$$f(1, 0, \tfrac{1}{2}) = (1 + 0 + \tfrac{1}{2})^{-1} \sin \pi(1 + 0 + \tfrac{1}{2})$$

$$= \left(\frac{3}{2}\right)^{-1} \sin \frac{3\pi}{2} = \left(\frac{2}{3}\right)(-1) = -\frac{2}{3}.$$

$$f(-1, 3, 2) = (-1 + 3 + 2)^{-1} \sin \pi(-1 + 3 + 2)$$

$$= 4^{-1} \sin 4\pi = (\tfrac{1}{4})(0) = 0.$$

$$f(1, 4, -5) = (1 + 4 - 5)^{-1} \sin \pi(1 + 4 - 5)$$

$$= 0^{-1} \sin 0 = \tfrac{0}{0} \quad \text{(undefined)}.$$

$f(x, y, z)$ is defined provided that the denominator $x + y + z$ does not vanish. Therefore its domain is

$$D = \{(x, y, z) \mid x + y + z \neq 0\}.$$

Clearly the point $(1, 4, -5)$ in the last case considered above does not belong to D since $1 + 4 - 5 = 0$.

When more than three independent variables occur it is common to use subscript notation to denote them rather than to introduce new letters. Thus if there are n independent variables we would denote them by x_1, x_2, x_3, and so on, up to x_n. Using z as the dependent variable we would denote a function of the n variables by $z = f(x_1, x_2, \ldots, x_n)$. Subscript notation is also frequently used for functions of two or three variables. For example, we might write $w = f(x_1, x_2, x_3)$ instead of $w = f(x, y, z)$.

433

EXAMPLE If

$$z = x_1^2 + e^{x_1 + x_2} + (2x_1 + x_4)^{-1} \sqrt{x_2^2 + x_3^2},$$

evaluate z at the points $(1, 2, 3, 4)$ and $(3, -3, 4, -5)$.

SOLUTION By substitution:
i. at $(1, 2, 3, 4)$,

$$z = 1^2 + e^{1+2} + [2(1) + 4]^{-1} \sqrt{2^2 + 3^2}$$

$$= 1 + e^3 + \frac{\sqrt{13}}{6};$$

ii. at $(3, -3, 4, -5)$,

$$z = 3^2 + e^{3+(-3)} + [2(3) + (-5)]^{-1} \sqrt{(-3)^2 + 4^2}$$

$$= 9 + e^0 + (6 - 5)^{-1} \sqrt{9 + 16}$$

$$= 9 + 1 + \tfrac{5}{1}$$

$$= 15.$$

EXERCISES 9.1

Calculate the values of the given functions at the indicated points.

1. $f(x, y) = x^2 - 2xy + y^2$; $(x, y) = (3, -2)$ and $(-4, -4)$.

2. $f(x, y) = \dfrac{(x - 1)(y - 1)}{(x + y)}$; $(x, y) = (1, -2)$, $(2, -2)$, and $(3, -2)$.

3. $f(x, t) = \dfrac{\cos \pi t}{x^2 + t^2}$; $(x, t) = (2, 1)$, $\left(3, \dfrac{1}{2}\right)$, and $\left(-\dfrac{1}{4}, \dfrac{3}{4}\right)$.

4. $f(u, v) = u + \ln |v|$; $(u, v) = (2, 1)$, $(-2, -e)$, and $(0, e^3)$.

5. $f(x, y, z) = x^2 + 2y^2 + 3z^2$; $(x, y, z) = (1, 2, 3)$ and $(-2, 1, -4)$.

6. $f(x, y, t) = \dfrac{x + y + t}{x + y - t}$; $(x, y, t) = \left(\dfrac{1}{2}, -\dfrac{1}{2}, 1\right)$ and $\left(\dfrac{1}{2}, \dfrac{1}{2}, -1\right)$.

7. $f(u, v, z) = \dfrac{\tan \pi u \sin \pi v}{z}$; $(u, v, z) = \left(\dfrac{1}{2}, 1, 1\right)$ and $\left(\dfrac{1}{4}, -\dfrac{1}{3}, 2\right)$.

8. $f(a, b, c) = \dfrac{2a^2 + b^2}{\sqrt{c^2 - 4}}$; $(a, b, c) = (1, 2, 3)$ and $(2, 2, -4)$.

Find the domains of the following functions and represent them graphically.

9. $f(x, y) = x^2 + 2xy + y^2$.

10. $f(x, y) = \sqrt{x^2 + y^2 - 9}$.

11. $f(x, t) = \sqrt{x + 2t - 2}$.

12. $f(u, v) = |u + v|$.

13. $f(x, y) = \dfrac{x^2}{y^2 - 1}$.

14. $f(x, y) = x \sqrt{y^2 - 1}$.

15. $f(x, y) = \sqrt{1 - (x + y)^2}$.

16. $f(z, w) = \ln (z - w)$.

434

17. $f(w, z) = \dfrac{1}{\ln(z - w)}.$ **18.** $f(x, y) = \sqrt{e^{x+y} - 1}.$

19. $f(x, y) = \tan(y - x).$ **20.** $f(x, y) = \operatorname{cosec}(xy).$

Find the domains of the following functions.

21. $f(x, y, z) = x + \sqrt{yz}.$ **22.** $f(x, y, z) = \ln\left[\dfrac{x}{y(z - 1)}\right].$

Write the following implicit relations as one or more explicit functions.

23. $x - 2y + 3z = 4.$ **24.** $x^2 - y^2 + z^2 = 1.$

25. $x^2 + y^2 + z^2 + 2xy + 2yz + 2zx = 1.$

26. $z + \sqrt{xyz} + 1 = 0.$

9.2 COORDINATES IN THREE DIMENSIONS

For functions $y = f(x)$ of a single variable, the graph provides an exceedingly useful means of visualizing the qualitative features of the function. Such properties of the function, as where it is increasing and where it is decreasing, where it is concave upwards or downwards, where its maxima and minima are located, the values of x at which the function becomes infinitely large, and the behavior of the function for large values of x, are all immediately apparent from its graph. In order to extend the idea of a graph to functions $z = f(x, y)$ of two independent variables we need three coordinate axes, one for each of the variables x, y, and z, and therefore we must concern ourselves with coordinate geometry in three rather than two dimensions.

In three dimensions the x-, y-, and z-axes are constructed at right angles to one another, as shown in Fig. 9.4. Each pair of axes determines a plane, for example, the

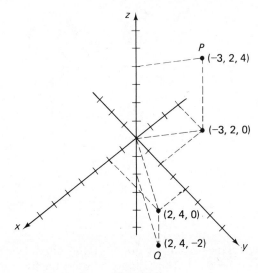

Figure 9.4

x-axis and *y*-axis determine the *xy*-plane, the *x*-axis and *z*-axis determine the *xz*-plane, and so on. On the *xy*-plane the third coordinate *z* is set equal to zero, and the coordinates *x* and *y* are used in the usual way to plot the positions of points in that plane. In Fig. 9.4 the points (2, 4, 0) and (−3, 2, 0) are plotted in order to demonstrate this procedure.

In order to plot the position of a general point (*x*, *y*, *z*) for which $z \neq 0$, we first plot the point (*x*, *y*, 0) in the *xy*-plane and then move from this point parallel to the *z*-axis according to the given value of the *z*-coordinate. For example, when plotting (−3, 2, 4) we first plot (−3, 2, 0), as in Fig. 9.4, and then move a distance of 4 units in the direction of the positive *z*-axis, to the point *P*. In plotting the point (2, 4, −2) we first plot (2, 4, 0) in the *xy*-plane and then move 2 units parallel to the negative *z*-axis, to the point *Q*.

It is often convenient to think of the *xy*-plane as being horizontal and the *z*-axis as pointing vertically upwards. The negative *z*-axis then points in the downwards direction.

All points in the *xy*-plane satisfy the condition $z = 0$. Correspondingly, all points in the *xz*-plane satisfy the condition $y = 0$ and all points in the *yz*-plane satisfy the condition $x = 0$. On the *z*-axis both *x* and *y* are zero. Correspondingly, on the *x*-axis, $y = z = 0$, and on the *y*-axis, $x = z = 0$.

EXAMPLE   Plot the points (0, 2, 4), (3, 0, −2), (0, 0, 5), and (0, −3, 0).

SOLUTION   The points are plotted in Fig. 9.5. Note that the four points lie respectively in the *yz*-plane, in the *xz*-plane, on the *z*-axis, and on the *y*-axis.

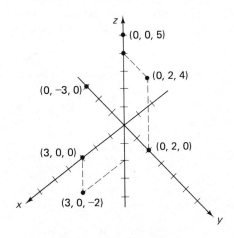

Figure 9.5

THEOREM 9.2.1

The distance from the origin of coordinates to the point (*x*, *y*, *z*) is equal to $\sqrt{x^2 + y^2 + z^2}$. The distance between the two points (x_1, y_1, z_1) and (x_2, y_2, z_2)

is equal to

$$\sqrt{(x_1 - x_2)^2 + (y_1 - y_2)^2 + (z_1 - z_2)^2}.$$

PROOF   Let O be the origin, P the point (x, y, z), and Q the point $(x, y, 0)$ (Fig. 9.6). Then Q lies in the xy-plane directly beneath P, and the distance $PQ = |z|$ (the absolute value sign is necessary in case z is negative).

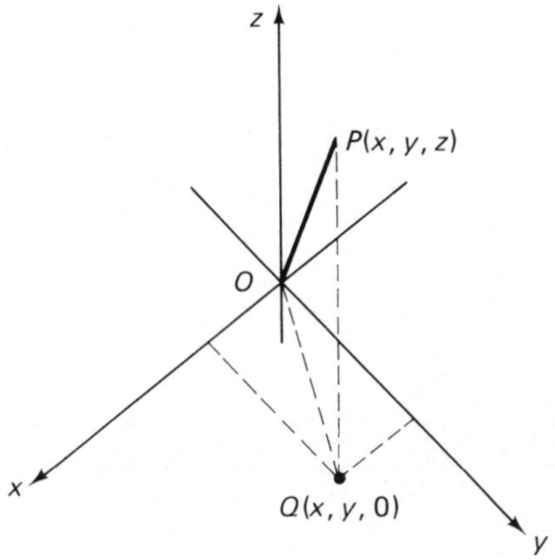

From the distance formula in two dimensions we know that $OQ = \sqrt{x^2 + y^2}$. But consider the triangle OPQ that has a right angle at the vertex Q. From Pythagoras's theorem,

$$OP^2 = OQ^2 + PQ^2$$
$$= (\sqrt{x^2 + y^2})^2 + |z|^2$$
$$= x^2 + y^2 + z^2.$$

Taking square roots, therefore, we obtain the required result,

$$OP = \sqrt{x^2 + y^2 + z^2}.$$

Figure 9.6

The proof of the second part of the theorem is given with reference to Fig. 9.7. Here P is the point (x_1, y_1, z_1) and Q is $(x_1, y_1, 0)$; R is (x_2, y_2, z_2) and S is $(x_2, y_2, 0)$. Then since Q and S are both in the xy-plane we can use the distance formula in the plane to show that $QS^2 = (x_1 - x_2)^2 + (y_1 - y_2)^2$.

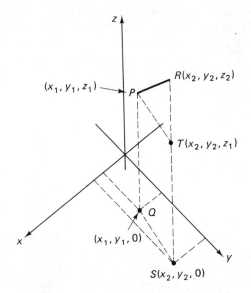

Figure 9.7

Now let T be the point (x_2, y_2, z_1), which lies on the same vertical line as R and S but is on the same horizontal level as P. Then PT is equal and parallel to QS so that $PT^2 = (x_1 - x_2)^2 + (y_1 - y_2)^2$. Furthermore, the distance $RT = |z_2 - z_1|$.

But the triangle PRT is right-angled at T since PT is horizontal and RT is vertical. Therefore

$$PR^2 = PT^2 + RT^2 = (x_1 - x_2)^2 + (y_1 - y_2)^2 + |z_2 - z_1|^2$$
$$= (x_1 - x_2)^2 + (y_1 - y_2)^2 + (z_1 - z_2)^2.$$

After taking square roots, the required formula for the distance PR is obtained.

EXAMPLE   Find the distance from $(1, 1, 1)$ to the origin.

SOLUTION   Distance $= \sqrt{1^2 + 1^2 + 1^2} = \sqrt{3}$.

EXAMPLE   Find the distance between $(2, -1, 3)$ and $(1, 2, -2)$.

SOLUTION
$$\text{Distance} = \sqrt{(2 - 1)^2 + [(-1) - (2)]^2 + [3 - (-2)]^2}$$
$$= \sqrt{1^2 + (-3)^2 + (5)^2}$$
$$= \sqrt{1 + 9 + 25}$$
$$= \sqrt{35}.$$

Consider the set of points for which the third coordinate has the fixed value $z = z_0$ while x and y are allowed to vary. Whatever the values of x and y, the point (x, y, z_0) lies at a height z_0 above the xy-plane (where z_0 is positive; if z_0 is negative the point lies below the xy-plane). In other words, each of the points (x, y, z_0) lies on the horizontal plane that lies a distance z_0 above the xy-plane if $z_0 > 0$ or a distance $|z_0|$ below the xy-plane if $z_0 < 0$ (Fig. 9.8). This plane intersects the z-axis at the point $(0, 0, z_0)$. It never intersects the x- and y-axes, being parallel to them. We can say that "$z = z_0$" is the *equation* of this horizontal plane.

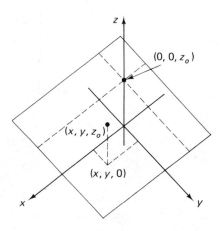

Figure 9.8

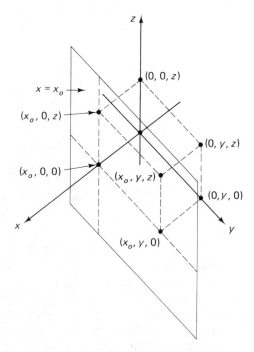

Figure 9.9

In a similar way, the set of points that satisfy the equation $x = x_0$ lie on a vertical plane that is parallel to the yz-plane. This is illustrated in Fig. 9.9. The plane in question meets the x-axis at the point $(x_0, 0, 0)$ and is parallel to the y- and z-axes.

Finally, the equation $y = y_0$ describes a vertical plane that is parallel to the xz-plane. All points (x, y_0, z) lie on the vertical plane that meets the y-axis at $(0, y_0, 0)$ and which is parallel to the x- and z-axes.

EXAMPLE   What is the equation of the vertical plane through the two points $(1, 2, 0)$ and $(-3, 2, 4)$?

SOLUTION   We observe that the two points in question both have the y-coordinate equal to 2. Hence they both lie on the vertical plane whose equation is $y = 2$.

We should like to extend this discussion of horizontal and vertical planes to planes that cut the coordinate axes at arbitrary angles. It turns out that any plane in three dimensions is determined by an equation of the form $ax + by + cz = d$, where a, b, c, and d are four constants. For example, the equation $x - y + 2z = 3$ is satisfied by all the points (x, y, z) on a certain plane. Here the coefficients are $a = 1$, $b = -1$, $c = 2$, and $d = 3$. The equation $2x + y - 3z = 0$, in which $a = 2$, $b = 1$, $c = -3$, and $d = 0$, is satisfied by the points on some other plane.

An equation of the form $ax + by + cz = d$ is called a *linear equation*. We shall show that any plane in xyz-space is described by a linear equation of this type.

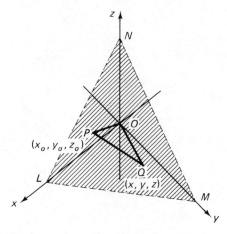

Figure 9.10

Figure 9.10 illustrates a plane in *xyz*-space, intersecting the three coordinate axes at *L*, *M*, and *N*. Let a perpendicular be dropped from the origin *O* onto the plane, meeting the plane at the point *P*, whose coordinates are taken to be (x_0, y_0, z_0). Let *Q* be a general point (x, y, z) lying in the plane. Then since *OP* was constructed to be perpendicular to the given plane, it follows that, in particular, *OP* must be perpendicular to the line *PQ*. In other words, the triangle *OPQ* is right-angled at *P*.

From Pythagoras's theorem, ther fore $OQ^2 = OP^2 + PQ^2$. Substituting the values of the three lengths, *OP*, *OQ*, and *PQ*, we obtain the equation

$$(x^2 + y^2 + z^2) = (x_0^2 + y_0^2 + z_0^2) + (x - x_0)^2 + (y - y_0)^2 + (z - z_0)^2$$

$$= x_0^2 + y_0^2 + z_0^2 + x^2 - 2xx_0 + x_0^2 + y^2 - 2yy_0 + y_0^2$$

$$+ z^2 - 2zz_0 + z_0^2$$

$$= x^2 + y^2 + z^2 + 2(x_0^2 + y_0^2 + z_0^2) - 2(xx_0 + yy_0 + zz_0).$$

After some simplification, therefore, we obtain

$$xx_0 + yy_0 + zz_0 = x_0^2 + y_0^2 + z_0^2. \tag{1}$$

This result actually completes the required proof, since we can take $a = x_0$, $b = y_0$, $c = z_0$, and $d = x_0^2 + y_0^2 + z_0^2$, in which case equation (1) can be rewritten in the form $ax + by + cz = d$.

NOTE. It is left as an exercise for the student to show that if the plane passes through the origin its equation is of the form $ax + by + cz = 0$ (in other words, the constant *d* is zero).

EXAMPLE   Find the equation of the plane that passes through the point *P* $(1, 2, -1)$ and that is perpendicular to *OP*.

SOLUTION   We simply take $(x_0, y_0, z_0) = (1, 2, -1)$ in equation (1). The required equation is then

$$x(1) + y(2) + z(-1) = (1)^2 + (2)^2 + (-1)^2 = 6,$$

that is,

$$x + 2y - z = 6.$$

In the equation $ax + by + cz = d$, describing a plane, the coefficients a, b, c, and d are not uniquely determined, since we can multiply them all by a constant without changing the given relation. For example, $x + 2y - z = 6$ and $2x + 4y - 2z = 12$ both describe the same plane, the latter equation having been obtained from the former simply by multiplying through by 2. In general, the equations $ax + by + cz = d$ and $(ka)x + (kb)y + (kc)z = (kd)$ describe the same plane, where k is any nonzero constant.

Let the plane whose equation is $ax + by + cz = d$ intersect the coordinate axes at the points L, M, and N, as illustrated in Fig. 9.10. At L, on the x-axis, both $y = 0$ and $z = 0$. Therefore, setting y and z to zero in the equation of the plane, we find

$$ax + b(0) + c(0) = d,$$

i.e.,

$$ax = d, \; x = \frac{d}{a}.$$

The coordinates of L are therefore $(d/a, 0, 0)$.

At M, both $x = 0$ and $z = 0$, and it follows that $y = d/b$. The coordinates of M are therefore $(0, d/b, 0)$. Similarly, the coordinates of N are seen to be $(0, 0, d/c)$.

EXAMPLE   The foot P of the perpendicular from the origin onto a plane is the point $(1, -1, 2)$. Calculate the points of intersection of the plane with the coordinate axes.

SOLUTION   We must first find the equation of the plane. If P is the point (x_0, y_0, z_0), then, according to our previous results, the equation has the form $ax + by + cz = d$ where $a = x_0$, $b = y_0$, $c = z_0$, and $d = x_0^2 + y_0^2 + z_0^2$. In this example, $x_0 = 1$, $y_0 = -1$, and $z_0 = 2$. Therefore $a = 1$, $b = -1$, $c = 2$, $d = (1)^2 + (-1)^2 + 2^2 = 6$, and so the equation of the plane is

$$x - y + 2z = 6.$$

To find the intersection with the x-axis we set $y = z = 0$ in the equation of the plane. Thus we obtain

$$x - 0 + 0 = 6, \quad \text{or} \quad x = 6.$$

The coordinates of the point of intersection are therefore $(6, 0, 0)$.

The intersection with the y-axis is obtained by setting $x = z = 0$, which gives $0 - y + 0 = 6$, or $y = -6$. The point is therefore $(0, -6, 0)$.

Finally, for the intersection with the z-axis we set $x = y = 0$, obtaining $0 - 0 + 2z = 6$, or $z = 3$. The coordinates are then $(0, 0, 3)$.

EXAMPLE Find the equation of the plane that meets the coordinate axes at the points $(2, 0, 0)$, $(0, -1, 0)$ and $(0, 0, 3)$.

SOLUTION Let us assume that the equation of the plane in standard form is $ax + by + cz = d$. This passes through the point $x = 2$, $y = 0$, and $z = 0$, so substituting these values of x, y, and z into the equation we obtain

$$a(2) + b(0) + c(0) = d,$$

that is,

$$2a = d, \quad \text{or} \quad a = \tfrac{1}{2}d.$$

Similarly, for the point of intersection with the y-axis, we can substitute $x = 0$, $y = -1$, and $z = 0$ into the equation, obtaining

$$a(0) + b(-1) + c(0) = d,$$

that is,

$$b = -d.$$

Finally, substituting $x = y = 0$ and $z = 3$ from the third point, we get

$$a(0) + b(0) + c(3) = d,$$

that is,

$$3c = d, \quad \text{or} \quad c = \tfrac{1}{3}d.$$

The equation of the plane is therefore

$$\tfrac{1}{2}dx + (-d)y + \tfrac{1}{3}dz = d$$

or, multiplying through by $6d^{-1}$,

$$3x - 6y + 2z = 6.$$

If in the general linear equation $ax + by + cz = d$ the coefficient c is zero, we obtain the equation $ax + by = d$, in which the variable z does not appear. Such an equation describes a plane that is vertical or, in other words, parallel to the z-axis. If $d \neq 0$, the plane never intersects the z-axis, while if $d = 0$, the whole z-axis lies in the plane.

EXAMPLE Find the points of intersection of the plane

$$2x - y = 3$$

with the three coordinate axes.

SOLUTION For the x-axis we set $y = z = 0$, thus obtaining $2x = 3$ or $x = \tfrac{3}{2}$. For the y-axis we set $x = z = 0$, thus obtaining $-y = 3$ or $y = -3$. The two points of intersection with these axes are therefore $(\tfrac{3}{2}, 0, 0)$ and $(0, -3, 0)$. For the z-axis we set $x = y = 0$, obtaining the equation $0 = 3$. This equation is nonsensical, and means that the plane cannot intersect the z-axis. This is because the plane is vertical and parallel to the z-axis.

In a similar way a plane parallel to the *x*-axis has an equation of the form $by + cz = d$ (i.e., $a = 0$), and a plane parallel to the *y*-axis has an equation of the form $ax + cz = d$ (i.e., $b = 0$).

EXERCISES 9.2

Plot the following points relative to a set of coordinate axes in three dimensions.

1. $(1, 1, 1)$.   **2.** $(-2, 3, 2)$.

3. $(1, 3, -1)$.   **4.** $(2, -4, -2)$.

5. $(-2, 2, -2)$.   **6.** $(-1, -1, -1)$.

Calculate the distances of the following points from the origin.

7. $(1, 2, 0)$.   **8.** $(-1, -3, -1)$.

9. $(2, 4, -4)$.   **10.** $(-3, 2, -1)$.

Calculate the distances between the following pairs of points.

11. $(0, 1, 2)$ and $(1, 2, 0)$.   **12.** $(-1, 1, -1)$ and $(2, 2, 2)$.

13. $(-3, 0, 1)$ and $(1, -2, 4)$.   **14.** $(5, -6, -3)$ and $(6, -5, 0)$.

Show that the following triangles PQR are right-angled and find which of the angles is the right angle.

15. $P(3, 1, 0)$; $Q(2, 0, -1)$; $R(-1, -1, 3)$.

16. $P(-1, 2, 0)$; $Q(-2, 0, 3)$; $R(-3, 2, 1)$.

Find the equations of the horizontal planes through the following points.

17. $(0, 0, 3)$.   **18.** $(1, 4, 1)$.

19. $(-2, -3, -1)$.   **20.** $(2, 0, -3)$.

Find the equations of the vertical planes through the following pairs of points.

21. $(0, 1, 3)$ and $(0, -2, 2)$.   **22.** $(3, -1, 0)$ and $(-1, -1, -1)$.

23. $(-1, 2, 2)$ and $(0, 2, 2)$.   **24.** $(2, -1, -1)$ and $(2, 1, -1)$.

Find the equations of the planes for which the foot P of the perpendicular from the origin has the following coordinates.

25. $(1, 1, 1)$.   **26.** $(-2, 0, 2)$.

27. $(0, 3, -1)$.   **28.** $(-2, -1, 4)$.

29–32. Find the coordinates of the points L, M, and N at which each of the planes in exercises 25–28 meets the coordinate axes.

Find the equations of the vertical planes that meet the *x*- and *y*-axes at the given points.

33. $(2, 0, 0)$ and $(0, 3, 0)$.   **34.** $(-1, 0, 0)$ and $(0, -4, 0)$.

35. Three breakfast foods (A, B, C) contain, respectively, 100 cal and 0.10 mg of thiamin per ounce (food A), 120 cal and 0.14 mg of thiamin per ounce (food B), and 90 cal and 0.08 mg of thiamin per ounce (food C). What combinations (x oz of A, y oz of B, and z oz of C) contain at most 400 cal and at least 0.4 mg of thiamin?

36. An area of at most 1000 acres is to be planted with three crops, x acres being planted with crop A, y acres with crop B, and z acres with crop C. In order to inhibit the spread of disease, no crop can exceed 60% of the total area planted. The protein yield from crop A is 100 kg/acre, from crop B is 80 kg/acre, and from crop C is 70 kg/acre. Write down the conditions on x, y, and z if the total protein yield is to be at least 80,000 kg.

Find the equations of the planes passing through the following sets of three points.

37. $(-1, 0, 0)$, $(0, 3, 0)$, and $(0, 0, 4)$.   **38.** $(2, 0, 0)$, $(0, -2, 0)$, and $(0, 0, -5)$.

9.3 GRAPHS IN THREE DIMENSIONS

Let $z = f(x, y)$ be a function of two variables. Its domain D consists of the set of points in the xy-plane at which the function is defined. For any point (x, y) in D, we can calculate the corresponding value of $z = f(x, y)$ and plot the point (x, y, z) using three-dimensional coordinates. By doing this for every point (x, y) in D we obtain a set of points (x, y, z), which form a surface in three dimensions. There is one point (x, y, z) on this surface lying above each point of the domain D [or below if $z = f(x, y)$ turns out to be negative]. This surface is said to be the *graph* of the function $z = f(x, y)$.

EXAMPLE   Find and sketch the graph of the function

$$z = \sqrt{4 - x^2 - y^2}.$$

SOLUTION   We saw in Section 9.1 that the domain D of this function consists of the circular region $\{(x, y, 0) \mid x^2 + y^2 \leq 4\}$ lying in the xy-plane, centered at the origin, and of radius equal to 2. The graph of the function must be a surface of some kind lying above or below this circle. Squaring both sides of the given equation we obtain

$$z^2 = 4 - x^2 - y^2$$

or

$$x^2 + y^2 + z^2 = 4.$$

But if P is the point (x, y, z) the distance OP from the origin to P is equal to $(x^2 + y^2 + z^2)^{1/2}$ so that, from the above equation, it follows that

$$OP = 2.$$

Therefore every point $P(x, y, z)$ on the graph of the given function lies a distance of 2 units from the origin. In other words, every such point P lies on a sphere of radius 2 units, centered at the origin. The graph does not consist of the whole of this sphere, however. Since from the original function, $z = \sqrt{4 - x^2 - y^2}$, z must be greater than or equal to zero, the graph consists of the hemisphere that lies above or in the xy-plane, that is, for which $z \geq 0$ (Fig. 9.11). For values of x and y that lie on the circle $x^2 + y^2 = 4$, which is the outer boundary of the domain D, the point (x, y, z) lies in the xy-plane. For values of x and y lying in the interior of D ($x^2 + y^2 < 4$), the point (x, y, z) lies above the xy-plane.

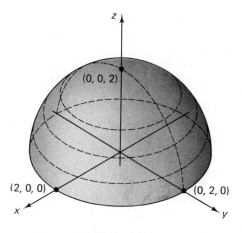

Figure 9.11

The task of sketching a surface in three dimensions such as the graph of a function $z = f(x, y)$ is by no means as easy as sketching the graph of a function $y = f(x)$ of a single variable. When faced with this task it is often helpful to examine what are called *sections* of the graph, which are slices made through the graph by certain planes. Usually for this purpose we use either slices by horizontal planes or slices by vertical planes parallel to the xz- and yz-planes. Any such plane slicing through the graph of a function $z = f(x, y)$ intersects the graph in a certain curve. By sketching a few such curves it is often possible to convey a good idea of the general shape of the surface of the graph itself.

Let us consider sections by horizontal planes. A horizontal plane (parallel to the xy-plane) satisfies an equation of the type $z = c$, where c is a constant that gives the height of the plane above the xy-plane (or below if $c < 0$). So the section of a graph by such a plane consists of points on the graph that lie at a constant height above (or below) the xy-plane. Such a horizontal section is often called a *contour line* on the graph.

Consider the preceding example, $z = \sqrt{4 - x^2 - y^2}$. The points on the graph of this function that also lie on the horizontal plane $z = c$ satisfy

$$c^2 = 4 - x^2 - y^2,$$

that is,

$$x^2 + y^2 = 4 - c^2.$$

This equation relating x and y is the equation of a circle centered at $x = y = 0$ and with radius equal to $\sqrt{4 - c^2}$. It follows, therefore, that the horizontal section of the graph by the plane $z = c$ consists of a circle centered on the z-axis, with radius $\sqrt{4 - c^2}$, and, of course, lying at a height c above the xy-plane.

For example, let us take $c = 1$ so that we are considering the horizontal slice through the graph by the plane lying one unit above the xy-plane. The section is then a horizontal circle whose radius is $\sqrt{4 - 1^2} = \sqrt{3}$ and that is centered at the point $(0, 0, 1)$. Similarly, if $c = \frac{1}{2}$, the section is a circle of radius $\sqrt{15}/2$ centered at $(0, 0, \frac{1}{2})$, while if $c = \frac{3}{2}$, the section is a circle of radius $\sqrt{7}/2$ centered at $(0, 0, \frac{3}{2})$.

The circles $x^2 + y^2 = 4 - c^2$ corresponding to these three values of c as well as the outer boundary $x^2 + y^2 = 4$ of the domain are shown in Fig. 9.12. These four horizontal circles were in fact drawn in Fig. 9.11, and it can be seen that they are of considerable assistance in representing the shape of the surface.

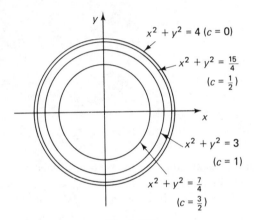

Figure 9.12

We can also make use of sections of a graph by vertical planes parallel to the xz- and yz-coordinate planes. We saw in Section 9.2 that a plane parallel to the xz-plane is one on which $y = $ constant, while a plane parallel to the yz-plane is one on which $x = $ constant. Thus these vertical sections can be found in a similar way to the above horizontal sections by setting x or y, rather than z, equal to a constant. We shall illustrate the procedure by means of an example.

EXAMPLE   Draw vertical sections of the graph of $z = \sqrt{4 - x^2 - y^2}$.

SOLUTION   We shall consider first the section on which $x = c$, a constant, that is, the section by the vertical plane parallel to the yz-plane that meets the x-axis at the point $(c, 0, 0)$. Substituting $x = c$ into the given form of the function we obtain $z = \sqrt{4 - c^2 - y^2}$. Squaring we get

$$z^2 = 4 - c^2 - y^2$$

or

$$y^2 + z^2 = 4 - c^2.$$

This equation describes a circle in the *yz*-plane with its center at $y = z = 0$ and a radius of $\sqrt{4 - c^2}$. Since at all points on the graph, $z \geq 0$, it follows that the section, in fact, consists of a semicircle that lies in the vertical plane $x = c$ and is centered on the *x*-axis at the point $(c, 0, 0)$ with a radius of $\sqrt{4 - c^2}$. When $c = 1$, for example, this semicircle has a radius equal to $\sqrt{3}$ and is centered at $(1, 0, 0)$. The semicircles corresponding to the sections $c = \frac{1}{2}$, $c = 1$, and $c = \frac{3}{2}$ are drawn in Fig. 9.13 (note that the semicircles all lie above the *xy*-plane, i.e., $z \geq 0$).

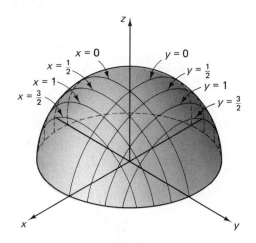

Figure 9.13

Next let us consider sections on which $y = c$, that is, vertical sections by planes parallel to the *xz*-plane. Setting $y = c$ we obtain $z = \sqrt{4 - x^2 - c^2}$, from which it follows that $x^2 + z^2 = 4 - c^2$ (and $z \geq 0$). Thus again the sections are semicircles, this time lying parallel to the *xz*-plane, being centered on the *y*-axis at $(0, c, 0)$, and having radius $\sqrt{4 - c^2}$. The sections corresponding to $c = \frac{1}{2}$, 1, and $\frac{3}{2}$ are drawn in Fig. 9.13.

It can be seen from this example that the two sets of vertical sections can again be a useful aid in representing the form of a surface.

EXAMPLE By drawing vertical sections, sketch the graph of the function $z = x^2 - y^2$.

SOLUTION We note first that z is defined for all values of x and y; the domain consists of the whole *xy*-plane.

Consider first a section for which $y = c$. Section by a plane parallel to the *xz*-plane. The equation of this section is $z = x^2 - c^2$. This equation in terms of x and z describes a parabola, opening upwards, with vertex at the point $x = 0$, $z = -c^2$. For example, if $c = 1$, the vertex of the parabola is $(0, 1, -1)$ and the parabola lies in the vertical plane $y = 1$.

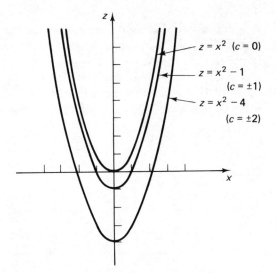

Figure 9.14

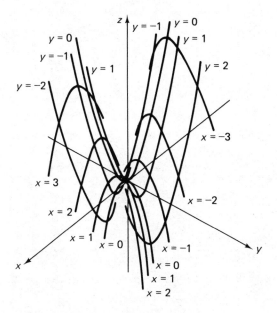

Figure 9.15

The parabolas corresponding to the values $c = 0$, ± 1, and ± 2 are drawn in Fig. 9.14, which shows them in the xz-plane. In Fig. 9.15 these same parabolas are drawn in a three-dimensional diagram showing vertical sections of the required graph.

Now consider a section for which $x = c$, that is, a slice of the graph by a plane parallel to the yz-plane. On this section

$$z = c^2 - y^2.$$

In terms of z and y, this equation describes a parabola that opens downwards, the vertex being at the point $y = 0$, $z = c^2$. For example, if $c = 2$, the vertex of the parabola is at the point $(2, 0, 4)$ and the parabola lies in the plane $x = 2$. These sections are sketched in Fig. 9.15 for the values $x = 0, \pm 1, \pm 2$, and ± 3 (the curves open downwards).

Again we see that the sections provide a good idea of the nature of the graph. In this example the overriding feature of the surface is its saddle-like shape, a feature that is readily apparent from Fig. 9.15.

In order to represent graphically a function of two variables, we can draw graphs of several vertical sections on the same sheet of graph paper. Any section parallel to the xz-plane of the function $z = f(x, y)$ satisfies the equation $y = c$, $z = f(x, c)$. We can plot the graph of the relation $z = f(x, c)$ using the xz-coordinate axes, and will obtain a curve in the usual way. This curve represents z as a function of x for a fixed value of y (namely $y = c$). By allowing c to have several different values, it is possible to plot a number of such curves on the same set of axes. This procedure is illustrated in the following example.

EXAMPLE   The surface area of the average human being, S, is given in square meters approximately by the formula

$$S = 2W^{0.4}H^{0.7},$$

where W is the weight in kilograms and H the height in meters. Draw graphs of S against W for $H = 1.6$, 1.7, and 1.8 m, and draw graphs of S against H for $W = 50$, 75, and 100 kg.

SOLUTION

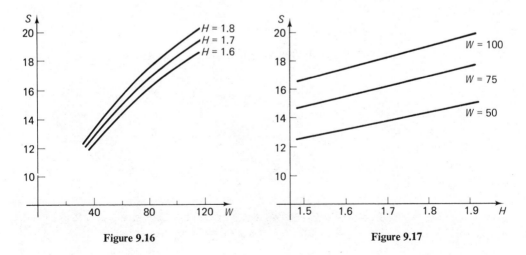

Figure 9.16   Figure 9.17

The required graphs are given in Figs. 9.16 and 9.17. For example, when $H = 1.6$ m,

$$S = 2(1.6)^{0.7}W^{0.4}$$

$$= 2.78W^{0.4},$$

and this section is plotted as a graph of S against W in the lower curve of Fig. 9.16. This curve provides the surface area as a function of weight for individuals whose heights are 1.6 m. Similarly, in Fig. 9.17, the middle one of the three curves, for example, is given by the function

$$S = 2(75)^{0.4}H^{0.7}$$

$$= 11.2H^{0.7}.$$

It gives the surface area as a function of height for individuals whose weights are 75 kg.

In the case of a function $w = f(x, y, z)$ of three variables, the independent variables themselves require three dimensions, and the graph of the function must be plotted in four dimensions, using (x, y, z, w) as the four coordinates. It is no longer possible to visualize the shape of the graph by means of a two-dimensional diagram, and one does not attempt to draw the surface in the way we have done above for functions of two variables. This is, if possible, even more true for functions $z = f(x_1, x_2, \ldots, x_n)$ of n variables. The graph of such a function is a "hypersurface" in an $(n + 1)$-dimensional space and would be impossible to draw in any useful way. In spite of this, many of the geometrical ideas can be usefully carried over from three dimensions to higher dimensions, and it is still useful to look upon a function of many variables as representing a surface in some space of higher dimension, even though the surface cannot be drawn in any convenient way.

EXERCISES 9.3

Find the equations of the horizontal sections of the graphs of the following functions.

1. $z = \sqrt{9 - x^2 - y^2}$.   **2.** $z = \sqrt{16 - x^2 - y^2}$.

3. $z = x^2 + y^2$.   **4.** $z = x^2 + 3xy + 2y^2$.

Sketch the horizontal sections of the following graphs corresponding to the given values of z.

5. $z = \sqrt{16 - x^2 - y^2}$; $z = 0, 1, 2, 3$.

6. $z = \sqrt{25 - x^2 - y^2}$; $z = 0, 1, 2, 3, 4, 5$.

7. $z = x^2 - y^2$; $z = 0, \pm 1, \pm 2$.   **8.** $z = 2x^2 - y^2$; $z = 0, \pm 1, \pm 2$.

Sketch the graphs of the following functions and relations by using appropriate vertical sections parallel to the coordinate planes.

9. $z = \sqrt{16 - x^2 - y^2}$.

10. $x^2 + y^2 + z^2 = 16$.

11. $z = x + 2y$.

12. $x + y - 2z = 2$.

13. $z = x^2 + y^2$.

14. $z = x^2 + 2y^2 + 1$.

9.4   PARTIAL DERIVATIVES

We shall now turn to the question of differentiating functions of several variables. In the present section we shall simply concern ourselves with the mechanics of differentiation, and shall in the following sections turn to matters of interpretation and use of the resulting derivatives.

Let $z = f(x, y)$ be a function of two independent variables. If the variable y is held fixed at a value $y = y_0$, then the relation $z = f(x, y_0)$ expresses z as a function of the one variable x. This function will have as its graph a curve in the xz-plane that is in fact the vertical section of the graph of $z = f(x, y)$ by the plane $y = y_0$.

Figure 9.18 illustrates a typical section by the plane $y = y_0$ of the graph of

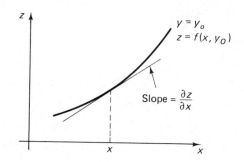

Figure 9.18

$z = f(x, y)$. The section here is drawn as a curve with respect to xz-coordinates and is described by the equation $z = f(x, y_0)$. At a general point on this curve, the tangent line can be constructed and its slope can be calculated by differentiating z with respect to x from the relation $z = f(x, y_0)$. This derivative is found in the usual way as a limit according to the following formula:

$$\frac{d}{dx} f(x, y_0) = \lim_{\Delta x \to 0} \frac{f(x + \Delta x, y_0) - f(x, y_0)}{\Delta x}.$$

It is called the *partial derivative* of z with respect to x, and is usually denoted by $\partial z / \partial x$. (Note the symbol ∂ (lower case delta) in this expression. Italic d is reserved for the derivative of a function of a single variable.)

DEFINITION   Let $z = f(x, y)$ be a function of x and y. Then the *partial derivative of z with respect to x* is defined to be

$$\frac{\partial z}{\partial x} = \lim_{\Delta x \to 0} \frac{f(x + \Delta x, y) - f(x, y)}{\Delta x}.$$

In writing this definition we have dropped the subscript from y_0: We must remember that when calculating $\partial z/\partial x$ the variable y is held constant. Correspondingly, the *partial derivative of z with respect to y* is defined to be

$$\frac{\partial z}{\partial y} = \lim_{\Delta y \to 0} \frac{f(x, y + \Delta y) - f(x, y)}{\Delta y}.$$

In calculating $\partial z/\partial y$ the variable x is held constant and the differentiation is carried out with respect to y only.

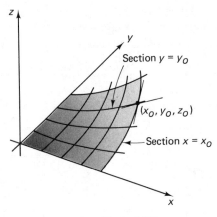

Figure 9.19

At a point (x_0, y_0) in the domain of the given function $z = f(x, y)$ the partial derivative $\partial z/\partial x$ provides the slope of the vertical section of the graph by the plane $y = y_0$. Correspondingly, the partial derivative $\partial z/\partial y$ provides the slope of the vertical section by the plane $x = x_0$. This vertical section has the equation $z = f(x_0, y)$, expressing z as a function of y, and its slope is obtained by differentiating z with respect to y, with x set equal to x_0 and held fixed.

Figure 9.19 shows a typical graph $z = f(x, y)$ with various vertical sections drawn. The tangent lines to the sections $x = x_0$ and $y = y_0$ are shown. The vertical slopes of these two tangent lines are given respectively by the partial derivatives $\partial z/\partial y$ and $\partial z/\partial x$ evaluated at (x_0, y_0).

EXAMPLE   Calculate $\partial z/\partial x$ and $\partial z/\partial y$ when $z = x^3 y$.

SOLUTION   When calculating $\partial z/\partial x$ we must treat y as a constant and simply differentiate the x^3 part of z with respect to x:

$$\begin{aligned}
\frac{\partial z}{\partial x} &= \frac{\partial}{\partial x}(x^3 y) = y \frac{\partial}{\partial x}(x^3) \\
&= y \cdot 3x^2 \\
&= 3x^2 y.
\end{aligned}$$

When calculating $\partial z/\partial y$ the x^3 factor is treated as a constant:

$$\frac{\partial z}{\partial y} = \frac{\partial}{\partial y}(x^3 y) = x^3 \frac{\partial}{\partial y}(y) = x^3 \cdot 1 = x^3.$$

EXAMPLE   Calculate $\partial z/\partial x$ for $z = \sqrt{x^2 + y^2}$.

SOLUTION   With y held constant we obtain from the chain rule that

$$\frac{\partial z}{\partial x} = \frac{\partial}{\partial x}(x^2 + y^2)^{1/2}$$

$$= \frac{1}{2}(x^2 + y^2)^{-1/2}\frac{\partial}{\partial x}(x^2 + y^2)$$

$$= \tfrac{1}{2}(x^2 + y^2)^{-1/2}(2x + 0)$$

since $\partial/\partial x\,(y^2) = 0$ as y^2 is held constant. Therefore

$$\frac{\partial z}{\partial x} = x(x^2 + y^2)^{-1/2} = \frac{x}{\sqrt{x^2 + y^2}}.$$

EXAMPLE   Calculate $\partial z/\partial y$ for $z = (x + y)\cos(x - y)$.

SOLUTION   The function z in this example is a product of the two functions $(x + y)$ and $\cos(x - y)$. When calculating $\partial z/\partial y$ we regard x as constant and differentiate with respect to y in the ordinary way. Therefore the product formula must continue to hold just as it does for ordinary differentiation:

$$\frac{\partial z}{\partial y} = \cos(x - y)\frac{\partial}{\partial y}(x + y) + (x + y)\frac{\partial}{\partial y}\cos(x - y)$$

$$= \cos(x - y)\,(0 + 1) + (x + y)[-\sin(x - y)]\,(0 - 1),$$

where the chain rule has been used in the second term. Consequently,

$$\frac{\partial z}{\partial y} = \cos(x - y) + (x + y)\sin(x - y).$$

EXAMPLE   Calculate $\partial z/\partial x$ and $\partial z/\partial y$ for $z = (x^2 + y^2)/\ln x$.

SOLUTION   Using the quotient formula we obtain that

$$\frac{\partial z}{\partial x} = \frac{\ln x\frac{\partial}{\partial x}(x^2 + y^2) - (x^2 + y^2)\frac{\partial}{\partial x}(\ln x)}{(\ln x)^2}$$

$$= \frac{\ln x \cdot (2x) - (x^2 + y^2) \cdot (1/x)}{(\ln x)^2}$$

$$= \frac{2x^2 \ln x - (x^2 + y^2)}{x(\ln x)^2}$$

after multiplying numerator and denominator by x.

We do not need to use the quotient formula in order to evaluate $\partial z/\partial y$ since the denominator of the given quotient is a function of x alone, and so is a constant as far as partial differentiation with respect to y is concerned.

$$\frac{\partial z}{\partial y} = \frac{\partial}{\partial y}\left(\frac{x^2 + y^2}{\ln x}\right) = \frac{1}{\ln x}\frac{\partial}{\partial y}(x^2 + y^2)$$

$$= \frac{1}{\ln x}(0 + 2y)$$

$$= \frac{2y}{\ln x}.$$

EXAMPLE   If $z = (ax + by)^n$ where a and b are constants, show that

$$x\frac{\partial z}{\partial x} + y\frac{\partial z}{\partial y} = nz.$$

SOLUTION   Using the chain rule we find that

$$\frac{\partial z}{\partial x} = n(ax + by)^{n-1}(a), \qquad \frac{\partial z}{\partial y} = n(ax + by)^{n-1}(b).$$

Therefore

$$x\frac{\partial z}{\partial x} + y\frac{\partial z}{\partial y} = n(ax + by)^{n-1}(xa + yb)$$

$$= n(ax + by)^n$$

$$= nz.$$

It can be seen from these examples that the calculation of partial derivatives of a function of two variables is essentially no different from differentiating a function of just one variable. We must simply remember that when finding the partial derivative with respect to one of the variables we treat the other variable as a constant, and then differentiate in the familiar way.

The derivative $\partial z/\partial x$ is found by differentiating with respect to x with y held constant. It therefore gives the rate of change of z with respect to x for a constant value of y. This interpretation is often useful in practical examples.

EXAMPLE   The surface area S (square meters) of the average human body of weight W(kg) and height H(m) is given approximately by $S = 2W^{0.4}H^{0.7}$. Then

$$\frac{\partial S}{\partial H} = 2W^{0.4}(0.7)H^{0.7-1} = 1.4W^{0.4}H^{-0.3},$$

$$\frac{\partial S}{\partial W} = 2(0.4)W^{0.4-1}H^{0.7} = 0.8W^{-0.6}H^{0.7}.$$

$\partial S/\partial H$ is the rate at which the surface area increases with respect to height for individuals whose weights are equal to the constant value W. Similarly, $\partial S/\partial W$ is the rate at which surface area increases with respect to weight for individuals of constant height H.

The partial derivative $\partial z/\partial x$ is itself a function of x and y, and therefore we can construct its partial derivatives with respect to both x and y. These are called *second-order partial derivatives* of z, and the following notation is used:

$$\frac{\partial^2 z}{\partial x^2} = \frac{\partial}{\partial x}\left(\frac{\partial z}{\partial x}\right), \qquad \frac{\partial^2 z}{\partial y\,\partial x} = \frac{\partial}{\partial y}\left(\frac{\partial z}{\partial x}\right).$$

Similarly, $\partial z/\partial y$ can be differentiated with respect to x and y, thus providing two more second-order partial derivatives:

$$\frac{\partial^2 z}{\partial y^2} = \frac{\partial}{\partial y}\left(\frac{\partial z}{\partial y}\right), \qquad \frac{\partial^2 z}{\partial x\,\partial y} = \frac{\partial}{\partial x}\left(\frac{\partial z}{\partial y}\right).$$

The two derivatives $\partial^2 z/\partial x\,\partial y$ and $\partial^2 z/\partial y\,\partial x$ are often called *mixed partial derivatives* of second order. It turns out that, provided all of the second-order partial derivatives are continuous functions of x and y, these two mixed partial derivatives are equal to one another,

$$\frac{\partial^2 z}{\partial x\,\partial y} = \frac{\partial^2 z}{\partial y\,\partial x}.$$

The proof of this result is outside the scope of this book, and we shall content ourselves with demonstrating its validity by the following examples.

EXAMPLE   Calculate the second derivatives of the function

$$z = x^3 y^4.$$

SOLUTION   The first-order derivatives are given by

$$\frac{\partial z}{\partial x} = 3x^2 y^4, \qquad \frac{\partial z}{\partial y} = 4x^3 y^3.$$

Therefore, differentiating again, we find

$$\frac{\partial^2 z}{\partial x^2} = \frac{\partial}{\partial x}(3x^2 y^4) = 3y^4 \frac{\partial}{\partial x}(x^2) = 3y^4 \cdot 2x$$

$$= 6xy^4.$$

$$\frac{\partial^2 z}{\partial y\,\partial x} = \frac{\partial}{\partial y}(3x^2 y^4) = 3x^2 \frac{\partial}{\partial y}(y^4) = 3x^2 \cdot 4y^3$$

$$= 12x^2 y^3.$$

$$\frac{\partial^2 z}{\partial y^2} = \frac{\partial}{\partial y}(4x^3 y^3) = 4x^3 \frac{\partial}{\partial y}(y^3) = 4x^3 \cdot 3y^2$$

$$= 12x^3 y^2.$$

$$\frac{\partial^2 z}{\partial x\,\partial y} = \frac{\partial}{\partial x}(4x^3 y^3) = 4y^3 \frac{\partial}{\partial x}(x^3) = 4y^3 \cdot 3x^2$$

$$= 12x^2 y^3.$$

The equality of the two mixed derivatives is apparent.

EXAMPLE   Calculate all the second-order derivatives of the function $z = \sqrt{x^2 + y^2}$.

SOLUTION   In an earlier example we showed that for this function

$$\frac{\partial z}{\partial x} = \frac{x}{\sqrt{x^2 + y^2}}.$$

It follows in a similar way that

$$\frac{\partial z}{\partial y} = \frac{y}{\sqrt{x^2 + y^2}}.$$

The two mixed derivatives of second order are obtained as follows:

$$\frac{\partial^2 z}{\partial y \, \partial x} = \frac{\partial}{\partial y}\left(\frac{x}{\sqrt{x^2 + y^2}}\right) = x\frac{\partial}{\partial y}[(x^2 + y^2)^{-1/2}]$$

$$= x(-\tfrac{1}{2})(x^2 + y^2)^{-3/2}(2y)$$

$$= -xy(x^2 + y^2)^{-3/2}$$

(the chain rule having been used in the intermediate step);

$$\frac{\partial^2 z}{\partial x \, \partial y} = \frac{\partial}{\partial x}\left(\frac{y}{\sqrt{x^2 + y^2}}\right) = y\frac{\partial}{\partial x}[(x^2 + y^2)^{-1/2}]$$

$$= y(-\tfrac{1}{2})(x^2 + y^2)^{-3/2}(2x)$$

$$= -xy(x^2 + y^2)^{-3/2}.$$

Once again we find that these two derivatives are equal to one another.
For the remaining two derivatives the quotient rule must be used:

$$\frac{\partial^2 z}{\partial x^2} = \frac{\partial}{\partial x}\left(\frac{x}{\sqrt{x^2 + y^2}}\right)$$

$$= \frac{\sqrt{x^2 + y^2} \cdot \frac{\partial}{\partial x}(x) - x\frac{\partial}{\partial x}(\sqrt{x^2 + y^2})}{(\sqrt{x^2 + y^2})^2}$$

$$= \frac{\sqrt{x^2 + y^2} \cdot (1) - x \cdot (x/\sqrt{x^2 + y^2})}{(x^2 + y^2)}.$$

In this expression, if we multiply numerator and denominator by $\sqrt{x^2 + y^2}$, we obtain

$$\frac{\partial^2 z}{\partial x^2} = \frac{\sqrt{x^2 + y^2} \cdot \sqrt{x^2 + y^2} - (x^2/\sqrt{x^2 + y^2}) \cdot \sqrt{x^2 + y^2}}{(x^2 + y^2)\sqrt{x^2 + y^2}}$$

$$= \frac{(x^2 + y^2) - x^2}{(x^2 + y^2)^{3/2}}$$

$$= \frac{y^2}{(x^2 + y^2)^{3/2}}.$$

In a similar way it is shown that

$$\frac{\partial^2 z}{\partial y^2} = \frac{x^2}{(x^2 + y^2)^{3/2}}.$$

We may continue this process and calculate partial derivatives of higher orders, for example,

$$\frac{\partial^3 z}{\partial x^3} = \frac{\partial}{\partial x}\left(\frac{\partial^2 z}{\partial x^2}\right), \qquad \frac{\partial^3 z}{\partial y\, \partial x^2} = \frac{\partial}{\partial y}\left(\frac{\partial^2 z}{\partial x^2}\right), \qquad \frac{\partial^3 z}{\partial x\, \partial y\, \partial x} = \frac{\partial}{\partial x}\left(\frac{\partial^2 z}{\partial y\, \partial x}\right),$$

and so on. Provided that all derivatives of the given order are continuous, the order in which the x and y differentiations are carried out is immaterial. Thus, for example, the following mixed derivatives are all equal:

$$\frac{\partial^3 z}{\partial y\, \partial x^2} = \frac{\partial^3 z}{\partial x\, \partial y\, \partial x} = \frac{\partial^3 z}{\partial x\, \partial x\, \partial y}.$$

They would be denoted by $\partial^3 z/\partial x^2 \partial y$, indicating two differentiations with respect to x and one with respect to y.

EXAMPLE   Calculate $\partial^3 z/\partial x^2\, \partial y$ and $\partial^4 z/\partial x\, \partial y^3$ for $z = x^3 y^4$.

SOLUTION $\dfrac{\partial z}{\partial y} = x^3 \cdot 4y^3 = 4x^3 y^3$

$$\frac{\partial^2 z}{\partial x\, \partial y} = \frac{\partial}{\partial x}(4x^3 y^3) = 4y^3 \cdot 3x^2 = 12x^2 y^3.$$

Therefore

$$\frac{\partial^3 z}{\partial x^2\, \partial y} = \frac{\partial}{\partial x}\left(\frac{\partial^2 z}{\partial x\, \partial y}\right) = \frac{\partial}{\partial x}(12x^2 y^3) = 24xy^3.$$

Also

$$\frac{\partial^3 z}{\partial x\, \partial y^2} = \frac{\partial}{\partial y}\left(\frac{\partial^2 z}{\partial x\, \partial y}\right) = \frac{\partial}{\partial y}(12x^2 y^3) = 36x^2 y^2,$$

and so

$$\frac{\partial^4 z}{\partial x\, \partial y^3} = \frac{\partial}{\partial y}\left(\frac{\partial^3 z}{\partial x\, \partial y^2}\right) = \frac{\partial}{\partial y}(36x^2 y^2) = 72x^2 y.$$

As with ordinary derivatives, there are several alternative notations that are used for partial derivatives. The most commonly encountered of these is the use of subscripts to indicate partial derivatives, and we shall use this notation from time to time. According to this notation

$$\frac{\partial z}{\partial x} \qquad \text{is denoted by } z_x \text{ or } f_x(x, y),$$

$$\frac{\partial z}{\partial y} \qquad \text{is denoted by } z_y \text{ or } f_y(x, y),$$

$$\frac{\partial^2 z}{\partial x^2} \qquad \text{is denoted by } z_{xx} \text{ or } f_{xx}(x, y),$$

$$\frac{\partial^2 z}{\partial x\, \partial y} \qquad \text{is denoted by } z_{xy} \text{ or } f_{xy}(x, y) \text{ (note } z_{xy} = z_{yx}),$$

$$\frac{\partial^4 z}{\partial x\, \partial y^3} \qquad \text{is denoted by } z_{xyyy} \text{ or } f_{xyyy}(x, y), \text{ and so on.}$$

The notion of partial derivatives extends in a straightforward way to functions $z = f(x_1, x_2, \ldots, x_n)$ of several variables. For example, $\partial z / \partial x_1$ is obtained by differentiating z with respect to x_1 keeping $x_2, \ldots, x_n$ all constant, and so on.

EXAMPLE   If $z = f(x_1, x_2, x_3) = x_1^2 + x_1 \sqrt{x_2^2 - x_3^2}$, find

$$\frac{\partial z}{\partial x_i} \ (i = 1, 2, 3).$$

SOLUTION

$$\frac{\partial z}{\partial x_1} = 2x_1 + \sqrt{x_2^2 - x_3^2}.$$

$$\frac{\partial z}{\partial x_2} = 0 + x_1 \left(\frac{1}{2}\right)(x_2^2 - x_3^2)^{-1/2}(2x_2)$$

$$= \frac{x_1 x_2}{\sqrt{x_2^2 - x_3^2}}.$$

$$\frac{\partial z}{\partial x_3} = 0 + x_1 \left(\frac{1}{2}\right)(x_2^2 - x_3^2)^{-1/2}(-2x_3)$$

$$= -\frac{x_1 x_3}{\sqrt{x_2^2 - x_3^2}}.$$

EXERCISES 9.4

Calculate $\partial z / \partial x$ and $\partial z / \partial y$ for the following functions.

1. $z = x^2 y^4$.

2. $z = \sqrt{xy^5}$.

3. $z = \sqrt{x - y}$.

4. $z = \sqrt{x^2 - y^2}$.

5. $z = e^{(x + 3y)}$.

6. $z = \ln(x^2 y^3)$.

7. $z = (x/y) \sin(x/y)$.

8. $z = \tan(x^2 + 2y^2)$.

9. $z = \ln \cos(x + y)$.

10. $z = \text{Sin}^{-1}(\sqrt{x^2 + y^2})$.

Calculate $\partial^2 z / \partial x^2$ and $\partial^2 z / \partial x \, \partial y$ for the following functions:

11. $z = x^{3/2} y^{-4}$.

12. $z = x^5 y^{-1/2}$.

13. $z = \ln(x + 2y)$.

14. $z = e^{(x^2 + y^2)}$.

15. If $z = \text{Sin}^{-1}(y/x)$ show that $xz_x + yz_y = 0$.

16. If $z = x \, \text{Tan}^{-1}(x/y)$ show that $xz_x + yz_y = z$.

17. If $z = x^3 + y^3$ show that $xz_x + yz_y = 3z$.

18. If $z = f(ax + by)$ show that $bz_x - az_y = 0$.

19. If a substance is injected into a vein, the concentration of the substance at any point in the vein will vary with time t and with the distance x from the point of injection. Under certain conditions the concentration can be described by a function of the form

$$C(x, t) = \frac{c}{\sqrt{t}} e^{[-x^2/(at)]},$$

where a and c are constants. Show that $C(x, t)$ satisfies the following equation

$$\frac{\partial C}{\partial t} = \left(\frac{a}{4}\right) \frac{\partial^2 C}{\partial x^2}.$$

(This equation is known as the diffusion equation.)

20. Show that the function $C = c e^{(kx+wt)}$ satisfies the diffusion equation (cf. exercise 19) provided that $w = ak^2/4$.

21. The model equation in exercise 19 for the diffusion of a substance through the bloodstream does not take account of the drift due to the motion of the blood. A better equation is

$$C(x, t) = \frac{c}{\sqrt{t}} e^{-[(x-vt)^2/at]},$$

where v is the velocity of the blood. Show that for this equation

$$\frac{\partial C}{\partial t} = \frac{a}{4} \frac{\partial^2 C}{\partial x^2} - v \frac{\partial C}{\partial x}.$$

(*Note:* This exercise is more difficult than the preceding ones.)

22. In the process of metabolism of a bacterium the rate M at which a chemical substance can be absorbed into the bacterium and distributed throughout its volume is given by $M = aS/V$, where S is the surface area, V is the volume of the bacterium, and a is a constant. For a cylindrical bacterium of radius r and length l, $V = \pi r^2 l$ and $S = 2\pi r l + 2\pi r^2$. Calculate $\partial M/\partial r$ and $\partial M/\partial l$, thus finding how an increase in radius or in length affects the rate of metabolism.

23. Repeat exercise 22 for a bacterium whose shape is a cylinder with hemispherical caps on each end.

24. The rate at which an animal's body loses heat by convection is given by $H = a(T - T_0)v^{1/3}$, where T and T_0 are the temperatures of the animal's body and of the surrounding air, v is the wind velocity, and a is a constant. Calculate $\partial H/\partial T$, $\partial H/\partial T_0$, and $\partial H/\partial v$, and interpret these quantities.

9.5 LINEAR APPROXIMATIONS

In this section we shall discuss the geometrical significance of the partial derivatives of a function of two variables. Using the ideas to be developed we shall show how functions of two variables can be approximated in small regions by linear functions.

Let (x_0, y_0, z_0) be a point on the graph of the function $z = f(x, y)$. We define the *tangent plane* at this point to be the plane that touches the graph at (x_0, y_0, z_0). The tangent plane to a surface in three dimensions is the natural analog to the tangent line to a graph in two dimensions.

If the graph of $z = f(x, y)$ has a corner or a crease at (x_0, y_0, z_0), the tangent plane will not exist. In order to have a well-defined tangent plane the surface must be *smooth* at this point. We shall therefore restrict our attention to functions whose graphs are smooth surfaces.

In Fig. 9.20 the surface of the graph $z = f(x, y)$ is indicated by various sections by planes $x = $ constant and $y = $ constant in the neighborhood of the point (x_0, y_0, z_0). The tangent plane at (x_0, y_0, z_0) is indicated by shading.

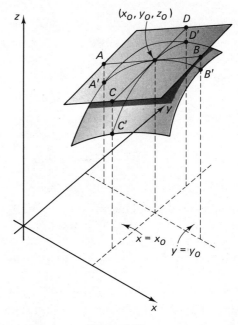

Figure 9.20

Let (x, y, z) be a general point P on the tangent plane. Since any plane is described by a linear equation, there must be certain constants a, b, c, and d such that

$$ax + by + cz = d$$

for all points P on the plane. But the point (x_0, y_0, z_0) itself lies on the tangent plane so that

$$ax_0 + by_0 + cz_0 = d.$$

Therefore

$$ax + by + cz = ax_0 + by_0 + cz_0.$$

Dividing through by c and rearranging the terms, we obtain

$$z - z_0 = p(x - x_0) + q(y - y_0),$$

where $p = -a/c$ and $q = -b/c$.

We wish to calculate the values of the constants p and q. Let AB be the line of intersection of the tangent plane and the vertical plane on which $y = y_0$ (Fig. 9.20). Any point (x, y, z) on AB must satisfy the equation

$$z - z_0 = p(x - x_0) + q(y - y_0) = p(x - x_0)$$

since $y - y_0 = 0$ on AB. That is,

$$z = px + (z_0 - px_0).$$

This equation describes a straight line in the xz-plane whose slope is equal to p. Therefore p is equal to the vertical slope of the line AB.

Consider the vertical section of the graph by the plane $y = y_0$ (the curved line $A'B'$ in Fig. 9.20). This curve must touch AB at (x_0, y_0, z_0). We observed at the beginning of Section 9.4 that the slope of the section $y = y_0$ at any point (x, y_0, z) is given by the partial derivative $\partial z / \partial x$ evaluated at that point. Therefore the slope p of AB must be equal to the value of $\partial z / \partial x$ at the point (x_0, y_0, z_0). That is,

$$p = f_x(x_0, y_0).$$

Similarly, we can consider the vertical section by the plane $x = x_0$, giving the straight line CD on the tangent plane and the curved line $C'D'$ on the surface. These two touch at (x_0, y_0, z_0). The points on CD satisfy the linear equation

$$z - z_0 = q(y - y_0)$$

so that q is equal to the vertical slope of CD. But this must be equal to the slope of the section $C'D'$ at (x_0, y_0, z_0), and so it follows that

$$q = f_y(x_0, y_0).$$

Substituting the values of p and q we obtain the result that the equation of the tangent plane at (x_0, y_0, z_0) is

$$\boxed{z - z_0 = f_x(x_0, y_0)(x - x_0) + f_y(x_0, y_0)(y - y_0).}$$

EXAMPLE   Find the equation of the tangent plane to the graph of the function $z = x^3 y$ at the point $(2, 1, 8)$.

SOLUTION   Differentiating the function $f(x, y) = x^3 y$ we obtain

$$f_x(x, y) = 3x^2 y, \qquad f_y(x, y) = x^3.$$

At the given point, $x_0 = 2$, $y_0 = 1$, $z_0 = 8$, and so

$$f_x(x_0, y_0) = 3(2^2)(1) = 12,$$
$$f_y(x_0, y_0) = 2^3 = 8.$$

The equation of the tangent plane is therefore

$$z - 8 = 12(x - 2) + 8(y - 1)$$
$$= 12x - 24 + 8y - 8,$$

and so

$$z = 12x + 8y - 24.$$

The equation of the tangent plane is therefore

$$12x + 8y - z = 24.$$

Observing that $z_0 = f(x_0, y_0)$, we can write the equation of the tangent plane at (x_0, y_0, z_0) in the form

$$z = f(x_0, y_0) + f_x(x_0, y_0)(x - x_0) + f_y(x_0, y_0)(y - y_0).$$

Now let us consider the case when the point (x, y) is very close to the point (x_0, y_0). Then the value of z on the tangent plane, given by the above equation, is approximately the same as the value of z on the graph, namely $z = f(x, y)$. Therefore, for (x, y) sufficiently close to (x_0, y_0), we have that

$$f(x, y) \simeq f(x_0, y_0) + f_x(x_0, y_0)(x - x_0) + f_y(x_0, y_0)(y - y_0).$$

We see then that the function $f(x, y)$ can be approximated by the linear function on the right-hand side of this equation provided that (x, y) is close enough to (x_0, y_0). This result is true for any function $f(x, y)$ whose graph is a smooth surface.

Let us set $x = x_0 + \Delta x$ and $y = y_0 + \Delta y$, where Δx and Δy are increments in the independent variables x and y. Then the above approximation formula takes the form

$$\boxed{\begin{aligned} f(x_0 + \Delta x, y_0 + \Delta y) &\simeq f(x_0, y_0) + f_x(x_0, y_0)\, \Delta x \\ &\quad + f_y(x_0, y_0)\, \Delta y. \end{aligned}}$$

Alternatively, we can introduce the increment Δz in the dependent variable

$$\Delta z = f(x_0 + \Delta x, y_0 + \Delta y) - f(x_0, y_0),$$

which corresponds to the increments Δx and Δy, in which case we have the approximate result that

$$\boxed{\Delta z \simeq f_x(x_0, y_0)\, \Delta x + f_y(x_0, y_0)\, \Delta y.}$$

These approximations hold provided that Δx and Δy are sufficiently small.

EXAMPLE For $f(x, y) = \sqrt{x + y} + \sqrt{x - y}$, it is readily seen that $f(10, 6) = 6$. Find the approximate value of $f(10.1, 5.8)$.

SOLUTION After partial differentiation we obtain that

$$f_x(x, y) = \tfrac{1}{2}(x + y)^{-1/2} + \tfrac{1}{2}(x - y)^{-1/2},$$
$$f_y(x, y) = \tfrac{1}{2}(x + y)^{-1/2} - \tfrac{1}{2}(x - y)^{-1/2}.$$

At the point $(x_0, y_0) = (10, 6)$, these partial derivatives have the values

$$f_x(10, 6) = \tfrac{1}{2}(10 + 6)^{-1/2} + \tfrac{1}{2}(10 - 6)^{-1/2}$$
$$= \frac{1}{2} \cdot \frac{1}{\sqrt{16}} + \frac{1}{2}\frac{1}{\sqrt{4}} = \frac{1}{8} + \frac{1}{4} = \frac{3}{8},$$
$$f_y(10, 6) = \tfrac{1}{2}(10 + 6)^{-1/2} - \tfrac{1}{2}(10 - 6)^{-1/2}$$
$$= \tfrac{1}{8} - \tfrac{1}{4} = -\tfrac{1}{8}.$$

The increments in the independent variables in this example are

$$\Delta x = 10.1 - 10 = 0.1,$$
$$\Delta y = 5.8 - 6 = -0.2.$$

Therefore

$$f(10.1, 5.8) = f(x_0 + \Delta x, y_0 + \Delta y)$$
$$\simeq f(x_0, y_0) + f_x(x_0, y_0) \Delta x + f_y(x_0, y_0) \Delta y$$
$$= f(10, 6) + f_x(10, 6)(0.1) + f_y(10, 6)(-0.2)$$
$$= 6 + (\tfrac{3}{8})(0.1) + (-\tfrac{1}{8})(-0.2)$$
$$= 6.0625.$$

[Note that the exact value of $f(10.1, 5.8)$ is equal to $\sqrt{15.9} + \sqrt{4.3} = 3.987 + 2.074 = 6.061$, so that the approximation is correct to the first two decimal places.]

EXAMPLE   The average human being's surface area S is given in terms of his height H and weight W approximately by the formula $S = 2W^{0.4}H^{0.7}$ (S in m², W in kg, H in m). A youth currently weighs 60 kg and is 155 cm in height. His weight is increasing at the rate of 5 kg/yr and his height at the rate of 4 cm/yr. What is the increment in surface area during one year?

SOLUTION $\dfrac{\partial S}{\partial W} = 0.8W^{-0.6}H^{0.7}, \qquad \dfrac{\partial S}{\partial H} = 1.4W^{0.4}H^{-0.3}.$

At the current values, $W = 60$ kg and $H = 1.55$ m, so that the values of these two partial derivatives are currently

$$\frac{\partial S}{\partial W} = 0.8(60)^{-0.6}(1.55)^{0.7} = 0.0932,$$

$$\frac{\partial S}{\partial H} = 1.4(60)^{0.4}(1.55)^{-0.3} = 6.31.$$

During one year the increments in W and H are as follows:

$$\Delta W = 5, \qquad \Delta H = 4 \text{ cm} = 0.04 \text{ m}.$$

Therefore the corresponding increment in S is given approximately by

$$\Delta S = \frac{\partial S}{\partial W} \Delta W + \frac{\partial S}{\partial H} \Delta H$$
$$= (0.0932)(5) + (6.31)(0.04)$$
$$= 0.466 + 0.252$$
$$= 0.718.$$

The approximate increase of surface area during 1 yr is therefore 0.72 m² (rounding off to two figures).

The reader may have noticed the similarity between the approximation introduced in this section and the approximation to functions of a single variable, which was introduced in Chapter 4 via the concept of differentials. In fact, these two approximations are completely analogous to one another, as the following definition makes clear.

DEFINITION For a given function $z = f(x, y)$ we define the differentials of the independent variables x and y to be arbitrary increments in these variables: $dx = \Delta x$ and $dy = \Delta y$. The differential of the dependent variable z is defined to be

$$dz = f_x(x, y)\, dx + f_y(x, y)\, dy.$$

The differential dz is in general different from the increment Δz. However, if dx and dy are sufficiently small, dz and Δz are almost equal to one another, and the preceding approximation formula for $f(x + dx, y + dy)$ is obtained by setting $dz \simeq \Delta z$.

EXERCISES 9.5

Find the equation of the tangent plane to the following functions at the given points.

1. $z = x^2y^2$; $x = 1$, $y = 1$.

2. $z = xy^{-1}$; $x = 6$, $y = 2$.

3. $z = \ln(x + y)$; $x = -1$, $y = 2$.

4. $z = xe^y$; $x = 2$, $y = \ln 2$.

5. $z = y^2 \sin x$; $x = \dfrac{\pi}{2}$, $y = -1$.

6. $z = \sqrt{x^2 + y^2}$; $x = 3$, $y = 4$.

If $f(x, y) = \sqrt{x^2 + y^2}$, find the approximate values of

7. $f(3.1, 4.1)$.

8. $f(5.1, 11.8)$.

If $f(x, y) = \sqrt{x^2 - y^2}$, find the approximate values of

9. $f(5.2, 2.9)$.

10. $f(25.1, 23.9)$.

If $f(x, y) = (x - y)/\sqrt{x + y}$, find the approximate values of

11. $f(2.1, 1.95)$.

12. $f(4.0, 5.1)$.

13. If $z = cx^ay^b$ where a, b, and c are constants, show that

$$\frac{\Delta z}{z} \simeq a\frac{\Delta x}{x} + b\frac{\Delta y}{y}$$

where Δx, Δy, and Δz are corresponding increments in x, y, and z.

14. A certain mass of gas occupies a volume V and is at temperature T (°K). Its pressure P is then given by

$$P = cTV^{-1} \quad (c = \text{constant}).$$

When $V = 1$ liter and $T = 200°$K, $P = 500$ dynes/cm². Use differentials to find the approximate value of P when $V = 950$ ml and $T = 220°$K. (*Hint:* exercise 13 can be used.)

15. At a time t after a patient is administered an amount x of a certain drug, his pulse rate rises by an amount y that is found to be given by

$$y = cxte^{-t},$$

where c is a constant. When $x = 1$ and $t = 2$, it is observed that $y = 24$. Use differentials to find the approximate value of y when $x = 0.9$ and $t = 2.2$.

16. The volume of fluid passing through a tube of radius a and length l is given by the formula

$$V = \frac{\pi p a^4}{8l\eta},$$

where p is the difference in pressure between the two ends of the tube and η is the viscosity of the fluid. By measuring V, p, a, and l, this equation is used to determine η. If there are percentage errors of $\pm 2\%$ in V, $\pm 3\%$ in p, and $\pm\frac{1}{3}\%$ in a and l, what is the maximum possible error in η?

17. The volume of a cylinder of radius r and length l is given by $V = \pi r^2 l$. If r and l are measured with errors respectively of Δr and Δl, find the resulting error ΔV in the volume.

9.6 MAXIMA AND MINIMA

We saw in Chapter 4 that one of the most important and widely applicable uses of the calculus of functions of a single variable is the calculation of the maximum and minimum values of functions. The corresponding problem, the calculation of the maxima and minima of functions of several variables, is equally important, and in this section we shall discuss it in the case of functions of two variables.

DEFINITION The function $f(x, y)$ has a *local maximum* at the point (x_0, y_0) if $f(x, y) < f(x_0, y_0)$ for all points (x, y) sufficiently close to (x_0, y_0), except for (x_0, y_0) itself.

The function $f(x, y)$ has a *local minimum* at the point (x_0, y_0) if $f(x, y) > f(x_0, y_0)$ for all points (x, y) sufficiently close to (x_0, y_0), except for (x_0, y_0) itself.

The corresponding value $f(x_0, y_0)$ is called the *local maximum value* (or *local minimum value* as the case may be) of the function f. The term *extremum* refers to both maxima and minima.

In the case of functions of one variable we discussed two types of extrema, one for which the derivative vanished and the other for which the derivative failed to exist, corresponding to a corner or a spike on the graph of the function. In this section we shall, for the sake of simplicity, restrict ourselves to the first type of extremum. We shall assume that the graph of $z = f(x, y)$ is "smooth" in the sense that a well-defined tangent plane exists at each point on the graph. The possibility that the graph has corners will not be considered. This restriction is not too serious since the vast majority of applications concern functions whose graphs are smooth.

Let the function $z = f(x, y)$ have a local maximum at (x_0, y_0). Let us construct the vertical section of the graph on which $y = y_0$, that is, the section through the local maximum point on the graph (Fig. 9.21). The equation of this section is $z = f(x, y_0)$.

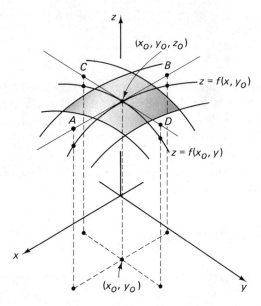

Figure 9.21

It is a curve lying in the vertical plane $y = y_0$, and along this curve, z has a local maximum value at $x = x_0$. Thus on the curve $z = f(x, y_0)$ the tangent line becomes horizontal when $x = x_0$ (the line AB in Fig. 9.21).

The slope of the section $z = f(x, y_0)$ is equal to the partial derivative $\partial z/\partial x = f_x(x, y_0)$ since this partial derivative is obtained precisely by holding y constant. Therefore the section $y = y_0$ has a horizontal tangent at (x_0, y_0) if and only if $f_x(x_0, y_0) = 0$.

Similarly the section $z = f(x_0, y)$, on which $x = x_0$, must also have a horizontal tangent at $y = y_0$ (CD in Fig. 9.21). The slope of this section is given by the partial derivative $f_y(x_0, y)$, and the fact that the slope must vanish at $y = y_0$ leads to the condition that $f_y(x_0, y_0) = 0$. This leads us to the following theorem.

THEOREM 9.6.1

If $f(x, y)$ has a local maximum or a local minimum at the point (x_0, y_0), then it is necessary that

$$f_x(x_0, y_0) = 0, \qquad f_y(x_0, y_0) = 0.$$

(The discussion for a local minimum proceeds in parallel with that given above for a local maximum.)

From Section 9.5 we know that the equation of the tangent plane at the point (x_0, y_0, z_0) to $z = f(x, y)$ is

$$z - z_0 = f_x(x_0, y_0)(x - x_0) + f_y(x_0, y_0)(y - y_0),$$

where (x, y, z) is a general point on the tangent plane. When $f_x(x_0, y_0) = f_y(x_0, y_0) = 0$, the equation of the tangent plane becomes simply $z = z_0$, that is, the plane is a horizontal one.

Another way of stating Theorem 9.6.1, therefore, is to say that a necessary condition for (x_0, y_0) to be an extremum is that the tangent plane at that point should be horizontal.

DEFINITION   A *critical point* of a smooth function $f(x, y)$ is a point (x_0, y_0) at which $f_x(x_0, y_0) = f_y(x_0, y_0) = 0$. Alternatively, we can say that a critical point is a point at which the tangent plane is horizontal.

It is clear from the preceding discussion that every local extremum of a smooth function must be a critical point. However every critical point is not an extremum, just as for functions of a single variable. We shall return to this question later.

EXAMPLE   Find the critical points of the function

$$f(x, y) = x^3 + x^2y + x - y.$$

SOLUTION   We must set the two partial derivatives f_x and f_y equal to zero:

$$f_x(x, y) = 3x^2 + 2xy + 1 = 0,$$
$$f_y(x, y) = x^2 - 1 = 0.$$

From the second of these equations it follows that $x^2 = 1$, or $x = \pm 1$. From the first equation it then follows that

$$2xy = -3x^2 - 1 = -3(1) - 1 = -4,$$

that is,

$$y = \frac{-4}{2x} = -\frac{2}{x}.$$

Thus, when $x = 1$, $y = -2$; and when $x = -1$, $y = +2$. There are therefore two critical points: $(1, -2)$ and $(-1, 2)$.

EXAMPLE   Find the critical points of the function

$$f(x, y) = x^2 - 2y + e^{x+y}.$$

SOLUTION

$$f_x(x, y) = 2x + e^{x+y} = 0,$$
$$f_y(x, y) = -2 + e^{x+y} = 0.$$

From the second equation, $e^{x+y} = 2$, and substituting this value of e^{x+y} into the first equation we obtain

$$2x = -e^{x+y} = -2, \quad \text{or} \quad x = -1.$$

Therefore

$$e^{x+y} = e^{-1+y} = 2,$$

and so, multiplying both sides by e, we get

$$e^y = 2e,$$

$$y = \ln(2e) = \ln e + \ln 2 = 1 + \ln 2.$$

In this case there is only one critical point, namely $(-1, 1 + \ln 2)$.

In the case of a function $f(x)$ of one variable, we saw in Chapter 4 that every critical point is not necessarily a local extremum. A critical point for which $f'(x) = 0$ can be either a local maximum or a local minimum, or else a point of inflection; in Chapter 4 we developed tests to distinguish between these possibilities. Similar tests are necessary in the case of a function $f(x,y)$ of two variables, since again it is true that not every critical point need be an extremum. This is illustrated by the function $z = x^2 - y^2$, whose graph was constructed as an example toward the end of Section 9.3. This function has a critical point at the origin that is neither a local maximum nor a local minimum. The vertical section of its graph by the plane $y = 0$ has a local minimum at the origin while the vertical section by the plane $x = 0$ has a local maximum at the origin. A critical point of this type is called a *saddle-point*.

If $f(x, y)$ has a local maximum at (x_0, y_0) then it is necessary that the section by the plane $y = y_0$ also have a local maximum at $y = y_0$. (This is clear from Fig. 9.21.) Thus it is necessary that $f_x(x, y_0)$ vanish at $x = x_0$, as we have seen; it is also necessary that the second partial derivative $f_{xx}(x, y_0)$ be nonpositive at $x = x_0$ since the section $y = y_0$ must be concave downwards at the local maximum point. Conversely, if $f_{xx}(x_0, y_0) < 0$, then the section by the plane $y = y_0$ has a local maximum at (x_0, y_0).

Similarly, we can see that if $f_{yy}(x_0, y_0) < 0$, then the section of the graph by the vertical plane $x = x_0$ is concave downwards and therefore has a local maximum at (x_0, y_0).

The two conditions $f_{xx} < 0$, $f_{yy} < 0$ at (x_0, y_0) are, however, not sufficient to guarantee that the surface itself has a local maximum at (x_0, y_0). They guarantee only that the vertical sections by the two coordinate planes $x = x_0$ and $y = y_0$ have local maxima at the point (x_0, y_0). It is quite possible for the sections of the graph to have local maxima on those vertical planes, yet to have a local minimum on some other vertical plane through (x_0, y_0).

EXAMPLE   Consider the function $f(x, y) = 2xy - x^2 - \frac{1}{2}y^2$.

It follows that $f_x = 2y - 2x$, $f_y = 2x - y$, and setting these derivatives equal to zero we find that the only critical point is at the origin; $x = 0$ and $y = 0$.

The second derivatives are $f_{xx} = -2 < 0$, $f_{yy} = -1 < 0$, so that the vertical sections of the graph by the planes $x = 0$ and $y = 0$ both have a local maxima at the origin. In fact, on $x = 0$, $f(0, y) = -\frac{1}{2}y^2$, so that this vertical section is a parabola opening downwards with its vertex at $y = 0$; and on $y = 0$, $f(x, 0) = -x^2$, so that this section is again a parabola opening downwards with its vertex at $x = 0$.

However, consider the values of the function on the vertical plane $x = y$: $f(x, x)$ $= 2x(x) - x^2 - \frac{1}{2}x^2 = \frac{1}{2}x^2$. Thus the section of the graph by the vertical plane $x = y$ is a parabola opening *upwards*, so that on this section the graph has a local minimum at the origin. The graph has a saddle-point at the origin, even though both f_{xx} and f_{yy} are negative. The graph of $z = 2xy - x^2 - \frac{1}{2}y^2$ is shown in Fig. 9.22. The saddle-shape at the origin is apparent.

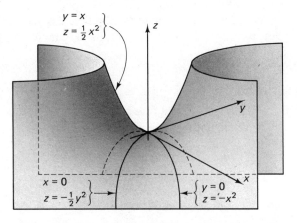

Figure 9.22

It is clear therefore that some extra condition is required in order to complete the test for a local maximum or minimum. This is provided by the following theorem, which we shall not prove here.

THEOREM 9.6.2

Let (x_0, y_0) be a critical point of the function $f(x, y)$ for which $f_x(x_0, y_0) = f_y(x_0, y_0) = 0$. Let

$$\Delta(x, y) = f_{xx}(x, y)f_{yy}(x, y) - [f_{xy}(x, y)]^2.$$

(a)   If $f_{xx}(x_0, y_0) < 0$, $f_{yy}(x_0, y_0) < 0$, and $\Delta(x_0, y_0) > 0$, then $f(x, y)$ has a local maximum at (x_0, y_0).

(b)   If $f_{xx}(x_0, y_0) > 0$, $f_{yy}(x_0, y_0) > 0$, and $\Delta(x_0, y_0) > 0$, then $f(x, y)$ has a local minimum at (x_0, y_0).

(c)   If $\Delta(x_0, y_0) < 0$, then (x_0, y_0) is not a local extremum of $f(x, y)$, but is a saddle-point.

NOTE. If $\Delta(x_0, y_0) = 0$ then this theorem cannot be used to test for a maximum or minimum.

EXAMPLE   Find the local extrema of the function

$$f(x, y) = x^2 + 2xy + 2y^2 + 2x - 2y.$$

SOLUTION   First let us find the critical points.

$$f_x = 2x + 2y + 2 = 0,$$
$$f_y = 2x + 4y - 2 = 0.$$

Solving these two simultaneous equations we find that $x = -3$ and $y = 2$. Thus $(-3, 2)$ is the only critical point.

Now let us apply the preceding theorem in order to test whether this critical point is a local maximum or a local minimum. We find upon differentiating a second time that

$$f_{xx} = 2, \qquad f_{yy} = 4, \qquad f_{xy} = 2.$$

Therefore $\Delta = f_{xx} f_{yy} - f_{xy}^2 = (2)(4) - 2^2 = 8 - 4 = 4$. So we see that $f_{xx} > 0$, $f_{yy} > 0$ and $\Delta > 0$, and thus the point $x = -3$, $y = 2$ is a local minimum of f. The local minimum value of f is

$$f(-3, 2) = (-3)^2 + 2(-3)(2) + 2(2)^2 + 2(-3) - 2(2)$$
$$= 9 - 12 + 8 - 6 - 4 = -5.$$

EXAMPLE   Find the local extrema of the function

$$f(x, y) = xe^y - 2y + \tfrac{1}{2}x^2 - 3x.$$

SOLUTION   At the critical points,

$$f_x = e^y + x - 3 = 0, \qquad f_y = xe^y - 2 = 0.$$

From the second of these equations it follows that $e^y = 2/x$, and substituting for e^y into the first equation we obtain

$$\frac{2}{x} + x - 3 = 0.$$

After multiplying through by x we obtain the quadratic equation

$$x^2 - 3x + 2 = 0,$$

whose roots are $x = 1$ and $x = 2$.

Since $e^y = 2/x$ it follows that when $x = 1$, $e^y = 2$, and therefore $y = \ln 2$; when $x = 2$, $e^y = 1$, and so $y = 0$. There are two critical points, therefore, namely $(1, \ln 2)$ and $(2, 0)$.

After differentiating a second time we find that

$$f_{xx} = 1, \qquad f_{yy} = xe^y, \qquad f_{xy} = e^y.$$

Therefore
$$\Delta = f_{xx}f_{yy} - f_{xy}^2 = (1)(xe^y) - (e^y)^2$$
$$= e^y(x - e^y).$$

When $x = 1$ and $e^y = 2$ it follows that $\Delta = -2$. Since $\Delta < 0$ the critical point $(1, \ln 2)$ must be a saddle-point. When $x = 2$ and $y = 0$ it follows that $\Delta = 1$. Since $\Delta > 0$ and also $f_{xx} = 1 > 0$ and $f_{yy} = 2 > 0$, the critical point $(2, 0)$ must be a local minimum for the given function.

EXAMPLE   The dollar value of a crop of tomatoes produced under artifical heat is given by
$$V = 25T(1 - e^{-x}) \quad \text{per unit area of ground.}$$

Here T is the maintained temperature in °C above 10°C and x the amount per unit area of fertilizer used. The cost of the fertilizer is $50x$ per unit area and the cost of heating is equal to T^2 per unit area. Find the values of x and T that maximize the profit from the crop.

SOLUTION   The profit is equal to the crop value minus the production costs involved, and therefore is equal to
$$P = 25T(1 - e^{-x}) - 50x - T^2.$$

We first find the critical points of P:
$$P_x = 25Te^{-x} - 50 = 0$$
$$P_T = 25(1 - e^{-x}) - 2T = 0.$$

From the second equation, $25e^{-x} = 25 - 2T$, so that the first equation reduces to
$$T(25 - 2T) - 50 = 0$$
or
$$2T^2 - 25T + 50 = 0.$$

The roots of this equation are $T = 2.5$ and $T = 10$. Corresponding to $T = 2.5$, $e^{-x} = (25 - 2T)/25 = \frac{4}{5}$, so $x = \ln(1.25)$. Corresponding to $T = 10$, $e^{-x} = \frac{1}{5}$, so $x = \ln 5$.

We now use Theorem 9.6.2 in order to test these two critical points. We have
$$P_{xx} = -25Te^{-x}, \qquad P_{TT} = -2, \qquad P_{xT} = 25e^{-x},$$
and so
$$\Delta = P_{xx}P_{TT} - P_{xT}^2 = -25Te^{-x}(-2) - (25e^{-x})^2$$
$$= 25e^{-x}(2T - 25e^{-x}).$$

At the critical points, therefore, since $25e^{-x} = 25 - 2T$,
$$\Delta = 25e^{-x}[2T - (25 - 2T)] = 25e^{-x}(4T - 25).$$

When $T = 2.5$ and $e^{-x} = \frac{4}{5}$, Δ is clearly negative. This critical point therefore does not correspond to an extremum. When $T = 10$ and $e^{-x} = \frac{1}{5}$, $\Delta =$

$(25)(\frac{1}{5})(40 - 25) = 75 > 0$. Since $P_{xx} < 0$ and $P_{TT} < 0$, it follows that P has a local maximum at this critical point.

The maximum value of P, attained when $T = 10$ and $x = \ln 5$, is given by

$$P = 25(10)(1 - \tfrac{1}{5}) - 50 \ln 5 - 10^2$$
$$= 100 - 50 \ln 5 = 19.6.$$

Problems also arise in which we are required to find the maximum and minimum values of a function $f(x_1, x_2, \ldots, x_n)$ of several variables. We again solve such problems by setting all of the first partial derivatives equal to zero:

$$\frac{\partial f}{\partial x_1} = \frac{\partial f}{\partial x_2} = \cdots = \frac{\partial f}{\partial x_n} = 0.$$

This provides n equations that must be solved for the variables $x_1, \ldots, x_n$. The resulting point is a critical point of f.

The test that must be applied in order to verify whether the critical point is a local maximum or local minimum or a saddle-point is more complicated than for functions of two variables, and we shall not explain it here.

EXERCISES 9.6

Find the critical values of the following functions and test whether each is a local maximum or a local minimum.

1. $f(x, y) = 2x^2 + xy + 2y^2$.

2. $f(x, y) = x^2 + 4xy + y^2$.

3. $f(x, y) = 2xy - x^2 - 3y^2 - x - 3y$.

4. $f(x, y) = x^2 + 2y^2 - 2x - 2y + 1$.

5. $f(x, y) = x^3 + 3x^2y + y^3 - y$.

6. $f(x, y) = 2xy(x + y) + x^2 + 2x$.

7. $f(x, y) = xy + \ln x + y^2$.

8. $f(x, y) = x^2 + y^2 - \ln(xy^2)$.

9. The reaction to an injection of x units of a certain drug when measured t hr after injection is given by $y = x^2(a - x)te^{-t}$. Calculate $\partial y/\partial x$ and $\partial y/\partial t$ and find the values of x and t that make the reaction maximum.

10. If in exercise 9 the reaction to the drug is given by the formula $y = x(a-x)t^{1/2}e^{-xt}$, calculate
    (a) the value of t which, for fixed x, makes y a maximum and
    (b) the values of x and t which together make y maximum.

11. Two drugs are used simultaneously in the treatment of a certain disease. The reaction R, measured in suitable units, to x units of the first drug and y units of the second is $R = x^2y^2(a - 2x - y)$. Find the values of x and y that make R a maximum.

12. A large area of land is to be sprayed by an airplane with x units of one insecticide and y units of another. Suppose that the number of pests killed is given by

$$K(x, y) = 10^7(1 - \tfrac{1}{2}e^{-x/50} - \tfrac{1}{2}e^{-y/100}).$$

Assuming that the plane can carry only 100 units total of pesticide, find the amounts of each pesticide that should be carried in order to maximize the number of pests killed.

13. The average number of apples produced per tree in an orchard in which there are n trees per acre is given by $(A - \alpha n + \beta x^{1/2})$, where A, α, and β are constants and x is the amount of fertilizer used per acre. The value of each apple is V and the cost per unit of fertilizer is F. Find the values of x and n that make the profit (i.e., value of the crop of apples less the cost of fertilizer) a maximum.

14. A lake is to be stocked with two species of fish. When there are x fish of the first species and y fish of the second species in the lake, the average weights of the fish in the two species at the end of the season are $(3 - \alpha x - \beta y)$ lb and $(4 - \beta x - 2\alpha y)$ lb respectively. Find the values of x and y that make the total weight of fish a maximum.

15. Repeat exercise 14 for the case when the average weights of the two species of fish are respectively $(5 - 2\alpha x - \beta y)$ lb and $(3 - 2\beta x - \alpha y)$ lb.

16. A fish tank is to be built with width x, length y, and depth z, sufficiently large to hold 256 ft³ of water. If the top is open, what dimensions will minimize the total area of the remaining five sides of the tank (and hence minimize the amount of material used in its construction)?

9.7 METHOD OF LEAST SQUARES

Let us suppose that an experiment is being conducted in which values of a variable y are measured for certain values of an independent variable x. Let the values of x that are used in the experiment be $x_1, x_2, x_3, \ldots, x_n$ and let the corresponding measured values of y be $y_1, y_2, y_3, \ldots, y_n$; that is, when x takes the value x_1 in the experiment, y turns out to have the value y_1, and when $x = x_2$, y is measured to be y_2, and so on. We shall assume that on theoretical grounds there is reason to believe that x and y are linearly related, $y = ax + b$, where the constants a and b are to be determined from the experiment. However, if the experimental points $(x_1, y_1), (x_2, y_2), \ldots, (x_n, y_n)$ are plotted on a graph, it is most likely that they will not fall on a straight line, due to the presence of small random errors that occur in all experimental measurements. What we must do is to draw the "best" straight line through the experimental points, that is, the straight line that passes closest to all of these points. The concept of "best" straight line has not yet been defined, and what we shall do in this section is to describe the most commonly used method of constructing this line, the so-called method of least squares.

We shall suppose that the independent variable x is measured precisely, essentially without error, while for each measured value of x the variable y involves substantial fluctuation from its theoretical value $ax + b$. Many experimental situations fall into this category. For example, let the experiment comprise measurements of the weights, y, of infants as a function of their ages, x. Then the values of x would be accurately measured while the values of y would exhibit random fluctuations from one infant to another (and also from one age to another of the same infant). Thus we shall suppose that each value x_i $(i = 1, 2, \ldots, n)$ is an accurate value of the variable x, but that the corresponding value y_i departs from its theoretical value, which would be $ax_i + b$, due either to random fluctuations or to errors in experimental measurement.

The *error* in the measurement y_i is equal to the difference $y_i - (ax_i + b)$ between the measured value and the theoretical value of y (Fig. 9.23). The *square error* is defined

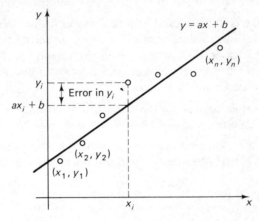

Figure 9.23

to be simply $(y_i - ax_i - b)^2$. The *mean square error* E is defined as the average over all the measurements of the various square errors. That is,

$$E = \frac{1}{n}[(y_1 - ax_1 - b)^2 + (y_2 - ax_2 - b)^2 + \ldots + (y_n - ax_n - b)^2]$$

$$= \frac{1}{n}\sum_{i=1}^{n}(y_i - ax_i - b)^2.$$

Thus, in calculating E, we form the square error from each individual measurement, add these together for all n measurements in the experiment, and then divide by n to form the average.

Of course we cannot calculate E yet, since we do not know the values of the constants a and b. According to the method of least squares, what we must do is to choose a and b in such a way that E is made minimum. Necessary conditions for this are that the two partial derivatives $\partial E/\partial a$ and $\partial E/\partial b$ be zero.

$$\frac{\partial E}{\partial a} = \frac{1}{n}\sum_{i=1}^{n}\frac{\partial}{\partial a}[(y_i - ax_i - b)^2]$$

$$= \frac{1}{n}\sum_{i=1}^{n}2(y_i - ax_i - b)(-x_i)$$

$$= \frac{2}{n}\sum_{i=1}^{n}(-x_iy_i + ax_i^2 + bx_i)$$

$$= \frac{2}{n}\left[-\sum_{i=1}^{n}x_iy_i + a\sum_{i=1}^{n}x_i^2 + b\sum_{i=1}^{n}x_i\right]$$

and

$$\frac{\partial E}{\partial b} = \frac{1}{n} \sum_{i=1}^{n} \frac{\partial}{\partial b} [(y_i - ax_i - b)^2]$$

$$= \frac{1}{n} \sum_{i=1}^{n} 2(y_i - ax_i - b)(-1)$$

$$= \frac{2}{n} \sum_{i=1}^{n} (-y_i + ax_i + b)$$

$$= \frac{2}{n} \left[-\sum_{i=1}^{n} y_i + a \sum_{i=1}^{n} x_i + nb \right].$$

Therefore, setting these two derivatives equal to zero, we obtain the two equations

$$\boxed{\begin{array}{c} a \sum x_i^2 + b \sum x_i = \sum x_i y_i \\ a \sum x_i + nb = \sum y_i \end{array}}$$

(the limits on the summation signs have been omitted for brevity).

These equations form a pair of simultaneous linear equations for the unknown constants a and b, and they can be solved in the usual way. Having calculated a and b, the best straight line through the given experimental points has equation $y = ax + b$.

EXAMPLE   The following experimental values of x and y have been obtained.

x:	1	2	3	4	5
y:	5.4	5.1	4.6	3.9	3.5

Find by the method of least squares the best straight line fitting these experimental data.

SOLUTION   When using this method it is convenient to set the calculation out in the form illustrated in the following table.

x_i	y_i	x_i^2	$x_i y_i$
1	5.4	1	5.4
2	5.1	4	10.2
3	4.6	9	13.8
4	3.9	16	15.6
5	3.5	25	17.5
$\sum$ 15	22.5	55	62.5

In the four columns of the table the values of x_i, y_i, x_i^2, and $x_i y_i$ are set out for each measurement in the experiment. The columns are also summed up. In this particular example the following values are obtained:

$$\sum x_i = 15, \qquad \sum y_i = 22.5, \qquad \sum x_i^2 = 55, \qquad \sum x_i y_i = 62.5.$$

In addition there are five measurements, so $n = 5$. The pair of simultaneous equations for a and b therefore take the form

$$55a + 15b = 62.5$$
$$15a + 5b = 22.5.$$

Eliminating b, we obtain that $10a = -5.0$, so that $a = -0.5$. Then $5b = 22.5 - 15a = 22.5 + 7.5 = 30.0$. Thus $b = 6.0$.

So the best straight line through the given points has the equation

$$y = -0.5x + 6.0.$$

It can be shown without too much difficulty that $\partial^2 E/\partial a^2 > 0$, $\partial^2 E/\partial b^2 > 0$, and (with a little more difficulty) that

$$\Delta = \frac{\partial^2 E}{\partial a^2} \cdot \frac{\partial^2 E}{\partial b^2} - \left(\frac{\partial^2 E}{\partial a \, \partial b}\right)^2 > 0,$$

so that the values of a and b found by setting $\partial E/\partial a = \partial E/\partial b = 0$ do correspond to a minimum of E.

EXERCISES 9.7

Using the method of least squares find the best straight line through the following sets of experimental data.

1.

x:	2	3	5	6	9	12
y:	3	4	6	5	7	8

2.

x:	3	4	5	6	7	8
y:	0.7	1.9	2.1	2.5	3.4	4.5

3.

x:	0	1	2	3
y:	1	1.5	2.5	3

4.

x:	2	3	4	5	6
y:	2	4	3.5	5	6.5

5. During the growth period of a certain cholera outbreak, the numbers of new cases (y) on successive days were as given in the following table (x denotes the day in question):

x:	1	2	3	4	5
y:	6	7	10	12	15

Find the best straight line through these data points and use it to "predict" how many new cases will arise on days 6 and 7. (This method of prediction is called *linear extrapolation*.)

6. A certain species of insect is gradually extending its habitat in a northerly direction. The following table gives the most northerly latitude y at which the insect was normally found during the year x.

x:	1955	1960	1965	1970	1975
y:	30°N	35°N	38°N	42°N	45°N

Find the best straight line through these data points and use it to predict when the insect will reach latitude 49°N.

9.8 CHAIN RULE

Let $z = f(u, v)$ be a function of the two variables u and v, and let u and v themselves be functions of x and y; $u = u(x, y)$, $v = v(x, y)$. Then, by substituting for u and v in $f(u, v)$, we can regard z as a function of x and y, i.e., a *composite function*:

$$z = f[u(x, y), v(x, y)].$$

The partial derivatives of z with respect to x and y can be calculated by means of a formula called the *chain rule*, which is the direct extension of the chain rule already encountered for functions of a single variable.

THEOREM 9.8.1 (Chain Rule)

If $z = f(u, v)$ and $u = u(x, y)$, $v = v(x, y)$, then

$$\frac{\partial z}{\partial x} = \frac{\partial z}{\partial u}\frac{\partial u}{\partial x} + \frac{\partial z}{\partial v}\frac{\partial v}{\partial x}$$

$$\frac{\partial z}{\partial y} = \frac{\partial z}{\partial u}\frac{\partial u}{\partial y} + \frac{\partial z}{\partial v}\frac{\partial v}{\partial y}.$$

We shall give a plausibility argument for these formulas that can be made rigorous. If Δx and Δy are small increments in x and y, then the corresponding increments in u and v (Section 9.5) are given approximately by

$$\Delta u \simeq \frac{\partial u}{\partial x}\Delta x + \frac{\partial u}{\partial y}\Delta y, \qquad \Delta v \simeq \frac{\partial v}{\partial x}\Delta x + \frac{\partial v}{\partial y}\Delta y.$$

These increments in u and v lead to an increment in z, which in turn is given approximately by

$$\Delta z \simeq \frac{\partial z}{\partial u}\Delta u + \frac{\partial z}{\partial v}\Delta v.$$

Substituting for Δu and Δv, therefore,

$$\Delta z \simeq \frac{\partial z}{\partial u}\left(\frac{\partial u}{\partial x}\Delta x + \frac{\partial u}{\partial y}\Delta y\right) + \frac{\partial z}{\partial v}\left(\frac{\partial v}{\partial x}\Delta x + \frac{\partial v}{\partial y}\Delta y\right)$$

or

$$\Delta z \simeq \left(\frac{\partial z}{\partial u}\frac{\partial u}{\partial x} + \frac{\partial z}{\partial v}\frac{\partial v}{\partial x}\right)\Delta x + \left(\frac{\partial z}{\partial u}\frac{\partial u}{\partial y} + \frac{\partial z}{\partial v}\frac{\partial v}{\partial y}\right)\Delta y.$$

But it is also true that, for small Δx and Δy, Δz must be given approximately by

$$\Delta z \simeq \frac{\partial z}{\partial x}\Delta x + \frac{\partial z}{\partial y}\Delta y.$$

Since Δx and Δy are independent increments, we must have equality of the coefficients of Δx and Δy in these two expressions for Δz. Therefore

$$\frac{\partial z}{\partial x} = \frac{\partial z}{\partial u}\frac{\partial u}{\partial x} + \frac{\partial z}{\partial v}\frac{\partial v}{\partial x}, \qquad \frac{\partial z}{\partial y} = \frac{\partial z}{\partial u}\frac{\partial u}{\partial y} + \frac{\partial z}{\partial v}\frac{\partial v}{\partial y},$$

as required.

EXAMPLE If $z = u^2 v^{1/2}$, $u = x + y$, and $v = x^2 + y^2$, calculate

$$\frac{\partial z}{\partial x} \quad \text{and} \quad \frac{\partial z}{\partial y}.$$

SOLUTION We have

$$\frac{\partial z}{\partial u} = 2uv^{1/2}, \qquad \frac{\partial z}{\partial v} = \frac{1}{2}u^2 v^{-1/2}$$

and

$$\frac{\partial u}{\partial x} = 1, \qquad \frac{\partial u}{\partial y} = 1, \qquad \frac{\partial v}{\partial x} = 2x, \qquad \frac{\partial v}{\partial y} = 2y.$$

Therefore, from the chain rule,

$$\begin{aligned}
\frac{\partial z}{\partial x} &= \frac{\partial z}{\partial u}\frac{\partial u}{\partial x} + \frac{\partial z}{\partial v}\frac{\partial v}{\partial x} \\
&= 2uv^{1/2}(1) + \tfrac{1}{2}u^2 v^{-1/2}(2x) \\
&= 2(x + y)(x^2 + y^2)^{1/2} + \tfrac{1}{2}(x + y)^2(x^2 + y^2)^{-1/2}(2x) \\
&= (x + y)(x^2 + y^2)^{-1/2}[2(x^2 + y^2) + x(x + y)] \\
&= (x + y)(x^2 + y^2)^{-1/2}(3x^2 + xy + 2y^2);
\end{aligned}$$

$$\begin{aligned}
\frac{\partial z}{\partial y} &= \frac{\partial z}{\partial u}\frac{\partial u}{\partial y} + \frac{\partial z}{\partial v}\frac{\partial v}{\partial y} \\
&= 2uv^{1/2}(1) + \tfrac{1}{2}u^2 v^{-1/2}(2y) \\
&= 2(x + y)(x^2 + y^2)^{1/2} + y(x + y)^2(x^2 + y^2)^{-1/2} \\
&= (x + y)(x^2 + y^2)^{-1/2}(2x^2 + xy + 3y^2).
\end{aligned}$$

EXAMPLE   If $z = u \ln v$ where $u = x^2 - y^2$ and $v = xy$, find

$$\frac{\partial z}{\partial x} \quad \text{and} \quad \frac{\partial z}{\partial y}.$$

SOLUTION   We have

$$\frac{\partial z}{\partial u} = \ln v, \qquad \frac{\partial z}{\partial v} = \frac{u}{v}$$

and

$$\frac{\partial u}{\partial x} = 2x, \qquad \frac{\partial u}{\partial y} = -2y, \qquad \frac{\partial v}{\partial x} = y, \qquad \frac{\partial v}{\partial y} = x.$$

Therefore

$$\frac{\partial z}{\partial x} = \ln v(2x) + \frac{u}{v}(y)$$

$$= 2x \ln(xy) + \frac{x^2 - y^2}{xy}y$$

$$= 2x \ln(xy) + \frac{x^2 - y^2}{x};$$

$$\frac{\partial z}{\partial y} = \ln v(-2y) + \frac{u}{v}(x)$$

$$= -2y \ln(xy) + \frac{x^2 - y^2}{y}.$$

EXAMPLE   In a certain geographical region, x and y are used as map coordinates measured from a certain origin O, the x-axis pointing due east and the y-axis due north (Fig. 9.24). The annual production of grain (bushels per acre) at the point $P(x, y)$ is given by

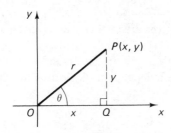

$$z = xe^{-(x^2 + 2y^2)}.$$

Let the distance OP be r and let OP make an angle θ with the x-axis. For a fixed value of θ, find the rate of change of the production z with respect to the distance r.

Figure 9.24

SOLUTION   In $\triangle OPQ$, $OQ = x$ and $PQ = y$, and therefore $x = r \cos \theta$ and $y = r \sin \theta$. (We note that r and θ are called the *polar coordinates* of the point P.) We wish to calculate the derivative $\partial z / \partial r$ (θ held fixed). From the chain rule this is given by

$$\frac{\partial z}{\partial r} = \frac{\partial z}{\partial x} \frac{\partial x}{\partial r} + \frac{\partial z}{\partial y} \frac{\partial y}{\partial r}.$$

Now

$$\frac{\partial z}{\partial x} = e^{-(x^2+2y^2)} + x(-2x)e^{-(x^2+2y^2)}$$

$$= (1 - 2x^2)e^{-(x^2+2y^2)},$$

$$\frac{\partial z}{\partial y} = x(-4y)e^{-(x^2+2y^2)}$$

$$= -4xye^{-(x^2+2y^2)},$$

$$\frac{\partial x}{\partial r} = \cos\theta, \qquad \frac{\partial y}{\partial r} = \sin\theta.$$

Therefore

$$\frac{\partial z}{\partial r} = (1 - 2x^2)e^{-(x^2+2y^2)}\cos\theta - 4xye^{-(x^2+2y^2)}\sin\theta$$

$$= e^{-(r^2\cos^2\theta + 2r^2\sin^2\theta)}[(1 - 2r^2\cos^2\theta)\cos\theta - 4(r\cos\theta)(r\sin\theta)\sin\theta]$$

$$= e^{-r^2(1+\sin^2\theta)}\cos\theta\,(1 - 2r^2 - 2r^2\sin^2\theta).$$

SPECIAL CASE. The following special case of the chain rule is often of use. Let $z = f(u, v)$ be a function of u and v as before, but now let u and v be functions of a single variable x. Then, after substitution for u and v, we obtain

$$z = f[u(x), v(x)],$$

expressing z as a composite function of the single variable x.

In this case the chain rule allows us to compute the ordinary derivative dz/dx:

$$\boxed{\frac{dz}{dx} = \frac{\partial z}{\partial u}\frac{du}{dx} + \frac{\partial z}{\partial v}\frac{dv}{dx}.}$$

(Note the italic d's here because these derivatives are ordinary derivatives, not partial derivatives.)

Let us introduce a purely fictitious variable y and regard u and v as functions of the two variables x and y. Of course they are independent of the second variable y so that

$$\frac{\partial u}{\partial y} = \frac{\partial v}{\partial y} = 0.$$

Furthermore, the partial derivatives $\partial u/\partial x$ and $\partial v/\partial x$ are identical with the ordinary derivatives du/dx and dv/dx respectively. The required result then follows directly from Theorem 9.8.1.

EXAMPLE   The pH of a certain chemical solution is given by

$$\mathrm{pH} = -\ln{(u + v)},$$

where u and v are related to the concentrations of two chemicals in the solution.

In the course of time a reaction proceeds that causes u and v to change according to the formulas

$$u = u_0 e^{-kt}, \qquad v = v_0(1 - e^{-kt}).$$

Calculate the rate of change of the pH.

SOLUTION   We have

$$\frac{\partial(\text{pH})}{\partial u} = -\frac{1}{u+v}, \qquad \frac{\partial(\text{pH})}{\partial v} = -\frac{1}{u+v}$$

and

$$\frac{du}{dt} = -ku_0 e^{-kt}, \qquad \frac{dv}{dt} = kv_0 e^{-kt}.$$

Therefore, from the chain rule,

$$\frac{d(\text{pH})}{dt} = \frac{\partial(\text{pH})}{\partial u}\frac{du}{dt} + \frac{\partial(\text{pH})}{\partial v}\frac{dv}{dt}$$

$$= \left(-\frac{1}{u+v}\right)(-ku_0 e^{-kt}) + \left(-\frac{1}{u+v}\right)(kv_0 e^{-kt})$$

$$= \frac{k(u_0 - v_0)e^{-kt}}{v_0 + (u_0 - v_0)e^{-kt}}.$$

EXAMPLE   The kinetic energy of an object moving in a two-dimensional plane is defined to be

$$K = \tfrac{1}{2}m(u^2 + v^2),$$

where m is a constant (the object's mass) and u and v are the velocities parallel to the x and y axes that are taken to lie in the plane. If at time t

$$u = \sqrt{t} + 1, \qquad v = t(\sqrt{t} + 1),$$

find the rate of change of kinetic energy.

SOLUTION   We have

$$\frac{\partial K}{\partial u} = mu, \qquad \frac{\partial K}{\partial v} = mv$$

and

$$\frac{du}{dt} = \frac{1}{2}t^{-1/2}, \qquad \frac{dv}{dt} = \frac{3}{2}t^{1/2} + 1.$$

Therefore

$$\frac{dK}{dt} = \frac{\partial K}{\partial u}\frac{du}{dt} + \frac{\partial K}{\partial v}\frac{dv}{dt}$$

$$= (mu)(\tfrac{1}{2}t^{-1/2}) + (mv)(\tfrac{3}{2}t^{1/2} + 1)$$

$$= \tfrac{1}{2}m[t^{-1/2}(\sqrt{t} + 1) + (3t^{1/2} + 2)t(\sqrt{t} + 1)]$$

$$= \tfrac{1}{2}m(\sqrt{t} + 1)(3t^{3/2} + 2t + t^{-1/2}).$$

Using the chain rule, calculate the indicated derivatives for the given functions.

1. $z = x^2 y^3$, $x = r \cos \theta$, $y = r \sin \theta$; calculate $\partial z/\partial r$ and $\partial z/\partial \theta$.

2. $z = x^2 - y^2$, $x = u(e^v + e^{-v})$, $y = u(e^v - e^{-v})$; calculate $\partial z/\partial u$ and $\partial z/\partial v$.

3. $z = x^3 \sqrt{y}$, $x = s^2$, $y = e^{s+t}$; calculate $\partial z/\partial s$ and $\partial z/\partial t$.

4. $z = x(e^y + 1)$, $x = r \cos \theta$, $y = \ln r - \theta$; calculate $\partial z/\partial r$ and $\partial z/\partial \theta$.

5. $z = x^2 + y^2$, $x = p \cos q$, $y = p \sin q$; calculate $\partial z/\partial p$ and $\partial z/\partial q$.

6. $z = u + v$, $u = x^2 y$, $v = \sqrt{x + y}$; calculate $\partial z/\partial x$ and $\partial z/\partial y$.

7. $K = \frac{1}{2}(u^2 + v^2)$, $u = 2t + 1$, $v = t^2$; calculate dK/dt.

8. $z = \sqrt{x^2 + y^2}$, $x = 2 \cos \theta$, $y = -2 \sin \theta$; calculate $dz/d\theta$.

9. $z = ye^{xy}$, $x = t - 1$, $y = t + 1$; calculate dz/dt.

10. $z = x/(x + y)$, $x = \sqrt{t}$, $y = \ln t$; calculate dz/dt.

11. By taking $z = uv$, $u = u(x)$, and $v = v(x)$, use the chain rule to show that dz/dx is given by the usual product formula of differential calculus.

12. Take $z = u/v$, $u = u(x)$, and $v = v(x)$, and use the chain rule to prove the usual quotient formula for dz/dx.

13. The trunk of a certain pine tree·can be regarded as a cylinder of radius r and height h. If the radius is equal to 4 in. and is increasing at the rate of 0.2 in./yr and if the height is 50 ft and is increasing at the rate of 4 in./yr, how fast is the volume increasing?

14. Repeat exercise 13 but with the assumption that the radius at the top of the trunk is always 1 in. less than at the bottom.

15. The cost per day to society of a certain epidemic is given by

$$c = 10g + 100m,$$

where g is the number of individuals currently infected and m is the current death rate (per day). If g and m are given as functions of time t (in days) by

$$m = kte^{-t/10}, \qquad g = 20kt^2 e^{-t/10},$$

find the value of dc/dt. Hence find the time t at which c is maximum.

16. The pressure p, volume V, and absolute temperature T of a fixed mass of gas are related according to the formula $pV = nRT$, where R is the gas constant and n the number of gram-molecules of gas. If p and T vary with time according to the formula

$$p = p_0(1 - e^{-kt}), \qquad T = T_0 + T_1 e^{-kt},$$

use the chain rule to calculate dV/dt.

REVIEW EXERCISES FOR CHAPTER 9

1. State whether the following statements are true or false. If false, give the corresponding correct statement.
   - (a) The range of a function $z = f(x, y)$ is the region in the xy-plane where the function takes its values.
   - (b) The domain of the function $z = x^2y^2$ is the set of all positive real numbers.
   - (c) The linear equation $Ax + By + Cz = 0$ has as its graph a plane that passes through the origin.
   - (d) The equation of any plane can be written in the form $z = \alpha x + \beta y + \gamma$, where α, β, and γ are constants.
   - (e) Any function $z = f(x, y)$ has as its graph a hemisphere in three dimensions that is centered at the origin.
   - (f) If $z = f(x + y)$ is any function in which x and y occur only in the combination $(x + y)$ [for example, $z = (x + y)^2 \cos(x + y)$], then $\partial z/\partial x = \partial z/\partial y$.
   - (g) If $z = f(x, y)$ and if $\partial z/\partial x = 0$, then z is independent of x, and we can write $z = f(y)$, a function of y only.
   - (h) If the second-order partial derivatives of $f(x, y)$ are all continuous, then $\partial^3 f/\partial x^2 \, \partial y = \partial^3 f/\partial y \, \partial x^2$.
   - (i) $(\partial/\partial x)(x^3y^2) = (x^4/4)y^2$.
   - (j) $(\partial/\partial y)(x^2/y) = x^2 \ln y$.
   - (k) The function $f(x, y)$ has a maximum whenever $f_x(x, y) = 0$, $f_y(x, y) = 0$, $f_{xx}(x, y) < 0$, and $f_{yy}(x, y) < 0$.
   - (l) If $z = f(x, y)$ then $\partial x/\partial z = (\partial z/\partial x)^{-1}$.

   Give the domains of the following functions:

2. $z = \sqrt{x(y + 1)}$.

3. $z = y^{-1} \ln(x + y)$.

4. $z = \sqrt{1 - x_1^2 - x_2^2 - x_3^2}$.

5. $z = \dfrac{\ln(x_1 + x_2 + x_3)}{\sqrt{x_1 - x_2}}$.

Find the equations of the planes such that the foot of the perpendicular from the origin onto the plane is as follows:

6. $(3, -1, -1)$.

7. $(-2, -1, 0)$.

Find the equations of the planes that meet the three coordinate axes in the following points:

8. $(2, 0, 0) \quad (0, -1, 0) \quad (0, 0, 1)$.

9. $(-3, 0, 0) \quad (0, 2, 0) \quad (0, 0, -4)$.

Evaluate $\partial z/\partial x$, $\partial z/\partial y$, $\partial^2 z/\partial x \, \partial y$, and $\partial^2 z/\partial y^2$ for the following functions:

10. $z = \sqrt{x}/y$.

11. $z = xy(x + y)$.

12. $z = xe^{x+y}$.

13. $z = \sin(x^2 + y^2)$.

Find the critical points of the following functions and verify whether they are maxima or minima:

14. $(x^2 - y^2)e^{-x}$.

15. $x^3 - 3xy^2 + 2y^3 - 3x^2 - 6xy + 6y^2$.

16. The surface area of the average human being of height H and weight W is $S = cH^{0.7}W^{0.4}$, where c is some constant. Show that increments $\Delta H, \Delta W$ lead to an increment ΔS where

$$\frac{\Delta S}{S} \approx 0.7\frac{\Delta H}{H} + 0.4\frac{\Delta W}{W}.$$

When $H = 1.75$ m and $W = 90$ kg, $S = 17.9$ m². Find the approximate value of S when $H = 1.65$ m and $W = 85$ kg.

17. For the function $z = (0.105)e^{x^2/y}$, we have that $z = 1$ when $x = 3$ and $y = 2$. Find the approximate value of z when $x = 3.06$ and $y = 1.92$.

18. Three spheres are to be made of materials of densities 1, 2, and 3 gm/cm³ such that their total weight is 10 gm. Find the radii of the spheres for which the sum of their three surface areas is a minimum.

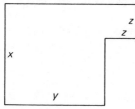

19. A field is to be enclosed by fencing in the shape shown in the figure. The total area to be enclosed is 8000 yd.² What dimensions for the field will use the minimum length of fencing for the perimeter?

20. Repeat exercise 19 for the case when the top side of the field as shown in the figure can make use of existing fencing and does not have to be constructed anew.

Use the chain rule to calculate $\partial z/\partial x$ and $\partial z/\partial y$ in the following cases:

21. $z = u^2 e^{u+v}$, $u = x^2 + y^2$, $v = 2xy$.

22. $z = re^{i\theta}$, $r = (x^2 + y^2)^{1/2}$, $\theta = \text{Tan}^{-1}(y/x)$.

23. $z = uv$, $x = u/v$, $y = u^2 + v^2$.

24. Find the least squares straight line through the following experimental data points:

x:	0	1	2	3	4
y:	9.9	8.2	6.1	3.7	2.4

25. It is desired to fit a set of experimental data $\{(x_i, y_i)\}$ with a quadratic function, $y = ax^2 + bx + c$. We define the mean square error to be

$$E = \frac{1}{n}\sum_{i=1}^{n}(y_i - ax_i^2 - bx_i - c)^2.$$

By setting $\partial E/\partial a$, $\partial E/\partial b$, and $\partial E/\partial c$ equal to zero, find three equations for a, b, and c. Evaluate a, b, and c from these equations for the experimental data given in exercise 24.

26. The maximum amount y of a certain substance that will dissolve in 1 liter of water depends on the temperature T. The following experimental results are obtained:

T (°C):	10	20	30	40	50	60
y (gm):	120	132	142	155	169	179

Find the best straight line fit to this data.

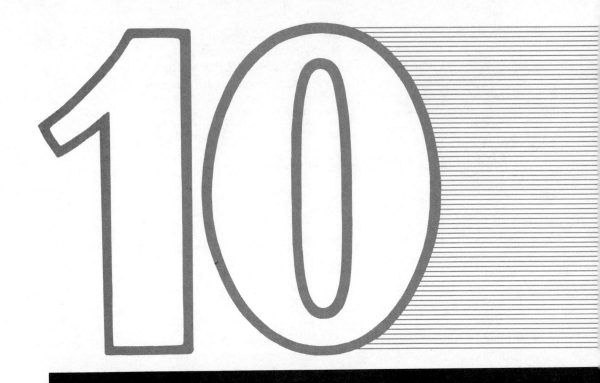

10

Differential Equations

10.1 INTRODUCTORY REMARKS

In this chapter we shall be concerned with certain mathematical techniques that can be used, among other things, to investigate biological or chemical processes that take place over time. The process might be, for example, the growth or decline of a biological population, the spread of an epidemic through a population, the progress of a chemical reaction, and so on. In such cases the independent variable is time, denoted by t, and the state of the system under consideration at any time is assumed to be described by a variable y that is a function of t. For example, y might be the size at time t of the population under consideration, or in the case of a chemical reaction, y might be the amount of some end-product that has been formed at time t.

In more complex phenomena it may be necessary to introduce more than one function of t in order to describe the state at time t. An example of this would be a chemical reaction or chain of reactions that involve the formation of several end-products. In such a case we may need to introduce as variables the amounts present at time t of each of these products. A second example arises when discussing the growth of two or more populations that interact with one another. In this example we need to use the sizes of all of the relevant populations as functions of time t in order to specify the state of the ecosystem as a whole. We shall introduce the student to such problems involving several variables in a later section, but in the initial sections of this chapter we shall stick to the case when only one function of time, $y = f(t)$, is needed in order to describe the process.

When investigating processes of the types mentioned it is often considered desirable to construct what is called a "mathematical model" of the process in question. This term is one that in recent years has come into vogue to mean what used to be called a "mathematical theory" of a process. A model for a particular process is obtained, first of all, by choosing the variable or variables that are to be used to describe the states of the system, and then by writing down equations that must be satisfied by these variables. Normally one would seek to base such equations on certain fundamental conceptual principles or empirical laws that underlie processes of the type under consideration. In order to have a complete mathematical model we must write down just enough equations so that all of the variables that describe the process are determined for each value of time.

Usually, when constructing a mathematical model, and this is especially true in the life sciences where phenomena are usually very complex, it is necessary to make a number of idealizations and simplifications. It is for this reason that the term "mathematical model" is now favored over the older term "mathematical theory," which endows the equations with a more respectable status than they commonly merit. When using a mathematical model it is necessary continually to bear in mind the assumptions and approximations that have gone into the construction of the equations and to be aware of the limitations on the applicability of those equations.

In an empirical science, the proof of any pudding must be in the eating, which is to say that any model is only good insofar as its predictions agree with the experiment.

No matter how aesthetically pleasing it may be, a set of equations must be discarded, or at least modified, if their solution does not match the corresponding experimental results. Having said that, however, it is equally true that a great deal is often gained by developing abstract principles and laws in terms of which empirical data can be "explained." This is so because these abstract concepts often underlie a range of different experimental situations, so that from just a few basic concepts it is often possible to bring together empirical results from a number of different areas of investigation, sometimes areas that are seemingly quite diverse.

Thus there is considerable value in attempting to construct mathematical models of biological phenomena, although one must always be questioning the validity of the application of the model to the phenomenon in question.

In this chapter on differential equations we shall take an approach partially oriented toward model-building. We shall usually develop each equation with reference to one or more particular biological problems, and often we shall discuss the reasonableness of the assumptions underlying the equation in the context of the problem or problems at hand. We hope that in this way we shall help the student to develop some model-building ability of his own as well, of course, as to learn how to solve the particular differential equations considered.

For a process described by a single variable y as a function of t, the rate at which the process is going on is given by the derivative dy/dt. Many mathematical models of processes lead to an equation expressing this rate as a function of time and of the current value of y, and possibly also of other external variables. For example, if y denotes the size of a population, such an equation specifies that the rate of growth of the population, dy/dt, is given by a certain function of its existing size y and also of time t. In addition the growth rate may also depend on external variables such as amount of food supply, prevalence of disease, weather conditions, and so on, and these can be included in the equation if necessary.

An equation of this type is a typical example of a *differential equation*, that is, an equation that involves the derivatives of the unknown function as well as the unknown function itself. Mathematical models of biological processes commonly lead to such differential equations, although sometimes the equations not only involve the first derivative dy/dt but also higher derivatives as well. Let us give some more formal definitions.

DEFINITION   Let $y = f(t)$ be a differentiable function of the independent variable t and let $y', y'', \ldots, y^{(n)}$ denote the derivatives of y with respect to t of orders up to and including n. Then a *differential equation of order n* for the function y is an equation relating the variables $t, y, y', \ldots, y^{(n)}$. The *order n* is the order of the highest derivative that occurs in the differential equation.

EXAMPLES   (i) $(dy/dt) + 2y = 0$ is a first-order differential equation.

(ii) $(d^2y/dt^2) - \sin(ty) = 0$ is a second-order differential equation.

(iii) $(d^4y/dt^4) - t^2(d^3y/dt^3) = \cos t$ is a fourth-order differential equation.

DEFINITION   A differential equation is said to be *linear* if the dependent variable y and all of its derivatives appear linearly in the equation. Otherwise it is said to be a *nonlinear* differential equation.

EXAMPLES   **(i)** Among the previous examples, the first and third differential equations are linear. The middle one is not linear, however, because y appears in the term $\sin(ty)$, which is not a linear function of y.

(ii) $(d^2y/dt^2) = 3(dy/dt)^2$ is a nonlinear, second-order differential equation.

(iii) $(d^2y/dt^2) = 3t^2(dy/dt)$ is a linear, second-order differential equation. Note that y and its derivatives appear linearly. The fact that the independent variable t occurs as a factor t^2 does not make the equation nonlinear.

DEFINITION   A linear differential equation is referred to as an *equation with constant coefficients* if the coefficients that multiply y and its derivatives are independent of the variable t.

EXAMPLE   **(i)** $2(dy/dt) + 5y = \sin t$ is a linear first-order differential equation with constant coefficients.

(ii) $(d^4y/dt^2) - 2(d^2y/dt^2) + 2y = e^t$ is a linear fourth-order differential equation with constant coefficients.

(iii) $(d^2y/dt^2) + t^2y = 1$ is a linear second-order differential equation, but *not* with constant coefficients since the coefficient of y is t^2, which varies with t.

DEFINITION   A function $y(t)$ is said to be a *solution* of a differential equation if, upon substituting $y(t)$ and its derivatives into the differential equation, this latter equation is satisfied for all values of t.

EXAMPLES   **(i)** The function $y = t^2$ is a solution of the differential equation $t(dy/dt) - 2y = 0$. This is so because $(dy/dt) = 2t$, and so

$$t(dy/dt) = t \cdot 2t = 2t^2 = 2y.$$

(ii) The function $y = \sin kt$, where k is a constant, is a solution of the differential equation $(d^2y/dt^2) + k^2y = 0$.

$$(dy/dt) = k \cos kt, \quad \text{and} \quad (d^2y/dt^2) = -k^2 \sin kt = -k^2y.$$

(iii) The function $y = 2 \ln t$ is a solution of the differential equation $(d^2y/dt^2) + \frac{1}{2}(dy/dt)^2 = 0$.

$$(dy/dt) = (2/t) \text{ and } (d^2y/dt^2) = -(2/t^2),$$

and so

$$(d^2y/dt^2) + \frac{1}{2}(dy/dt)^2 = -(2/t^2) + \frac{1}{2}(2/t)^2 = 0.$$

Usually, when presented with a certain differential equation, we wish to find all of its solutions. In the remainder of this first section we shall indicate how this can be done for the simplest type of differential equation, that is, for equations of the form

$$\frac{dy}{dt} = f(t),$$

where $f(t)$ is a given function of t.

By regarding the derivative dy/dt as the ratio of the differentials dy and dt, the above differential equation can be written in the form

$$dy = f(t)\,dt.$$

Therefore, integrating, we find

$$\int dy = \int f(t)\,dt$$

or

$$y = \int f(t)\,dt + C,$$

where $\int f(t)\,dt$ denotes any antiderivative of $f(t)$ and C is a constant of integration.

EXAMPLE   Find the solutions of the differential equations

(i) $(dy/dt) = t^2$.   **(ii)** $(dy/dt) = \ln t$.

SOLUTION   **(i)** $y = \int t^2\,dt + C = \tfrac{1}{3}t^3 + C$.

(ii) $y = \int \ln t\,dt + C = t \ln t - t + C$,

where we have used formula 120 in Appendix II. It can readily be verified that in each of these two cases, the function determined does form a solution of the differential equation given.

We observe from these examples that the solution of the differential equations involved is not completely determined, but involves a constant of integration C, which can have any arbitrary value. This is in fact a general feature that is found for all first-order differential equations that are in a certain sense well-behaved. The solution of such equations always involves one undetermined constant. This constant must in practice be fixed by specifying the value of the function y for one specific value of t, in addition to giving the differential equation itself.

EXAMPLE   The growth rate of a certain population of bacteria is given as a function of t by $2/(t + 2)$ mg/h. If the size of the population is 3 mg when $t = 0$, calculate the weight at a general time t.

SOLUTION   Let y be the size of the population in milligrams. Then we are given that the rate of growth is

$$\frac{dy}{dt} = \frac{2}{t + 2}.$$

Therefore

$$y = \int \frac{2}{t+2} dt + C$$
$$= 2 \ln (t + 2) + C,$$

after using formula 6 in Appendix II. The value of the constant C can be determined by using the additional information that when $t = 0$, $y = 3$. Setting $t = 0$ in the above solution, we find that

$$y = 2 \ln (0 + 2) + C = 2 \ln 2 + C = 3.$$

Therefore

$$C = 3 - 2 \ln 2 = 3 - 2(0.693) = 1.614.$$

The solution is therefore

$$y = 2 \ln (t + 2) + 1.614.$$

EXERCISES 10.1

Classify the following differential equations as to their order, linearity or non-linearity, and as to whether they have constant coefficients.

1. $\frac{dy}{dt} - 3y = t^2$.

2. $2\frac{dy}{dt} + 5y - e^t = 0$.

3. $\frac{d^2y}{dt^2} + y = t^3$.

4. $\frac{d^2y}{dt^2} + \frac{dy}{dt} - 4y = 0$.

5. $\frac{d^2y}{dt^2} + t\frac{dy}{dt} = 1$.

6. $t\frac{d^2y}{dt^2} + 5y = t - 3$.

7. $\frac{dy}{dt} + \frac{t}{y} = -y$.

8. $\frac{dy}{dt} - \frac{y}{t} = -t$.

9. $\frac{d^3y}{dt^3} - \frac{1}{t}\frac{dy}{dt} + y = 0$.

10. $\frac{d^4y}{dt^4} + t^3y = 0$.

11. $t = \ln \left(\frac{dy}{dt} - 3y\right)$.

12. $t = \sin \left(\frac{d^4y}{dt^4} + \frac{d^3y}{dt^3}\right)$.

Show that the functions given below satisfy the stated differential equations.

13. $y = t^{-4}$; $t\frac{dy}{dt} + 4y = 0$.

14. $y = \cos 3t$; $\frac{d^2y}{dt^2} + 9y = 0$.

15. $y = \tan t$; $\frac{dy}{dt} - y^2 = 1$.

16. $y = t^3 + 2t^{1/2}$; $2t^2\frac{d^2y}{dt^2} - 5t\frac{dy}{dt} + 3y = 0$.

Find the solutions of the following differential equations that satisfy the given additional conditions.

17. $\dfrac{dy}{dt} = t^{1/2}$; $y = 1$ when $t = 0$.   **18.** $\dfrac{dy}{dt} = te^t$; $y = 1$ when $t = 0$.

19. $\dfrac{dy}{dt} = t \cos t$; $y = 3$ when $t = \dfrac{\pi}{2}$.   **20.** $\dfrac{dy}{dt} = t^2 + t^{-1}$; $y = 2$ when $t = 1$.

21. A bacteria population is growing in such a way that the growth rate at time t (measured in hours) is equal to $1000\,(1 + 3t)^{-1}$. If the population size is 1000 at $t = 0$, what will be its size after 4 hr?

22. A population of insects grows from an initial size of 3000 to a size $p(t)$ after time t (measured in days). If the growth rate at time t is $5(t + 2t^2)$, determine $p(t)$ and $p(10)$.

23. The rate of change of the concentration $c(t)$ at time t of a radioactive tracer is $c'(t) = 5^{-t/2}$, where t is measured in hours and c in micrograms per liter. If the initial concentration is 1 μgm/liter, determine $c(t)$.

24. The rate of change of a population of rabbits was observed to be given by $dy/dt = 2^{(t/2)}$, where t is measured in months. Find the change in the population size between $t = 4$ and $t = 5$.

25. The population of a city is growing at the rate of $5000e^{t/10}$ (t is measured in years). If at $t = 0$ the size of the population is 200,000, find the size for any subsequent time.

26. During the course of an epidemic the rate of arrival of new cases at a certain hospital is equal to $5te^{-(t/10)}$, where t is measured in days, $t = 0$ being the start of the epidemic. How many cases has the hospital handled in total when $t = 5$ and when $t = 10$?

10.2 EXPONENTIAL FUNCTIONS REVISITED

In this section we shall study the two differential equations

$$\frac{dy}{dt} = ky \quad \text{and} \quad \frac{dy}{dt} = -ky,$$

where k is some positive constant. We shall begin by finding the general solution of the first of these equations, by which we mean the most general function $y = f(t)$ that satisfies this differential equation.

As in the last section, let us regard dy/dt as the ratio of the differentials dy and dt. Then, multiplying through the differential equation by dt, we can write it in the form

$$dy = ky\, dt$$

or

$$\frac{dy}{y} = k\, dt.$$

Let us now integrate both sides of this equation:

$$\int \frac{dy}{y} = \int k \; dt.$$

After using standard antiderivatives we get

$$\ln y = kt + C,$$

where C is the constant of integration. [Here it is assumed that y is positive. If y is negative, the left-hand side should be replaced by $\ln (-y)$.]

If we set $C = \ln c$, the two logarithms can be combined in the form

$$\ln \frac{y}{c} = kt.$$

Taking exponentials of both sides we therefore obtain

$$\frac{y}{c} = e^{kt}, \qquad y = ce^{kt}.$$

We note that the constant of integration C is an arbitrary constant. Since c is defined by the equation $C = \ln c$, it follows that c is also an arbitrary constant, but c must be positive. Thus the solution of the given differential equation has turned out not to be completely determined, but rather to contain an unknown and arbitrary constant. (If y is negative, it is readily seen that the solution can be written in the same form, $y = ce^{kt}$, but now $c = -e^C$ so that c is an arbitrary negative constant.)

As we remarked in Section 10.1, this is in fact a common feature of the solution of all first-order differential equations, with few exceptions. Apart from these special situations, the general solution of a first-order equation is always found to involve an undetermined constant. In order to fix this constant we need further information about the solution. As a rule this extra information consists of the value of y being given for one particular value of t.

Consider the above solution $y = ce^{kt}$, for example, and let us suppose that when $t = t_0$, the value of y is given to be equal to y_0. Then it must be true that

$$y_0 = ce^{kt_0}.$$

Therefore

$$c = y_0 e^{-kt_0},$$

and we see that the value of c is determined.

The simplest case occurs when the value of y is given equal to y_0 at $t = 0$ (that is, $t_0 = 0$). Then simply $c = y_0$, and the solution becomes $y = y_0 e^{kt}$.

EXAMPLE   Find the solution of the differential equation

$$\frac{dy}{dt} = 3y$$

that satisfies the condition that $y = 2$ when $t = 1$.

SOLUTION   In this equation k has the value of 3. Therefore the general solution has the form $y = ce^{3t}$. Setting $t = 1$ we know that y has the value 2, and so

$$2 = ce^{3(1)} = ce^3.$$

Therefore

$$c = 2e^{-3},$$

and the solution takes the form

$$y = 2e^{-3}e^{3t} = 2e^{3(t-1)}.$$

In the second of these expressions, we have used the usual rules of exponents in order to combine the two exponentials.

The differential equation $(dy/dt) = ky$ arises in simple models of population growth. Consider for example a population of bacteria growing in the laboratory in an environment in which there is a plentiful supply of nutrient. During a small interval of time from t to $t + \Delta t$ the population size increases, let us say from y to $y + \Delta y$, The ratio $\Delta y/y$ represents the proportion of the population of bacterial cells that have divided during the time interval Δt. It is reasonable to expect that the cells of such a population divide at an even rate, and therefore the proportion that divides during a small interval Δt should be proportional to Δt. That is,

$$\frac{\Delta y}{y} = k \, \Delta t$$

where k is the constant of proportionality. k is called the *specific growth rate* of the population. Thus we have

$$\frac{\Delta y}{\Delta t} = ky.$$

Strictly speaking, this equation will only be approximately true for finite intervals Δt. In order for it to be exactly true we must take Δt to be infinitesimally small. In the limit as $\Delta t \rightarrow 0$, we therefore obtain the differential equation

$$\lim_{\Delta t \to 0} \frac{\Delta y}{\Delta t} = ky, \qquad \frac{dy}{dt} = ky.$$

The solution of this equation is the exponential function, and we have seen in Sections 3.1 and 3.5 that exponential functions are applicable to a number of problems involving population growth. It can be seen now that the basic reason for this lies in the fact that the growth of such populations follows the differential equation $(dy/dt) = ky$. We should emphasize however that this differential equation is too simplistic to apply to any but the most elementary situations as regards the growth of real populations. As a rule, more complex differential equations are required, and in later sections of this chapter we shall see how more complicated phenomena can be incorporated into the mathematical model.

One such phenomenon that is relatively easy to include is the combination of the birth and death processes. Consider a population of a certain species of animal such that during a small time interval from t to $t + \Delta t$ a fraction $B \, \Delta t$ of the population give birth to young animals. If the number of surviving young produced by each of these adults is, on the average, equal to n, then we see that the total increase in the population during the given time interval is equal to $nyB \, \Delta t$.

Furthermore, let us assume that during this interval a fraction $D \, \Delta t$ of the population dies, where D is a constant called the *specific death rate*. The total number of deaths is then $D \, \Delta t \cdot y$. Therefore we have the equation

$$\text{Increase in population} = \text{Number of births} - \text{Number of deaths},$$

i.e., $\quad \Delta y = nyB \, \Delta t - yD \, \Delta t.$

Therefore

$$\frac{\Delta y}{\Delta t} = (nB - D)y.$$

Taking the limit as $\Delta t \longrightarrow 0$ we arrive at the differential equation $(dy/dt) = ky$, where $k = nB - D$.

Therefore we see that the inclusion of both birth and death processes does not change the type of differential equation. The solution is again of simple exponential type.

A differential equation that is closely related to the one we have just been studying is the equation

$$\frac{dy}{dt} = -ky.$$

In fact this is exactly the same type of equation as before, with the constant k simply being replaced by $-k$. The solution therefore is given by

$$y = ce^{-kt},$$

where c is an unknown and arbitrary constant. The value of c is determined from some given value of y, for example, if $y = y_0$ when $t = 0$, then $c = y_0$.

The solution of this differential equation has the form of a decaying exponential. As mentioned in Section 3.5, a number of processes found in nature exhibit a decaying exponential behavior, perhaps the most important of which is the phenomenon of radioactivity. Let us consider further the question of why radioactive decay should follow an exponential decay law.

Suppose that at time t an amount y of radioactive material remains undecayed. Let t increase by a small increment to $t + \Delta t$ and let y change at the same time to $y + \Delta y$. (Note that Δy is in fact negative since y decreases as time increases.) Now radioactive decay is a random process, each nucleus of radioactive material having the same probability of decaying during the interval Δt. The more radioactive material there is, the more of it will decay. In other words, the amount of decay $(-\Delta y)$ must be proportional to the amount of material there is left to decay: $(-\Delta y) \propto y$. Also, for small increments Δt, we can expect that the amount of decay is proportional to the length of the time interval: $(-\Delta y) \propto \Delta t$. Thus, bringing these two proportionalities together, we have

$$-\Delta y = ky \, \Delta t,$$

where k is a certain constant, called the decay constant.

Thus $(\Delta y/\Delta t) = -ky$. Strictly speaking, this equation holds only for an infinitesimally small increment Δt, so we should take the limit as $\Delta t \to 0$:

$$\lim_{\Delta t \to 0} \frac{\Delta y}{\Delta t} = -ky, \qquad \frac{dy}{dt} = -ky.$$

It is clear then that the reason why radioactive decay obeys the law of exponential decay derives from the more fundamental fact that this process satisfies the differential equation $(dy/dt) = -ky$.

Another wide class of problems in which this differential equation finds application arises in investigations of the survival of populations that are subjected to stress. Let us consider the radiation treatment of cancer as an example. In this case the population under investigation is the set of living cancer cells in the patient who is undergoing treatment, while the stress that is imposed on this population is, of course, the radiation.

Let x denote the amount of radiation to which the population is subjected, measured in rads. Let y denote the number of cancer cells that survive the treatment. Then, of course, y is a function of x; moreover, it must be a decreasing function since the more radiation the more cells will be killed. When $x = 0$ (no radiation), $y = y_0$, the initial number of cancer cells in the patient.

Let us assume that each additional small increment Δx of radiation kills a certain fraction of the remaining cancer cells, this fraction being proportional to the increment of radiation. Then if y changes to $y + \Delta y$ (again $\Delta y < 0$) when x increases to $x + \Delta x$, the proportion of the remaining cells that are killed is equal to $(-\Delta y)/y$, and by hypothesis this is proportional to Δx:

$$\frac{(-\Delta y)}{y} \propto \Delta x, \qquad \frac{-\Delta y}{y} = k\,\Delta x,$$

where k is the constant of proportionality. Hence again we have $\Delta y/\Delta x = -ky$ or, in the limit as $\Delta x \to 0$, we obtain the differential equation

$$\frac{dy}{dx} = -ky.$$

The solution is

$$y = ce^{-kx}.$$

The constant c is in fact given by $c = y_0$ since $y = y_0$ when $x = 0$.

EXAMPLE   If a dose of 0.1 rads is sufficient to kill 50 % of the cells of a certain type of cancer, how much radiation must be given to kill 99 % of the cells?

SOLUTION   When $x = 0.1$, then $y = 0.5\, y_0$ since half of the cells survive a radiation dose of 0.1 rads. Thus, since

$$y = y_0 e^{-kx},$$

it follows that

$$0.5\, y_0 = y_0 e^{-k(0.1)}$$

$$0.5 = e^{-k(0.1)}$$

$$-k(0.1) = \ln{(0.5)} = -\ln 2 = -0.6931.$$

Therefore

$$k = 6.931.$$

If 99 % of the cells are to be killed, the amount surviving would be $y = 0.01y_0$. Therefore the required dose of radiation x is given by the equation

$$y = 0.01y_0 = y_0 e^{-kx}.$$

Thus

$$e^{-kx} = 0.01$$

$$kx = -\ln(0.01) = \ln(100) = 4.605.$$

Therefore

$$x = \frac{4.605}{6.931} = 0.664 \text{ rads.}$$

Questions of survival of populations arise in many other contexts. For example, the survival of a population of insect pests that are subjected to insecticide is an example, like the treatment of cancer, in which the desired aim of the investigation is to make the proportion of surviving population sufficiently small. Other examples fall into the opposite category. For example, we may be concerned with the proportion of a deer population that survive the winter as a function of the harshness of the winter, or with the proportion of a fish population that will survive varying degrees of environmental stress brought about by certain types of industrial development, or with the survival of a human population that is subject to stress of some kind, such as an influenza epidemic or, more drastically, a famine.

Many of these examples, and others not mentioned, are rather complex phenomena, and the simple exponential model does not apply. One problem that immediately arises in many cases is the question of how to measure stress in a quantitative way. Consider for instance a population of deer subjected to the rigors of winter. There are many variables that must be considered as having a bearing on the overall harshness of the winter, such as the average temperature, the length of the winter, the average snowfall, the number of severe storms, and so on. To extract from these various factors a single measure of harshness that correctly weights the significance of each factor is no easy matter.

Even in cases where a clear-cut definition of a quantitative measure of stress can be given, the exponential law may not apply, or may only apply within limits. The basic point of view of the exponential model is that individuals in the population are either completely healthy or are dead, and that each increment of stress kills off a definite proportion of the remaining healthy individuals. But this would quite obviously not apply in many cases since, besides killing off a certain number $(y_0 - y)$ of the original population, the existing stress level x will often have had a weakening effect on the remaining y members. Thus an increment Δx of stress can be expected to have a more disastrous effect than it would have had on the initial healthy population. We can incorporate this kind of effect into the model by allowing the "constant" of proportionality k to increase with the existing stress level. Then the fraction $(-\Delta y)/y$ killed by an increment of stress Δx is equal to $k(x)\,\Delta x$, still proportional to Δx but with an increasing constant of proportionality $k(x)$. Thus we obtain the more general differential equation

$$\frac{dy}{dx} = -k(x)y.$$

We shall examine differential equations of this type in Section 10.3.

Another assumption of the exponential model is that all individuals in the original population are equally susceptible to the effects of stress. This may be a reasonable assumption in the case of a population of cancer cells but would not, for example, be reasonable in the case of a human population subjected to the stress of an epidemic. In this latter case there would be considerable variability among members of the population as regards their susceptibility to the disease.

We mention these points as an indication of some of the pitfalls in model-building. One must be continually aware of the assumptions that go into a particular model and must be prepared to question the applicability of those assumptions to the situation under investigation. The advantage of pursuing models such as exponential growth and decay back to the differential equations upon which they are based is that the assumptions are usually more evident from the differential equation. And further, it is usually easier to see how to modify the assumptions at the differential equation level in cases where they turn out to be inadequate.

A number of other applications of the differential equations $dy/dx = \pm kx$ to problems in biological areas will be found among the following exercises.

EXERCISES 10.2

Find the solutions of the following differential equations that satisfy the given conditions.

1. $\dfrac{dy}{dt} - y = 0$; $y = 2$ when $t = 0$.   2. $\dfrac{dy}{dt} - 4y = 0$; $y = \dfrac{1}{2}$ when $t = \dfrac{1}{2}$.

3. $\dfrac{dy}{dt} + 2y = 0$; $y = 1$ when $t = 1$.   4. $2\dfrac{dy}{dt} + y = 0$; $y = 3$ when $t = 4$.

5. $2\dfrac{dy}{dt} - y = 0$; $y = 3$ when $t = \dfrac{1}{4}$.   6. $\dfrac{dy}{dt} - 3y = 0$; $y = -1$ when $t = 0$.

7. A bacteria population grows in such a way that its growth rate at time t is equal to its population divided by 5. Describe the growth process by a differential equation and find the solution for the population size at time t if at $t = 0$ it is equal to 10^4.

8. Yeast is growing in a sugar solution in such a way that the rate of increase in the weight of yeast is equal to one-third of the weight already formed (when t is measured in hours). Describe the change in weight by a differential equation and find the weight at time t if at $t = 0$, the weight is 1 gm.

9. If the specific growth rate of a population is equal to 0.5 when time is measured in hours, obtain an expression for the population size at time t in terms of its value at time $t = 0$.

10. The specific growth rate of a population is 0.05/yr. If at $t = 0$ the population size is 1 million, what will be its size 20 years later?

11. Assume that the proportional growth rate $y'(t)/y(t)$ of the human population of the earth is constant. The population in 1930 was 2 billion and in 1975 was 4 billion. Taking 1930 to be $t = 0$, determine the population $y(t)$ of the earth at time t. According to this model, what should the population have been in 1960?

12. The population of the United States was 75 million in 1900 and in 1950 it was 150 million. Assuming that the growth rate is at any time proportional to the population size, find the population size at a general time t. (Take 1900 as $t = 0$). What is the projected population in 1980 on the basis of this solution?

13. Deep-sea sediment can be dated using radioactive beryllium. Beryllium decay satisfies the differential equation $dy/dt = -1.5 \times 10^{-7} \, y$ when t is measured in years. What is the half-life of beryllium?

14. Thorium is used to date coral and shells. Its decay satisfies the differential equation $(dy/dt) = -9.2 \times 10^{-6} y$ when t is measured in years. What is the half-life of radioactive thorium?

15. A population of insects hatched in the spring ($t = 0$) has an initial size of 20 million and a specific death rate of 0.5/month(mo). What will be the population size after 4 mo?

16. If a radiation dosage of 0.2 rads is sufficient to kill 40% of a population of cancer cells, how big a dosage is required to kill 95% of the population?

17. When subjected to 1.0 rads of X-radiation only 3000 of a certain bacterium survive out of an original population of 10^4. How much radiation would be necessary to kill only 2000 of the original population?

18. It is found that for aerial spraying against mosquitoes with a certain insecticide, an amount of 2000 lb/mi² of insecticide will kill 40% of the insects. How much insecticide is needed per square mile to kill 90% of the mosquitoes? (Assume that the exponential law of survival applies.)

19. A body whose temperature is an amount $T°$ above that of its surroundings loses heat by conduction and convection. Its temperature falls according to the differential equation $(dT/dt) = -kT$, where k is certain constant. (This equation is called *Newton's law of cooling*.) The body is originally 100°C hotter than its surroundings and after 15 min this temperature difference has fallen to 60°C. How long will it take the body to reach a temperature 10°C above that of its surroundings?

20. In exercise 19, what will be the temperature of the body after 25 min?

21. Consider a simple chemical reaction in which a molecule A dissociates into two or more simpler molecules B, C, ... Let $y(t)$ be the concentration of A molecules at time t. Explain why it may be reasonable to use the differential equation $dy/dt = -ky$ to model this process. Express y as a function of t.

22. The amount of sunlight that passes through a transparent or translucent material is given by $y = ce^{-kx}$, where x is the thickness of the material and k is a constant for each material. Construct a differential equation model that provides a reasonable explanation of this result. What is the physical meaning of c?

23. Zones in which there is a high level of radioactivity are shielded by means of some dense material, usually lead. Construct a differential equation for the amount of radiation y that penetrates through the shielding as a function of the thickness x of the shielding. Obtain an explicit expression for y as a function of x. (*Hint:* Compare with exercise 22.)

24 Consider the spread of a disease that has the property that once an individual becomes infected he always remains infectious. Show that while only a small proportion of the population is infected with the disease, its spread can reasonably be modeled by the differential equation $dy/dt = ky$ (where y is the number of infected individuals at time t). Obtain y as a function of t assuming that at time $t = 0$ there are 587 infected individuals and at time $t = 1$ yr there are 831 infected individuals in the population.

10.3   SEPARABLE EQUATIONS

We have seen that for the simple exponential models of population growth the change Δy in population size during a small interval of time Δt is given by $ky \, \Delta t$, where k is a constant. Such an equation applies to populations of microorganisms growing by cell division as well as to populations changing via a birth and death process (provided that suitable simplifying assumptions are made). Let us now suppose that, as well as these natural processes, the population is changing also by virtue of a constant rate b of immigration. Then during the time interval Δt the total change in population would be

$$\Delta y = ky \, \Delta t + b \, \Delta t,$$

where the last term gives the increase in population during Δt due to immigration at the rate b. Dividing through by Δt and proceeding to the limit $\Delta t \to 0$, we obtain the differential equation

$$\frac{dy}{dt} = ky + b. \tag{i}$$

The situation in which the population is being reduced by emigration rather than increased by immigration is also covered by this differential equation. In this case the constant b simply has a negative value equal to minus the rate of emigration. Another case is that of a population that is being harvested at a constant rate; for example, many fish populations fall into this category. b would again be negative in this case, equal to minus the rate at which the population is harvested.

In order to solve the above differential equation we again treat dy/dt as the ratio of differentials, and write

$$dy = (ky + b)\, dt = \left(y + \frac{b}{k}\right) k\, dt.$$

Therefore

$$\frac{dy}{y + (b/k)} = k\, dt.$$

Integrating this equation we obtain

$$\int \frac{dy}{y + (b/k)} = \int k\, dt$$

$$\ln\left|y + \frac{b}{k}\right| = kt + C,$$

where C is an arbitrary constant of integration. Let us set $C = \ln|c|$ with the sign of c chosen to be the same as that of $y + (b/k)$. Then we can combine the two logarithms, obtaining

$$\ln\left|\frac{y + (b/k)}{c}\right| = kt.$$

But the quantity inside the absolute value sign is always positive by virtue of the choice of the sign of c. Therefore, after exponentiation, we obtain

$$y + \frac{b}{k} = ce^{kt}.$$

Again this solution is seen to involve an unknown constant c, which must be determined by specifying the value of y at one particular value of t. For example, if $y = y_0$ when $t = 0$, we find, upon putting $t = 0$ into the solution, that

$$y_0 + \frac{b}{k} = ce^0 = c.$$

Therefore the solution takes the form

$$y = \left(y_0 + \frac{b}{k}\right)e^{kt} - \frac{b}{k}$$

when the value of c is substituted back.

EXAMPLE   A population has an initial size of 1.1 million and a specific growth rate of 0.05 when time is measured in years. If the population is losing members due to harvesting at a steady rate of 50,000/yr, what will be the population size after 10 yr?

SOLUTION   In the preceding differential equation we have $k = 0.05$ and $b = -50,000$. (Note that b is negative since harvesting, rather than immigration, is taking place.) Then the differential equation becomes

$$\frac{dy}{dt} = 0.05y - 50,000 = 0.05(y - 10^6).$$

This can be written

$$\frac{dy}{y - 10^6} = 0.05 \, dt,$$

and so, after integration,

$$\ln (y - 10^6) = 0.05t + C.$$

In order to determine the constant of integration C, let us set $t = 0$ since we know that at that value of t, $y = 1.1 \times 10^6$. Therefore

$$\ln (1.1 \times 10^6 - 10^6) = 0.05(0) + C$$

and so

$$C = \ln (0.1 \times 10^6) = \ln (10^5).$$

The solution therefore takes the form

$$\ln (y - 10^6) = 0.05t + \ln (10^5).$$

Upon taking exponentials of both sides we obtain

$$y - 10^6 = e^{[0.05t + \ln (10^5)]} = e^{0.05t} \cdot e^{\ln (10^5)}$$

$$= 10^5 \times e^{0.05t}$$

since $e^{\ln a} = a$. After 10 yr, therefore (setting $t = 10$),

$$y = 10^6 + 10^5 \times e^{0.05(10)}$$

$$= 10^6 + 10^5 \times e^{0.5}$$

$$= 10^6 + 1.65 \times 10^5$$

$$= 1.165 \times 10^6.$$

The population has therefore increased to 1.165 million after 10 yr.

In the exponential model of population growth it is assumed that the population can continue to grow indefinitely. However in practice this can never be the case, and eventually certain factors come to bear that limit further growth of the population. The most significant of such factors is simply the limitation in food supply, which would usually prevent any population from growing beyond a certain size. Other factors, for example, are the activity of predators that might be sufficiently powerful to keep the population down to a certain level in some cases, or simply physical overcrowding which, even in the presence of a plentiful food supply, can act as a limiting factor on the population density.

We see then that for any population in a given environment, there is a maximum population, say y_m, which the environment can support. As the population size approaches y_m, the rate of growth can be expected to slow down, and if y is actually equal to y_m, the growth rate dy/dt should be zero. The simplest mathematical model that has these properties is the differential equation

$$\frac{dy}{dt} = p(y_m - y), \tag{ii}$$

where p is some constant. We observe that as $y \to y_m$, the maximum supportable population, the growth rate dy/dt tends to zero.

It can be seen also that if $y > y_m$, then $dy/dt < 0$. Thus if by some chance the population size exceeds y_m (which might happen, for example, if there is a sudden worsening in the environment that causes a reduction in y_m), then the population size will decrease. This, again, is an expected type of behavior.

The differential equation (ii) becomes identical with equation (i) studied in the earlier part of this section if we set

$$py_m = b, \qquad -p = k.$$

We can therefore take over the previous solution, simply substituting into it these values of b and k:

$$y = \left(y_0 + \frac{b}{k}\right)e^{kt} - \frac{b}{k}$$

$$= (y_0 - y_m)e^{-pt} + y_m$$

$$= y_0 e^{-pt} + y_m(1 - e^{-pt}).$$

The graph of this solution has one of the forms shown in Fig. 10.1. If the initial population size $y_0 < y_m$, the solution steadily increases toward the limiting value y_m; if $y_0 > y_m$, the size steadily decreases and again approaches the limiting value y_m as $t \to \infty$. We can speak of y_m as the *equilibrium population size* for this model.

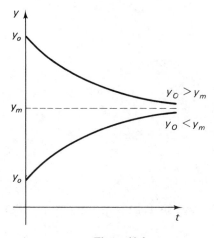

Figure 10.1

The differential equation (ii) represents the simplest model of limited growth, and indeed it is a model that finds application in certain areas, but it is clear that it is too simple an equation to serve as a model except within certain limits. We can see this by examining the right-hand side of (ii) for very small values of y. As $y \to 0$, this righthand side approaches the constant value py_m, so that, according to such a model, the growth rate of a very small population would have the practically constant value py_m. But this is absurd since for any real population, the growth rate must tend to zero as the population size tends to zero. Thus the differential equation (ii) cannot model correctly the growth of a very small population; it can only apply when the population size is reasonably close to its equilibrium value y_m.

A more sophisticated model that overcomes this criticism is based on the differential equation

$$\frac{dy}{dt} = py(y_m - y). \tag{iii}$$

Again the growth rate approaches zero as $y \to y_m$, and if $y > y_m$ the growth rate becomes negative, so that this differential equation, like the previous one, satisfies the appropriate conditions for y_m to be the maximum supportable population. But, in addition, for very small values of y, $y_m - y \simeq y_m$, and so (iii) becomes approximately equivalent to the equation

$$\frac{dy}{dt} = (py_m)y.$$

This equation is the standard equation for exponential growth studied in Section 10.2 (with $k = py_m$). Thus we see that the model equation (iii) combines the property of limited growth toward the maximum population y_m together with the property of exponential growth when the population size is very small.

Because it is the simplest differential equation to combine these two properties, equation (iii) is widely used to model population growth. It is called the *logistic differential equation*. We shall now proceed to obtain its solution.

We have, from (iii), that

$$dy = py(y_m - y)\, dt$$

or

$$\frac{dy}{y(y_m - y)} = p\, dt.$$

In order to integrate the left-hand side we must split it into partial fractions (cf. Section 6.4):

$$\frac{1}{y(y_m - y)} = \frac{1}{y_m} \left\{ \frac{1}{y} + \frac{1}{y_m - y} \right\}.$$

Therefore, multiplying by y_m and integrating, we obtain

$$\int \left\{ \frac{1}{y} + \frac{1}{y_m - y} \right\} dy = \int k\, dt, \qquad (k = py_m).$$

So

$$\ln|y| - \ln|y_m - y| = kt + C,$$

where C is the constant of integration. Therefore

$$\ln \left| \frac{y}{y_m - y} \right| = kt + C.$$

If we now set $C = \ln|c|$, with c having the same sign as the quantity $y/(y_m - y)$, then this solution takes the form

$$\ln \left\{ \frac{y}{c(y_m - y)} \right\} = kt$$

or

$$\frac{y}{y_m - y} = ce^{kt}. \tag{a}$$

This can be written in the form

$$y = (y_m - y)ce^{kt},$$

that is,

$$y(1 + ce^{kt}) = y_m ce^{kt}$$

$$y = \frac{y_m ce^{kt}}{ce^{kt} + 1}$$

(b)

$$= \frac{y_m}{1 + c^{-1}e^{-kt}}.$$

The unknown constant c appearing in this solution can be found in terms of some given value of y. If $y = y_0$ when $t = 0$, we find from equation (a) above that

$$\frac{y_0}{y_m - y_0} = ce^0 = c.$$

Substituting this value of c into (b) we obtain the solution in the final form

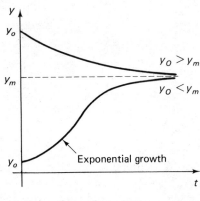

Figure 10.2

$$y = \frac{y_m}{1 + \left(\dfrac{y_m - y_0}{y_0}\right)e^{-kt}}$$

$$= \frac{y_0 y_m}{y_0 + (y_m - y_0)e^{-kt}}.$$

(c)

The graph of this solution is shown in Fig. 10.2 for the two cases $y_0 > y_m$ and $y_0 < y_m$. In either case, as $t \to \infty$, $e^{-kt} \to 0$ and the denominator in the above solution (c) approaches y_0. Thus y approaches the limiting value $y_0 y_m / y_0 = y_m$; y_m represents the asymptotic value of the population no matter what its initial value.

However, in the case where the initial population is small, the graph of y against t shows a phase of approximately exponential growth before it begins to level off. When y is much smaller than y_m the solution is given approximately by $y \simeq y_0 e^{kt}$.

EXAMPLE Find the solution of the differential equation

$$\frac{dy}{dt} = y(2 - y)$$

for which $y = \tfrac{1}{2}$ when $t = 0$.

SOLUTION We write the differential equation in the form

$$\frac{dy}{y(2 - y)} = dt.$$

Splitting the left-hand side into partial fractions we find that

$$\frac{1}{y(2 - y)} = \frac{1}{2}\left(\frac{1}{y} + \frac{1}{2 - y}\right)$$

and so, upon integration, we get

$$\int \left(\frac{1}{y} + \frac{1}{2-y} \right) dy = \int 2 \, dt.$$

Therefore

$$\ln y - \ln (2-y) = 2t + C.$$

We can determine the constant of integration C immediately by using the fact that $y = \frac{1}{2}$ when $t = 0$:

$$\ln \left(\tfrac{1}{2} \right) - \ln \left(\tfrac{3}{2} \right) = 2(0) + C$$

so that

$$C = \ln \left(\tfrac{1}{2} \div \tfrac{3}{2} \right) = \ln \left(\tfrac{1}{3} \right) = -\ln 3.$$

The solution then becomes

$$2t = \ln y - \ln (2-y) + \ln 3$$

$$= \ln \left(\frac{3y}{2-y} \right).$$

Therefore

$$\frac{3y}{2-y} = e^{2t}$$

and so

$$y = \frac{2e^{2t}}{3 + e^{2t}} = \frac{2}{1 + 3e^{-2t}}.$$

EXAMPLE   For a certain rabbit population the growth of population follows the logistic equation (iii) with the constant $k = py_m$ having the value 0.25 when time is measured in months. The population is suddenly reduced from its equilibrium size y_m to a size equal to 1% of y_m by an epidemic of myxamatosis. How many months does it take the population to climb back up to a level of 50% of its equilibrium size?

SOLUTION   The population size $y(t)$ changes according to the differential equation

$$\frac{dy}{dt} = py(y_m - y) = \frac{0.25}{y_m} y(y_m - y)$$

since

$$p = \frac{k}{y_m} = \frac{0.25}{y_m}.$$

Therefore

$$\frac{y_m}{y(y_m - y)} \, dy = 0.25 \, dt.$$

Splitting the left-hand side into its partial fractions and integrating, we find that

$$\int \left\{ \frac{1}{y} + \frac{1}{y_m - y} \right\} dy = \int 0.25 \, dt$$

or

$$\ln y - \ln (y_m - y) = 0.25t + C,$$

where C is the constant of integration. In order to determine C we use the fact that at $t = 0$, y has a value equal to 1% of y_m, i.e., $y = 0.01 y_m$. Therefore, setting $t = 0$, we get

$$\ln (0.01y_m) - \ln (y_m - 0.01y_m) = 0.25(0) + C$$

or

$$\ln \left(\frac{0.01y_m}{0.99y_m}\right) = C.$$

Therefore $C = -\ln 99$ and the solution takes the form

$$\ln y - \ln (y_m - y) = 0.25t - \ln 99.$$

From this equation we can find an explicit expression for y in terms of t, corresponding to equation (c) above. However in this example we wish to find the value of t at which $y = 0.5y_m$, and so we write this solution in the form

$$0.25t = \ln y - \ln (y_m - y) + \ln 99$$

$$= \ln \left(\frac{99y}{y_m - y}\right).$$

Substituting $y = 0.5y_m$ we find that

$$0.25t = \ln \left(\frac{99(0.5y_m)}{y_m - 0.5y_m}\right) = \ln 99 = 4.60.$$

Therefore

$$t = \frac{4.60}{0.25} = 18.4.$$

Thus it takes the population 18.4 mo to recover to half its equilibrium size.

The logistic equation finds application to a number of different areas, apart from population models, some of which will be mentioned in the exercises at the end of this section.

All of the differential equations considered in this section fall into the category of equations of the type

$$\frac{dy}{dt} = P(y)Q(t),$$

that is, equations in which the derivative dy/dt is given as the product of some function of y multiplied by some function of t. Such equations are said to be *separable*. They can be solved by "separating" the variables y and t,

$$\frac{dy}{P(y)} = Q(t)\, dt,$$

and then simply integrating:

$$\int \frac{dy}{P(y)} = \int Q(t)\, dt + C.$$

Of course it may happen, if $P(y)$ and $Q(t)$ are sufficiently complicated functions, that it is not possible to evaluate one or both of these integrals. However, apart from this possible difficulty, separable differential equations can always be solved by the above method.

EXAMPLE   When subjected to a stress level x, the proportion y of a population that survives varies according to the differential equation

$$\frac{dy}{dx} = -k\sqrt{x}\,y,$$

where k is a constant. When $x = 1$ unit, 50% of the population survives. What proportion will survive when the stress level equals 2 units?

SOLUTION   After separating the variables in the above differential equation and then integrating, we obtain

$$\int \frac{dy}{y} = -\int kx^{1/2}\, dx$$

or

$$\ln y = -\tfrac{2}{3}kx^{3/2} + C.$$

Setting $C = \ln c$, as usual, we get

$$y = ce^{-(2/3)kx^{3/2}}.$$

We note that when $x = 0$, $y = c$, so that the constant c is equal to the initial population size before any stress is placed upon it.

We are given that when $x = 1$, $y = \tfrac{1}{2}c$, so that

$$\tfrac{1}{2}c = ce^{-(2/3)k(1)^{3/2}} = ce^{-(2/3)k}.$$

Dividing through by c and taking logs we get

$$\tfrac{2}{3}k = -\ln\left(\tfrac{1}{2}\right) = \ln 2$$

$$k = \tfrac{3}{2}\ln 2 = \tfrac{3}{2}(0.693) = 1.04.$$

When $x = 2$ the proportion of the population that survives, y/c, is given by

$$\frac{y}{c} = e^{-(2/3)k(2)^{3/2}} = e^{-(2/3)(1.04)(2.83)} = 0.141.$$

Thus 14.1% of the population survive when $x = 2$ units.

EXERCISES 10.3

Find the general solutions of the following differential equations by separating variables.

1. $\dfrac{dy}{dt} = 2y + 1.$ **2.** $\dfrac{dy}{dt} + y = 2.$

3. $2\dfrac{dy}{dt} + y = 3$.

4. $3\dfrac{dy}{dt} = 2y - 4$.

5. $\dfrac{dy}{dt} = y(y - 1)$.

6. $\dfrac{dy}{dt} = y(y + 1)$.

7. $\dfrac{dy}{dt} = 6y(y + 3)$.

8. $\dfrac{dy}{dt} = 4y(y - 8)$.

9. $\dfrac{dy}{dt} = y \cos t$.

10. $\dfrac{dy}{dt} = y(1 + \sin t)$.

11. $\dfrac{dy}{dx} = xy$.

12. $\dfrac{dy}{dx} = xe^{x+y}$.

13. $\dfrac{dy}{dt} + y^2 = 4$.

14. $\dfrac{dy}{dt} = 1 + y^2$.

15. A population has an initial size of 10,000 and a specific growth rate of 0.04 (time measured in years). If the population increases due to immigration at the rate of 100/yr, what will be the population size after 10 yr?

16. Repeat exercise 15 in the case when the population is losing members at a rate of 150/yr due to emigration.

17. An infectious disease spreads slowly through a large population. Let $p(t)$ be the proportion of the population that has been exposed to the disease within t yr of its introduction. If $p'(t) = \frac{1}{5}[1 - p(t)]$ and $p(0) = 0$, find $p(t)$ for all $t > 0$. After how many years has the proportion increased to 75%?

18. According to Fick's law, the diffusion of a solute across the wall membrane of a cell is governed by the differential equation $c'(t) = \kappa[c_s - c(t)]$, where $c(t)$ is the concentration of solute in the cell, c_s is the concentration of solute in the surrounding medium, and κ is a constant depending on the size of the cell and on the membrane properties. If $c(0) = c_0$, show that when $c_0 < c_s$

$$c(t) = c_s + (c_0 - c_s)e^{-\kappa t}.$$

19. A population is growing according to the differential equation $dy/dt = 0.1\, y \times [1 - 10^{-6}y]$, when t is measured in years. How many years does it take the population to increase from an initial size of 10^5 to a size 5×10^5?

20. It has been estimated that for the global population of blue whales the population grows according to the logistic model (iii) with $y_m = 200{,}000$ and $p = 6 \times 10^{-7}$ (time measured in years). The population size was estimated in 1972 to be 6000 whales. Assuming that the present ban on harvesting these whales is maintained, how long will it take for the population to recover to 100,000?

21. Consider the chemical reaction $A + B \rightarrow C + D$ in which the compounds A and B are converted into compounds C and D. Let $a(t)$ be the number of gram-molecules of A and $b(t)$ the number of gram-molecules of B present at time t. The numbers of gram-molecules initially present are $a(0)$ and $b(0)$ respectively. Then $a(0) - a(t) = b(0) - b(t)$ since each of these quantities is the number of gram-molecules of A or B that have been used up in the reaction up to time t. The rate of reaction, $-(da/dt)$, satisfies the differential equation

$$-\frac{da}{dt} = ka(t)b(t),$$

where k is a constant. By eliminating $b(t)$ solve this equation for $a(t)$.

22. We can construct a simple model of the spread of an infection through a population in the following way. Let n be the total number of susceptible (i.e., non-immune) individuals in the original population. Let $y(t)$ be the number of infected individuals at time t. Then $n - y(t)$ gives the number of remaining uninfected susceptibles. The model consists of setting

$$\frac{dy}{dt} = ky(n - y),$$

where k is a constant. (Note that dy/dt is the rate of spread of the infection.) Find the solution for y as a function of t, and sketch its graph. Discuss the limitations of this model.

23. In a town whose population is 2000, the spread of an epidemic of influenza follows the differential equation

$$\frac{dy}{dt} = py(2000 - y),$$

where y is the number of people infected at time t (t measured in weeks) and $p = 0.002$. If there are initially two people infected, find y as a function of t. How long is it before three-quarters of the population are infected.

24. The fluctuation of the size of a population throughout the seasons of the year can be modeled by the equation

$$\frac{dy}{dt} = k(t)y,$$

where $k(t)$ is a periodic function with period 1 yr. Show that the population size does not change from one year to the next provided that $\int_0^1 k(t)\, dt = 0$. Find the solution for y when $k(t) = k_1 \cos 2\pi t + k_2 \cos 4\pi t$, k_1 and k_2 being constants.

25. An alternative model to the logistic model of limited growth is based on the Gompertz differential equation

$$\frac{dy}{dt} = ky \ln \left(\frac{y_m}{y}\right).$$

Examine the sign of dy/dt accordingly as $y > y_m$ or $y < y_m$. What conclusions do you draw from this? Find the solution y as a function of t. (*Hint:* Substitute $y = y_m e^u$ in order to evaluate the integral.)

26. The limited growth model of von Bertalanffy can be obtained on the basis of the differential equation

$$\frac{dy}{dt} = 3ky^{2/3}(y_m^{1/3} - y^{1/3}).$$

Find an expression for y as a function of t. (*Hint:* Substitute $y^{1/3} = u$ in the integral to be evaluated.)

27. According to the Michaelis–Menton equation, the rate at which an enzyme reaction takes place is given by

$$\frac{dy}{dt} = -\frac{My}{K + y},$$

where M and K are constant and y is the amount present at time t of the substrate that is being transformed by the enzyme. Find an implicit equation expressing y as a function of t.

10.4 LINEAR FIRST-ORDER EQUATIONS

Let us consider again the problem with which we began Section 10.3 in which a population grows by a combination of natural exponential expansion and immigration, but now let us suppose that the rate of immigration varies with time. The differential equation will be analogous to equation (i) of Section 10.3, namely

$$\frac{dy}{dt} = ky + b(t), \tag{i}$$

where k is the specific growth rate for natural expansion and $b(t)$ is the rate of immigration. The only difference between this and the earlier equation is that b is a function of t rather than simply a constant.

We note that the possibility of depletion of the population by emigration or other type of removal of members such as harvesting can be accommodated within this model by allowing $b(t)$ to take negative values.

The above differential equation cannot be solved by separation of the variables. Instead we use a technique known as *variation of parameters* in order to find the solution. This technique is quite widely used for solving linear differential equations, and we shall encounter its use again later.

The method of procedure is as follows. First we set $b(t) = 0$ in the given equation, obtaining the simpler equation

$$\frac{dy}{dt} = ky,$$

in which all of the terms are multiples of either y or its derivatives. This equation is called the *homogeneous equation* corresponding to the given differential equation (i). Its solution, as seen in Section 10.2, is simply $y = ce^{kt}$, where c is an arbitrary constant.

We now seek a solution of the given equation (i) above in the form $y = c(t)e^{kt}$, where the constant c in the solution of the homogeneous equation is allowed to vary as a function of t. Using the product rule we find that

$$\frac{dy}{dt} = \frac{dc}{dt}e^{kt} + kce^{kt}.$$

But from (i),

$$\frac{dy}{dt} = ky + b(t) = kce^{kt} + b(t).$$

Comparing these two expressions for dy/dt we see that two of the terms are identical, so that

$$\frac{dc}{dt}e^{kt} = b(t).$$

Thus

$$\frac{dc}{dt} = b(t)e^{-kt}$$

and so, after integration,

$$c(t) = \int b(t)e^{-kt}\,dt + C,$$

where C is a constant of integration.

Thus the solution of (i) is given by $y = c(t)e^{kt}$, or

$$y = e^{kt}\int b(t)e^{-kt}\,dt + Ce^{kt}.$$

EXAMPLE   Find the solution of the differential equation

$$2\frac{dy}{dt} + y = e^{-t/2}$$

that satisfies the additional condition that $y = 1$ when $t = 0$.

SOLUTION   Dividing through the given equation by 2 we obtain

$$\frac{dy}{dt} + \frac{1}{2}y = \frac{1}{2}e^{-t/2},$$

which has the same form as the general equation (i) with $k = -\frac{1}{2}$ and $b(t) = \frac{1}{2}e^{-t/2}$. The corresponding homogeneous equation is

$$\frac{dy}{dt} + \frac{1}{2}y = 0,$$

whose solution is $y = ce^{-(1/2)t}$. We therefore seek a solution of the given equation in the form $y = c(t)e^{-(1/2)t}$; then

$$\frac{dy}{dt} + \frac{1}{2}y = \left[\frac{dc}{dt}e^{-(1/2)t} + c\left(-\frac{1}{2}\right)e^{-(1/2)t}\right] + \frac{1}{2}ce^{-(1/2)t}$$

$$= \frac{dc}{dt}e^{-(1/2)t} = \frac{1}{2}e^{-t/2}.$$

Therefore $(dc/dt) = \frac{1}{2}$, and so

$$c(t) = \int \tfrac{1}{2}\,dt = \tfrac{1}{2}t + C,$$

where C is the constant of integration. The solution is then given by

$$y = c(t)e^{kt}$$
$$= (\tfrac{1}{2}t + C)e^{-(1/2)t}.$$

In order to find C we make use of the given initial value of y. Setting $t = 0$ we find that

$$y = 1 = (\tfrac{1}{2}0 + C)e^0 = C,$$

so $C = 1$ and therefore

$$y = (\tfrac{1}{2}t + 1)e^{-(1/2)t}.$$

EXAMPLE A population has an initial size of 1.1 million and a specific growth rate of 0.05 when time is measured in years. The population is losing members due to emigration at the rate of 50,000 $(1 + \cos \tfrac{1}{5}\pi t)$ at time t. What will be the population size after 5 yr?

SOLUTION The size of the population is governed by the differential equation

$$\frac{dy}{dt} = 0.05y - 50{,}000\left(1 + \cos \frac{1}{5}\pi t\right).$$

Therefore in the preceding general solution we must put $b(t) = -50{,}000(1 + \cos \tfrac{1}{5}\pi t)$ and $k = 0.05$. We obtain

$$c(t) = C - 50{,}000 \int (1 + \cos \tfrac{1}{5}\pi t)e^{-kt}\,dt.$$

Making use of formula 117 in the table of integrals in Appendix II, we find that

$$\int e^{-kt} \cos \frac{1}{5}\pi t\,dt = \frac{1}{k^2 + \frac{1}{25}\pi^2}e^{-kt}\left[\frac{1}{5}\pi \sin \frac{1}{5}\pi t - k \cos \frac{1}{5}\pi t\right].$$

Therefore

$$c(t) = C - 50{,}000\left\{-\frac{1}{k}e^{-kt} + \frac{1}{k^2 + \frac{1}{25}\pi^2}e^{-kt}\left[\frac{1}{5}\pi \sin \frac{1}{5}\pi t - k \cos \frac{1}{5}\pi t\right]\right\}$$

and so, since $y = c(t)e^{kt}$, we obtain the solution

$$y = Ce^{kt} + 50{,}000\left\{\frac{1}{k} - \frac{1}{k^2 + \frac{1}{25}\pi^2}\left[\frac{1}{5}\pi \sin \frac{1}{5}\pi t - k \cos \frac{1}{5}\pi t\right]\right\}.$$

The value of the unknown constant C is determined from the initial size of the population. Since $y = 1.1$ million when $t = 0$, we have

$$1.1 \times 10^6 = C + 50{,}000\left\{\frac{1}{k} + \frac{25k}{25k^2 + \pi^2}\right\}$$

$$= C + 50{,}000\,\frac{50k^2 + \pi^2}{k(25k^2 + \pi^2)}$$

$$= C + 50{,}000\,\frac{50(0.05)^2 + \pi^2}{0.05[25(0.05)^2 + \pi^2]}$$

$$= C + 1.006 \times 10^6.$$

Therefore

$$C = 0.094 \times 10^6 = 9.4 \times 10^4.$$

The population size after 5 yr is obtained by setting $t = 5$, and we find, since $\sin \pi = 0$ and $\cos \pi = -1$,

$$y = 9.4 \times 10^4 e^{5k} + 50{,}000\left\{\frac{1}{k} - \frac{25k}{25k^2 + \pi^2}\right\}$$

$$= 9.4 \times 10^4 e^{0.25} + \frac{50{,}000\pi^2}{25(0.05)^2 + \pi^2}$$

$$= 1.21 \times 10^5 + 9.94 \times 10^5$$

$$= 1.115 \times 10^6.$$

We can generalize the preceding problem and consider a population for which the specific growth rate as well as the immigration rate varies with time. The differential equation that models such a population is then

$$\frac{dy}{dt} = k(t)y + b(t),$$

where $k(t)$ and $b(t)$ are both functions of t. This differential equation is in fact the most general kind of linear first-order equation that is possible. Its solution can be found by extending the above method, but we shall not provide the details here.

Another application of first-order linear differential equations is to the operation of the aorta in controlling bloodflow. The aorta is the large blood vessel into which the arterial blood first flows upon leaving the heart. During the systolic phase of the heartbeat cycle, blood is pumped under pressure from the heart into one end of the aorta, and the walls of the aorta stretch in order to accommodate the blood. During the diastolic phase, there is no flow of blood into the aorta and the walls of the aorta elastically contract, thus squeezing the blood out of the aorta and around the circulatory system of the body.

Figure 10.3 illustrates the system of heart, aorta, and circulatory system. Let V be the volume of the aorta and P the pressure of blood within it at time t. Then since the aorta expands under the blood pressure inside it, it is reasonable to assume that V is a linear function of P: $V = V_0 + kP$, where V_0 and k are two constants.

The rate of change of volume, dV/dt, is equal to the rate at which blood is pumped into the aorta by the heart minus the rate at which blood flows out of the aorta into

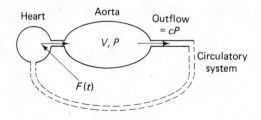

Figure 10.3

the circulatory system. It is as a result of fluid mechanics that this latter rate is proportional to the pressure driving the blood. Therefore we can write

$$\frac{dV}{dt} = F(t) - cP,$$

where c is a constant and $F(t)$ is the rate at which the heart pumps blood into the aorta.

$F(t)$ can be assumed to be a known function of t, and has a graph of the form shown in Fig. 10.4. During the diastolic phase, $F(t) = 0$. During the systolic phase, $F(t)$ can reasonably be taken to be given by $F(t) = A \sin \omega t$ when the duration of the systolic phase is $0 < t < \pi/\omega$.

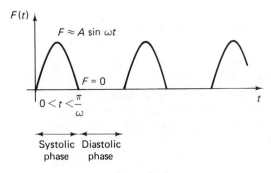

Figure 10.4

Noting that $dV/dt = k(dP/dt)$, we obtain the following differential equation for P:

$$k\frac{dP}{dt} + cP = F(t) = \begin{cases} 0 & \text{during diastolic phase} \\ A \sin \omega t & \text{during systolic phase, } 0 < t < \pi/\omega. \end{cases}$$

Using the techniques developed earlier in the section we find that the solution for P during the systolic phase, $0 < t < \pi/\omega$, is

$$P = P_0 e^{-ct/k} + A(c^2 + \omega^2 k^2)^{-1}[\omega k e^{-ct/k} + c \sin \omega t - \omega k \cos \omega t],$$

where P_0 is the value of P at $t = 0$, that is, at the beginning of the systolic phase.

During the diastolic phase the solution for P is a simple exponential decay:

$$P = P_1 e^{-ct/k},$$

where P_1 is the value of P at the end of the systolic phase.

EXERCISES 10.4

Find the general solutions of the following differential equations.

1. $\dfrac{dy}{dt} + y = t.$

2. $\dfrac{dy}{dt} = y + t^2.$

3. $2\dfrac{dy}{dt} - y = e^t.$

4. $\dfrac{dy}{dt} + 3y = \cos t.$

5. $3\dfrac{dy}{dt} + y = t^2.$

6. $\dfrac{dy}{dt} + 3y = t^2.$

7. $\dfrac{dy}{dt} - y = \cos t.$

8. $\dfrac{dy}{dt} + y = \sin 2t.$

Find the solutions of the following differential equations that satisfy the given additional conditions.

9. $\dfrac{dy}{dt} - 3y = 1; \; y = 0$ when $t = 1.$

10. $3\dfrac{dy}{dt} - y = 1; \; y = 0$ when $t = 1.$

11. $\dfrac{dy}{dt} + 2y = e^{-2t}; \; y = 3$ when $t = 0.$

12. $\dfrac{dy}{dt} - 2y = e^{-2t}; \; y = 3$ when $t = 0.$

13. A population has an initial size of 1000 and a specific growth rate of 0.5 when t is measured in months. The population is harvested at the rate of $h(2 + \cos \frac{1}{6}\pi t)/$mo throughout the year, where h is given constant. Find the population size as a function of t.

14. In exercise 13, what value of h would lead to a zero change in population over the 12-month period from $t = 0$ to $t = 12$?

15. A population has a specific growth rate at time t equal to $\frac{1}{20}$. If the population has a size y_0 at $t = 0$ and is harvested at a rate h per unit time, find the size y as a function of t.

16. A population has an initial size y_0 and a specific growth rate at time t given by k. If the population is emigrating at the rate he^{kt} per unit time, find the population size at time t.

17. A population has specific growth rate k and is being harvested at the rate $b(t) = b_0(1 - \sin 2\pi t)$ (where time is measured in years). Find the population size as a function of t and find the value of b_0 that produces no change in the population size over a period of 1 yr.

10.5 SECOND-ORDER EQUATIONS

In this section we shall investigate the solution of differential equations of the type

$$\frac{d^2y}{dt^2} + M\frac{dy}{dt} + Ny = 0, \qquad \text{(i)}$$

in which M and N are two given constants. Such equations belong to the class of linear second-order differential equations with constant coefficients.

In the case of the first-order equations with constant coefficients, we found in Section 10.2 that the solution is an exponential function of t. Let us try to find a solution of similar type for equation (i) above, that is, let us seek a solution of (i) in the form $y = e^{kt}$, where k is a certain constant. For such a solution we would have

$$\frac{dy}{dt} = ke^{kt}, \qquad \frac{d^2y}{dt^2} = k^2e^{kt},$$

and upon substituting these into (i) we find

$$k^2e^{kt} + Mke^{kt} + Ne^{kt} = 0,$$

and so

$$k^2 + Mk + N = 0. \tag{a}$$

In the last step we have divided through by the factor e^{kt}.

Equation (a) is known as the *auxiliary equation* of the given second-order differential equation (i). It is a quadratic equation for the unknown constant k, and from the usual quadratic formula we find that

$$k = \tfrac{1}{2}\{-M \pm \sqrt{M^2 - 4N}\}.$$

In the event that $M^2 - 4N > 0$, there are two distinct real roots that we denote by k_1 and k_2. For definiteness we denote by k_1 the root with the positive square root and by k_2 the root with the negative square root; in other words, k_1 is the larger of the two roots. Thus in this case we have two distinct solutions of the form

$$y = e^{k_1 t} \quad \text{and} \quad y = e^{k_2 t}.$$

It is easily seen that by combining these two solutions to obtain a function given by

$$y = c_1 e^{k_1 t} + c_2 e^{k_2 t}, \tag{b}$$

we still have a solution of the given differential equation (i), where c_1 and c_2 are arbitrary constants. Equation (b) in fact represents the most general solution of the given differential equation, as is shown by the following theorem, which we shall not prove.

THEOREM 10.5.1

If $y = f(t)$ and $y = g(t)$ are any two solutions of the differential equation (i), where $f(t)$ is not a constant multiple of $g(t)$, then the general solution of (i) is given by

$$y = c_1 f(t) + c_2 g(t),$$

where c_1 and c_2 are arbitrary constants.

This general solution involves two constants c_1 and c_2 that are not determined by the differential equation itself, just as we found that the solution of a first-order differential equation involves a single arbitrary constant. In this case, as earlier, c_1 and c_2 must be fixed by specifying extra conditions. Usually these conditions consist of the values of y and dy/dt being given for some particular value of t, say for $t = 0$.

EXAMPLE   Find the solution of the differential equation

$$\frac{d^2y}{dt^2} - 2\frac{dy}{dt} - 3y = 0$$

that satisfies the conditions that $y = 2$ and $dy/dt = 14$ when $t = 0$.

SOLUTION   The given equation is of the type (i) discussed above in which $M = -2$ and $N = -3$. The auxiliary equation is therefore

$$k^2 - 2k - 3 = 0,$$

or

$$(k - 3)(k + 1) = 0.$$

The roots of this equation are $k_1 = 3$ and $k_2 = -1$. It follows therefore that the solution of the differential equation is

$$y = c_1e^{3t} + c_2e^{-t}.$$

In order to determine c_1 and c_2 we must make use of the given values of y and dy/dt at $t = 0$. For y itself, setting $t = 0$, we find

$$y = c_1e^{3(0)} + c_2e^{-(0)} = c_1 + c_2 = 2.$$

The derivative dy/dt is given for a general value of t by

$$\frac{dy}{dt} = 3c_1e^{3t} - c_2e^{-t},$$

and so at $t = 0$

$$\frac{dy}{dt} = 3c_1 - c_2 = 14.$$

Thus we have the two simultaneous equations

$$c_1 + c_2 = 2, \qquad 3c_1 - c_2 = 14,$$

which can be solved for c_1 and c_2. The solution is $c_1 = 4$ and $c_2 = -2$. Therefore the solution is

$$y = 4e^{3t} - 2e^{-t}.$$

Next let us consider the case when the coefficients appearing in the differential equation (i) satisfy the condition that $M^2 - 4N = 0$. In this case the auxiliary equation (a) has only a single real root, which is given by $k = -M/2$. It follows therefore that we can, in this case, find only the solution $y = ce^{kt}$ of exponential type. In order to get the general solution we use something like the method of variation of parameters, which was encountered in Section 10.4.

We know that one particular solution is $y = ce^{kt}$, where c is an arbitrary constant and $k = -M/2$. Let us seek the general solution in this same form but with the constant c replaced by a function of t:

518

$$y = c(t)e^{kt}.$$

Then

$$\frac{dy}{dt} = \frac{dc}{dt}e^{kt} + kce^{kt}$$

and

$$\frac{d^2y}{dt^2} = \frac{d^2c}{dt^2}e^{kt} + 2k\frac{dc}{dt}e^{kt} + k^2ce^{kt}.$$

So if these are all substituted into (i) we find

$$\left(\frac{d^2c}{dt^2} + 2k\frac{dc}{dt} + k^2c\right)e^{kt} + M\left(\frac{dc}{dt} + kc\right)e^{kt} + Nce^{kt} = 0.$$

The factor e^{kt} cancels through this equation, and after rearrangement it becomes

$$\frac{d^2c}{dt^2} + (2k + M)\frac{dc}{dt} + (k^2 + Mk + N)c = 0.$$

But the coefficient of c vanishes because k satisfies the auxiliary equation (a); and furthermore the coefficient of dc/dt also vanishes since $k = -M/2$ in the case being considered ($M^2 = 4N$). Therefore we obtain

$$\frac{d^2c}{dt^2} = 0.$$

It follows that the first derivative of c must be constant,

$$\frac{dc}{dt} = c_1,$$

where c_1 is some constant. Integrating a second time then we find that

$$c(t) = c_1t + c_2,$$

where c_2 is a second constant of integration. Therefore, after substitution of this value of $c(t)$,

$$y = (c_1t + c_2)e^{kt}.$$

Again we have found a solution that involves two unknown constants, c_1 and c_2. These must be determined from conditions specified in addition to the differential equation itself.

EXAMPLE   Find the solution of the differential equation

$$\frac{d^2y}{dt^2} - 4\frac{dy}{dt} + 4y = 0$$

that satisfies the conditions that $y = 1$ and $dy/dt = 4$ when $t = 0$.

SOLUTION   The auxiliary equation obtained from the given differential equation is

$$k^2 - 4k + 4 = 0$$

or

$$(k - 2)^2 = 0.$$

In this case there is only one distinct value of k, namely $k = 2$, and so we must take the solution in the form

$$y = (c_1 t + c_2)e^{2t}.$$

Setting $t = 0$ we find

$$y = (c_1 0 + c_2)e^{2(0)} = c_2 = 1$$

and so the value of c_2 is determined. In order to determine c_1 we must use the given value of dy/dt at $t = 0$. For general t

$$\frac{dy}{dt} = c_1 e^{2t} + (c_1 t + c_2)2e^{2t},$$

and so at $t = 0$

$$\frac{dy}{dt} = c_1 e^{2(0)} + (c_1 0 + c_2)2e^{2(0)} = c_1 + 2c_2 = 4.$$

Therefore

$$c_1 = 4 - 2c_2 = 4 - 2 = 2,$$

and so the solution is finally

$$y = (2t + 1)e^{2t}.$$

We see that in both the cases $M^2 - 4N > 0$ and $M^2 - 4N = 0$ the solution of the given differential equation (i) involves exponential functions. In the case $M^2 - 4N < 0$, which we shall discuss in the next section, it turns out that the solution involves trigonometric functions that give it an oscillatory nature.

For completeness we summarize the solutions of the differential equation (i) in the following table. Certain of the results in this table will appear unfamiliar to the student until the next section has been read.

Differential equation	$\dfrac{d^2 y}{dt^2} + M\dfrac{dy}{dt} + Ny = 0$	
Auxiliary equation	$k^2 + Mk + N = 0$	with roots k_1, k_2
k_1, k_2 real and distinct	$y = c_1 e^{k_1 t} + c_2 e^{k_2 t}$	c_1, c_2 arbitrary
$k_1 = k_2$	$y = (c_1 t + c_2)e^{k_1 t}$	c_1, c_2 arbitrary
k_1, k_2 not real $(k_1, k_2 = k \pm i\omega)$ where $i = \sqrt{-1}$	$y = ce^{kt}\cos(\omega t + \alpha)$ where $k = -M/2$ $\omega = \frac{1}{2}\sqrt{4N - M^2}$	c, α arbitrary

EXERCISES 10.5

Find the general solutions of the following differential equations.

1. $\dfrac{d^2 y}{dt^2} - 5\dfrac{dy}{dt} + 4y = 0.$

2. $\dfrac{d^2 y}{dt^2} + 4\dfrac{dy}{dt} - 5y = 0.$

3. $\dfrac{d^2y}{dt^2} + 2\dfrac{dy}{dt} + y = 0.$ $\qquad\qquad$ **4.** $4\dfrac{d^2y}{dt^2} - 4\dfrac{dy}{dt} + y = 0.$

Find the solutions of the following differential equations that satisfy the given conditions.

5. $\dfrac{d^2y}{dt^2} - y = 0$; $y = 2$ and $\dfrac{dy}{dt} = 4$ when $t = 0$.

6. $\dfrac{d^2y}{dt^2} - 9y = 0$; $y = 3$ and $\dfrac{dy}{dt} = 3$ when $t = 0$.

7. $\dfrac{d^2y}{dt^2} + \dfrac{dy}{dt} = 0$; $y = 1$ and $\dfrac{dy}{dt} = -2$ when $t = 0$.

8. $\dfrac{d^2y}{dt^2} - 3\dfrac{dy}{dt} = 0$; $y = 0$ and $\dfrac{dy}{dt} = 3$ when $t = 0$.

9. $\dfrac{d^2y}{dt^2} - 5\dfrac{dy}{dt} + 6y = 0$; $y = 2$ and $\dfrac{dy}{dt} = 5$ when $t = 0$.

10. $4\dfrac{d^2y}{dt^2} + 3\dfrac{dy}{dt} - y = 0$; $y = 9$ and $\dfrac{dy}{dt} = 1$ when $t = 0$.

11. $9\dfrac{d^2y}{dt^2} + 6\dfrac{dy}{dt} + y = 0$; $y = 6$ and $\dfrac{dy}{dt} = 1$ when $t = 0$.

12. $9\dfrac{d^2y}{dt^2} + 12\dfrac{dy}{dt} + 4y = 0$; $y = 6$ and $\dfrac{dy}{dt} = 4$ when $t = 0$.

10.6 OSCILLATORY SOLUTIONS

In this section we shall consider the differential equation

$$\frac{d^2y}{dt^2} + M\frac{dy}{dt} + Ny = 0 \qquad\qquad\text{(i)}$$

for the case when $M^2 - 4N < 0$. It will turn out that the solutions are oscillatory functions of time in this case rather than the exponential functions that we found in Section 10.5.

We shall begin by considering the function

$$y = c \cos(\omega t + \alpha),$$

where c, ω, and α are constants. As we saw in Section 5.5, this function describes a sinusoidal oscillation with angular frequency ω and period equal to $2\pi/\omega$. The variable y oscillates between the minimum value $-c$ and maximum value $+c$, so that c is called the *amplitude* of oscillation (Fig. 10.5). The constant α is called the *phase* of y, and defines the location of the oscillation with respect to the time origin $t = 0$; a maximum of y occurs at $t = -\alpha/\omega$.

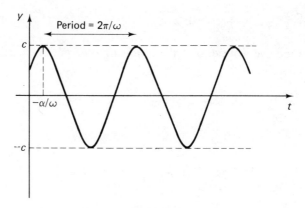

Figure 10.5

For this function we find, upon successive differentiation, that

$$\frac{dy}{dt} = -\omega c \sin(\omega t + \alpha)$$

$$\frac{d^2 y}{dt^2} = -\omega^2 c \cos(\omega t + \alpha).$$

It follows therefore that

$$\frac{d^2 y}{dt^2} + \omega^2 y = 0. \tag{ii}$$

We see from this result that the function $y = c \cos(\omega t + \alpha)$ satisfies the differential equation (ii), which is a special case of equation (i) in which $M = 0$ and $N = \omega^2$. It is clear that in this special case, $M^2 - 4N < 0$.

Equation (ii) is a very important differential equation that is known in physics as the *equation of simple harmonic motion*. There are many different problems in the fields of mechanical and electrical vibrations in which some variable oscillates in time in a sinusoidal manner, and this differential equation is of immense significance for such problems. As a simple example we mention the pendulum, which consists of a weight swinging from side to side on the end of a string. If y denotes the angle that the string makes with the vertical at time t (Fig. 10.6), then y satisfies equation (ii) with $\omega^2 = (g/l)$, where l is the length of the pendulum and g is the acceleration due to gravity, which has the value $g = 981$ cm/sec². (In actual fact this statement is only true for small amplitudes of vibration of the pendulum.)

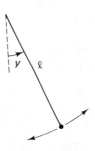

Figure 10.6

The general solution of the equation of simple harmonic motion (ii) is the function $y = c \cos(\omega t + \alpha)$. The two constants c and α are arbitrary and must be determined from additional conditions on the solution, which would as a rule consist of the values of y and dy/dt being specified for a certain value of t.

EXAMPLE   A variable y satisfies the differential equation

$$\frac{d^2y}{dt^2} + 9y = 0.$$

At $t = 0$, $y = 2$ and $dy/dt = -6$. Find the solution for y as a function of t.

SOLUTION   The given differential equation is of the type (ii) with $\omega = 3$. Therefore the general solution is

$$y = c \cos (3t + \alpha),$$

where c and α are constants yet to be determined. Setting $t = 0$ we find that

$$y = c \cos \alpha = 2.$$

Further,

$$\frac{dy}{dt} = -3c \sin (3t + \alpha)$$

and at $t = 0$,

$$\frac{dy}{dt} = -3c \sin \alpha = -6.$$

We see therefore that in order to obtain the values of c and α we must solve the two equations

$$c \cos \alpha = 2, \qquad c \sin \alpha = 2.$$

We proceed as follows:

$$(c \cos \alpha)^2 + (c \sin \alpha)^2 = (2)^2 + (2)^2 = 8$$
$$= c^2 \cos^2 \alpha + c^2 \sin^2 \alpha$$
$$= c^2 (\cos^2 \alpha + \sin^2 \alpha) = c^2$$

since $\cos^2 \alpha + \sin^2 \alpha = 1$. It follows therefore that $c = \sqrt{8} = 2\sqrt{2}$. Consequently, $\cos \alpha = 2/c = 1/\sqrt{2}$ and $\sin \alpha = 2/c = 1/\sqrt{2}$. Therefore $\alpha = \pi/4$. (Note that α must lie in the first quadrant, that is, between 0 and $\pi/2$, since its cosine and sine are both positive.) The solution therefore takes the final form

$$y = 2\sqrt{2} \cos \left(3t + \frac{\pi}{4}\right).$$

Now let us turn to the general equation (i) in the case $M^2 - 4N < 0$. It turns out that the general solution is

$$y = ce^{kt} \cos (\omega t + \alpha),$$

where $k = -M/2$ and $\omega = \frac{1}{2}\sqrt{4N - M^2}$. The constants c and α are arbitrary and must be determined from extra conditions, and not from the differential equation itself.

Let us verify that this function is a solution of the differential equation (i). Differentiating y twice, we obtain

$$\frac{dy}{dt} = kce^{kt} \cos(\omega t + \alpha) - \omega ce^{kt} \sin(\omega t + \alpha)$$

$$\frac{d^2y}{dt^2} = k^2 ce^{kt} \cos(\omega t + \alpha) - 2k\omega ce^{kt} \sin(\omega t + \alpha) - \omega^2 ce^{kt} \cos(\omega t + \alpha).$$

Therefore

$$\frac{d^2y}{dt^2} + M\frac{dy}{dt} + Ny = ce^{kt}\{k^2 \cos(\omega t + \alpha) - 2\omega k \sin(\omega t + \alpha) - \omega^2 \cos(\omega t + \alpha)$$

$$+ M[k \cos(\omega t + \alpha) - \omega \sin(\omega t + \alpha)] + N \cos(\omega t + \alpha)\}$$

$$= ce^{kt}\{(k^2 - \omega^2 + Mk + N) \cos(\omega t + \alpha)$$

$$- (2\omega k + M\omega) \sin(\omega t + \alpha)\}.$$

But since $k = -M/2$ and $\omega^2 = N - \frac{1}{4}M^2$, it is easily seen that the coefficients of $\cos(\omega t + \alpha)$ and $\sin(\omega t + \alpha)$ inside the braces are both zero:

$$k^2 - \omega^2 + Mk + N = \left(-\frac{M}{2}\right)^2 - N + \frac{1}{4}M^2 + M\left(-\frac{1}{2}M\right) + N = 0$$

$$2\omega k + M\omega = \omega(2k + M) = \omega(-M + M) = 0.$$

Consequently the given function y satisfies the differential equation (i) as required.

EXAMPLE   Find the general solution of the differential equation

$$\frac{d^2y}{dt^2} + Ny = 0$$

for $N > 0$.

SOLUTION   In this case the coefficient M is zero, and, since N is given to be positive, $M^2 - 4N = -4N$ is negative. Hence we must use the solution $y = ce^{kt} \cos(\omega t + \alpha)$. However since $M = 0$, then $k = 0$ also, and the solution reduces to

$$y = c \cos(\omega t + \alpha).$$

The constants c and α are not determined by the differential equation, and

$$\omega = \frac{1}{2}\sqrt{4N - M^2} = \frac{1}{2}\sqrt{4N} = \sqrt{N}.$$

The equation in this example is that for simple harmonic motion, and we have simply recovered our earlier solution for this case.

EXAMPLE   Find the solution of the differential equation

$$\frac{d^2y}{dt^2} - 2\frac{dy}{dt} + 2y = 0$$

for which $y = 4$ and $dy/dt = 7$ when $t = 0$.

Since $M = -2$ and $N = 2$ for the given equation,

$$M^2 - 4N = (-2)^2 - 4(2) = 4 - 8 = -4 < 0.$$

Hence the general solution is $y = ce^{kt} \cos{(\omega t + \alpha)}$. In this particular solution,

$$k = -\tfrac{1}{2}M = 1, \quad \text{and} \quad \omega = \tfrac{1}{2}\sqrt{4N - M^2} = \tfrac{1}{2}\sqrt{4} = 1.$$

Therefore the solution takes the form

$$y = ce^t \cos{(t + \alpha)}$$

when the values of k and ω are substituted. Now setting $t = 0$ we find that

$$y = ce^0 \cos{(0 + \alpha)} = c \cos{\alpha} = 4.$$

For a general t,

$$\frac{dy}{dt} = \frac{d}{dt}[ce^t \cos{(t + \alpha)}]$$

$$= ce^t \cos{(t + \alpha)} - ce^t \sin{(t + \alpha)}.$$

Therefore, setting $t = 0$,

$$\frac{dy}{dt} = c \cos{\alpha} - c \sin{\alpha} = 7.$$

Consequently, since $c \cos{\alpha} = 4$, it follows that

$$c \sin{\alpha} = c \cos{\alpha} - 7 = 4 - 7 = -3.$$

We must therefore determine the values of c and α, knowing that $c \cos{\alpha} = 4$ and $c \sin{\alpha} = -3$. We proceed as before:

$$(c \cos{\alpha})^2 + (c \sin{\alpha})^2 = 4^2 + (-3)^2 = 16 + 9 = 25$$

$$= c^2 (\sin^2{\alpha} + \cos^2{\alpha}) = c^2$$

since $\sin^2{\alpha} + \cos^2{\alpha} = 1$. So we conclude that $c^2 = 25$ or $c = 5$. Finally $\cos{\alpha} = 4/c = \tfrac{4}{5} = 0.8$ and $\sin{\alpha} = -3/c = -\tfrac{3}{5} = -0.6$. Hence $\alpha = -0.64$ radians. (Note; Since $\cos{\alpha}$ is positive and $\sin{\alpha}$ negative, α must lie in the fourth quadrant, that is, $-\pi/2 < \alpha < 0$.) Thus the final solution is

$$y = 5e^t \cos{(t - 0.64)}.$$

We shall conclude this section by describing briefly the nature of the solution

$$y = ce^{kt} \cos{(\omega t + \alpha)}.$$

The solution is still oscillatory in the sense that y takes alternately positive and negative values, but the presence of the exponential factor e^{kt} makes the oscillation no longer sinusoidal. In fact, if $k > 0$, the amplitude of oscillation grows exponentially in time, as shown in Fig. 10.7. The dashed curves in this figure represent the two exponential functions $y = \pm ce^{kt}$, and the solution of the differential equation oscillates between these two curves with growing amplitude. The solution is no longer periodic, but the time interval between successive instants at which y crosses from negative to positive values is still given by $2\pi/\omega$, as shown.

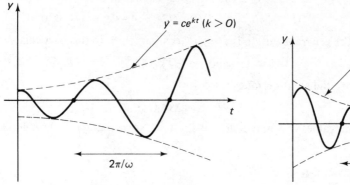

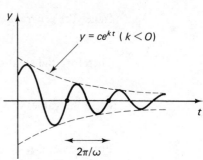

| Figure 10.7 | Figure 10.8 |

When $k < 0$ the exponential factor e^{kt} makes the amplitude of oscillation decrease as t increases (Fig. 10.8). This type of solution is called a *damped oscillation*. The amplitude in this case decays to zero at $t \to \infty$.

EXERCISES 10.6

Find the general solutions of the following differential equations:

1. $\dfrac{d^2y}{dt^2} + y = 0.$ **2.** $\dfrac{d^2y}{dt^2} + 3y = 0.$

3. $3\dfrac{d^2y}{dt^2} + y = 0.$ **4.** $3\dfrac{d^2y}{dt^2} + 5y = 0.$

5. $\dfrac{d^2y}{dt^2} + 2\dfrac{dy}{dt} + 5y = 0.$ **6.** $\dfrac{d^2y}{dt^2} - 4\dfrac{dy}{dt} + 20y = 0.$

Find the solutions of the following differential equations that satisfy the given conditions:

7. $\dfrac{d^2y}{dt^2} + y = 0; y = -3, \dfrac{dy}{dt} = -3$ when $t = 0.$

8. $\dfrac{d^2y}{dt^2} + 3y = 0; y = -1, \dfrac{dy}{dt} = -3$ when $t = 0.$

9. $\dfrac{d^2y}{dt^2} + 4y = 0; y = 0, \dfrac{dy}{dt} = 2$ at $t = 0.$

10. $4\dfrac{d^2y}{dt^2} + y = 0; y = 4, \dfrac{dy}{dt} = 0$ at $t = 0.$

11. $\dfrac{d^2y}{dt^2} + 4\dfrac{dy}{dt} + 13y = 0; y = 1, \dfrac{dy}{dt} = -5$ at $t = 0.$

12. $2\dfrac{d^2y}{dt^2} + 2\dfrac{dy}{dt} + y = 0;\ y = -1,\ \dfrac{dy}{dt} = 1$ at $t = 0$.

13. $5\dfrac{d^2y}{dt^2} - 2\dfrac{dy}{dt} + y = 0;\ y = -3,\ \dfrac{dy}{dt} = 1$ at $t = 0$.

14. $\dfrac{d^2y}{dt^2} - 6\dfrac{dy}{dt} + 10y = 0;\ y = -5,\ \dfrac{dy}{dt} = -27$ at $t = 0$.

10.7 SYSTEMS OF DIFFERENTIAL EQUATIONS

So far in our models of population growth, we have supposed that the population of each species grows in isolation and is not influenced by any other species. While this is sometimes a reasonable supposition it also often happens that populations interact with one another to a considerable degree. Consider for example the two populations of deer and wolves in a certain region where the deer provide the main food supply for the wolf population. Then we can expect that the larger the deer population, the faster the wolf population will expand, that is, the deer population exerts a positive influence on the growth of the wolf population. On the other hand, the more wolves there are, the slower the deer population will grow; in fact with a large enough wolf population, the deer population will expand at a large negative rate. Thus the wolf population exerts a negative influence on the growth of the deer population.

This example is typical of the so-called *predator–prey* relationship between two species in which one species provides the food source for the other. Another commonly occurring type of interaction between species is the *competition* relationship in which the two species are competing for some resource, typically for the same limited supply of food. In this case each population exerts a negative influence on the other since the larger the population of the one species, the smaller the amount of food left over for the other species will be, hence the growth of this second population is inhibited. A third type of interaction is that found in *symbiosis*, in which each population exerts a positive influence on the other. For example, a population of trees may aid the expansion of a bird population by providing food and/or shelter while the birds may further the growth of the tree population by spreading the seeds.

While these examples illustrate some simple types of interaction between populations, there are many more complex interactions that occur in nature involving perhaps several different populations. We shall in this section restrict ourselves to some simple models that deal with interactions between only two populations.

Let us consider two species A and B that interact with each other, and let $x(t)$ be the size of the population of species A and $y(t)$ the size of the population of species B at time t. The rates of growth of these two populations are given by the derivatives dx/dt and dy/dt. Then, since the two populations influence one another, we can expect that the growth rate dx/dt for species A will depend on y, the size of population B, as well as on x; and similarly, dy/dt, the growth rate for species B, should depend on x as well as on y.

The simplest mathematical model of the growth of two interacting populations is obtained by assuming that the growth rates are given by linear functions of the population sizes,

$$\frac{dx}{dt} = Px + Qy, \qquad \frac{dy}{dt} = Rx + Sy, \tag{i}$$

where P, Q, R, and S are four given constants. In the first of these equations, if y is set equal to zero, we obtain the differential equation

$$\frac{dx}{dt} = Px,$$

whose solution is an exponential growth function. Thus the coefficient P is equal to the specific growth rate that the A species would have if the B species were absent. The term Qy in the first equation represents the influence of the B-species on the growth of the A-species. If the coefficient Q is positive, then this term Qy represents a positive contribution to the growth of species A, so that species B exerts a beneficial influence on species A in this case. On the other hand, if $Q < 0$, species B exerts a negative influence on the growth of species A.

Similarly, in the second differential equation, the term Sy corresponds to the natural exponential growth of population B while the term Rx accounts for the effect of species A on the growth of species B. Again, if $R > 0$, this effect is beneficial, whereas if $R < 0$, species A exerts a negative influence on species B.

The pair of differential equations (i) are said to form a *system of first-order differential equations*. We note that in this case the system contains two differential equations for the two unknown functions $x(t)$ and $y(t)$, and that each of the equations involves both x and y. Furthermore, in this case the system is *linear* since x and y and their derivatives appear linearly in each equation.

Our first task in this section will be to investigate the solution of the system (i). We shall do so by transforming the system into a single second-order differential equation of the type considered in the preceding sections.

The second differential equation in the system (i) can be written as

$$Rx = \frac{dy}{dt} - Sy, \tag{a}$$

thus expressing x in terms of y and its first derivative. Differentiating we obtain

$$R\frac{dx}{dt} = \frac{d^2y}{dt^2} - S\frac{dy}{dt}.$$

Now if the first differential equation in the system (i) is multiplied through by R it becomes

$$R\frac{dx}{dt} = PRx + QRy,$$

and so, after substitution for Rx and $R\,(dx/dt)$, we obtain

$$\frac{d^2y}{dt^2} - S\frac{dy}{dt} = P\left(\frac{dy}{dt} - Sy\right) + QRy.$$

Reorganizing, therefore, we get

$$\frac{d^2y}{dt^2} - (P + S)\frac{dy}{dt} + (PS - QR)y = 0,$$

which is an equation of the type solved in Sections 10.5 and 10.6.

Using the techniques developed in the preceding two sections, this equation can be solved for y as a function of t. Having found y, we can immediately determine x from equation (a) above.

EXAMPLE   Find the general solution of the system

$$\frac{dx}{dt} = x + y \qquad \frac{dy}{dt} = 4x + y.$$

SOLUTION   Proceeding as above we solve the second differential equation for x:

$$4x = \frac{dy}{dt} - y, \qquad 4\frac{dx}{dt} = \frac{d^2y}{dt^2} - \frac{dy}{dt}.$$

Therefore, substituting into the first equation, we find

$$4\frac{dx}{dt} = 4x + 4y,$$

i.e.,

$$\frac{d^2y}{dt^2} - \frac{dy}{dt} = \frac{dy}{dt} - y + 4y,$$

i.e.,

$$\frac{d^2y}{dt^2} - 2\frac{dy}{dt} - 3y = 0.$$

The auxiliary equation for this second-order differential equation is

$$k^2 - 2k - 3 = 0$$

or

$$(k - 3)(k + 1) = 0.$$

So there are two real roots, $k_1 = 3$ and $k_2 = -1$, and the general solution is

$$y = c_1 e^{k_1 t} + c_2 e^{k_2 t} = c_1 e^{3t} + c_2 e^{-t}.$$

Note that two unknown constants, c_1 and c_2, enter the solution. To find x we proceed as follows:

$$4x = \frac{dy}{dt} - y = \frac{d}{dt}(c_1 e^{3t} + c_2 e^{-t}) - (c_1 e^{3t} + c_2 e^{-t})$$

$$= 3c_1 e^{3t} - c_2 e^{-t} - (c_1 e^{3t} + c_2 e^{-t})$$

$$= 2c_1 e^{3t} - 2c_2 e^{-t}.$$

Therefore the solution is obtained in the form

$$x = \tfrac{1}{2}c_1 e^{3t} - \tfrac{1}{2}c_2 e^{-t}, \qquad y = c_1 e^{3t} + c_2 e^{-t}.$$

The two unknown constants c_1 and c_2 are not determined by the system of differential equations, but require additional conditions in order that their values be fixed. Usually these conditions would consist of the values of x and y being given for $t = 0$.

EXAMPLE   Find the solution of the system

$$\frac{dx}{dt} = x + y, \qquad \frac{dy}{dt} = -x + y$$

that satisfies the additional conditions that $x = 1$ and $y = 1$ when $t = 0$.

SOLUTION   From the second differential equation we find that

$$x = -\frac{dy}{dt} + y, \quad \text{and so} \quad \frac{dx}{dt} = -\frac{d^2 y}{dt^2} + \frac{dy}{dt}.$$

Substituting for x and dx/dt into the first differential equation, we find

$$-\frac{d^2 y}{dt^2} + \frac{dy}{dt} = -\frac{dy}{dt} + y + y.$$

Therefore

$$\frac{d^2 y}{dt^2} - 2\frac{dy}{dt} + 2y = 0.$$

The auxiliary equation is

$$k^2 - 2k + 2 = 0,$$

and it is readily seen that this equation has no real roots. In fact the above differential equation for y was solved in Section 10.6, and we found there the general solution

$$y = ce^t \cos (t + \alpha),$$

where c and α are two unknown constants. From this solution we can determine x:

$$x = -\frac{dy}{dt} + y$$

$$= -[ce^t \cos (t + \alpha) - ce^t \sin (t + \alpha)] + ce^t \cos (t + \alpha)$$

$$= ce^t \sin (t + \alpha).$$

In order to determine c and α we make use of the given initial values of x and y. Setting $t = 0$,

$$x = c \sin \alpha = 1, \qquad y = c \cos \alpha = 1.$$

Therefore

$$(c \sin \alpha)^2 + (c \cos \alpha)^2 = 1^2 + 1^2 = 2$$
$$= c^2 (\sin^2 \alpha + \cos^2 \alpha) = c^2$$

and so $c = \sqrt{2}$. Then

$$\sin \alpha = \cos \alpha = \frac{1}{c} = \frac{1}{\sqrt{2}}$$

so that $\alpha = \pi/4$. The final form of the solution is therefore

$$x = \sqrt{2} \, e^t \sin \left(t + \frac{\pi}{4} \right), \qquad y = \sqrt{2} \, e^t \cos \left(t + \frac{\pi}{4} \right).$$

As mentioned at the beginning of this section, the system of differential equations (i) can be used as a simple model of two interacting species. Let us consider in a little more detail the three particular types of interactions mentioned earlier—the predator–prey relationship, the competition relationship, and the symbiotic relationship between two species.

In the first case let species A with population x be the predator species and species B with population y be the prey species. Then we would expect that the larger the value of y, that is, the more prey there are available, then the greater should be the expansion of the predator population. Thus the coefficient Q in (i) should be positive, so that any increase in y produces an increase in dx/dt. On the other hand we would expect that the coefficient R should be negative since the bigger the x value, the more predators there are, and the slower should be expansion of the prey population. In fact if x is large enough then the prey population would decrease in size, that is, dy/dt would be negative.

The other two coefficients, P and S, would usually be positive since they represent the specific growth rates of each separate population. However if either of the species is overpopulated, so that the species is competing with itself, it could be more appropriate to use a model in which the corresponding coefficient P or S is negative.

Next consider the case in which the two species A and B are in competition for some limited resource. In this case we must use a model in which both Q and R are negative. For example, the bigger the value of y, the smaller will be the rate of growth of x since there will be less of the resource available to species A. Hence Q must be negative. Similarly, the bigger the population of species A, the less resource will be available to species B, hence the smaller dy/dt will be. Thus $R < 0$ also. As in the predator–prey case, P and S would usually be positive, though they could in certain cases be taken to be negative.

In the third case, that in which the two species enjoy a symbiotic relationship, both Q and R are positive since each population enhances the growth of the other. Again P and S would usually be positive. In this connection we should also mention the case when $Q = 0$ and $R > 0$. This corresponds to the situation in which species A enhances the growth of species B but species B has no effect on the growth of population of species A.

Two species in competition for the same food supply satisfy the following system of differential equations:

$$\frac{dx}{dt} = 3x - 2y, \qquad \frac{dy}{dt} = -2x + 3y.$$

The initial sizes of the populations are $x = 2000$, $y = 1600$ at $t = 0$. Find the population sizes at time t.

SOLUTION Proceeding as usual to eliminate one of the variables, we obtain from the second equation that

$$2x = -\frac{dy}{dt} + 3y$$

and hence

$$2\frac{dx}{dt} = -\frac{d^2y}{dt^2} + 3\frac{dy}{dt}$$

$$= 6x - 4y \qquad \text{(from the first differential equation)}$$

$$= 3\left(-\frac{dy}{dt} + 3y\right) - 4y.$$

Therefore we obtain

$$\frac{d^2y}{dt^2} - 6\frac{dy}{dt} + 5y = 0.$$

The auxiliary equation for this is

$$k^2 - 6k + 5 = 0$$

$$(k - 5)(k - 1) = 0.$$

The two roots are $k_1 = 5$ and $k_2 = 1$ so that the solution for y takes the form

$$y = c_1 e^{k_1 t} + c_2 e^{k_2 t} = c_1 e^{5t} + c_2 e^{t}.$$

Then for x we find that

$$x = -\frac{1}{2}\frac{dy}{dt} + \frac{3}{2}y = -\frac{1}{2}(5c_1 e^{5t} + c_2 e^{t}) + \frac{3}{2}(c_1 e^{5t} + c_2 e^{t})$$

$$= -c_1 e^{5t} + c_2 e^{t}.$$

The values of c_1 and c_2 are determined by setting $t = 0$. We find

$$x = -c_1 + c_2 = 2000, \qquad y = c_1 + c_2 = 1600$$

and therefore $c_1 = -200$, $c_2 = 1800$. The solution therefore finally takes the form

$$x = 200e^{5t} + 1800e^{t}, \qquad y = -200e^{5t} + 1800e^{t}.$$

The graphs of these solutions are shown in Fig. 10.9. It can be seen that $x(t)$ increases for all values of t. However $y(t)$, after increasing slowly for small values of t, reaches a maximum value and then decreases. It becomes zero at a

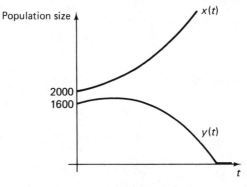

Figure 10.9

finite value of t, given by

$$-200e^{5t} + 1800e^t = 0,$$

i.e.,

$$e^{4t} = \tfrac{1800}{200} = 9$$

or

$$t = \tfrac{1}{4}\ln 9 = 0.549.$$

Therefore at $t = 0.549$, the population of species B becomes extinct. For $t > 0.549$, y remains identically zero since, once extinct, the species cannot revive. Of course it makes no sense to consider negative values of x or y in a population problem.

The methods of solution of systems of the type (i) given in this section can very readily be extended to systems of the type

$$\frac{dx}{dt} = Px + Qy + u, \qquad \frac{dy}{dt} = Rx + Sy + v, \tag{ii}$$

where u and v are constants. This system can be reduced to a system of type (i) by a simple trick. What we do is to write

$$x = x_1 + X(t), \qquad y = y_1 + Y(t),$$

where x_1 and y_1 are two constants that will be determined and $X(t)$ and $Y(t)$ are two new variables. Then, since $dx/dt = dX/dt$ and $dy/dt = dY/dt$, system (ii) becomes

$$\frac{dX}{dt} = P(x_1 + X) + Q(y_1 + Y) + u, \qquad \frac{dY}{dt} = R(x_1 + X) + S(y_1 + Y) + v.$$

If we now choose x_1 and y_1 in such a way that

$$Px_1 + Qy_1 + u = 0, \qquad Rx_1 + Sy_1 + v = 0, \tag{b}$$

then these two differential equations become

$$\frac{dX}{dt} = PX + QY, \qquad \frac{dY}{dt} = RX + SY.$$

Thus for the two new functions $X(t)$ and $Y(t)$ we obtain a system of differential equations of the same type as (i).

The above equations (b) form a pair of simultaneous algebraic equations for x_1 and y_1 that can be solved in the usual way.

EXAMPLE   Reduce the system

$$\frac{dx}{dt} = x + y + 1, \qquad \frac{dy}{dt} = -x + y - 2$$

to the form (i). Find its general solution.

SOLUTION   Writing $x = x_1 + X$ and $y = y_1 + Y$, we obtain the differential equations in this example in the form

$$\frac{dX}{dt} = X + Y + x_1 + y_1 + 1, \qquad \frac{dY}{dt} = -X + Y - x_1 + y_1 - 2.$$

Therefore we must choose

$$x_1 + y_1 + 1 = 0, \qquad -x_1 + y_1 - 2 = 0.$$

Adding these two equations together we get $0x_1 + 2y_1 - 1 = 0$, i.e., $2y_1 = 1$, $y_1 = \frac{1}{2}$. Then, from the first equation,

$$x_1 = -y_1 - 1 = -\tfrac{1}{2} - 1 = -\tfrac{3}{2}.$$

Thus if we set $x = -\frac{3}{2} + X(t)$ and $y = \frac{1}{2} + Y(t)$, the differential equations become

$$\frac{dX}{dt} = X + Y, \qquad \frac{dY}{dt} = -X + Y.$$

The solution of this system was actually found in an earlier example. Making use of it we see that the solution of the present example is

$$x = -\tfrac{3}{2} + ce^t \sin(t + \alpha),$$
$$y = \tfrac{1}{2} + ce^t \cos(t + \alpha),$$

where c and α are arbitrary constants.

EXAMPLE   Two species A and B coexist in a predator–prey relationship, A being the predator species with population $x(t)$, and B the prey species with population $y(t)$. These populations vary according to the differential equations

$$\frac{dx}{dt} = q(y - y_1), \qquad \frac{dy}{dt} = -r(x - x_1),$$

in which q, r, y_1, and x_1 are four positive constants. Find the solution for x and y.

SOLUTION   Note that when $x = x_1$ and $y = y_1$ the derivatives dx/dt and dy/dt are zero, so that x and y remain constant. Thus the population levels x_1 and y_1 represent an equilibrium situation for the two species. If, however, either of the populations at any time differs from these equilibrium sizes, then the population sizes will not remain constant but both of them will vary with time.

The given system is not quite in the form (i) because of the presence of the x_1 and y_1 terms in the two differential equations. This difficulty is minor, however, and is easily overcome by writing

$$x = x_1 + X, \qquad y = y_1 + Y.$$

Then

$$\frac{dx}{dt} = \frac{dX}{dt} = q(y - y_1) = qY$$

and

$$\frac{dy}{dt} = \frac{dY}{dt} = -r(x - x_1) = -rX.$$

Thus, in terms of X and Y, the given system takes the form

$$\frac{dX}{dt} = qY, \qquad \frac{dY}{dt} = -rX.$$

This is in the standard form (i) with coefficients given by $P = S = 0$, $Q = q$, and $R = -r$.

Eliminating X, we obtain the equivalent second-order equation

$$\frac{d^2Y}{dt^2} = -r\frac{dX}{dt} = -rqY,$$

that is,

$$\frac{d^2Y}{dt^2} + \omega^2 Y = 0,$$

where $\omega^2 = qr$. Thus we obtain for Y the equation of simple harmonic motion whose solution, as we saw in Section 10.6, is given by

$$Y = c \cos(\omega t + \alpha).$$

The corresponding value of X is found to be

$$X = -r^{-1}\frac{dY}{dt} = \omega r^{-1} c \sin(\omega t + \alpha).$$

The values of the two constants, c and α, can be found as usual in terms of the initial sizes of the populations. The actual sizes of the populations are given by $x = x_1 + X$ and $y = y_1 + Y$, and so are equal to

$$x = x_1 + \omega r^{-1} c \sin(\omega t + \alpha),$$

$$y = y_1 + c \cos(\omega t + \alpha).$$

Graphs of these solutions are shown in Fig. 10.10 for a typical case.

It can be observed that both x and y fluctuate sinusoidally in time with angular frequency ω (see Section 5.5). x fluctuates between a maximum value of $x_1 + \omega r^{-1} c$ and a minumum value of $x_1 - \omega r^{-1} c$ while y fluctuates between a maximum value of $y_1 + c$ and a minimum value of $y_1 - c$. If either of the minimum values $x_1 - \omega r^{-1} c$ or $y_1 - c$ is negative or zero, then the respective

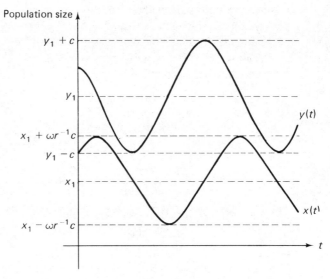

Figure 10.10

population will become extinct, and the predator–prey relationship will, of course, be destroyed. Otherwise the two populations will continue to oscillate up and down in size as indicated in Fig. 10.10.

It can also be observed that $x(t)$ reaches its maximum values later than $y(t)$ does. This is actually a quite realistic feature of a predator–prey situation, namely that increases and decreases in the predator population lag behind the corresponding increases and decreases in the prey population. For example, when the prey species reaches its maximum population size and begins to decrease, there will, for a time, still be enough prey around to permit the predator population to continue to increase. It will not be until the prey population has shrunk to some extent that the predator population is forced to begin to decrease by the shortage of prey. Similarly, when the prey population reaches its minimum value and starts to increase, there will, for a time, continue to be a shortage of prey so that the predator population will continue to decrease until the prey population has increased sufficiently. This qualitative behavior can be observed in the graphs shown in Fig. 10.10.

NOTE: In solving the system of differential equations (i), we eliminated x and so obtained a single differential equation of second order for y. Rather than eliminating x, we could eliminate y instead and get a single differential equation for x. The solution we would obtain this way would be the same as that found by eliminating x. When eliminating y, we must express y in terms of x and dx/dt from one of the differential equations, and then substitute for y and dy/dt into the second of the equations.

Find the general solutions of the following systems.

1. $\dfrac{dx}{dt} = y, \dfrac{dy}{dt} = -x.$　　　　　　　　　2. $\dfrac{dx}{dt} = y, \dfrac{dy}{dt} = x.$

3. $\dfrac{dx}{dt} = 4x - y, \dfrac{dy}{dt} = -4x + 4y.$　　　　4. $\dfrac{dy}{dt} = x, \dfrac{dx}{dt} = 6x - 8y.$

5. $\dfrac{dy}{dt} = 2y + x, \dfrac{dx}{dt} = -4y + 2x.$　　　　6. $\dfrac{dx}{dt} = y, \dfrac{dy}{dt} = 2x - 2y.$

7.–12. Find the particular solutions of exercises 1–6 that satisfy the conditions that $x = 1$ and $y = -1$ when $t = 0$.

13. Two species coexist in a symbiotic relationship, their populations $x(t)$ and $y(t)$ satisfying the system of differential equations

$$\frac{dx}{dt} = -2x + 4y, \qquad \frac{dy}{dt} = x - 2y.$$

Find the general solution for x and y. If at $t = 0$, the sizes of the two populations are $x = 1000$ and $y = 2000$, find the solution for x and y and sketch their graphs as functions of t.

14. In exercise 13 show that whatever the initial values of x and y, the quantity $(x + 2y)$ is constant, independent of t. Show also that as $t \to \infty$, x and y approach constant values, with $x \to 2y$.

15. A predator–prey relationship between two species is modeled by the pair of differential equations

$$\frac{dx}{dt} = x + y, \qquad \frac{dy}{dt} = y - 9x$$

(x being the population of the predator species, and y of the prey species). If the initial population sizes are $x = 100$ and $y = 1000$, find the solution for x and y. Show that the prey become extinct after a certain time, which should be found.

16. The development of two species in competition for a common resource is represented by the two equations

$$\frac{dx}{dt} = x - ky, \qquad \frac{dy}{dt} = -kx + y,$$

where k is a certain postive constant. Find the solution given that initially $x = 100$ and $y = 90$.

17. For the system of exercise 16, show that if $x > y$ at $t = 0$ then $x > y$ for all values of t and that the y-population becomes extinct after a certain finite time. What would be the corresponding properties if $x < y$ at $t = 0$? And what if $x = y$ at $t = 0$?

18. x is the population of a predator species, y the population of the prey, satisfying the differential equations

$$\frac{dx}{dt} = 2y - 1000, \qquad \frac{dy}{dt} = 800 - 8x.$$

Find the solutions for x and y as functions of t given that at $t = 0$, $x = 100$ and $y = 400$.

19. In exercise 18, find the solution if initially $x = 100$ and $y = 200$. Show that in this case x becomes zero after a certain finite time, which should be found.

20. The exchange of some resource between different component parts of a complex biological system can often be represented by a system of differential equations. Consider the exchange of some substance, for example, a drug, between the blood and the tissue of a human body. Let x be the concentration of the substance in the bloodstream and y the concentration in the body tissue; then the system of differential equations

$$\frac{dx}{dt} = k_1(y - x) - px, \qquad \frac{dy}{dt} = k_2(x - y)$$

can be used to describe the exchange process. The terms proportional to k_1 and k_2 refer to the rate of transfer between blood and tissue and the term px arises from the extraction of the drug from the bloodstream by the kidneys. If $k_1 = k_2 = 1$ and $p = \frac{1}{2}$, find the solution for x and y as functions of t. Show that x and y tend to zero as $t \longrightarrow \infty$.

10.8   SOME MORE ADVANCED MODELS

In this last section we shall discuss a few mathematical models based on differential equations that are of a somewhat more complex nature than those dealt with hitherto.

Maximum Sustained Yield

Let us suppose that a certain population, if left on its own would develop according to the logistic model,

$$\frac{dy}{dt} = py(y_m - y),$$

with p and y_m being constants. If this population is harvested at a constant rate h per unit time, then this differential equation must be modified by subtracting a term from the right-hand side equal to the rate of harvesting. We get the differential equation

$$\frac{dy}{dt} = py(y_m - y) - h. \tag{i}$$

We know, from the results in Section 10.3, that when h is zero, the population size always approaches the limiting value y_m as $t \to \infty$. Thus in this case y_m represents the stable equilibrium size of the population. Now we would expect intuitively that the effect of harvesting would be to reduce the equilibrium size of population below y_m, but as long as the rate of harvesting is not too large, presumably there would continue to exist a stable equilibrium size for the population. On the other hand, if the rate of harvesting becomes too great, then we would expect that no equilibrium would exist because a sufficiently large amount of harvesting would wipe out the population. The question is, what is the largest rate of harvesting that does not make the population become extinct?

Let us suppose that an equilibrium size of the population exists, and denote it by y_E. Then when $y = y_E$ we must have $dy/dt = 0$, since by definition the population size does not change when it equals y_E. From the differential equation therefore, when $y = y_E$,

$$\frac{dy}{dt} = py_E(y_m - y_E) - h = 0.$$

That is,

$$py_E^2 - py_m y_E + h = 0.$$

This equation is a quadratic equation for y_E, and the value of y_E is obtained from the usual quadratic formula:

$$y_E = \frac{py_m \pm \sqrt{(py_m)^2 - 4ph}}{2p}$$

$$= \frac{1}{2}\left[y_m \pm \sqrt{y_m^2 - \frac{4h}{p}}\right].$$

We see then that there are two possible equilibrium sizes of the population corresponding to the $\pm$ signs before the square root. We denote these two values of y_E by $y_+(h)$ and $y_-(h)$. The graphs of $y_\pm(h)$ as functions of

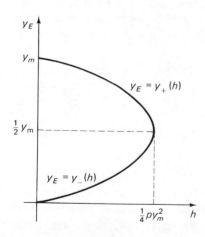

Figure 10.11

h are shown in Fig. 10.11. When $h = 0$, the two roots are $y_+(0) = y_m$ and $y_-(0) = 0$. As the harvesting rate h is increased from zero, $y_+(h)$ decreases from y_m while $y_-(h)$ increases from its initial value of zero. Eventually the two roots $y_\pm$ become equal to each other at the value of h, which makes the square root term equal to zero:

$$y_m^2 - \frac{4h}{p} = 0$$

$$h = \tfrac{1}{4}py_m^2.$$

When the rate of harvesting exceeds this value there is no real root for y_E. Hence in such a case there is no equilibrium size for the population. The value $h = \tfrac{1}{4}py_m^2$ is called the *maximum sustained yield* from the population, and represents the maximum rate at which the population can be harvested without being caused to decline toward extinction.

EXAMPLE It is estimated that for the global population of blue whales, $y_m = 200,000$ and $p = 6 \times 10^{-7}$ when time is measured in years. The maximum sustained yield of these whales is therefore

$$h = \tfrac{1}{4}(6 \times 10^{-7})(2 \times 10^5)^2 = 6000 \qquad \text{(per year).}$$

It is because many more than this number of blue whales were harvested prior to the 1950s that this population has been driven to the verge of extinction.

Stable and Unstable Equilibria

Let us examine the differential equation (i) in a little more detail. First of all let us consider the case when $h > \tfrac{1}{4}py_m^2$. Then (i) can be written

$$\frac{dy}{dt} = -p\left(y - \frac{1}{2}y_m\right)^2 - \left(h - \frac{1}{4}py_m^2\right),$$

as can easily be seen by expanding the square on the right-hand side of this last equation and cancelling out two of the terms. Now the term $-p(y - \tfrac{1}{2}y_m)^2$ cannot be positive, whatever the value of y. Therefore

$$\frac{dy}{dt} \leq -\left(h - \frac{1}{4}py_m^2\right).$$

It follows from this that when $h > \tfrac{1}{4}py_m^2$, the rate of change of the population is always negative, so that the population always decreases in size, and furthermore, the rate of decrease is at least equal to $(h - \tfrac{1}{4}py_m^2)$. In such a case the population will always become extinct, no matter what its initial size.

Now let us examine the more interesting case when $h < \tfrac{1}{4}py_m^2$. In this case there are two values of y, namely the values denoted above by $y_\pm$, at which the right-hand side of (i) is equal to zero. This differential equation can therefore be written

$$\frac{dy}{dt} = -p\left[\left(y - \frac{1}{2}y_m\right)^2 - \left(\frac{h}{p} - \frac{1}{4}y_m^2\right)\right]$$

$$= -p\left[y - \frac{1}{2}y_m + \sqrt{\frac{h}{p} - \frac{1}{4}y_m^2}\right]\left[y - \frac{1}{2}y_m - \sqrt{\frac{h}{p} - \frac{1}{4}y_m^2}\right],$$

after using the formula for the difference of two squares. Therefore we see that

$$\frac{dy}{dt} = -p(y - y_-)(y - y_+). \tag{ii}$$

Consider the case in which $y > y_+$. Then $y > y_-$ also and so both of the factors $(y - y_-)$ and $(y - y_+)$ on the right-hand side are positive. Therefore $dy/dt < 0$, and the population decreases in size. As y decreases toward y_+, the factor $(y - y_+)$ becomes small, so that the rate at which the population decreases slows down. In fact it turns out that y approaches y_+ asymptotically in the limit as $t \to \infty$. Thus y_+

represents the limiting stable size of population toward which any larger population will tend as $t \rightarrow \infty$. This type of solution is illustrated by the uppermost curve in Fig. 10.12.

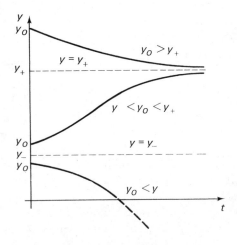

Figure 10.12

Next consider the case when $y_- < y < y_+$. Then the factor $(y - y_-)$ is positive but $(y - y_+)$ is negative, so that the right-hand side of the differential equation is positive. Consequently, $dy/dt > 0$, and the population size increases. Again as y approaches y_+, the rate of increase of y tends to zero and $y \rightarrow y_+$ as $t \rightarrow \infty$. This type of solution is represented by the middle curve in Fig. 10.12.

Finally, if $y < y_-$, then both $(y - y_-)$ and $(y - y_+)$ are negative, and so $dy/dt < 0$. In this case the population decreases, and in fact it passes through zero after a finite length of time.

We see then that as long as the initial size of the population exceeds y_-, the population eventually settles down to an equilibrium size equal to y_+. We speak of y_+ as representing the *stable equilibrium size* of the population. On the other hand, if the initial population size, y_0, is less than y_-, the population will decrease in size and will eventually become extinct. This happens if

$$y_0 < y_-, \quad \text{i.e.,} \quad y_0 < \frac{1}{2}\left[y_m - \sqrt{y_m^2 - \frac{4h}{p}}\right]$$

i.e.,

$$\sqrt{y_m^2 - \frac{4h}{p}} < y_m - 2y_0.$$

(The right-hand side here must be positive, so $y_0 < \frac{1}{2}y_m$.) Squaring both sides we find

$$y_m^2 - \frac{4h}{p} < (y_m - 2y_0)^2 = y_m^2 - 4y_m y_0 + 4y_0^2$$

and so

$$\frac{4h}{p} > 4y_m y_0 - 4y_0^2 = 4y_0(y_m - y_0).$$

We conclude therefore that if the initial population size y_0 is less than $\frac{1}{2}y_m$, the population will be made extinct by a rate of harvesting that exceeds the value $py_0(y_m - y_0)$.

EXAMPLE   For the blue whale population of the preceding example, the actual population in 1972 was estimated as 6000. The maximum rate of harvesting that would not cause the population to become extinct is therefore equal to

$$py_0(y_m - y_0) = (6 \times 10^{-7})(6000)[200,000 - 6000] = 698.$$

If more than 698 whales per year are harvested, the population according to this model will shrink even further.

The value $y = y_-$ is also an equilibrium size for the population in the sense that when $y = y_-$ then $(dy/dt) = 0$. However we observe from the preceding figure that solutions $y(t)$ that differ slightly from y_- do not approach y_- as $t \to \infty$, but actually diverge away from the value $y = y_-$. For this reason, y_- is termed an *unstable equilibrium* value of y.

The above conclusions regarding the behavior of various types of solutions of the differential equation (i) were obtained from a qualitative study of the differential equation itself, without having recourse to the solution. This type of method is very useful, since often we encounter differential equations whose solutions are not known. In such cases we can usually extract many useful properties of the solution by studies such as those we made above.

In the case of the present differential equation (ii) however, we are in a position to find the solution in closed form. The variables can be separated, and we can write the equation in the form

$$\frac{dy}{(y - y_-)(y - y_+)} = -p\,dt.$$

By partial fractions,

$$\frac{1}{(y - y_-)(y - y_+)} = \frac{1}{(y_+ - y_-)}\left\{\frac{1}{y - y_+} - \frac{1}{y - y_-}\right\}.$$

Noting that $y_+ - y_- = \sqrt{y_m^2 - (4h/p)}$, we see that the differential equation can be written in the form

$$\left\{\frac{1}{y - y_+} - \frac{1}{y - y_-}\right\}dy = -k\,dt,$$

where $k = \sqrt{p^2 y_m^2 - 4hp}$. Integrating, therefore, we get

$$\ln\left|\frac{y - y_+}{y - y_-}\right| = -kt + C.$$

If as usual we write $C = \ln|c|$, where the sign of c is chosen to be the same as the sign of $(y - y_+)/(y - y_-)$, the two logarithms can be combined in the form

$$\ln\left\{\frac{(y - y_+)}{c(y - y_-)}\right\} = -kt,$$

and therefore

$$\frac{y - y_+}{y - y_-} = ce^{-kt}. \tag{a}$$

Thus

$$y - y_+ = (y - y_-)ce^{-kt}$$

$$y(1 - ce^{-kt}) = y_+ - y_-ce^{-kt},$$

so that the solution is finally obtained in the form

$$y = \frac{y_+ - y_-ce^{-kt}}{1 - ce^{-kt}}.$$

The value of the constant c is as usual found from the initial value of y. Setting $t = 0$ and $y = y_0$ in (a) above we see immediately that

$$c = \frac{y_0 - y_+}{y_0 - y_-}.$$

The qualitative properties of the solution that were derived earlier can all be proved starting from this form for the solution. For example, if $y_0 < y_-$, the population decreases in size and becomes extinct after a finite interval of time. The value of t at which extinction occurs is found by setting $y = 0$, and is given by

$$y_+ - y_-ce^{-kt} = 0,$$

i.e.,

$$e^{kt} = \frac{y_-c}{y_+} = \frac{y_-(y_+ - y_0)}{y_+(y_- - y_0)},$$

so that

$$t = \frac{1}{k} \ln \left\{ \frac{y_-(y_+ - y_0)}{y_+(y_- - y_0)} \right\},$$

Lotka–Volterra Equations

In Section 10.7 we discussed several models of interacting populations. All of the models consisted of systems of differential equations, each equation in the system being linear in the variables x and y, which represent the two population sizes. Models of this type, based on linear equations, can only represent realistic behavior at best within a limited range of population sizes. The situation is analogous to the exponential model of the growth of a single population: Within a limited range of values of the population size, the exponential model can be used, but outside this range a nonlinear differential equation such as the logistic equation must be used. Similarly for interacting populations we must expect to require a nonlinear system of differential equations in order to provide a realistic model.

The earliest of such models was devised in the 1920s by Lotka and Volterra as a model of the predator–prey relationship between two species. This type of model has since been modified and extended to other kinds of interaction. The original Lotka–Volterra equations can be derived in the following way.

Let $x(t)$ be the size of the predator population and $y(t)$ the size of the prey population at time t. Then we can write

$$\frac{dx}{dt} = (B_1 - D_1)x, \qquad \frac{dy}{dt} = (B_2 - D_2)y,$$

where B_1 and B_2 are the specific birth rates and D_1 and D_2 are the specific death rates of the predator and prey populations. We can reasonably assume that B_1 and B_2 are constant, since the birth rates should not be very strongly influenced by population sizes. However the death rates can be expected to be quite strongly dependent on population sizes. For example, the specific death rate D_1 of the predator species will decrease as the population y of prey increases. We can assume as the simplest guess that

$$D_1 = M - Py,$$

where M and P are two positive constants; that is, the death rate of predators decreases linearly with the size of prey population. Similarly the specific death rate D_2 of the prey population will increase as the size x of the predator population increases, and the simplest model is again to take a linear function:

$$D_2 = N + Qx,$$

where N and Q are positive constants.

Substituting these two models for D_1 and D_2 into the differential equations, we find

$$\frac{dx}{dt} = [Py - (M - B_1)]x, \qquad \frac{dy}{dt} = [(B_2 - N) - Qx]y.$$

If for brevity we denote $M - B_1 = C$ and $B_2 - N = D$, we have

$$\frac{dx}{dt} = (Py - C)x, \qquad \frac{dy}{dt} = (D - Qx)y.$$

These differential equations are known as the Lotka–Volterra equations.

NOTE: If $y = 0$, that is, there is no prey population, then $dx/dt = -Cx$. Since it can reasonably be expected that the predator population would decrease in such a situation, we can expect that $C > 0$. Similarly, if $x = 0$, that is, there are no predators, then $dy/dt = Dy$. Therefore we can reasonably take $D > 0$ also, since the prey population should expand in the absence of predators.

The Lotka–Volterra equations cannot be solved in closed form. However a partial solution can be found. Let us begin by establishing certain qualitative properties of the solution.

We note first of all that the equations can be written

$$\frac{dx}{dt} = Px\left(y - \frac{C}{P}\right), \qquad \frac{dy}{dt} = Qy\left(\frac{D}{Q} - x\right).$$

Thus $dx/dt = 0$ when $y = C/P$ and $dy/dt = 0$ when $x = D/Q$. These values represent an equilibrium point for the two populations: If $x = D/Q$ and $y = C/P$ then the population sizes remain constant as functions of time. However if either of the population sizes differs from its equilibrium value, then both x and y change with time.

Let us examine the case where x and y are close to their equilibrium values. Initially we set

$$x = \frac{D}{Q} + X, \qquad y = \frac{C}{P} + Y,$$

where X and Y represent the differences between the population sizes at any time and their equilibrium values. Then $dx/dt = dX/dt$ and $dy/dt = dY/dt$, so that the differential equations take the form

$$\frac{dX}{dt} = P\left(\frac{D}{Q} + X\right)Y, \qquad \frac{dY}{dt} = Q\left(\frac{C}{P} + Y\right)(-X).$$

Now by assumption X and Y are small since the populations are close to equilibrium; therefore the terms involving the product XY in these differential equations are doubly small, being the product of two small quantities, and these terms can be neglected as a first approximation. Thus we have approximately

$$\frac{dX}{dt} = \frac{PD}{Q}Y, \qquad \frac{dY}{dt} = -\frac{QC}{P}X.$$

The solution of this system can be obtained straightforwardly using the methods of Section 10.7, since it is linear in X and Y. The solution can be written in the form

$$X = P\sqrt{D}\,a\sin(\omega t + \alpha), \qquad Y = Q\sqrt{C}\,a\cos(\omega t + \alpha),$$

where a and α are the two arbitrary constants of integration and $\omega = \sqrt{CD}$. Therefore the population sizes themselves are given by

$$x = \frac{D}{Q} + P\sqrt{D}\,a\sin(\omega t + \alpha), \qquad y = \frac{C}{P} + Q\sqrt{C}\,a\cos(\omega t + \alpha).$$

In Fig. 10.13 we plot the curve in the xy-plane that is traced out by this solution. In order to do this, we regard the above equations as parametric equations for x and y

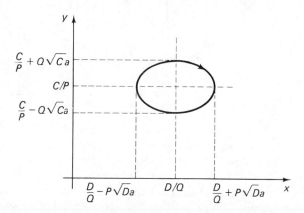

Figure 10.13

in terms of the "parameter" t. The curve is actually an ellipse with center at the equilibrium point $x = D/Q$, $y = C/P$. As t increases, the point (x, y) corresponding to the solution moves around the ellipse in the direction shown in Fig. 10.13. The motion around the ellipse is periodic in time, with period equal to $2\pi/\omega$. The path of the solution in the xy-plane is called its *trajectory*.

We note that this approximate solution of the Lotka–Volterra equations actually corresponds to the predator–prey model described in the last worked example of Section 10.7.

Now let us return to the general Lotka–Volterra equations. Dividing one equation by the other we can eliminate t, obtaining

$$\frac{dy}{dx} = \frac{dy/dt}{dx/dt} = \frac{y(D - Qx)}{x(Py - C)}.$$

In this form the variables can be separated, and we get

$$\left(P - \frac{C}{y}\right) dy = \left(\frac{D}{x} - Q\right) dx.$$

Therefore, integrating, we find

$$Py - C \ln y = D \ln x - Qx + k,$$

k being the constant of integration.

This relation between x and y has as its graph in the xy-plane an oval-shaped curve that forms the trajectory of the solution, shown in Fig. 10.14. As t increases,

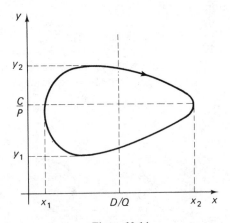

Figure 10.14

the point (x, y) corresponding to the solution of the Lotka–Volterra equations moves around the oval in the direction indicated. Thus again x and y vary periodically in time, x ranging between a minimum value x_1 and a maximum value x_2 and y varying between a minimum value y_1 and a maximum value y_2. We note that x, the predator population, reaches its peak size x_2 after y, the prey population has passed its peak size y_2. And similarly x reaches its minimum value x_1 after y has passed its minimum value y_1.

The oval always encloses the equilibrium point $(D/Q, C/P)$. In general the periodic motion of x and y is not sinusoidal, but if the oval is a very small one, then the motion is approximately sinusoidal, and the oval itself becomes approximately elliptical in shape, with the equilibrium point at its center.

Numerical Solutions

Almost all of the differential equations in this chapter have had solutions that could be expressed in closed form by means of some formula involving algebraic or transcendental functions. The main exception to this has been the Lotka–Volterra equations. In this latter case, although the equation of the trajectory in the xy-plane can be obtained in closed form, it is not possible to obtain closed-form expressions for x and y as functions of time.

Many differential equations fall into this category; in fact, more often than not, a model of a complex process will lead to a differential equation or system of equations for which the solution cannot be found in closed form. In such a situation it is necessary to resort to what are called numerical methods of solution by means of which approximate solutions in numerical form can be calculated. By using a modern computer, the solution of almost any differential equation can be generated quickly and accurately with such methods. (Even some currently available programmable pocket calculators can be used to solve first-order differential equations of considerable complexity.)

The subject of computer methods of solution of differential equations is a very extensive one and we do not wish to pursue a discussion of it here. For further reading we refer the interested student to more specialized books on numerical analysis or else to the manuals available at his/her computer center.

EXERCISES 10.8

Solve the following differential equations:

1. $\dfrac{dy}{dt} = (y - 1)(2 - y)$.

2. $\dfrac{dy}{dt} = \dfrac{1}{2}y(6 - y) - 4$.

3. $\dfrac{dy}{dt} = y(2 - y) - 2$.

4. $\dfrac{dy}{dt} = y(1 - y) - 1$.

Find the equilibrium values of y for the following differential equations. By examining the sign of dy/dt for values of y close to the equilibrium, show whether each equilibrium is stable or unstable.

5. $\dfrac{dy}{dt} = y(1 - y)$.

6. $\dfrac{dy}{dt} = y(y - 1)$.

7. $\dfrac{dy}{dt} = y^5 - y^3$.

8. $\dfrac{dy}{dt} = 2y(y-1)(y-2)$.

9. $\dfrac{dy}{dt} = (y-1)^2(2-y)$.

10. $\dfrac{dy}{dt} = (1-y)(2-y)^2$.

Find the equations of the trajectories for the following systems:

11. $\dfrac{dx}{dt} = x(a-y), \dfrac{dy}{dt} = xy$.

12. $\dfrac{dx}{dt} = 1 - y, \dfrac{dy}{dt} = xy^3$.

13. A population whose natural growth follows the logistic model and that is also gaining members through immigration at the rate b per unit time would be governed by the differential equation

$$\frac{dy}{dt} = py(y_m - y) + b.$$

Find the solution of this equation for which $y = y_0$ when $t = 0$. Show that as $t \rightarrow \infty$, y approaches a limiting value that is greater than the natural equilibrium level y_m.

14. Suppose that in exercise 13, $y_0 = 10{,}000$, $y_m = 20{,}000$, $py_m = 0.04$, and $b = 100$, time being measured in years. What is the population size after 10 yr? What is the limiting population size as $t \rightarrow \infty$?

15. Find the solution of the differential equation (i) in this section, namely

$$\frac{dy}{dt} = py(y_m - y) - h,$$

for the case where $h > \frac{1}{4}py_m^2$. Show that the population always becomes extinct and find the time at which extinction occurs.

16. One extension of the Lotka–Volterra equations is to the more general system

$$\frac{dx}{dt} = x(A - Px - Qy), \qquad \frac{dy}{dt} = y(B - Rx - Sy).$$

If all of the coefficients A, B, P, Q, R, and S are positive, we can regard this as a model of two competing species.
(a) Show that in the absence of one of the species, the other species grows according to the logistic equation.
(b) Find the equilibrium point for this model at which x and y are both nonzero.
(c) Obtain the solution $x(t)$, $y(t)$ for the case when x and y always remain close to this equilibrium point. Draw the trajectory.
(d) The two lines $Px + Qy = A$ and $Rx + Sy = B$ divide the first quadrant of the xy-plane into four regions. Discuss the signs of dx/dt and dy/dt in each of these regions.

17. Repeat parts (b), (c), and (d) of the preceding exercise for the following system:

$$\frac{dx}{dt} = x(2 - x - y), \qquad \frac{dy}{dt} = y(3 - 2x - y).$$

18. The differential equations

$$\frac{dx}{dt} = x(1 + y - x), \qquad \frac{dy}{dt} = y(1 + x - 2y)$$

form a particular model for the growth of two cooperating species growing in symbiosis. Find the equilibrium sizes of each of the two populations in the absence of the other. Show that $x = 3$, $y = 2$ is an equilibrium point for the interacting populations. By setting $x = 3 + X$ and $y = 2 + Y$ where X and Y are small, obtain the approximate solutions for X and Y and show that X and Y both approach zero as $t \to \infty$. What conclusion can be drawn regarding the point $(3, 2)$?

19. Modify the Lotka–Volterra equations for the case in which both predator and prey populations are harvested at constant rates h and k respectively. Show that equilibrium points continue to exist for the modified equations no matter what the sizes of h and k. Is this a realistic feature of the model?

20. A simple model of the spread of an infectious disease through a population can be constructed as follows. Let $x(t)$ be the number of uninfected but susceptible individuals in the population at time t and let $y(t)$ be the number of infected individuals at time t. Then we suppose that

$$\frac{dx}{dt} = -axy, \qquad \frac{dy}{dt} = axy - by,$$

where a and b are constant. (The term axy is the rate at which susceptible individuals become infected, proportional to both x and y. The term by is the rate at which infected individuals become non-infectious, as a result of cure, death, or some other form of removal.) Find the equation of the trajectory in the xy-plane. Show that y achieves its maximum value when $x = b/a$ regardless of the initial values of x and y. What conclusion can you draw if the initial number of susceptibles in the population is smaller than b/a?

REVIEW EXERCISES FOR CHAPTER 10

1. State whether the following statements are true or false. If false, give the corresponding correct statement.
   (a) The function $y = ce^{kt}$, where c is an arbitrary constant, is the most general solution of the differential equation $dy/dt = ky$.
   (b) The function $y = e^{kt}$ is a solution of the differential equation $y(d^2y/dt^2) = (dy/dt)^2$ for all values of k.
   (c) The equation $dy/dt = y + t$ can be solved by separation of variables.
   (d) The function $y = c \sin(\omega t + \alpha)$ satisfies the differential equation $(d^2y/dt^2) + \omega^2 y = 0$ for all values of c and α.
   (e) Any population for which growth is limited must obey the logistic model.

(f)   The solution of the differential equation $dy/dt = yt^2$ can be obtained as follows:

$$y = \int yt^2 \, dt = y \int t^2 \, dt = \tfrac{1}{3}yt^3 + C.$$

Therefore

$$y(1 - \tfrac{1}{3}t^3) = C, \quad \text{or} \quad y = C(1 - \tfrac{1}{3}t^3)^{-1}.$$

(g)   Any system of first-order linear differential equations can be reduced to a single differential equation whose order is 2.

(h)   The differential equation $2(dy/dt)^2 = t^2(dy/dt)$ is of second order.

(i)   The differential equation $(d^2y/dt^2) - [1 + (dy/dt)]\sin t = 0$ is of first order.

Find the general solutions of the following differential equations:

2. $3\dfrac{dy}{dt} + 4y = 0.$
   3. $\dfrac{dy}{dt} + 2y = t.$

4. $\dfrac{dy}{dt} + y(1 - y) = 0.$
   5. $\dfrac{dy}{dt} = y^2 t^2.$

6. $\dfrac{d^2y}{dt^2} + 5\dfrac{dy}{dt^2} - 6y = 0.$
   7. $\dfrac{d^2y}{dt^2} + 4\dfrac{dy}{dt} + 5y = 0.$

Find the solutions of the following differential equations that satisfy the given conditions:

8. $\dfrac{dy}{dt} = y + \sin t; \; y = 1$ when $t = 0.$
   9. $\dfrac{dy}{dt} = \dfrac{y}{1 - y}; \; y = \dfrac{1}{2}$ when $t = 0.$

10. $2\dfrac{d^2y}{dt^2} + \dfrac{dy}{dt} + y = 0; \; y = 1, \dfrac{dy}{dt} = 0$ when $t = 0.$

11. $2\dfrac{d^2y}{dt^2} - 6\dfrac{dy}{dt} + 5y = 0; \; y = 0, \dfrac{dy}{dt} = 1$ when $t = 1.$

12. $\dfrac{dx}{dt} = 2x + 2y, \dfrac{dy}{dt} = 8x + 2y; \; x = 1, y = 2$ when $t = 0.$

13. $\dfrac{dx}{dt} = x - 4y, \dfrac{dy}{dt} = x + 5y; \; x = 3, y = 2$ when $t = 0.$

14. $\dfrac{dx}{dt} = x + y, \dfrac{dy}{dt} = 2y; \; x = 0, y = 1$ when $t = 0.$

15. Strontium-90 has a decay constant equal to 0.028 when time is measured in years. Write a differential equation for the decay of strontium-90. What is the half-life? How many years does it take for 90% of a specimen of strontium-90 to decay?

16. Let y be the yield of a certain crop per acre and x the amount of fertilizer used. Then the change in y as a function of changes in x is sometimes modeled by the differential equation

$$\frac{dy}{dx} = k(y_m - y).$$

Find y as a function of x, given that the yield is equal to y_0 when no fertilizer is used. What is the physical meaning of y_m?

17. The logistic equation is sometimes used to describe the spread of information through a group of individuals, for example, the spread of a rumor through a society. Let $p(t)$ be the proportion of individuals in the group who have heard the information. Then according to the logistic model,

$$\frac{dp}{dt} = kp(1 - p),$$

where k is a constant. Find p as a function of t. If at $t = 0$, 10% of the group have heard the information, find the value of t at which half the group has heard it.

18. A radioactive tracer (C^{14}) is used to investigate the rate of exchange of carbon between a plant leaf and the atmosphere. The leaf is made to absorb C^{14} so that the initial amount of C^{14} in the leaf as a proportion of the total carbon content is equal to s_0 (the so-called *specific activity*). As a result of exchange of carbon between leaf and atmosphere, the proportion of C^{14} changes as time progresses. If this proportion is equal to $s(t)$ at time t, show that $ds/dt = -ks$, where k is the proportion of carbon atoms exchanged per unit time. Find s as a function of t. How could your result be used to measure k experimentally?

19. The diagram represents a biochemical reaction in which an intermediate product I (amino acid) is transformed into a protein M and also into waste products X. The amino acid is supplied to the system at a steady rate from a "food source" F. The food source is labeled by means of a radioactive tracer so that as time develops, proportions $s_I(t)$ and $s_M(t)$ of the products I and M involve radioactive atoms. These quantities satisfy the differential equations

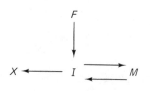

$$\frac{ds_M}{dt} = k_1(s_I - s_M), \qquad \frac{ds_I}{dt} = k_2(s_M - s_I) + k_3(s_F - s_I),$$

where k_1, k_2, and k_3 are constants related to the reaction rates for the parts of the reaction. $s_F(t)$ is the proportion of radioactive tracer in the food supply. Obtain the solution for $s_I(t)$ and $s_F(t)$ in the cases:
(a) $s_F(t) = 0$, $s_I(0) = s_0$, $s_M(0) = 0$;
(b) $s_F(t) = $ constant, $s_I(0) = s_M(0) = 0$.

20. Let $y(t)$ be the weight of a culture of yeast growing in sugar solution. As the culture grows it produces alcohol, $x(t)$ denoting the amount of alcohol produced at time t. The alcohol inhibits the growth of the yeast. We can represent this situation by the differential equations

$$\frac{dx}{dt} = Qy, \qquad \frac{dy}{dt} = R(x_m - x)y,$$

where Q, R, and x_m are certain positive constants. Note that when x exceeds x_m, the yeast culture decreases in size. Find the solution for x and y as functions of t given that x and y are very small when $t = 0$. Show that $y \to 0$ and $x \to 2x_m$ as $t \to \infty$. [*Hint:* by dividing one of the differential equations by the other, show that $y = (R/2Q)(2x_m - x)x$].

21. Show that the differential equation $t(dy/dt) - 2y = 0$ has the solution $y = c_1 t^2$ in the region $t > 0$. Explain why the method of separation fails to produce a solution at $t = 0$. Show that the function

$$y = \begin{cases} c_1 t^2 & t \geq 0 \\ c_2 t^2 & t \leq 0, \end{cases}$$

where c_1 and c_2 are two arbitrary constants, satisfies the differential equation for all values of t. (Note that this solution involves two arbitrary constants even though the differential equation is of first order. The point $t = 0$ is a singular point for this equation.)

Difference
Equations

11.1 DISCRETE TIME MODELS

In Chapter 10 we discussed mathematical models of a number of biological processes taking place in time. In these models the time variable t was treated as a continuous variable taking all real values in some interval. The process in question was modeled by one or more functions that vary smoothly as functions of t. However in certain instances it makes more sense, when building a mathematical model, to treat time as a discrete variable taking only integer values.

For example, let us suppose that we wish to devise a model of the growth of a certain deer population. Through the year the size of such a population would vary in a smooth manner except during the spring birth season when, within a very short space of time, the population increases greatly in size. The rate of increase during this period is so large as to make the change appear to be almost discontinuous. Instead of treating the population as a continuous function of t, in such a situation it is more appropriate to consider the population size only at certain discrete intervals of time. For example, we might consider the population at the end of each birth season and build a model that takes account of the change of population from one year to the next. Such a model would ignore the detailed fluctuation of the population during the course of the year from the end of one birth season to the end of the next.

In fact a model of this kind makes sense from another point of view. Usually in the case of wild animal populations, counts of population size are conducted only at regular time intervals, typically once per year. Thus in such cases there is only empirical data governing the change in population from one year to the next. In such a situation it is more appropriate to take the discrete view of time rather than the continuous view, in other words, to construct a model that involves as variables the population at the time of each count.

A typical model of this kind would lead to an equation relating the population size at any count to the population sizes at previous counts. An equation of this type is called a *difference equation*, and we shall study such equations briefly in this chapter. We shall find that there is a close connection between the solutions (and methods of solution also) of difference equations and of differential equations.

Let us begin by considering a simple example related to the deer population mentioned earlier. Suppose that the growth of the population from one year to the next is proportional to the population size in the first of these two years, the constant of proportionality being k. So we have the following equation:

[Population after n years] — [Population after $(n-1)$ years]

$$= k[\text{Population after } (n-1) \text{ years}].$$

The left-hand side here gives the increase in the population between the $(n-1)$th and the nth years, and we have supposed that this is equal to k times the population in the $(n-1)$th year.

Let x_n denote the size of the population after n years; the size after $(n-1)$ years will then of course be x_{n-1}. The above equation takes the form

$$x_n - x_{n-1} = kx_{n-1}$$

or

$$x_n = (1 + \mathrm{k})x_{n-1}.$$

We can expect this equation to provide a reasonable model of the growth of a population that is not subject to limiting pressures from its environment. In this case, the number of births in the nth year should be a fixed proportion of x_{n-1}, the population after the previous birth season; the number of deaths during the intervening year should likewise be in proportion to x_{n-1}. Thus the total increase in population (number of births minus number of deaths) can reasonably be taken proportional to x_{n-1}.

The above equation relating x_n to x_{n-1} is an example of a first-order difference equation.

DEFINITION   Let n be a variable taking integer values and let x_n denote the value of a variable x as a function of n. Then a *difference equation* is an equation that relates the values of x_n for different values of n. If P and Q are the largest and smallest values respectively of n that occur in the difference equation then the *order* is equal to $P - Q$. The difference equation is said to be *linear* if each of the x_n's appearing in the equation enters it linearly. An equation that is not linear is called a *nonlinear* difference equation.

EXAMPLES   (a) $x_{n+1} - 2x_n = 1$.
Here $P = n + 1$ and $Q = n$ are the highest and lowest values of n, so the order $(P - Q)$ is equal to 1. The equation is linear since x_{n+1} and x_n enter linearly.

(b) $x_{n+1} + x_n = e^{x_n}$.
Again this is a first-order equation. It is nonlinear, however, because of the e^{x_n} term.

(c) $x_{n+2} - 2x_{n+1} = \dfrac{1}{x_{n-1} + 1}$.
Here $P = n + 2$ and $Q = n - 1$ so that the order is equal to $(n + 2) - (n - 1) = 3$. The left-hand side is linear, but the reciprocal on the right is not, so that the equation is a nonlinear, third-order difference equation.

(d) $x_{n+1} - 2nx_{n-1} = n^2$.
This equation is of second order $(P = n + 1, Q = n - 1)$ and is linear.

DEFINITION   A *solution* of a difference equation is a set of values for x_n as a function of n such that when these values are substituted into the difference equation, this latter equation is satisfied for all n.

EXAMPLES   (a) $x_n = \frac{1}{2}n(n + 1)$ is a solution of the difference equation $x_n - x_{n-1} = n$. For if $x_n = \frac{1}{2}n(n + 1)$ then $x_{n-1} = \frac{1}{2}(n - 1)(n - 1 + 1) = \frac{1}{2}n(n - 1)$, and so

$$\begin{aligned}
x_n - x_{n-1} &= \tfrac{1}{2}n(n + 1) - \tfrac{1}{2}n(n - 1) \\
&= \tfrac{1}{2}n[(n + 1) - (n - 1)] \\
&= \tfrac{1}{2}n(2) = n.
\end{aligned}$$

Therefore the given function $x_n = \frac{1}{2}n(n + 1)$ satisfies the difference equation.

(b) $x_n = 1/(c + n)$ ($c = $ constant) is a solution of the difference equation

$$x_{n+1}(1 + x_n) = x_n.$$

For if $x_n = 1/(c + n)$ then $x_{n+1} = 1/(c + n + 1)$. Therefore

$$x_{n+1}(1 + x_n) = \frac{1}{c + n + 1}\left(1 + \frac{1}{c + n}\right)$$

$$= \frac{1}{c + n + 1} \cdot \frac{c + n + 1}{c + n}$$

$$= \frac{1}{c + n} = x_n.$$

Thus the given difference equation is satisfied by this expression for x_n.

In these examples the solution x_n is given as a function of n by means of an algebraic formula. However, another way which can be used to state the solution of any difference equation is to give a table of values of x_n for a series of values of n. In fact, such a form of solution *must* be used in many cases when an algebraic expression for the solution cannot be found. In our initial discussion of difference equations we shall concentrate mainly on this form of solution. In later sections of the chapter we shall study certain linear difference equations with the aim of producing algebraic expressions for the solutions.

EXAMPLE   A certain population has an initial size of 1000 and grows by 25% each year. Write down a difference equation for the size x_n of this population after n years. Give a table of values of x_n for $n = 0, 1, 2, \ldots, 10$.

SOLUTION   The growth of population between the $(n - 1)$th and the nth years is $x_n - x_{n-1}$. This is given to be equal to 25% of x_{n-1}. Therefore

$$x_n - x_{n-1} = 0.25x_{n-1},$$

i.e.,

$$x_n = 1.25x_{n-1}.$$

This is the required difference equation. Since this equation holds for each value of n, we can set $n = 1$ in particular, and obtain

$$x_1 = 1.25x_0.$$

Now x_0 is the initial population and is given to be equal to 1000. Therefore

$$x_1 = 1.25(1000) = 1250.$$

Thus after one year the population has increased in size to 1250. Clearly, the increase of 250 represents 25% of the initial population of 1000. Now let us put $n = 2$ in the difference equation. We get

$$x_2 = 1.25x_1$$

$$= 1.25(1250) = 1562.5.$$

Thus, rounding off the decimal, we see that after 2 yr the population has increased in size to 1563. We can continue in this way; for example,

$$x_3 = 1.25x_2 = 1.25(1562.5) = 1953.125,$$

$$x_4 = 1.25x_3 = 2441.4,$$

and so on. After rounding off to the nearest whole number, the values of x_n for $n = 0, 1, 2, \ldots, 10$ are given in the following table.

n	x_n
0	1000
1	1250
2	1563
3	1953
4	2441
5	3052
6	3815
7	4768
8	5960
9	7451
10	9313

The graph of the solution is shown in Fig. 11.1, where x_n is plotted as a function of n. Since n takes only integer values, the graph consists of a number of discrete points. The solution in this case represents a typical example of *exponential growth*.

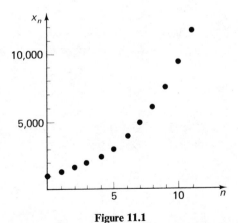

Figure 11.1

In this particular example it is also possible to contruct a formula for x_n as a function of n. Putting $n = 1, 2, 3$, etc. successively in the difference equation, we obtain the sequence:

$$x_1 = 1.25x_0, \quad x_2 = 1.25x_1, \quad x_3 = 1.25x_2, \quad \text{and so on.}$$

Thus

$$x_2 = 1.25x_1 = 1.25(1.25x_0) = (1.25)^2 x_0$$
$$x_3 = 1.25x_2 = 1.25[(1.25)^2 x_0] = (1.25)^3 x_0$$
$$x_4 = 1.25x_3 = 1.25[(1.25)^3 x_0] = (1.25)^4 x_0, \text{ etc.}$$

It should be pretty clear now what the formula is going to be for x_n. It is

$$x_n = (1.25)^n x_0.$$

Since $x_0 = 1000$, this becomes

$$x_n = 1000(1.25)^n,$$

a formula that allows us to calculate x_n for any value of n. For example, setting $n = 8$, we find

$$x_8 = 1000(1.25)^8 = 1000(5.96046) = 5960.46.$$

EXAMPLE   A certain animal population is increasing by 10% each year. If the initial population is 100, how many years will it take the population to exceed 200 individuals?

SOLUTION   If x_n is the population after n yr, the increase during the nth year is the difference $x_n - x_{n-1}$. This is given to be equal to 10% of x_{n-1}, so that

$$x_n - x_{n-1} = 0.1x_{n-1}.$$

Therefore

$$x_n = x_{n-1} + 0.1x_{n-1} \quad \text{or} \quad x_n = 1.1x_{n-1}.$$

Setting $n = 1$ we find that

$$x_1 = 1.1x_0 = 1.1(100)$$
$$= 110,$$

where we have used the fact that the initial population size x_0 is 100. Next, setting $n = 2$ in the difference equation, we get

$$x_2 = 1.1x_1 = 1.1(110)$$
$$= 121.$$

Continuing in this way we obtain the sequence of values of x_n given in the table. It is clear from these values that x_7 is less than 200 while x_8 is greater. Therefore it takes 8 yr before the population size exceeds 200.

n	x_n
0	100
1	110
2	121
3	133
4	146
5	161
6	177
7	195
8	214

This example can also be solved by developing a formula for the solution. From the difference equation $x_n = 1.1x_{n-1}$, by putting n successively equal to 1, 2, 3, and so on, we obtain

$$x_1 = 1.1x_0$$
$$x_2 = 1.1x_1 = (1.1)^2 x_0$$
$$x_3 = 1.1x_2 = (1.1)^3 x_0 \quad \text{and so on.}$$

In general it can be seen that

$$x_n = (1.1)^n x_0 = 100(1.1)^n.$$

It is then easy enough to verify that x_7 is less that 200 but x_8 is greater than 200.

The difference equations which have arisen in the examples so far have been of the type $x_n = ax_{n-1}$ where a is a certain constant. We shall now move on to an example which leads to another type of difference equation, $x_n = ax_{n-1} + b$, where a and b are two constants.

EXAMPLE   A certain system of lakes contains an initial population of one million fish. Due to natural growth, this population would increase by one-third each year. If current fishing regulations permit the harvesting of 400,000 fish per year, find the size of the fish stock after n yr and calculate how many years it will take before the population is depleted to less than half a million, and how many years it will take before the population is wiped out.

SOLUTION   Let x_n denote the number of millions of fish in the nth year. Then the initial population of one million corresponds to $x_0 = 1$.

The difference $x_n - x_{n-1}$ measures the increase in millions of the fish population during the nth year. This has two components. First of all, there is the natural growth of the population that gives an increase equal to $\frac{1}{3}x_{n-1}$ during the nth year. Then secondly, harvesting leads to a decrease in population of 0.4 million fish per year. Therefore

Increase in population = Natural growth — Amount of harvesting,

or

$$x_n - x_{n-1} = \tfrac{1}{3}x_{n-1} - 0.4.$$

Thus we obtain the difference equation

$$x_n = \tfrac{4}{3}x_{n-1} - 0.4.$$

n	x_n
0	1.0
1	0.933
2	0.844
3	0.726
4	0.568
5	0.357
6	0.076
7	−0.298

The table of values of x_n is constructed in the usual way by giving n the successive values 1, 2, 3, ...Since $x_0 = 1$,

$$x_1 = \tfrac{4}{3}x_0 - 0.4 = 0.933.$$

Again

$$x_2 = \tfrac{4}{3}x_1 - 0.4 = \tfrac{4}{3}(0.933) - 0.4 = 0.844.$$

Continuing, we observe that x_4 is still greater than 0.5 (half a million fish) while x_5 is less than 0.5, and therefore it takes 5 yr before the population shrinks below the half-million level. Thereafter the decline is very swift. After six years the population has shrunk to a mere 76,000 (0.076 million), while in year 7, x_7 is negative. This indicates that the population is extinct in the 7th year (and in that year there are insufficient fish to permit the full quota of 400,000 to be harvested). The decline of the fish stock in this example is illustrated in Fig. 11.2, which shows x_n as a function of n for $n = 0, 1, 2, \ldots, 6$.

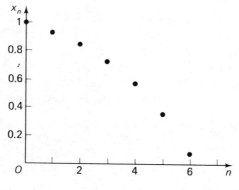

Figure 11.2

Difference equations similar to the one in this example arise in situations where a population is increasing or decreasing by virtue of immigration or emigration. Suppose that the natural growth of the population in any year is equal to k times the population at the beginning of the year and that new members are immigrating into the population at the rate of b/yr. Then the change in population size, $x_n - x_{n-1}$, during the nth year will be equal to the natural growth, kx_{n-1}, plus the amount b of immigration:

$$x_n - x_{n-1} = kx_{n-1} + b.$$

Therefore we obtain the difference equation

$$x_n = (1 + k)x_{n-1} + b.$$

This type of difference equation is also of considerable importance in finance. The growth of a regular savings program is governed by an equation of this type as also is the repayment by regular installments of a loan (for example, the repayment of a house mortgage). A further application is given in the following example.

EXAMPLE Among the North American population, a number of people suffer from a certain debilitating disease. Each year 1000 new cases of the disease occur and half of the existing cases are cured. At the end of 1970 there were 1200 cases of the disease. Calculate the number of cases at the end of each subsequent year through 1978.

SOLUTION Let us denote the end of 1970 by $n = 0$ and subsequent years by $n = 1, 2, 3$, etc. Then $n = 8$ corresponds to the end of 1978. Let x_n denote the number of cases of the disease at the end of year n. Then we are given that $x_0 = 1200$. We can write the difference equation

$$x_n = 0.5x_{n-1} + 1000$$

since the number x_n consists of half of the x_{n-1} cases that remain uncured from the previous year together with 1000 new cases. Putting $n = 1$, we find

$$x_1 = 0.5x_0 + 1000 = 0.5(1200) + 1000 = 1600.$$

Then putting $n = 2$, we obtain

$$x_2 = 0.5x_1 + 1000 = 0.5(1600) + 1000 = 1800.$$

Continuing in this way, we find the number of cases in each of the years 1970–1978 to be as given in the table.

n	Year	x_n
0	1970	1200
1	1971	1600
2	1972	1800
3	1973	1900
4	1974	1950
5	1975	1975
6	1976	1988
7	1977	1994
8	1978	1997

The graph of this solution is shown in Fig. 11.3. It is clear that as n increases,

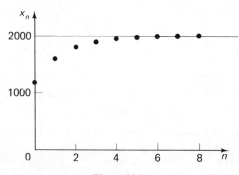

Figure 11.3

the number of cases approaches closer and closer to 2000. We can speak of the number 2000 as being the *equilibrium value* of the number of cases since if x_n were actually equal to 2000, it would remain unchanged from one year to the next (i.e., $x_n = x_{n-1}$). The solution whose graph is shown approaches the equilibrium value as n approaches ∞.

The reader is encouraged to reconsider this example for the case where x_0 is greater than 2000, say $x_0 = 3000$. The solution should be found once again to approach the equilibrium value of 2000 as n gets larger and larger.

EXERCISES 11.1

Give the orders of the following difference equations and state whether they are linear or not:

1. $x_{n+1} + 2x_{n-1} = \dfrac{1}{n}$.

2. $x_{n+2} + x_n^2 = 0$.

3. $x_{n+3} + \dfrac{1}{x_{n+2}} = nx_{n+2}$.

4. $x_{n-1} + \sin x_n = 1$.

5. $x_{n+4} + 2x_{n-1} = (n - 2)^3$.

6. $nx_{n-2} + n^2 x_{n-3} = 2$.

Show that the following functions are solutions of the given difference equations (a, b, and c are constants):

7. $x_n = n(n + 2)$; $x_{n+1} - x_{n-1} = 4(n + 1)$.

8. $x_n = c(-1)^n$; $x_n + x_{n-1} = 0$.

9. $x_n = (-1)^n(n + c)$; $x_n + x_{n-1} = (-1)^n$.

10. $x_n = \dfrac{1}{(n + c)^2}$; $x_{n+1} = \dfrac{x_n}{(1 + \sqrt{x_n})^2}$.

11. $x_n = a + b(-1)^n$; $x_{n+2} = x_n$.

12. $x_n = a + b(2^n) + c(3^n)$; $x_n - 6x_{n-1} + 11x_{n-2} - 6x_{n-3} = 0$.

13. Assuming that the increase in the global population of blue whales in any year equals 20% of the existing population, calculate how many years it would take the population to increase from 1000 to more than 2000 animals.

14. The average weight of cattle in a feed-lot increases by 5%/wk. Write down a difference equation for w_n, the average weight after n weeks, and find its solution for $n = 1, 2, \ldots, 8$ given that $w_0 = 500$ lb.

15. During a period of fasting, an individual was observed to lose weight each day by an amount equal to 1% of his weight at the beginning of that day. If his initial weight was 100 kg, how many days did it take for his weight to decrease below 90 kg?

16. Each severe storm in winter kills 5% of a certain animal population. Find an expression for the proportion of the original population that remains alive after n storms. If an average winter involves 10 such storms, what proportion of the population would survive, on the average?

17. A population of birds on an island increases by $20\%/\text{yr}$ by natural growth and by 20 birds per year because of immigration. Write down a difference equation for the population size x_n after n years. If the initial population size is 100, find the size of the population in each of the first 12 yr.

18. A population has an initial value of 1000 and grows by $50\%/\text{yr}$. If the population is harvested at the rate of 400/yr, find the population size in each of the first 8 yr.

19. Repeat exercise 18 for the situation in which the rate of harvesting is increased by 10% each year (440 the second year, 484 the third year, and so on). What is the difference equation in this case?

20. A disease spreads through a population in such a way that the number of individuals who become infected during any year is equal to 100, and 25% of those infected at the beginning of a year die before the end of the year. Write a difference equation for x_n, the number of people infected after n yr. Find its solution for $n = 1, 2, \ldots, 10$ in the two cases: (a) $x_0 = 200$; (b) $x_0 = 500$. What conclusion can you draw?

11.2   THE LOGISTIC MODEL

In the preceding section we considered a number of population models in which the natural growth of the population in any time unit was proportional to the size of the population at the beginning of the period in question. Such growth models lead to a difference equation of the form

$$x_n - x_{n-1} = kx_{n-1}, \tag{1}$$

or perhaps with an extra constant term on the right-hand side in the event that immigration into the population or harvesting of it is taking place. The constant k is the *specific growth rate* for the population; it represents the increase in the population during one unit of time as a proportion of the existing population. The increase kx_{n-1} represents the number of births minus the number of deaths.

Now as we have seen, the solution of the difference equation (1) grows exponentially as n increases. A population that grows according to this equation would continue to increase indefinitely and its size would eventually reach arbitrarily large values. For example, in Section 11.1 we considered the case of a population whose initial size was 1000 and that increased by $25\%/\text{yr}$ ($k = 0.25$). After n yr the population size was found to be given by $x_n = 1000(1.25)^n$. After 10 yr this population will have increased from 1000 to 9313; after 20 yr it will have increased to 86,700, and after 50 yr to 88 million.

Clearly a population growing like this is eventually going to exhaust the capacity of its habitat to support the species in question. Resources such as food and shelter, which were abundant when the population was a mere 1000, may be in short supply 20 yr later when the population has climbed to 86,700. Then, with member competing against member for the limited supply, the death rate will be higher than it was earlier and the birth rate will be lower. The specific growth rate k, which represents the birth rate minus the death rate, will therefore be smaller than it was when the population was small.

In other words, as the population increases, the specific growth rate k can be expected to decrease due to pressure on the population from its environment. The simplest model we can take is that k decreases linearly with population size; that is,

$$k = a - bx_{n-1},$$

where a and b are two constants. In such a case we obtain the difference equation

$$x_n - x_{n-1} = (a - bx_{n-1})x_{n-1} \qquad (2)$$

in place of equation (1). This difference equation is called the *logistic difference equation*. It is the analogue for discrete time models of the logistic differential equation, which was studied in Section 10.3.

EXAMPLE   When a certain population has size x, the birth rate during the next year (that is, the number of births as a proportion of the population) is equal to $0.7 - 0.00005x$ and the death rate per year is equal to $0.2 + 0.00015x$. Then during the year between time $(n - 1)$ and time n the number of births will be

$$(0.7 - 0.00005x_{n-1})x_{n-1}$$

and the number of deaths will be

$$(0.2 + 0.00015x_{n-1})x_{n-1}.$$

We can write the following equation for the change in population during that year:

$$x_n - x_{n-1} = \text{Number of births} - \text{Number of deaths}$$
$$= (0.7 - 0.00005x_{n-1})x_{n-1} - (0.2 + 0.00015x_{n-1})x_{n-1} \qquad (3)$$
$$= (0.5 - 0.0002x_{n-1})x_{n-1}.$$

Thus we obtain a logistic difference equation of the type (2) with the constants having the values $a = 0.5$ and $b = 0.0002$.

Let us examine the solution of this equation for the case of a population that starts out with a size of 1000 (that is, $x_0 = 1000$). Setting $n = 1$ in (3), we get

$$x_1 - x_0 = (0.5 - 0.0002x_0)x_0,$$

and so, since $x_0 = 1000$,

$$x_1 = 1000 + [0.5 - (0.0002)(1000)](1000)$$
$$= 1000 + (0.3)(1000) = 1300.$$

Then, setting $n = 2$ in (3), we obtain

$$x_2 - x_1 = (0.5 - 0.0002x_1)x_1$$
$$= [0.5 - (0.0002)(1300)](1300)$$
$$= (0.24)(1300) = 312.$$

Therefore

$$x_2 = 1300 + 312 = 1612.$$

In this way we can generate the solution, setting n successively equal to 1, 2, 3, ... in the difference equation and thus obtaining in turn the values of x_1, x_2, x_3, and so on. These values are set out in the table for values of n up to 16, and the graph of the solution is shown in Fig. 11.4. The values of x_n in the table have been rounded off to the nearest whole number. The form of solution

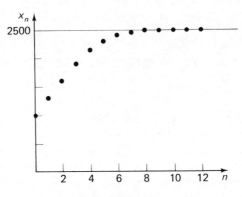

Figure 11.4

n	x_n	n	x_n
0	1000	9	2484
1	1300	10	2492
2	1612	11	2496
3	1898	12	2498
4	2127	13	2499
5	2286	14	2500
6	2384	15	2500
7	2439	16	2500
8	2469		

shown in Fig. 11.4 is exactly what one would expect in a situation in which the expansion of the population is limited by the environment. The population grows rapidly at first while it is small, but the rate of growth slows down and eventually the population settles down to a constant level representing the size that can be maintained by the available resources. In this example the maximum supportable population is 2500.

In the preceding example we can speak of the size 2500 as being the *equilibrium size* for the population, since once the population reaches this size it does not change. We can see this result from a different point of view. If the population does not change from one year to the next, then $x_n = x_{n-1}$. From the difference equation (3) then,

$$x_n - x_{n-1} = (0.5 - 0.0002x_{n-1})x_{n-1} = 0.$$

Therefore either $x_{n-1} = 0$ (which does not represent a very interesting case) or else

$$0.5 - 0.0002x_{n-1} = 0.$$

From this,

$$x_{n-1} = 0.5/0.0002 = 2500.$$

Thus again we see that if the population is 2500 in the year $n-1$ it has this same size in the next year and, in fact, in all subsequent years.

For the general logistic equation (2), we similarly obtain the equilibrium population size by setting $x_n = x_{n-1}$. We get

$$x_n - x_{n-1} = (a - bx_{n-1})x_{n-1} = 0.$$

From this, apart from the trivial solution $x_{n-1} = 0$, we obtain the result that

$$x_{n-1} = \frac{a}{b},$$

a value that represents the equilibrium population size.

The question arises whether the solution of the logistic difference equation always approaches the equilibrium population size as n gets larger and larger. Let us begin to answer this question by considering the difference equation in the above example, but in the case when the initial population size exceeds the equilibrium size of 2500. We shall in fact consider three cases: (a) $x_0 = 4000$, (b) $x_0 = 5000$, and (c) $x_0 = 6000$. The solutions obtained for $n = 1, 2, \ldots, 13$ in these three cases are given in the table, and their graphs are plotted in Fig. 11.5 for cases (a) and (c).

	(a)	(b)	(c)
n	x_n	x_n	x_n
0	4000	5000	6000
1	2800	2500	1800
2	2632	2500	2052
3	2563	.	2236
4	2530	.	2354
5	2515	.	2423
6	2507	.	2460
7	2504		2480
8	2502		2490
9	2501		2495
10	2500		2497
11	2500		2499
12	2500		2499
13	2500		2500

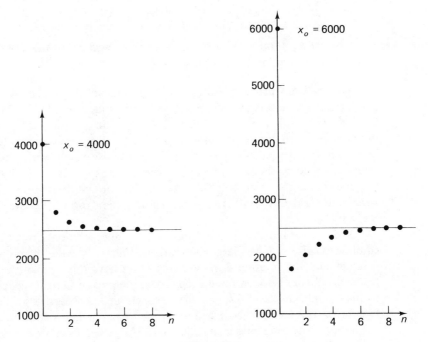

Figure 11.5

We see that in case (a) the solution steadily decreases, getting closer and closer to the equilibrium value 2500 as n increases. In case (b), when the initial population size is 5000, it drops immediately in the first year to the equilibrium value of 2500, and thereafter remains constant at that level. In case (c) the initial population size exceeds 5000, but drops in the first year to below the equilibrium level. Thereafter the solution increases steadily toward the equilibrium level of 2500.

These various examples illustrate general properties of the logistic difference equation, which are summarized in the following theorem.

THEOREM 11.2.1

Let x_n ($n = 0, 1, 2, 3, \ldots$) satisfy the logistic difference equation

$$x_n - x_{n-1} = (a - bx_{n-1})x_{n-1},$$

where a and b are positive constants and $a < 1$. Then:

(a) If $x_0 < a/b$, the x_n form an increasing sequence (that is, $x_n > x_{n-1}$ for each n) that approaches the limiting value a/b as n gets larger.

(b) If $a/b < x_0 < 1/b$, the x_n form a decreasing sequence (that is, $x_n < x_{n-1}$ for each n) that approaches the limiting value a/b as n gets larger.

(c) If $x_0 = 1/b$ then $x_n = a/b$ for all $n \geq 1$.

(d) If $1/b < x_0 < (a + 1)/b$ then $x_1 < a/b$ and for $n \geq 1$ the x_n form an increasing sequence that again approaches the limiting value a/b as n increases.

(e) If $x_0 \geq (a + 1)/b$ then x_1 is zero or negative, so this case leads to immediate extinction of the population.

We shall not prove this theorem. Note that in the preceding examples whose solutions were evaluated explicity, $a = 0.5$ and $b = 0.0002$. Therefore $a/b = 0.5/0.0002 = 2500$ and $1/b = 1/0.0002 = 5000$. So the theorem states that if the initial population size x_0 is below 2500, the population size steadily increases and approaches the level 2500 as $n \rightarrow \infty$; if the initial size is between 2500 and 5000, the size steadily decreases and again approaches 2500 as $n \rightarrow \infty$; and if x_0 exceeds 5000, but does not equal or exceed 7500, the population first drops below 2500, then rises steadily toward this level as $n \rightarrow \infty$. These general properties are consistent with the solutions we found.

We see then that if the constant $a < 1$, the solution always approaches the equilibrium size as n gets larger and larger. When $a > 1$ the situation is much more complex and the solution does not necessarily have this simple type of asymptotic behavior. If a lies between 1 and 2 then the sequence x_n eventually get closer and closer to the equilibrium level a/b, but the members of the sequence oscillate first above, then below this level. When $a > 2$ the sequence never settles down to the equilibrium value, but always oscillates up and down with large swings in population size from one year to the next. These types of behavior are in marked contrast to the behavior of solutions of the logistic differential equation in Section 10.3.

We illustrate these kinds of solutions in the following examples.

EXAMPLE   Investigate the solutions of the logistic difference equation

$$x_n - x_{n-1} = (1.5 - 0.001x_{n-1})x_{n-1}$$

for the cases (a) $x_0 = 600$, (b) $x_0 = 1800$, and (c) $x_0 = 2300$.

SOLUTION   The solutions are found in the usual way by setting n successively equal to 1, 2, 3, etc. in the difference equation. In the case when $x_0 = 600$ we find

$$x_1 = x_0 + (1.5 - 0.001x_0)x_0$$
$$= 600 + [1.5 - 0.001(600)](600)$$
$$= 600 + (0.9)(600) = 1140;$$
$$x_2 = x_1 + (1.5 - 0.001x_1)x_1$$
$$= 1140 + [1.5 - 0.001(1140)](1140)$$
$$= 1140 + (0.36)(1140) = 1550,   \text{and so on.}$$

The solutions in the three cases are set out in the table. It can be seen that in each case the solution settles down to a pattern of oscillation about the 1500 level. This is illustrated by Fig. 11.6, which shows a graph of the solution in case (a).

n	(a) x_n	(b) x_n	(c) x_n
0	600	1800	2300
1	1140	1260	460
2	1550	1562	938
3	1472	1465	1465
4	1513	1516	1516
5	1493	1492	1492
6	1503	1504	1504
7	1498	1498	1498
8	1501	1501	1501
9	1500	1499	1499
10	1500	1500	1500

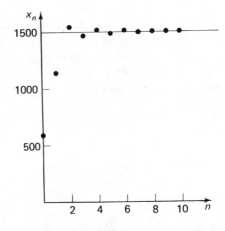

Figure 11.6

For the difference equation in this example the equilibrium population size is $a/b = 1.5/0.001 = 1500$. No matter what the initial size of the population, it eventually settles down to this equilibrium level. However, when the population gets close to 1500, it always oscillates first above, then below this level, the magnitude of the oscillations getting smaller and smaller as n increases. (Note: If $x_0 \geq 2500$ then $x_1 \leq 0$ so that in this case immediate extinction of the population occurs.)

The results obtained in this example are special cases of the following general theorem.

THEOREM 11.2.2

Let x_n $(n = 0, 1, 2, 3, \ldots)$ satisfy the logistic difference equation

$$x_n - x_{n-1} = (a - bx_{n-1})x_{n-1},$$

where $b > 0$ and $1 < a < 2$. Then:

(a) If $1/b < x_0 < 2/b$ the x_n get closer and closer to a/b as n increases, but oscillate from one side of this limiting value to the other as n goes from one value to the next.

(b) If $x_0 < 1/b$ the x_n at first increase steadily until they exceed $1/b$ and thereafter they follow the same type of oscillatory behavior as in case (a), approaching the limit a/b as $n \longrightarrow \infty$.

(c) If $2/b < x_0 < (a + 1)/b$ then the sequence immediately falls below the equilibrium level (i.e., $x_1 < a/b$) and thereafter the x_n follow the same pattern as in case (a) or (b), depending on whether $x_1 > 1/b$ or $x_1 < 1/b$.

(d) If $x_0 \geq (a + 1)/b$ then $x_1 \leq 0$ and the population immediately becomes extinct.

The unusual behavior found in the solution when $a > 2$ is illustrated by the following example. It will be seen that in this example, the solution eventually settles down to a regular pattern that consists of finite oscillations up and down between four levels. A population varying according to this equation would show regular variations in size on a four-year cycle.

EXAMPLE   Find the solutions of the difference equation

$$x_n - x_{n-1} = (2.5 - 0.001x_{n-1})x_{n-1}$$

when (a) $x_0 = 200$; (b) $x_0 = 3400$.

SOLUTION   Proceeding in the usual way, setting n successively equal to $1, 2, 3, \ldots$, the sequences given in the table are found for the two cases. The graph of the solution in case (b) is shown in Fig. 11.7. We can see that eventually the solution reaches a state of oscillation between the four levels 2894, 1753, 3062, and 1340.

	(a)	(b)		(a)	(b)
n	x_n	x_n	n	x_n	x_n
0	200	3400	12	1341	1715
1	660	340	13	2895	3061
2	1874	1074	14	1751	1343
3	3047	2606	15	3062	2897
4	1380	2330	16	1340	1747
5	2926	2727	17	2894	3062
6	1680	2109	18	1753	1340
7	3058	2934	19	3062	2894
8	1353	1661	20	1340	1753
9	2905	3055	21	2894	3062
10	1729	1360	22	1753	1340
11	3062	2911			

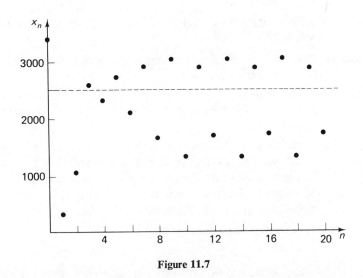

Figure 11.7

In this example the equilibrium level is $a/b = 2.5/0.001 = 2500$. The solution does not approach this size as $n \rightarrow \infty$. In fact if the solution is close to 2500 for any value of n, it subsequently moves further away from this equilibrium level. In such a case we say that the equilibrium level is *unstable*. For any logistic equation for which $a > 2$ the equilibrium level a/b is always unstable.

EXERCISES 11.2

Construct a table of values of the solution of each of the following difference equations for the indicated values of n.

1. $x_n - x_{n-1} = (0.6 - 0.01x_{n-1})x_{n-1}$; $x_0 = 10$, $0 \le n \le 10$.

2. $x_n - x_{n-1} = (0.6 - 0.01x_{n-1})x_{n-1}$; $x_0 = 140$, $0 \le n \le 10$.

3. $x_n - x_{n-1} = (1.6 - 0.01x_{n-1})x_{n-1}$; $x_0 = 10$, $0 \le n \le 10$.

4. $x_n - x_{n-1} = (1.6 - 0.01x_{n-1})x_{n-1}$; $x_0 = 180$, $0 \le n \le 10$.

5. $x_n - x_{n-1} = (2.8 - 0.01x_{n-1})x_{n-1}$; $x_0 = 20$, $0 \le n \le 20$.

6. $x_n - x_{n-1} = (2.8 - 0.01x_{n-1})x_{n-1}$; $x_0 = 175$, $0 \le n \le 20$.

7. $x_n - x_{n-1} = (2.45 - 0.01x_{n-1})x_{n-1}$; $x_0 = 20$, $0 \le n \le 16$.

8. $x_n - x_{n-1} = (2.45 - 0.01x_{n-1})x_{n-1}$; $x_0 = 246$, $0 \le n \le 16$.

9. $x_n - x_{n-1} = (2.35 - 0.01x_{n-1})x_{n-1}$; $x_0 = 236$; $0 \le n \le 16$.

10. The specific growth rate for any 1-yr period for a certain population is equal to $(0.2 - 0.0001x)$ when the population at the beginning of the year is x. Write down a difference equation for the growth of the population and construct a table of its solution for $n = 0, 1, 2, \ldots, 12$ given that initially the population size is 200.

11. For a given population of size p the birth rate during a 1-yr period is equal to $(0.5 - 0.0005p)$ and the death rate is equal to $(0.2 + 0.005p)$. Write down a difference equation for the growth of this population. If p_0, the population in year 0, is 100, construct a table of values of the population for $n = 0, 1, 2, \ldots, 20$.

12. A certain population grows from year to year according to the logistic difference equation, the equilibrium size of the population being 2000. If the population size approaches zero the specific growth becomes equal to 1.6. Construct a table of solutions for the population size for a 15-yr period, given that the initial population size is equal to 2800.

13. In the difference equation

$$x_n - x_{n-1} = (a - bx_{n-1})x_{n-1}$$

set $y_n = bx_n$ and find the resulting difference equation for y_n. What is the equilibrium value of y_n?

14. For the logistic difference equation

$$x_n - x_{n-1} = (a - bx_{n-1})x_{n-1}$$

show that $x_n \leq 0$ if and only if $x_{n-1} \geq (1 + a)/b$. [Hence a population that exceeds $(1 + a)/b$ in size becomes extinct in the next year.] Show that if $a < 3$, x_n is always less than $(1 + a)/b$ when x_{n-1} is less than $(1 + a)/b$. [Hence when $a < 3$, the population never becomes extinct. When $a > 3$ it is possible for the population to become extinct at some stage.]

15. A population grows according to the difference equation

$$x_n - x_{n-1} = (0.5 - 0.003x_{n-1} - 0.00002x_{n-1}^2)x_{n-1}.$$

What is the equilibrium level for the population? Construct a table of values of x_n for $0 \leq n \leq 12$ given that $x_0 = 10$.

16. Find a table of values of x_n for $0 \leq n \leq 10$ in exercise 15 given that $x_0 = 120$.

11.3   LINEAR FIRST-ORDER EQUATIONS

In this section we shall study the solution of first-order linear difference equations. The general equation of this type is of the form

$$x_n - a_n x_{n-1} = b_n, \tag{1}$$

in which a_n and b_n are given functions of n, and x_n is the unknown function whose solution is required. For example, the difference equations

$$x_n - nx_{n-1} = 1 \quad (a_n = n, b_n = 1)$$

$$x_n - \frac{1}{n}x_{n-1} = e^n \quad (a_n = n^{-1}, b_n = e^n)$$

are both of this type. However in this section we shall only give the solution for the case in which $a_n = a$ is a constant, independent of n.

We shall begin by considering the case when $b_n = 0$, in which event the difference equation reduces to

$$x_n - ax_{n-1} = 0.$$

In Section 11.1 several examples were considered of problems that led to difference equations of this type. In each of those examples a general formula could be found expressing x_n as an explicit algebraic function of n. This is in fact generally true, as shown by the following theorem.

THEOREM 11.3.1

The difference equation

$$x_n - ax_{n-1} = 0$$

has the solution

$$x_n = ca^n,$$

where c is a certain constant. The value of c is determined if we know the value of x_n for some n; if x_p is known, then $c = x_p a^{-p}$.

PROOF

Writing down the difference equation successively for $n = 1, 2, 3, \ldots$ we find:

$$x_1 = ax_0,$$
$$x_2 = ax_1 = a(ax_0) = a^2 x_0,$$
$$x_3 = ax_2 = a(a^2 x_0) = a^3 x_0,$$

.

.

.

It is clear that in general $x_n = a^n x_0$. Thus the theorem is established, the constant c being equal to x_0.

If x_p is known, then since $x_p = ca^p$, $c = x_p/a^p = x_p a^{-p}$. This equation allows us to calculate c in terms of the known value of x_p.

EXAMPLE

(a)   The solution of the difference equation

$$x_n = 1.25x_{n-1}$$

is given by

$$x_n = c(1.25)^n,$$

where $c = x_0$.

(b)   The solution of the difference equation

$$x_n = 1.5x_{n-1}$$

is given by

$$x_n = c(1.5)^n.$$

If we are given that $x_2 = 2$ then the value of the constant c is obtained from the equation

$$x_2 = c(1.5)^2 = 2.$$

Consequently $c = 2(1.5)^{-2}$. The solution then takes the form

$$x_n = [2(1.5)^{-2}](1.5)^n$$
$$= 2(1.5)^{n-2}.$$

The solution $x_n = ca^n$ of the difference equation $x_n = ax_{n-1}$ expresses x_n as an exponential function of n. If $a > 1$, the solution represents an exponential growth, while if $a < 1$, it represents exponential decay.

For the next case we shall consider the difference equation (1) when $a_n = 1$. The right-hand side, b_n, is assumed to be nonzero and to depend in general on n. The solution in this case is provided by the following theorem.

THEOREM 11.3.2

The difference equation

$$x_n - x_{n-1} = b_n,$$

where b_n is given for each n, has the solution

$$\boxed{x_n = x_0 + \sum_{p=1}^{n} b_p.}$$

PROOF   Writing down the given difference equation for successive values of the index n yields the following series of equations:

$$x_1 - x_0 = b_1$$
$$x_2 - x_1 = b_2$$
$$x_3 - x_2 = b_3$$
$$\cdot$$
$$\cdot$$
$$\cdot$$
$$x_n - x_{n-1} = b_n.$$

If all these equations are added together, most of the terms on the left-hand side cancel out, and we are simply left with

$$x_n - x_0 = b_1 + b_2 + \ldots + b_n$$
$$= \sum_{p=1}^{n} b_p.$$

After moving x_0 across to the right-hand side we obtain the formula given in the statement of the theorem.

EXAMPLE   Solve the difference equation:

$$x_n - x_{n-1} = 2n + 1$$

for which $x_1 = 2$.

In Theorem 11.3.2 we can set $b_n = 2n + 1$. Therefore

$$x_n = x_0 + \sum_{p=1}^{n} b_p$$

$$= x_0 + \sum_{p=1}^{n} (2p + 1)$$

$$= x_0 + 2 \sum_{p=1}^{n} p + \sum_{p=1}^{n} 1$$

$$= x_0 + 2 \cdot \tfrac{1}{2}n(n + 1) + n$$

after using the standard sum in Theorem 7.1.2. Therefore

$$x_n = x_0 + n(n + 2).$$

We do not know the value of x_0, but we do know that $x_1 = 2$. Therefore, setting $n = 1$, we obtain that

$$x_1 = x_0 + 1(1 + 2)$$

$$= x_0 + 3$$

$$= 2$$

and therefore $x_0 = -1$. Consequently the solution for x_n is given by

$$x_n = n(n + 2) - 1.$$

THEOREM 11.3.3

The difference equation

$$x_n - ax_{n-1} = b_n,$$

where a and b_n are given, has the solution

$$\boxed{x_n = x_0 a^n + \sum_{p=1}^{n} a^{n-p} b_p.}$$

PROOF   When $b_n = 0$ we know from Theorem 11.3.1 that the solution has the form $x_n = ca^n$, where c is some constant. In order to find the solution when $b_n \neq 0$, we assume that the solution has a similar form but that c, instead of being constant, depends on n. That is, we set

$$x_n = c_n a^n. \tag{i}$$

(Note the analogy with the method of variation of parameters in Section 10.4.) Then, of course, $x_{n-1} = c_{n-1}a^{n-1}$, and substituting for x_n and x_{n-1} in the given difference equation we obtain

$$c_n a^n - ac_{n-1}a^{n-1} = b_n.$$

Therefore

$$a^n(c_n - c_{n-1}) = b_n,$$

i.e.,

$$c_n - c_{n-1} = a^{-n}b_n.$$

Thus we have obtained a new difference equation for the variables c_n. But this new equation is much simpler than the original difference equation; in fact it is of the type whose solution is given in Theorem 11.3.2. The only difference is that instead of just b_n on the right-hand side we have $a^{-n}b_n$. Therefore the solution is given by

$$c_n = c_0 + \sum_{p=1}^{n} a^{-p}b_p.$$

Putting $n = 0$ in equation (i) above we get

$$x_0 = c_0 a^0 = c_0.$$

Therefore

$$c_n = x_0 + \sum_{p=1}^{n} a^{-p}b_p.$$

After substituting this value of c_n into (i) we finally obtain

$$x_n = \left[x_0 + \sum_{p=1}^{n} a^{-p}b_p \right]a^n$$

$$= x_0 a^n + \sum_{p=1}^{n} a^{n-p}b_p.$$

EXAMPLE   Find the solution of the equation

$$x_n - 3x_{n-1} = 2.$$

SOLUTION   We can use the result of Theorem 11.3.3, setting $a = 3$ and $b_n = 2$ (b_n is independent of n in this example).
Then

$$x_n = x_0 3^n + \sum_{p=1}^{n} 3^{n-p} \cdot 2$$

$$= x_0 3^n + 2[3^{n-1} + 3^{n-2} + \ldots + 3^1 + 3^0],$$

where we have expanded the sum. The term in the square bracket here is a geometrical progression whose common ratio is 3 and whose lowest term is $3^0 = 1$. Its sum is given by the standard formula (see below)

$$1 + 3^1 + \ldots + 3^{n-1} = \frac{3^n - 1}{3 - 1} = \frac{3^n - 1}{2}.$$

Therefore

$$x_n = x_0 3^n + (3^n - 1)$$

$$= (x_0 + 1)3^n - 1.$$

In this example the solution involved a geometrical progression. Such progressions and related ones occur quite commonly with this type of difference equation, and we take this opportunity to list two useful formulas. For geometrical progressions themselves we recall that

$$1 + r + r^2 + \ldots + r^{n-1} = \frac{r^n - 1}{r - 1}.$$

Using $\sum$-notation this sum can be written in either of the two forms

$$\sum_{p=1}^{n} r^{n-p} = \sum_{q=0}^{n-1} r^q = 1 + r + \ldots + r^{n-1}.$$

A second sum that is useful is the following:

$$\sum_{q=1}^{n} q r^{q-1} = 1 + 2r + 3r^2 + \ldots + nr^{n-1} = \frac{nr^{n+1} - (n+1)r^n + 1}{(r-1)^2}.$$

EXAMPLE   A population of animals increases through natural growth by 25%/yr. Its size in the year $n = 0$ is 1100. If every year 300 of the animals are harvested, find an expression for the size of the population after n yr. After how many years is the population size reduced to below 500?

SOLUTION   An example similar to this was solved in Section 11.1 by explicitly constructing a table of values of the population size in each year. We shall solve the present example by finding an algebraic formula for x_n, the population size in the nth year. The change $x_n - x_{n-1}$ in population between the $(n-1)$th and nth years consists of a natural increase equal to $0.25x_{n-1}$ and a decrease of 300 due to harvesting. Therefore

$$x_n - x_{n-1} = 0.25x_{n-1} - 300,$$

or

$$x_n - 1.25x_{n-1} = -300.$$

This difference equation is of the type covered by Theorem 11.3.3, with $a = 1.25$ and $b_n = -300$. Therefore

$$x_n = x_0(1.25)^n + \sum_{p=1}^{n} (-300)(1.25)^{n-p}$$

$$= x_0(1.25)^n - 300[(1.25)^{n-1} + (1.25)^{n-2} + \ldots + (1.25) + 1]$$

$$= x_0(1.25)^n - 300\frac{(1.25)^n - 1}{1.25 - 1}$$

after using the formula for the sum of a geometrical progression. With some simplification we then obtain that

$$x_n = 1200 + (x_0 - 1200)(1.25)^n.$$

Since $x_0 = 1100$, the population in the nth year is then

$$x_n = 1200 - 100(1.25)^n.$$

In order to find when the population is depleted below 500 we set $x_n = 500$, obtaining

$$1200 - 100(1.25)^n = 500;$$

therefore

$$(1.25)^n = 7.$$

Taking logs of both sides of this expression we get

$$n \log (1.25) = \log 7$$

or

$$n = \frac{\log 7}{\log 1.25} = \frac{0.8451}{0.0969} = 8.72.$$

This means that when $n = 8$, x_n is still greater than 500, but when $n = 9$, x_n has fallen below 500. It therefore takes 9 yr for the population to be reduced to below the level of 500.

A further example of the use of first-order difference equations occurs in one of the methods used to measure the CO_2 content of blood entering the lungs. In this method the subject is required to breath repeatedly in and out of a closed plastic bag. After each breath the partial pressure of CO_2 in the plastic bag is measured. Let p_n denote this partial pressure after n breaths; then of course p_n increases as n increases since the CO_2 content in the bag rises as the subject continues to breathe into it. It is found that if a graph of p_n against p_{n-1} is plotted, a straight line is obtained, so that p_n is related to p_{n-1} by a linear relation of the form

$$p_n = ap_{n-1} + b,$$

where a and b are two constants that can be determined from the graph ($a < 1$). Thus the partial pressure of CO_2 in the bag after each breath is a linear function of the partial pressure after the previous breath.

The above relation between p_n and p_{n-1} is a linear difference equation that can be solved using Theorem 11.3.3. The solution is easily seen to be

$$p_n = ca^n + b\frac{1 - a^{n+1}}{1 - a},$$

where c is an arbitrary constant.

Now since $a < 1$, we must have that $a^n \to 0$ as $n \to \infty$. Thus as $n \to \infty$, the partial pressure p_n approaches the limiting value

$$\lim_{n \to \infty} p_n = c(0) + b\frac{1 - 0}{1 - a} = \frac{b}{1 - a}.$$

But eventually the partial pressure of CO_2 in the bag must equilibrate with the partial pressure of CO_2 in the venous blood entering the lung. Hence this latter partial pressure must be equal to $b/(1 - a)$.

So by measuring a and b from a graph of p_n plotted against p_{n-1} we are able to determine the partial pressure of CO_2 in the blood entering the lung.

The solutions of the difference equations studied in this section are set out in the following table for ease of reference.

DIFFERENCE EQUATION	SOLUTION
$x_n - ax_{n-1} = 0$	$x_n = x_0 a^n$
$x_n - x_{n-1} = b_n$	$x_n = x_0 + \sum_{p=1}^{n} b_p$
$x_n - ax_{n-1} = b_n$	$x_n = x_0 a^n + \sum_{p=1}^{n} a^{n-p} b_p$

EXERCISES 11.3

Solve the following difference equations:

1. $x_n - 2x_{n-1} = 0.$

2. $x_n - \frac{1}{3}x_{n-1} = 0.$

3. $x_{n+1} + \frac{1}{4}x_n = 0.$

4. $2x_{n+1} + 3x_n = 0.$

5. $x_n - x_{n-1} = 3.$

6. $x_n - x_{n-1} = -1.$

7. $x_n - x_{n-1} = n + 2.$

8. $x_{n+1} - x_n = n + 2.$

9. $x_{n+1} - x_n = 3n^2 + n.$

10. $x_n - x_{n-1} = n^2 - n.$

11. $x_n - 2x_{n-1} = 1.$

12. $x_n - \frac{1}{2}x_{n-1} = 3.$

13. $x_n + x_{n-1} = n.$

14. $x_n + 2x_{n-1} = 2.$

15. $x_n - 2x_{n-1} = n.$

16. $x_n + \frac{1}{2}x_{n-1} = 3n.$

17. During a period of fasting an individual loses weight each day by an amount equal to 1% of his weight at the beginning of that day. If his initial weight was 100 kg, find an expression for his weight after n days of fasting. How many days does it take him to reduce to 80 kg?

18. A population has an initial size of 1000 and increases by 20%/yr. Find an expression for the size of the population after n yrs. How many years is it before the population exceeds 2000 in size?

19. A certain animal population has an initial size of 1000 and increases by 20%/yr. If each year 100 of the population are harvested, how many years is it before the population exceeds 2000 in size?

20. Repeat exercise 19 in the case when the number of animals harvested increases by 10/yr (110 harvested in the second year, 120 in the third year, and so on).

21. A herd of animals increases each year by 10% through normal birth and death processes and in addition receives 20 animals per year through immigration from neighboring areas. If the initial size of the herd is 500, calculate the size after n yr.

11.4 SECOND-ORDER EQUATIONS

We shall in this section consider the second-order linear difference equation

$$Ax_n + Bx_{n-1} + Cx_{n-2} = 0, \tag{i}$$

in which A, B, and C are constants.

This equation is a generalization of the first-order difference equation whose solution was given in Theorem 11.3.1. In that theorem we found that the solution for x_n turned out to be a certain exponential function of n, namely $x_n = ca^n$. Let us try to find a solution of the present second-order difference equation of a similar exponential form, that is, let us suppose that $x_n = a^n$, where a is a constant that is to be determined.

Substituting $x_n = a^n$, $x_{n-1} = a^{n-1}$, and $x_{n-2} = a^{n-2}$ into the given difference equation we obtain

$$Aa^n + Ba^{n-1} + Ca^{n-2} = 0.$$

Then, after dividing through by a^{n-2}, we are left with the equation

$$Aa^2 + Ba + C = 0.$$

This equation is a quadratic equation for the constant a, which is called the *auxiliary equation* of the difference equation (i). Hence there are two possible roots for a, which we denote by $a = a_1$ and $a = a_2$, where

$$a_1 = \frac{-B + \sqrt{B^2 - 4AC}}{2A}, \qquad a_2 = \frac{-B - \sqrt{B^2 - 4AC}}{2A}.$$

Thus we have found that the given difference equation (i) has two different solutions of the assumed type, namely $x_n = a_1^n$ and $x_n = a_2^n$, where a_1 and a_2 are given above. From these two solutions the following theorem allows us to construct the general solution of the difference equation (i) in the form

$$\boxed{x_n = c_1 a_1^n + c_2 a_2^n,}$$

where c_1 and c_2 are arbitrary constants. Any solution of (i) must be expressible in this form for certain values of c_1 and c_2.

THEOREM 11.4.1

If $x_n = f(n)$ and $x_n = g(n)$ are two particular solutions of the difference equation $Ax_n + Bx_{n-1} + Cx_{n-2} = 0$, where $f(n)$ is not a constant multiple of $g(n)$, then the general solution is given by $x_n = c_1 f(n) + c_2 g(n)$, where c_1 and c_2 are two arbitrary constants.

For the first-order difference equation discussed in Theorem 11.3.1, the constant c appearing in the solution was determined by making use of the value of x_n for one value of n assumed given. As a rule we assume that x_0 is known, but in fact any one of the x_n's would do just as well to determine c. In the present case, with a second-

580

order difference equation, we have found a solution that involves two constants, c_1 and c_2. Clearly then, in order to determine these constants, we must be given the values of x_n for two values of n. Usually it is assumed that x_0 and x_1 are given. In this case,

$$x_0 = c_1 a_1^0 + c_2 a_2^0 = c_1 + c_2$$
$$x_1 = c_1 a_1^1 + c_2 a_2^1 = c_1 a_1 + c_2 a_2.$$

These two equations can be solved for c_1 and c_2. For example,

$$x_1 - a_1 x_0 = c_1 a_1 + c_2 a_2 - a_1(c_1 + c_2)$$
$$= c_2(a_2 - a_1).$$

Therefore

$$c_2 = \frac{x_1 - a_1 x_0}{a_2 - a_1}.$$

In a similar way we obtain

$$c_1 = \frac{x_1 - a_2 x_0}{a_1 - a_2}.$$

EXAMPLE   Find the solution of the difference equation

$$x_n - 2x_{n-1} - 3x_{n-2} = 0,$$

for which $x_0 = 1$ and $x_1 = 2$.

SOLUTION   The coefficients of this difference equation corresponding to the general form (i) are $A = 1$, $B = -2$, and $C = -3$. Therefore the auxiliary equation takes the form

$$a^2 - 2a - 3 = 0,$$

i.e.,

$$(a + 1)(a - 3) = 0.$$

So the two roots are $a_1 = -1$ and $a_2 = 3$. The general solution is therefore

$$x_n = c_1(-1)^n + c_2 3^n.$$

In order to determine c_1 and c_2, we must set $n = 0$ and $n = 1$ and make use of the given values of x_0 and x_1:

$$x_0 = c_1 + c_2 = 1,$$
$$x_1 = c_1(-1) + c_2(3) = -c_1 + 3c_2 = 2.$$

The solution of these two simultaneous equations is readily found to be $c_1 = \frac{1}{4}$, $c_2 = \frac{3}{4}$. Therefore the final solution for any value of n is given by

$$x_n = \tfrac{1}{4}(-1)^n + (\tfrac{3}{4})3^n.$$

For example, putting $n = 4$ we see that

$$x_4 = \tfrac{1}{4}(-1)^4 + \tfrac{3}{4}3^4 = \tfrac{1}{4} + \tfrac{243}{4} = 61.$$

EXAMPLE A biological population is growing in such a way that its growth in any year is equal to twice its growth in the previous year. If the size of the population at year 0 is 1000 and the size at year 1 is 1050, what is the size in year n?

SOLUTION Let x_n denote the size of the population at year n. Then the growth between year $(n - 1)$ and year n is equal to $x_n - x_{n-1}$. This is equal to twice the population growth in the preceding year, that is,

$$x_n - x_{n-1} = 2(x_{n-1} - x_{n-2}).$$

Bringing all the terms to the left hand side we get

$$x_n - 3x_{n-1} + 2x_{n-2} = 0.$$

This is a difference equation of the type (i) with coefficients $A = 1$, $B = -3$, $C = 2$. Therefore the auxiliary equation is

$$a^2 - 3a + 2 = 0$$

or

$$(a - 1)(a - 2) = 0.$$

So we obtain the roots $a_1 = 1$, $a_2 = 2$, and hence the general solution of the difference equation is

$$x_n = c_1(1)^n + c_2 2^n = c_1 + c_2 2^n.$$

Putting $n = 0$ and $n = 1$, we find that

$$x_0 = c_1 + c_2 = 1000,$$
$$x_1 = c_1 + 2c_2 = 1050.$$

Solving for c_1 and c_2 we get $c_1 = 950$ and $c_2 = 50$. Therefore finally the population after n yr is given by

$$x_n = 950 + (50)2^n.$$

We see then that the first-order linear difference equations studied in Sections 11.1 to 11.3 require one of the x_n's, usually x_0, to be specified in order that the solution should be completely determined, while the second-order equation (i) requires two of the x_n's to be specified, say x_0 and x_1. We can see that this has to be the case by solving (i) for x_n:

$$x_n = -A^{-1}(Bx_{n-1} + Cx_{n-2}).$$

Putting $n = 2$ we obtain

$$x_2 = -A^{-1}(Bx_1 + Cx_0),$$

so that if x_0 and x_1 are specified, we are able to determine x_2. Putting $n = 3$, then, we get

$$x_3 = -A^{-1}(Bx_2 + Cx_1).$$

Since x_1 is specified and x_2 has been determined we are now able to determine the value of x_3.

We can clearly continue this process; putting n successively equal to $4, 5, 6, \ldots$ we are able to determine in turn $x_4, x_5, x_6, \ldots$. Thus it is clear that any x_n can be found by means of this process of repeated solution, once we are given the values of x_0 and x_1 to enable the process to start.

As a practical method of solution this process is usually inferior to the use of the formulas developed earlier in this section unless a computer is used to do the calculations. It is introduced here in order to explain why, for a second-order difference equation, we need two starting values x_0 and x_1 in order to determine the solution.

The formula developed above for solving the difference equation (i) works provided the coefficients A, B, and C satisfy the condition $B^2 - 4AC > 0$. If $B^2 - 4AC \leq 0$ some changes must be made.

Let us consider first the case in which $B^2 - 4AC = 0$. Then the two roots a_1 and a_2 of the auxiliary equation are equal, and we denote them simply by a:

$$a_1 = a_2 = a = -\frac{B}{2A}.$$

The solution we found before for x_n reduces to

$$x_n = c_1 a_1^n + c_2 a_2^n = (c_1 + c_2)a^n = ca^n,$$

where we have set $c = c_1 + c_2$. It is clear that this solution involves only a single constant c. Therefore it cannot possibly be the most general solution, since we know that this general solution has to involve two constants in order to satisfy the given values of x_0 and x_1.

Therefore we must look a little further. Let us set $B = -2Aa$ and $C = B^2/4A = Aa^2$ in the difference equation (i). After dividing through by A this equation reduces to

$$x_n - 2ax_{n-1} + a^2 x_{n-2} = 0,$$

that is,

$$x_n - ax_{n-1} - a(x_{n-1} - ax_{n-2}) = 0.$$

Therefore if we introduce new quantities v_n by means of the definition

$$v_n = x_n - ax_{n-1}, \qquad v_{n-1} = x_{n-1} - ax_{n-2},$$

we see that these v_n's satisfy the difference equation

$$v_n - av_{n-1} = 0.$$

But the solution of this difference equation can be obtained from Theorem 11.3.1; the result is $v_n = c_1 a^n$, where c_1 is a certain constant. In fact the value of c_1 can be determined in terms of the given initial values x_0 and x_1, since on the one hand $v_1 = c_1 a^1 = c_1 a$ and on the other hand, from the definition of v_n, it follows that $v_1 = x_1 - ax_0$. Comparing these two values of v_1 we find that

$$c_1 = a^{-1}x_1 - x_0.$$

Since v_n is now known, the following equation remains to be solved for x_n:

$$x_n - ax_{n-1} = v_n.$$

But from Theorem 11.3.3, replacing b_n in the theorem by $v_n = c_1 a^n$, we find that

$$x_n = c_2 a^n + \sum_{p=1}^{n} a^{n-p} v_p,$$

where $c_2 = x_0$. Substituting for $v_p = c_1 a^p$ we obtain therefore

$$x_n = c_2 a^n + \sum_{p=1}^{n} a^{n-p} c_1 a^p$$

$$= c_2 a^n + \sum_{p=1}^{n} c_1 a^n$$

$$= c_2 a^n + n c_1 a^n$$

since there are n terms in the sum and these terms are all equal to $c_1 a^n$, independent of the summation index p.

Consequently we have proved that, when $B^2 - 4AC = 0$, the solution of the difference equation (i) is given by

$$\boxed{x_n = (c_1 n + c_2) a^n,}$$

where c_1 and c_2 are two constants. Their values are in fact

$$c_1 = a^{-1} x_1 - x_0, \qquad c_2 = x_0.$$

EXAMPLE   As a result of competition for limited resources, a certain biological population is decreasing. The decrease in population size between the $(n-1)$th year and the nth year is equal to one-quarter of the size at the $(n-2)$th year. If the original population sizes were 20,000 in year 0 and 19,500 in year 1, what is the population size x_n in the nth year? Evaluate the size in the 8th year.

SOLUTION   The decrease in population referred to in the example is $(x_{n-1} - x_n)$. This is equal to $\frac{1}{4} x_{n-2}$, so that the following difference equation results:

$$x_{n-1} - x_n = \tfrac{1}{4} x_{n-2},$$

i.e.,

$$x_n - x_{n-1} + \tfrac{1}{4} x_{n-2} = 0.$$

The coefficients are therefore $A = 1$, $B = -1$, $C = \frac{1}{4}$; we see that $B^2 - 4AC = 0$. Thus $a = -B/2A = \frac{1}{2}$, and the solution of the difference equation has the general form

$$x_n = (c_1 n + c_2)(\tfrac{1}{2})^n.$$

Setting $n = 0$ and $n = 1$, we find

$$x_0 = (c_1 0 + c_2)(\tfrac{1}{2})^0 = c_2 = 20{,}000,$$
$$x_1 = (c_1 1 + c_2)(\tfrac{1}{2})^1 = \tfrac{1}{2}(c_1 + c_2) = 19{,}500.$$

Therefore $c_1 = 19{,}000$, and so

$$x_n = (19{,}000 n + 20{,}000) 2^{-n}.$$

In particular, in the 8th year, the population size is

$$x_8 = [(19,000)(8) + 20,000]2^{-8} = \frac{172,000}{256} = 672.$$

Finally we must consider the case in which $B^2 - 4AC < 0$. Here the auxiliary equation has no root that is a real number. It turns out that the solution of the difference equation (i) can be expressed in the form

$$\boxed{x_n = cb^n \cos (n\theta + \alpha),}$$

where b and θ are two constants whose values we shall derive in terms of the coefficients A, B, and C. The other two constants c and α are analogous to c_1 and c_2 in the earlier solutions and must be determined from the starting values x_0 and x_1.

If x_n is given by the above formula then

$$\begin{aligned}
x_{n-1} &= cb^{n-1} \cos [(n - 1)\theta + \alpha] \\
&= cb^{n-1} \cos (n\theta + \alpha - \theta) \\
&= cb^{n-1}[\cos (n\theta + \alpha) \cos \theta + \sin (n\theta + \alpha) \sin \theta],
\end{aligned}$$

where we have used the standard formula for the cosine of the difference between two angles (see p. 231). Similarly,

$$\begin{aligned}
x_{n-2} &= cb^{n-2} \cos [(n - 2)\theta + \alpha] \\
&= cb^{n-2}[\cos (n\theta + \alpha) \cos 2\theta + \sin (n\theta + \alpha) \sin 2\theta].
\end{aligned}$$

Substituting these expressions for x_n, x_{n-1}, and x_{n-2} into the difference equation (i), therefore, we obtain

$$\begin{aligned}
Acb^n \cos (n\theta + \alpha) &+ Bcb^{n-1}[\cos (n\theta + \alpha) \cos \theta + \sin (n\theta + \alpha) \sin \theta] \\
&+ Ccb^{n-2}[\cos (n\theta + \alpha) \cos 2\theta + \sin (n\theta + \alpha) \sin 2\theta] = 0,
\end{aligned}$$

and so, after dividing through by the common factor cb^{n-2} and grouping the coefficients of $\cos (n\theta + \alpha)$ and $\sin (n\theta + \alpha)$, we find

$$(Ab^2 + Bb \cos \theta + C \cos 2\theta) \cos (n\theta + \alpha) + (Bb \sin \theta + C \sin 2\theta) \sin (n\theta + \alpha) = 0.$$

This equation will be satisfied for all values of n if we choose b and θ in such a way that

$$Ab^2 + Bb \cos \theta + C \cos 2\theta = 0$$
$$Bb \sin \theta + C \sin 2\theta = 0.$$

Using the fact that $\sin 2\theta = 2 \sin \theta \cos \theta$, we obtain from the second of the two above equations the result that

$$Bb + 2C \cos \theta = 0$$

or

$$\cos \theta = -\frac{Bb}{2C}. \tag{ii}$$

We now substitute this into the first equation, observing that $\cos 2\theta = 2\cos^2\theta - 1$:

$$Ab^2 + Bb\left(-\frac{Bb}{2C}\right) + C\left[2\left(-\frac{Bb}{2C}\right)^2 - 1\right] = 0,$$

and this reduces simply to

$$Ab^2 - C = 0.$$

Hence $b = \sqrt{C/A}$. Substituting this into (ii) above, we obtain that

$$\cos\theta = -\frac{B}{2C}\left(\frac{C}{A}\right)^{1/2} = -\frac{B}{2\sqrt{AC}}.$$

EXAMPLE   Find the solution of the difference equation

$$3x_n - 6x_{n-1} + 4x_{n-2} = 0$$

for which $x_0 = 1$ and $x_1 = 2$.

SOLUTION   The coefficients are $A = 3$, $B = -6$, and $C = 4$, and so

$$B^2 - 4AC = (-6)^2 - 4(3)(4) = 36 - 48 = -12 < 0.$$

Consequently we must use the solution involving trigonometric functions. We have:

$$b = \sqrt{C/A} = \sqrt{4/3} = 2/\sqrt{3}$$

and

$$\cos\theta = -\frac{B}{2\sqrt{AC}} = -\frac{(-6)}{2\sqrt{(3)(4)}} = \frac{\sqrt{3}}{2}.$$

It follows therefore that we can take $\theta = \pi/6$. Consequently the solution of the given difference equation is:

$$x_n = cb^n \cos(n\theta + \alpha)$$

$$= c\left(\frac{2}{\sqrt{3}}\right)^n \cos\left(\frac{n\pi}{6} + \alpha\right).$$

In order to determine c and α we must make use of the given initial values x_0 and x_1. Setting $n = 0$ and $n = 1$ in the solution, therefore, we obtain

$$x_0 = c\cos\alpha = 1,$$

$$x_1 = c\frac{2}{\sqrt{3}}\cos\left(\frac{\pi}{6} + \alpha\right) = 2.$$

Thus

$$c\cos\left(\frac{\pi}{6} + \alpha\right) = \sqrt{3},$$

and expanding this cosine using the sum formula we get

$$c\left(\cos\frac{\pi}{6}\cos\alpha - \sin\frac{\pi}{6}\sin\alpha\right) = \sqrt{3}.$$

Inserting the values $\cos\pi/6 = \sqrt{3}/2$, $\sin\pi/6 = \frac{1}{2}$ and using the fact that c

$\cos \alpha = 1$, we obtain

$$\frac{\sqrt{3}}{2}(1) - \frac{1}{2}c \sin \alpha = \sqrt{3}$$

and therefore

$$c \sin \alpha = 2\left(\frac{\sqrt{3}}{2} - \sqrt{3}\right) = -\sqrt{3}.$$

Knowing the values of $c \sin \alpha$ and $c \cos \alpha$, we can determine c in the usual way by squaring and adding:

$$c^2 = (c \sin \alpha)^2 + (c \cos \alpha)^2 = (-\sqrt{3})^2 + (1)^2 = 3 + 1 = 4,$$

and so $c = 2$. Therefore

$$\cos \alpha = \frac{1}{c} = \frac{1}{2} \qquad \text{and} \qquad \sin \alpha = -\frac{\sqrt{3}}{c} = -\frac{\sqrt{3}}{2}.$$

Consequently $\alpha = -\pi/3$, and so the solution is finally obtained in the form

$$x_n = 2\left(\frac{2}{\sqrt{3}}\right)^n \cos\left(\frac{n\pi}{6} - \frac{\pi}{3}\right).$$

It is interesting to examine the graph of this solution, which is plotted in Fig. 11.8. We see that the solution oscillates between positive and negative values and also

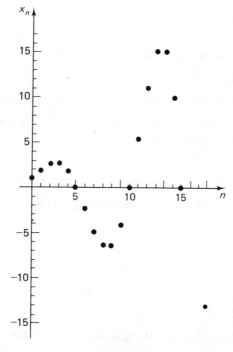

Figure 11.8

that, as n increases, the oscillations grow larger in amplitude. We can see that the origin of the oscillatory nature of the solution rests with the cosine term that appears in the solution. The exponential factor b^n, or $(2/\sqrt{3})^n$ in the present case, is responsible for the growth in amplitude as n increases.

This type of behavior is quite typical of the solution of the present class of difference equations when $B^2 - 4AC < 0$. The solution always oscillates between positive and negative values because of the factor $\cos(n\theta + \alpha)$ that appears in the solution. The other factor cb^n makes the amplitude of this oscillation either grow exponentially as n increases if $b > 1$ or decay exponentially to zero as n increases if $b < 1$.

The solutions of the second-order difference equations studied in this section are set out in the following table.

Difference equation	$Ax_n + Bx_{n-1} + Cx_{n-2} = 0$	
Auxiliary equation	$Aa^2 + Ba + C = 0$	with roots a_1, a_2
a_1, a_2 real and distinct	$x_n = c_1 a_1^n + c_2 a_2^n$	c_1, c_2 arbitrary
$a_1 = a_2 (= a)$	$x_n = (c_1 n + c_2) a^n$	c_1, c_2 arbitrary
a_1, a_2 not real	$x_n = cb^n \cos(n\theta + \alpha)$ where $b = \sqrt{C/A}$, $\cos\theta = -B/2\sqrt{AC}$	c, α arbitrary

EXERCISES 11.4

Solve the following difference equations:

1. $x_n - x_{n-2} = 0.$ **2.** $x_n - 4x_{n-2} = 0.$

3. $x_n + x_{n-1} - 6x_{n-2} = 0.$ **4.** $x_{n+2} - 2x_{n+1} - 3x_n = 0.$

Find the solutions of the following difference equations that satisfy the initial conditions $x_0 = 1$, $x_1 = 2$:

5. $x_n + 2x_{n-1} + 2x_{n-2} = 0.$ **6.** $x_{n+2} - 3x_{n+1} - 4x_n = 0.$

7. $x_{n+2} - 6x_{n+1} + 9x_n = 0.$ **8.** $4x_n + 4x_{n-1} + x_{n-2} = 0.$

Find the solutions of the following difference equations that satisfy the given initial conditions:

9. $x_{n+1} + 4x_n + 4x_{n-1} = 0;$ $x_0 = 2$, $x_1 = 1.$

10. $x_n - 8x_{n-1} + 16x_{n-2} = 0;$ $x_0 = 1$, $x_1 = 1.$

11. $x_n - 3x_{n-1} + 3x_{n-2} = 0;$ $x_0 = 1$, $x_1 = 2.$

12. $x_n - 2x_{n-1} + 2x_{n-2} = 0;$ $x_0 = 1$, $x_1 = -1.$

13. $x_{n+1} + x_n + x_{n-1} = 0;$ $x_0 = 2$, $x_1 = 4.$

14. $8x_{n+2} + 4x_{n+1} + x_n = 0;$ $x_0 = 1$, $x_1 = 0.$

15. A population is increasing in such a way that its growth in any year is three times its growth in the previous year. Find the population size x_n after n yr in terms of its sizes x_0 and x_1 in year 0 and year 1.

16. Repeat exercise 15 for the case in which the increase in any year is k times the increase the previous year, where k is some given constant.

17. Assume that in a rabbit colony each pair of adult rabbits produces a pair of young rabbits every month and that the rabbits start to produce young when they are 2 mo old. Set up a difference equation for x_n, the number of pairs of adult rabbits after n mo. If the colony starts at $n = 0$ with one adult pair, find x_n. (This problem was first investigated by Fibonacci in the thirteenth century. The values of x_n are said to form a *Fibonacci sequence*.)

18. Repeat exercise 17 on the assumption that each month every adult pair of rabbits produces two pairs of young rabbits.

19. During the spread of an epidemic the number of new cases occurring during the nth week is equal to twice the number of cases x_{n-2} that existed at the end of the $(n - 2)$th week. Write down a difference equation for x_n and solve it on the assumption that $x_0 = x_1 = 1$.

20. In exercise 19 assume that during the nth week half of the cases that existed at the end of the previous week are cured. In addition, as in exercise 19, assume that the number of new cases during that week is $2x_{n-2}$. Construct the new difference equation and solve it with the initial condition $x_0 = 1$, $x_1 = 4$.

21. The figure illustrates the conduction of an electrical impulse along a myelinated

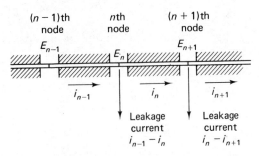

nerve fiber. If E_n is the electrical potential of the nth node along the fiber and i_n the current in the fiber between the nth and $(n + 1)$th nodes, then $i_n = (E_n - E_{n+1})/R$, where R is the resistance of the segment of fiber between these two nodes. Current can leak through each node, and the amount of leakage at the nth node is given by $(i_{n-1} - i_n) = E_n/r$, where r is the leakage resistance of the nodes. Show that

$$i_{n+1} - \left(2 + \frac{R}{r}\right)i_n + i_{n-1} = 0$$

and hence obtain an expression for i_n as a function of n.

REVIEW EXERCISES FOR CHAPTER 11

1. State whether each of the following statements is true or false. If false, give the corresponding correct statement:
   (a) The difference equation $x_{n+1} - n(x_n)^2 = 1$ is of second order.
   (b) The difference equation $x_{n+1} = \sqrt{x_{n-1}} + 1$ is of second order.
   (c) The difference equation $x_n + (x_{n-1})^2 - 3x_{n-1} = 0$ is of second order.
   (d) $x_n = (c + n)^2$, where c is an arbitrary constant, is a solution of the difference equation

   $$x_n - x_{n-1} - 2\sqrt{x_{n-1}} = 1.$$

   (e) $x_n = \tfrac{1}{2}(n + 1) + cn^{-1}$, where c is an arbitrary constant, is a solution of the difference equation

   $$nx_n - (n - 1)x_{n-1} = n.$$

   (f) The difference equation $x_n(x_{n-2})^p = (x_{n-1})^{p+1}$ has as its solution $x_n = cd^n$, where c and d are arbitrary constants.

   Find a table of values of the solution of each of the following difference equations for $n = 0, 1, 2, \ldots, 10$.

2. $x_n = 3x_{n-1}/2$, $x_0 = 4$.

3. $x_n = 2x_{n-1} - 1$, $x_0 = 1.5$.

4. $x_n = [2.2 - 0.1x_{n-1}]x_{n-1}$, $x_0 = 30$.

5. $x_n = \dfrac{1}{1 + x_{n-1}}$, $x_0 = 0$.   What do you observe about the solution in this case?

   Find the solutions of the following difference equations that satisfy the given conditions:

6. $x_{n+1} - 3x_n = 0$, $x_1 = 1$.

7. $2x_n + x_{n-1} = 0$, $x_0 = 3$.

8. $x_n + x_{n-1} = n$, $x_0 = 0$.

9. $x_n = x_{n-1} + 2n^2$, $x_1 = 1$.

10. $x_n = 3x_{n-1} + n - 1$, $x_0 = 0$.

11. $9x_{n+2} + 6x_{n+1} + x_n = 0$, $x_0 = 1$, $x_1 = -1$.

12. $x_{n+1} - 6x_n + 10x_{n-1} = 0$, $x_1 = 2$, $x_2 = 3$.

   Just as in the case of differential equations, we can consider systems of difference equations. By eliminating x_n, reduce the following systems to second-order difference equations for y_n, and hence find their solutions:

13. $y_n = x_{n-1}$, $x_n = 6x_{n-1} - 5y_{n-1}$; $x_0 = 1$, $y_0 = 2$.

14. $x_n = -4y_{n-1} + 2x_{n-1}$, $y_n = 2y_{n-1} + x_{n-1}$; $x_0 = -3$, $y_0 = 0$.

15. By writing $x_n = e^{u_n}$ and solving the resulting difference equation for u_n, find the solution of the equation

$$\frac{x_n}{x_{n-1}} = \left(\frac{x_{n-1}}{x_{n-2}}\right)^p.$$

12

Vectors
and
Matrices

12.1 VECTORS IN TWO DIMENSIONS

It is useful to distinguish between two types of physical quantities called *scalars* and *vectors*, which occur in many problems. A *scalar* is a quantity that is completely characterized by a single real number, the magnitude of the quantity concerned. Quantities such as length, mass, time, temperature, volume, and so on are scalars. For example, the length of a newborn rattlesnake is a scalar quantity, inasmuch as it is specified by one real number, namely the appropriate number of inches.

On the other hand, a *vector* is a quantity that is characterized by two things, a magnitude and a direction. An example of a quantity that is a vector is the velocity of a moving object. Such a velocity has a magnitude, namely the number of miles per hour, and it also has a direction, namely the direction in which the object is traveling. Another example of a vector quantity that we shall discuss at greater length in Section 12.4 is force: Forces exerted on one another by different parts of a system are vector quantities, having magnitude—the size of the force—and direction. A third example of a vector quantity is the magnetic field at a point, for instance, the earth's magnetic field at a given point on its surface. The magnitude of this vector is the field strength and its direction is the direction indicated by a compass needle.

Perhaps the simplest example of a quantity that has both size and direction is what is called a *directed line segment*. Let P and Q be any two points in a given plane. The directed line segment from P to Q, denoted by $\overrightarrow{PQ}$, is the straight line segment that starts from P and ends at Q. P is called the *initial point* and Q the *terminal point* of the segment. Geometrically the segment $\overrightarrow{PQ}$ is represented by an arrow based at P and directed toward Q. Figure 12.1 illustrates the segments $\overrightarrow{PQ}$ and $\overrightarrow{QP}$. The length of the directed line segment $\overrightarrow{PQ}$ is denoted by $|\overrightarrow{PQ}|$.

Two directed line segments $\overrightarrow{PQ}$ and $\overrightarrow{RS}$ are said to be *equivalent* to each other if they have equal lengths and are parallel to each other. In Fig. 12.2 $\overrightarrow{PQ}$, $\overrightarrow{RS}$, and $\overrightarrow{TU}$ are all equivalent to one another. If two directed line segments are equivalent to each other, then one of them can be moved, all the time remaining parallel to its original direction, until it lies exactly on top of the other.

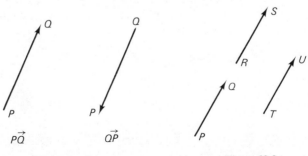

Figure 12.1   Figure 12.2

Any vector can be represented geometrically by a directed line segment, the length and direction of the segment being respectively the magnitude and direction of the vector. In making this representation it is necessary to decide upon a length scale for the magnitudes of the vectors concerned. For example, in representing a velocity vector by a directed line segment, one might choose a scale of 1 in. to represent a velocity of 10 mph. Having chosen this scale, any velocity vector can then be represented by a directed line segment of the appropriate length and direction.

A given vector can in this way be represented by many different directed line segments since the location of the line segment in the plane is not restricted, but only its length and direction. Any two equivalent directed line segments represent the same vector, and conversely, if a given vector can be represented by a segment $\overrightarrow{PQ}$ then it can equally well be represented by any directed line segment $\overrightarrow{RS}$ that is equivalent to $\overrightarrow{PQ}$. Because of this flexibility, a vector when represented by a directed line segment is often called a *free vector*.

Ordinarily a single boldface letter is used to denote a vector, for example **A**, **B**, and **C**, etc. This symbol is written next to the corresponding directed line segment when the vector is represented geometrically.

The symbol **0** is used to denote the so-called *zero vector*, which is the vector whose magnitude is zero. It is not necessary to specify a direction for the zero vector. The

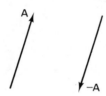

symbol −**A** is used to denote the vector that has the same magnitude as the vector **A** but whose direction is diametrically opposite to that of **A**. This is illustrated in Fig. 12.3. This symbol, −**A**, is read either "minus **A**" or "the negative of vector **A**."

A vector whose magnitude is one unit is called a *unit vector*. A unit vector having the same direction as **A** is denoted by **Â**.

Figure 12.3

DEFINITION   *Scalar Multiplication of a Vector*

Let c be a scalar, i.e., a real number, and **A** a nonzero vector. Then we define $c\mathbf{A}$ to be the vector whose magnitude is $|c|$ times the magnitude of **A** and whose direction is the same as that of **A** if $c > 0$ and opposite to that of **A** if $c < 0$. Some illustrations of scalar multiplication of the vector **A** are shown in Fig. 12.4.

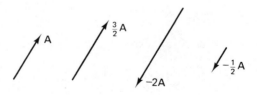

Figure 12.4

Addition of Vectors

Imagine that the directed line segment $\overrightarrow{PQ}$ represents the displacement of a point along a straight line as it move from P to Q. Then, as is clear from Fig. 12.5, a displacement $\overrightarrow{PQ}$ followed by another displacement $\overrightarrow{QR}$ is equivalent to a single displacement $\overrightarrow{PR}$. The directed line segment $\overrightarrow{PR}$ is called the *sum* of the first two segments $\overrightarrow{PQ}$ and $\overrightarrow{QR}$. We write this as

$$\overrightarrow{PR} = \overrightarrow{PQ} + \overrightarrow{QR}.$$

Since any vector can be represented by directed line segments of appropriate length and direction, we can define the sum of two vectors **A** and **B** by placing the initial point of **B** at the terminal point of **A**. The sum **A** + **B** is then the vector that runs from the initial point of **A** to the terminal point of **B** (Fig. 12.6).

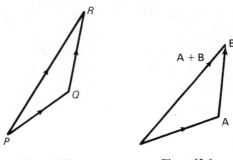

Figure 12.5   **Figure 12.6**

Another way of finding the sum of two vectors **A** and **B** is to take two directed line segments (**A** and **B** in Fig. 12.7) that are equivalent to the two given vectors but have the same initial point. If we complete the parallelogram of which two adjacent sides are formed by these vectors, then **A** + **B** is the vector that runs from the common initial point of **A** and **B** to the opposite vertex of the parallelogram.

The *difference* of two vectors **A** and **B** is defined as

$$\mathbf{A} - \mathbf{B} = \mathbf{A} + (-1)\mathbf{B}.$$

Geometrically, **A** − **B** can be viewed as a vector that, when added to **B**, gives the vector **A**. Thus if we place **A** and **B** with a common initial point, then the vector **A** − **B** runs from the terminal point of **B** to the terminal point of **A**, as shown in Fig. 12.8.

Let us now suppose that all the vectors under consideration lie in some fixed plane. Then it is possible to assign coordinates to the initial and final points of any free vectors; statements about the vectors can be interpreted algebraically in terms of these coordinates. This leads us to the algebraic approach to vectors as opposed to the above geometric approach.

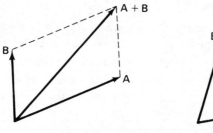

Figure 12.7

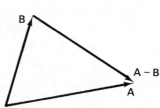

Figure 12.8

To use rectangular coordinates in dealing with vectors it is convenient to introduce unit vectors **i** and **j** along each of the coordinate axes. The unit vector **i** is taken to point in the positive direction of the *x*-axis and **j** is taken to point in the positive direction of the *y*-axis, as shown in Fig. 12.9.

If $P(x, y)$ is any point in the plane and O the origin of coordinates, then $\overrightarrow{OP}$ is called the *position vector* of *P*. If the vector **A** is equivalent to the segment $\overrightarrow{OP}$, then we can write

$$\mathbf{A} = \overrightarrow{OP} = x\mathbf{i} + y\mathbf{j}.$$

This result follows from the fact that the vectors $x\mathbf{i}$ and $y\mathbf{j}$ run from the origin to the points $(x, 0)$ and $(0, y)$ on the coordinate axes, whereas $\mathbf{A} = \overrightarrow{OP}$ is the diagonal of the rectangle, two sides of which are formed by these two vectors (Fig. 12.10).

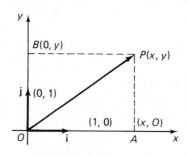

Figure 12.9

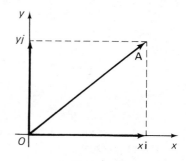

Figure 12.10

The numbers x and y in the expansion $\mathbf{A} = x\mathbf{i} + y\mathbf{j}$ are respectively called the *x*- *and y-components* of the vector **A**. Commonly, the notation (x, y) is used to denote the vector $x\mathbf{i} + y\mathbf{j}$, so that

$$(x, y) = x\mathbf{i} + y\mathbf{j}.$$

In particular, $\mathbf{i} = (1, 0)$ and $\mathbf{j} = (0, 1)$. The zero vector is $\mathbf{0} = (0, 0)$ and the negative of a vector $\mathbf{A} = (x, y)$ is $-\mathbf{A} = (-x, -y)$.

THEOREM 12.1.1

If $P(x_1, y_1)$ and $Q(x_2, y_2)$ are any two points in a plane, then the directed line segment $\overrightarrow{PQ} = (x_2 - x_1, y_2 - y_1)$.

The position vectors $\overrightarrow{OP}$ and $\overrightarrow{OQ}$ in this case (Fig. 12.11) are given by

$$\overrightarrow{OP} = (x_1, y_1) = x_1\mathbf{i} + y_1\mathbf{j},$$
$$\overrightarrow{OQ} = (x_2, y_2) = x_2\mathbf{i} + y_2\mathbf{j}.$$

Since by the law of vector addition, we have

$$\overrightarrow{OP} + \overrightarrow{PQ} = \overrightarrow{OQ},$$

it follows that

$$\begin{aligned}\overrightarrow{PQ} &= \overrightarrow{OQ} - \overrightarrow{OP} \\ &= x_2\mathbf{i} + y_2\mathbf{j} - (x_1\mathbf{i} + y_1\mathbf{j}) \\ &= (x_2 - x_1)\mathbf{i} + (y_2 - y_1)\mathbf{j} \\ &= (x_2 - x_1, y_2 - y_1).\end{aligned}$$

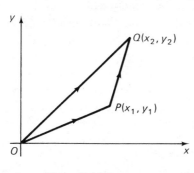

Figure 12.11

Thus the *components of any directed line segment are the coordinates of the terminal point minus the coordinates of the initial point.*

Consider a position vector $\mathbf{A} = \overrightarrow{OP} = (x, y)$. Since it is possible to move $\mathbf{A}$ parallel to itself to any position in the plane, we can represent $\mathbf{A}$ equivalently by another directed line segment with any arbitrary point as the initial point. Let us find the line segment with initial point $Q(x_0, y_0)$ that is equivalent to $\mathbf{A}$. Let $R(x_1, y_1)$ be the terminal point of this segment. Then by the above theorem, this new segment $\overrightarrow{QR}$ is in component form $(x_1 - x_0, y_1 - y_0)$. Since this is equivalent to the original vector (x, y), we must have

$$x_1 - x_0 = x, \qquad y_1 - y_0 = y.$$

Therefore

$$x_1 = x + x_0, \qquad y_1 = y + y_0.$$

Thus the vector $\mathbf{A} = (x, y)$ can also be represented by $\overrightarrow{QR}$, where Q is the point (x_0, y_0) and R is $(x + x_0, y + y_0)$. Some of the geometric representations of $\mathbf{A} = (x, y)$ are shown in Fig. 12.12. It can be remarked that although $\mathbf{A} = (x, y)$ can be

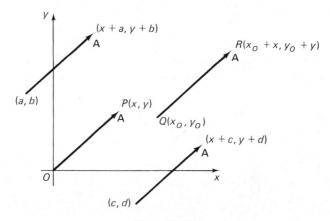

Figure 12.12

represented geometrically in many ways (because we can choose any point as the initial point), there is precisely one *position vector* corresponding to **A**.

Two vectors can be added by simply adding their components:

$$(x_1, y_1) + (x_2, y_2) = (x_1\mathbf{i} + y_1\mathbf{j}) + (x_2\mathbf{i} + y_2\mathbf{j})$$
$$= (x_1 + x_2)\mathbf{i} + (y_1 + y_2)\mathbf{j}.$$
$$= (x_1 + x_2, y_1 + y_2).$$

Similarly, multiplying a vector by a scalar is equivalent to multiplying the components of the vector by the scalar:

$$c(x, y) = c(x\mathbf{i} + y\mathbf{j})$$
$$= cx\mathbf{i} + cy\mathbf{j}$$
$$= (cx, cy).$$

Two vectors can be subtracted by subtracting their components, that is,

$$(x_1, y_1) - (x_2, y_2) = (x_1 - x_2, y_1 - y_2).$$

EXAMPLE   If $\mathbf{A} = (2, -3)$, $\mathbf{B} = (3, 4)$ compute:

(a) $3\mathbf{A} + 4\mathbf{B}$.
(b) $2\mathbf{B} - \mathbf{A}$.

SOLUTION

(a) $3\mathbf{A} + 4\mathbf{B} = 3(2, -3) + 4(3, 4)$
$$= (3(2), 3(-3)) + (4(3), 4(4))$$
$$= (6, -9) + (12, 16)$$
$$= (6 + 12, -9 + 16)$$
$$= (18, 7).$$

(b) $2\mathbf{B} - \mathbf{A} = 2(3, 4) - (2, -3)$
$$= (6, 8) - (2, -3)$$
$$= (6 - 2, 8 - (-3))$$
$$= (4, 11).$$

EXAMPLE   If $P(1, -2)$, $Q(-3, 1)$, $R(2, 4)$, and $S(-1, 6)$ are four points in a plane, then compute the vector $\overrightarrow{PQ} + 2\overrightarrow{RS}$.

SOLUTION   Since the components of any vector are the coordinates of the terminal point *minus* the coordinates of the initial point,

$$\overrightarrow{PQ} = (-3 - 1, 1 - (-2)) = (-4, 3)$$

and

$$\overrightarrow{RS} = (-1 - 2, 6 - 4) = (-3, 2).$$

Therefore

$$\overrightarrow{PQ} + 2\overrightarrow{RS} = (-4, 3) + 2(-3, 2)$$
$$= (-4, 3) + (-6, 4)$$
$$= (-4 + (-6), 3 + 4)$$
$$= (-10, 7).$$

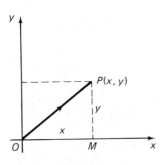

Figure 12.13

Let $\overrightarrow{OP} = (x, y)$ and let PM be the perpendicular from P onto the x-axis, as shown in Fig. 12.13. Then from Pythagoras's theorem,

$$OP^2 = OM^2 + MP^2 = x^2 + y^2,$$

and hence it follows that the *length of the vector* (x, y) is given by

$$|(x, y)| = \sqrt{x^2 + y^2}.$$

If $\mathbf{A} = (x, y)$ is any nonzero vector, we can find a unit vector $\hat{\mathbf{A}}$ with the same direction as $\mathbf{A}$ by dividing $\mathbf{A}$ by its length $|\mathbf{A}|$, that is,

$$\hat{\mathbf{A}} = \frac{\mathbf{A}}{|\mathbf{A}|} = \frac{1}{\sqrt{x^2 + y^2}}(x. y)$$
$$= \left(\frac{x}{\sqrt{x^2 + y^2}}, \frac{y}{\sqrt{x^2 + y^2}}\right).$$

EXAMPLE   If $P(1, 2)$ and $Q(4, 6)$ are two points in a plane, find the unit vector that has the same direction as the vector $\overrightarrow{PQ}$.

SOLUTION   Using Theorem 12.1.1, the components of the vector $\overrightarrow{PQ}$ are given by

$$\overrightarrow{PQ} = (4 - 1, 6 - 2) = (3, 4).$$

The length of the vector $\overrightarrow{PQ}$ is $|\overrightarrow{PQ}| = \sqrt{3^2 + 4^2} = 5$. Therefore the unit vector in the direction of $\overrightarrow{PQ}$ is given by

$$\frac{\overrightarrow{PQ}}{|\overrightarrow{PQ}|} = \frac{1}{5}(3, 4) = \left(\frac{3}{5}, \frac{4}{5}\right).$$

The *direction* of a nonzero vector $\mathbf{A} = (x, y)$ is defined to be the smallest non-negative angle θ from the positive x-axis to the position vector $\overrightarrow{OP}$ corresponding to $\mathbf{A}$. The directions of some typical vectors are as shown in Fig. 12.14. Note that $0 \leq \theta < 2\pi$.

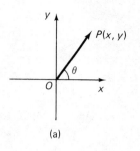
(a)

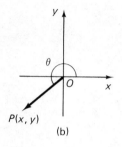
(b)

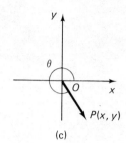
(c)

Figure 12.14

Consider the vector $\mathbf{A} = (x, y)$ that makes an angle θ with the positive x-axis, as shown in Fig. 12.15. Then from the right-angle triangle OMP we have

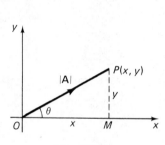

Figure 12.15

$$\cos \theta = \frac{OM}{OP} = \frac{x}{|\mathbf{A}|},$$

$$\sin \theta = \frac{MP}{OP} = \frac{y}{|\mathbf{A}|};$$

i.e.,

$$x = |\mathbf{A}| \cos \theta, \qquad y = |\mathbf{A}| \sin \theta.$$

Thus the vector $\mathbf{A} = (x, y)$ can be expressed in terms of its magnitude and direction as

$$\mathbf{A} = (|\mathbf{A}| \cos \theta, |\mathbf{A}| \sin \theta)$$
$$= |\mathbf{A}| (\cos \theta, \sin \theta).$$

This is known as the *polar form* of the vector $\mathbf{A}$.

EXAMPLE   Give the x- and y-components of a vector $\mathbf{A}$ that has a magnitude of five units and makes an angle of $\pi/6$ with the positive x-axis.

SOLUTION   In this case we have

$$|\mathbf{A}| = 5 \quad \text{and} \quad \theta = \frac{\pi}{6}.$$

Therefore

$$\mathbf{A} = |\mathbf{A}| (\cos \theta, \sin \theta)$$
$$= 5\left(\cos \frac{\pi}{6}, \sin \frac{\pi}{6}\right)$$
$$= 5\left(\frac{\sqrt{3}}{2}, \frac{1}{2}\right)$$
$$= \left(\frac{5\sqrt{3}}{2}, \frac{5}{2}\right).$$

Thus the x-component of $\mathbf{A}$ is $5\sqrt{3}/2$ and the y-component is $\frac{5}{2}$.

Note that, in terms of the polar form $\mathbf{A} = |\mathbf{A}|(\cos\theta, \sin\theta)$ of the vector $\mathbf{A}$, the unit vector in the same direction as $\mathbf{A}$ is given simply by

$$\hat{\mathbf{A}} = \frac{\mathbf{A}}{|\mathbf{A}|} = (\cos\theta, \sin\theta).$$

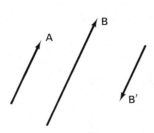

If the two vectors are parallel then they have either the same direction or opposite directions (Fig. 12.16). Their magnitudes could be different. All this implies that one vector is a scalar multiple of the other. Thus if $\mathbf{A}$ is parallel to $\mathbf{B}$ then $\mathbf{B} = c\mathbf{A}$ for some scalar c. Conversely, if $\mathbf{B} = c\mathbf{A}$ then $\mathbf{A}$ and $\mathbf{B}$ are parallel. The vectors $\mathbf{A}$ and $\mathbf{B}$ have the same or opposite directions depending on whether the real number c is positive or negative. The magnitude of $\mathbf{B}$ is $|c|$ times the magnitude of $\mathbf{A}$.

Figure 12.16

EXAMPLE   Given that the two vectors $\mathbf{A} = (2, 3)$ and $\mathbf{B} = (k, -4)$ are parallel, determine k.

SOLUTION   Since $\mathbf{A}$ and $\mathbf{B}$ are parallel, there exists a real number c such that

$$\mathbf{B} = c\mathbf{A}$$

or

$$(k, -4) = c(2, 3) = (2c, 3c).$$

This implies that $k = 2c$ and $3c = -4$. Eliminating c from the two equations we get $k = -\frac{8}{3}$.

Let us now give some simple geometric applications of vectors in two dimensions.

EXAMPLE   Show that the coordinates of the middle point of the line joining $P(x_1, y_1)$ and $Q(x_2, y_2)$ are $[(x_1 + x_2)/2, (y_1 + y_2)/2]$.

SOLUTION   The position vectors of the points P and Q are

$$\overrightarrow{OP} = (x_1, y_1) \quad \text{and} \quad \overrightarrow{OQ} = (x_2, y_2).$$

Also, the vector $\overrightarrow{PQ}$ is given by

$$\overrightarrow{PQ} = (x_2 - x_1, y_2 - y_1).$$

Let R be the midpoint of the line joining P and Q (Fig. 12.17). Then the vector $\overrightarrow{PR}$ has the same direction as $\overrightarrow{PQ}$ but is half the size of $\overrightarrow{PQ}$. Thus

$$\overrightarrow{PR} = \tfrac{1}{2}\overrightarrow{PQ}$$
$$= \tfrac{1}{2}(x_2 - x_1, y_2 - y_1)$$
$$= \left(\frac{x_2 - x_1}{2}, \frac{y_2 - y_1}{2}\right).$$

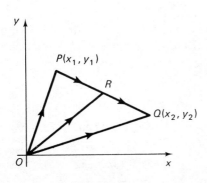

Figure 12.17

Now from $\triangle OPR$, by the rule of addition of vectors, we have

$$\overrightarrow{OR} = \overrightarrow{OP} + \overrightarrow{PR}$$

$$= (x_1, y_1) + \left(\frac{x_2 - x_1}{2}, \frac{y_2 - y_1}{2}\right)$$

$$= \left(x_1 + \frac{x_2 - x_1}{2}, y_1 + \frac{y_2 - y_1}{2}\right)$$

$$= \left(\frac{x_1 + x_2}{2}, \frac{y_1 + y_2}{2}\right),$$

which implies that the coordinates of the point R are

$$\left(\frac{x_1 + x_2}{2}, \frac{y_1 + y_2}{2}\right).$$

EXAMPLE Use vectors to show that the line joining the midpoints of the two sides of a triangle is parallel to the third side and is half as long.

SOLUTION Let P and Q be the midpoints of the sides AB and AC of the triangle ABC (Fig. 12.18). Then

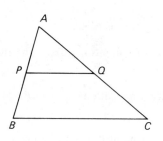

Figure 12.18

$$\overrightarrow{PA} = \tfrac{1}{2}\overrightarrow{BA} \quad \text{and} \quad \overrightarrow{AQ} = \tfrac{1}{2}\overrightarrow{AC}.$$

Now in $\triangle APQ$, by the law of vector addition,

$$\overrightarrow{PQ} = \overrightarrow{PA} + \overrightarrow{AQ}$$

$$= \tfrac{1}{2}\overrightarrow{BA} + \tfrac{1}{2}\overrightarrow{AC}$$

$$= \tfrac{1}{2}(\overrightarrow{BA} + \overrightarrow{AC}) = \tfrac{1}{2}\overrightarrow{BC}.$$

This shows that $\overrightarrow{PQ}$ is parallel to $\overrightarrow{BC}$ and is equal to half its size, which proves the required result.

EXERCISES 12.1

In the following exercises determine the magnitude of **A**, draw a position vector corresponding to **A**, and draw a particular representation of **A** through the given point Q.

1. $\mathbf{A} = (2, -3)$; $Q(1, 2)$. **2.** $\mathbf{A} = (3, 2)$; $Q(-1, -2)$.

3. $\mathbf{A} = (0, -3)$; $Q(2, 5)$. **4.** $\mathbf{A} = (4, 0)$; $Q(-2, 3)$.

In the following exercises find the vector **B** having $\overrightarrow{PQ}$ as one of its representations and draw the position vector corresponding to **B**.

5. $P(2, -3)$; $Q(3, 1)$. **6.** $P(-3, 2)$; $Q(2, -5)$.

7. $P(1, 5)$; $Q(0, -2)$. **8.** $P(-1, 0)$; $Q(3, 2)$.

9. If $\mathbf{A} = (2, -3)$ and $\mathbf{B} = (3, 7)$, find $\mathbf{A} + 2\mathbf{B}$, $3\mathbf{A} - 2\mathbf{B}$.

10. If $\mathbf{P} = (1, 3)$, $\mathbf{Q} = (0, -2)$, and $\mathbf{R} = (2, 5)$, find $2\mathbf{P} + 3\mathbf{R}$, $3\mathbf{Q} - \mathbf{P} + 2\mathbf{R}$.

11. If $\mathbf{A} = (5, -2)$, determine c so that $\mathbf{B} = (10, c)$ is parallel to $\mathbf{A}$.

12. If $\mathbf{A} = (2, -1)$, $\mathbf{B} = (3, 4)$, and $\mathbf{C} = (p, 15)$, determine p so that $\mathbf{C}$ is parallel to $\mathbf{B} - \mathbf{A}$.

13. If $\mathbf{A} = (1, 3)$ and $\mathbf{B} = (2, -1)$, find $|\mathbf{A} - 2\mathbf{B}|$.

14. If $\mathbf{A} = (5, -2)$, $\mathbf{B} = (-3, 2)$, and $\mathbf{C} = (0, 1)$, determine $|2\mathbf{A} - 3\mathbf{B} + 4\mathbf{C}|$.

15. Given $\mathbf{A} = (1, 4)$, $\mathbf{C} = (2, 1)$, and $\mathbf{A} + \mathbf{B} = \mathbf{C}$, find $|\mathbf{B}|$.

16. Given $\mathbf{A} = (5, -2)$, $\mathbf{C} = (7, 0)$, and $3\mathbf{A} + 2\mathbf{B} = \mathbf{C}$, determine $|\mathbf{B}|$.

In the following exercises find a unit vector having (a) the same direction as $\mathbf{A}$, and (b) the opposite direction to $\mathbf{A}$.

17. $\mathbf{A} = (3, -4)$.   **18.** $\mathbf{A} = (-5, 12)$.

19. $\mathbf{A} = (-6, 8)$.   **20.** $\mathbf{A} = (0, 0)$.

21. Give two unit vectors that are parallel to the line $2x - 3y + 6 = 0$.

22. Write down two unit vectors that are parallel to the tangent line to $y = x^2 + 1$ at $x = 1$.

Find the numbers a and b such that:

23. $a(2, 3) + b(-3, 1) = (-4, 5)$.   **24.** $a(1, -4) - b(5, 2) = (7, -6)$.

Give the x- and y-components of the vector:

25. whose magnitude is 5 units and which makes an angle of $\pi/6$ with the positive x-axis;

26. whose magnitude is 6 units and which makes an angle of $\pi/4$ with the direction of $\mathbf{i}$.

27. Find the coordinates of the point that is one-third of the way from $(2, -1)$ toward $(5, 8)$.

28. Find the coordinates of the point that is two-fifths of the way from $(1, -3)$ toward $(6, 7)$.

29. Use vectors to show that if in a quadrilateral one pair of opposite sides are equal and parallel, then so are the other pair; i.e., then the quadrilateral is a parallelogram.

30. Use vectors to show that the line joining the midpoints of two opposite sides of a parallelogram is parallel and equal to the other two sides.

31. Prove that the four lines joining the middle points of adjacent sides of a quadrilateral form a parallelogram.

32. Prove that the line joining the middle points of two nonparallel sides of a trapezoid is parallel to the two parallel sides.

12.2 THE INNER PRODUCT

If $A = (x_1, y_1)$ and $B = (x_2, y_2)$ are any two vectors then the *inner product* of A and B denoted by $A \cdot B$ is defined by

$$A \cdot B = x_1 x_2 + y_1 y_2.$$

The inner product is also known as the *dot product* or *scalar product*. It should be observed that $A \cdot B$ is a scalar (real number) and not a vector.

We see then that in forming the inner product of two vectors A and B, the x-components of A and B are multiplied together, the y-components are multiplied together, and then these two products are added together.

EXAMPLE If $A = (2, -3)$ and $B = (-1, -2)$, then

$$A \cdot B = (2)(-1) + (-3)(-2) = -2 + 6 = 4.$$

Note that, in particular,

$$i \cdot i = (1, 0) \cdot (1, 0) = (1)(1) + (0)(0) = 1,$$
$$j \cdot j = (0, 1) \cdot (0, 1) = 1,$$

and

$$i \cdot j = (1, 0) \cdot (0, 1) = 0.$$

If $A = (x, y)$ is any vector then the inner product of A with itself is the square of the length of A:

$$A \cdot A = (x, y) \cdot (x, y) = (x)(x) + (y)(y) = x^2 + y^2$$
$$= |A|^2.$$

DEFINITION *Angle Between Two Vectors*

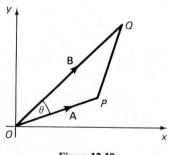

Figure 12.19

Let A and B be two nonzero vectors and let $\overrightarrow{OP}$ and $\overrightarrow{OQ}$ be the position vectors corresponding to A and B respectively (Fig. 12.19). Then the angle θ between A and B is defined to be the angle POQ of the triangle formed by the points O, P, and Q.

The inner product is a useful tool to determine the angle between two nonzero vectors, as is shown by the following theorem.

THEOREM 12.2.1

The angle θ between any two nonzero vectors A and B is given by

$$A \cdot B = |A| |B| \cos \theta.$$

604

PROOF Let $\mathbf{A} = (x_1, y_1)$ and $\mathbf{B} = (x_2, y_2)$ be two nonzero vectors. Let $\overrightarrow{OP}$ and $\overrightarrow{OQ}$ be the position vectors corresponding to $\mathbf{A}$ and $\mathbf{B}$ (Fig. 12.20). Then the points P and Q have the coordinates (x_1, y_1) and (x_2, y_2) respectively. By Theorem 12.1.1 the vector $\overrightarrow{PQ}$ is given by $\overrightarrow{PQ} = (x_2 - x_1, y_2 - y_1)$.

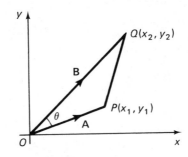

Figure 12.20

Applying the law of cosines to triangle OPQ, we have

$$|\overrightarrow{PQ}|^2 = |\overrightarrow{OP}|^2 + |\overrightarrow{OQ}|^2 - 2|\overrightarrow{OP}||\overrightarrow{OQ}| \cos\theta$$
$$= |\mathbf{A}|^2 + |\mathbf{B}|^2 - 2|\mathbf{A}||\mathbf{B}| \cos\theta. \qquad \text{(i)}$$

Now

$$|\overrightarrow{PQ}|^2 = |(x_2 - x_1, y_2 - y_1)|^2$$
$$= (x_2 - x_1)^2 + (y_2 - y_1)^2$$
$$= (x_2^2 - 2x_1 x_2 + x_1^2) + (y_2^2 - 2y_1 y_2 + y_1^2)$$
$$= (x_1^2 + y_1^2) - 2(x_1 x_2 + y_1 y_2) + (x_2^2 + y_2^2)$$
$$= |\mathbf{A}|^2 - 2\mathbf{A} \cdot \mathbf{B} + |\mathbf{B}|^2.$$

Using this in (i) above we get

$$|\mathbf{A}|^2 - 2\mathbf{A} \cdot \mathbf{B} + |\mathbf{B}|^2 = |\mathbf{A}|^2 + |\mathbf{B}|^2 - 2|\mathbf{A}||\mathbf{B}| \cos\theta,$$

from which, after cancelling terms and dividing both sides by -2, we get the required result:

$$\mathbf{A} \cdot \mathbf{B} = |\mathbf{A}||\mathbf{B}| \cos\theta.$$

COROLLARY 1 If the two vectors $\mathbf{A}$ and $\mathbf{B}$ are at right angles then $\theta = 90°$ and $\cos\theta = \cos 90° = 0$. Therefore from the above theorem it follows that

$$\mathbf{A} \cdot \mathbf{B} = 0.$$

COROLLARY 2 If the vectors $\mathbf{A}$ and $\mathbf{B}$ are parallel then $\theta = 0$ or π and therefore $\cos\theta = 1$ or -1. Consequently

$$\mathbf{A} \cdot \mathbf{B} = \pm|\mathbf{A}||\mathbf{B}|.$$

In the last section we showed that the two vectors **A** and **B** are parallel if $\mathbf{B} = c\mathbf{A}$ for some scalar c. We leave it as an exercise for the reader to show that these two conditions of parallelism are equivalent to each other.

EXAMPLE   Find the cosine of the angle between the vectors $\mathbf{A} = (2, -3)$ and $\mathbf{B} = (3, 1)$.

SOLUTION   If θ is the angle between **A** and **B** then

$$\mathbf{A} \cdot \mathbf{B} = |\mathbf{A}||\mathbf{B}| \cos \theta. \tag{1}$$

Now

$$\mathbf{A} \cdot \mathbf{B} = 2(3) + (-3)(1) = 6 - 3 = 3,$$
$$|\mathbf{A}| = \sqrt{2^2 + (-3)^2} = \sqrt{13},$$

and

$$|\mathbf{B}| = \sqrt{3^2 + 1^2} = \sqrt{10}.$$

Therefore from (1) we have

$$3 = \sqrt{13}\sqrt{10} \cos \theta \quad \text{or} \quad \cos \theta = \frac{3}{\sqrt{130}}.$$

EXAMPLE   Given $\mathbf{A} = (1, -2)$ and $\mathbf{B} = (3, p)$. Determine the constant p so that:

(a)   **A** and **B** are perpendicular to each other.

(b)   **A** and **B** make an angle of $\pi/3$ with each other.

SOLUTION   We have $\mathbf{A} = (1, -2)$ and $\mathbf{B} = (3, p)$.

(a)   Now if **A** and **B** are perpendicular to each other, then

$$\mathbf{A} \cdot \mathbf{B} = 0$$

or

$$3(1) + p(-2) = 0,$$
$$3 - 2p = 0,$$

which gives $p = \frac{3}{2}$. Thus the vector $\mathbf{B} = (3, \frac{3}{2})$ is perpendicular to **A**.

(b)   Using Theorem 12.2.1 we have

$$\mathbf{A} \cdot \mathbf{B} = |\mathbf{A}||\mathbf{B}| \cos \theta,$$

which in this case becomes

$$3 - 2p = \sqrt{5}\sqrt{9 + p^2}(\tfrac{1}{2})$$

since $\cos \theta = \cos \pi/3 = \frac{1}{2}$. Squaring both sides of this equation we find

$$(6 - 4p)^2 = 5(9 + p^2)$$

or

$$36 - 48p + 16p^2 = 45 + 5p^2;$$

i.e.,

$$11p^2 - 48p - 9 = 0,$$

which on solving gives

$$p = \frac{24 \pm 15\sqrt{3}}{11}.$$

So there are two vectors **B** that make an angle of $\pi/3$ with **A**, namely

$$\left(3, \frac{24 + 15\sqrt{3}}{11}\right) \quad \text{and} \quad \left(3, \frac{24 - 15\sqrt{3}}{11}\right).$$

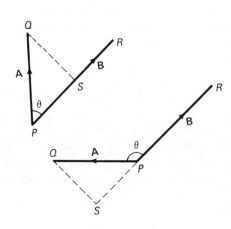

Figure 12.21

Let **A** and **B** be any two vectors. Then it is possible to represent them by directed line segments, $\overrightarrow{PQ}$ and $\overrightarrow{PR}$ say, with a common initial point P. Let a perpendicular be dropped from the terminal point Q of **A** onto the line of the segment $\overrightarrow{PR}$, meeting this line at the point S. In Fig. 12.21 are shown the two cases when the angle θ between **A** and **B** is either acute or obtuse. In the first case

$$PS = |\overrightarrow{PQ}| \cos \theta = |\mathbf{A}| \cos \theta,$$

while in the second case

$$-PS = |\overrightarrow{PQ}| \cos \theta = |\mathbf{A}| \cos \theta.$$

The quantity $|\mathbf{A}| \cos \theta$ that appears in these expressions is called the *component of* **A** *in the direction of* **B**. Note that this component is positive or negative according as the angle θ is acute or obtuse.

Since $\mathbf{A} \cdot \mathbf{B} = |\mathbf{A}||\mathbf{B}| \cos \theta$, it follows that the component of **A** in the direction of **B** is given by

$$|\mathbf{A}| \cos \theta = \frac{\mathbf{A} \cdot \mathbf{B}}{|\mathbf{B}|} \quad \text{(provided } |\mathbf{B}| \neq 0).$$

In this way this component is expressed in terms of the inner product of **A** and **B**.

EXAMPLE   Find the component of $\mathbf{A} = 2\mathbf{i} - 3\mathbf{j}$ in the direction of $\mathbf{B} = 5\mathbf{i} + 2\mathbf{j}$.

SOLUTION   Here $\mathbf{A} = (2, -3)$ and $\mathbf{B} = (5, 2)$. The component of **A** in the direction of **B** equals

$$\frac{\mathbf{A} \cdot \mathbf{B}}{|\mathbf{B}|} = \frac{(2, -3) \cdot (5, 2)}{|(5, 2)|}$$

$$\frac{2(5) + (-3)(2)}{\sqrt{5^2 + 2^2}} = \frac{10 - 6}{\sqrt{29}} = \frac{4}{\sqrt{29}}.$$

When the component of **A** in the direction of **B** is multiplied by a unit vector in the direction of **B**, we obtain a vector that is known as the *projection of* **A** *in the direction of* **B**. Thus, since the unit vector along the direction of **B** is given by **B**/|**B**|, the projection of **A** in the direction of **B** is given by

$$\frac{\mathbf{A} \cdot \mathbf{B}}{|\mathbf{B}|} \frac{\mathbf{B}}{|\mathbf{B}|} = \frac{\mathbf{A} \cdot \mathbf{B}}{|\mathbf{B}|^2} \mathbf{B}.$$

The projection of **A** along the direction **B** corresponds to the directed line segment $\overrightarrow{PS}$ in Fig. 12.22.

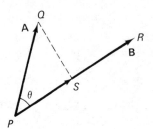

The directed line segment $\overrightarrow{SQ}$ corresponds to what is called the *projection of* **A** *perpendicular to* **B**. Since $\overrightarrow{PQ} = \overrightarrow{PS} + \overrightarrow{SQ}$, it follows that this projection is given by

$$\overrightarrow{SQ} = \overrightarrow{PQ} - \overrightarrow{PS}$$

$$= \mathbf{A} - \frac{\mathbf{A} \cdot \mathbf{B}}{|\mathbf{B}|^2} \mathbf{B}.$$

Figure 12.22

The vector **A** can be written as the sum of its two projections along and perpendicular to **B**, since we can write $\mathbf{A} = \overrightarrow{PS} + \overrightarrow{SQ}$. When decomposed into these two projections, the vector **A** is often said to be *resolved* along and perpendicular to **B**.

EXAMPLE   If $\mathbf{A} = (3, -1)$ and $\mathbf{B} = (1, 2)$, find the projections of **A** along and perpendicular to **B**.

SOLUTION   The projection of **A**-along **B** is

$$\frac{\mathbf{A} \cdot \mathbf{B}}{|\mathbf{B}|^2} \mathbf{B} = \frac{(3, -1) \cdot (1, 2)}{|(1, 2)|^2}(1, 2)$$

$$= \frac{(3)(1) + (-1)(2)}{1^2 + 2^2}(1, 2)$$

$$= \tfrac{1}{5}(1, 2) = (\tfrac{1}{5}, \tfrac{2}{5}).$$

The projection perpendicular to **B** is therefore

$$\mathbf{A} - \text{projection of } \mathbf{A} \text{ along } \mathbf{B} = (3, -1) - (\tfrac{1}{5}, \tfrac{2}{5})$$

$$= (\tfrac{14}{5}, -\tfrac{7}{5}).$$

One important feature of the algebraic approach to vectors is that we can readily extend the definitions and properties of vectors in two dimensions to vectors in n dimensions for any integer $n > 2$. An ordered set of n real numbers $(x_1, x_2, \ldots, x_n)$ is called an *n-dimensional vector*, analogous to the ordered pair (x, y), which represents a two-dimensional vector. The numbers $x_1, x_2, \ldots, x_n$ are called the *components* of the vector. The vector all of whose components are zero is called the *zero vector* and as before is denoted by **0**. Thus

$$\mathbf{0} = (0, 0, \ldots, 0).$$

Addition of Two Vectors

Two vectors can be added provided they are of the same dimension. The sum is obtained by adding the corresponding components of the two vectors. Thus if

$$\mathbf{A} = (x_1, x_2, \ldots, x_n) \quad \text{and} \quad \mathbf{B} = (y_1, y_2, \ldots, y_n)$$

then

$$\mathbf{A} + \mathbf{B} = (x_1 + y_1, x_2 + y_2, \ldots, x_n + y_n).$$

Scalar Multiplication of a Vector

If c is any scalar, i.e., a real number, and $\mathbf{A} = (x_1, x_2, \ldots, x_n)$ then

$$c\mathbf{A} = (cx_1, cx_2, \ldots, cx_n).$$

Magnitude of a Vector

The magnitude of an n-dimensional vector $\mathbf{A} = (x_1, x_2, \ldots, x_n)$ is given by

$$|\mathbf{A}| = \sqrt{x_1^2 + x_2^2 + \ldots + x_n^2}.$$

Inner Product of Two Vectors

If $\mathbf{A} = (x_1, x_2, \ldots, x_n)$ and $\mathbf{B} = (y_1, y_2, \ldots, y_n)$ then the inner product of $\mathbf{A}$ and $\mathbf{B}$ is defined as

$$\mathbf{A} \cdot \mathbf{B} = x_1 y_1 + x_2 y_2 + \ldots + x_n y_n.$$

The angle θ between two n-dimensional vectors $\mathbf{A}$ and $\mathbf{B}$ is determined by the equation

$$\mathbf{A} \cdot \mathbf{B} = |\mathbf{A}| |\mathbf{B}| \cos \theta$$

and the component of $\mathbf{A}$ in the direction of $\mathbf{B}$ is

$$\frac{\mathbf{A} \cdot \mathbf{B}}{|\mathbf{B}|}.$$

The vectors $(1, 0, 0, \ldots, 0), (0, 1, 0, \ldots, 0), \ldots, (0, 0, 0, \ldots, 1)$ are the n unit vectors corresponding to two unit vectors $\mathbf{i} = (1, 0)$ and $\mathbf{j} = (0, 1)$ in two dimensions.

EXAMPLE   If the four-dimensional vectors $\mathbf{A}$ and $\mathbf{B}$ are given by $\mathbf{A} = (1, -2, 3, 7)$ and $\mathbf{B} = (2, 3, 0, -4)$, determine:

(a) $\mathbf{A} + 2\mathbf{B}$.

(b) $|\mathbf{A}|$ and $|-3\mathbf{B}|$.

(c) $\mathbf{A} \cdot \mathbf{B}$.

(d)   the cosine of the angle between $\mathbf{A}$ and $\mathbf{B}$.

SOLUTION (a) $\mathbf{A} + 2\mathbf{B} = (1, -2, 3, 7) + 2(2, 3, 0, -4)$
$$= (1, -2, 3, 7) + (4, 6, 0, -8)$$
$$= (1 + 4, -2 + 6, 3 + 0, 7 - 8)$$
$$= (5, 4, 3, -1).$$

(b) $|\mathbf{A}| = \sqrt{1^2 + (-2)^2 + 3^2 + 7^2} = \sqrt{63} = 3\sqrt{7}.$

$-3\mathbf{B} = -3(2, 3, 0, -4) = (-6, -9, 0, 12)$

and therefore

$$|-3\mathbf{B}| = \sqrt{(-6)^2 + (-9)^2 + 0^2 + 12^2}$$
$$= \sqrt{36 + 81 + 0 + 144}$$
$$= \sqrt{261} = 3\sqrt{29}.$$

(c) $\mathbf{A} \cdot \mathbf{B} = (1, -2, 3, 7) \cdot (2, 3, 0, -4)$
$$= 1(2) + (-2)(3) + 3(0) + 7(-4)$$
$$= 2 - 6 + 0 - 28 = -32.$$

(d) If θ is the angle between $\mathbf{A}$ and $\mathbf{B}$ then
$$\mathbf{A} \cdot \mathbf{B} = |\mathbf{A}||\mathbf{B}| \cos \theta$$
or
$$-32 = 3\sqrt{7}\sqrt{2^2 + 3^2 + 0^2 + (-4)^2} \cos \theta$$
$$= 3\sqrt{7}\sqrt{29} \cos \theta,$$

which gives
$$\cos \theta = \frac{-32}{3\sqrt{203}}.$$

EXAMPLE Given that $\mathbf{A} = (1, -2, 3)$ and $\mathbf{B} = (3, 1, x)$ are two perpendicular vectors, determine x.

SOLUTION If $\mathbf{A}$ is perpendicular to $\mathbf{B}$, then $\mathbf{A} \cdot \mathbf{B} = 0$, or
$$1(3) + (-2)(1) + 3(x) = 0$$
$$3 - 2 + 3x = 0,$$

which gives $x = -\frac{1}{3}$.

EXERCISES 12.2

1. If $\mathbf{A} = (2, 3)$ and $\mathbf{B} = (-1, 2)$, evaluate $\mathbf{A} \cdot \mathbf{B}$.

2. If $\mathbf{A} = 5\mathbf{i} - 3\mathbf{j}$ and $\mathbf{B} = 2\mathbf{i} + 3\mathbf{j}$, compute $\mathbf{A} \cdot \mathbf{B}$.

610 3. If $\mathbf{A} = (2, 3, -1)$ and $\mathbf{B} = (1, -2, 5)$, evaluate $(\mathbf{A} + 2\mathbf{B}) \cdot (\mathbf{B} - 2\mathbf{A})$.

4. If $\mathbf{P} = (3, -1, 0, 4)$ and $\mathbf{Q} = (2, 3, 4, 5)$, evaluate $(\mathbf{P} + \mathbf{Q}) \cdot (\mathbf{P} - 2\mathbf{Q})$.

If θ denotes the angle between $\mathbf{A}$ and $\mathbf{B}$, find $\cos \theta$ if

5. $\mathbf{A} = (2, -1)$, $\mathbf{B} = (1, 2)$.   **6.** $\mathbf{A} = 5\mathbf{i} + 12\mathbf{j}$, $\mathbf{B} = 3\mathbf{i} - 4\mathbf{j}$.

In the following exercises find the number k (if possible) such that given conditions are satisfied:

7. $\mathbf{A} = (2, k)$ and $\mathbf{B} = (3, -1)$; $\mathbf{A}$ and $\mathbf{B}$ are perpendicular.

8. $\mathbf{A} = (k, -2)$ and $\mathbf{B} = k\mathbf{i} + 8\mathbf{j}$; $\mathbf{A}$ and $\mathbf{B}$ are perpendicular.

9. $\mathbf{A} = (2, 3, k)$ and $\mathbf{B} = (3, -2k, 5)$; $\mathbf{A}$ and $\mathbf{B}$ are perpendicular.

10. $\mathbf{A} = (2, 1, 3, k)$ and $\mathbf{B} = (1, -2, 5, 0)$; $\mathbf{A} + 2\mathbf{B}$ and $\mathbf{B} - 2\mathbf{A}$ are perpendicular.

11. $\mathbf{A} = (3, -1)$ and $\mathbf{B} = (k, 2)$; $\mathbf{A}$ and $\mathbf{B}$ are parallel.

12. $\mathbf{A} = (2, 5)$ and $\mathbf{B} = (0, k)$; $\mathbf{A}$ and $\mathbf{B}$ are parallel.

13. $\mathbf{A} = (-k, 2)$ and $\mathbf{B} = (1, k)$; $\mathbf{A}$ and $\mathbf{B}$ are parallel.

14. $\mathbf{A} = (2, -1, k)$ and $\mathbf{B} = (6, -3, 0)$; $\mathbf{A}$ and $\mathbf{B}$ are parallel.

15. $\mathbf{A} = 3\mathbf{i} + 4\mathbf{j}$ and $\mathbf{B} = \mathbf{i} + k\mathbf{j}$; $\mathbf{A}$ and $\mathbf{B}$ make an angle of $\pi/3$.

16. $\mathbf{A} = 12\mathbf{i} - 5\mathbf{j}$ and $\mathbf{B} = (k, -1)$; $\mathbf{A}$ and $\mathbf{B}$ make an angle of $\pi/4$.

17. $\mathbf{A} = (1, 0, -1)$ and $\mathbf{B} = (-1, k, 2)$, $\mathbf{A} + \mathbf{B}$ and $\mathbf{B} - \mathbf{A}$ make an angle of $\pi/6$.

In the following exercises find the component of $\mathbf{A}$ in the direction of $\mathbf{B}$ and also the projection of $\mathbf{A}$ in the direction of $\mathbf{B}$.

18. $\mathbf{A} = (2, -1)$, $\mathbf{B} = (4, 3)$.   **19.** $\mathbf{A} = (3, 5)$, $\mathbf{B} = (1, 2)$.

20. $\mathbf{A} = (1, 2, 3)$, $\mathbf{B} = (0, -3, 1)$.

Under what circumstances do the following hold?

21. $|\mathbf{A} \cdot \mathbf{B}| = |\mathbf{A}||\mathbf{B}|$.   **22.** $|\mathbf{A} + \mathbf{B}| = |\mathbf{A}| + |\mathbf{B}|$.

23. Given $\mathbf{A} \cdot \mathbf{B} = \pm|\mathbf{A}||\mathbf{B}|$, show that $\mathbf{B} = c\mathbf{A}$ for some nonzero scalar c.

24. Given $\mathbf{B} = c\mathbf{A}$ where c is some scalar, show that $\mathbf{A} \cdot \mathbf{B} = \pm|\mathbf{A}||\mathbf{B}|$.

(Exercises 23 and 24 prove that the two conditions given in this section for the parallelism of $\mathbf{A}$ and $\mathbf{B}$ are equivalent.)

12.3 RELATIVE VELOCITIES

An important application of the theory of vectors is to the velocities of moving objects or moving animals. To be specific, let us consider a migrating goose that is flying with a constant speed V in a fixed direction, so that its path is a straight line. We define the *velocity* of the goose to be a vector $\mathbf{V}$ with a magnitude equal to the speed V and a direction along the given direction of flight.

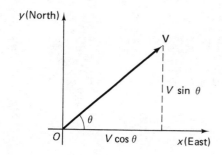

Figure 12.23

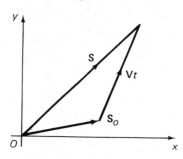

Figure 12.24

In problems of this kind it is often convenient to choose axes in such a way that the x-axis points due east and the y-axis points due north (as in Fig. 12.23). If the velocity vector **V** makes an angle θ with the x-axis as shown, then we can resolve **V** into an easterly velocity component of $V\cos\theta$ and a northerly velocity component of $V\sin\theta$. We can write, in the usual way, $\mathbf{V} = (V\cos\theta,\ V\sin\theta)$.

During a time interval t, the goose covers a distance of Vt along the direction of flight. We can represent this change in position by the vector **V**t since this vector has the requisite magnitude, namely Vt, and the requisite direction, namely the direction of **V**.

Let the initial position of the goose be represented by the position vector $\mathbf{S}_0$ with respect to some origin of position O, and let **S** be the position at a time t later. Then **S** can be found by the vector addition of the original position $\mathbf{S}_0$ and the change in position **V**t, as in Fig. 12.24:

$$\mathbf{S} = \mathbf{S}_0 + \mathbf{V}t.$$

EXAMPLE   With respect to a certain set of coordinate axes a bird has an initial position $(-5, 2)$ (units of distance in miles). Its easterly velocity component is 5 mph and northerly velocity component -8 mph. What is its position after 2 hr?

SOLUTION   The velocity vector is $\mathbf{V} = (5, -8)$. The initial position is $(-5, 12)$, and during 2 hr the change in position is

$$\mathbf{V}t = 2(5, -8) = (10, -16).$$

Therefore the new position vector is

$$\mathbf{S} = (-5, 12) + (10, -16) = (5, -4) \quad \text{(Fig. 12.25)}.$$

It is a common observation that a person riding on a moving vehicle feels the wind to be blowing differently from the wind that is felt by a person who is standing still. For example, if there is no wind at all, and if you ride a bicycle at 10 mph, then you experience a wind blowing at 10 mph in your face, whereas a person standing still would notice no wind. This "apparent wind" is caused by the fact that you on your bicycle are moving through the air, creating the same feeling as if a real wind were blowing against you. The apparent wind can be represented by a vector that is equal and opposite to the velocity vector of the moving person; if the velocity is 10 mph from left to right then the apparent wind blows at 10 mph from right to left (Fig. 12.26).

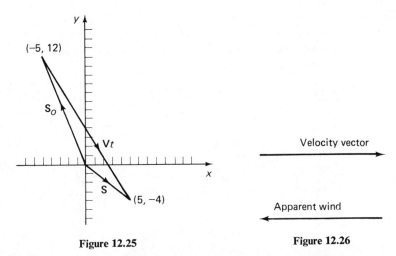

Figure 12.25

Figure 12.26

Now consider the case when there is a real wind blowing as well. Suppose first of all that the real wind is blowing from right to left at 5 mph and you are cycling from left to right at 10 mph. Since you are cycling into the wind you will feel an apparent wind of 15 mph. This is illustrated in Fig. 12.27; the total apparent wind is found by adding the real wind of 5 mph to the artificial wind of 10 mph that is created by your motion.

Suppose on the other hand that you are cycling at 10 mph from right to left, so that now the real wind is behind you. Then this real wind will cancel out part of the artificial 10-mph wind caused by your motion, and you will feel a total apparent wind of only 5 mph blowing in your face (Fig. 12.28).

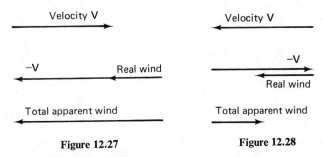

Figure 12.27

Figure 12.28

In general, we can conclude that if the real wind has a speed W then when you travel into the wind with a speed V you experience an apparent wind of $V + W$ in your face, and when you travel downwind with a speed V you experience an apparent wind of $V - W$ in your face.

EXAMPLE   A wind blows from the west at 15 mph. An individual travels toward the east at 10 mph. What apparent wind does he feel?

SOLUTION The situation is illustrated in Fig. 12.29. In this case $V = 10$ and $W = 15$. The wind is behind the individual, so the apparent wind is $V - W = 10 - 15 = -5$ mph in his face. Since this is negative, it means that the apparent wind is 5 mph blowing from behind him.

Now let us turn to the situation in which the wind is blowing from some general direction, not necessarily from dead ahead or behind the moving individual. We let V denote the velocity vector of the individual and W denote the velocity vector of the real wind. Then by moving with velocity V through the air, the individual creates an artificial wind whose velocity is $-V$. The total apparent wind is obtained by adding this artificial wind to the real wind velocity W, using the parallelogram law of vector addition (Fig. 12.30). If we use W_A to denote the velocity vector of the apparent wind then we can write

$$W_A = W - V.$$

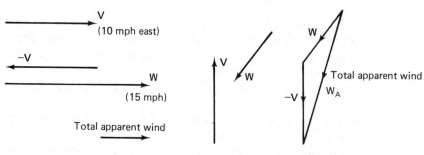

Figure 12.29   Figure 12.30

EXAMPLE An individual cycles due north at 10 mph, the wind blowing from the east at 5 mph. Find the magnitude and direction of the apparent wind.

SOLUTION With the x-axis pointing eastward and the y-axis pointing northward, we have, in terms of components,

$$V = (0, 10), \qquad W = (-5, 0).$$

Therefore, as illustrated in Fig. 12.31, the apparent wind velocity is given by

$$W_A = W - V = (-5, 0) - (0, 10)$$
$$= (-5, -10).$$

The magnitude of this vector is

$$|W_A| = \sqrt{(-5)^2 + (-10)^2} = \sqrt{125} = 11.18.$$

The apparent wind blows in a direction that makes an angle θ with due east (as shown in Fig. 12.31) where

$$\cos \theta = \frac{-5}{\sqrt{125}} = -0.4472, \qquad \sin \theta = \frac{-10}{\sqrt{125}} = -0.8944.$$

614

It follows that $\theta = 243° \, 26'$. Another way of saying this is that the apparent wind has a speed of 11.18 mph, blowing from a direction 26° 34' east of north (Fig. 12.32).

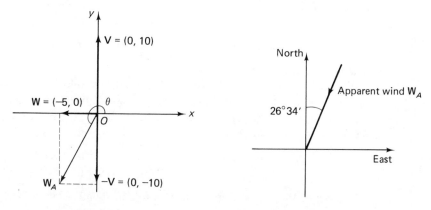

Figure 12.31   **Figure 12.32**

If A and B are any two moving objects then the term *velocity of A relative to B* is used to denote the velocity vector that A appears to have to an observer who sits on the object B and moves with it. Thus, for example, another way of describing the apparent wind is to say that it is the velocity of the air *relative* to the moving individual. In the same terms, the velocity of the real wind, $\mathbf{W}$, can be described as the velocity of the air relative to the ground, while the vector $\mathbf{V}$ is the velocity of the individual relative to the ground. The equation $\mathbf{W}_A = \mathbf{W} - \mathbf{V}$ can be rewritten in the form

$$\mathbf{W} = \mathbf{W}_A + \mathbf{V},$$

and as such can be expressed in words as follows:

Velocity of air relative to ground = Velocity of air relative to traveler
                          + Velocity of traveler relative to ground.

In this form the apparent velocity equation is an example of the general rule of relative velocities, which can be stated in the following way. If A, B, and C are any three objects moving relative to one another then

Velocity vector of A relative to B = Velocity vector of A relative to C
                          + Velocity vector of C relative to B.   (1)

If we take A to be the air, B the ground, and C the traveling individual, then this equation of relative velocities simply becomes our earlier equation $\mathbf{W} = \mathbf{W}_A + \mathbf{V}$.

Before looking at some further examples of relative velocities, let us note the following rule, which is often useful:

Velocity of A relative to B = −Velocity of B relative to A.

For example, the velocity of the ground relative to a traveling individual is the negative of the velocity of the individual relative to the ground. If the individual is traveling at 10 mph due east, then relative to him the ground appears to be moving at 10 mph due west.

   This second rule can in fact be deduced from the first rule (1) of relative velocities by replacing B with A. Then the left-hand side of (1) becomes the velocity of A relative to itself, which is the zero vector. The rule (1) then reduces to the statement

$$\text{Velocity of } A \text{ relative to } C + \text{Velocity of } C \text{ relative to } A = \mathbf{0}.$$

This is just the second rule if we replace C by B.

   We now turn to some further examples of relative velocities. First let us consider a second example related to real and apparent wind, but this time as it relates to birds flying in the wind. In the general rule let us take A to be the bird, B to be the ground, and C to be the air. We obtain

$$\text{Velocity of bird relative to the ground} = \text{Velocity of bird relative to the air}$$
$$+ \text{ Velocity of air relative to the ground.}$$

   The last of the three velocities here is just the wind velocity $\mathbf{W}$ again. The other velocity on the right, the velocity of the bird relative to the air, is the velocity with which the bird feels he is flying; it depends primarily on how fast he flaps his wings. We denote this velocity by $\mathbf{U}$. The velocity on the left-hand side is the velocity with which the bird covers the ground. Calling this $\mathbf{V}$, as before, we can write the above equation in the form

$$\mathbf{V} = \mathbf{U} + \mathbf{W}.$$

EXAMPLE   A homing pigeon is released from a point 100 mi due south of its loft. It can fly with a velocity of 20 mph. If there is a wind blowing at 10 mph from the north-east, what is the shortest time within which the pigeon can reach home?

SOLUTION   Relative to the air, let the pigeon fly in a direction that makes an angle θ with the x-axis (which points east). Then since the relative velocity $\mathbf{U}$ has a magnitude of 20 mph, we can write $\mathbf{U} = (20 \cos \theta, 20 \sin \theta)$, as illustrated in Fig. 12.33. The wind velocity $\mathbf{W}$ has a magnitude of 10 mph, blowing from a direction that makes

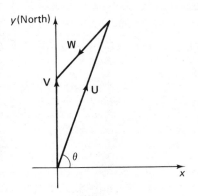

Figure 12.33

an angle of 45° with the x-axis. Hence

$$\mathbf{W} = \left(-\frac{10}{\sqrt{2}}, -\frac{10}{\sqrt{2}}\right) = (-7.071, -7.071).$$

Therefore the velocity of the pigeon over the ground is

$$\mathbf{V} = \mathbf{U} + \mathbf{W}$$

$$= (20 \cos \theta - 7.071, 20 \sin \theta - 7.071).$$

Now, in order to reach home as fast as possible, the velocity $\mathbf{V}$ should be due north. That is, $\mathbf{V} = (0, V)$. Therefore, comparing components, we obtain

$$20 \cos \theta - 7.071 = 0, \qquad 20 \sin \theta - 7.071 = V.$$

From the first of these equations we find that

$$\cos \theta = 0.3536$$

and hence $\theta = 69° 18'$. Then from the second equation we obtain

$$V = 20 \sin (69° 18) - 7.071 = 20(0.9354) - 7.071$$

$$= 11.64.$$

The quantity V is the northerly speed of the pigeon over the ground. Since the distance it has to cover is 100 mi, the time taken to do this is

$$\text{Time} = \frac{\text{Distance}}{\text{Speed}} = \frac{100}{11.64} = 8.59 \text{ hr.}$$

This kind of application to the flight of birds relative to the wind applies equally well to the motion of airplanes. The selection of the course to fly and the calculation of the estimated time of arrival for an airplane flying through the wind are quite similar to the preceding example of the homing pigeon. In a similar way the motion of objects, such as fish or other swimming animals as well as ships, that are moving through water that is flowing with some nonzero current follows the same pattern. All we must do in such examples is to replace the wind velocity with the velocity of the current. We can then write

Velocity of object relative to the ground = Velocity of object relative to the water
+ Velocity of the current relative to the ground.

The following example provides a somewhat different illustration of relative velocities.

EXAMPLE   Two people start from a certain point O and walk away from it with velocity vectors $(3, 1)$ and $(-3, 2)$ with respect to some coordinate axes. How fast is the distance between them increasing?

SOLUTION   Calling the people A and B we have: Velocity of A relative to $O = (3, 1)$; velocity of B relative to $O = (-3, 2)$ (Fig. 12.34). Therefore the velocity of O relative to $B = -(-3, 2) = (3, -2)$. It follows therefore that

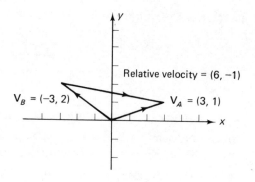

Figure 12.34

Velocity of A relative to B = Velocity of A relative to O
$$+ \text{ Velocity of } O \text{ relative to } B = (3, 1) + (3, -2)$$
$$= (6, -1).$$

The magnitude of this relative velocity is the rate at which the distance between A and B is increasing, namely,

$$\sqrt{6^2 + (-1)^2} = \sqrt{37} = 6.08 \text{ mph.}$$

EXERCISES 12.3

In the following exercises $\mathbf{V}_{XY}$ is used to denote the velocity of the object X relative to the object Y.

1. If $\mathbf{V}_{YX} = (1, 2)$, what is $\mathbf{V}_{XY}$?

2. If $\mathbf{V}_{XY} = (1, 2)$ and $\mathbf{V}_{YZ} = (2, -3)$, what is $\mathbf{V}_{XZ}$?

3. If $\mathbf{V}_{XY} = (2, 3)$ and $\mathbf{V}_{ZY} = (2, 1)$, what is $\mathbf{V}_{ZX}$?

4. If $\mathbf{V}_{BA} = (-1, 1)$ and $\mathbf{V}_{BC} = (2, 2)$, what is $\mathbf{V}_{AC}$?

5. If $\mathbf{V}_{AB} = (1, 1)$, $\mathbf{V}_{BC} = (2, -2)$, and $\mathbf{V}_{CD} = (3, 0)$, what is $\mathbf{V}_{AD}$?

6. If $\mathbf{V}_{AB} = (2, 0)$, $\mathbf{V}_{CB} = (-1, 1)$, and $\mathbf{V}_{AD} = (1, 3)$, what is $\mathbf{V}_{DC}$?

7. A is traveling east at 10 mph and B is traveling north at 20 mph. Find the velocity of A relative to B.

8. A is traveling north at 5 mph and B is traveling southwest at 8 mph. Find the velocity of B relative to A.

9. An automobile is traveling at 30 mph. The driver throws a cigarette butt out of the window with a velocity of 20 mph at a right angle to the car. What is the speed and direction of the cigarette butt relative to the ground?

10. Repeat exercise 9 in the case when the cigarette butt is thrown in a direction that makes an angle of $60°$ with the direction of motion of the car.

11. A car travels north at 50 mph. If the wind is blowing at 30 mph from the north-west, what is the strength and direction of the apparent wind?

12. A ship is sailing due east at 15 mph. A man standing on deck feels that the wind is blowing from the northeast at 20 mph. What is the strength and direction of the actual wind?

13. A homing pigeon is released a distance of 50 mi to the southwest of its loft. The wind is blowing from the west at 20 mph. If the pigeon can fly at 20 mph, what is the shortest time within which it can reach home, and in what direction must it fly?

14. A salmon swims at 4 mph in a westerly direction relative to ocean water that is flowing toward the northeast at 1 mph. What is the speed and direction of the salmon over the bottom of the ocean?

12.4 FORCES

Different components of the universe around us interact with one another by exerting forces on each other. For example, standing on the ground, a person exerts a force on the ground equal to his weight, the force acting vertically downward. Or, pushing a jalopy to the roadside, his hands exert a force on the automobile; in this case the direction of the force is horizontal. Similarly, a man pulling on rope exerts a force on the rope, the direction of the force being along the rope in this case.

A force is a vector quantity. It has a magnitude equal to the size of the force (for example, the weight of a body pressing down on the ground) and it has a direction.

When several forces act on a particular object, we can define the *total force* or *resultant force* to be the sum of all the separate forces, where these latter are added together by the usual law of vector addition.

EXAMPLE   Three ropes are fastened to an object A, as in Fig. 12.35, one rope being along the x-direction, the other two making angles of 60° to each side. Forces of F, $2F$, and $3F$ are exerted along the ropes, as shown. Find the resultant force on A.

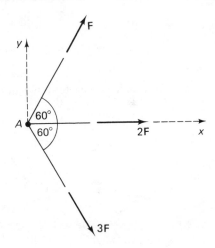

Figure 12.35

SOLUTION The components of the three forces are respectively

$$(F \cos 60°, F \sin 60°) = \left(\frac{F}{2}, \frac{F\sqrt{3}}{2}\right),$$

$$(2F, 0),$$

$$(3F \cos 60°, -3F \sin 60°) = \left(\frac{3F}{2}, \frac{-3F\sqrt{3}}{2}\right).$$

Adding these three together we obtain the resultant

$$\left(\frac{F}{2} + 2F + \frac{3F}{2}, \frac{F\sqrt{3}}{2} + 0 - \frac{3F\sqrt{3}}{2}\right) = (4F, -\sqrt{3}\,F).$$

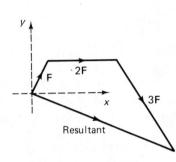

Alternatively, the three forces can be added geometrically using the parallelogram law, as in Fig. 12.36. The resultant has components $4F$ in the x-direction and $\sqrt{3}\,F$ in the negative y-direction.

An object is said to be in *equilibrium* if the various forces acting on it do not cause it to accelerate and develop velocity. It is one of the laws of mechanics that in order for a body to be in equilibrium it is necessary that the resultant force on it be zero. If the resultant force on any body is non-zero, then the body will not remain stationary but

Figure 12.36

will always accelerate and therefore begin to move.

EXAMPLE Three forces, F_1, F_2, and F_3, act on a body. F_1 and F_2 have components $F_1 = (2, 3)$, $F_2 = (3, -3)$. What must be the components of F_3 in order that the body will be in equilibrium?

SOLUTION In order for the body to be in equilibrium the resultant force must be zero:

$$F_1 + F_2 + F_3 = 0.$$

That is,

$$F_3 = -F_1 - F_2 = -(2, 3) - (3, -3) = (-2 - 3, -3 + 3)$$
$$= (-5, 0).$$

So F_3 must be a force of 5 units acting along the negative x-direction (Fig. 12.37).

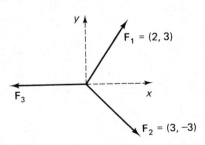

Figure 12.37

EXAMPLE A pulley is used to raise a weight of 100 kg, as shown in Fig. 12.38. The tail end of the rope is pulled at an angle of 30° below the horizontal. What is the force that its support must exert on the pulley?

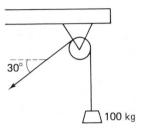

30°

100 kg

Figure 12.38

SOLUTION The force with which the tail end of the rope is pulled must be equal to 100 kg, since this force is transmitted via the tension in the rope and is exactly sufficient

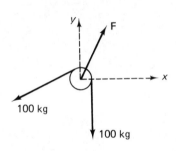

F

100 kg

100 kg

Figure 12.39

to raise the load of 100 kg. The forces on the pulley are therefore as follows: a force of 100 kg acting vertically downwards; a force of 100 kg acting in a direction 30° to the horizontal; and a certain force **F** provided by the pulley support. These three forces must add to zero if the pulley is to be in equilibrium. Using axes as shown in Fig. 12.39 we can resolve the first two forces into the components $(0, -100)$ and $(-100 \cos 30°, -100 \sin 30°) = (-50\sqrt{3}, -50)$, respectively. Therefore, for equilibrium,

$$(0, -100) + (-50\sqrt{3}, -50) + \mathbf{F} = 0.$$

Therefore $\mathbf{F} = (50\sqrt{3}, 150) = (86.6, 150)$. So the force on the pulley has a horizontal component of 86.6 kg and a vertical component of 150 kg.

If a part A of the universe exerts a certain force **F** on a part B, then it is one of the fundamental laws of mechanics that part B exerts a force on part A equal to $-\mathbf{F}$. This law is called Newton's third law, and is usually stated laconically as "Action and reaction are equal and opposite." Thus, for example, if by standing on the floor I exert a vertical force of 75 kg downward on the floor, then it follows that the floor must be exerting a corresponding upward force of 75 kg on me. This upwards force is, of course, what balances my weight and keeps me in equilibrium. Or, in the last example, the pulley support exerts a force $\mathbf{F} = (86.6, 150)$ on the pulley, and so it follows that the force on the pulley support exerted by the pulley is equal to $-\mathbf{F} = (-86.6, -150)$.

That the resultant force on a body should be zero is a necessary condition for the body to be in equilibrium. However it is not a sufficient condition. This is clear from the example illustrated in Fig. 12.40. In this example two equal and opposite forces, **F** and −**F**, are applied one at each end of a rod, in a direction perpendicular to the rod. Clearly the resultant force on the rod, the sum of **F** and −**F**, is zero; and equally clearly, the rod will not be in equilibrium; the two forces will cause the rod to rotate in a counterclockwise direction.

The basic reason why the rod is not in equilibrium is that the two forces acting on it do not act along the same line. In general, if a body is acted on by several forces, **F**$_1$, **F**$_2$, ..., **F**$_n$, which are such that the lines of action of all the forces pass through a certain point P, then the body is in equilibrium if and only if the sum of all the force vectors is zero (Fig. 12.41). We refer to such a situation as one of *concurrent* forces, which means that the forces "run together" at a single point. If, however, the forces are not concurrent (as in the above example of the rod) then the fact that the resultant force is zero does not guarantee equilibrium. A further condition is necessary.

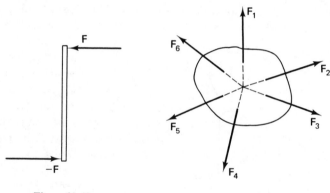

Figure 12.40   Figure 12.41

DEFINITION   Let a force **F** act at a certain point P of a body and let Q be any other point. Let QR be drawn perpendicular to the line of action of **F** (Fig. 12.42). Then if the length of QR is d and if $|\mathbf{F}|$ denotes the magnitude of **F**, then the product $|\mathbf{F}|d$ is called the *moment* of the force **F** about the point Q.

In Fig. 12.42, the moments are both in the clockwise sense. In Fig. 12.43 an example is shown in which the moment of **F** about Q is in the counterclockwise sense. We give clockwise moments a negative sign and counterclockwise moments a positive sign.

We can think of the moment of **F** about Q as measuring the tendency of this force to make the body rotate about the point Q.

When a body is acted upon by several forces, we define the total moment about Q to be the algebraic sum of the moments of all the separate forces, counting counterclockwise moments positive and clockwise moments negative. Then we have the following law of mechanics: A body acted on by several forces is in equilibrium if and only if the resultant force vanishes and the total moment about a point Q is zero.

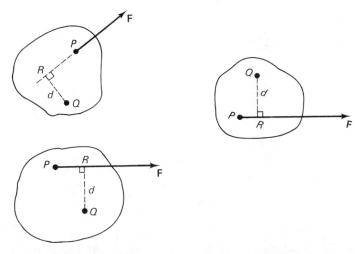

Figure 12.42   **Figure 12.43**

Note that if the total moment is zero about any one point, it is zero about every point. Therefore it does not matter which point we choose for Q in stating this condition for equilibrium.

EXAMPLE   Consider the rod of the preceding example, and let the force $\mathbf{F}$ act at P and the force $-\mathbf{F}$ act at Q, as shown in Fig. 12.44. Let the length PQ be ℓ and let R be a point on the line PQ, situated a distance x from P and a distance $(\ell - x)$ from Q. The moment of the force $\mathbf{F}$ about R is $-|\mathbf{F}|x$ (negative because it is clockwise). The moment of the force $-\mathbf{F}$ about R is $-|-\mathbf{F}|(\ell - x) = -|\mathbf{F}|(\ell - x)$, again clockwise and negative. Therefore the total moment about R is equal to

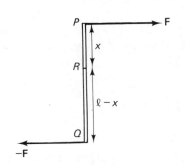

Figure 12.44

$$-|\mathbf{F}|x - |\mathbf{F}|(\ell - x) = -|\mathbf{F}|\ell.$$

Observe that this total moment is independent of x; it is the same for all points R on the line PQ. It is also always nonzero as long as ℓ, the distance between P and Q, is not equal to zero.

The analysis of the action of muscles in the human body leads to many problems that involve the use of resultant forces and moments. As an example, we shall consider the action of the biceps muscle when the arm is used to raise a weight. Figure 12.45 shows the case when the upper arm bone remains in a vertically downward position and the weight is lifted by the rotation of the lower arm about the elbow. At a general position let θ denote the angle that the lower arm makes with the vertical.

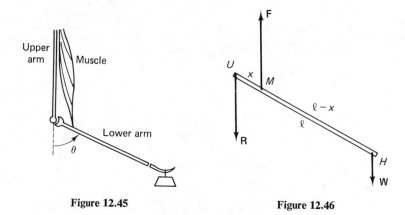

Figure 12.45   **Figure 12.46**

The forces acting on the lower arm are shown schematically in Fig. 12.46: the weight $\mathbf{W}$ acting on the hand in a direction vertically downward; the muscle force $\mathbf{F}$, which we can take to act vertically upward as a first approximation although its direction is in actual fact not quite parallel to the upper arm bone; and a force $\mathbf{R}$, which is caused by the upper arm bone. The muscle force acts at a certain point M where the ligament attaches to the lower arm and we suppose that the distance $UM = x$. The length UH of the lower arm is denoted by ℓ.

The resultant force on the lower arm is $\mathbf{R} + \mathbf{F} + \mathbf{W}$, which, for equilibrium, must vanish. Since $\mathbf{F}$ and $\mathbf{W}$ both have vertical directions, it follows that $\mathbf{R} = -\mathbf{F} - \mathbf{W}$ must also be a vertical force. We can see in fact that $\mathbf{R}$ acts vertically downward and has a magnitude equal to $|\mathbf{F}| - |\mathbf{W}|$.

The moment of $\mathbf{F}$ about the point U is equal to $|\mathbf{F}| x \sin \theta$. The moment of $\mathbf{W}$ about U is $-|\mathbf{W}| \ell \sin \theta$ (negative because it is clockwise). Since $\mathbf{R}$ acts through U, the moment of this third force about U is zero. The total moment about U is therefore $|\mathbf{F}| x \sin \theta - |\mathbf{W}| \ell \sin \theta$, which must be zero. It follows therefore that

$$|\mathbf{F}| = \left(\frac{\ell}{x}\right)|\mathbf{W}|,$$

which gives the magnitude of the muscle force.

Since x is much smaller than ℓ, the ratio (ℓ/x) is a large number, so it follows that the magnitude of the muscle force $|\mathbf{F}|$ is many times larger than the magnitude of the weight being lifted. Note also that the muscle force $|\mathbf{F}|$ does not depend on the angle θ: $|\mathbf{F}|$ remains constant as the weight is raised.

EXERCISES 12.4

What are the resultants of the following sets of forces?

1. $\mathbf{F}_1 = (2, 2)$, $\mathbf{F}_2 = (-3, 1)$, $\mathbf{F}_3 = (1, 2)$.

2. $F_1 = (-1, 4)$, $F_2 = (1, 2)$, $F_3 = (2, -2)$, $F_4 = (3, -3)$.

3.

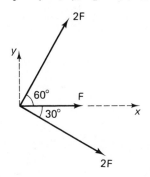

4.

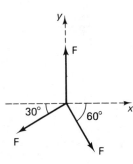

Find the force **G** in the following examples such that the given system of forces is in equilibrium.

5.

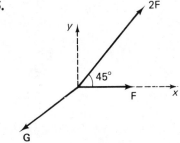

6.

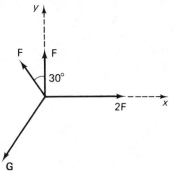

7.

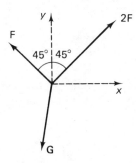

8.

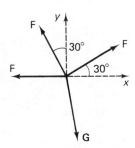

Evaluate the moments of the following forces and systems of forces about the origin.

9.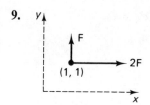

10.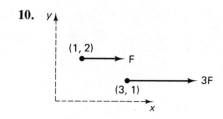

11.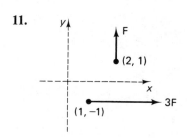

12.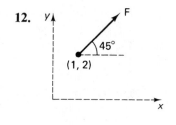

13.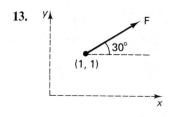

14.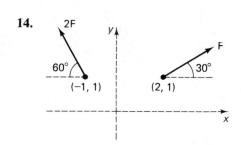

(*Hint:* In exercises 12–14, resolve the given forces into horizontal and vertical components and then evaluate the moment as for exercises 9–11.)

15–20. In Exercises 9–14, a certain force **G** acts at a point $(a, 0)$ on the x-axis, in addition to the forces shown. Find **G** and a in each case in order that the total system of forces will be in equilibrium.

21. The arm is used to raise a weight in such a way that the upper arm and lower arm maintain the same angle θ with the vertical. Show that the force F in the biceps muscle is given by

$$F = \frac{W\ell}{2x \cos \theta}.$$

(Assume that the muscle remains parallel with the upper arm.)

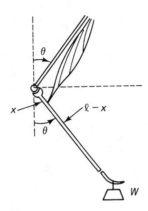

12.5 MATRICES

DEFINITION   An $m \times n$ *matrix* is a rectangular array of real numbers with m rows and n columns. When written down, the array is enclosed within a pair of large brackets.

EXAMPLES

(i)

$$\begin{pmatrix} 1 & 2 \\ 4 & -1 \end{pmatrix}$$

is a 2×2 matrix; it has 2 rows and 2 columns.

(ii)

$$\begin{pmatrix} -1 & 2 & 6 \\ 3 & 0 & -2 \end{pmatrix}$$

is a 2×3 matrix, having 2 rows and 3 columns.

(iii)

$$\begin{pmatrix} -1 & 3 \\ 2 & 0 \\ 6 & -2 \end{pmatrix}$$

is a 3×2 matrix, having 3 rows and 2 columns.

(iv)

$$\begin{pmatrix} a & b & c \\ d & e & f \\ g & h & i \end{pmatrix}$$

is a 3×3 matrix in which the values of the different entries are left unspecified, being denoted by letters.

It is often convenient, as in example (iv) above, to use symbols for the various entries in the matrix rather than to give the entries particular numerical values. However, it is usually inconvenient to use a different letter for each entry, as in this example, if for no other reason than that with a large matrix we would run out of letters. So we use a double-subscript notation instead. For example, a 2×2 matrix might be taken in the general form

$$\mathbf{A} = \begin{pmatrix} A_{11} & A_{12} \\ A_{21} & A_{22} \end{pmatrix}$$

and a general $m \times n$ matrix might be

$$\mathbf{B} = \begin{pmatrix} B_{11} & B_{12} & B_{13} & \dots & B_{1n} \\ B_{21} & B_{22} & B_{23} & \dots & B_{2n} \\ \cdot & & & & \\ \cdot & & & & \\ \cdot & & & & \\ B_{m1} & B_{m2} & B_{m3} & \dots & B_{mn} \end{pmatrix}.$$

In the second example, B_{ij} is used to denote the entry that is situated in the ith row and the jth column. For example, B_{23} is the entry in the second row and third column; B_{m2} is the entry in the mth row and second column.

Note also in these examples the way in which a single boldface letter ($\mathbf{A}$ or $\mathbf{B}$) is used to stand for the whole matrix.

EXAMPLE   If the 2×2 matrix $\mathbf{A}$ is equal to

$$\begin{pmatrix} 1 & 2 \\ 4 & -1 \end{pmatrix}$$

then in terms of subscript notation,

$$A_{11} = 1, \quad A_{12} = 2, \quad A_{21} = 4, \quad A_{22} = -1.$$

We speak of A_{ij} as being the *ij-element* of the matrix $\mathbf{A}$. Instead of writing the matrix out in full, we often write simply (A_{ij}) to denote the matrix whose *ij*-element is A_{ij}. Of course the size of the matrix must be already understood when such a notation is used.

DEFINITION   If $\mathbf{A}$ is the matrix (A_{ij}) and c is any real number, then the product $c\mathbf{A}$ is used to denote the matrix whose size is the same as that of $\mathbf{A}$ and whose *ij*-element is cA_{ij}. That is, each element of $\mathbf{A}$ is multiplied by c to give $c\mathbf{A}$.

EXAMPLE   If

$$\mathbf{A} = \begin{pmatrix} 1 & 0 & -1 \\ 0 & -2 & 4 \end{pmatrix}$$

then

$$2\mathbf{A} = \begin{pmatrix} 2 & 0 & -2 \\ 0 & -4 & 8 \end{pmatrix} \quad \text{and} \quad -3\mathbf{A} = \begin{pmatrix} -3 & 0 & 3 \\ 0 & 6 & -12 \end{pmatrix}.$$

DEFINITION   If $A = (A_{ij})$ and $B = (B_{ij})$ are two matrices of the same size, $m \times n$ say, then the sum $A + B$ is used to denote the $m \times n$ matrix whose ij-element is $A_{ij} + B_{ij}$.

EXAMPLE   If

$$A = \begin{pmatrix} 1 & 0 & -1 \\ 0 & -2 & 4 \end{pmatrix} \quad \text{and} \quad B = \begin{pmatrix} 3 & 2 & 1 \\ -2 & 1 & -2 \end{pmatrix}$$

then

$$A + B = \begin{pmatrix} 1+3 & 0+2 & -1+1 \\ 0-2 & -2+1 & 4-2 \end{pmatrix} = \begin{pmatrix} 4 & 2 & 0 \\ -2 & -1 & 2 \end{pmatrix}.$$

A matrix of size $1 \times n$ has only one row, and, when written out, resembles the component form of a vector. As a result, such a matrix is called a *row-vector*. In a similar way a matrix of size $m \times 1$ has only one column containing m elements, and is called a *column-vector*.

EXAMPLES

(i)   The 1×4 matrix $(0 \quad 5 \quad -2 \quad 1)$ is a row-vector with 4 elements.

(ii)   The 3×1 matrix

$$\begin{pmatrix} -2 \\ 3 \\ -3 \end{pmatrix}$$

is a column-vector with 3 elements.

When using subscript notation to specify the elements of a row- or column-vector it is not necessary to use a double subscript since the elements all lie in a single row or column. For example, we can denote the elements of a 1×3 row-vector by $(x_1 \quad x_2 \quad x_3)$, using x_i as the ith element in the row. Similarly, the ith element in an $m \times 1$ column-vector can be denoted simply by y_i, and the vector itself is then

$$\begin{pmatrix} y_1 \\ y_2 \\ \cdot \\ \cdot \\ \cdot \\ y_m \end{pmatrix}$$

The above definitions of the operations of addition of matrices and their multiplication by scalars are quite straightforward. In certain cases matrices can also be multiplied together, but the rule governing the multiplication of matrices is more complicated. We shall begin by considering the multiplication of a row-vector and a column-vector.

DEFINITION   Let X be a $1 \times n$ row-vector and Y be an $n \times 1$ column-vector. Then the *matrix product* XY is defined to be equal to the inner product of X and Y when X and Y are regarded as ordinary vectors.

In terms of components, if $\mathbf{X} = (x_1 \quad x_2 \quad \ldots \quad x_n)$ and

$$\mathbf{Y} = \begin{pmatrix} y_1 \\ y_2 \\ \cdot \\ \cdot \\ \cdot \\ y_n \end{pmatrix},$$

then

$$\mathbf{XY} = (x_1 \quad x_2 \quad \ldots \quad x_n) \begin{pmatrix} y_1 \\ y_2 \\ \cdot \\ \cdot \\ \cdot \\ y_n \end{pmatrix} = x_1 y_1 + x_2 y_2 + \ldots + x_n y_n.$$

So, in forming the product of a row-vector and a column-vector we multiply corresponding elements in the two vectors together (for example, x_1 times y_1, x_2 times y_2, etc.) and then add the results together.

EXAMPLES

(i) $(2 \quad 3) \begin{pmatrix} -1 \\ 4 \end{pmatrix} = 2(-1) + 3(4) = -2 + 12 = 10.$

(ii) $(4 \quad 5 \quad 6) \begin{pmatrix} 2 \\ 0 \\ -1 \end{pmatrix} = 4(2) + 5(0) + 6(-1) = 8 + 0 - 6 = 2.$

NOTE. If $\mathbf{X}$ is a $1 \times n$ matrix and $\mathbf{Y}$ is an $m \times 1$ matrix then the product $\mathbf{XY}$ is *not defined* when $n \neq m$. The row-vector and column-vector *must* have the same number of elements.

Now let us turn to the definition of the product $\mathbf{AB}$ of two matrices $\mathbf{A}$ and $\mathbf{B}$. This product is only defined when the number of columns in the first matrix $\mathbf{A}$ is equal to the number of rows in the second matrix $\mathbf{B}$.

DEFINITION   Let $\mathbf{A}$ be an $m \times n$ matrix and $\mathbf{B}$ an $n \times p$ matrix. Then the product $\mathbf{AB}$ is an $m \times p$ matrix whose *ij*-element is obtained by taking the product of the *i*th row in $\mathbf{A}$ (regarded as a row-vector) and the *j*th column in $\mathbf{B}$ (regarded as a column-vector).

EXAMPLE   Let us consider the product of the two 2×2 matrices

$$\mathbf{A} = \begin{pmatrix} 1 & 2 \\ 3 & 4 \end{pmatrix}, \quad \mathbf{B} = \begin{pmatrix} 5 & 6 \\ 7 & 8 \end{pmatrix}$$

The product will also be a 2×2 matrix. Its 11-element is obtained by multiplying the first row of **A** by the first column of **B**:

$$(\mathbf{AB})_{11} = (1 \quad 2)\binom{5}{7} = (1)(5) + (2)(7) = 19.$$

Its 12-element is obtained by multiplying the first row of **A** by the second column of **B**:

$$(\mathbf{AB})_{12} = (1 \quad 2)\binom{6}{8} = (1)(6) + (2)(8) = 22.$$

Its 21-element is obtained by multiplying the second row of **A** by the first column of **B**:

$$(\mathbf{AB})_{21} = (3 \quad 4)\binom{5}{7} = (3)(5) + (4)(7) = 43.$$

And finally, the 22-element is obtained by multiplying the second row of **A** by the second column of **B**:

$$(\mathbf{AB})_{22} = (3 \quad 4)\binom{6}{8} = (3)(6) + (4)(8) = 50.$$

Therefore

$$\mathbf{AB} = \begin{pmatrix} 19 & 22 \\ 43 & 50 \end{pmatrix}.$$

EXAMPLE   The 2×3 matrix **A** and the 3×3 matrix **B** are as follows:

$$\mathbf{A} = \begin{pmatrix} 1 & 2 & 3 \\ 4 & 5. & 6 \end{pmatrix}, \qquad \mathbf{B} = \begin{pmatrix} 1 & 2 & 3 \\ 4 & 5 & 6 \\ 7 & 8 & 9 \end{pmatrix}.$$

The product **AB** is a 2×3 matrix. Its 23-element, for example, is obtained by multiplying the second row of **A** and the third column of **B**:

$$(\mathbf{AB})_{23} = (4 \quad 5 \quad 6)\begin{pmatrix} 3 \\ 6 \\ 9 \end{pmatrix} = (4)(3) + (5)(6) + (6)(9)$$

$$= 12 + 30 + 54 = 96.$$

The complete matrix **AB** is as follows:

$$\mathbf{AB} = \begin{pmatrix} (1)(1) + (2)(4) & (1)(2) + (2)(5) & (1)(3) + (2)(6) \\ \quad + (3)(7) & + (3)(8) & + (3)(9) \\ (4)(1) + (5)(4) & (4)(2) + (5)(5) & (4)(3) + (5)(6) \\ \quad + (6)(7) & + (6)(8) & + (6)(9) \end{pmatrix}$$

$$= \begin{pmatrix} 1 + 8 + 21 & 2 + 10 + 24 & 3 + 12 + 27 \\ 4 + 20 + 42 & 8 + 25 + 48 & 12 + 30 + 54 \end{pmatrix}$$

$$= \begin{pmatrix} 30 & 36 & 42 \\ 66 & 81 & 96 \end{pmatrix}.$$

If **A** is an $m \times n$ matrix and **B** is a $q \times p$ matrix, then the product **AB** is defined only when the second dimension of **A** is equal to the first dimension of **B**, that is, when $n = q$. If $n \neq q$, no meaning can be given to the product **AB**.

We observe that matrix multiplication does not obey the commutative law—that is, **AB** is not equal to **BA** in general. In fact, it often happens that one of these products is not even defined while the other is. In the preceding example, in which **A** was 2×3 and **B** was 3×3, the product **AB** is defined but the product **BA** is not, since the number of columns of **B** is not equal to the number of rows of **A**.

If **A** is $m \times n$ and **B** is $n \times p$, the product **AB** is an $m \times p$ matrix. If **C** is a third matrix of size $p \times q$, then the product $(\mathbf{AB})\mathbf{C}$ is defined and is of size $m \times q$. Now also we can form the product $(\mathbf{BC})$, which will be $n \times q$, and then multiply this by **A** to get $\mathbf{A}(\mathbf{BC})$, an $m \times q$ matrix. It turns out always that these two final $m \times q$ matrices are equal:

$$\mathbf{A}(\mathbf{BC}) = (\mathbf{AB})\mathbf{C}. \tag{i}$$

That is, matrix multiplication is associative. Because of this we do not need to keep the brackets in such products, and we can simply write **ABC** to stand for either of the two products in (i) above.

The rule for multiplying matrices can be written in a compact form by employing the summation notation introduced in Section 7.1. First of all let **x** be a $1 \times n$ row-vector and **y** an $n \times 1$ column vector. Then the matrix product **xy** can be written

$$\mathbf{xy} = x_1 y_1 + x_2 y_2 + \ldots + x_n y_n$$
$$= \sum_{k=1}^{n} x_k y_k,$$

using the $\sum$-notation.

Now let **A** be an $m \times n$ matrix with elements (A_{ij}) and **B** an $n \times p$ matrix with elements (B_{ij}). The product **AB** is an $m \times p$ matrix, and its ij-element is obtained by taking the product of the ith row of **A** and the jth column of **B**. Now the ith row of **A** consists of the row of elements $[A_{i1} \quad A_{i2} \quad A_{i3} \quad \ldots \quad A_{in}]$; for example, the second row of **A** is $[A_{21} \quad A_{22} \quad A_{23} \quad \ldots \quad A_{2n}]$, where we have set $i = 2$. Similarly, the jth column of **B** consists of the column of elements

$$\begin{pmatrix} B_{1j} \\ B_{2j} \\ \cdot \\ \cdot \\ \cdot \\ B_{nj} \end{pmatrix}.$$

Therefore, when we form the product of this row and column, we obtain that

$$(\mathbf{AB})_{ij} = A_{i1}B_{1j} + A_{i2}B_{2j} + \ldots + A_{in}B_{nj}$$
$$= \sum_{k=1}^{n} A_{ik}B_{kj}.$$

DEFINITION A matrix of size $n \times n$, that is, for which the numbers of rows and columns are equal, is called a *square matrix*.

If A is an $n \times n$ square matrix, then the product AA is defined, and is again of size $n \times n$. It is denoted by A^2. Similarly, the product AAA is denoted by A^3, and in general, the product of r A's is denoted by A^r. All of these matrices are $n \times n$.

An important square matrix is the *identity matrix*, usually denoted by I, all of whose elements are zero except those that lie on the diagonal of the matrix, which are equal to 1. For example, the 2×2 identity matrix is

$$\begin{pmatrix} 1 & 0 \\ 0 & 1 \end{pmatrix}$$

and the 3×3 identity matrix is

$$\begin{pmatrix} 1 & 0 & 0 \\ 0 & 1 & 0 \\ 0 & 0 & 1 \end{pmatrix}.$$

In general, if I_{ij} denotes the ij-element of the identity matrix, then $I_{ij} = 0$ whenever $i \neq j$, and $I_{ij} = 1$ whenever $i = j$.

EXERCISES 12.5

1. What are the sizes of the following matrices?

$$A = \begin{pmatrix} 1 & 0 \\ -1 & -2 \end{pmatrix}, \quad B = \begin{pmatrix} 2 & 3 & 1 \\ 3 & 1 & 2 \end{pmatrix}, \quad C = \begin{pmatrix} -1 \\ 2 \\ 2 \end{pmatrix},$$

$$D = \begin{pmatrix} 1 & 2 & 0 \\ 0 & 1 & 2 \\ -1 & 3 & 4 \end{pmatrix}, \quad E = \begin{pmatrix} 2 & 1 & -1 \\ -2 & 3 & 0 \end{pmatrix}, \quad F = \begin{pmatrix} -3 & 2 \\ 1 & 1 \end{pmatrix},$$

$$G = (4 \quad 1), \quad H = (2 \quad -2 \quad 1).$$

For the matrices in exercise 1, evaluate the following:

2. 2D.

3. A + B.

4. A − F.

5. 2B + 3E.

6. C + 2H.

7. BC.

8. GA.

9. HDC.

10. GF − 2G.

11. AF − FA.

12. ABC.

13. DE.

14. BD.

15. ED.

16. FB — 2AE.

17. Show that $(\mathbf{AB})\mathbf{D} = \mathbf{A}(\mathbf{BD})$.

18. Let $\mathbf{I}$ be the $n \times n$ identity matrix and $\mathbf{A}$ any $n \times p$ matrix. Show that $\mathbf{IA} = \mathbf{A}$.

19. For the matrix $\mathbf{A}$ in exercise 1, find a matrix $\mathbf{P}$ such that $\mathbf{AP} = \mathbf{I}$, where $\mathbf{I}$ is the 2×2 identity matrix. Evaluate $\mathbf{PA}$.

12.6 APPLICATIONS OF MATRICES

Data Storage

Let us suppose that we are conducting tests to measure the yields of different varieties of wheat subjected to various fertilizer treatments. Suppose that there are four varieties of wheat to be tested and three different fertilizers to be used for each. Then there are clearly 12 combinations of varieties of wheat and fertilizers to be tested, and measurements of the yield obtained from each of these combinations must be made. The results of these measurements can be conveniently stored in a matrix of size 3×4. Each column in the matrix will correspond to one of the four varieties of wheat and each row will correspond to one of the three fertilizers.

For example, the results of the experiment might give the following matrix:

$$\mathbf{Y} = \begin{pmatrix} 10 & 13 & 11 & 8 \\ 12 & 14 & 11 & 9 \\ 11 & 11 & 13 & 12 \end{pmatrix}.$$

In this example the first variety of wheat when treated with the second fertilizer gave a yield of 12 units; the third variety when treated with the third fertilizer gave a yield of 13 units; and so on.

If Y_{ij} denotes the ij-element in the above matrix, then

$Y_{ij} =$ yield of the jth variety of wheat when treated with the ith fertilizer.

This type of usage of matrices for the storage of experimental data is frequently employed for measurements conducted for different values of two parameters. The two parameters in the preceding example are the variety of wheat and the type of fertilizer, and a measurement is taken for each pair of values of these two parameters. It is a particularly convenient means of storing the data when the two parameters take a large number of different values, in which case the number of experimental results can be very large. For example, we may be concerned with data on the incidence of lung cancer among individuals as a function of their age and of the number of cigarettes per day that they consume. In such a case the number of categories into which the individuals are divided may be very large—we may wish to consider upwards of 50 age groups for instance. The data matrix will be a large one, of course, but nevertheless it is the most convenient way as a rule of storing the results.

To return to the wheat-yield example, suppose that the experiment is rerun using the same combinations of variety and fertilizer, giving the results contained in the following matrix:

$$\mathbf{Y}' = \begin{pmatrix} 11 & 13 & 12 & 9 \\ 11 & 13 & 12 & 9 \\ 10 & 12 & 12 & 11 \end{pmatrix}.$$

Then the data from the two experiments can be *pooled* by adding the two matrices $\mathbf{Y} + \mathbf{Y}'$. Since in this type of experiment, the yields would be measured in volume of grain per unit area of land (e.g., bushels per acre), the data would be most appropriately pooled by considering the average of the two experiments,

$$\tfrac{1}{2}(\mathbf{Y} + \mathbf{Y}') = \begin{pmatrix} 10.5 & 13 & 11.5 & 8.5 \\ 11.5 & 13.5 & 11.5 & 9 \\ 10.5 & 11.5 & 12.5 & 11.5 \end{pmatrix}.$$

Linear Systems

Let us suppose that two species of animals live in the same environment and eat the same two types of food, F_1 and F_2. Each animal of species 1 requires 2 units of F_1 and 3 units of F_2 per day and each animal of species 2 requires 2 units of F_1 and 1 unit of F_2 per day. If there are x_1 animals of species 1 and x_2 of species 2 present in the environment, then we can work out the total consumption of the two foods as follows.

The x_1 animals of species 1 consume $2x_1$ units of F_1 and $3x_1$ units of F_2; the x_2 animals of species 2 consume $2x_2$ units of F_1 and x_2 units of F_2. Therefore the total daily consumption consists of $(2x_1 + 2x_2)$ units of F_1 and $(3x_1 + x_2)$ units of F_2. If C_1 and C_2 are used to denote the daily consumption of units of F_1 and F_2 respectively, then we can write

$$C_1 = 2x_1 + 2x_2, \qquad C_2 = 3x_1 + x_2. \tag{i}$$

This pair of equations can be written as a matrix equation. In order to do this we observe the matrix product

$$\begin{pmatrix} 2 & 2 \\ 3 & 1 \end{pmatrix}\begin{pmatrix} x_1 \\ x_2 \end{pmatrix} = \begin{pmatrix} 2x_1 + 2x_2 \\ 3x_1 + x_2 \end{pmatrix}.$$

Therefore the two equations (i) above take the form

$$\begin{pmatrix} 2 & 2 \\ 3 & 1 \end{pmatrix}\begin{pmatrix} x_1 \\ x_2 \end{pmatrix} = \begin{pmatrix} C_1 \\ C_2 \end{pmatrix}.$$

We can abbreviate this matrix equation considerably by introducing single symbols for the matrices involved. We set

$$\mathbf{X} = \begin{pmatrix} x_1 \\ x_2 \end{pmatrix}, \qquad \mathbf{A} = \begin{pmatrix} 2 & 2 \\ 3 & 1 \end{pmatrix}, \quad \text{and} \quad \mathbf{C} = \begin{pmatrix} C_1 \\ C_2 \end{pmatrix}.$$

The column vector $\mathbf{X}$ is called the *population vector* for the given environment since its different entries provide the sizes of the populations of the species. The matrix $\mathbf{A}$ is called the *consumption matrix*. Its *ij*-element, A_{ij}, gives the amount of the *i*th food F_i that is consumed by each animal of the *j*th species. For example, taking $i = 2$ and $j = 1$, $A_{21} = 3$ is the amount of F_2 consumed by each animal of species 1.

With this notation, the above matrix equation can be written in the very simple form

$$\mathbf{AX} = \mathbf{C}. \tag{ii}$$

Now let us suppose that each day 1000 units of F_1 and 1200 units of F_2 are available, and let us ask the question: What sizes of the two animal populations will consume exactly all of the available supplies of the two foods? This question is answered by setting the consumption of each of the two foods equal to the supply, that is $C_1 = 1000$ and $C_2 = 1200$. The two equations (i) now become

$$2x_1 + 2x_2 = 1000, \qquad 3x_1 + x_2 = 1200, \tag{iii}$$

which comprise two simultaneous equations for x_1 and x_2. The solution of these simultaneous equations provides the required sizes of the two populations.

The solution is obtained in the usual way. For example, multiplying the second equation by 2 and subtracting the first equation from it, we obtain

$$(6x_1 + 2x_2) - (2x_1 + 2x_2) = 2400 - 1000$$

or

$$4x_1 = 1400.$$

Hence $x_1 = 350$ and therefore $x_2 = 150$. So the required population sizes are 350 of species 1 and 150 of species 2.

We note that the two simultaneous equations (iii) are equivalent to the matrix equation (ii), with the components of $\mathbf{C}$ having the given values. That is,

$$\mathbf{AX} = \mathbf{C} = \begin{pmatrix} 1000 \\ 1200 \end{pmatrix}.$$

Note that in this matrix equation, the components of $\mathbf{A}$ and $\mathbf{C}$ are known quantities having given numerical values. The components of the column vector $\mathbf{X}$ are, however, not known; they are in fact the unknown population sizes. So the task of solving the pair of equations (iii) can be rephrased as that of finding the column vector $\mathbf{X}$ that satisfies the matrix equation $\mathbf{AX} = \mathbf{C}$.

Our ability to rewrite systems of simultaneous linear equations as matrix equations does not apply only to systems of two equations, as the following example shows.

EXAMPLE   The system

$$\begin{aligned} x_1 - 2x_2 + x_3 + x_4 &= 1, \\ 2x_1 + x_2 - 3x_3 - x_4 &= 2, \\ 3x_2 + x_3 + 2x_4 &= 3 \end{aligned}$$

consists of three simultaneous equations in four unknown quantities, x_1, x_2, x_3, and x_4. It can be written as a matrix equation as follows:

$$
\begin{pmatrix} 1 & -2 & 1 & 1 \\ 2 & 1 & -3 & -1 \\ 0 & 3 & 1 & 2 \end{pmatrix} \begin{pmatrix} x_1 \\ x_2 \\ x_3 \\ x_4 \end{pmatrix} = \begin{pmatrix} x_1 - 2x_2 + x_3 + x_4 \\ 2x_1 + x_2 - 3x_3 - x_4 \\ 3x_2 + x_3 + 2x_4 \end{pmatrix} = \begin{pmatrix} 1 \\ 2 \\ 3 \end{pmatrix}.
$$

Therefore if we introduce the definitions

$$
\mathbf{X} = \begin{pmatrix} x_1 \\ x_2 \\ x_3 \\ x_4 \end{pmatrix}, \qquad \mathbf{A} = \begin{pmatrix} 1 & -2 & 1 & 1 \\ 2 & 1 & -3 & -1 \\ 0 & 3 & 1 & 2 \end{pmatrix}, \qquad \mathbf{C} = \begin{pmatrix} 1 \\ 2 \\ 3 \end{pmatrix},
$$

then the equation takes the form $\mathbf{AX} = \mathbf{C}$. Observe again that $\mathbf{A}$ and $\mathbf{C}$ are matrices whose elements are known while $\mathbf{X}$ is a column vector whose elements are not known.

The fact that systems of simultaneous linear equations can be written in matrix form is very important. There are a number of theorems in the theory of matrices that can be applied to matrix equations of the type $\mathbf{AX} = \mathbf{C}$ encountered above and that lead to immensely useful consequences for systems of simultaneous linear equations. Further development of these results is beyond the scope of this book and we refer the interested student to books on what is called linear algebra for further reading.

Age Distribution in a Population

In Chapters 10 and 11, a number of mathematical models were introduced for the growth of populations under various conditions. In each of these models the population in question was treated as homogeneous and no account was taken of the differing ages of members of the population. We shall now discuss a simple model that includes the age structure of the population.

Let us begin by focusing our attention on the population at some particular instant of time. The population consists of individuals of various ages, so let us divide the population up into groups, each group comprising individuals of the same age. Suppose that there are y_0 individuals who are less than 1 year old, y_1 individuals who are more than 1 and less than 2 years old, and so on. In general, y_k is the number of individuals who have passed their kth birthday but have not yet passed their $(k + 1)$th. Of course the sum of all these numbers is the total population size y:

$$
y_0 + y_1 + y_2 + \ldots = \sum_k y_k = y.
$$

The sizes y_k of these age groups can be arranged in a column-vector $\mathbf{Y}$, which is called the *population vector*:

$$\mathbf{Y} = \begin{pmatrix} y_0 \\ y_1 \\ y_2 \\ \cdot \\ \cdot \\ \cdot \end{pmatrix}.$$

The size of this vector depends on the number of age categories present in the population. In the case of a human population, the population vector might involve a hundred or more entries.

We are interested in the changes that occur in the population from one year to the next. The population vector varies as a function of time, and we let $\mathbf{Y}_n$ denote this vector as applies to the nth year. The elements of $\mathbf{Y}_n$ are denoted by $y_{0n}, y_{1n}, y_{2n}, \ldots$. Thus, for example, y_{01} denotes the number of individuals present in the first year who are less than 1 year old; y_{73} denotes the number of individuals present in the third year who are between 7 and 8 years old, and so on. In general, y_{kn} is the number present in the year n who are between their kth and $(k + 1)$th birthdays.

Consider the group of y_{01} individuals who are less than 1 year old in year 1. The following year these individuals will be between 1 and 2 years old, and hence will comprise the group of y_{12} individuals who are in this age category in year 2. However we cannot set $y_{12} = y_{01}$ since some of the individuals may have perished during the intervening year. What we must do is to write $y_{12} = S_0 y_{01}$, where S_0 is the proportion of these individuals who survive to the following year.

In a similar way we can write $y_{22} = S_1 y_{11}$, where S_1 is the proportion of individuals of age 1 who survive to the following year. So the number of individuals who are age 2 in the second year is equal to the surviving fraction of those who were age 1 in the first year. The proportion of individuals who survive will in general vary with their ages, so that S_1 will not necessarily be equal to S_0.

The number of individuals who are age $(k + 1)$ in year 2, $y_{k+1, 2}$, will equal the surviving proportion of those individuals who were age k the previous year. Hence in general we can write

$$y_{k+1, 2} = S_k y_{k1} \qquad (k = 0, 1, 2, \ldots), \tag{i}$$

where S_k is the proportion of individuals of age k who survive to the following year.

This set of equations determines all of the group sizes $y_{12}, y_{22}, y_{32}, \ldots$ in year 2, assuming that the survival proportions S_0, S_1, etc., are known. However, the group size y_{02} is not determined. This group consists of all the new individuals who were born into the population between the first and second years. We can model this group as follows.

Within the existing population in year 1, each age group produces a certain number of offspring during the following year. We can reasonably assume that this number of offspring is proportional to the number of individuals in the group. For example, we can assume that the number of offspring produced by the y_{71} individuals in the group of seven-year-olds is equal to $B_7 y_{71}$, where B_7 is a certain constant equal to the average number of surviving offspring produced by each seven-year-old individual. In general, the number of offspring produced by the group of k-year-olds can be assumed to be $B_k y_{k1}$, where B_k is a constant. Thus we can write

$$y_{02} = B_0 y_{01} + B_1 y_{11} + \ldots = \sum_k B_k y_{k1}. \tag{ii}$$

The set of equations (i) and (ii) can be written as a matrix equation as follows:

$$\begin{pmatrix} y_{02} \\ y_{12} \\ y_{22} \\ \cdot \\ \cdot \\ \cdot \end{pmatrix} = \begin{pmatrix} B_0 & B_1 & B_2 & B_3 & \ldots \\ S_0 & 0 & 0 & 0 & \ldots \\ 0 & S_1 & 0 & 0 & \ldots \\ 0 & 0 & S_2 & 0 & \ldots \\ \cdot & \cdot & \cdot & \cdot & \\ \cdot & \cdot & \cdot & \cdot & \end{pmatrix} \begin{pmatrix} y_{01} \\ y_{11} \\ y_{21} \\ \cdot \\ \cdot \\ \cdot \end{pmatrix}.$$

Introducing the symbol $\mathbf{P}$ to stand for the large matrix here, we obtain the relation

$$\mathbf{Y}_2 = \mathbf{P}\mathbf{Y}_1$$

between the population vectors in the first and second years.

But, between the second and third years a similar equation holds:

$$\mathbf{Y}_3 = \mathbf{P}\mathbf{Y}_2;$$

hence it follows that

$$\mathbf{Y}_3 = \mathbf{P}(\mathbf{P}\mathbf{Y}_1) = \mathbf{P}^2\mathbf{Y}_1.$$

In general, the same relation holds between the population vectors in the nth and $(n + 1)$th years, and we can write

$$\mathbf{Y}_{n+1} = \mathbf{P}\mathbf{Y}_n.$$

Therefore

$$\mathbf{Y}_{n+1} = \mathbf{P}^n\mathbf{Y}_1.$$

This last equation provides the age structure of the population in the year $(n + 1)$ in terms of the structure in the year 1. In order to make use of it, we must calculate the matrix $\mathbf{P}$ raised to the power n, which may in practice be a quite difficult task.

As an example, let us consider a population of organisms whose lifespan is 3 yr, so that the population vector $\mathbf{y}_n$ has three elements, y_{0n}, y_{1n}, and y_{2n}, and $\mathbf{P}$ is a 3×3 matrix. We shall further assume that the organisms reproduce only in their third year of life, so that $B_0 = B_1 = 0$. The matrix $\mathbf{P}$ is then

$$\mathbf{P} = \begin{pmatrix} 0 & 0 & B_2 \\ S_0 & 0 & 0 \\ 0 & S_1 & 0 \end{pmatrix}.$$

It follows that

$$\mathbf{P}^2 = \begin{pmatrix} 0 & 0 & B_2 \\ S_0 & 0 & 0 \\ 0 & S_1 & 0 \end{pmatrix} \begin{pmatrix} 0 & 0 & B_2 \\ S_0 & 0 & 0 \\ 0 & S_1 & 0 \end{pmatrix} = \begin{pmatrix} 0 & B_2 S_1 & 0 \\ 0 & 0 & B_2 S_0 \\ S_0 S_1 & 0 & 0 \end{pmatrix}$$

and

$$\mathbf{P}^3 = \begin{pmatrix} 0 & B_2S_1 & 0 \\ 0 & 0 & B_2S_0 \\ S_0S_1 & 0 & 0 \end{pmatrix} \begin{pmatrix} 0 & 0 & B_2 \\ S_0 & 0 & 0 \\ 0 & S_1 & 0 \end{pmatrix} = \begin{pmatrix} B_2S_0S_1 & 0 & 0 \\ 0 & B_2S_0S_1 & 0 \\ 0 & 0 & B_2S_0S_1 \end{pmatrix}$$

$$= B_2S_0S_1 \begin{pmatrix} 1 & 0 & 0 \\ 0 & 1 & 0 \\ 0 & 0 & 1 \end{pmatrix}.$$

That is,

$$\mathbf{P}^3 = (B_2S_0S_1)\mathbf{I} = k\mathbf{I},$$

I being the identity matrix and $k = B_2S_0S_1$. Thus the population vector in year 4 is given by

$$\mathbf{Y}_4 = \mathbf{P}^3\mathbf{Y}_1 = (k)\mathbf{I}\mathbf{Y}_1 = k\mathbf{Y}_1.$$

It follows therefore that the age structure in the fourth year is the same as it was in the first year except that the sizes of all of the age groups are multiplied by the factor k. That is,

$$y_{04} = ky_{01}, \qquad y_{14} = ky_{11}, \qquad y_{24} = ky_{21}.$$

The relative sizes of the different age groups remain unchanged and the sizes themselves are changed by a factor k.

Furthermore,

$$\mathbf{Y}_7 = \mathbf{P}^6\mathbf{Y}_1 = \mathbf{P}^3\mathbf{P}^3\mathbf{Y}_1 = k^2\mathbf{Y}_1.$$

Thus in year 7, the relative sizes of the different age groups are again the same as in year 1, the sizes themselves being changed by a factor k^2. Continuing in this way we see that the population progresses in a 3-yr cycle, and every third year the relative sizes of the three age groups return to their initial values. The sizes of the age groups themselves change by a factor k every 3 yr.

An Example from Genetics (More Difficult)

At the end of Section 8.4 we discussed the distribution of sex-linked genes throughout a population. We considered the case of a gene with two alleles A and a, and we denoted by p and P the gene frequencies among, respectively, the females and males in the population for the A allele. That is, for example, the proportion of the genes in question among females in the population that are A type is equal to p, and the proportion that are a type is therefore $(1 - p)$. P and $(1 - P)$ apply likewise to the males.

If p_n and P_n are the corresponding gene frequencies after n generations, then we showed in Section 8.4 that

$$p_n = \tfrac{1}{2}(p_{n-1} + P_{n-1}), \qquad P_n = p_{n-1}.$$

This pair of equations can be written in matrix form. We introduce a gene-frequency vector for the nth generation, defined as

$$\mathbf{F}_n = \begin{pmatrix} p_n \\ P_n \end{pmatrix}.$$

Then for the previous generation, the corresponding vector is, of course,

$$\mathbf{F}_{n-1} = \begin{pmatrix} p_{n-1} \\ P_{n-1} \end{pmatrix}.$$

Then, noting that

$$\begin{pmatrix} \frac{1}{2} & \frac{1}{2} \\ 1 & 0 \end{pmatrix} \begin{pmatrix} p_{n-1} \\ P_{n-1} \end{pmatrix} = \begin{pmatrix} \frac{1}{2}p_{n-1} + \frac{1}{2}P_{n-1} \\ p_{n-1} \end{pmatrix} = \begin{pmatrix} p_n \\ P_n \end{pmatrix},$$

we see that the given pair of equations can be written as

$$\mathbf{F}_n = \mathbf{A}\mathbf{F}_{n-1}, \tag{i}$$

where $\mathbf{A}$ is the matrix

$$\mathbf{A} = \begin{pmatrix} \frac{1}{2} & \frac{1}{2} \\ 1 & 0 \end{pmatrix}.$$

Equation (i), when n is replaced by $n - 1$, becomes

$$\mathbf{F}_{n-1} = \mathbf{A}\mathbf{F}_{n-2}.$$

Therefore

$$\mathbf{F}_n = \mathbf{A}\mathbf{F}_{n-1} = \mathbf{A}(\mathbf{A}\mathbf{F}_{n-2}) = \mathbf{A}^2\mathbf{F}_{n-2}.$$

Similarly, $\mathbf{F}_{n-2} = \mathbf{A}\mathbf{F}_{n-3}$, and so

$$\mathbf{F}_n = \mathbf{A}^2(\mathbf{A}\mathbf{F}_{n-3}) = \mathbf{A}^3\mathbf{F}_{n-3}.$$

Continuing in this way we can see that

$$\mathbf{F}_n = \mathbf{A}^n\mathbf{F}_0.$$

But $\mathbf{F}_0$ is the gene-frequency vector for the initial population, that is,

$$\mathbf{F}_0 = \begin{pmatrix} p \\ P \end{pmatrix},$$

where p and P are certain given frequencies. Therefore the gene-frequency vector $\mathbf{F}_n$ for the nth generation is given explicitly by the equation $\mathbf{F}_n = \mathbf{A}^n\mathbf{F}_0$. In order to compute $\mathbf{F}_n$, all we need to do is to calculate the matrix $\mathbf{A}$ raised to the nth power and then multiply this by the column vector $\mathbf{F}_0$.

For $n = 2$:

$$\mathbf{A}^2 = \begin{pmatrix} \frac{1}{2} & \frac{1}{2} \\ 1 & 0 \end{pmatrix} \begin{pmatrix} \frac{1}{2} & \frac{1}{2} \\ 1 & 0 \end{pmatrix} = \begin{pmatrix} \frac{3}{4} & \frac{1}{4} \\ \frac{1}{2} & \frac{1}{2} \end{pmatrix}$$

and therefore

$$\mathbf{F}_2 = \begin{pmatrix} p_2 \\ P_2 \end{pmatrix} = \begin{pmatrix} \frac{3}{4} & \frac{1}{4} \\ \frac{1}{2} & \frac{1}{2} \end{pmatrix} \begin{pmatrix} p \\ P \end{pmatrix} = \begin{pmatrix} \frac{3}{4}p + \frac{1}{4}P \\ \frac{1}{2}p + \frac{1}{2}P \end{pmatrix}.$$

Hence the gene-frequencies in the second generation are $p_2 = \frac{3}{4}p + \frac{1}{4}P$ for the males and $P_2 = \frac{1}{2}p + \frac{1}{2}P$ for the females.

For $n = 4$,

$$\mathbf{A}^4 = \mathbf{A}^2\mathbf{A}^2 = \begin{pmatrix} \frac{3}{4} & \frac{1}{4} \\ \frac{1}{2} & \frac{1}{2} \end{pmatrix}\begin{pmatrix} \frac{3}{4} & \frac{1}{4} \\ \frac{1}{2} & \frac{1}{2} \end{pmatrix} = \begin{pmatrix} \frac{11}{16} & \frac{5}{16} \\ \frac{5}{8} & \frac{3}{8} \end{pmatrix}.$$

Since $\mathbf{F}_4 = \mathbf{A}^4\mathbf{F}_0$, the gene frequencies on the fourth generation are

$$p_4 = \frac{11}{16}p + \frac{5}{16}P, \qquad P_4 = \frac{5}{8}p + \frac{3}{8}P.$$

We can continue in this way to compute $\mathbf{A}^n$ for any generation n. However it turns out that an explicit formula can be found for the matrix $\mathbf{A}^n$, and hence the solution for $\mathbf{F}_n$ can be obtained as an algebraic formula for all n. The matrix $\mathbf{A}^n$ is in fact given by

$$\mathbf{A}^n = (\tfrac{1}{3})\begin{pmatrix} 2 + k^n & 1 - k^n \\ 2 - 2k^n & 1 + 2k^n \end{pmatrix},$$

where $k = -\frac{1}{2}$.

Let us verify this. We denote the above matrix by $\mathbf{M}_n$ and show that $\mathbf{M}_n = \mathbf{A}^n$. First of all, putting $n = 1$ in the matrix $\mathbf{M}_n$ we get

$$\mathbf{M}_1 = (\tfrac{1}{3})\begin{pmatrix} 2 + k & 1 - k \\ 2 - 2k & 1 + 2k \end{pmatrix} = (\tfrac{1}{3})\begin{pmatrix} \frac{3}{2} & \frac{3}{2} \\ 3 & 0 \end{pmatrix} = \begin{pmatrix} \frac{1}{2} & \frac{1}{2} \\ 1 & 0 \end{pmatrix} = \mathbf{A}.$$

Thus the quoted formula agrees with $\mathbf{A}$ when $n = 1$: $\mathbf{M}_1 = \mathbf{A}^1$.

Next let us multiply the given matrix by $\mathbf{A}$:

$$\mathbf{M}_n\mathbf{A} = (\tfrac{1}{3})\begin{pmatrix} 2 + k^n & 1 - k^n \\ 2 - 2k^n & 1 + 2k^n \end{pmatrix}\begin{pmatrix} \frac{1}{2} & \frac{1}{2} \\ 1 & 0 \end{pmatrix} = (\tfrac{1}{3})\begin{pmatrix} 2 - \frac{1}{2}k^n & 1 + \frac{1}{2}k^n \\ 2 + k^n & 1 - k^n \end{pmatrix}$$

$$= (\tfrac{1}{3})\begin{pmatrix} 2 + k^{n+1} & 1 - k^{n+1} \\ 2 - 2k^{n+1} & 1 + 2k^{n+1} \end{pmatrix}$$

since $k = -\frac{1}{2}$. Thus we see that $\mathbf{M}_n\mathbf{A} = \mathbf{M}_{n+1}$.

This holds for any value of n, and hence we can proceed inductively to show that $\mathbf{M}_n = \mathbf{A}^n$ for any n. For, setting $n = 1, 2, 3, \ldots$, in turn, we obtain

$$\mathbf{M}_2 = \mathbf{M}_1\mathbf{A} = \mathbf{A}\mathbf{A} = \mathbf{A}^2,$$

$$\mathbf{M}_3 = \mathbf{M}_2\mathbf{A} = \mathbf{A}^2\mathbf{A} = \mathbf{A}^3,$$

and so on.

Having found a formula for $\mathbf{A}^n$, we can write the solution for the gene frequency after n generations as

$$\mathbf{F}_n = \begin{pmatrix} p_n \\ P_n \end{pmatrix} = \mathbf{A}^n\mathbf{F}_0 = (\tfrac{1}{3})\begin{pmatrix} 2 + k^n & 1 - k^n \\ 2 - 2k^n & 1 + 2k^n \end{pmatrix}\begin{pmatrix} p \\ P \end{pmatrix}.$$

Therefore

$$p_n = \tfrac{1}{3}[(2 + k^n)p + (1 - k^n)P],$$

$$P_n = \tfrac{1}{3}[(2 - 2k^n)p + (1 + 2k^n)P].$$

We note that as $n \to \infty$, $k^n = (-\frac{1}{2})^n \to 0$. Therefore, after a larger number of generations, the gene frequencies approach the limiting values

$$p_n \to \tfrac{1}{3}[2p + P], \qquad P_n \to \tfrac{1}{3}[2p + P].$$

The conclusion that p_n and P_n approach equality as the generations proceed was drawn in Section 8.4 by a qualitative (but nevertheless sound!) argument. We see here that this conclusion also follows from the algebraic solution.

EXERCISES 12.6

Write the following systems of equations as matrix equations.

1. $x_1 + x_2 = 2$
  $x_1 - x_2 = 3.$

2. $2x_1 - 3x_2 = 1$
  $x_1 + 4x_2 = -3.$

3. $x_1 + 2x_2 - x_3 = 0$
  $x_1 - 3x_2 = 2.$

4. $2x_1 + x_2 = 2$
  $-x_1 - x_2 + 3x_3 = 1$
  $3x_1 + 2x_2 - x_3 = 1.$

5. $x_1 - x_2 = 1$
  $2x_1 + x_2 = 2$
  $-x_1 + 3x_2 = 2.$

6. $x_1 + x_2 + x_3 + x_4 = 1$
  $2x_1 - x_3 - x_4 = 0$
  $x_2 + 3x_4 = 2.$

For the matrices $\mathbf{A}$ and $\mathbf{B}$ given in the following exercises, solve the matrix equation $\mathbf{AX} = \mathbf{B}$ for the matrix $\mathbf{X}$ by transforming the matrix equation into a system of algebraic equations.

7. $\mathbf{A} = \begin{pmatrix} 1 & 1 \\ 1 & -1 \end{pmatrix}$, $\quad \mathbf{B} = \begin{pmatrix} 1 \\ 2 \end{pmatrix}.$

8. $\mathbf{A} = \begin{pmatrix} 2 & 1 \\ 1 & 3 \end{pmatrix}$, $\quad \mathbf{B} = \begin{pmatrix} 4 \\ 7 \end{pmatrix}.$

9. $\mathbf{A} = \begin{pmatrix} 1 & 1 \\ 1 & -1 \end{pmatrix}$, $\quad \mathbf{B} = \begin{pmatrix} 2 & 3 \\ 0 & 1 \end{pmatrix}.$

10. $\mathbf{A} = \begin{pmatrix} 1 & 2 & 3 \\ 2 & 1 & 0 \\ 1 & -1 & 0 \end{pmatrix}$, $\quad \mathbf{B} = \begin{pmatrix} 3 \\ 3 \\ 3 \end{pmatrix}.$

11. A fish of species 1 consumes 10 gm of food 1 and 5 gm of food 2 per day. A fish of species 2 consumes 6 gm of food 1 and 4 gm of food 2 per day. If a given environment has 2.2 kg of food 1 and 1.3 kg of food 2 available daily, what population sizes of the two species will consume exactly all of the available food?

12. Three species of birds eat aphids from different parts of trees. Species 1 feed half of the time on the top levels and half of the time on the middle levels of the trees. Species 2 feed half on the middle levels and half on the lower levels. Species 3 feed entirely on the lower levels. There are equal numbers of aphids available on the middle and lower levels, but only half this number available on the upper levels. What should be the relative sizes of the populations of the three species in order that the supply of aphids will be entirely consumed?

13. Two species, labeled 1 and 2, consume two foods, F_1 and F_2. Let C_{ij} denote the amount of food F_i consumed daily by each individual of the species j. If S_1 and S_2

are the daily supplies of the two foods, determine the sizes x_1 and x_2 of the two species populations that consume exactly all of the available food supply. What happens when the two species consume the foods in exactly the same proportions?

14. Repeat exercise 13 in the case when $C_{11} = \alpha c$, $C_{21} = (1 - \alpha)c$, $C_{12} = \beta c$, $C_{22} = (1 - \beta)c$, and $S_1 = S_2 = s$ where α, β, c, and s are constants. What is the physical significance of α and β? For a fixed value of α ($\alpha < \frac{1}{2}$), determine the value of β that makes the size of population 2 a maximum.

15. Several species in an ecosystem provide the food sources for one another. The element C_{ij} of a consumption matrix equals the number of units of species j consumed daily by an individual of species i. Construct the matrix (C_{ij}) for the following simple ecosystems consisting of just 3 species.
    (a) Each species consumes on the average 1 unit of each of the other two species.
    (b) Species 1 consumes one unit of species 2; species 2 consumes $\frac{1}{2}$ unit each of species 1 and 3; species 3 consumes 2 units of species 1.
    (c) Species 1 consumes 2 units of species 3; species 2 consumes 1 unit of species 1; species 3 does not consume any other species.

16. Consider a population of biannual organisms that reproduce only in their second year of life. Show that the relative age distribution in the population repeats every 2 yr, but that the actual sizes of the two age groups change by a certain factor over each 2-yr period.

REVIEW EXERCISES FOR CHAPTER 12

1. State whether the following are true or false. If false, give the corresponding correct statement.
    (a) Any vector can be represented by one and only one directed line segment.
    (b) A vector in two dimensions has two components.
    (c) All vectors have two components.
    (d) The vectors $\mathbf{X}$ and $\mathbf{Y}$ are parallel if and only if $\mathbf{X} \cdot \mathbf{Y} = |\mathbf{X}| \, |\mathbf{Y}|$.
    (e) $\mathbf{X} \cdot \mathbf{Y}$ cannot be greater than $|\mathbf{X}||\mathbf{Y}|$ for any pair of vectors.
    (f) $\mathbf{X}$ and $\mathbf{Y}$ make an angle of $60°$ with each other if $\mathbf{X} \cdot \mathbf{Y} = \frac{1}{2}$.
    (g) Three points P, Q, and R lie on a straight line if and only if $(\overrightarrow{OP} - \overrightarrow{OQ})$ is parallel to $(\overrightarrow{OR} - \overrightarrow{OQ})$.
    (h) If $\mathbf{A}$ is $m \times n$ and $\mathbf{B}$ is $n \times m$ then the matrix products $\mathbf{AB}$ and $\mathbf{BA}$ both exist.
    (i) If $\mathbf{A}$ and $\mathbf{B}$ are $n \times n$ square matrices then $\mathbf{AB} = \mathbf{BA}$.
    (j) If $\mathbf{A}$ is an $m \times n$ and $\mathbf{B}$ an $n \times p$ matrix then the matrix $\mathbf{A} + \mathbf{B}$ is $m \times p$.

2. Given $\mathbf{A} = (2, -3, 4)$ and $\mathbf{A} + 2\mathbf{B} = \mathbf{0}$, find $\mathbf{B}$.

3. Given $\mathbf{A} = (4, 5)$, $\mathbf{B} = (2, -3)$, find $\mathbf{C}$ if $\mathbf{A} + 3\mathbf{B} = 2\mathbf{C}$.

4. If $x(2, 3) + y(-1, 4) = (0, -11)$, find x and y.

   In the following exercises, determine a, b, and c.

5. $\begin{pmatrix} 2 & 3 \\ 1 & c \end{pmatrix} + \begin{pmatrix} a & -1 \\ 0 & b \end{pmatrix} = \begin{pmatrix} 3 & b \\ 1 & 0 \end{pmatrix}.$

6. $\begin{pmatrix} a & -1 \\ 2 & b \end{pmatrix} + \begin{pmatrix} 1 & b \\ c & -4 \end{pmatrix} = \begin{pmatrix} 2 & a \\ b & c \end{pmatrix}.$

7. $\begin{pmatrix} a^2 & 1 & b^2 \\ 0 & -2 & 3 \end{pmatrix} - 2 \begin{pmatrix} a & 2 & 1 \\ -1 & c & 4 \end{pmatrix} = \begin{pmatrix} 15 & -3 & 7 \\ 2 & -8 & -5 \end{pmatrix}.$

8. $\begin{pmatrix} a & b & 3 \\ 2 & -1 & c \\ 0 & 1 & 5 \end{pmatrix} + 3 \begin{pmatrix} 1 & -2 & a \\ b & 0 & 3 \\ -1 & c & 2 \end{pmatrix} = \begin{pmatrix} 4 & 4c & 6 \\ 8 & -1 & 8 \\ -3 & -2 & 11 \end{pmatrix}.$

9. If

$$\mathbf{A} = \begin{pmatrix} 1 & -2 & 3 \\ 4 & 1 & -5 \\ -2 & 0 & 3 \end{pmatrix} \quad \text{and} \quad \mathbf{B} = \begin{pmatrix} 0 & 1 & -3 \\ 2 & 0 & 5 \\ -1 & 4 & 0 \end{pmatrix}.$$

find $\mathbf{A} + 2\mathbf{B}$ and $\mathbf{AB}$.

10. If

$$\mathbf{A} = \begin{pmatrix} 2 & -1 & 4 \\ 1 & 0 & 3 \end{pmatrix} \quad \text{and} \quad \mathbf{B} = \begin{pmatrix} -1 & 2 & 4 \\ 3 & 0 & 1 \\ 0 & -1 & 2 \end{pmatrix},$$

compute $\mathbf{AB}$. Can we compute $\mathbf{A} + \mathbf{B}$? If not, explain why. If so, compute it.

Express the following sets of linear equations in matrix form: $\mathbf{AX} = \mathbf{B}$.

11. $x + 2y - z = 4$
$\qquad y + 3z = 5$
$\quad 2x - y + z = 1.$

12. $-2x_1 + x_2 - 3x_3 = 4$
$\qquad 3x_1 - x_2 + 2x_3 = 0$
$\quad -2x_1 + 3x_2 - x_3 = 1.$

Evaluate x, y, and z from the following:

13. $\begin{pmatrix} 1 & 1 & 1 \\ 2 & -1 & 1 \\ 3 & 1 & 5 \end{pmatrix} \begin{pmatrix} x \\ y \\ z \end{pmatrix} = \begin{pmatrix} 5 \\ -2 \\ 6 \end{pmatrix}.$

14. $\begin{pmatrix} 1 & -2 & 3 \\ 0 & 1 & 2 \\ -1 & 3 & 0 \end{pmatrix} \begin{pmatrix} x \\ y \\ z \end{pmatrix} = \begin{pmatrix} 14 \\ 4 \\ 8 \end{pmatrix}.$

15. $\begin{pmatrix} x & 0 & 1 \\ 2 & y & 0 \\ -1 & 0 & z \end{pmatrix} \begin{pmatrix} 0 & 1 \\ -1 & 0 \\ 3 & -2 \end{pmatrix} = \begin{pmatrix} 3 & 1 \\ 1 & 2 \\ x & -3 \end{pmatrix}.$

16. $\begin{pmatrix} 1 & x & 0 \\ 3 & 1 & y \end{pmatrix} \begin{pmatrix} z & 0 \\ 0 & -1 \\ 1 & 2 \end{pmatrix} = \begin{pmatrix} 1 & 2 \\ 4 & 1 \end{pmatrix}.$

Appendices

Table of Standard Derivatives

I. BASIC FORMULAS

1. The derivative of a constant is zero.

2. For any constant c, $\dfrac{d}{dx}[cf(x)] = cf'(x)$.

3. $\dfrac{d}{dx}[f(x) \cdot g(x)] = f(x)g'(x) + g(x)f'(x)$.   —Product Formula

4. $\dfrac{d}{dx}\left(\dfrac{f(x)}{g(x)}\right) = \dfrac{g(x)f'(x) - f(x)g'(x)}{[g(x)]^2}$.   —Quotient Formula

5. If $y = f(u)$ and $u = g(x)$ then

$$\frac{dy}{dx} = \frac{dy}{du} \cdot \frac{du}{dx},$$ —Chain Rule

or

$$\frac{d}{dx}(f[g(x)]) = f'[g(x)] \cdot g'(x)$$ —Chain Rule

6. $\dfrac{d}{dx}(x^n) = nx^{n-1}$ —Power Formula

7. If $x = f(t)$, $y = g(t)$ then $\dfrac{dy}{dx} = \dfrac{g'(t)}{f'(t)}$ or $\dfrac{dy}{dx} = \dfrac{dy/dt}{dx/dt}$.

II. EXPONENTIAL AND LOGARITHMIC FUNCTIONS

$$\frac{d}{dx}(e^x) = e^x.$$

$$\frac{d}{dx}(a^x) = a^x \ln a.$$

$$\frac{d}{dx}(\ln x) = \frac{1}{x}.$$

$$\frac{d}{dx}(\log_a x) = \frac{1}{x} \log_a e = \frac{1}{x \ln a}.$$

III. TRIGONOMETRIC FUNCTIONS

$$\frac{d}{dx}(\sin x) = \cos x; \quad \frac{d}{dx}(\cos x) = -\sin x.$$

$$\frac{d}{dx}(\tan x) = \sec^2 x; \quad \frac{d}{dx}(\cot x) = -\operatorname{cosec}^2 x.$$

$$\frac{d}{dx}(\sec x) = \sec x \tan x; \quad \frac{d}{dx}(\operatorname{cosec} x) = -\operatorname{cosec} x \cot x.$$

NOTE. The derivatives of all cofunctions ($\cos x$, $\cot x$, $\operatorname{cosec} x$) are negative.

IV. INVERSE TRIGONOMETRIC FUNCTIONS

$$\frac{d}{dx}(\text{Sin}^{-1} x) = \frac{1}{\sqrt{1 - x^2}}.$$

$$\frac{d}{dx}(\text{Cos}^{-1} x) = -\frac{1}{\sqrt{1 - x^2}}.$$

$$\frac{d}{dx}(\text{Tan}^{-1} x) = -\frac{d}{dx}(\text{Cot}^{-1} x) = \frac{1}{1 + x^2}.$$

$$\frac{d}{dx}(\text{Sec}^{-1} x) = -\frac{d}{dx}(\text{Cosec}^{-1} x) = \frac{1}{|x|\sqrt{x^2 - 1}}.$$

Table of Integrals

$$\frac{dN}{dt} = rN$$

$$dN = rN\,dt$$

Appendix

NOTE. In every integral the constant of integration is omitted and should be supplied by the reader himself.

SOME FUNDAMENTAL FORMULAS

1. $\int [f(x) \pm g(x)]\, dx = \int f(x)\, dx \pm \int g(x)\, dx.$

2. $\int cf(x)\, dx = c \int f(x)\, dx.$

3. $\int f[g(x)]g'(x)\, dx = \int f(u)\, du \quad \text{where} \quad u = g(x).$

4. $\int f(x)g(x)\, dx = f(x) \int g(x)\, dx - \int f'(x)\left\{\int g(x)\, dx\right\} dx.$

RATIONAL INTEGRANDS INVOLVING $(ax + b)$

5. $\displaystyle\int (ax + b)^n\, dx = \frac{(ax + b)^{n+1}}{a(n + 1)} \quad (n \neq -1).$

6. $\displaystyle\int (ax + b)^{-1}\, dx = \frac{1}{a} \ln|ax + b|.$

7. $\displaystyle\int x(ax + b)^n\, dx = \frac{1}{a^2}(ax + b)^{n+1}\left[\frac{ax + b}{n + 2} - \frac{b}{n + 1}\right] \quad (n \neq -1, -2).$

8. $\displaystyle\int x(ax + b)^{-1}\, dx = \frac{x}{a} - \frac{b}{a^2} \ln|ax + b|.$

9. $\displaystyle\int x(ax + b)^{-2}\, dx = \frac{1}{a^2}\left[\ln|ax + b| + \frac{b}{ax + b}\right].$

10. $\displaystyle\int \frac{x^2}{ax + b}\, dx = \frac{1}{a^3}\left[\frac{1}{2}(ax + b)^2 - 2b(ax + b) + b^2 \ln|ax + b|\right].$

11. $\displaystyle\int \frac{x^2}{(ax + b)^2}\, dx = \frac{1}{a^3}\left[ax + b - \frac{b^2}{ax + b} - 2b \ln|ax + b|\right].$

12. $\displaystyle\int \frac{1}{x(ax + b)}\, dx = \frac{1}{b} \ln\left|\frac{x}{ax + b}\right| \quad (b \neq 0).$

13. $\displaystyle\int \frac{1}{x^2(ax + b)}\, dx = -\frac{1}{bx} + \frac{a}{b^2} \ln\left|\frac{ax + b}{x}\right| \quad (b \neq 0).$

14. $\displaystyle\int \frac{1}{x(ax + b)^2}\, dx = \frac{1}{b(ax + b)} - \frac{1}{b^2} \ln\left|\frac{ax + b}{x}\right| \quad (b \neq 0).$

15. $\displaystyle\int \frac{1}{(ax + b)(cx + d)}\, dx = \frac{1}{bc - ad} \ln\left|\frac{cx + d}{ax + b}\right| \quad (bc - ad \neq 0).$

16. $\int \dfrac{x}{(ax + b)(cx + d)} \, dx = \dfrac{1}{bc - ad} \left\{ \dfrac{b}{a} \ln |ax + b| - \dfrac{d}{c} \ln |cx + d| \right\}$

$$(bc - ad \neq 0).$$

17. $\int \dfrac{1}{(ax + b)^2(cx + d)} \, dx = \dfrac{1}{bc - ad} \left\{ \dfrac{1}{ax + b} + \dfrac{c}{bc - ad} \ln \left| \dfrac{cx + d}{ax + b} \right| \right\}$

$$(bc - ad \neq 0).$$

18. $\int \dfrac{x}{(ax + b)^2(cx + d)} \, dx = -\dfrac{1}{bc - ad} \left\{ \dfrac{b}{a(ax + b)} + \dfrac{d}{bc - ad} \ln \left| \dfrac{cx + d}{ax + b} \right| \right\}$

$$(bc - ad \neq 0).$$

INTEGRALS CONTAINING $\sqrt{ax + b}$

19. $\int x\sqrt{ax + b} \, dx = \dfrac{2}{a^2} \left[\dfrac{(ax + b)^{5/2}}{5} - \dfrac{b(ax + b)^{3/2}}{3} \right].$

20. $\int x^2\sqrt{ax + b} \, dx = \dfrac{2}{a^3} \left[\dfrac{(ax + b)^{7/2}}{7} - \dfrac{2b(ax + b)^{5/2}}{5} + \dfrac{b^2(ax + b)^{3/2}}{3} \right].$

21. $\int \dfrac{x}{\sqrt{ax + b}} \, dx = \dfrac{2ax - 4b}{3a^2} \sqrt{ax + b}.$

22. $\int \dfrac{1}{x\sqrt{ax + b}} \, dx = \dfrac{1}{\sqrt{b}} \ln \left| \dfrac{\sqrt{ax + b} - \sqrt{b}}{\sqrt{ax + b} + \sqrt{b}} \right| \quad (b > 0).$

$$= \dfrac{2}{\sqrt{-b}} \, \text{Tan}^{-1} \sqrt{\dfrac{ax + b}{-b}} \quad (b < 0).$$

23. $\int \dfrac{1}{x^n\sqrt{ax + b}} \, dx = -\dfrac{1}{b(n - 1)} \dfrac{\sqrt{ax + b}}{x^{n-1}} - \dfrac{(2n - 3)a}{(2n - 2)b} \int \dfrac{1}{x^{n-1}\sqrt{ax + b}} \, dx \quad (n \neq 1).$

24. $\int \dfrac{\sqrt{ax + b}}{x} \, dx = 2\sqrt{ax + b} + b \int \dfrac{1}{x\sqrt{ax + b}} \, dx \quad \text{(see 22)}.$

25. $\int \dfrac{\sqrt{ax + b}}{x^2} \, dx = -\dfrac{\sqrt{ax + b}}{x} + \dfrac{a}{2} \int \dfrac{1}{x\sqrt{ax + b}} \, dx \quad \text{(see 22)}.$

INTEGRALS CONTAINING $a^2 \pm x^2$

26. $\int \dfrac{1}{a^2 + x^2} \, dx = \dfrac{1}{a} \, \text{Tan}^{-1} \dfrac{x}{a}.$

27. $\int \dfrac{1}{(a^2 + x^2)^2} \, dx = \dfrac{x}{2a^2(a^2 + x^2)} + \dfrac{1}{2a^3} \, \text{Tan}^{-1} \dfrac{x}{a}.$

28. $\int \dfrac{1}{a^2 - x^2} \, dx = \dfrac{1}{2a} \ln \left| \dfrac{x + a}{x - a} \right|.$

29. $\int \dfrac{1}{(a^2 - x^2)^2} \, dx = \dfrac{x}{2a^2(a^2 - x^2)} + \dfrac{1}{4a^3} \ln \left| \dfrac{x + a}{x - a} \right|.$

30. $\int \dfrac{x}{(a^2 \pm x^2)}\, dx = \pm \dfrac{1}{2} \ln |a^2 \pm x^2|.$

31. $\int \dfrac{1}{x(a^2 \pm x^2)}\, dx = \dfrac{1}{2a^2} \ln \left| \dfrac{x^2}{a^2 \pm x^2} \right|.$

INTEGRALS CONTAINING $\sqrt{a^2 - x^2}$

32. $\int \dfrac{1}{\sqrt{a^2 - x^2}}\, dx = \operatorname{Sin}^{-1} \dfrac{x}{a} \quad (a > 0).$

33. $\int \dfrac{x}{\sqrt{a^2 - x^2}}\, dx = -\sqrt{a^2 - x^2}.$

34. $\int \dfrac{x^2}{\sqrt{a^2 - x^2}}\, dx = -\dfrac{x\sqrt{a^2 - x^2}}{2} + \dfrac{a^2}{2} \operatorname{Sin}^{-1} \dfrac{x}{a} \quad (a > 0).$

35. $\int \dfrac{1}{x\sqrt{a^2 - x^2}}\, dx = -\dfrac{1}{a} \ln \left| \dfrac{a + \sqrt{a^2 - x^2}}{x} \right|.$

36. $\int \dfrac{1}{x^2\sqrt{a^2 - x^2}}\, dx = -\dfrac{\sqrt{a^2 - x^2}}{a^2 x}.$

37. $\int \dfrac{1}{(a^2 - x^2)^{3/2}}\, dx = \dfrac{1}{a^2} \dfrac{x}{\sqrt{a^2 - x^2}}.$

38. $\int \dfrac{x}{(a^2 - x^2)^{3/2}}\, dx = \dfrac{1}{\sqrt{a^2 - x^2}}.$

39. $\int \dfrac{x^2}{(a^2 - x^2)^{3/2}}\, dx = \dfrac{x}{\sqrt{a^2 - x^2}} - \operatorname{Sin}^{-1} \dfrac{x}{a} \quad (a > 0).$

40. $\int \dfrac{1}{x(a^2 - x^2)^{3/2}}\, dx = \dfrac{1}{a^2\sqrt{a^2 - x^2}} - \dfrac{1}{a^3} \ln \left| \dfrac{a + \sqrt{a^2 - x^2}}{x} \right|.$

41. $\int \dfrac{1}{x^2(a^2 - x^2)^{3/2}}\, dx = \dfrac{1}{a^4} \left[-\dfrac{\sqrt{a^2 - x^2}}{x} + \dfrac{x}{\sqrt{a^2 - x^2}} \right].$

42. $\int \sqrt{a^2 - x^2}\, dx = \dfrac{1}{2} a^2 \operatorname{Sin}^{-1} \dfrac{x}{a} + \dfrac{1}{2} x\sqrt{a^2 - x^2} \quad (a > 0).$

43. $\int x\sqrt{a^2 - x^2}\, dx = -\dfrac{1}{3}(a^2 - x^2)^{3/2}.$

44. $\int x^2\sqrt{a^2 - x^2}\, dx = -\dfrac{x(a^2 - x^2)^{3/2}}{4} + \dfrac{a^2 x\sqrt{a^2 - x^2}}{8} + \dfrac{a^4}{8} \operatorname{Sin}^{-1} \dfrac{x}{a} \quad (a > 0).$

45. $\int \dfrac{\sqrt{a^2 - x^2}}{x}\, dx = \sqrt{a^2 - x^2} - a \ln \left| \dfrac{a + \sqrt{a^2 - x^2}}{x} \right|.$

46. $\int \dfrac{\sqrt{a^2 - x^2}}{x^2}\, dx = -\dfrac{\sqrt{a^2 - x^2}}{x} - \operatorname{Sin}^{-1} \dfrac{x}{a} \quad (a > 0).$

47. $\int (a^2 - x^2)^{3/2}\, dx = \dfrac{x}{4}(a^2 - x^2)^{3/2} + \dfrac{3}{8} a^2 x\sqrt{a^2 - x^2} + \dfrac{3}{8} a^4 \operatorname{Sin}^{-1} \dfrac{x}{a} \quad (a > 0).$

48. $\int x(a^2 - x^2)^{3/2} \, dx = -\frac{1}{5}(a^2 - x^2)^{5/2}.$

49. $\int \frac{(a^2 - x^2)^{3/2}}{x} \, dx = \frac{(a^2 - x^2)^{3/2}}{3} + a^2\sqrt{a^2 - x^2} - a^3 \ln\left|\frac{a + \sqrt{a^2 - x^2}}{x}\right|.$

50. $\int x^n(a^2 - x^2)^{3/2} \, dx = \frac{1}{n+1}x^{n+1}(a^2 - x^2)^{3/2} + \frac{3}{n+1}\int x^{n+2}\sqrt{a^2 - x^2} \, dx$

$\qquad\qquad\qquad\qquad\qquad\qquad\qquad\qquad\qquad\qquad\qquad (n \neq -1).$

51. $\int x^n\sqrt{a^2 - x^2} \, dx = -\frac{1}{n+2}x^{n-1}(a^2 - x^2)^{3/2} + \frac{a^2(n-1)}{n+2}\int x^{n-2}\sqrt{a^2 - x^2} \, dx$

$\qquad\qquad\qquad\qquad\qquad\qquad\qquad\qquad\qquad\qquad\qquad (n \neq -2).$

INTEGRALS CONTAINING $\sqrt{x^2 \pm a^2}$

52. $\int \frac{1}{\sqrt{x^2 \pm a^2}} \, dx = \ln|x + \sqrt{x^2 \pm a^2}|.$

53. $\int \frac{x}{\sqrt{x^2 \pm a^2}} \, dx = \sqrt{x^2 \pm a^2}.$

54. $\int \frac{x^2}{\sqrt{x^2 \pm a^2}} \, dx = \frac{1}{2}x\sqrt{x^2 \pm a^2} \mp \frac{1}{2}a^2 \ln|x + \sqrt{x^2 \pm a^2}|.$

55. $\int \frac{1}{x\sqrt{x^2 + a^2}} \, dx = -\frac{1}{a} \ln\left|\frac{a + \sqrt{x^2 + a^2}}{x}\right|.$

56. $\int \frac{1}{x\sqrt{x^2 - a^2}} \, dx = \frac{1}{a} \text{Sec}^{-1} \frac{x}{a} \quad (a > 0).$

57. $\int \frac{1}{x^2\sqrt{x^2 \pm a^2}} \, dx = \mp \frac{\sqrt{x^2 \pm a^2}}{a^2 x}.$

58. $\int \frac{1}{(x^2 \pm a^2)^{3/2}} \, dx = \pm \frac{1}{a^2}\frac{x}{\sqrt{x^2 \pm a^2}}.$

59. $\int \frac{x}{(x^2 \pm a^2)^{3/2}} \, dx = -\frac{1}{\sqrt{x^2 \pm a^2}}.$

60. $\int \frac{x^2}{(x^2 \pm a^2)^{3/2}} \, dx = -\frac{x}{\sqrt{x^2 \pm a^2}} + \ln|x + \sqrt{x^2 \pm a^2}|.$

61. $\int \frac{1}{x(x^2 \pm a^2)^{3/2}} \, dx = \pm\frac{1}{a^2}\left\{\frac{1}{\sqrt{x^2 \pm a^2}} + \int \frac{1}{x\sqrt{x^2 \pm a^2}} \, dx\right\} \quad \text{(see 55 or 56)}.$

62. $\int \frac{1}{x^2(x^2 \pm a^2)^{3/2}} \, dx = -\frac{1}{a^4}\left\{\frac{\sqrt{x^2 \pm a^2}}{x} + \frac{x}{\sqrt{x^2 \pm a^2}}\right\}.$

63. $\int \frac{1}{(x^2 \pm a^2)^{5/2}} \, dx = \frac{1}{a^4}\left\{\frac{x}{\sqrt{x^2 \pm a^2}} - \frac{1}{3}\frac{x^3}{(x^2 \pm a^2)^{3/2}}\right\}.$

64. $\int \frac{x}{(x^2 \pm a^2)^{5/2}} \, dx = -\frac{1}{3}\frac{1}{(x^2 \pm a^2)^{3/2}}.$

65. $\int \frac{x^2}{(x^2 \pm a^2)^{5/2}} \, dx = \pm\frac{1}{3a^2}\frac{x^3}{(x^2 \pm a^2)^{3/2}}.$

66. $\int \sqrt{x^2 \pm a^2} \, dx = \pm \frac{1}{2} x \sqrt{x^2 \pm a^2} \pm \frac{1}{2} a^2 \ln |x + \sqrt{x^2 \pm a^2}|.$

67. $\int x\sqrt{x^2 \pm a^2} \, dx = \frac{1}{3}(x^2 \pm a^2)^{3/2}.$

68. $\int x^2 \sqrt{x^2 \pm a^2} \, dx = \frac{x}{4}(x^2 \pm a^2)^{3/2} \mp \frac{1}{8} a^2 x \sqrt{x^2 \pm a^2} - \frac{1}{8} a^4 \ln |x + \sqrt{x^2 \pm a^2}|.$

69. $\int \frac{\sqrt{x^2 + a^2}}{x} \, dx = \sqrt{x^2 + a^2} - a \ln \left| \frac{a + \sqrt{x^2 + a^2}}{x} \right|.$

70. $\int \frac{\sqrt{x^2 - a^2}}{x} \, dx = \sqrt{x^2 - a^2} - a \, \text{Sec}^{-1} \frac{x}{a} \quad (a > 0).$

71. $\int \frac{\sqrt{x^2 \pm a^2}}{x^2} \, dx = -\frac{\sqrt{x^2 \pm a^2}}{x} + \ln |x + \sqrt{x^2 \pm a^2}|.$

72. $\int (x^2 \pm a^2)^{3/2} \, dx = \frac{x}{4}(x^2 \pm a^2)^{3/2} \pm \frac{3}{8} a^2 x \sqrt{x^2 \pm a^2} + \frac{3}{8} a^4 \ln |x + \sqrt{x^2 \pm a^2}|.$

73. $\int x(x^2 \pm a^2)^{3/2} \, dx = \frac{1}{5}(x^2 \pm a^2)^{5/2}.$

74. $\int \frac{(x^2 \pm a^2)^{3/2}}{x} \, dx = \frac{1}{3}(x^2 \pm a^2)^{3/2} \pm a^2 \int \frac{\sqrt{x^2 \pm a^2}}{x} \, dx \quad \text{(see 69 or 70)}.$

75. $\int x^n (x^2 \pm a^2)^{3/2} \, dx = \frac{1}{n+1} x^{n+1}(x^2 \pm a^2)^{3/2} - \frac{3}{n+1} \int x^{n+2}\sqrt{x^2 \pm a^2} \, dx$

$$(n \neq -1).$$

76. $\int x^n \sqrt{x^2 \pm a^2} \, dx = \frac{1}{n+2} x^{n-1}(x^2 \pm a^2)^{3/2} \mp \frac{a^2(n-1)}{n+2} \int x^{n-2}\sqrt{x^2 \pm a^2} \, dx$

$$(n \neq -2).$$

INTEGRALS CONTAINING $ax^2 + bx + c$

77. $\int \frac{1}{ax^2 + bx + c} \, dx = \frac{2}{\sqrt{4ac - b^2}} \, \text{Tan}^{-1} \frac{2ax + b}{\sqrt{4ac - b^2}} \quad (b^2 - 4ac < 0)$

$$= \frac{1}{\sqrt{b^2 - 4ac}} \ln \left| \frac{2ax + b - \sqrt{b^2 - 4ac}}{2ax + b + \sqrt{b^2 - 4ac}} \right| \quad (b^2 - 4ac > 0).$$

78. $\int \frac{1}{(ax^2 + bx + c)^2} \, dx = \frac{2ax + b}{(4ac - b^2)(ax^2 + bx + c)} + \frac{2a}{4ac - b^2} \int \frac{1}{ax^2 + bx + c} \, dx$

$$\text{(see 77).}$$

TRIGONOMETRIC INTEGRALS

79. $\int \sin x \, dx = -\cos x.$

80. $\int \cos x \, dx = \sin x.$

81. $\int \sec^2 x \, dx = \tan x.$

82. $\int \csc^2 x \, dx = -\cot x.$

83. $\int \sec x \tan x \, dx = \sec x.$

84. $\int \csc x \cot x \, dx = -\csc x.$

85. $\int \sin^2 x \, dx = -\frac{1}{2} \sin x \cos x + \frac{x}{2}.$

86. $\int \cos^2 x \, dx = \frac{1}{2} \sin x \cos x + \frac{x}{2}.$

87. $\int \sin^n x \, dx = -\frac{1}{n} \sin^{n-1} x \cos x + \frac{n-1}{n} \int \sin^{n-2} x \, dx \quad (n \geq 2).$

88. $\int \cos^n x \, dx = \frac{1}{n} \cos^{n-1} x \sin x + \frac{n-1}{n} \int \cos^{n-2} x \, dx \quad (n \geq 2).$

89. $\int \sin x \cos x \, dx = \frac{1}{2} \sin^2 x \text{ or } -\frac{1}{2} \cos^2 x.$

90. $\int \sin^2 x \cos^2 x \, dx = -\frac{1}{4} \sin x \cos^3 x + \frac{1}{8} \sin x \cos x + \frac{x}{8}.$

91. $\int \sin^m x \cos^n x \, dx = \frac{1}{m+n} \left\{ -\sin^{m-1} x \cos^{n+1} x + (m-1) \int \sin^{m-2} x \cos^n x \, dx \right\}$

$$= \frac{1}{m+n} \left\{ \sin^{m+1} x \cos^{n-1} x + (n-1) \int \sin^m x \cos^{n-2} x \, dx \right\}.$$

92. $\int x \sin x \, dx = \sin x - x \cos x.$

93. $\int x \cos x \, dx = \cos x + x \sin x.$

94. $\int x^n \sin x \, dx = -x^n \cos x + nx^{n-1} \sin x - n(n-1) \int x^{n-2} \sin x \, dx.$

95. $\int x^n \cos x \, dx = x^n \sin x + nx^{n-1} \sin x - n(n-1) \int x^{n-2} \cos x \, dx.$

96. $\int x \sin^2 x \, dx = \frac{x^2}{4} - \frac{x \sin 2x}{4} - \frac{\cos 2x}{8}.$

97. $\int x^2 \sin^2 x \, dx = \frac{x^3}{6} - \left(\frac{x^2}{4} - \frac{1}{8}\right) \sin 2x - \frac{x \cos 2x}{4}.$

98. $\int x \cos^2 x \, dx = \frac{x^2}{4} + \frac{x \sin 2x}{4} + \frac{\cos 2x}{8}.$

99. $\int x^2 \cos^2 x \, dx = \frac{x^3}{6} + \left(\frac{x^2}{4} - \frac{1}{8}\right) \sin 2x + \frac{x \cos 2x}{4}.$

100. $\int \tan x \, dx = -\ln|\cos x|.$

101. $\int \tan^2 x \, dx = \tan x - x.$

102. $\int \tan^n x \, dx = \dfrac{1}{n-1} \tan^{n-1} x - \int \tan^{n-2} x \, dx \quad (n \geq 2).$

103. $\int \cot x \, dx = \ln|\sin x|.$

104. $\int \cot^2 x \, dx = -\cot x - x.$

105. $\int \cot^n x \, dx = -\dfrac{1}{n-1} \cot^{n-1} x - \int \cot^{n-2} x \, dx \quad (n \geq 2).$

106. $\int \sec x \, dx = \ln|\sec x + \tan x|.$

107. $\int \sec^n x \, dx = \dfrac{1}{n-1}\left\{\sec^{n-2} x \tan x + (n-2)\int \sec^{n-2} x \, dx\right\} \quad (n \geq 2).$

108. $\int \operatorname{cosec} x \, dx = \ln|\operatorname{cosec} x - \cot x|.$

109. $\int \operatorname{cosec}^n x \, dx = \dfrac{1}{n-1}\left\{-\operatorname{cosec}^{n-2} x \cot x + (n-2)\int \operatorname{cosec}^{n-2} x \, dx\right\} \quad (n \geq 2).$

INTEGRALS CONTAINING EXPONENTIALS AND LOGARITHMS

110. $\int e^{ax} \, dx = \dfrac{1}{a} e^{ax}.$

111. $\int a^x \, dx = \dfrac{1}{\ln a} a^x.$

112. $\int x e^{ax} \, dx = \dfrac{1}{a^2}(ax - 1)e^{ax}.$

113. $\int x^n e^{ax} \, dx = \dfrac{1}{a} x^n e^{ax} - \dfrac{n}{a} \int x^{n-1} e^{ax} \, dx.$

114. $\int \dfrac{1}{b + ce^{ax}} \, dx = \dfrac{1}{ab}[ax - \ln(b + ce^{ax})] \quad (ab \neq 0).$

115. $\int \dfrac{1}{be^{ax} + ce^{-ax}} \, dx = \dfrac{1}{a\sqrt{bc}} \operatorname{Tan}^{-1}\left(e^{ax}\sqrt{\dfrac{b}{c}}\right) \quad (bc > 0, \, a \neq 0).$

116. $\int e^{ax} \sin bx \, dx = \dfrac{1}{a^2 + b^2} e^{ax}(a \sin bx - b \cos bx).$

117. $\int e^{ax} \cos bx \, dx = \dfrac{1}{a^2 + b^2} e^{ax}(a \cos bx + b \sin bx).$

118. $\displaystyle\int e^{ax} \sin^n bx\, dx = \frac{1}{a^2 + n^2 b^2}\Big\{e^{ax} \sin^{n-1} bx\, (a \sin bx - nb \cos bx)$

$$+ n(n-1)b^2 \int e^{ax} \sin^{n-2} bx\, dx\Big\}.$$

119. $\displaystyle\int e^{ax} \cos^n bx\, dx = \frac{1}{a^2 + n^2 b^2}\Big\{e^{ax} \cos^{n-1} bx\, (a \cos bx + nb \sin bx)$

$$+ n(n-1)b^2 \int e^{ax} \cos^{n-2} bx\, dx\Big\}.$$

120. $\displaystyle\int \ln|x|\, dx = x \ln|x| - x.$

121. $\displaystyle\int x \ln|x|\, dx = \tfrac{1}{2}x^2 \ln|x| - \tfrac{1}{4}x^2.$

122. $\displaystyle\int x^n \ln|x|\, dx = \frac{1}{n+1}x^{n+1}\Big[\ln|x| - \frac{1}{n+1}\Big] \quad (n \ne -1).$

123. $\displaystyle\int \frac{\ln|x|}{x}\, dx = \ln|\ln|x||.$

124. $\displaystyle\int \ln^n|x|\, dx = x \ln^n|x| - n \int \ln^{n-1}|x|\, dx.$

125. $\displaystyle\int x^m \ln^n|x|\, dx = \frac{1}{m+1}\Big\{x^{m+1} \ln^n|x| - n \int x^m \ln^{n-1}|x|\, dx\Big\} \quad (m \ne -1).$

126. $\displaystyle\int \frac{\ln^n|x|}{x}\, dx = \frac{1}{n+1} \ln^{n+1}|x|.$

INTEGRALS INVOLVING INVERSE TRIGONOMETRIC FUNCTIONS

127. $\displaystyle\int \mathrm{Sin}^{-1} x\, dx = x\, \mathrm{Sin}^{-1} x + \sqrt{1 - x^2}.$

128. $\displaystyle\int \mathrm{Cos}^{-1} x\, dx = x\, \mathrm{Cos}^{-1} x - \sqrt{1 - x^2}.$

129. $\displaystyle\int x^n \mathrm{Sin}^{-1} x\, dx = \frac{1}{n+1}\Big\{x^{n+1} \mathrm{Sin}^{-1} x - \int \frac{x^{n+1}}{\sqrt{1 - x^2}}\, dx\Big\} \quad (n \ne -1).$

130. $\displaystyle\int x^n \mathrm{Cos}^{-1} x\, dx = \frac{1}{n+1}\Big\{x^{n+1} \mathrm{Cos}^{-1} x + \int \frac{x^{n+1}}{\sqrt{1 - x^2}}\, dx\Big\} \quad (n \ne -1).$

131. $\displaystyle\int \mathrm{Tan}^{-1} x\, dx = x\, \mathrm{Tan}^{-1} x - \tfrac{1}{2} \ln(x^2 + 1).$

132. $\displaystyle\int x^n \mathrm{Tan}^{-1} x\, dx = \frac{1}{n+1}\Big\{x^{n+1} \mathrm{Tan}^{-1} x - \int \frac{x^{n+1}}{x^2 + 1}\, dx\Big\}.$

133. $\displaystyle\int \mathrm{Sec}^{-1} x\, dx = x\, \mathrm{Sec}^{-1} x - \ln|x + \sqrt{x^2 - 1}|.$

134. $\displaystyle\int \frac{1}{x(ax^n + b)}\, dx = \frac{1}{nb} \ln \left| \frac{x^n}{ax^n + b} \right| \quad (n \neq 0,\, b \neq 0).$

135. $\displaystyle\int \frac{1}{x\sqrt{ax^n + b}}\, dx = \frac{1}{n\sqrt{b}} \ln \left| \frac{\sqrt{ax^n + b} - \sqrt{b}}{\sqrt{ax^n + b} + \sqrt{b}} \right| \quad (b > 0)$

$\displaystyle\qquad\qquad\qquad\quad = \frac{2}{n\sqrt{-b}} \operatorname{Sec}^{-1} \sqrt{-\frac{ax^n}{b}} \quad (b < 0).$

136. $\displaystyle\int \sqrt{\frac{x + a}{x + b}}\, dx = \sqrt{x + b}\,\sqrt{x + a} + (a - b) \ln |\sqrt{x + b} + \sqrt{x + a}|.$

137. $\displaystyle\int \sqrt{\frac{a + x}{a - x}}\, dx = a \operatorname{Sin}^{-1} \frac{x}{a} - \sqrt{a^2 - x^2}.$

Numerical Tables

A.3.1 Natural Logarithms

	0.00	0.01	0.02	0.03	0.04	0.05	0.06	0.07	0.08	0.09
1.0	0.0000	0.0100	0.0198	0.0296	0.0392	0.0488	0.0583	0.0677	0.0770	0.0862
1.1	0.0953	0.1044	0.1133	0.1222	0.1310	0.1398	0.1484	0.1570	0.1655	0.1740
1.2	0.1823	0.1906	0.1989	0.2070	0.2151	0.2231	0.2311	0.2390	0.2469	0.2546
1.3	0.2624	0.2700	0.2776	0.2852	0.2927	0.3001	0.3075	0.3148	0.3221	0.3293
1.4	0.3365	0.3436	0.3507	0.3577	0.3646	0.3716	0.3784	0.3853	0.3920	0.3988
1.5	0.4055	0.4121	0.4187	0.4253	0.4318	0.4383	0.4447	0.4511	0.4574	0.4637
1.6	0.4700	0.4762	0.4824	0.4886	0.4947	0.5008	0.5068	0.5128	0.5188	0.5247
1.7	0.5306	0.5365	0.5423	0.5481	0.5539	0.5596	0.5653	0.5710	0.5766	0.5822
1.8	0.5878	0.5933	0.5988	0.6043	0.6098	0.6152	0.6206	0.6259	0.6313	0.6366
1.9	0.6419	0.6471	0.6523	0.6575	0.6627	0.6678	0.6729	0.6780	0.6831	0.6881
2.0	0.6931	0.6981	0.7031	0.7080	0.7130	0.7178	0.7227	0.7275	0.7324	0.7372
2.1	0.7419	0.7467	0.7514	0.7561	0.7608	0.7655	0.7701	0.7747	0.7793	0.7839
2.2	0.7885	0.7930	0.7975	0.8020	0.8065	0.8109	0.8154	0.8198	0.8242	0.8286
2.3	0.8329	0.8372	0.8416	0.8459	0.8502	0.8544	0.8587	0.8629	0.8671	0.8713
2.4	0.8755	0.8796	0.8838	0.8879	0.8920	0.8961	0.9002	0.9042	0.9083	0.9123
2.5	0.9163	0.9203	0.9243	0.9282	0.9322	0.9361	0.9400	0.9439	0.9478	0.9517
2.6	0.9555	0.9594	0.9632	0.9670	0.9708	0.9746	0.9783	0.9821	0.9858	0.9895
2.7	0.9933	0.9969	1.0006	1.0043	1.0080	1.0116	1.0152	1.0188	1.0225	1.0260
2.8	1.0296	1.0332	1.0367	1.0403	1.0438	1.0473	1.0508	1.0543	1.0578	1.0613
2.9	1.0647	1.0682	1.0716	1.0750	1.0784	1.0818	1.0852	1.0886	1.0919	1.0953
3.0	1.0986	1.1019	1.1053	1.1086	1.1119	1.1151	1.1184	1.1217	1.1249	1.1282
3.1	1.1314	1.1346	1.1378	1.1410	1.1442	1.1474	1.1506	1.1537	1.1569	1.1600
3.2	1.1632	1.1663	1.1694	1.1725	1.1756	1.1787	1.1817	1.1848	1.1878	1.1909
3.3	1.1939	1.1970	1.2000	1.2030	1.2060	1.2090	1.2119	1.2149	1.2179	1.2208
3.4	1.2238	1.2267	1.2296	1.2326	1.2355	1.2384	1.2413	1.2442	1.2470	1.2499
3.5	1.2528	1.2556	1.2585	1.2613	1.2641	1.2669	1.2698	1.2726	1.2754	1.2782
3.6	1.2809	1.2837	1.2865	1.2892	1.2920	1.2947	1.2975	1.3002	1.3029	1.3056
3.7	1.3083	1.3110	1.3137	1.3164	1.3191	1.3218	1.3244	1.3271	1.3297	1.3324
3.8	1.3350	1.3376	1.3403	1.3429	1.3455	1.3481	1.3507	1.3533	1.3558	1.3584
3.9	1.3610	1.3635	1.3661	1.3686	1.3712	1.3737	1.3762	1.3788	1.3813	1.3838
4.0	1.3863	1.3888	1.3913	1.3938	1.3962	1.3987	1.4012	1.4036	1.4061	1.4085
4.1	1.4110	1.4134	1.4159	1.4183	1.4207	1.4231	1.4255	1.4279	1.4303	1.4327
4.2	1.4351	1.4375	1.4398	1.4422	1.4446	1.4469	1.4493	1.4516	1.4540	1.4563
4.3	1.4586	1.4609	1.4633	1.4656	1.4679	1.4702	1.4725	1.4748	1.4770	1.4793
4.4	1.4816	1.4839	1.4861	1.4884	1.4907	1.4929	1.4952	1.4974	1.4996	1.5019
4.5	1.5041	1.5063	1.5085	1.5107	1.5129	1.5151	1.5173	1.5195	1.5217	1.5239
4.6	1.5261	1.5282	1.5304	1.5326	1.5347	1.5369	1.5390	1.5412	1.5433	1.5454
4.7	1.5476	1.5497	1.5518	1.5539	1.5560	1.5581	1.5602	1.5623	1.5644	1.5665
4.8	1.5686	1.5707	1.5728	1.5748	1.5769	1.5790	1.5810	1.5831	1.5851	1.5872
4.9	1.5892	1.5913	1.5933	1.5953	1.5974	1.5994	1.6014	1.6034	1.6054	1.6074
5.0	1.6094	1.6114	1.6134	1.6154	1.6174	1.6194	1.6214	1.6233	1.6253	1.6273
5.1	1.6292	1.6312	1.6332	1.6351	1.6371	1.6390	1.6409	1.6429	1.6448	1.6467
5.2	1.6487	1.6506	1.6525	1.6544	1.6563	1.6582	1.6601	1.6620	1.6639	1.6658
5.3	1.6677	1.6696	1.6715	1.6734	1.6752	1.6771	1.6790	1.6808	1.6827	1.6845
5.4	1.6864	1.6882	1.6901	1.6919	1.6938	1.6956	1.6974	1.6993	1.7011	1.7029

$$\ln (N \cdot 10^m) = \ln N + m \ln 10, \quad \ln 10 = 2.3026$$

A.3.1 Natural Logarithms (*continued*)

	0.00	0.01	0.02	0.03	0.04	0.05	0.06	0.07	0.08	0.09
5.5	1.7047	1.7066	1.7084	1.7102	1.7120	1.7138	1.7156	1.7174	1.7192	1.7210
5.6	1.7228	1.7246	1.7263	1.7281	1.7299	1.7317	1.7334	1.7352	1.7370	1.7387
5.7	1.7405	1.7422	1.7440	1.7457	1.7475	1.7492	1.7509	1.7527	1.7544	1.7561
5.8	1.7579	1.7596	1.7613	1.7630	1.7647	1.7664	1.7682	1.7699	1.7716	1.7733
5.9	1.7750	1.7766	1.7783	1.7800	1.7817	1.7834	1.7851	1.7867	1.7884	1.7901
6.0	1.7918	1.7934	1.7951	1.7967	1.7984	1.8001	1.8017	1.8034	1.8050	1.8066
6.1	1.8083	1.8099	1.8116	1.8132	1.8148	1.8165	1.8181	1.8197	1.8213	1.8229
6.2	1.8245	1.8262	1.8278	1.8294	1.8310	1.8326	1.8342	1.8358	1.8374	1.8390
6.3	1.8406	1.8421	1.8437	1.8453	1.8469	1.8485	1.8500	1.8516	1.8532	1.8547
6.4	1.8563	1.8579	1.8594	1.8610	1.8625	1.8641	1.8656	1.8672	1.8687	1.8703
6.5	1.8718	1.8733	1.8749	1.8764	1.8779	1.8795	1.8810	1.8825	1.8840	1.8856
6.6	1.8871	1.8886	1.8901	1.8916	1.8931	1.8946	1.8961	1.8976	1.8991	1.9006
6.7	1.9021	1.9036	1.9051	1.9066	1.9081	1.9095	1.9110	1.9125	1.9140	1.9155
6.8	1.9169	1.9184	1.9199	1.9213	1.9228	1.9242	1.9257	1.9272	1.9286	1.9301
6.9	1.9315	1.9330	1.9344	1.9359	1.9373	1.9387	1.9402	1.9416	1.9430	1.9445
7.0	1.9459	1.9473	1.9488	1.9502	1.9516	1.9530	1.9544	1.9559	1.9573	1.9587
7.1	1.9601	1.9615	1.9629	1.9643	1.9657	1.9671	1.9685	1.9699	1.9713	1.9727
7.2	1.9741	1.9755	1.9769	1.9782	1.9796	1.9810	1.9824	1.9838	1.9851	1.9865
7.3	1.9879	1.9892	1.9906	1.9920	1.9933	1.9947	1.9961	1.9974	1.9988	2.0001
7.4	2.0015	2.0028	2.0042	2.0055	2.0069	2.0082	2.0096	2.0109	2.0122	2.0136
7.5	2.0149	2.0162	2.0176	2.0189	2.0202	2.0215	2.0229	2.0242	2.0255	2.0268
7.6	2.0282	2.0295	2.0308	2.0321	2.0334	2.0347	2.0360	2.0373	2.0386	2.0399
7.7	2.0412	2.0425	2.0438	2.0451	2.0464	2.0477	2.0490	2.0503	2.0516	2.0528
7.8	2.0541	2.0554	2.0567	2.0580	2.0592	2.0605	2.0618	2.0631	2.0643	2.0656
7.9	2.0669	2.0681	2.0694	2.0707	2.0719	2.0732	2.0744	2.0757	2.0769	2.0782
8.0	2.0794	2.0807	2.0819.	2.0832	2.0844	2.0857	2.0869	2.0882	2.0894	2.0906
8.1	2.0919	2.0931	2.0943	2.0956	2.0968	2.0980	2.0992	2.1005	2.1017	2.1029
8.2	2.1041	2.1054	2.1066	2.1078	2.1090	2.1102	2.1114	2.1126	2.1138	2.1150
8.3	2.1163	2.1175	2.1187	2.1190	2.1211	2.1223	2.1235	2.1247	2.1258	2.1270
8.4	2.1282	2.1294	2.1306	2.1318	2.1330	2.1342	2.1353	2.1365	2.1377	2.1389
8.5	2.1401	2.1412	2.1424	2.1436	2.1448	2.1459	2.1471	2.1483	2.1494	2.1506
8.6	2.1518	2.1529	2.1541	2.1552	2.1564	2.1576	2.1587	2.1599	2.1610	2.1622
8.7	2.1633	2.1645	2.1656	2.1668	2.1679	2.1691	2.1702	2.1713	2.1725	2.1736
8.8	2.1748	2.1759	2.1770	2.1782	2.1793	2.1804	2.1815	2.1827	2.1838	2.1849
8.9	2.1861	2.1872	2.1883	2.1894	2.1905	2.1917	2.1928	2.1939	2.1950	2.1961
9.0	2.1972	2.1983	2.1994	2.2006	2.2017	2.2028	2.2039	2.2050	2.2061	2.2072
9.1	2.2083	2.2094	2.2105	2.2116	2.2127	2.2138	2.2148	2.2159	2.2170	2.2181
9.2	2.2192	2.2203	2.2214	2.2225	2.2235	2.2246	2.2257	2.2268	2.2279	2.2289
9.3	2.2300	2.2311	2.2322	2.2332	2.2343	2.2354	2.2364	2.2375	2.2386	2.2396
9.4	2.2407	2.2418	2.2428	2.2439	2.2450	2.2460	2.2471	2.2481	2.2492	2.2502
9.5	2.2513	2.2523	2.2534	2.2544	2.2555	2.2565	2.2576	2.2586	2.2597	2.2607
9.6	2.2618	2.2628	2.2638	2.2649	2.2659	2.2670	2.2680	2.2690	2.2701	2.2711
9.7	2.2721	2.2732	2.2742	2.2752	2.2762	2.2773	2.2783	2.2793	2.2803	2.2814
9.8	2.2824	2.2834	2.2844	2.2854	2.2865	2.2875	2.2885	2.2895	2.2905	2.2915
9.9	2.2925	2.2935	2.2946	2.2956	2.2966	2.2976	2.2986	2.2996	2.3006	2.3016

COMMON LOGARITHMS

A.3.2

n	0	1	2	3	4	5	6	7	8	9
1.0	.0000	.0043	.0086	.0128	.0170	.0212	.0253	.0294	.0334	.0374
1.1	.0414	.0453	.0492	.0531	.0569	.0607	.0645	.0682	.0719	.0755
1.2	.0792	.0828	.0864	.0899	.0934	.0969	.1004	.1038	.1072	.1106
1.3	.1139	.1173	.1206	.1239	.1271	.1303	.1335	.1367	.1399	.1430
1.4	.1461	.1492	.1523	.1553	.1584	.1614	.1644	.1673	.1703	.1732
1.5	.1761	.1790	.1818	.1847	.1875	.1903	.1931	.1959	.1987	.2014
1.6	.2041	.2068	.2095	.2122	.2148	.2175	.2201	.2227	.2253	.2279
1.7	.2304	.2330	.2355	.2380	.2405	.2430	.2455	.2480	.2504	.2529
1.8	.2553	.2577	.2601	.2625	.2648	.2672	.2695	.2718	.2742	.2765
1.9	.2788	.2810	.2833	.2856	.2878	.2900	.2923	.2945	.2967	.2989
2.0	.3010	.3032	.3054	.3075	.3096	.3118	.3139	.3160	.3181	.3201
2.1	.3222	.3243	.3263	.3284	.3304	.3324	.3345	.3365	.3385	.3404
2.2	.3424	.3444	.3464	.3483	.3502	.3522	.3541	.3560	.3579	.5398
2.3	.3617	.3636	.3655	.3674	.3692	.3711	.3729	.3747	.3766	.3784
2.4	.3802	.3820	.3838	.3856	.3774	.3892	.3909	.3927	.3945	.3962
2.5	.3979	.3997	.4014	.4031	.4048	.4065	.6082	.4099	.4116	.4133
2.6	.4150	.4166	.4183	.4200	.4216	.4232	.4249	.4265	.4281	.4298
2.7	.4314	.4330	.4346	.4362	.4378	.4393	.4409	.4425	.4440	.4456
2.8	.4472	.4487	.4502	.4518	.4533	.4548	.4564	.4579	.4594	.4609
2.9	.4624	.4639	.4654	.4669	.4683	.4698	.4713	.4728	.4742	.4757
3.0	.4771	.4786	.4800	.4814	.4829	.4843	.4857	.4871	.4886	.4900
3.1	.4914	.4928	.4942	.4955	.4969	.4983	.4997	.5011	.5024	.5038
3.2	.5051	.5065	.5079	.5092	.5105	.5119	.5132	.5145	.5159	.5172
3.3	.5185	.5198	.5211	.5224	.5237	.5250	.5263	.5276	.5289	.5302
3.4	.5315	.5328	.5340	.5353	.5366	.5378	.5391	.5403	.5416	.5428
3.5	.5441	.5453	.5465	.5478	.5490	.5502	.5514	.5527	.5539	.5551
3.6	.5563	.5575	.5587	.5599	.5611	.5623	.5635	.5647	.5658	.5670
3.7	.5682	.5694	.5705	.5717	.5729	.5740	.5752	.5763	.5775	.5786
3.8	.5798	.5809	.5821	.5832	.5843	.5855	.5866	.5877	.5888	.5899
3.9	.5911	.5922	.5933	.5944	.5955	.5966	.5977	.5988	.5999	.6010
4.0	.6021	.6031	.6042	.6053	.6064	.6075	.6085	.6096	.6107	.6117
4.1	.6128	.6138	.6149	.6160	.6170	.6180	.6191	.6201	.6212	.6222
4.2	.6232	.6243	.6253	.6263	.6274	.6284	.6294	.6304	.6314	.6325
4.3	.6335	.6345	.6355	.6365	.6375	.6385	.6395	.6405	.6415	.6425
4.4	.6435	.6444	.6454	.6464	.6474	.6484	.6493	.6503	.6513	.6522
4.5	.6532	.6542	.6551	.6561	.6571	.6580	.6590	.6599	.6609	.6618
4.6	.6628	.6637	.6646	.6656	.6665	.6675	.6684	.6693	.6702	.6712
4.7	.6721	.6730	.6739	.6749	.6758	.6767	.6776	.6785	.6794	.6803
4.8	.6812	.6821	.6830	.6839	.6848	.6857	.6866	.6875	.6884	.6893
4.9	.6902	.6911	.6920	.6928	.6937	.6946	.6955	.6964	.6972	.6981
5.0	.6990	.6998	.7007	.7016	.7024	.7033	.7042	.7050	.7059	.7067
5.1	.7076	.7084	.7093	.7101	.7110	.7118	.7126	.7135	.7143	.7152
5.2	.7160	.7168	.7177	.7185	.7193	.7202	.7210	.7218	.7226	.7235
5.3	.7243	.7251	.7259	.7267	.7275	.7284	.7292	.7300	.7308	.7316
5.4	.7324	.7332	.7340	.7348	.7356	.7364	.7372	.7380	.7388	.7396

COMMON LOGARITHMS (continued)

n	0	1	2	3	4	5	6	7	8	9
5.5	.7404	.7412	.7419	.7427	.7435	.7443	.7451	.7459	.7466	.7474
5.6	.7482	.7490	.7497	.7505	.7513	.7520	.7528	.7536	.7543	.7551
5.7	.7559	.7566	.7574	.7582	.7589	.7597	.7604	.7612	.7619	.7627
5.8	.7634	.7642	.7649	.7657	.7664	.7672	.7679	.7686	.7694	.7701
5.9	.7709	.7716	.7723	.7731	.7738	.7745	.7752	.7760	.7767	.7774
6.0	.7782	.7789	.7796	.7803	.7810	.7818	.7825	.7832	.7839	.7846
6.1	.7853	.7860	.7868	.7875	.7882	.7889	.7896	.7903	.7910	.7917
6.2	.7924	.7931	.7938	.7945	.7952	.7959	.7966	.7973	.7980	.7987
6.3	.7993	.8000	.8007	.8014	.8021	.8028	.8035	.8041	.8048	.8055
6.4	.8062	.8069	.8075	.8082	.8089	.8096	.8102	.8109	.8116	.8122
6.5	.8129	.8136	.8142	.8149	.8156	.8162	.8169	.8176	.8182	.8189
6.6	.8195	.8202	.8209	.8215	.8222	.8228	.8235	.8241	.8248	.8254
6.7	.8261	.8267	.8274	.8280	.8287	.8293	.8299	.8306	.8312	.8319
6.8	.8325	.8331	.8338	.8344	.8351	.8357	.8363	.8370	.8376	.8382
6.9	.8388	.8395	.8401	.8407	.8414	.8420	.8426	.8432	.8439	.8445
7.0	.8451	.8457	.8463	.8470	.8476	.8482	.8488	.8494	.8500	.8506
7.1	.8513	.8519	.8525	.8531	.8537	.8543	.8549	.8555	.8561	.8567
7.2	.8573	.8579	.8585	.8591	.8597	.8603	.8609	.8615	.8621	.8627
7.3	.8633	.8639	.8645	.8651	.8657	.8663	.8669	.8675	.8681	.8686
7.4	.8692	.8698	.8704	.8710	.8716	.8722	.8727	.8733	.8739	.8745
7.5	.8751	.8756	.8762	.8768	.8774	.8779	.8785	.8791	.8797	.8802
7.6	.8808	.8814	.8820	.8825	.8831	.8837	.8842	.8848	.8854	.8859
7.7	.8865	.8871	.8876	.8882	.8887	.8893	.8899	.8904	.8910	.8915
7.8	.8921	.8927	.8932	.8938	.8943	.8949	.8954	.8960	.8965	.8971
7.9	.8976	.8982	.8987	.8993	.8998	.9004	.9009	.9015	.9020	.9025
8.0	.9031	.9036	.9042	.9047	.9053	.9058	.9063	.9069	.9074	.9079
8.1	.9085	.9090	.9096	.9101	.9106	.9112	.9117	.9122	.9128	.9133
8.2	.9138	.9143	.9149	.9154	.9159	.9165	.9170	.9175	.9180	.9186
8.3	.9191	.9196	.9201	.9206	.9212	.9217	.9222	.9227	.9232	.9238
8.4	.9243	.9248	.9253	.9258	.9263	.9269	.9274	.9279	.9284	.9289
8.5	.9294	.9299	.9304	.9309	.9315	.9320	.9325	.9330	.9335	.9340
8.6	.9345	.9350	.9355	.9360	.9365	.9370	.9375	.9380	.9385	.9390
8.7	.9395	.9400	.9405	.9410	.9415	.9420	.9425	.9430	.9435	.9440
8.8	.9445	.9450	.9455	.9460	.9465	.9469	.9474	.9479	.9484	.9489
8.9	.9494	.9499	.9504	.9509	.9513	.9518	.9523	.9528	.9533	.9538
9.0	.9542	.9547	.9552	.9557	.9562	.9566	.9571	.9576	.9581	.9586
9.1	.9590	.9595	.9600	.9605	.9609	.9614	.9619	.9624	.9628	.9633
9.2	.9638	.9643	.9647	.9652	.9657	.9661	.9666	.9671	.9675	.9680
9.3	.9685	.9689	.9694	.9699	.9703	.9708	.9713	.9717	.9722	.9727
9.4	.9731	.9736	.9741	.9745	.9750	.9754	.9759	.9763	.9768	.9773
9.5	.9777	.9782	.9786	.9791	.9795	.9800	.9805	.9809	.9814	.9818
9.6	.9823	.9827	.9832	.9836	.9841	.9845	.9850	.9854	.9859	.9863
9.7	.9868	.9872	.9877	.9881	.9886	.9890	.9894	.9899	.9903	.9908
9.8	.9912	.9917	.9921	.9926	.9930	.9934	.9939	.9943	.9948	.9952
9.9	.9956	.9961	.9965	.9969	.9974	.9978	.9983	.9987	.9991	.9996

From: ALGEBRA & TRIGONOMETRY — FLEMING & VARBERG.

(PRENTICE-HALL)

A.3.2 Exponential Function

x	e^x	e^{-x}
0.00	1.0000	1.0000
0.01	1.0101	0.9900
0.02	1.0202	0.9802
0.03	1.0305	0.9704
0.04	1.0408	0.9608
0.05	1.0513	0.9512
0.06	1.0618	0.9418
0.07	1.0725	0.9324
0.08	1.0833	0.9231
0.09	1.0942	0.9139
0.10	1.1052	0.9048
0.11	1.1163	0.8958
0.12	1.1275	0.8869
0.13	1.1388	0.8781
0.14	1.1503	0.9694
0.15	1.1618	0.8607
0.16	1.1735	0.8521
0.17	1.1853	0.8437
0.18	1.1972	0.8353
0.19	1.2092	0.8270
0.20	1.2214	0.8187
0.21	1.2337	0.8106
0.22	1.2461	0.8025
0.23	1.2586	0.7945
0.24	1.2712	0.7866
0.25	1.2840	0.7788
0.26	1.2969	0.7711
0.27	1.3100	0.7634
0.28	1.3231	0.7558
0.29	1.3364	0.7483
0.30	1.3499	0.7408
0.31	1.3634	0.7334
0.32	1.3771	0.7261
0.33	1.3910	0.7189
0.34	1.4049	0.7118
0.35	1.4191	0.7047
0.36	1.4333	0.6977
0.37	1.4477	0.6907
0.38	1.4623	0.6839
0.39	1.4770	0.6771
0.40	1.4918	0.6703
0.41	1.5068	0.6637
0.42	1.5220	0.6570
0.43	1.5373	0.6505
0.44	1.5527	0.6440

x	e^x	e^{-x}
0.45	1.5683	0.6376
0.46	1.5841	0.6313
0.47	1.6000	0.6250
0.48	1.6161	0.6188
0.49	1.6323	0.6126
0.50	1.6487	0.6065
0.51	1.6653	0.6005
0.52	1.6820	0.5945
0.53	1.6989	0.5886
0.54	1.7160	0.5827
0.55	1.7333	0.5769
0.56	1.7507	0.5712
0.57	1.7683	0.5655
0.58	1.7860	0.5599
0.59	1.8040	0.5543
0.60	1.8221	0.5488
0.61	1.8044	0.5434
0.62	1.8589	0.5379
0.63	1.8776	0.5326
0.64	1.8965	0.5273
0.65	1.9155	0.5220
0.66	1.9348	0.5169
0.67	1.9542	0.5117
0.68	1.9739	0.5066
0.69	1.9937	0.5016
0.70	2.0138	0.4966
0.71	2.0340	0.4916
0.72	2.0544	0.4868
0.73	2.0751	0.4819
0.74	2.0959	0.4771
0.75	2.1170	0.4724
0.76	2.1383	0.4677
0.77	2.1598	0.4630
0.78	2.1815	0.4584
0.79	2.2034	0.4538
0.80	2.2255	0.4493
0.81	2.2479	0.4449
0.82	2.2705	0.4404
0.83	2.2933	0.4360
0.84	2.3164	0.4317
0.85	2.3396	0.4274
0.86	2.3632	0.4232
0.87	2.3869	0.4190
0.88	2.4109	0.4148
0.89	2.4351	0.4107

A.3.2 Exponential Function (*continued*)

x	e^x	e^{-x}		x	e^x	e^{-x}
0.90	2.4596	0.4066		2.75	15.643	0.0639
0.91	2.4843	0.4025		2.80	16.445	0.0608
0.92	2.5093	0.3985		2.85	17.288	0.0578
0.93	2.5345	0.3946		2.90	18.174	0.0550
0.94	2.5600	0.3906		2.95	19.106	0.0523
0.95	2.5857	0.3867		3.00	20.086	0.0498
0.96	2.6117	0.3829		3.05	21.115	0.0474
0.97	2.6379	0.3791		3.10	22.198	0.0450
0.98	2.6645	0.3753		3.15	23.336	0.0429
0.99	2.6912	0.3716		3.20	24.533	0.0408
1.00	2.7183	0.3679		3.25	25.790	0.0388
1.05	2.8577	0.3499		3.30	27.113	0.0369
1.10	3.0042	0.3329		3.35	28.503	0.0351
1.15	3.1582	0.3166		3.40	29.964	0.0334
1.20	3.3201	0.3012		3.45	31.500	0.0317
1.25	3.4903	0.2865		3.50	33.115	0.0302
1.30	3.6693	0.2725		3.55	34.813	0.0287
1.35	3.8574	0.2592		3.60	36.598	0.0273
1.40	4.0552	0.2466		3.65	38.475	0.0260
1.45	4.2631	0.2346		3.70	40.447	0.0247
1.50	4.4817	0.2231		3.75	42.521	0.0235
1.55	4.7115	0.2122		3.80	44.701	0.0224
1.60	4.9530	0.2019		3.85	46.993	0.0213
1.65	5.2070	0.1920		3.90	49.402	0.0202
1.70	5.4739	0.1827		3.95	51.935	0.0193
1.75	5.7546	0.1738		4.00	54.598	0.0183
1.80	6.0496	0.1653		4.10	60.340	0.0166
1.85	6.3598	0.1572		4.20	66.686	0.0150
1.90	6.6859	0.1496		4.30	73.700	0.0136
1.95	7.0287	0.1423		4.40	81.451	0.0123
2.00	7.3891	0.1353		4.50	90.017	0.0111
2.05	7.7679	0.1287		4.60	99.484	0.0101
2.10	8.1662	0.1225		4.70	109.95	0.0091
2.15	8.5849	0.1165		4.80	121.51	0.0082
2.20	9.0250	0.1108		4.90	134.29	0.0074
2.25	9.4877	0.1054		5.00	148.41	0.0067
2.30	9.9742	0.1003		5.20	181.27	0.0055
2.35	10.486	0.0954		5.40	221.41	0.0045
2.40	11.023	0.0907		5.60	270.43	0.0037
2.45	11.588	0.0863		5.80	330.30	0.0030
2.50	12.182	0.0821		6.00	403.43	0.0025
2.55	12.807	0.0781		7.00	1096.6	0.0009
2.60	13.464	0.0743		8.00	2981.0	0.0003
2.65	14.154	0.0707		9.00	8103.1	0.0001
2.70	14.880	0.0672		10.00	22026.	0.00005

A.3.4 Trigonometric Ratios

ANGLE	SIN	COS	TAN	ANGLE	SIN	COS	TAN
0°	.0000	1.000	.0000	45°	.7071	.7071	1.0000
1°	.0175	.9998	.0175	46°	.7193	.6947	1.0355
2°	.0349	.9994	.0349	47°	.7314	.6820	1.0724
3°	.0523	.9986	.0524	48°	.7341	.6691	1.1106
4°	.0698	.9976	.0699	49°	.7547	.6561	1.1504
5°	.0872	.9962	.0875	50°	.7660	.6428	1.1918
6°	.1045	.9945	.1051	51°	.7771	.6293	1.2349
7°	.1219	.9925	.1228	52°	.7880	.6157	1.2799
8°	.1392	.9903	.1405	53°	.7986	.6108	1.3270
9°	.1564	.9877	.1584	54°	.8090	.5878	1.3764
10°	.1736	.9848	.1763	55°	.8192	.5736	1.4281
11°	.1908	.9816	.1944	56°	.8290	.5592	1.4826
12°	.2079	.9781	.2126	57°	.8387	.5446	1.5399
13°	.2250	.9744	.2309	58°	.8480	.5299	1.6003
14°	.2419	.9703	.2493	59°	.8572	.5150	1.6643
15°	.2588	.9659	.2679	60°	.8660	.5000	1.7321
16°	.2756	.9613	.2867	61°	.8746	.4848	1.8040
17°	.2924	.9563	.3057	62°	.8829	.4695	1.8807
18°	.3090	.9511	.3249	63°	.8910	.4540	1.9626
19°	.3256	.9455	.3443	64°	.8988	.4384	2.0503
20°	.3420	.9397	.3640	65°	.9063	.4226	2.1445
21°	.3584	.9336	.3839	66°	.9135	.4067	2.2460
22°	.3746	.9272	.4040	67°	.9205	.3907	2.3559
23°	.3907	.9205	.4245	68°	.9272	.3746	2.4751
24°	.0467	.9135	.4452	69°	.9336	.3584	2.6051
25°	.4226	.9063	.4663	70°	.9397	.3420	2.7475
26°	4384	.8988	.4877	71°	.9455	.3256	2.9042
27°	.4540	.8910	.5095	72°	.9511	.3090	3.0777
28°	.4695	.8829	.5317	73°	.9563	.2924	3.2709
29°	.4848	.8746	.5543	74°	.9613	.2756	3.4874
30°	.5000	.8660	.5774	75°	.9659	.2588	3.7321
31°	.5150	.8572	.6009	76°	.9703	.2419	4.0108
32°	.5299	.8480	.6249	77°	.9744	.2250	4.3315
33°	.5446	.8387	.6494	78°	.9781	.2079	4.7046
34°	.5592	.8290	.6745	79°	.9816	.1908	5.1446
35°	.5736	.8192	.7002	80°	.9848	.1736	5.6713
36°	.5878	.8090	.7265	81°	.9877	.1564	6.3138
37°	.6018	.7986	.7536	82°	.9903	.1392	7.1154
38°	.6157	.7880	.7813	83°	.9925	.1219	8.1443
39°	.6293	.7771	.8098	84°	.9945	.1045	9.5144
40°	.6428	.7660	.8391	85°	.9962	.0872	11.4330
41°	.6561	.7547	.8693	86°	.9976	.0698	14.3010
42°	.6691	.7431	.9004	87°	.9986	.0523	19.0810
43°	.6820	.7314	.9325	88°	.9994	.0349	28.6360
44°	.6947	.7193	.9657	89°	.9998	.0175	57.2900
45°	.7071	.7071	1.0000				

From: ELEMENTARY MATH FOR THE TECHNICIAN—EDWIN M. HEMMERLING
Published by *McGRAW HILL*

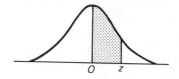

A.3.3 Normal Curve Areas

z	.00	.01	.02	.03	.04	.05	.06	.07	.08	.09
0.0	.0000	.0040	.0080	.0120	.0160	.0199	.0239	.0279	.0319	.0359
0.1	.0398	.0438	.0478	.0517	.0557	.0596	.0636	.0675	.0714	.0753
0.2	.0793	.0832	.0871	.0910	.0948	.0987	.1026	.1064	.1103	.1141
0.3	.1179	.1217	.1255	.1293	.1331	.1368	.1406	.1443	.1480	.1517
0.4	.1554	.1591	.1628	.1664	.1700	.1736	.1772	.1808	.1844	.1879
0.5	.1915	.1950	.1985	.2019	.2054	.2088	.2123	.2157	.2190	.2224
0.6	.2257	.2291	.2324	.2357	.2389	.2422	.2454	.2486	.2517	.2549
0.7	.2580	.2611	.2642	.2673	.2704	.2734	.2764	.2794	.2823	.2852
0.8	.2881	.2910	.2939	.2967	.2995	.3023	.3051	.3078	.3106	.3133
0.9	.3159	.3186	.3212	.3238	.3264	.3289	.3315	.3340	.3365	.3389
1.0	.3413	.3438	.3461	.3485	.3508	.3531	.3554	.3577	.3599	.3621
1.1	.3643	.3665	.3686	.3708	.3729	.3749	.3770	.3790	.3810	.3830
1.2	.3849	.3869	.3888	.3907	.3925	.3944	.3962	.3980	.3997	.4015
1.3	.4032	.4049	.4066	.4082	.4099	.4115	.4131	.4147	.4162	.4177
1.4	.4192	.4207	.4222	.4236	.4251	.4265	.4279	.4292	.4306	.4319
1.5	.4332	.4345	.4357	.4370	.4382	.4394	.4406	.4418	.4429	.4441
1.6	.4452	.4463	.4474	.4484	.4495	.4505	.4515	.4525	.4535	.4545
1.7	.4554	.4564	.4573	.4582	.4591	.4599	.4608	.4616	.4625	.4633
1.8	.4641	.4649	.4656	.4664	.4671	.4678	.4686	.4693	.4699	.4706
1.9	.4713	.4719	.4726	.4732	.4738	.4744	.4750	.4756	.4761	.4767
2.0	.4772	.4778	.4783	.4788	.4793	.4798	.4803	.4808	.4812	.4817
2.1	.4821	.4826	.4830	.4834	.4838	.4842	.4846	.4850	.4854	.4857
2.2	.4861	.4864	.4868	.4871	.4875	.4878	.4881	.4884	.4887	.4890
2.3	.4893	.4896	.4898	.4901	.4904	.4906	.4909	.4911	.4913	.4916
2.4	.4918	.4920	.4922	.4925	.4927	.4929	.4931	.4932	.4934	.4936
2.5	.4938	.4940	.4941	.4943	.4945	.4946	.4948	.4949	.4951	.4952
2.6	.4953	.4955	.4956	.4957	.4959	.4960	.4961	.4962	.4963	.4964
2.7	.4965	.4966	.4967	.4968	.4969	.4970	.4971	.4972	.4973	.4974
2.8	.4974	.4975	.4976	.4977	.4977	.4978	.4979	.4979	.4980	.4981
2.9	.4981	.4982	.4982	.4983	.4984	.4984	.4985	.4985	.4986	.4986
3.0	.4987	.4987	.4987	.4988	.4988	.4989	.4989	.4989	.4990	.4990

*(From J. E. Freund and F. J. Williams, *Elementary Business Statistics,* 2nd ed. Englewood Cliffs, N.J.: Prentice-Hall, Inc., 1972, p. 473.) Reprinted by permission.

Answers to Odd-Numbered Exercises

CHAPTER 1

EXERCISES 1.1

1. $x \leq 1$.   **3.** $x < 1$.   **5.** $x > 1$.
7. $x \geq -5$.   **9.** $-2 \leq x \leq 1$.   **11.** $x < -1$ or $x > 2$.
13. $1 < x < 3$.   **15.** $x \leq -\sqrt{3}$ or $x \geq \sqrt{3}$.

EXERCISES 1.2

1. $\{-1, 0, 1, 2, 3, 4\}$.   **3.** $\{2, 3, 5, 7, 11, 13, 17, 19\}$.   **5.** $\{2, 3\}$.
7. $\{x \mid x$ is an even number, $0 < x < 100\}$
     or $\{x \mid x = 2n, n$ is a natural number, and $1 \leq n \leq 49\}$.
9. $\{x \mid x$ is an odd number, $0 < x < 20\}$
     or $\{x \mid x = 2n + 1, n$ is an integer, and $0 \leq n \leq 9\}$.
11. $\{x \mid x$ is a natural number divisible by 3$\}$
     or $\{x \mid x = 3n, n$ is a natural number$\}$.
13. True.   **15.** True.   **17.** False.
19. True.   **21.** True.
23. $\varnothing, \{a\}, \{b\}, \{c\}, \{a, b\}, \{b, c\}, \{a, c\}, \{a, b, c\}$; $2^3 = 8$.
25. 2^N.

EXERCISES 1.3

1. (a) 3; (b) 27; (c) 15; (d) $\frac{31}{4}$; (e) $5a^2 - 7a + 3$; (f) $5(a + h)^2 - 7(a + h) + 3$;
     (g) $10a + 5h - 7$.
3. (a) -3; (b) 20.
5. $D_f = $ set of all real numbers; $R_f = $ set of all real numbers greater than or equal to 2.
7. $D_f = R_f = \{x \mid x \neq 1\}$. (It is understood that x is a real number.)
9. $D_F = \{x \mid x < 1\}$; $R_f = \{y \mid y > 0\}$.
11. $D_F = \{x \mid -1 \leq x \leq 1\}$; $R_f = \{y \mid 1 \leq y \leq 2\}$.
13. $p(0) = 1000, p(1) = 2000, p(2) = 2600$.
15. (a) 3.9; (b) 4.0; (c) 3.775 ($= 3.8$ rounded off to one decimal place).
17. No.   **19.** Yes.   **21.** $D_f = \{x \mid 1 \leq x \leq 2\}$.
23. $D_f = \{x \mid x \leq -4$ or $x \geq 4\}$.

25. $F(x) = \begin{cases} 79x \text{ if } x < 5 \\ 69x \text{ if } x > 5 \end{cases}$

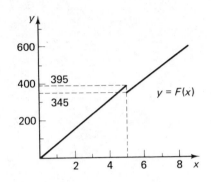

EXERCISES 1.4

1. 3.
7. $y = -3x + 1$.

3. -5.
9. $y = 4x - 11$.

5. 0.
11. $y = -2x + 3$.

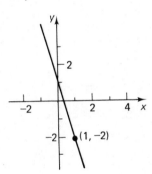

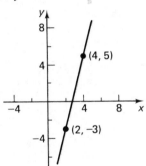

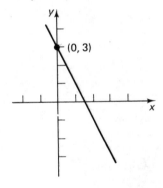

13. Slope $= -\frac{3}{5}$, y-intercept $= 3$.
17. $5x + 8y = 100$.

15. Slope $= -2$, y-intercept $= -6$.
19. Proportion of A to B should be $2 : 1$.

EXERCISES 1.5

1.

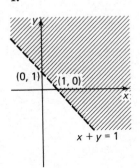

$x + y = 1$

3.

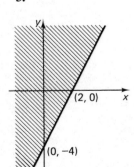

5.

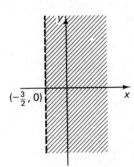

7.   **9.** $5x + 6y \geq 20.$

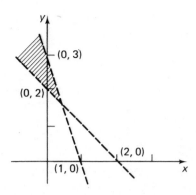

11. Let $x =$ number of beakers of first size, and $y =$ number of beakers of second size. Then $x \geq 300$, $y \geq 400$, and $x + y \leq 1200$.

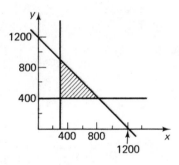

13. If x gm of meat and y gm of soybeans are consumed per day, then $7x + 3y \geq 50$.

15. If the numbers of S and T are denoted respectively by x and y, then $2x + 3y \leq 600$, $3x + y \leq 300$, $3x + 2y \geq 400$ (and $x \geq 0$, $y \geq 0$ also).

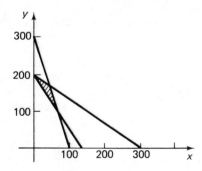

1. $(-\frac{3}{4}, -\frac{1}{8})$.

3. $(-\frac{1}{6}, \frac{37}{12})$.

5. $D_f = \{x \mid -2 \le x \le 2\}$; $R_f = \{y \mid 0 \le y \le 2\}$.

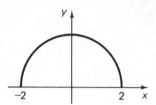

7. $D_f = \{x \mid x \le 3\}$; $R_f = \{y \mid y \le 0\}$.

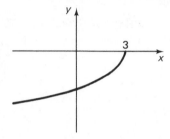

9. $D_f = \{x \mid x \ne 0\}$; $R_f = \{y \mid y \ne 0\}$.

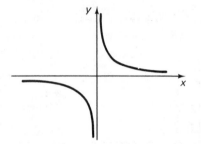

11. $D_f = R_f =$ set of all real numbers.

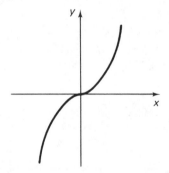

13. $D_f =$ set of all real numbers; $R_f = \{y \mid y \le 2\}$.

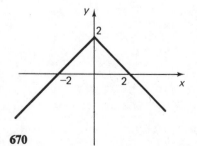

15. D_f = set of all real numbers; $R_f = \{y \mid y \geq 0\}$.

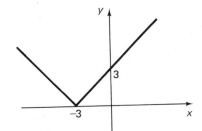

17. $D_f = \{x \mid x \neq 3\}$; $R_f = \{1, -1\}$.

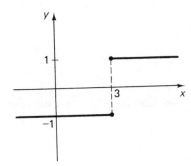

19. (a) Yes; $y = \sqrt{9 - x^2}$;   (b) No;   (c) No;   (d) Yes; $y = -\sqrt{4 - x^2}$.
21. $x^2 + y^2 - 4x - 10y + 20 = 0$.   **23.** $x^2 + y^2 + 6x - 7 = 0$.
25. $x^2 + y^2 - 6x + 6y + 9 = 0$.   **27.** 10,000 yd².

29. $x = 10$.

31. $g(x) = \begin{cases} 200 - x & \text{if } 0 < x \leq 200 \\ 700 - x & \text{if } 200 < x \leq 700 \\ 1000 - x & \text{if } 700 < x \leq 1000. \end{cases}$

EXERCISES 1.7

1. $(f \pm g)(x) = x^2 \pm \dfrac{1}{x - 1}$; $(fg)(x) = \dfrac{x^2}{x - 1}$; $\left(\dfrac{f}{g}\right)(x) = x^2(x - 1)$; $\left(\dfrac{g}{f}\right)(x) = \dfrac{1}{x^2(x - 1)}$.
$D_{f+g} = D_{f-g} = D_{fg} = D_{f/g} = \{x \mid x \neq 1\}$; $D_{g/f} = \{x \mid x \neq 0, 1\}$.

3. $(f \pm g)(x) = \sqrt{x - 1} \pm \dfrac{1}{x + 2}$; $(fg)(x) = \dfrac{\sqrt{x - 1}}{x + 2}$;
$\left(\dfrac{f}{g}\right)(x) = \sqrt{x - 1}(x + 2)$; $\left(\dfrac{g}{f}\right)(x) = \dfrac{1}{\sqrt{x - 1}(x + 2)}$.
$D_{f+g} = D_{f-g} = D_{fg} = D_{f/g} = \{x \mid x \geq 1\}$; $D_{g/f} = \{x \mid x > 1\}$.

5. $(f \pm g)(x) = (x + 1)^2 \pm \dfrac{1}{x^2 - 1}$; $(fg)(x) = \dfrac{x + 1}{x - 1}$;
$\left(\dfrac{f}{g}\right)(x) = (x + 1)^3(x - 1)$; $\left(\dfrac{g}{f}\right)(x) = \dfrac{1}{(x + 1)^3(x - 1)}$.
$D_{f+g} = D_{f-g} = D_{fg} = D_{f/g} = D_{g/f} = \{x \mid x \neq \pm 1\}$.

7. $\sqrt{8}$.   **9.** $\sqrt{3}$.
11. Not defined.   **13.** 0.
15. $f \circ g(x) = |x| + 1; g \circ f(x) = (\sqrt{x} + 1)^2$. $D_{f \circ g} = \{x \mid x \text{ real}\}; D_{g \circ f} = \{x \mid x \geq 0\}$.
17. $f \circ g(x) = 2 + |x - 2|; g \circ f(x) = x$. $D_{f \circ g} = \{x \mid x \text{ real}\}; D_{g \circ f} = \{x \mid x \geq 0\}$.
19. $g(x) = 1 + x^4$.   **21.** $g(x) = x - 1$.
23. $f(x) = x^3, g(x) = x^2 + 1$ is the "simplest" answer.

EXERCISES 1.8

1. 3.   **3.** 0.   **5.** ∞.
7. $\frac{1}{2}$.   **9.** 2.   **11.** 0.
13. ∞.   **15.** 2.

REVIEW EXERCISES FOR CHAPTER 1

1. (a) False; for example, $\sqrt{2}$ is a real number that is not rational; (b) True; (c) True;
  (d) False; a curve is the graph of a function if any vertical line meets the curve in *at most* one point;
  (e) False; in general $f \circ g \neq g \circ f$;
  (f) False; for example, the equation $x^2 + y^2 = 4$ does not express y as a function of x;
  (g) False; the statement is true provided that a and b are not both zero;
  (h) False; if $b < 0$ the graph is the half-plane below the line $ax + by + c = 0$;
  (i) False; D_f is the set of all real numbers;
  (j) False; $(x^2 - 9)/(x - 3) = x + 3$ only for $x \neq 3$;
  (k) False; the domain of f/g may differ from that of $f + g$ and fg;
  (l) False; $-2y > 4x - 6$ is equivalent to $y < -2x + 3$. (Dividing an inequality by a negative number changes the direction of the inequality);
  (m) True.
3. $f \circ g(x) = g \circ f(x) = x^2$. The domain of $f \circ g$ and $g \circ f$ are both the set of all real numbers.
5. (a) $y = 3x + 4$; (b) $y = 3x - 16$.
7. $x^2 + y^2 + 2x - 4y - 11 = 0$.   **9.** $(-\frac{1}{3}, \frac{2}{3})$.
11.   **13.** $1/\sqrt{2}$.

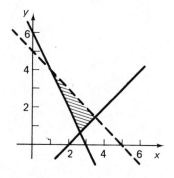

15. $g \circ f(x) = -(1 + x)$. $D_{g \circ f} = \{x \mid x \neq -1\}$.

CHAPTER 2

EXERCISES 2.1

1. 0.40.

3. $g(x)$ is not defined over the whole interval from x to $x + \Delta x$.

5. -50.

7. $\Delta x - \dfrac{2\Delta x}{x(x + \Delta x)}$.

9. -7.

11. 1.

13. 0.1613.

15. $3a^2 + 3ah + h^2 + 1$.

17. (a) 40; (b) 160; (c) 220; (d) 250; (e) $1000 - 240t - 120\Delta t$.

19. 7.

EXERCISES 2.2

1. 25.

3. 4.

5. 0.

7. 4.

9. 1.

11. Limit does not exist.

13. $\frac{1}{4}$.

15. $\frac{1}{8}$.

17. 2.

19. 2.

21. 7.

23. 0.

25. $4x + 5$.

EXERCISES 2.3

1. 0.

3. Limit does not exist.

5. 1.

7. $\frac{1}{6}$.

9. ∞.

11. ∞.

13. Limit does not exist.

15. Limit does not exist.

EXERCISES 2.4

1. Discontinuous at $x = 0$.

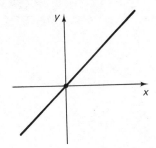

3. Discontinuous at $x = 0$.

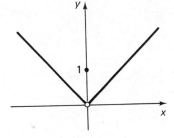

5. Discontinuous at $x = 0$.

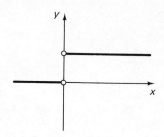

7. Continuous at $x = 1$.

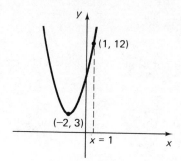

9. Continuous at $x = 2$.

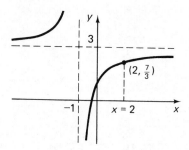

11. Continuous at $x = 2$.

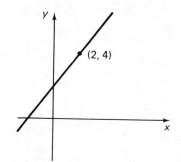

13. $h = 2$.

15. $h = -1$.

EXERCISES 2.5

1. 2.

3. $\dfrac{-1}{(t + 1)^2}$.

5. (a) $-4x$; (b) 3.

7. (a) $\dfrac{1}{2\sqrt{y}}$; (b) $-\dfrac{2}{y^3}$.

9. 10.

11. 1.

13. Slope $= -1$; $y = 3 - x$.

15. Slope $= -2$; $y = 7 - 2x$.

17. $p(0) = 10{,}000$; $p'(t) = 1000 - 600t$.

EXERCISES 2.6

1. $12x^2 - 6x$.

3. $6u - \dfrac{6}{u^3}$.

5. $\dfrac{1}{2\sqrt{y}} - \dfrac{1}{2y^{3/2}}$.

7. $4x - 23$.

9. $-3 + 8t + \dfrac{7}{t^2}$.

11. $3x^2 - \dfrac{3}{x^4}$.

13. $3u^2 - 10u - \dfrac{14}{3u^3}$.

17. (a) $6t^2 - \dfrac{1}{2\sqrt{t}}$; (b) $95\frac{3}{4}$.

19. $5000(2 + t)$; 11,250; 20,000.

21. $\gamma c \rho^{\gamma - 1}$ or $\dfrac{\gamma p}{\rho}$.

EXERCISES 2.7

1. $4x^3 + 3x^2 + 3$.

3. $11 - 42x$.

5. $6x^2 - 14x - 13$.

7. $24x^3 - 33x^2 + 18x - 11$.

9. $-\dfrac{3}{(x-1)^2}$.

11. $\dfrac{t^2 - 10t + 35}{(t-5)^2}$.

13. $\dfrac{-1}{\sqrt{u}\,(\sqrt{u}-1)^2}$.

15. $\dfrac{-2x}{(x^2+1)^2}$.

17. $5 - \dfrac{6t}{(t^2+1)^2}$.

19. $\dfrac{dT}{dt} = c\left(3 + 6t^2 - \dfrac{1}{t^2}\right)$.

21. $\dfrac{3t^2 + 16t - 12}{(t^2+4)^2}$.

EXERCISES 2.8

1. $21(3x+5)^6$.

3. $6x(2x^2+1)^{1/2}$.

5. $\dfrac{-8x}{(x^2+1)^5}$.

7. $\dfrac{t}{\sqrt{t^2+a^2}}$.

9. $2(u^2+1)^2(7u^2+3u+1)$.

11. $-t^2(t^3+1)^{-4/3}$.

13. $\dfrac{(x^2+1)(3x^2+4x-1)}{(x+1)^2}$.

15. $\dfrac{t^3 + 8t}{(t^2+4)^{3/2}}$.

17. $\dfrac{5}{2(t+2)^{3/2}\sqrt{3t+1}}$.

19. $7(x^2+3)^6(x^2+6x-3)(x+3)^{-8}$.

EXERCISES 2.9

1. $\dfrac{dy}{dx} = 15x^4 + 21x^2 - 8x$; $\dfrac{d^2y}{dx^2} = 60x^3 + 42x - 8$; $\dfrac{d^3y}{dx^3} = 180x^2 + 42$; $\dfrac{d^4y}{dx^4} = 360x$;

$\dfrac{d^5y}{dx^5} = 360$; $\dfrac{d^n y}{dx^n} = 0$ for $n \geq 6$.

3. $f'(x) = 3x^2 - 12x + 9$; $f''(x) = 6x - 12$; $f'''(x) = 6$; $f^{(n)}(x) = 0$ for $n \geq 4$.

5. $y'' = \dfrac{2(1 - 3x^2)}{(1+x^2)^3}$.

7. $g^{(iv)}(u) = \dfrac{1944}{(3u+1)^5}$.

9. $\dfrac{2(3x^2-1)}{(x^2+1)^3}$.

11. (a) vel. $= 9 + 32t$; acc. $= 32$; (b) vel. $= 9t^2 + 14t - 5$; acc. $= 18t + 14$.

REVIEW EXERCISES FOR CHAPTER 2

1. (a) True; (b) False; $\dfrac{d}{dx}(uv) = uv' + vu'$; (c) False; $\dfrac{d}{dx}\left(\dfrac{u}{v}\right) = \dfrac{vu' - uv'}{v^2}$;

(d) False; $\dfrac{d}{dx}[f(x)]^n = n[f(x)]^{n-1} \cdot f'(x)$;

(e) False; the derivative of y w.r.t. x represents the instantaneous rate of change of y w.r.t. x;

(f) False; the derivative of $f(y)$ w.r.t. u is $f'(y) \cdot \dfrac{dy}{du}$;

(g) False; a function continuous at a point need not be differentiable at that point. For example, $f(x) = |x|$ is continuous at $x = 0$ but is not differentiable at $x = 0$;

(h) False; if the derivative of a function does not exist at a point, the function may or may not be defined at that point. For example, $f(x) = \sqrt{x}$ is defined at $x = 0$ but is not differentiable at $x = 0$, whereas $f(x) = 1/\sqrt{x}$ is neither defined nor differentiable at $x = 0$;

(i) False; $\lim\limits_{x \to c} f(x)$ may exist without $f(x)$ being continuous at $x = c$. For example, $f(x) = \dfrac{x^2 - 9}{x - 3}$ is not continuous at $x = 3$ because $f(x)$ is not defined at $x = 3$. But $\lim\limits_{x \to 3} f(x)$ exists and is

equal to 6;

(j) False; for example, $f(x) = \dfrac{x^2 - 4}{x - 2}$ is *not* defined at $x = 2$ but $\displaystyle\lim_{x \to 2} f(x)$ exists and is equal to 4;

(k) False; consider $f(x) = \begin{cases} \dfrac{x^2 - 4}{x - 2} & \text{for } x \neq 2 \\ 7 & \text{for } x = 2; \end{cases}$

In this case $\displaystyle\lim_{x \to 2} f(x) = 4 \neq f(2)$.

(l) False; if acceleration is zero then velocity is constant, not necessarily zero.

3. $\dfrac{1}{2\sqrt{a}}$.

5. 2.

7. Discontinuous at $x = 3$.

9. -2.

11. $\dfrac{2x^2 + 4}{\sqrt{x^2 + 4}}$.

13. $6(2x + 1)^2(3x - 1)^3(7x + 1)$.

15. $\frac{1}{6}(x + 1)^{-5/6}$.

17. $2xa^3(x^3 + a^3)^{-5/3}$.

19. $\dfrac{dv}{dx} = \dfrac{VM}{(x + M)^2}$.

CHAPTER 3

EXERCISES 3.1

1. $3\sqrt{2}$.

3. 4.

5. $\sqrt{3}$.

7. 1.62×10^{-3}.

9. 3.0×10^2.

11. 1.2×10^{11}.

13. $40(1 + 20 + 20^2 + 20^3 + 20^4) = 6736840$.

15. $10^4 \times 2^4$; $10^4 \times 2^8$; $10^4 \times 2^{17/3} = 5.08 \times 10^5$; 19.9 hr.

17. 28.98 billion; 46.27 yr after 1976.

19. 25.01 yr after 1970.

21. 23.32 yr after 1970.

EXERCISES 3.2

1. $x = f^{-1}(y) = -\frac{1}{3}(y + 4)$.

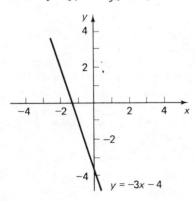

$y = -3x - 4$

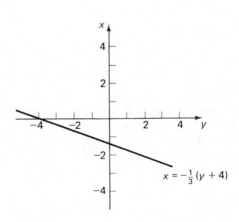

$x = -\frac{1}{3}(y + 4)$

3. $x = f^{-1}(y) = \frac{1}{3}(y^2 + 4)$.

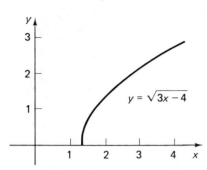

$y = \sqrt{3x - 4}$

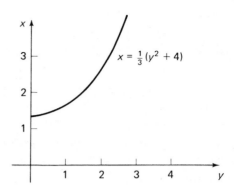

$x = \frac{1}{3}(y^2 + 4)$

5. $x = f^{-1}(y) = y^{1/5}$.

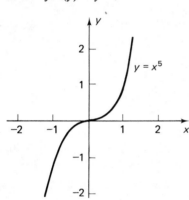

$y = x^5$

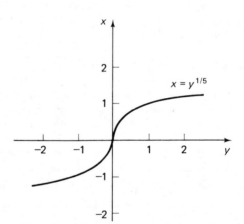

$x = y^{1/5}$

7. $x = f^{-1}(y) = -1 + \sqrt{y}$, $x \geq -1$;
$x = f^{-1}(y) = -1 - \sqrt{y}$, $x \leq -1$.

9. $x = f^{-1}(y) = y^{3/2}$ if $x \geq 0$;
$x = f^{-1}(y) = -y^{3/2}$ if $x \leq 0$.

11. 9.   **13.** 8.   **15.** -3.

17. 100.   **19.** $\dfrac{P}{2}$.   **21.** $\frac{1}{3}$.

23. 3.51.   **25.** $x = -\frac{13}{8}$.   **27.** 1.8702.

EXERCISES 3.3

1. $m = .0083$, $b = -.083$.   **3.** $a = 114$, $b = 0.33$.

5. $a = 131$, $b = 0.3$.   **7.** $y = 10^7 e^{-(.115)t}$ (t in minutes).

EXERCISES 3.4

1. $(x + 1)e^x$.

3. $2xe^{x^2}$.

5. $\left(\dfrac{1}{2\sqrt{x}} - 1\right)e^{\sqrt{x}-x}$.

7. $\dfrac{2}{x}$.

9. $-\dfrac{1}{x(\ln x)^2}$.

11. $\dfrac{1}{\ln(10)}$.

13. $\ln x$.

15. $\dfrac{1 - \ln x}{x^2}$.

17. $\dfrac{1}{x + 2} - \dfrac{x}{x^2 + 1}$.

19. $\dfrac{1}{2x\sqrt{\ln x}}$.

21. 2.1972.

23. 8.0064.

25. $\bar{4}.7811 = -3.2189$.

27. 1.7918.

29. $3pcke^{-kt}(1 - ce^{-kt})^2$.

31. $w = \dfrac{A}{1 + e^{-B(t-c)}}$.

EXERCISES 3.5

1. $y = e^{(.6931)t}$.

3. $y = 5e^{(.0392)t}$.

5. $y = 4e^{(.0198)t}$ billion.

7. 0.66%/yr.

9. 11,178 yr.

11. 6.47 gm.

13. 512 yr.

15. 5×10^6; 4×10^7; 2×10^8.

REVIEW EXERCISES FOR CHAPTER 3

1. (a) False; $a^m \cdot a^n = a^{m+n}$;   (b) False; $a^m/a^n = a^{m-n}$;   (c) True;
   (d) False; $\log m + \log n = \log(mn)$, for $m, n > 0$;
   (e) False; $\log m - \log n = \log(m/n)$ for $m, n > 0$;
   (f) False; $\log(mn) = \log m + \log n$ for $m, n > 0$;
   (g) False; $\log(m/n) = \log m - \log n$, for $m > 0, n > 0$;
   (h) False; $\log(m^n) = n \log m$, for $m > 0$;   (i) True;   (j) False; $\log_a a = 1$, for all $a > 0$;
   (k) False; domain is the set of all real numbers;   (l) False; domain of $\ln x$ is $\{x \mid x > 0\}$;
   (m) False; $\dfrac{d}{dx}[e^{f(x)}] = e^{f(x)} \cdot f'(x)$;   (n) False; $\dfrac{d}{dx}(e^2) = 0$, because e^2 is a constant;
   (o) False; $\dfrac{d}{dx}[\ln f(x)] = \dfrac{f'(x)}{f(x)}$;   (p) False; $\dfrac{d}{dx}(\ln a) = 0$, because $\ln a$ is a constant;   (q) True.

3. $2x$. (Note: $e^{\ln u} = u$.)

5. $\ln 2 + \dfrac{1}{2(x - 1)}$.

7. $\frac{1}{2}x(2 \ln x + 1)$.

9. $1 + \dfrac{2}{t}$.

11. $2t \ln 2$.

13. $e^{(x+e^x)}$.

15. $f'(t) = 0$ when $t = p$. $f(t)$ is maximum at $t = p$.

17. 37,138 yr.

19. 127 min.

21. $p = \dfrac{1}{1 + e^{-(kt+c)}}$; $p \to 1$ as $t \to \infty$.

CHAPTER 4

EXERCISES 4.1

1. (a) $x > 3$;   (b) $x < 3$;   (c) All x;   (d) No x. No point of inflection.
3. (a) $x > 1$ or $x < -1$;   (b) $-1 < x < 1$;   (c) $x > 0$;
   (d) $x < 0$; $x = 0$ is the point of inflection.

5. (a) $x > 1$ or $x < -1$;   (b) $-1 < x < 1$;   (c) $x > 0$;   (d) $x < 0$; No point of inflection.
7. (a) All $x \neq -1$;   (b) No x;   (c) $x < -1$;   (d) $x > -1$; No point of inflection.
9. (a) $x > 0$;   (b) No. x;   (c) No x;
    (d) All $x > 0$; (Note: y is not defined for $x < 0$.) No point of inflection.
11. (a) $x > \dfrac{1}{e}$;   (b) $0 < x < \dfrac{1}{e}$;   (c) $x > 0$;
    (d) No x; No point of inflection. (Note: y is not defined for $x < 0$.)
13. (a) $x < 0$ or $x > 4$;   (b) $0 < x < 4$;   (c) $x > 3$;   (d) $x < 3$; Point of inflection at $x = 3$.
15.   **17.**

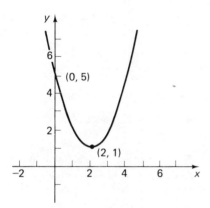

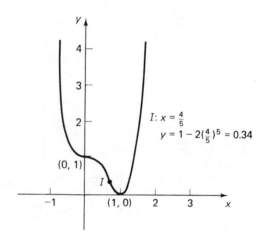

19.   **21.**

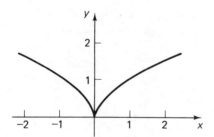

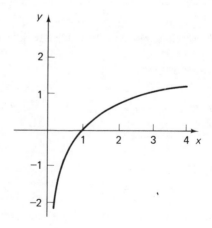

23.

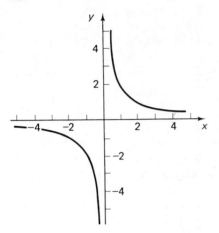

EXERCISES 4.2

1. Minimum at $x = 6$.
3. Minimum at $x = 4$; maximum at $x = 0$.
5. Maximum at $x = 1$; minimum at $x = 2$.
7. Maximum at $x = 4$; minimum at $x = 8$.
9. Maximum at $x = 1$; minimum at $x = 3$; point of inflection at $x = 0$.
11. Maximum at $x = \frac{3}{5}$; minimum at $x = 1$; point of inflection at $x = 0$.
13. Minimum at $x = 0$.
15. Minimum at $x = \dfrac{1}{e}$.
17. Maximum value 5 when $x = -2$; minimum value -22 when $x = 1$.
19. Minimum value $-\dfrac{1}{e}$ occurs at $x = -1$.
23. Absolute minimum -249 occurs at $x = 5$; absolute maximum 75 occurs at $x = -1$.
25. Absolute minimum 1 occurs at $x = 1$; absolute maximum 3 occurs when $x \longrightarrow \infty$.

EXERCISES 4.3

1. 5; 5.
3. 50; 25.
7. Maximum area a^2 occurs when the length of rectangle is twice the width $(= a/\sqrt{2})$.
9. 1250 yd².
11. 6 ft $\times$ 6 ft $\times$ 9 ft.
13. $n = 45$.
15. $y = \dfrac{p}{2}$.
17. $\dfrac{1}{r-s} \ln \dfrac{r}{s}$.
19. $t = 2$.
21. 6,500 at $t = 10$.
23. $x = 1$.
25. $x = \dfrac{1}{k} \ln\left(\dfrac{apk}{c}\right)$.
27. $t = \dfrac{1}{4b}(\sqrt{1 + 16b^2k} - 1)$.
29. Maximum value 6000 when $t = 0$; minimum value 2000 when $t = 20$.

EXERCISES 4.4

1. 1.414.
3. 1.817.
5. 1.316.
7. 0.76 and 5.24.
9. 2.17.
11. -0.70, 1.24, and 3.46.
13. -2.95 and 1.51.
15. 1.56.
17. $t = 10.74$.

EXERCISES 4.5

1. $dy = (2x + 7)\, dx$.

3. $(1 + \ln t)\, dt$.

5. $\dfrac{2z}{1 + z^2}\, dz$.

7. $-\left(\dfrac{x^3 + 3}{x^2}\right) dx$.

9. $\dfrac{z - 1}{z^2}\, e^{1+z}$.

11. 0.12.

13. 0.003.

15. $dy = 0.12;\ \Delta y = 0.1203$.

17. $dy = 0.02;\ \Delta y = 0.01980$.

19. 2.083.

21. 1.9875.

23. $\pm 0.512\pi/\text{cm}^3$.

25. $\tfrac{2}{3}\%$.

EXERCISES 4.6

1. $y = \dfrac{x}{1 - x}$.

3. $y = 2 - x$ and $y = -2 - x$.

5. $y = \dfrac{1}{2x}(-x^2 \pm \sqrt{x^4 + 24x})$.

7. $\dfrac{-x}{y + 1}$.

9. $-\dfrac{x^2}{y^2}$.

11. $\dfrac{4x - y}{x + 2y}$.

13. $-\dfrac{y(2x + y)}{x(2y + x)}$.

15. $\dfrac{x^4 - y}{x - y^4}$.

17. $\dfrac{-y}{e^y + x}$.

19. $-\dfrac{y}{x}$.

21. $\dfrac{-5t}{3x}$.

23. $\dfrac{x - 3y^2}{3x^2 - y}$.

25. $y = \dfrac{1}{3}x + \dfrac{5}{3}$.

27. 0.

29. $-\dfrac{4xy}{(x^2 - y)^3}$ (use $x^3 + y^3 - 3xy = 1$ to simplify the answer).

31. $-\dfrac{(b^2 + a^2t^2)x + (b^2 - a^2)ty + (a^2tk - b^2h)}{(b^2 - a^2)tx + (a^2 + b^2t^2)y - (a^2k + b^2th)}$.

EXERCISES 4.7

1. $y = \dfrac{1}{x^2};\ D = \{t \mid t \neq 0\}$.

3. $y^2 = x^3;\ D = $ all real numbers.

5. $y = 3x - 5$.

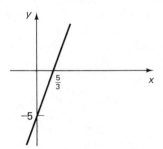

7. $-\tfrac{1}{2}$.

9. $\dfrac{3bt}{2a}$.

11. $\tfrac{2}{3}t^{-1/6}$.

13. $\dfrac{1}{2u}$.

15. $\tfrac{1}{2}e^t$.

17. $y = -2x + 3$.

19. $y = -\frac{2}{3}x - \frac{11}{3}.$ **21.** $y = -\frac{2}{3}x + \frac{5}{3}.$ **23.** $-\frac{1}{2at^3}.$

25. $-\frac{(1 + t^2)^3}{8t^3}.$ **27.** $\frac{dy}{dx} = 2\theta^2; \frac{d^2y}{dx^2} = 4\theta^2.$

29. 10 ft/sec; slope of path $= -\frac{8}{3}$ upon arrival.

31. slope $= \sqrt{3} - 1.6t; t = 5\sqrt{3}/8$ sec.

REVIEW EXERCISES FOR CHAPTER 4

1. (a) False; for example, $f(x) = x^{2/3}$ has a minimum at $x = 0$ and $f'(0) \neq 0$. (In fact $f'(0)$ does not exist);

(b) False; when $f'(c) = 0$ then $f(x)$ may have a point of inflection at $x = c$;

(c) False; the graph may have a corner at the extremum;

(d) False; the tangent need not be horizontal at a point of inflection. See Figure 4.17;

(e) False; only $f''(x) > 0$; $f'(x)$ may be > 0 or ≤ 0;

(f) False; a local maximum value of a function f may be less than a local minimum value of f;

(g) True; (h) False; the differential of x^3 is $3x^2 \, dx$; (i) True; (j) True, in general;

(k) False; to find dy for $y = f(x), f(x)$ must be differentiable with respect to x;

(l) False; the function $f(x)$ is increasing or decreasing in $a \leq x \leq b$ according as $f'(x) > 0$ or $f'(x) < 0$ respectively;

(m) False; in an implicit function of the form $F(x, y) = 0$, if x is an independent variable then y is a dependent variable or vice versa. x and y are not both independent variables. One is independent and the other is dependent;

(n) False; the curve may or may not represent the graph of a function. For example, the graph of parametric equations $x = t^2, y = t$ does not represent the graph of a function, because on eliminating t, we have $y^2 = x$ or $y = \pm\sqrt{x}$. Thus, to a value of x, the value of y is *not* unique.

3. (a) $-1 < x < 1$; (b) $x < -1$ or $x > -1$; (c) $-\sqrt{3} < x < 0, x > \sqrt{3}$;

(d) $x < -\sqrt{3}, 0 < x < \sqrt{3}$. The points of inflection are at $x = 0, \pm\sqrt{3}$.

5. (a) all x; (b) No x; (c) $x > 0$; (d) $x < 0$; $x = 0$ is a point of inflection.

7. Absolute maximum $g(3) = \sqrt{5}$, absolute minimum $g(2) = 0$.

9. (a) $K = \frac{3}{2}$; (b) $K = -48$; (c) $K = -32$.

11. $A > 0.$ **13.** $r = \frac{A}{B}.$ **15.** $x = \frac{1}{\sqrt{2}}; x = 0.83.$

17. 19.2. **19.** $-\left(\frac{y}{x}\right)^{1/3}.$ **21.** $\frac{2t(1 + t)^2}{(1 + t^2)^2}.$

23. 0. **25.** $\frac{2y(3 - y^2)(y^2 + 1)^3}{(1 - y^2)^3}.$ **27.** $-\frac{2}{9t^4}.$

CHAPTER 5

EXERCISES 5.1

1. $\frac{\pi}{12}.$ **3.** $\frac{5\pi}{6}.$ **5.** $\frac{11\pi}{5}.$

7. 135°. **9.** 330°.

11. $L = 2.705$ cm, $A = 6.763$ cm². **13.** $L = 6$ in., $A = 9$ in.²

15. $\cos\theta = \pm\dfrac{\sqrt{15}}{4}$; $\tan\theta = \pm\dfrac{1}{\sqrt{15}}$; $\sec\theta = \pm\dfrac{4}{\sqrt{15}}$.

17. $\tan\theta = \pm 2\sqrt{6}$; $\sin\theta = \pm\dfrac{2\sqrt{6}}{5}$.   **19.** $-\sin\theta$.

21. $-\sin\theta$.   **23.** $\cos\theta$.   **25.** $-\cot\theta$.

27. (a) -0.0523; (b) -0.3420; (c) -5.1446.

31. 120.363 ft.   **33.** 49.872 ft.   **35.** 1.8737 m.

EXERCISES 5.2

1. $\dfrac{2}{\pi}$.   **3.** Does not exist.   **5.** Does not exist.

7. 1.   **9.** 2.   **11.** 0.

13. Does not exist.   **15.** 9.143×10^4 ft $= 17.3$ mi.

EXERCISES 5.3

1. $4\cos 4x$.   **3.** $\cos x - x\sin x$.   **5.** $\cos^2 x - \sin^2 x$.

7. $\sin^2 x \cos^4 x\,(3\cos^2 x - 5\sin^2 x)$.   **9.** $\dfrac{1}{1 + \cos x}$.

11. $\dfrac{1 - \tan x + 2x\sec^2 x}{2\sqrt{x}\,(1 - \tan x)^2}$.   **13.** $-2\operatorname{cosec}^2 x \cot x$.   **15.** $a\cos x\, e^{a\sin x}$.

17. $-\dfrac{1}{x}\sin(\ln x)$.   **19.** $-\tan x$.   **21.** $\dfrac{1}{x} + \cot x$.

23. $-3\cos x$.   **25.** $(\cos^2 x - \sin x)e^{\sin x}$.

27. $\dfrac{dy}{dx} = -\dfrac{\sin\theta}{2\cos 2\theta}$; $\dfrac{d^2y}{dx^2} = -\dfrac{(\cos\theta\cos 2\theta + 2\sin\theta\sin 2\theta)}{4\cos^3 2\theta}$.

31. 1; $3e^{\pi/2}$.   **33.** 1 ft.   **35.** 6.125 ft.

EXERCISES 5.4

1. $\dfrac{\pi}{2}$.   **3.** $\dfrac{\pi}{2}$.   **5.** $2 - \pi$.

7. 0.   **9.** $\dfrac{3\pi}{4}$.   **11.** $\dfrac{2\pi}{3}$.

13. $-\dfrac{\pi}{4}$.   **15.** 0.   **17.** $\dfrac{7}{\sqrt{1 - 49x^2}}$.

19. $\dfrac{\left(1 + \dfrac{1}{2\sqrt{x}}\right)}{1 + (x + \sqrt{x})^2}$.   **21.** $2x\operatorname{Tan}^{-1}\left(\dfrac{1}{2}x\right) + \dfrac{2x^2}{(4 + x^2)}$.

23. $\dfrac{1}{2\sqrt{x - x^2}}$.   **25.** $\dfrac{-1}{2\sqrt{(1 - x^2)}\operatorname{Cos}^{-1} x}$.

27. $\dfrac{dy}{dx} = \dfrac{1 + t^2}{\sqrt{1 - t^2}}$; $\dfrac{d^2y}{dx^2} = \dfrac{t(1 + t^2)(3 - t^2)}{(1 - t^2)^{3/2}}$.

29. $\operatorname*{Lim}_{t\to\infty} N = A\left[\dfrac{\pi}{2} + \operatorname{Tan}^{-1} kt_0\right]$; point of inflection at $t = t_0$, $N = A\operatorname{Tan}^{-1} kt_0$.

1. $\dfrac{2\pi}{5}$.   **3.** $\dfrac{\pi}{3}$.   **5.** 3.

7. $\sqrt{10}$.

9. $\sqrt{2+\sqrt{2}}$.

11. $y = 1 + 2\cos \pi t$ (t in seconds).

13. $y = 70\sin 10\pi t$.

15. 3; 2π.

17. 4; 2.

REVIEW EXERCISES FOR CHAPTER 5

1. (a) True; (b) False; $\tan\theta$ is not defined for $\theta = \pm\dfrac{\pi}{2}, \pm\dfrac{3\pi}{2}$, etc;

    (c) False; the relation $\sin^2\theta + \cos^2\theta = 1$ can be proved for all values of θ;   (d) True;

    (e) False; $\cos\left(\dfrac{\pi}{2} + x\right) = -\sin x$;   (f) True;   (g) False; $\dfrac{d}{dx}\cos f(x) = -[\sin f(x)]f'(x)$;

    (h) False; the statement is true for $\text{Sin}^{-1} x$ and $\text{Tan}^{-1} x$, but the values of $\text{Cos}^{-1} x$ lie between 0 and π (inclusive);

    (i) True;   (j) False; $\tan x = \dfrac{\sin x}{\cos x}$. There is no such relation between the inverse functions.

3. $\frac{1}{2}$.   **5.** $\frac{1}{4}$.   **7.** $\frac{1}{3}$.

9. $2x$.   **11.** $\sec x \operatorname{cosec} x$.   **13.** $\cot x - x\cot x \operatorname{cosec} x$.

15. He must jump at an angle of $\text{Sin}^{-1}\frac{1}{4}(1 + \sqrt{5})$ to the horizontal.

17. $\dfrac{5\pi}{6}$.   **19.** $z = 7 + 2\cos\dfrac{\pi}{12}(t - 4)$.

CHAPTER 6

EXERCISES 6.1

(Note that the constant of integration has been omitted from the answers given below.)

1. (a) $\dfrac{x^8}{8}$; (b) $\frac{2}{3}x^{3/2}$; (c) $2\sqrt{x}$; (d) $7x$.   **3.** $\dfrac{x^8}{8} + \dfrac{7x^2}{2} + \dfrac{x^2}{14} + 7\ln|x|$.

5. $\dfrac{x^4}{4} + \dfrac{4}{5}x^{5/2}$.   **7.** $x^4 + x^3 + x^2 + x + \ln|x| - \frac{1}{2}x^2$.

9. $\frac{2}{7}u^{7/2} + \frac{6}{5}u^{5/2} + \frac{14}{3}u^{3/2}$.   **11.** $\frac{4}{7}x^{7/2} + \frac{2}{3}x^{5/2} - \frac{2}{3}x^{3/2}$.

13. $-\dfrac{1}{x} + 3\ln|x| + 7x - x^2$.   **15.** $\theta^3 - 3\theta^2 + 9\ln|\theta| + 4e^\theta$.

17. $\tan x - x$.   **19.** 250.5.

EXERCISES 6.2

1. $\frac{1}{16}(2x + 1)^8$.   **3.** $\frac{1}{2}\ln|2y - 1|$.   **5.** $\frac{1}{3}e^{3x+2}$.

7. $-\frac{1}{2}\cos(2x + 1)$.   **9.** $2\sqrt{t + 1}$.   **11.** $\frac{1}{5}(x^2 + 7x + 3)^5$.

13. $\dfrac{-1}{2(x^2 + 3x + 1)^2}$.   **15.** $\ln|x^2 + 3x + 1|$.   **17.** $\frac{1}{2}e^{t^2}$.

19. $e^{\text{Sin}^{-1} x}$.   **21.** $\frac{1}{4}(\ln x)^4$.   **23.** $\ln|1 + \ln x|$.

25. $-\cos(\ln x)$.

27. $\frac{1}{3}\tan(x^3)$.

29. $-\dfrac{1}{1+e^x}$.

31. $\ln|t^3+t|$.

33. $\frac{1}{8}\sin^4 2t$.

35. $\frac{1}{6}\tan^6\theta$.

37. $\frac{1}{3}\sec^3 x$.

39. $\ln|1+\tan x|$.

41. $-\ln(\cos x)$.

EXERCISES 6.3

1. $\dfrac{1}{\sqrt5}\ln\left|\dfrac{2x-3-\sqrt5}{2x-3+\sqrt5}\right|$ (from formula 77 in Appendix II).

3. $\dfrac{1}{4}\left[\ln|2x-3|-\dfrac{3}{2x-3}\right]$ (formula 9).

5. $2\sqrt{3x+1}+\ln\left|\dfrac{\sqrt{3x+1}-1}{\sqrt{3x+1}+1}\right|$ (formulas 22 and 24).

7. $-\dfrac{1}{4}\ln\left|\dfrac{4+\sqrt{t^2+16}}{t}\right|$ (formula 55).

9. $\frac{1}{2}x\sqrt{x^2-9}+\frac{9}{2}\ln|x+\sqrt{x^2-9}|$ (formula 54).

11. $\dfrac{t}{4}(t^2-4)^{3/2}-\dfrac{3}{2}t\sqrt{t^2-4}+6\ln|t+\sqrt{t^2-4}|$ (formula 72).

13. $\frac{1}{56}(21x-3\sin 7x\cos 7x-2\sin^3 7x\cos 7x)$ (formulas 85 and 87).

15. $\frac{1}{15}[2\sin x+\cos^2 x\sin x-3\cos^4 x\sin x]$ (formulas 88 and 91).

17. $-x^2\cos x+2x\sin x+2\cos x$ (formula 94).

19. $\frac{1}{4}\tan^4\theta-\frac{1}{2}\tan^2\theta-\ln|\cos\theta|$ (formulas 100 and 102).

21. $(\frac{1}{2}x^3-\frac{3}{4}x^2+\frac{3}{4}x-\frac{3}{8})e^{2x}$ (formula 113).

23. $\frac{1}{2}(1+\theta^2)\,\mathrm{Tan}^{-1}\theta-\frac{1}{2}\theta$.   (formulas 26 and 132).

25. $\dfrac{\sqrt{x-1}}{x}+\mathrm{Tan}^{-1}\sqrt{x-1}$ (formulas 22 and 23).

27. $\frac{1}{13}e^{2x}(2\sin 3x-3\cos 3x)$ (formula 116).

29. $\ln\left|\dfrac{1-e^x}{2-3e^x}\right|$ (formula 15).

EXERCISES 6.4

1. $\ln\left|\dfrac{x}{x+1}\right|$.

3. $2\ln|y-1|-\ln|y+1|$.

5. $2\ln|z-2|+\ln|z-1|$.

7. $\frac{5}{3}\ln|y+2|+\frac{1}{3}\ln|y-1|-\ln|y+1|$.

9. $\ln|t|-\frac{1}{2}\ln|1-t|-\frac{1}{2}\ln|1+t|$.

11. $x+\ln|x-4|+\ln|x+1|$.

13. $\dfrac{1}{3}\ln\left|\dfrac{1+\sin x}{2-\sin x}\right|$.

15. $\dfrac{1}{5}\ln\left|\dfrac{2+\tan x}{3-\tan x}\right|$.

EXERCISES 6.5

1. $a\,\mathrm{Tan}^{-1}\left(\dfrac{x}{a}\right)$.

3. $\mathrm{Sin}^{-1}\left(\dfrac{y}{3}\right)$.

5. $\dfrac{1}{4}\mathrm{Sec}^{-1}\left(\dfrac{x}{4}\right)$.

7. $\dfrac{t}{9\sqrt{t^2+9}}$.

9. $-\dfrac{x}{a^2\sqrt{x^2-a^2}}$.

11. $\dfrac{1}{\sqrt{a^2-x^2}}$.

13. $\frac{1}{2}\ln(\theta^2+3)$.

EXERCISES 6.6

1. $\dfrac{x^2}{2}\ln x - \dfrac{x^2}{4}.$

3. $\dfrac{x^{n+1}}{n+1}\ln x - \dfrac{x^{n+1}}{(n+1)^2}.$

5. $x\ln x - x.$

7. $(x-1)e^x.$

9. $\dfrac{1}{m^2}(mx-1)e^{mx}.$

11. $x\,\mathrm{Cos}^{-1}\,x - \sqrt{1-x^2}.$

13. $\dfrac{x^2}{2}\ln x - \dfrac{x^2}{4}.$

15. $-\dfrac{1}{2}x\cos 2x + \dfrac{1}{4}\sin 2x.$

17. $\dfrac{1}{4}(2x^2-1)\sin 2x + \dfrac{1}{2}x\cos 2x.$

19. $\dfrac{1}{4}(2x^2+2x+1)e^{2x}.$

REVIEW EXERCISES FOR CHAPTER 6

1. (a) False; the antiderivative contains an arbitrary constant; (b) True;
 (c) False; the integral of a product of two function can often be obtained by integration by parts;

 (d) False; $\displaystyle\int \dfrac{d}{dx}[f(x)]\,dx = f(x) + C$, where C is an arbitrary constant;

 (e) False; $\dfrac{d}{dx}\left[\displaystyle\int f(x)\,dx\right] = f(x);$

 (f) False; if $f'(x) = g'(x)$ then $f(x) - g(x)$ is constant, not necessarily zero;

 (g) False; $\displaystyle\int \dfrac{1}{x}\,dx = \ln|x| + C$; (h) False; $\displaystyle\int e^x\,dx = e^x + C$;

 (i) False; $\displaystyle\int \sec^2 x\,dx = \tan x + C$;

 (j) False; $\displaystyle\int [f(x)]^n\,f'(x)\,dx = \dfrac{[f(x)]^{n+1}}{n+1} + C,\quad (n \neq -1)$;

 (k) False; the statement is true for all n except $n = -1$;

 (l) False; integration by parts gives $\displaystyle\int x\,f(x)\,dx = x\displaystyle\int f(x)\,dx - \displaystyle\int \left\{\displaystyle\int f(x)\,dx\right\}dx$;

 (m) False; $\displaystyle\int \dfrac{1}{x^2}\,dx = -\dfrac{1}{x} + C$;

 (n) False; $\displaystyle\int e^{x^2}\,dx$ cannot be expressed in terms of elementary functions;

 (o) False; $\displaystyle\int e^t\,dt = e^t + C.$

3. $\ln|1 + \ln x|.$

5. $-\cos t + \tfrac{2}{3}\cos^3 t - \tfrac{1}{5}\cos^5 t$ (substitute $\cos t = x$)
  or $-\tfrac{1}{5}\cos t\,(\sin^4 t + \tfrac{4}{3}\sin^2 t + \tfrac{8}{3})$ (formula 87).

7. $\tfrac{1}{2}\tan^2 u + \tfrac{1}{4}\tan^4 u$ $(\tan u = x)$ or $\tfrac{1}{4}\sec^4 u$ $(\sec u\tan u = x)$.

9. $\tfrac{1}{3}(\mathrm{Tan}^{-1}\,u)^3$ $(x = \mathrm{Tan}^{-1}\,u)$.

11. $-\dfrac{1}{x}\sqrt{2x+1} - \ln\left|\dfrac{\sqrt{2x+1}-1}{\sqrt{2x+1}+1}\right|$ (formulas 22 and 23).

13. $-\tfrac{1}{3}\ln|x-1| - \tfrac{7}{15}\ln|x+2| + \tfrac{4}{5}\ln|x-3|$ (partial fractions).

15. $\tfrac{3}{2}x\sqrt{x^2+4} + 6\ln|x + \sqrt{x^2+4}|$ (formula 66).

17. $\tfrac{1}{4}t(25t^2-9)^{3/2} - \tfrac{27}{8}t\sqrt{25\,t^2-9} + \tfrac{243}{40}\ln|5t + \sqrt{25t^2-9}|$ (formula 72).

19. $\tfrac{1}{6}x^6[\ln x - \tfrac{1}{6}]$ (formula 122; x must be positive).

21. $\dfrac{1}{3}\ln\left|\dfrac{1 + \ln t}{2 - \ln t}\right|$ (substitute $\ln t = x$ and use formula 15).

23. $\frac{1}{5}\tan^5 x - \frac{1}{3}\tan^3 x + \tan x - x$ (formulas 101 and 102).
25. $\frac{1}{13}e^{-3x}(-3\cos 2x + 2\sin 2x)$ (formula 117).

CHAPTER 7

EXERCISES 7.1

1. 8.
3. 64.
5. $\frac{29}{6}$.
7. $\frac{3}{4}$.
9. n^2.
11. $\frac{1}{3}n(n^2 + 3n + 5)$.
13. $\frac{1}{6}n(4n^2 + 9n - 1)$.
15. $\frac{1}{4}n(n^3 + 2n^2 + 15n + 10)$.
17. $\frac{1}{3}(20)(20^2 + 12 \cdot 20 - 7) = 4220$.
19. $\frac{1}{6}(25)(2 \cdot 25^2 + 15 \cdot 25 + 31) = 6900$.
21. 42540.
25. (a) 11;   (b) 151.
27. (a) 9;   (b) 496.

EXERCISES 7.2

1. $A_4 = 40$.
3. $A_8 = \frac{51}{2}$.
5. $A_4 = \frac{31}{4}$.
7. $A_4 = \dfrac{\pi(1 + \sqrt{2})}{4}$.
9. $A_5 = \ln(\frac{7}{2})$.
11. 16.
13. $\frac{16}{3}$.
15. $\frac{81}{4}$.

EXERCISES 7.3

1. $\frac{1}{3}$.
3. $\frac{13}{2}$.
5. 1.
7. $\frac{3}{2}$.
9. $e^{\pi/4} - 1$.
11. 0.
13. 0.
15. $\dfrac{e + 1}{3 + \ln 2}$.
17. 16.
19. $\frac{16}{3}$.
21. $\frac{81}{4}$.
23. $\dfrac{e^x \ln x}{1 + x^2}$.
25. $-\dfrac{\sin x}{1 + x}$.
27. 0.

EXERCISES 7.4

1. 9.
3. 3.
5. $\frac{23}{3}$.
7. 6.
9. $\frac{13}{6}$.
11. $\frac{1}{3}$.
13. $e - \frac{4}{3}$.
15. $\frac{8}{3}$.
17. $\frac{1}{12}$.
19. $\frac{16}{3}$.

EXERCISES 7.5

1. $\dfrac{32\pi}{5}$.
3. $\dfrac{\pi^2}{2}$.
5. $\dfrac{32\pi}{3}$.
7. $\dfrac{96\pi}{5}$.
9. 6π.
11. $\dfrac{2\pi}{15}$.
13. $\frac{4}{3}\pi r^3$.
15. $\dfrac{175\pi}{6}$ ft³.

EXERCISES 7.6

1. $M_x = 9$, $M_y = 7$; $\bar{x} = \frac{7}{10}$, $\bar{y} = \frac{9}{10}$. **3.** $M_x = -3$, $M_y = 31$; $\bar{x} = \frac{31}{10}$, $\bar{y} = -\frac{3}{10}$.

5. $(\frac{27}{5}, \frac{16}{5})$. **7.** $(\frac{17}{5}, \frac{16}{5})$. **9.** $(\frac{29}{34}, \frac{5}{2})$.

11. $(\frac{9}{4}, \frac{27}{10})$. **13.** $(\frac{4}{5}, \frac{2}{7})$. **15.** $\left(\dfrac{4a}{3\pi}, \dfrac{4a}{3\pi}\right)$.

17. $\left(\dfrac{\pi}{2}, \dfrac{\pi}{8}\right)$. **19.** $\left[\dfrac{8\ln 2 - 3}{8\ln 2 - 4}, \dfrac{(\ln 2 - 1)^2}{(2\ln 2 - 1)}\right]$.

REVIEW EXERCISES FOR CHAPTER 7

1. (a) False; $\left(\sum\limits_{k=1}^{n} x_k\right)^2 = (x_1 + x_2 + \cdots + x_n)^2$ and $\sum\limits_{k=1}^{n} x_k^2 = x_1^2 + x_2^2 + \cdots x_n^2$, which are unequal;

(b) True, provided $C = nc$; (c) True, provided that $f(x) \geq 0$ for $a \leq x \leq b$; (d) True;

(e) False; $\dfrac{d}{dx}\displaystyle\int_a^x f(t)\,dt = f(x)$;

(f) False; $\dfrac{d}{dx}\displaystyle\int_a^b f(x)\,dx = 0$ and $\displaystyle\int_a^b \dfrac{d}{dx}[f(x)]\,dx = [f(x)]_a^b = f(b) - f(a)$;

(g) True; (h) True; (i) False; $\displaystyle\int_a^b f(x)\,dx = \int_a^b f(t)\,dt$; (j) True; (k) True;

(l) False; $\ln 2$ is the area under the curve $y = 1/x$ bounded by the x-axis and the lines $x = 1$ and $x = 2$;

(m) True; (n) False; $\ln x \longrightarrow -\infty$ as $x \longrightarrow 0^+$;

(o) True; $\ln(1 + 2 + 3) = \ln 6$; $\ln 1 + \ln 2 + \ln 3 = \ln(1\cdot 2\cdot 3) = \ln 6$.

5. $\dfrac{3\pi}{2}$. **7.** $2\ln 2 - 1$. **9.** $\dfrac{2\sqrt{2\pi}}{3}$.

11. $(\frac{3}{2}\ln 3, \frac{13}{54})$.

CHAPTER 8

EXERCISES 8.1

1. $\{0, 1, 2, 3, 4, 5\}$.

3. $\{H1, H2, H3, H4, H5, H6, T1, T2, T3, T4, T5, T6\}$.

5. $\{MER, MRE, RME, REM, EMR, ERM\}$ where M = mouse, E = elk and R = rabbit.

7. The sample space consists of four sample points, each sample point representing one of the aces.

9. The sample space consists of 26 sample points, each point representing one of the red cards in the deck.

11. $\{(6, 4), (5, 5), (4, 6)\}$.

13. $\{(1, 6), (2, 5), (3, 4), (4, 3), (5, 2), (6, 1)\}$

15. $E_1 \cup E_2 = \{x \mid x$ is a black card or a heart$\}$ is the event that the card drawn is either a heart or a black card.

17. $E_1 \cup E_3 = \{x \mid x$ is a heart or is of denomination less than 7$\}$ is the event that the card drawn is either a heart or of denomination less than 7.

19. $E_3 \cap E_4 = \{x \mid x$ is an ace$\} = E_4$ is the event that the card drawn is an ace.

21. $E_2' = \{x \mid x$ is a red card$\}$ is the event that the card drawn is not a black card.

23. The following pairs of events are mutually exclusive: E_1 and E_4; E_3 and E_4; E_3 and E_5; E_6 and E_k, where $k = 1, 2, 3, 4, 5$.

25. The given two events *are not* mutually exclusive.

27. The given two events *are* mutually exclusive.

EXERCISES 8.2

1. $\frac{1}{4}$.

3. $\frac{3}{4}$.

5. $\frac{19}{52}$.

7. $\frac{3}{8}$.

9. $\frac{7}{8}$.

11. 1.

13. $\frac{5}{18}$.

15. (a) $\frac{7}{10}$;   (b) $\frac{3}{10}$.

17. (a) $\frac{7}{10}$;   (b) $\frac{3}{10}$;   (c) 0;   (d) 0.

19. No; smoking and cancer disorders are *not* independent events: $P(S \cap C) = \frac{1}{25} > P(S) \cdot P(C) = (\frac{3}{5})(\frac{1}{20})$.

21. $P(E_1) = \frac{6}{13}, P(E_2) = \frac{1}{13}, P(E_3) = \frac{1}{4}, P(E_4) = \frac{1}{13}$; E_1 and E_3, E_2 and E_3, and E_3 and E_4 are independent pairs.

EXERCISES 8.3

1. 90.

3. 120.

5. 435.

7. 380.

9. 7.

11. $\dfrac{1}{13^3} = \dfrac{1}{2197}$

13. (a) $\frac{2}{21}$;   (b) $\frac{10}{21}$.

15. $\frac{969}{2639} = 0.3672$.

17. 4^{mf}.

19. $4^{(m-s)f} 2^{sf} = 2^{(2m-s)f}$.

EXERCISES 8.4

1. $2^3 = 8$ germ cells; $2^6 = 64$ offspring.

3. Denote the two chromosomes by C_1 and C_2. There are two possible germ cells and three possible offspring: $(C_1 C_1)$, $(C_1 C_2)$, and $(C_2 C_2)$. If C_1 and C_2 are identical only one offspring is possible.

5. 3^n.

7. $3^5 = 243$.

9. $4^2 = 16$.

11. The brown hair allele A is dominant over the blonde hair allele a. The man is type Aa (brown), and the woman is type aa (blonde). Half the offspring are Aa and half are aa.

13. The tall plant is type AA, the short plant type aa. All the offspring are Aa and are tall if A is dominant over a.

15. Genotypes AA and Aa are tall, and aa is short. One of the initial pairs of plants was AA and the other was Aa. Of the offspring, half were AA and half were Aa, so the first generation were all tall. One quarter of their germ cells were allele a, so the proportion of aa plants in the second generation was $(1/4)^2 = 1/16$.

17. 16% red, 48% pink, 36% white.

19. $p(p + 2q)$ tall, q^2 short.

21. Three pure types and three hybrid types.

23. $\frac{1}{2}(f + f') + \frac{1}{4}(g + g') = p$; $\frac{1}{2}(h + h') + \frac{1}{4}(g + g') = q$.

EXERCISES 8.5

1. $\frac{1}{4}$

3. $\frac{10 \cdot 5^3}{6^5} = \frac{625}{3888}$.

5. $\frac{128}{625}$.

7. $1 - (\frac{4}{5})^6 = \frac{11529}{15625}$.

9. $\frac{27}{128}$.

11. (a) $\frac{135}{4096}$; (b) $\frac{1}{4096}$.

13. $\frac{1904229}{1953125} = .975$.

15. $45(.02)^2(.98)^8 = 0.0153$.

17. (a) 0.251; (b) 0.215; (c) 0.201.

EXERCISES 8.6

1. $C = \frac{1}{2}$; $F(x) = \frac{1}{2}(1 - \cos x)$.

3. $C = 2$; $F(x) = 1 - \frac{1}{(1 + x)^2}$.

5. $C = \frac{1}{\pi}$; $F(x) = \frac{1}{\pi}\left(\frac{\pi}{2} + \text{Tan}^{-1} x\right)$.

7. $C = 3$; $F(x) = 1 - e^{-x^3}$.

9. (a) $F(x) = 1 - e^{-x/100}$; (b) 100 days;
   (c) $P(0 \le x \le 50) = F(50) - F(0) = 1 - e^{-0.5} = 0.3935$.

11. $P(x < 1) = F(1) = \frac{3}{4} = 0.75$; $P(x = 2) = 0$; $P(1 < x < 3) = \frac{3}{16} = .1875$.

13. $F(x) = 1 - (3x + 1)e^{-3x}$; $P(0 \le x \le 2) = 1 - 7e^{-6} = 0.9826$; $P(x > 3) = 0.0012$;   Average
   time $= \frac{2}{3}$ hr.

EXERCISES 8.7

1. 0.0375.

3. 0.8849.

5. 0.0827.

7. 0.8064.

9. 0.0808.

11. 0.6976.

13. (a) 0.8944; (b) 0.0401.

15. $\sigma = 3.5$.

17. $\mu = 28$.

19. (a) 0.9821; (b) 0.9811.

21. (a) 0.0073; (b) 0.8787; (c) 0.2877.

23. 0.0228.

REVIEW EXERCISES FOR CHAPTER 8

1. (a) False; for example, any event E and its complement E' are mutually exclusive but never
   independent unless $p(E)$ is 0 or 1;
   (b) True; (c) False; $p(E_1 \cap E_2) = p(E_1)p(E_2)$ for independent events; (d) True;
   (e) True; (f) True;
   (g) False; the mean and standard deviation of a *standard* normal distribution are 0 and 1 respectively;
   (h) False; x is a normal random variable with mean 12 and standard deviation 3;
   (i) True;
   (j) False; two *mutually exclusive* events cannot occur simultaneously;
   (k) False; the probability is approximately 0.27.

3. (a) $\frac{1}{19}$; (b) $\frac{10}{19}$.

5. 0.50; The events $E_1 = $ "excessive weight" and $E_2 = $ "high blood pressure" are *not* independent
   because $P(E_1 \cap E_2) \ne P(E_1)P(E_2)$.

7. (a) $\frac{1}{20}$; (b) $\frac{2}{3}$; (c) $\frac{1}{5}$.

11. $k = \frac{3}{20}$; $P(2 \le x \le 3) = \frac{11}{40}$.

13. (a) 64; (b) 6.

15. 0.0069.

17. 0.0025.

CHAPTER 9

EXERCISES 9.1

1. $f(3, -2) = 25; f(-4, -4) = 0.$

3. $f(2, 1) = -\dfrac{1}{5}; f\left(3, \dfrac{1}{2}\right) = 0; f\left(-\dfrac{1}{4}, \dfrac{3}{4}\right) = -\dfrac{4\sqrt{2}}{5}.$

5. $f(1, 2, 3) = 36; f(-2, 1, -4) = 54.$

7. $f\left(\dfrac{1}{2}, 1, 1\right)$ is not defined; $f\left(\dfrac{1}{4}, -\dfrac{1}{3}, 2\right) = -\dfrac{\sqrt{3}}{4}.$

9. $D = $ whole xy-plane.

11. $D = \{(x, t) \,|\, x + 2t \geq 2\}.$

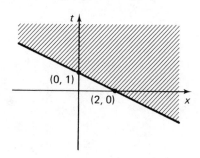

13. $D = \{(x, y) \,|\, y \neq \pm 1\}.$

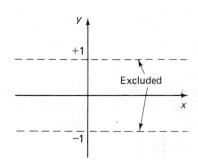

15. $D = \{(x, y) \,|\, -1 \leq x + y \leq 1\}.$

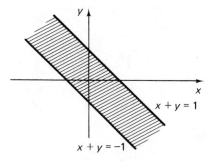

17. $D = \{(z, w) \,|\, z - w > 0 \text{ and } z - w \neq 1\}.$

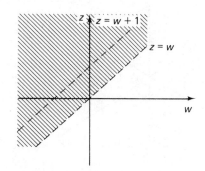

19. $D = \left\{(x, y)\,|\,y - x \neq \pm\dfrac{\pi}{2}, \pm\dfrac{3\pi}{2}, \ldots\right\}.$

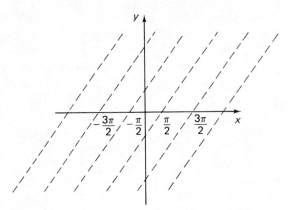

21. $D = \{(x, y, z)\,|\,yz \geq 0\}.$

23. $z = -\dfrac{1}{3}x + \dfrac{2}{3}y + \dfrac{4}{3}.$

25. $z = -x - y \pm 1.$

EXERCISES 9.2

1, 3, 5.

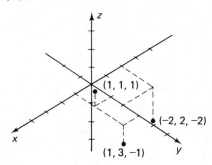

7. $\sqrt{5}.$

9. 6.

11. $\sqrt{6}.$

13. $\sqrt{29}.$

15. $PR^2 = PQ^2 + QR^2$, so PR is the hypotenuse and the right angle is at Q.

17. $z = 3.$

19. $z = -1.$

21. $x = 0.$

23. $y = 2.$

25. $x + y + z = 3.$

27. $3y - z = 10.$

29. $(3, 0, 0), (0, 3, 0), (0, 0, 3).$

31. Does not intersect x-axis; $(0, \frac{10}{3}, 0), (0, 0, -10).$

33. $3x + 2y = 6.$

35. $10x + 12y + 9z \leq 40$ and $10x + 14y + 8z \geq 40.$

37. $12x - 4y - 3z + 12 = 0.$

EXERCISES 9.3

1. $z = c, x^2 + y^2 = 9 - c^2 \quad (-3 < c < 3$.

3. $z = c, x^2 + y^2 = c \quad (c > 0)$.

5.

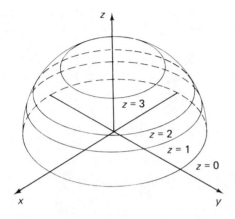

7.

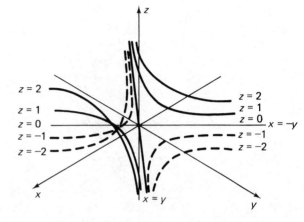

9.

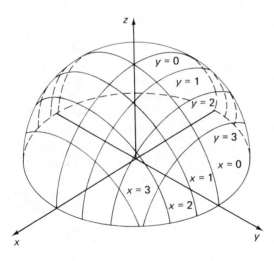

11.

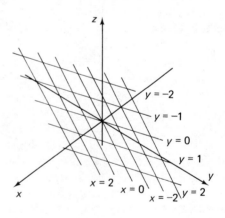

13.

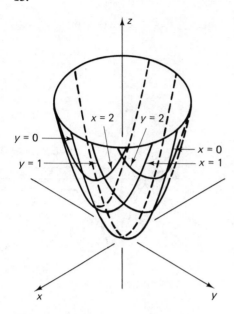

EXERCISES 9.4

1. $2xy^4$, $4x^2y^3$.   **3.** $\dfrac{1}{2\sqrt{x-y}}$, $-\dfrac{1}{2\sqrt{x-y}}$.   **5.** e^{x+3y}, $3e^{x+3y}$.

7. $\dfrac{1}{y}\sin\left(\dfrac{x}{y}\right) + \dfrac{x}{y^2}\cos\left(\dfrac{x}{y}\right)$, $-\dfrac{x}{y^2}\sin\left(\dfrac{x}{y}\right) - \dfrac{x^2}{y^3}\cos\left(\dfrac{x}{y}\right)$.

9. $-\tan(x+y)$, $-\tan(x+y)$.   **11.** $\frac{3}{4}x^{-1/2}y^{-4}$, $-6x^{1/2}y^{-5}$.

13. $-\dfrac{1}{(x+2y)^2}$, $-\dfrac{2}{(x+2y)^2}$.

23. $\dfrac{\partial M}{\partial r} = -6a\dfrac{(3l^2+8rl+8r^2)}{r^2(3l+4r)^2}$, $\dfrac{\partial M}{\partial l} = \dfrac{-12a}{(3l+4r)^2}$.

EXERCISES 9.5

1. $z - 2x - 2y = -3$.   **3.** $z - x - y = -1$.   **5.** $z + 2y = -1$.

7. 5.14.   **9.** 4.325.   **11.** 0.075.

15. 19.2.   **17.** $\Delta V \simeq \pi r(2l\Delta r + r\Delta l)$.

EXERCISES 9.6

1. Local minimum at $(0, 0)$.   **3.** Local maximum at $\left(-\frac{3}{2}, -1\right)$.

5. Local minimum at $\left(0, \dfrac{1}{\sqrt{3}}\right)$; local maximum at $\left(0, -\dfrac{1}{\sqrt{3}}\right)$; saddle points at $\left(\pm\dfrac{2}{\sqrt{15}}, \mp\dfrac{1}{\sqrt{15}}\right)$.

7. Saddle points at $\left(\sqrt{2}, -\dfrac{1}{\sqrt{2}}\right)$, no extrema. (Note that $f(x, y)$ is on'y defined for $x > 0$.)

9. $\partial y/\partial x = x(2a - 3x)\, te^{-t}$, $\partial y/\partial t = x^2(a - x)(1 - t)e^{-t}$. The reaction is maximum when $x = 2a/3$ and $t = 1$.

11. R is maximum when $x = a/5$ and $y = 2a/5$.

13. $n = 2FA/D$, $x = (\beta VA/D)^2$, where $D = 4\alpha F - \beta^2 V$ (D must be > 0).

15. $x = (10\alpha - 9\beta)/(8\alpha^2 - 9\beta^2)$, $y = (12\alpha - 15\beta)/(8\alpha^2 - 9\beta^2)$ gives maximum provided that $8\alpha^2 - 9\beta^2 > 0$.

EXERCISES 9.7

1. $y = 0.47x + 2.58$.   **3.** $y = 0.7x + 0.95$.

5. $y = 2.3x + 3.1$. Predictions are therefore $y = 16.9$ (or 17) when $x = 6$ and $y = 19.2$ (or 19) when $x = 7$.

EXERCISES 9.8

1. $\dfrac{\partial z}{\partial r} = 5r^4 \cos^2 \theta \sin^3 \theta$; $\dfrac{\partial z}{\partial \theta} = r^5 \sin^2 \theta \cos \theta (3 \cos^2 \theta - 2 \sin^2 \theta)$.

3. $\dfrac{\partial z}{\partial s} = s^5\left(6s + \dfrac{1}{2}\right)e^{(s+t)/2}$; $\dfrac{\partial z}{\partial t} = \dfrac{1}{2}s^6 e^{(s+t)/2}$.

5. $\dfrac{\partial z}{\partial p} = 2p$, $\dfrac{\partial z}{\partial q} = 0$.   **7.** $2(t^3 + 2t + 1)$.   **9.** $(2t^2 + 2t + 1)e^{(t^2-1)}$.

13. $\dfrac{dV}{dt} = 2\pi r h \dfrac{dr}{dt} + \pi r^2 \dfrac{dh}{dt} = 1024\pi$ in.^{3}/yr.

15. $\dfrac{dc}{dt} = 10k(10 + 39t - 2t^2)e^{-t/10}$;

c is maximum when $t = \tfrac{1}{4}(39 + \sqrt{1601}) = 19.8$.

REVIEW EXERCISES FOR CHAPTER 9

1. (a) False; the range of a function $f(x, y)$ is the set of values that the function takes. It is a subset of the real numbers, not of the xy-plane;

    (b) False; $D = \{(x, y)\,|\,x$ and y cover all real numbers$\}$;   (c) True;

    (d) False; the equation $ax + by + cz = d$ can be written in the given form only if $c \neq 0$. Then $\alpha = -a/c$, $\beta = -b/c$, and $\gamma = d/c$;

    (e) False; only a function of the form $z = \sqrt{a^2 - x^2 - y^2}$ has as its graph a hemisphere centered at the origin, where a is some constant;

    (f) True;   (g) True;

    (h) False; if the *third*-order partial derivatives of f are continuous, then it follows that $\partial^3 f/\partial x^2 \partial y = \partial^3 f/\partial y \partial x^2$;

    (i) False; $\dfrac{\partial}{\partial x}(x^3 y^2) = 3x^2 y^2$;   (j) False; $\dfrac{\partial}{\partial y}\left(\dfrac{x^2}{y}\right) = -\dfrac{x^2}{y^2}$;

    (k) False; the conditions stated are necessary in order that $f(x, y)$ should have a maximum, but are not sufficient. Sufficient conditions are obtained by adding the condition $f_{xx}f_{yy} - f_{xy}^2 > 0$;

    (l) True.

3. $D = \{(x, y)\,|\,x + y > 0, y \neq 0\}$.

5. $D = \{(x_1, x_2, x_3)\,|\,x_1 + x_2 + x_3 > 0, x_1 > x_2\}$.

7. $2x + y = -5$.

9. $4x - 6y + 3z = -12$.

11. $\dfrac{\partial z}{\partial x} = y(2x + y); \dfrac{\partial z}{\partial y} = x(x + 2y); \dfrac{\partial^2 z}{\partial x \partial y} = 2(x + y); \dfrac{\partial^2 z}{\partial y^2} = 2x$.

13. $\dfrac{\partial z}{\partial x} = 2x \cos(x^2 + y^2); \dfrac{\partial z}{\partial y} = 2y \cos(x^2 + y^2); \dfrac{\partial^2 z}{\partial x \partial y} = -4xy \sin(x^2 + y^2)$;

$\dfrac{\partial^2 z}{\partial y^2} = 2 \cos(x^2 + y^2) - 4y^2 \sin(x^2 + y^2)$.

15. Local maximum at $(0, -2)$; saddle points at $(0, 0)$ and $(3/2, -3/2)$.

17. 1.36.

19. $x = y = 80$ yd, $z = 40$ yd.

21. $z_x = 2(x^2 + y^2)[2x + (x^2 + y^2)(x + y)]e^{(x+y)^2}$;

$z_y = 2(x^2 + y^2)[2y + (x^2 + y^2)(x + y)]e^{(x+y)^2}$.

23. $z_x = \dfrac{v(v^2 - u^2)}{xu + v} = \dfrac{y(1 - x^2)}{(1 + x^2)^2}; z_y = \dfrac{xv + u}{2(xu + v)} = \dfrac{x}{x^2 + 1}$.

25. $\left. \begin{array}{l} a \sum x_i^4 + b \sum x_i^3 + c \sum x_i^2 = \sum x_i^2 y_i \\ a \sum x_i^3 + b \sum x_i^2 + c \sum x_i = \sum x_i y_i \\ a \sum x_i^2 + b \sum x_i + cn = \sum y_i \end{array} \right\}$ all sums run from 1 to n.

For exercise **24**, these equations reduce to:

$354a + 100b + 30c = 104.3$

$100a + 30b + 10c = 41.1$

$30a + 10b + 5c = 30.3$

Then $a = 0.036$; $b = -2.09$; $c = 10.03$.

CHAPTER 10

EXERCISES 10.1

1. First-order, linear, with constant coefficients.

3. Second-order, linear, with constant coefficients.

5. Second-order, linear, but not with constant coefficients.

7. First-order, nonlinear.

9. Third-order, linear, but not with constant coefficients.

11. If rewritten in the form $dy/dt - 3y = e^t$, the equation is first-order, linear, with constant coefficients.

17. $y = \frac{2}{3}t^{3/2} + 1$.

19. $y = t \sin t + \cos t + 3 - (\pi/2)$.

21. $1000 + \frac{1000}{3} \ln(13)$.

23. $c = 1 + \dfrac{2}{\ln 5}(1 - 5^{-t/2})$.

25. $50{,}000(3 + e^{t/10})$.

EXERCISES 10.2

1. $y = 2e^t$.

3. $y = e^{-2(t-1)}$.

5. $y = 3e^{-1/8}e^{t/2} = 3e^{(4t-1)/8}$.

7. $\dfrac{dy}{dt} = \dfrac{1}{5}y, y = 10^4 e^{t/5}$.

9. $y = y_0 e^{t/2}$.

11. $y = 2e^{kt}$ billion, where $k = (\ln 2)/45 = 0.0154$. In 1960, $y = 2e^{30k} = 3.17$ billion.

13. 4.62×10^6 yr.

15. 2.71 million.

17. 0.185 rads.

19. 67.6 min.

21. The rate at which A dissociates is proportional to the amount of substance A that remains, i.e., $-dy/dt \propto y$ at each time t. (The argument is similar to that for radioactive decay.) If $dy/dt = -ky$, $y = y_0 e^{-kt}$.

23. $dy/dx = -ky$, where k is a constant for each material. Therefore $y = y_0 e^{-kx}$ gives the amount of radiation that penetrates shielding of thickness x. y_0 is the incident radiation level.

EXERCISES 10.3

1. $y = -\frac{1}{2} + ce^{2t}$.

3. $y = 3 + ce^{-t/2}$.

5. $y = [1 - ce^t]^{-1}$.

7. $3[ce^{-18t} - 1]^{-1}$.

9. $y = ce^{\sin t}$.

11. $y = ce^{x^2/2}$.

13. $y = 2(ce^{4t} - 1)/(ce^{4t} + 1)$ (assuming $y < 2$).

15. 16,148.

17. $p(t) = 1 - e^{-t/5}$; $p = 0.75$ when $t = 6.9$ yr.

19. 22.0 yr.

21. $a(t) = \dfrac{(a_0 - b_0)a_0}{[a_0 - b_0 e^{-k(a_0 - b_0)t}]}$ $(a_0 \neq b_0)$.

23. $y = \dfrac{4000}{2 + 1998e^{-4t}}$; 2.0 wk.

25. $dy/dt < 0$ for $y > y_m$; $dy/dt > 0$ for $y < y_m$. A population larger than y_m decreases in size and a population smaller than y_m increases in size. $y = y_m e^{ce^{-kt}}$, where the constant c is determined from the equation $y_0 = y_m e^c$.

27. $y + K \ln y + Mt = $ constant.

EXERCISES 10.4

1. $y = t - 1 + ce^{-t}$.

3. $y = e^t + ce^{t/2}$.

5. $y = t^2 - 6t + 18 + ce^{-t/3}$.

7. $y = \frac{1}{2}(\sin t - \cos t) + ce^t$.

9. $y = \frac{1}{3}(e^{3t-3} - 1)$.

11. $y = (t + 3)e^{-2t}$.

13. $y = 4h + (1000 - 4h)e^{t/2} - \dfrac{6h}{9 + \pi^2}\left[3e^{t/2} + \pi \sin \dfrac{\pi t}{6} - 3 \cos \dfrac{\pi t}{6}\right]$.

15. $y = 20h + (y_0 - 20h)e^{t/20}$.

17. $y = ce^{kt} + \dfrac{b_0}{k} - \dfrac{b_0}{k^2 + 4\pi^2}[k \sin 2\pi t + 2\pi \cos 2\pi t]$.

If $b_0 = y_0 k(k^2 + 4\pi^2)/(k^2 + 4\pi^2 - 2\pi k)$ then $y(1) = y(0)$.

EXERCISES 10.5

1. $y = c_1 e^{4t} + c_2 e^t$.

3. $y = (c_1 + c_2 t)e^{-t}$.

5. $y = 3e^t - e^{-t}$.

7. $y = 2e^{-t} - 1$.

9. $y = e^{2t} + e^{3t}$.

11. $y = 3(t + 2)e^{-t/3}$.

EXERCISES 10.6

1. $y = c \cos (t + \alpha)$.

3. $y = c \cos \left(\dfrac{t}{\sqrt{3}} + \alpha\right)$.

5. $y = ce^{-t} \cos (2t + \alpha)$.

7. $y = 3\sqrt{2} \cos \left(t + \dfrac{3\pi}{4}\right)$.

9. $y = \sin 2t$.

11. $y = \sqrt{2}\, e^{-2t} \cos\left(3t + \dfrac{\pi}{4}\right)$.

13. $y = 5e^{t/5} \cos \left(\dfrac{2t}{5} + \pi + \text{Tan}^{-1} \dfrac{4}{3}\right)$.

1. $x = c \sin(t + \alpha)$; $y = c \cos(t + \alpha)$.   3. $x = -\tfrac{1}{2}c_1 e^{6t} + \tfrac{1}{2}c_2 e^{2t}$; $y = c_1 e^{6t} + c_2 e^{2t}$.

5. $x = -2ce^{2t} \sin(2t + \alpha)$; $y = ce^{2t} \cos(2t + \alpha)$.

7. $x = \sqrt{2} \sin\left(t + \dfrac{3\pi}{4}\right)$; $y = \sqrt{2} \cos\left(t + \dfrac{3\pi}{4}\right)$.

9. $x = \tfrac{3}{4}e^{6t} + \tfrac{1}{4}e^{2t}$; $y = -\tfrac{3}{2}e^{6t} + \tfrac{1}{2}e^{2t}$.

11. $x = \sqrt{5}\, e^{2t} \sin(2t + 0.46)$; $y = -\dfrac{\sqrt{5}}{2} e^{2t} \cos(2t + 0.46)$.

13. $x = 2500 - 1500e^{-4t}$
    $y = 1250 + 750e^{-4t}$.

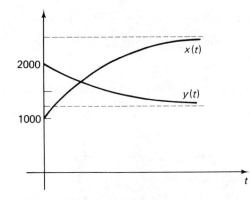

15. $x = 348e^t \sin(3t + 0.29)$; $y = 1044e^t \cos(3t + 0.29)$; $y = 0$ when $3t + 0.29 = \pi/2$ or $t = 0.43$.

17. If $x < y$ at $t = 0$ then $x < y$ for all t; if $x = y$ at $t = 0$ then $x = y$ for all t.

19. $x = 100 - 150 \sin 4t$; $y = 500 - 300 \cos 4t$; $x = 0$ when $t = \tfrac{1}{4} \operatorname{Sin}^{-1}(2/3) = 0.18$.

1. $y = 2 - \dfrac{1}{1 + ce^t}$.   3. $y = 1 - \tan(t + c)$.

5. $y = 0$, unstable; $y = 1$, stable.

7. $y = 0$, stable; $y = \pm 1$, both unstable.

9. $y = 2$, stable; the equilibrium at $y = 1$ is stable from the side $y < 1$, but is unstable from the side $y > 1$ since when $1 < y < 2$, $\dfrac{dy}{dt} > 0$ and y increases away from 1 and toward 2.

11. $x + y - a \ln y = c$.

13. $y = \dfrac{y_+ - y_- ce^{-kt}}{1 - ce^{-kt}}$ when $k = py_m$ and $y_\pm = \dfrac{1}{2}\left[y_m \pm \sqrt{y_m^2 + \dfrac{4b}{p}}\right]$; also, $c = (y_0 - y_+)/(y_0 - y_-)$. As $t \to \infty$, $y \to y_+$ and the limiting value $y_+ > y_m$.

15. $y = \tfrac{1}{2}y_m - a \tan(apt + c)$, where $a = \sqrt{h - \tfrac{1}{4}py_m^2}$. Extinction occurs when $t = (ap)^{-1} \times [\operatorname{Tan}^{-1}(y_m/2a) - c]$.

17. (b) $x = 1$; $y = 1$;
    (c) $x = 1 + c_1 e^{(\sqrt{2}-1)t} + c_2 e^{-(\sqrt{2}+1)t}$; $y = 1 - \sqrt{2}c_1 e^{(\sqrt{2}-1)t} + \sqrt{2}c_2 e^{-(\sqrt{2}+1)t}$;

(d)

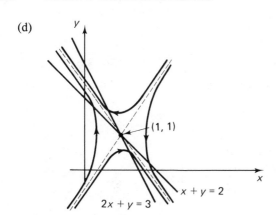

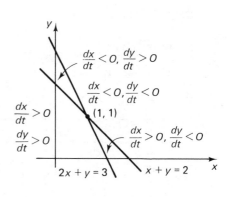

19. $\frac{dx}{dt} = (Py - C)x - h; \frac{dy}{dt} = (D - Qx)y - k.$ Equilibrium points $x = \frac{1}{2CQ}\{CD - Pk - Qh +$

$\sqrt{\Delta}; y = \frac{1}{2DP}\{CD + Pk + Qh + \sqrt{\Delta}\}$, where $\Delta = (CD + Qh + Pk)^2 - 4CDPk.$ This equi-

librium point always lies in the quadrant $x > 0, y > 0$ provided that $h > 0$.

REVIEW EXERCISES FOR CHAPTER 10

1. (a) True;   (b) True;
   (c) False; the given equation must be treated as a linear first-order differential equation (Section 10.4);
   (d) True;
   (e) False; there are several models of limited growth besides the logistic model (see for example exercises 25 and 26 of Section 10.3);
   (f) False; it is incorrect to write $\int yt^2\, dt = y \int t^2\, dt$ since y is a function of t and may not be taken outside the integral;
   (g) Almost true; the exception is the system $dx/dt = Px, dy/dt = Sy$, which simply consists of two independent first-order equations;
   (h) False; the given equation is of first order;
   (i) False; the given equation is a second-order differential equation for y.

3. $y = \frac{1}{2}t - \frac{1}{4} + ce^{-2t}.$ **5.** $y = [c - \frac{1}{3}t^3]^{-1}.$ **7.** $y = ce^{-2t} \cos(t + \alpha).$

9. $\ln(2y) - y = t - \frac{1}{2}.$ **11.** $y = 2e^{(3/2)(t-1)} \sin \frac{1}{2}(t - 1).$

13. $x = (3 - 14t)e^{3t}; y = (2 + 7t)e^{3t}.$ **15.** $\frac{dy}{dt} = -0.028y;$ 24.8 yr; 82.2 yr.

17. $p = (1 + ce^{-kt})^{-1}.$ If $p = 0.1$ at $t = 0$ then $c = 9$ and so $p = 0.5$ when $t = k^{-1} \ln 9.$

19. General solution for $s_F = $ constant is:
   $s_I = (q_1 - k_1 - k_2)c_1 e^{-q_1 t} + (q_2 - k_1 - k_2)c_2 e^{-q_2 t} + s_F$
   $s_M = -q_2 c_1 e^{-q_1 t} - q_1 c_2 e^{-q_2 t} + s_F$, where $q_1 = \frac{1}{2}\{k_1 + k_2 + k_3 - \sqrt{\Delta}$, $q_2 = \frac{1}{2}\{k_1 + k_2 + k_3$
   $+ \sqrt{\Delta}\}$ and $\Delta = (k_1 + k_2 + k_3)^2 - 4k_1 k_3.$
   (a) $s_F = 0; c_1 = -q_1 s_0/(q_2 - q_1); c_2 = q_2 s_0/(q_2 - q_1);$
   (b) $s_F \neq 0; c_1 = -c_2 = s_F/(q_2 - q_1).$

CHAPTER 11

EXERCISES 11.1

1. Second-order, linear.   **3.** First-order, nonlinear.   **5.** Fifth-order, linear.

13. 4 yr.   **15.** 11 days.

17. $x_n = 1.2x_{n-1} + 20$; $x_0 = 100$.

n	x_n	n	x_n	n	x_n
0	100	5	398	10	1138
1	140	6	497	11	1386
2	188	7	617	12	1683
3	246	8	760		
4	315	9	932		

19. $x_n = 1.5\, x_{n-1} - 400(1.1)^{n-1}$.

n	x_n	n	x_n
0	1000	5	1611
1	1100	6	1772
2	1210	7	1949
3	1331	8	2144
4	1464	9	2358

EXERCISES 11.2

1.

n	x_n
0	10
1	15
2	22
3	30
4	39
5	47
6	53
7	57
8	59
9	59
10	60

3.

n	x_n
0	10
1	25
2	59
3	118
4	168
5	155
6	163
7	158
8	161
9	159
10	160

5.

n	x_n
0	20
1	72
2	222
3	351
4	102
5	284
6	273
7	292
8	258
9	315
10	204
11	360
12	75
13	229
14	346
15	118
16	309
17	220
18	352
19	98
20	275

7.

n	x_n
0	20
1	65
2	182
3	297
4	143
5	289
6	162
7	296
8	144
9	290
10	161
11	296
12	145
13	290
14	160
15	296
16	145

9.

n	x_n
0	238
1	231
2	240
3	227
4	245
5	221
6	252
7	209
8	263
9	189
10	276
11	163
12	280
13	153
14	279
15	157
16	280
17	155
18	279
19	156
20	279

11. $p_n = p_{n-1} + (0.3 - 0.001p_{n-1})p_{n-1}$.

n	p_n	n	p_n	n	p_n	n	p_n
0	100	5	208	10	277	15	296
1	120	6	227	11	283	16	297
2	142	7	243	12	288	17	298
3	164	8	257	13	291	18	299
4	186	9	268	14	294	19	299

13. $y_n - y_{n-1} = (a - y_{n-1})y_{n-1}$; equilibrium value of y_n is a.

15. Equilibrium value of x_n is 100.

n	x_n	n	x_n	n	x_n
0	10	5	57	10	99
1	15	6	72	11	100
2	21	7	85	12	100
3	30	8	93		
4	42	9	98		

EXERCISES 11.3

1. $x_n = c2^n$, $c = $ constant.

3. $x_n = c(-1/4)^n$, $c = $ constant.

5. $x_n = x_0 + 3n$.

7. $x_n = x_0 + \frac{1}{2}n(n + 5)$.

9. $x_n = n^2(n - 1)$.

11. $x_n = (1 + x_0)2^n - 1$.

13. $x_n = (x_0 - \frac{1}{4})(-1)^n + \frac{1}{4}(2n + 1)$.

15. $x_n = (x_0 + 2)2^n - (n + 2)$.

17. $100(.99)^n$; 23 days.

19. 7 yr.

21. $x_n = 700(1.1)^n - 200$.

EXERCISES 11.4

1. $x_n = c_1 + c_2(-1)^n$.

3. $x_n = c_1(-3)^n + c_2(2^n)$.

5. $x_n = \sqrt{10} \cdot 2^{n/2} \cos(\frac{3}{4}n\pi - \text{Tan}^{-1} 3)$.

7. $x_n = \left(1 - \frac{n}{3}\right)3^n$.

9. $x_n = \left(2 - \frac{5n}{2}\right)(-2)^n$.

11. $x_n = 2 \cdot 3^{(n-1)/2} \cos(n - 1)\frac{\pi}{6}$.

13. $x = 4\sqrt{\frac{7}{3}} \cos\left(\frac{2n\pi}{3} - \text{Tan}^{-1}\frac{5}{\sqrt{3}}\right)$.

15. $x_n = \frac{3x_0 - x_1}{2} + \frac{x_1 - x_0}{2} \cdot 3^n$.

17. $x_n = x_{n-1} + x_{n-2}$; if $x_0 = x_1 = 1$ then $x_n = \frac{1}{\sqrt{5}}\left(\frac{\sqrt{5} + 1}{2}\right)^{n+1} - \frac{1}{\sqrt{5}}\left(\frac{1 - \sqrt{5}}{2}\right)^{n+1}$.

19. $x_n - x_{n-1} = 2x_{n-2}$; $x_n = \frac{1}{3}[2^{n+1} + (-1)^n]$.

21. $i_n = c_1\left[\frac{R + 2r + \sqrt{R(R + 4r)}}{2r}\right]^n + c_2\left[\frac{R + 2r - \sqrt{R(R + 4r)}}{2r}\right]^n$.

REVIEW EXERCISES FOR CHAPTER 11

1. (a) False; the given equation is of first order; (b) True;
   (c) False; the given equation is of first order; (d) True; (e) True;
   (f) False unless $p = 1$. See exercise 15 for $p \neq 1$;

3.

n	x_n	n	x_n
0	1.5	6	33
1	2	7	65
2	3	8	129
3	5	9	257
4	9	10	513
5	17		

5.

n	x_n	n	x_n
0	0	6	0.6154
1	1	7	0.6190
2	0.5	8	0.6176
3	0.667	9	0.6182
4	0.6	10	0.6180
·5	0.625		

Observe that the sequence x_n approaches closer and closer to a limiting value as n gets larger.

7. $x_n = 3(-\frac{1}{2})^n$.

9. $x_n = -1 + \frac{1}{3}n(n + 1)(2n + 1)$.

11. $x_n = (1 + 2n)(-\frac{1}{3})^n$.

13. $x_n = \frac{9}{4} - \frac{5}{4} \cdot 5^n$; $y = \frac{9}{4} - \frac{1}{4}(5^n)$.

15. $x_n = e^{(c_1 + c_2 p^n)}$ $(p \neq 1)$; $x_n = e^{(c_1 + c_2 n)}$ $(p = 1)$.

CHAPTER 12

EXERCISES 12.1

1.

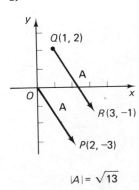

$$|A| = \sqrt{13}$$

3.

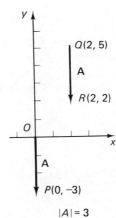

$$|A| = 3$$

5.

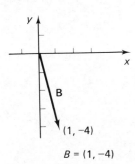

$$B = (1, -4)$$

7.

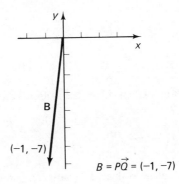

$$B = \vec{PQ} = (-1, -7)$$

9. $A + 2B = (8, 11)$; $3A - 2B = (0, -23)$.   **11.** $c = -4$.

13. $|A - 2B| = \sqrt{34}$.   **15.** $|B| = \sqrt{10}$.   **17.** $(\frac{3}{5}, -\frac{4}{5})$ and $(-\frac{3}{5}, \frac{4}{5})$.

19. $(-\frac{3}{5}, \frac{4}{5})$ and $(\frac{3}{5}, -\frac{4}{5})$.   **21.** $\left(\frac{3}{\sqrt{13}}, \frac{2}{\sqrt{13}}\right)$ and $\left(-\frac{3}{\sqrt{13}}, -\frac{2}{\sqrt{13}}\right)$.

23. $a = 1$; $b = 2$.   **25.** $\left(\frac{5\sqrt{3}}{2}, \frac{5}{2}\right)$.   **27.** $(3, 2)$.

EXERCISES 12.2

1. 4.   **3.** 59.   **5.** 0.

7. $k = 6$.   **9.** $k = 6$.   **11.** $k = -6$.

13. No value of k makes **A** and **B** parallel.   **15.** $k = -\dfrac{48 \pm 25\sqrt{3}}{39}$.

17. $k = \pm\sqrt{9 + 2\sqrt{21}}$.

19. Component $= \dfrac{13}{\sqrt{5}}$; Projection $= \left(\dfrac{13}{5}, \dfrac{26}{5}\right)$.

21. When **A** and **B** are parallel.

EXERCISES 12.3

1. $V_{XY} = (-1, -2)$.   **3.** $V_{ZX} = (0, -2)$.   **5.** $V_{AD} = (6, -1)$.
7. $V_{AB} = (10, -20)$ (x-axis east, y-axis north).
9. Speed $= 10\sqrt{13}$ mph at an angle $\text{Tan}^{-1}\left(\frac{2}{3}\right)$ with direction of automobile.
11. Speed $= 10\sqrt{34 + 15\sqrt{2}}$ mph in direction $\text{Tan}^{-1}\,[3/(3 + 5\sqrt{2})]$ W of N.
13. Pigeon flies due north; arrives home in $5\sqrt{2}/4 = 1.77$ hr.

EXERCISES 12.4

1. $(0, 5)$.

3. $((2 + \sqrt{3})F, (\sqrt{3} - 1)F)$.

5. $G = (-(1 + \sqrt{2})F, -\sqrt{2}F)$.

7. $G = \left(-\dfrac{1}{\sqrt{2}}F, -\dfrac{3}{\sqrt{2}}F\right)$.

9. $-F$.

11. $5F$.

13. $\frac{1}{2}(1 - \sqrt{3})F$.

15. $G = (-2F, -F)$; $a = -1$.

17. $G = (-3F, -F)$; $a = 5$.

19. $G = \left(-\dfrac{\sqrt{3}}{2}F, -\dfrac{1}{2}F\right)$; $a = \sqrt{3} - 1)$.

EXERCISES 12.5

1. Matrix: **A**   **B**   **C**   **D**   **E**   **F**   **G**   **H**
    Size: 2×2 2×3 3×1 3×3 2×3 2×2 1×2 1×3.
3. Cannot be evaluated because **A** and **B** are not of the same size.

5. $\begin{pmatrix} 10 & 9 & -1 \\ 0 & 11 & 4 \end{pmatrix}$.   **7.** $\begin{pmatrix} 6 \\ 3 \end{pmatrix}$.   **9.** (9)

11. $\begin{pmatrix} 2 & 6 \\ 1 & -2 \end{pmatrix}$.

13. **DE** cannot be evaluated because the number of columns in **D** is not equal to the number of rows in **E**.

15. $\begin{pmatrix} 3 & 2 & -2 \\ -2 & -1 & 6 \end{pmatrix}$.

19. $P = \begin{pmatrix} 1 & 0 \\ -\frac{1}{2} & -\frac{1}{2} \end{pmatrix}$, $PA = \begin{pmatrix} 1 & 0 \\ 0 & 1 \end{pmatrix} = I$.

EXERCISES 12.6

1. $\begin{pmatrix} 1 & 1 \\ 1 & -1 \end{pmatrix}\begin{pmatrix} x_1 \\ x_2 \end{pmatrix} = \begin{pmatrix} 2 \\ 3 \end{pmatrix}$.

3. $\begin{pmatrix} 1 & 2 & -1 \\ 1 & -3 & 0 \end{pmatrix}\begin{pmatrix} x_1 \\ x_2 \\ x_3 \end{pmatrix} = \begin{pmatrix} 0 \\ 2 \end{pmatrix}$.

5. $\begin{pmatrix} 1 & -1 \\ 2 & 1 \\ -1 & 3 \end{pmatrix}\begin{pmatrix} x_1 \\ x_2 \end{pmatrix} = \begin{pmatrix} 1 \\ 2 \\ 2 \end{pmatrix}$.   **7.** $X = \begin{pmatrix} \frac{3}{2} \\ -\frac{1}{2} \end{pmatrix}$.   **9.** $X = \begin{pmatrix} 1 & 2 \\ 1 & 1 \end{pmatrix}$.

11. 100 of species 1 and 200 of species 2.

13. $x_1 = \dfrac{s_1 c_{22} - s_2 c_{12}}{c_{11} c_{22} - c_{21} c_{12}}; \; x_2 = \dfrac{s_2 c_{11} - s_1 c_{21}}{c_{11} c_{22} - c_{21} c_{12}}.$

When two species consume the foods in exactly the same proportions then $c_{12} : c_{22} = c_{11} : c_{21}$. No solution exists unless $s_1 : s_2 = c_{12} : c_{22}$. If this condition is satisfied there exist infinitely many solutions for x_1 and x_2.

15. (a) $\begin{pmatrix} 0 & 1 & 1 \\ 1 & 0 & 1 \\ 1 & 1 & 0 \end{pmatrix}$; (b) $\begin{pmatrix} 0 & 1 & 0 \\ \frac{1}{2} & 0 & \frac{1}{2} \\ 2 & 0 & 0 \end{pmatrix}$; (c) $\begin{pmatrix} 0 & 0 & 2 \\ 1 & 0 & 0 \\ 0 & 0 & 0 \end{pmatrix}$.

REVIEW EXERCISES FOR CHAPTER 12

1. (a) False; a vector can be represented by an infinite number of directed line segments;
(b) True;
(c) False; a vector in n dimensions has n components, where n is any positive integer;
(d) False; $\mathbf{X}$ and $\mathbf{Y}$ are parallel if $\mathbf{X} \cdot \mathbf{Y} = \pm |\mathbf{X}||\mathbf{Y}|$; (e) True;
(f) False; $\mathbf{X}$ and $\mathbf{Y}$ make an angle of 60° with each oth r if $\mathbf{X} \cdot \mathbf{Y} = \frac{1}{2} |\mathbf{X}||\mathbf{Y}|$; (g) True;
(h) True; (i) False; $\mathbf{AB} \neq \mathbf{BA}$ in general;
(j) False; if $\mathbf{A}$ is $m \times n$ and $\mathbf{B}$ is $n \times p$ then $\mathbf{AB}$ is an $m \times p$ matrix;

3. $\mathbf{C} = (5, -2)$.   **5.** $a = 1, b = 2, c = -2$.

7. $a = 5, -3; b = \pm 3; c = 3$.

9. $\mathbf{A} + 2\mathbf{B} = \begin{pmatrix} 1 & 0 & -3 \\ 8 & 1 & 5 \\ -4 & 8 & 3 \end{pmatrix}; \; \mathbf{AB} = \begin{pmatrix} -7 & 13 & -13 \\ 7 & -16 & -7 \\ -3 & 10 & 6 \end{pmatrix}.$

11. $\begin{pmatrix} 1 & 2 & -1 \\ 0 & 1 & 3 \\ 2 & -1 & 1 \end{pmatrix} \begin{pmatrix} x \\ y \\ z \end{pmatrix} = \begin{pmatrix} 4 \\ 5 \\ 1 \end{pmatrix}.$

13. $x = \frac{5}{4}, y = \frac{33}{8}, z = -\frac{3}{8}$.   **15.** $x = 3, y = -1, z = 1$.

Index

BASIC DIFFERENTIATION FORMULAS

1. The derivative of a constant is zero.

2. For any constant c, $\dfrac{d}{dx}[cf(x)] = cf'(x)$

3. $\dfrac{d}{dx}[f(x) \cdot g(x)] = f(x)g'(x) + g(x)f'(x)$ Product Formula

4. $\dfrac{d}{dx}\left(\dfrac{f(x)}{g(x)}\right) = \dfrac{g(x)f'(x) - f(x)g'(x)}{[g(x)]^2}$ Quotient Formula

5. If $y = f(u)$ and $u = g(x)$, then $\dfrac{dy}{dx} = \dfrac{dy}{du} \cdot \dfrac{du}{dx}$ Chain Rule

     or $\dfrac{d}{dx}(f[g(x)]) = f'[g(x)] \cdot g'(x)$ Chain Rule

6. $\dfrac{d}{dx}(x^n) = nx^{n-1}$ Power Formula

7. If $x = f(t)$, $y = g(t)$, then $\dfrac{dy}{dx} = \dfrac{g'(t)}{f'(t)}$

     or $\dfrac{dy}{dx} = \dfrac{dy/dt}{dx/dt}$

8. $\dfrac{d}{dx}(e^x) = e^x$

9. $\dfrac{d}{dx}(\ln x) = \dfrac{1}{x}$

10. $\dfrac{d}{dx}(\sin x) = \cos x$

11. $\dfrac{d}{dx}(\cos x) = -\sin x$

12. $\dfrac{d}{dx}(\tan x) = \sec^2 x$

13. $\dfrac{d}{dx}(\sec x) = \sec x \tan x$

14. $\dfrac{d}{dx}(\text{Sin}^{-1} x) = -\dfrac{d}{dx}(\text{Cos}^{-1} x) = \dfrac{1}{\sqrt{1 - x^2}}$

15. $\dfrac{d}{dx}(\text{Tan}^{-1} x) = \dfrac{1}{1 + x^2}$